Conceptual Foundations of
MODERN PARTICLE PHYSICS

0 1856994

PHYSICS

Published by

World Scientific Publishing Co. Pte. Ltd.

P O Box 128, Farrer Road, Singapore 9128

USA office: Suite 1B. 1060 Main Street, River Edge, NJ 07661

UK office: 73 Lynton Mead, Totteridge, London N20 8DH

Cover illustration: Artist's rendering of a computer simulation of a proton-proton collision at SSC, courtesy of the Superconducting Super Collider Laboratory, Dallas, Texas.

Cover design: Jack W. Davis Graphic Design

Library of Congress Cataloging-in-Publication Data

Marshak, Robert Eugene, 1916–1992
   Conceptual foundations of modern particle physics / Robert E.
Marshak.
    p.   cm.
   Includes bibliographical references and index.
    ISBN 9810210981 : $86.00 – – ISBN 9810211066 (pbk.) : $46.00
    1. Particle (Nuclear physics)   I. Title.
   QC793.2.M375   1993
   539.7'2--dc20   93–16444
                                                   CIP

Printed in Singapore by Continental Press Pte Ltd

Conceptual Foundations of
MODERN PARTICLE PHYSICS

Robert E. Marshak

Virginia Polytechnic Institute and State University

World Scientific
Singapore • New Jersey • London • Hong Kong

To Ruth, my wife of fifty years

TABLE OF CONTENTS

LIST OF FIGURES

LIST OF TABLES

FOREWORD

The first half of this century witnessed the creation of relativity and quantum mechanics. The two critical paradoxes of the last century — the absence of an absolute inertial frame and the conflict of the dual wave-corpuscular behavior of particles — were thereby resolved. Their solutions will serve forever as a testimonial to the highest intellectual achievement of humanity.

By 1925, the theoretical foundations of these two milestones were firmly established. All the modern scientific and technological developments — nuclear energy, atomic physics, molecular beams, lasers, x-ray technology, semiconductors, superconductors, supercomputers — only exist because we have relativity and quantum mechanics. To our society and to our understanding of nature, these are all-encompassing.

After 1945 the world of physics again encountered a series of crises and successes. The complexity of the subnuclear world offers challenges that are at least as great as our predecessors faced. At present our theoretical framework, based on QCD for strong interactions, the standard model for the electroweak forces and general relativity for gravity, appears to be phenomenologically successful. Yet, there are deeply disturbing features which suggest that we are still in the transition period. In order to apply the present theories, we need about seventeen *ad hoc* parameters. All these theories are based on symmetry considerations, yet most of the symmetry quantum numbers do not appear to be conserved. All hadrons are made of quarks and yet no single quark can be individually observed. Now, fifty years after the beginning of modern particle physics, our successes have brought us to the deeper problems. We are in a serious dilemma about how to make the next giant step. Because the challenge is related to the very foundation of the totality of physics, a breakthrough is bound to bring us a profound change in basic science. If we are successful, this will be our legacy to the civilization of the next century.

Just before Robert Marshak left on a trip to Mexico with his family to celebrate his and Ruth's fiftieth wedding anniversary, a trip that ended tragically with Bob's death, he completed the final manuscript of this book on modern particle physics. He was proud and happy about what he had achieved in tracing and analyzing its conceptual foundations.

Only a true master and a creative participant could record with accuracy the genesis of the field from its early days through the major events culminating in the formulation of the standard model of strong and electroweak interactions, and then move on to the experimental test of the theory and today's speculations that go beyond it. All the important ideas and the related mathematical techniques were treated from a fresh perspective that is a delight to the reader, and will be extremely helpful to researchers and students.

Indeed Robert Marshak grew up with the field. His life was closely interwoven with the history of modern particle physics. When I first met Bob in 1947 at the American Physical Society meeting being held at the University of Chicago, I was very much impressed by his presentation of the two-meson theory that he and Bethe had just proposed. Later on, my work on white dwarfs was inspired by his pioneering papers. Nearly a decade later, his formulation of $V - A$ theory gave the important next step to the idea of parity nonconservation. When he came to New York to be the President of the City College, we quickly became good friends. Among people of high intellectual power, it was rare to find someone with Bob's warm feeling towards people and straightforward approach to science as well as life.

The whole scientific community owed much of its international unity to Bob. In 1950 he started the first Rochester meeting on High Energy Physics. This was so successful that now the major bi-annual international symposium in the field is always called the Rochester Conference, even though it has not been held there for three decades. Appropriately, just before his untimely death, the American Association for the Advancement of Science named Bob as the first recipient of a new prize to be given for contribution to international scientific cooperation.

It was an honor to know Bob. While we will always miss him, through this book he leaves us a legacy that will impact on the course of the future development of particle physics.

T. D. Lee

PREFACE

Forty years ago — in 1952 — I published my first high energy physics book on "Meson Physics" (the name 'particle physics' for our discipline was still some years off) based on a series of lectures delivered at Nevis Laboratory (Columbia University). This was five years after the two-meson theory and the Bristol discovery of $\pi \rightarrow \mu$ decays, and four years after the observation of the copious production of pions on the first synchrocyclotron ever built (Berkeley). Although the sub-discipline of meson physics had barely reached its maturity, I was moved to say in the preface of my book:

> "...the body of experimental knowledge in the field will no doubt grow rapidly in the years ahead as new meson-producing accelerators swing into action and as cosmic-ray workers take increasing advantage of the latest experimental techniques...".

Since theory was then in a state of befuddlement as to how to deal with the rapidly increasing body of experimental information concerning the Yukawa meson (pion) and the second-generation lepton (muon), it did not require an act of courage to devote the last chapter of "Meson Physics" to the newly-discovered V particles (soon to acquire the name 'strange particles') in which the concluding paragraph reads:

> "Quite a few theoretical attempts have been made to explain the longevity of the V particles — either by invoking new kinds of selection rules, requiring production in pairs or by postulating complex structures for the new unstable particles. However, none of these theories has shown itself capable of coping with the rapidly changing experi-

mental situation, and discussion of the relative merits of the various theories is postponed to a more auspicious occasion".

The next two decades (the 1950s and 1960s) saw a number of valiant attempts — in which I happily participated — to develop serious dynamical theories of the strong and weak interactions (QED was already in good shape) but progress was slow. It was not apparent in 1970 that all the essential ingredients of the standard model were in hand and no one told me — I guess no one knew — that a "heroic decade" in particle physics was ready to be launched. And so, just as the "heroic decade" with its exciting parade of experimental discoveries and theoretical triumphs began, I accepted the City College presidency as an act of public service. I took the CCNY job very seriously — trying to develop on many fronts what I called the "urban educational model" — and there was very little time for physics. Thanks to my friendly physics colleagues at City College, Bunji Sakita and Rabi Mohapatra, I received a weekly one-hour briefing (through seminar or private discussion) on the latest developments in particle physics, but this was hardly sufficient to keep abreast of the swift-moving developments in our field. In 1979, I resigned the City College presidency, eager to learn what had actually transpired during the "heroic decade" in particle physics and to join the ranks of those who wished to push beyond the standard model. But, again, I called the shots incorrectly: the resiliency of the standard model was phenomenal and soon transformed the "heroic decade" into the more deliberate "period of consolidation and speculation" in which we presently live.

There is clearly much to learn from the fluctuating fortunes of particle physics and, upon my retirement in 1987, I decided to write a book that would record my personal assessment of what had taken place in the conceptual evolution of modern particle physics during my own scientific lifetime, especially during the "heroic period". The writing of the book turned out to be a major "voyage of discovery", of brilliant theoretical insights and ingenious experiments, that completely transformed particle physics by 1980. I hope that some of the reflections on my retrospective journey will help to prepare the younger generation for the next "heroic decade" in "super-modern" particle physics.

It is always a pleasure to thank those persons who have taken the trouble to critique portions (large or small) of an evolving manuscript; such advice has been especially valuable in the present instance by eliminating a number of omissions and misinterpretations, and I acknowledge with thanks the assistance of L.N. Chang, L. Mo, R.N. Mohapatra, Y. Okamoto, P. Orland, J. Pasupathy, Riazuddin, S. Samuel and B. Sakita. Two former Virginia Tech students, Drs. Chao-Qiang Geng and Chopin Soo, have been most generous with their time over an extended period, and their assistance is gratefully acknowledged. I should like to take this opportunity to thank Professors D. Dubbers and P. Nozières for the invitation to deliver a series of lectures on the basic concepts of particle physics to a mixed physics audience at the I.L.L. (Grenoble) in 1986; the Grenoble lecture notes were expanded into this book. My secretary, Janet Manning, receives warm thanks for the out-standing job she performed in producing the electronic manuscript of the book. Finally, I must express my appreciation to my daughter-in-law Kathy and my son Steve for their assistance during the final stages of production; the completion of this book truly became a family affair with the benevolent support of World Scientific.

Robert E. Marshak

Chapter 1

GENESIS OF MODERN PARTICLE PHYSICS

§1.1. Historical introduction (prior to 1945)

We do not propose to review in this brief historical introduction — covering the period prior to 1945 (from whence we date the birth of "modern particle physics") — the multiplicity of partial insights into the nature of matter and forces advanced by the early Greeks and other curious "natural philosophers" during ancient and medieval times. We prefer to think of Isaac Newton as the father of "early" particle physics — as of three centuries ago — because he not only had much to say about one of the basic forces of nature but also because he tried to deal with the equally challenging question of the fundamental constituents of matter. With regard to the composition of matter, the empirical knowledge base at the time was severely limited and Newton's reasoning was fairly traditional and still teleological in character [1.1]:

> "All these Things being considered, it seems probable to me that God in the beginning formed Matter in solid, massy, hard, impenetrable, moveable Particles, of such Sizes and Figures, and with other Properties, and in such Proportion to Space, as most conduced to the End for which He formed them; and that these primitive Particles being Solids are incomparably harder than any porous Bodies compounded of them; even so very hard, as never to wear or break in Pieces; no ordinary Power being able to divide what God himself made in the first Creation....And therefore that Nature may be lasting, the Changes of corporeal Things are to be placed only in the various Separations and new Associations and Motions of these permanent Particles."

1

Since much more was known about the macroscopic gravity phenomenon, Newton did much better scientifically with regard to the problem of basic forces when he came up — in 1687 — with his universal law of gravitation [1.2]. Newton's solution of the gravity problem was only superceded by Einstein's theory of general relativity in 1915 [1.3] more than two centuries later. It is amusing that the first basic force identified by man is the last holdout in modern particle physics! The "theory of everything" (TOE) [1.4] — which aims to achieve a grand (or, should I say, supergrand) unification of all known interactions including gravity — uses Einstein's formulation of gravitational theory — not Newton's — as the model for TOE building. But the fact remains that the essential properties of gravitation (e.g. that all masses in the universe attract each other, thereby excluding a vector boson-mediated interaction) were already identified by Newton back in 1687 and that it is gravity (Einsteinian, to be sure) which is proving to be so intractable in efforts to quantize it and to "unify" it with the three other gauge-mediated interactions (strong, weak and electromagnetic).

Progress in our understanding of the fundamental constituents of matter and the basic forces of nature came in mutually reinforcing steps after Newton. Maxwell brought together the key results of Coulomb, Ampère and Faraday into his "unified" theory of electromagnetism [1.5] in 1864. The Maxwell equations not only placed electromagnetism on a firm footing but they were such an inspired step forward into the future that they have survived the later requirements of special relativity (Lorentz invariance) and gauge invariance. As implied by the last statement, the important role played by the Maxwell equations in the development of modern particle physics was made possible by Einstein's special theory of relativity (1905) [1.6] and the discovery of the photon (by Planck [1.7] and Einstein [1.8]). Stimulated by the nuclear model of the atom (Rutherford [1.9] and Bohr [1.10]), as well as the wave-particle duality for matter (DeBroglie [1.11]), quantum mechanics received its rigorous formulation in the mid-1920s (by Heisenberg [1.12] and Schrodinger [1.13]), followed by Dirac's relativistic theory of the electron [1.14] in 1928. The success of Dirac's equation for the electron, with its theory of "holes" and its prediction of the positron, led, with the help of second quantization, to the development of quantum field theory.

Thus, quantum field theory resulted from the merging of the two seminal concepts of special relativity and quantum mechanics and did provide a suitable language — despite the renormalization obstacles that had to be surmounted — to describe the properties of fields and the interactions

among particles. The initial application of quantum field theory to the interaction of electrons with the massless gauge quanta of the electromagnetic field (photons) created quantum electrodynamics in the early 1930s [1.15] which matured by the late 1940s [1.16] — with the help of renormalization techniques — into the only Lorentz-invariant, Abelian, gauge-invariant and renormalizable theory that we know as QED. Indeed, QED is not only a beautiful example of Einstein's famous precept that "symmetry dictates dynamics" [1.17] but it has also served as a paradigm — with certain novel and deeply significant modifications — for the gauging of the electroweak and strong interactions.

While the universal (classical) law of a second basic force in nature — electromagnetism — was being established during the mid-nineteenth century, progress was also being made in identifying the atomic constituents of matter. By the end of the 19th century, the structure of the atom itself began to be unraveled — with the measurement of the electron charge (by J.J. Thomson [1.18] in 1899) — and soon Rutherford's alpha particle scattering experiments [1.9] established the existence of the positively-charged nuclear core of the atom. The nuclear model of the atom, that was inspired by Rutherford's experiments, was not only a major bulwark of the quantum mechanics that followed, but the nucleus itself eventually became the source of major discoveries. While work in the domain of the atomic nucleus was interrupted by World War I, the pace picked up after the war and soon low energy particle accelerators (e.g. the Cockroft-Walton machine [1.19] and the cyclotron [1.20]) were being constructed to study artificially-produced nuclear transmutations. By 1927, the problem of the continuous beta ray spectrum from radioactive nuclei [1.21] also had to be confronted. The first intimations of the strong (nuclear) and weak (nuclear) interactions as basic forces of nature were beginning to manifest themselves.

The early 1930s were revolutionary years in the development of nuclear physics. By 1932, the neutronic constituent of the atomic nucleus was identified (by Chadwick [1.22]), which led quickly to spelling out the key properties of the two-nucleon interaction (by Heisenberg [1.23] and others) and then to the proposal by Yukawa [1.24] — in 1935 — of a quantum field theory of nuclear forces based on quanta of finite mass (mesons). The next three years (1932-35) of the 1930s' decade also saw the postulation of the neutrino (Pauli [1.25]) and Fermi's theory of beta decay [1.26], based on the fermionic quanta (electron and neutrino) of the weak field. Thus, by the mid-1930s, the two short-range strong (nuclear) and weak (nuclear) interactions [we have inserted the word "nuclear" after "strong" and "weak" to indicate that these two new forces were first discovered in nuclear processes but the word "nuclear" will be dropped henceforth] joined the long-range gravitational and electromagnetic interactions to complete the roster — at least until now — of the four basic forces of nature.

As far as the fundamental constituents of matter were concerned, it appeared, by 1939, that the proton (p), neutron (n), electron (e) and neutrino (ν) — plus their antiparticles — were the chosen candidates for the fermionic constituents (particles) of matter, supplemented by the photon and the hypothesized Yukawa meson as the boson-mediating particles of the electromagnetic and strong interaction respectively; there appeared to be no great need for a boson-mediating particle for the weak interaction [since Fermi's theory of beta decay — involving the direct interaction of n, p, e and ν (a local four-fermion interaction) — seemed quite adequate] and the graviton hardly entered the picture. It should be pointed out that the Lorentz structure of the Fermi beta decay interaction was poorly known and the Yukawa meson still had to be identified. Actually, particles of intermediate mass were discovered in the cosmic radiation (by Anderson and Neddermeyer [1.27]) in 1937 but they turned out to be the μ mesons — not the π mesons predicted by Yukawa — and it took another decade — extending beyond World War II — before the confusion was straightened out [1.28].

The onset of World War II brought a halt to most basic physics research activities, including the nascent sub-discipline of high energy particle physics. Surprisingly enough, two important experiments were carried out and one crucial accelerator design principle was uncovered before the end of the war. The experiments and accelerator principle were: (1) the measurement of a finite $2s_{\frac{1}{2}} \rightarrow 2p_{\frac{1}{2}}$ spectral shift in atomic hydrogen (Lamb shift [1.29]), disagreeing with Dirac's relativistic theory of the electron; (2) the Italian cosmic ray experiment at sea level [1.30], demonstrating a very different behaviour of positively and negatively charged (sea-level) mesons in interaction with light nuclei and thereby contradicting Yukawa's meson theory; and (3) the principle of phase stability in high energy particle accelerators [1.31]. As will be seen, the explanation of the Lamb shift gave rise to renormalization and QED [1.16], the Italian cosmic ray experiment gave rise to the two-meson theory [1.32] with its strongly interacting Yukawa meson (pion) and weakly interacting "second-generation" charged lepton (muon) [1.33] and, finally, the phase stability principle gave rise to the first generation of high-energy proton synchrocyclotrons and electron synchrotrons. In view of all this, it seems reasonable to consider the end of World War II, to wit 1945, as the date for the "birth" of modern particle physics.

With a "running start" at war's end, it is not surprising that the tempo of experimental discovery and theoretical understanding in particle physics accelerated rapidly after 1945, and that the next fifteen years (1945–60) witnessed a series of major experimental and theoretical developments that began to define the conceptual contours of the modern gauge theory of particle interactions; we call this fifteen year-period the "Startup Period" in particle physics. [Fifteen-year periods appear to be reasonable intervals for assessing progress in particle physics since 1945 and we label the three fifteen-year periods since 1945 successively as: the "Startup Period" (1945–60), the "Heroic Period" (1960–75), and the "Period of Consolidation and Specu-lation" (1975–90).] It took another fifteen years (1960–75) — the "Heroic Period" — to unravel the physical content and complete the mathematical formulation of the standard gauge theory of strong and electroweak interac-tions of the three generations of quarks and leptons. Finally, the last fifteen years (1975–90) — the "Period of Consolidation and Speculation" — have seen impressive precision tests of the standard model (all testimony to the re-silience of the model), further theoretical consolidation of the standard gauge theory, and a wide range of theoretical speculations (using a full panoply of topological concepts) purporting to go beyond the standard model and even to move in the direction of TOE.

We plan, in the remainder of this introductory historical chapter, to sum-marize (in§1.2–§1.2e) the early contributions to the basic concepts and prin-ciples of modern particle physics during the "Startup Period" of rapid growth in the field. A more detailed discussion of these concepts and principles will be given in Chapters 2 — 3 and throughout the book. We then proceed (in §1.3–§1.3d) — still within a historical context — to review the key steps in the formulation of the standard model of strong and electroweak inter-actions during the "Heroic Period"; this section should provide a preview of Chapters 4–7. Finally, (in §1.4–§1.4c), we offer a brief overview of the accomplishments and frustrations of the "Period of Consolidation and Spec-ulation" during the course of which we explain the rationale for the last three chapters (Chapters 8-10) of the book.

§1.2. Startup period (1945–60): early contributions to the basic concepts of modern particle physics

Particle physics experienced an explosive growth soon after the termination of World War II for at least two technological reasons: (1) the ever-expanding capability to construct particle accelerators of increasingly higher energy and intensity; and (2) the ability to build particle detectors of increasing complexity and sensitivity. With the TeV accelerators in operation now and in the planning stage, it is difficult to believe that the first pion-producing accelerator that began to operate in 1948 was the 340 MeV synchrocyclotron (at Berkeley), to be followed in the next dozen years (prior to 1960), by more than a score of proton synchrocyclotrons and electron synchrotrons throughout the world (the majority in the United States) — highlighted by the 3 GeV proton Cosmotron (at the Brookhaven National Laboratory in 1953) and the 6 GeV proton Bevatron (at Berkeley in 1956). However, it should be pointed out that, great as were the contributions of accelerator experiments to the progress of particle physics during the 1945–60 period, this period was marked as well by the pioneering contributions of cosmic ray experimentalists (chiefly to pion and strange particle physics) and by the important achievements of (non-accelerator) nuclear physics experimentalists (particularly, in weak interaction physics). Finally, the close interplay between theory and experiment — facilitated by the Shelter Island and Rochester conferences (see below) — made a vital contribution to the forward movement of particle physics during the 1950s' decade so that 1960 served — in a variety of ways to be spelled out — as the "jumping-off point" for the remarkable achievements that transpired after that watershed year.

Our review of the Startup Period traces the conceptual development of particle physics from the first Shelter Island Conference (1947) [1.34] to the Tenth Rochester Conference (1960) [1.35], highlighting those advances that contributed to the edifice now known as the standard model of strong and electroweak interactions. We present this review under the following five headings:

§1.2a. quantum electrodynamics and the gauge principle

§1.2b. pion physics and dispersion theory

§1.2c. strange particle physics and global internal symmetries in strong interactions

§1.2d. weak interactions and baryon-lepton symmetry

§1.2e. Yang-Mills fields and spontaneous symmetry breaking

§1.2a. Quantum electrodynamics and the gauge principle

The rapid maturation of the modern renormalized theory of quantum electrodynamics (QED) during the course of the three "Shelter Island" conferences (1947-49) [1.34] owed nothing to the appearance of high energy accelerators but owed much to the multi-pronged theoretical attack on understanding the measured Lamb shift in atomic hydrogen (and the observed anomalous magnetic moment of the electron [1.36]), that excited the first Shelter Island conference. Judicious exploitation of Lorentz invariance and gauge invariance of the Lagrangian for the photon-electron system permitted the unique subtraction of the finite number of ultra-violet divergences in the usually formulated quantum field theory of electromagnetism. A major key to success in the renormalization program lay in the application of the gauge principle to derive the vector "Ward-Takahashi identities" [1.37] in QED and then to show that consistent use of these identities makes it possible to subtract all the ultraviolet divergences and to secure uniquely finite answers; fortunately, vector-like QED is not plagued by anomalies (see §7.2a) (in contrast to the later chiral gauge theory of the electroweak interaction) and so this obstacle did not have to be surmounted when modern QED was created. Indeed, quite early in the "Startup Period", renormalized QED was so carefully framed that the residual finite predictions for electromagnetic processes (involving charged leptons) could, in principle, be calculated to arbitrary order in the renormalized fine structure constant; ambitious attempts in this direction have met with singular success in recent years (see §1.4a).

The importance of gauge invariance in defining the renormalized and phenomenologically correct QED led to attempts to probe more deeply into the significance of the gauge principle and, during the period under review (in 1959), Aharonov and Bohm [1.38] demonstrated — through the so-called Aharonov-Bohm effect — that the vector gauge field A_μ in electromagnetism (when A_μ is taken around a closed path) gives rise to a measurable phase shift in a two-slit electron interference experiment even when the electrons do not experience an intervening magnetic field. The Aharonov-Bohm effect is a quantum mechanical effect and is absent in classical approximation. Actually, Dirac, in his 1931 paper [1.39] (better known for its invention of the magnetic monopole), introduced the concept of "non-integrable phase" — to give unique physical meaning to quantum mechanical phase — and noted that a field of force, defined by a (vector) gauge field (i.e. gauge invariance), is required to achieve this result. Dirac's "non-integrable phase" is discussed in §4.2a, and its role in non-Abelian gauge theories is further examined in §4.3. As the example of Abelian QED illustrates, the gauging [in the case of QED, the global internal symmetry is Abelian, namely that of the (single) electric charge] converts the global internal symmetry of the quantum field theory into a fully dynamical gauge theory. The extension of the gauge principle to global non-Abelian internal symmetries, implicit in some of the prevailing ideas on strong and weak interactions towards the end of the "Startup Period", was a major step in the development of modern particle physics (see §4.1).

Another theoretical idea, put forth during the mid-1950s, that made an important contribution to the conceptual development of modern particle physics, was the idea of the renormalization group. It was realized, once renormalized QED was formulated, that despite its reliability, the perturbation-theoretic renormalization approach is clumsy to implement in higher order — even when the coupling constant is as small as in electromagnetism; for asymptotic purposes, one may take advantage of a "residual" scale invariance (see §2.3b) in QED by introducing the concept of the renormalization group (which governs the energy dependence of the renormalized coupling constant). The existence of the renormalization group for Abelian QED was first proved in 1953 [1.40] and it clarified the status of the "Landau zero" problem (see §4.5a) that arose when Landau and co-workers [1.41] summed up the leading logarithmic singularities in the perturbative expansion for QED. But it is for non-Abelian gauge theories, such as QCD, that renormalization group analysis is most useful and reveals novel and critical features such as asymptotic freedom; the renormalization group for non-Abelian (as well as Abelian) gauge groups is discussed in §5.2a.

While QED essentially achieved its definitive formulation during the "Startup Period" in particle physics, the theory of strong interactions experienced severe "growing pains" and only reached its highly successful QCD stage two decades after the renormalization group was applied to QED. The protracted evolution of strong interaction theory into QCD becomes comprehensible if one examines the contradictory trends in this field during the "Startup Period" in particle physics. We deal with this early history in the next two sections under the separate categories of "pion physics and dispersion theory" and "strange particle physics and global internal symmetries in strong interactions".

§1.2b. Pion physics and dispersion theory

Yukawa's hypothesis of a boson of intermediate mass mediating the strong short-range nuclear force was proposed in 1935 [1.24] and seemed to be spectacularly confirmed just two years later when a particle of several hundred electron masses was discovered in the cosmic radiation [1.27]; however, the confirmation of the Yukawa hypothesis was more apparent than real. If the observed cosmic-ray meson at sea level was the Yukawa meson, theory predicted a scattering cross-section of the order of 10^{-26} cm^2 per nucleon at energies of several GeV. By 1940, cosmic ray experiments at sea level, using cloud chambers and metal plates, seemed to give cross-sections lower by at least a factor of 10^2 [1.42]. This discrepancy was troublesome but the onset of World War II halted further experimentation on this problem (but not theoretical speculation [1.43]) until the war ended. Early in 1947 — four months before the first Shelter Island Conference [1.34], Conversi, Pancini and Piccioni [1.30] published a paper, in which they reported that a substantial fraction of negative "sea-level" mesons decayed in a carbon plate but were captured in an iron plate; all positively charged mesons decayed in both carbon and iron plates. According to theory [1.44], "Yukawa" mesons carrying positive charge should always decay in carbon plates (in agreement with experiment) and those carrying negative charge should never decay (in disagreement with experiment). Analysis of this experiment by Fermi, Teller and Weisskopf [1.45] led to the startling conclusion that:

> "the time of capture from the lowest orbit of carbon is not less than the time of natural decay, that is, about 10^{-6}s. This is in disagreement with the previous estimate by a factor of about 10^{12}. Changes in the spin of the meson or the interaction form may reduce this disagreement to 10^{10}...".

This tremendous discrepancy — by a factor of 10^{10-12} — between theory and experiment in the expected behaviour of a "Yukawa" meson, as deduced from the decay of sea level mesons in carbon, replaced the earlier discrepancy of a factor of 10^2 deduced from meson scattering.

The mystery of the "Yukawa" meson thereby assumed crisis proportions in early 1947 and Oppenheimer sent to each participant in the forthcoming Shelter Island conference (to be held in June) a memo entitled "The Foundations of Quantum Mechanics: Outline of Topics for Discussions"; this memo said in part [1.46]:

"There is an apparent difficulty in reconciling, on the basis of the usual quantum mechanical formalism, the high rate of production of mesons in the upper atmosphere with the small interactions which these mesons subsequently manifest in traversing matter. To date, no completely satisfactory understanding of this discrepancy exists, nor is it clear to what extent it indicates a breakdown in the customary formalism of quantum mechanics. It would appear profitable to discuss this and related questions in some detail... One [reason for these difficulties] is that in all current theory there is a formal correspondence between the creation of a particle and the absorption of an anti-particle...The question that we should attempt to answer is whether, perhaps along the lines of an S matrix formulation, this condition must be abandoned to accord with the experimental facts...".

Not surprisingly, the discussion of the Italian experiment became very animated at the conference, but there was very little inclination to support Oppenheimer's suggestion that one should surrender "microscopic reversibility" to solve the puzzle. In the midst of this discussion, the two-meson hypothesis was suggested by me [1.47], namely that two kinds of mesons exist in nature: the heavier one is produced copiously in the upper atmosphere and is responsible for nuclear forces (i.e. it is the "Yukawa" meson), and the lighter meson is a (weak) decay product of the heavy meson and the one normally observed to interact weakly with matter at sea level (i.e. it is the one observed in the Italian experiment). In the published paper [1.32], the decay lifetime from the heavy to the light meson is estimated — of the order of 10^{-8} sec — by relating the observed decay time of the light meson in carbon [1.45] to the strength of the heavy (Yukawa) meson-nucleon interaction. The authors of the two-meson theory [1.32] did not directly state that the light meson is "a second-generation" lepton; this equivalence was suggested by Pontecorvo [1.33] within the framework of a "one-meson" (not a "two-meson") theory when he observed that:

> "The probability (10^6 s^{-1}) of capture of a bound negative meson is of the order of the probability of ordinary K capture processes, when allowance is made for the difference in the disintegration energy and the volumes of the K shell and of the meson orbit. We assume that this is significant and wish to discuss the possibility of a fundamental analogy between β processes and processes of emission or absorption of charged mesons...".

The combination of the two-meson hypothesis and Pontecorvo's observation immediately led to the choice of spin 0 for the (heavy) Yukawa meson and spin $\frac{1}{2}$ for the (light) cosmic ray meson (i.e. a "second-generation" lepton).

Unbeknownst to the participants in the first Shelter Island conference and to Pontecorvo, the Bristol group was refining the nuclear emulsion technique and was soon rewarded with the discovery [1.28] of $\pi \to \mu$ decays confirming the two-meson theory. Within a year (1948), pions were produced artificially by the Berkeley 340 MeV synchrocyclotron and soon the combined efforts of the cosmic ray and accelerator experimentalists determined the pion spin, its dominant decay mode: $\pi \to \mu + \nu$ (with ν some form of neutrino initially identified with the neutrino in β decay), and several other properties that clearly distinguished the strongly interacting pion from the weakly interacting muon. In 1950, it seemed that the world of strong interactions — in which the pion appeared to occupy a pivotal position — and the world of weak interactions — into which the muon had entered as an unexpected intruder — were very distinct worlds and were only linked by the tenuous thread of a massless, chargeless fermion called the neutrino.

The above scenario was substantially modified by later developments in particle physics but, during the "Startup Period", the pion played a key role in the search for a theory of strong interactions. As a start — in the early 1950s — the rapid increase in the number of pion-producing accelerators quickly led to the determination of the basic properties of the pion: its three charge states, their masses, spins, parities and lifetimes for decay [1.48]. The two charged pions (charge conjugates of each other) and the neutral pion were found to be pseudoscalar particles with almost equal masses (both small compared to the nucleon mass); these properties qualified the three pions, as we will see later, as Nambu-Goldstone (N-G) bosons associated with the spontaneous breaking of global chiral quark flavor symmetry in QCD (see §3.4 and §5.1). The three charge states required the assignment of the $I = 1$ (**3**) representation of the isospin group $SU(2)_I$ to the pion [in contrast to the $I = 1/2$ (**2**) representation for the nucleon] and opened the door to the application of the very fruitful concept of isospin invariance [i.e. invariance under the global internal symmetry $SU(2)_I$ group] to the πN system. When a series of charged pion-proton elastic scattering experiments was performed during the early 1950s in the several hundred MeV region, it was also established [1.49] that the first excited state of N carries the quantum numbers $I = 3/2$ and $J = 3/2$ (Δ resonance) — in accord with the then existing isospin-invariant strong coupling calculations for the πN system [1.50]. While the pion-nucleon strong-coupling theory is only valid in the static approximation for the nucleon, its success in predicting the Δ resonance [furnishing four baryon states for the $J = \frac{3}{2}^+$ $SU(3)_F$ decuplet — see §3.2c] later provided motivation for the $SU(6)$ model (see §3.2d) and the recent revival of interest in the Skyrme model (see §10.5).

The very success in predicting the Δ resonance within the framework of strong coupling theory with its large pseudoscalar πN coupling constant [in contrast to the small vector ("gauge") coupling constant of electromagnetism] soon raised doubts about applying the renormalization strategy of

QED to the pion-nucleon system. Some new hope was engendered that quantum field-theoretic methods could be employed to develop a theory of strong interactions when — in 1954 — Yang and Mills [1.51] showed how the global non-Abelian $SU(2)$ internal symmetry of the pion-nucleon system could be converted into a gauged non-Abelian quantum field theory. However, the Yang-Mills proposal replaced the phenomenologically attractive pion (whose Compton wavelength is approximately equal to the range of the two-nucleon interaction) by massless gauge fields which — it was thought — would give rise to a phenomenologically incorrect infinite-range (Coulombic-type) nuclear force. Ultimately, non-Abelian Yang-Mills gauge fields became the salvation of strong interaction theory but, in 1954, particle physics was not prepared to accept the Yang-Mills version of quantum field theory. Instead — trying to cast aside quantum field theory to the maximum extent possible — particle theorists embarked with increasing vigor on extending the Heisenberg S matrix approach to strong interactions.

The disenchantment with the application of perturbative quantum field theory to pion physics in the mid-1950s was clinched after several semi-phenomenological successes in the application of dispersion relations to forward and, then, non-forward scattering of pions by nucleons [1.52]. The non-forward dispersion relations were interesting because they brought dispersion theory closer to quantum field theory by clarifying the role of three general properties of quantum field theory in momentum-space scattering calculations, namely: (1) crossing symmetry — related to charge conjugation

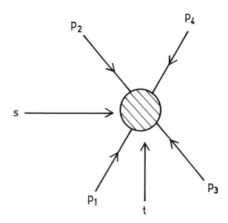

Fig. 1.1. A four-particle amplitude, using Mandelstam variables $s = (p_1 + p_2)^2$, $t = (p_2 + p_3)^2$, $u = (p_1 + p_4)^3$, $s + t + u = \sum_{i=1}^{4} m_i^2$.

invariance; (2) unitarity — guaranteeing conservation of probability; and (3) analyticity — equivalent to microcausality in space-time. Dispersion theory can yield useful predictions in strong interaction physics provided one is willing to accept phenomenological input (i.e. the known internal symmetries, values of the "renormalized" coupling constants, etc.). At the beginning, the dispersion relations were only "single-variable" dispersion relations [depending on the energy — the s channel (see Fig. 1.1)] but they were sufficient to lead to the static Chew-Low equations [1.53] which required phenomenological input of only two constants, "renormalized" coupling constant and mass), and gave excellent results for the p wave pion-nucleon scattering phase shifts. The Chew-Low theory [1.53] showed — for the static (non-relativistic) model of the pion-nucleon interaction — that: (1) the S matrix is an analytic function of the energy; (2) the "forces" can be associated with the singularities of the scattering amplitude in unphysical regions; and (3) a knowledge of the location and strength of these singularities is sufficient to determine the S matrix.

The implementation of the Chew-Low program for the low-energy pion-nucleon system was very encouraging. It was soon realized, however, that the relativistic single-variable (in s) dispersion-theoretic generalization of the Chew-Low Approach [1.54] failed to give the full analytical structure of the S matrix. This difficulty was supposed to be overcome by Mandelstam's "double-variable" dispersion representation [1.55] — in terms of the s and t channels (energy and momentum transfer) — thereby tying together the dependence on s and t into a single Lorentz-invariant analytic function. The Mandelstam representation gave impetus to the "bootstrap" idea [1.56], since it was now possible for the hadron to appear as a resonance pole in the s channel and to also represent the "exchange" hadron in the "crossed" t channel; the "exchange" hadron, in turn, could generate the interaction between the initial hadrons which could then give rise to the resonance pole to complete the "bootstrap" cycle. For example, two initial pions in the s channel could give rise to the ρ resonance pole, which could serve as the "exchange" particle in the t channel responsible for the 2π interaction generating the ρ resonance.

While the "bootstrap" idea was very attractive (as was the concomitant idea of "nuclear democracy" (that no hadron particle is more "elementary" than any other), it was still necessary to prove — within the framework of the Mandlestam representation — that analyticity, unitarity and crossing are sufficient to guarantee the "closure" of the "bootstrap" argument. Many attempts were made — invoking quantum field theory — to place the Mandelstam represntation on a firm mathematical foundation; however, "anomalous thresholds" and more complex singularities continued to surface and the extension of the Mandelstam representation to processes more complicated than simple two-particle scattering became more and more onerous. Despite these obstacles, the proponents of S matrix theory — as the entire dispersion-theoretic program was called [1.57] — began to claim that S matrix theory had superceded quantum field theory as the correct approach to the strong interactions — a claim that became moot with the advent of QCD.

In a burst of enthusiasm for S matrix theory, two of the chief protagonists — Chew and Landau — led the attack on quantum field theory but for somewhat different reasons. It is worth quoting statements by Landau and Chew made about 1960 (the end of the "Startup Period") to illustrate the strong convictions held at that time. In the mid-1950s, Landau and co-workers, who had computed the next-to-the-leading singularities in QED by summing all the Feynman graphs [1.41], were dismayed by the singularity contained in the equation:

$$\alpha_e(E) = \alpha_e(\mu)/[1 - \frac{3\alpha_e(\mu)}{16\pi^2} \ln (E/\mu)] \tag{1.1}$$

where $\alpha_e(E)$ and $\alpha_e(\mu)$ are the renormalized coupling constants at the energy E and the reference scale of renormalization μ respectively. The so-called "Landau singularity" occurs at $\ln(E/\mu) = 16\pi^2/3\alpha_e(\mu) \simeq 720$, a huge number. Landau judged this behavior to be intolerable and argued that $\alpha_e(\mu)$ must be zero and hence that $\alpha_e(E)$ must also be zero, i.e. QED is a "free field theory". This conclusion, of course, did not make any sense since electrons evidently interact with photons and Landau decided that the quantum field-theoretic mode of describing electromagnetic processes was at fault and that one must invoke "graph techniques" — equivalent to S matrix theory — to describe electromagnetic phenomena. This "conversion" in the mid-1950s led Landau to write [1.58]:

> "Since a rigorous theory which makes use of the Hamiltonian reduces the interaction to zero, the sole completely rigorous dispersion relation in this theory is 0 = 0 [this became known as the "Moscow Zero"]. By posing the problem of analytic properties of quantum field values, we actually go beyond the framework of the current theory. An assumption is thereby automatically made that there exists a non-vanishing theory in which quantum field operators and Hamiltonians are not employed, yet graph techniques are retained...".

Landau was properly concerned about the divergent behavior of QED for very large energy (or very small distance) but his outright rejection of all quantum field theories was premature [1.59]. QED (like the scalar Higgs field theory) possesses — as we will see (in §5.2a) — a stable "infrared" fixed point at the "origin" and, in the one-loop approximation, exhibits the "Landau singularity"; however, it is not clear what happens at very high energies in higher-order approximations in "pure" QED although we do know that the strong and weak interactions already affect QED in order α_e^4 (see §4.5a). What is clear is that non-Abelian QCD , the gauge theory of the strong interaction, possesses a stable "ultraviolet" fixed point at the "origin", i.e. is asymptotically free (so that the interaction definitely vanishes at very high energies) and the "Landau singularity" is gone.

As we have indicated, by the end of the Startup Period, Landau's strongly-voiced negative opinion of quantum field theory was shared by Chew who expressed his views in somewhat different terms; Chew stated his abandonment of quantum field theory as follows [1.56]:

> "So that there can be no misunderstanding, let me say at once that I believe the conventional association of fields with strongly interacting particles to be empty. I do not have firm convictions about leptons or photons, but it seems to me that no aspect of strong interactions has been clarified by the field concept. Whatever success theory has achieved in this area, is based on the unitarity of the analytically continued S matrix plus symmetry principles...The general goal then is, given the strong-interaction symmetry principles, to make a maximum number of predictions about physical singularities in terms of a minimum amount of information about unphysical singularities...We have absolutely no ideas as to the origin of the strong-interaction symmetries, but we expect that promising developments here can be incorporated directly into the S matrix without reference to the field concept...".

While Landau's and Chew's views and prophecies about the triumph of S matrix theory over quantum field theory in the strong interaction arena — expressed towards the end of the Startup Period — were found wanting by the early 1970s, it must be stated that the decade of the 1960s saw the development of a host of fresh ideas (Regge pole theory, dual resonance models, string theories, etc.) within the context of the S matrix program that have enriched modern particle theory. We will say more about the post-1960 S matrix work when we discuss the "Heroic Period" in §1.3b. Meanwhile, we note that Chew's quoted statement above is neutral as regards the applicability of quantum field theory to the weak and electromagnetic interactions and acknowledges that S matrix theory can not explain the origin of the symmetry principles governing the strong interactions. But it was precisely symmetry principles that became increasingly cogent for particle theory as experimental progress was made in strange particle physics during the 1950s. The search for strange particle resonances (both mesonic and baryonic) was actively pursued during the entire 1950s and the identification of their spins and parities, as well as their classification under new global internal symmetry groups, finally led to the important quark flavor model of hadrons. We review the historical highlights in strange particle physics prior to 1960 in the next section.

§1.2c. Strange particle physics and global internal symmetries in strong interactions

The discovery and consolidation of the empirical data concerning strange particles was an important undertaking during the 1950s and led, first, to a further demonstration of the power of symmetry principles to dictate the classification of hadronic states and properties and, ultimately, to a revival of quantum field theory as the correct approach to the strong interaction (QCD). The first consequence of the strange particle discoveries was the introduction of a new global $U(1)$ charge, the strangeness quantum number S [or, more useful from the group-theoretic point of view, the hypercharge $Y = B+S$ (B is the baryon charge)], leading to the enlargement of the isospin group $SU(2)_I$ to $SU(2)_I \times U(1)_Y$ with its associated Gell-Mann-Nakano-Nishijima (GNN) relation $Q = I_3 + Y/2$ [1.60]. The "strong" GNN relation [we call it the "strong" GNN relation in anticipation of the "weak" GNN relation to be mentioned later — see §1.2d], connecting one of the $SU(2)_I$ charges (I_3) and the U_Y charge to the electric charge Q, follows automatically when the "strong" GNN group $[SU(2)_I \times U(1)_Y]$ is enlarged to $SU(3)_F$ (as it was by 1960 — see §1.3b). From 1953 (when strangeness was first introduced) to the end of the 1950s, many strange particle experiments were performed and all of them favored the "strong" GNN group $SU(2)_I \times U(1)_Y$.

With invariance of the strong interactions under the "strong" GNN group established by many strange particle experiments, it was possible to make rapid progress in the further enlargement of the global strong flavor symmetry group. A useful step was taken by Sakata [1.61] when he used the baryon triplet (p, n, Λ) (Λ is the Λ hyperon) [i.e. a combination of the $I = 1/2$, $Y = 1$ nucleon and the $I = 0$, $Y = 0$ hyperon] to replicate the isospin and hypercharge quantum numbers of all hadronic resonances known at the time. This success and the observation of the small mass difference between n and Λ — compared to the masses of p, n and Λ — encouraged the Nagoya group [1.62] to enlarge the "strong" GNN group to the $SU(3)$ flavor group [then called the "unitary group" but which we denote by $SU(3)_F$ ($SU(3)$ flavor), consistent with later notation]. The $SU(3)_F$ group, in the Sakata-type formulation, could duplicate the isospin and hypercharge quantum numbers of all known hadronic resonances [regarded as composites of members of the baryon triplet (p, n, Λ) and their anti-particles] — as of 1960 — and could successfully predict the pseudoscalar meson octet (the η meson was unknown in 1960!).

However, the Sakata-inspired $SU(3)_F$ group soon failed with the $J = \frac{1}{2}^+$ baryon octet, where it predicted a **15** representation for the $J = \frac{1}{2}^+$ "composite" baryons instead of the observed octet. This difficulty was overcome in the early 1960s with the "eightfold way" [1.63] by recognizing that the fundamental **3** representation of $SU(3)_F$ must be assigned fractional charges (the details will be given in §3.2c). But the fact remains that the potential significance of a global hadronic $SU(3)$ flavor symmetry in the strong interactions, i.e. $SU(3)_F$, was established by 1960 and that the path was cleared for the quark flavor model of hadrons and global $SU(3)$ "color". The development of the quark flavor model approach to strong interaction physics will be discussed in the section on the "Heroic Period" in particle physics (§1.3b) but, before doing so, we continue our remarks about the "Startup Period" with a brief description of the extent to which the conceptual foundations of weak interactions were in hand by 1960 for the creation of present-day electroweak theory (QFD).

§1.2d. Weak interactions and baryon-lepton symmetry

Before the sharp distinction between the strongly interacting pion and the weakly interacting muon was confirmed in 1947, the only handle on the weak interactions was to study various features of β decay by means of: lifetimes, allowed and forbidden β spectra, spin and parity selection rules in nuclear transitions, K captures and electron-neutrino correlations — all parity-conserving experiments, since no one expected the sacrosanct principle of parity conservation to fail in those early days. Beta decay experiments are notoriously difficult to perform but, by 1947, it was generally felt that Fermi's theory of β decay [no derivatives in the four-fermion interaction, massless neutrino, implicit baryon and lepton conservation, etc.] was basically correct. It was clear that the Lorentz structure of the four-fermion β interaction could not be purely vector (as Fermi had hypothesized) because of the occurrence of Gamow-Teller [1.64] in addition to Fermi β transitions. Once $\pi \to \mu$ decays were seen in 1947, the number of weak interaction processes that could be studied increased from one (β decay) to three (muon decay and muon capture by the nucleon — in addition to β decay). The weak interactions joined the strong interactions as a major field of investigation by experimentalists and the rate of progress in all branches of particle physics greatly accelerated. Within one year of the discovery of $\pi \to \mu$ decay, the famous "triangle" of weak interactions (see Fig. 1.2) was used to test the concept of a universal Fermi interaction (UFI), i.e. the hypothesis that all weak interaction processes possess the same structure and strength.

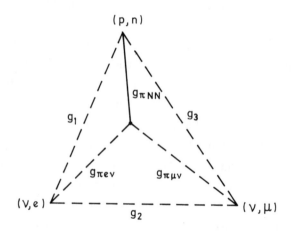

Fig. 1.2. Diagrammatic sketch showing the weak interactions (dotted lines) and the strong π interaction (solid line).

A number of authors, notably Tiomno and Wheeler [1.65], examined the relationship of the various weak processes implied by Fig. 1.2, with and without the mediation of the strong pion-nucleon interaction. Since the Lorentz structure of the weak interaction was not really known for any weak process, the analysis had to be "order of magnitude" and, within this limitation, Tiomno and Wheeler's comprehensive calculations did suggest approximate equality for the strengths of muon decay, beta decay and muon capture, provided that the Lorentz structure of the β decay interaction is not predominantly pseudoscalar. The "universal Fermi interaction" (UFI) hypothesis gave impetus to the calculation of the complete muon decay spectrum by Michel [1.66] and the "Michel parameter" became a simple test for various UFI theories. One very useful observation resulting from the early UFI discussion was that the ratio $\mathcal{R}$ of decay rates of the pseudoscalar π, i.e. $\mathcal{R} = \Gamma(\pi \to e\nu)/\Gamma(\pi \to \mu\nu)$, is independent of the strong pion-nucleon interaction [1.67] and that, if one assumes electron-muon universality, $\mathcal{R}$ depends only on the Lorentz structure of the weak interaction in the following fashion: $\mathcal{R} = 1.2 \times 10^{-4}$ for the axial vector (A), $\mathcal{R} = 5.4$ for the pseudoscalar (P) and $\mathcal{R} = 0$ for the scalar (S), vector (V) or tensor (T) form of the weak interaction. Hence, the measurement of $\mathcal{R}$ could — and later did — play a decisive role in pinning down the Lorentz structure of UFI. However, experimentalists were not prepared to decide the UFI question in the late 1940s and it was not until 1957 — after parity violation was discovered in both β decay and muon decay (in 1956–57 — see below) — that the universal V-A Lorentz structure was established for all weak processes and UFI received definite confirmation.

Two other papers appeared shortly before the watershed (parity-violation) year of 1956 that should be mentioned: the first, a crucial new measurement (unfortunately mistaken) of the electron-neutrino angular correlation coefficient λ in the pure Gamow-Teller transition $\mathrm{He}^6 \to \mathrm{Li}^6 + e^- + \bar{\nu}_e$, which gave $\lambda = 0.33 \pm 0.08$ [1.68] (theory predicts $\lambda = \frac{1}{3}$ for a T interaction and $\lambda = -\frac{1}{3}$ for an A interaction), thereby giving strong preference to the T interaction as the Gamow-Teller part of the total β interaction. The other paper, a theoretical one, by Gershtein and Zeldovich [1.69], noted a peculiar property of the V interaction — i.e. that the strength of the vector interaction at low energies is not affected by the strong interaction — and dismissed the idea; to quote the authors [1.69]:

"It is of no practical significance but only of theoretical interest that, in the case of the vector interaction type V, we should expect the equality [$g_F(V)$ and $g'_F(V)$ are the renormalized and unrenormalized coupling constants respectively]:

$$g_{F(V)} \equiv g'_{F(V)}$$

to any order of the meson-nucleon coupling constant, taking nucleon recoil into account and allowing also for interaction of the nucleon with the electromagnetic field, etc. This result might be foreseen by analogy with Ward's identity for the interaction of a charged particle with the electromagnetic field; in this case, virtual processes involving particles (self-energy and vertex parts) do not lead to charge renormalization of the particle...".

In their paper, Gershtein and Zeldovich rejected their very original idea of a conserved vector current (CVC) in weak interactions as being of "no practical significance" because they accepted the conclusion of an earlier experimental paper [1.70] (reporting on a measurement of λ in Ne^{19} — a mixed Fermi-Gamow-Teller transition) — that the β interaction is a combination of S and T (T being taken from the He^6 experiment!). We will come back to CVC within the context of the universal V-A theory.

Without question, the big breakthrough in weak interactions came in 1956 when careful high energy accelerator experiments established the full magnitude of the $\theta - \tau$ dilemma in strange particle physics (i.e. the existence of $\theta \to 2\pi$ and $\tau \to 3\pi$ decay modes with equal θ and τ masses and lifetimes [1.71]) and led — thanks to Lee and Yang's detailed analysis of the consequences of parity violation in weak interactions [1.72] — to the discovery of maximal parity violation in beta decay [1.73] and muon decay [1.74]. The two-component (Weyl) equation for the massless neutrino was vindicated [1.75] and, within one year, the universal V-A theory of weak interactions with a lefthanded (massless) neutrino was announced [1.76]. The universal V-A theory was supposed to be an "effective" charged current-current (four-fermion) theory in the low energy region for all weak interactions with: (1) the purely leptonic weak processes resulting from the product of two charged lepton currents; (2) the semi-leptonic weak processes resulting from the product of a charged lepton current and a hypercharge-conserving ($\Delta Y = 0$) plus a hypercharge-violating ($\Delta Y = 1$) charged baryon current; and (3) the non-leptonic weak processes resulting from the product of two charged baryon currents. The charged leptons were treated as structureless (point-like) particles in the V-A current whereas the finite structure of the baryons in the V-A current was represented by form factors (later, the baryon currents were replaced by quark currents — with the quarks treated on a par with the point-like leptons).

When the universal V-A theory was put forth in 1957 [1.76], it could explain almost all parity-conserving and parity-violating experiments in β decay, muon decay, and even the hypercharge-changing weak decay (both semi-leptonic and non-leptonic). There were, however, several experiments that contradicted the universal V-A theory; the two published experiments in disagreement with V-A (there were two unpublished experiments — see the first of ref. [1.76]) were: the wrong sign of the electron-neutrino angular correlation coefficient in the Gamow-Teller transition: He^6 $\to$Li^6 $+ e^- + \bar{\nu}_e$ [1.68] (supporting T — not A — for the Gamow-Teller part of the β interaction) and too low a value [1.77] for $\mathcal{R}$ corresponding to the A interaction (see above). Here is where the UFI hypothesis came to the rescue in strongly favoring the V-A Lorentz structure of weak currents: within the framework of UFI, only a V-A form for all charged lepton and baryon currents — with left-handed neutrinos — could explain all parity-conserving and parity-violating experiments (including muon decay [1.74]), provided the four contradictory experiments mentioned above were mistaken. All four experiments disagreeing with the universal V-A theory were redone within the next two years (1957–59) [1.78] and the new results were in complete agreement with theory. In addition, a direct measurement of the helicity of the neutrino — by means of an ingenious experiment that measured the helicity of the outgoing γ ray following K capture by Eu^{151} [1.79] — established the lefthandedness of the neutrino. As a consequence, the universal V-A charged current theory became the "standard" low-energy theory of all weak interactions until the "true"' standard model came along!

The overwhelming phenomenological support for the universal V-A theory drew attention to the original formulation of the V-A theory [1.80], which exploited the principle of "chirality invariance". The principle of "chirality (γ_5) invariance" was modeled on the two-component theory of the neutrino [1.75], which affirms that a massless neutrino satisfies the γ_5-invariant Weyl equation. This "neutrino paradigm" — so to speak — was generalized to massive fermions in the weak interaction Lagrangian [1.80], with the new "chirality invariance" principle stating that all baryon and lepton weak charged currents should be invariant under the separate replacement of each spinor wavefunction ψ_i by its "chirality transform" $\gamma_5\psi_i$ (i.e. $\psi_i \to \gamma_5\psi_i$). More precisely, "chirality invariance" requires that, if one writes $\bar{\psi}_i\mathcal{O}\psi_j$ ($i \neq j$) for the general Lorentz structure of a charged (two-flavor) current [where $\mathcal{O}$ represents the five possible relativistically covariant combinations of the Dirac γ_μ operators (choosing γ_0 to be Hermitian and γ_i anti-Hermitian so that $\gamma_5 = -i\gamma_0\gamma_1\gamma_2\gamma_3$): $1(S), \gamma_5(PS), \gamma_\mu(V), \gamma_5\gamma_\mu(A), \frac{i}{2}[\gamma_\mu, \gamma_\nu](T)$, one is compelled to choose $\mathcal{O} = \gamma_\mu(1 - \gamma_5)$ for the Lorentz structure of the charged fermion (baryon or lepton) current, i.e. only lefthanded chiral fermions are involved in charged currents. What is equally significant is that the same argument of "chirality invariance", applied to a general (one-flavor) neutral current (weak or electromagnetic), does not require chiral fermions and only mandates a linear combination of the V and A currents, i.e. $(V + \alpha A)$ (α a constant); this result reflects the fact that only one fermion flavor is involved in a neutral current but two flavors are needed to describe a charged current. Hence, the principle of "chirality invariance" (for neutral currents) is consistent with the parity-conserving (V) neutral electromagnetic current as well as with the form of the neutral current in the standard electroweak theory (see §6.2).

Because of the crucial role played by chiral fermions in the standard electroweak theory, it is worth quoting from the original formulation of the universal V-A theory [1.80] [$(1 + \gamma_5)$ implies negative helicity in [1.80]]:

> "...One can rewrite the (V-A) interaction of the four [Dirac] fields A, B, C, D, in the form:
>
> $$g \bar{A}' \gamma_\mu B' \bar{C}' \gamma_\mu D'$$
>
> where A', B', C', D' are the "two-component" fields: $A' = (1/\sqrt{2})$ $(1+\gamma_5)A$, $\bar{A}' = (1/\sqrt{2})\bar{A}(1-\gamma_5)$, etc. Now the "two-component" field $(1/\sqrt{2})(1\pm\gamma_5)A$ is an eigenstate of the chirality operation with eigenvalues ± 1. Thus, the universal Fermi interaction, while not preserving parity, preserves chirality and the maximal violation of parity is brought about by the requirement of "chirality invariance". This is an elegant formal principle, which can now replace the Lee-Yang requirement of a two-component neutrino field coupling...Thus our scheme of Fermi interactions is such that, if one switches off all mesonic interactions, the gauge-invariant electromagnetic interactions (with Pauli couplings omitted) and Fermi couplings retain chirality as a good quantum number...".

It is the sharing of "chirality invariance" by the chiral (parity-violating) $SU(2)_L$ weak group and the non-chiral (parity-conserving) electromagnetic group $U(1)_{\rm EM}$, that justifies their combination into the standard chiral $SU(2)_L \times U(1)_{Y_W}$ electroweak group (Y_W is the "weak" chiral hypercharge); and it is the spontaneous breaking of chiral $SU(2)_L \times U(1)_{Y_W}$ into non-chiral $U(1)_{\rm EM}$ that leads to the "weak" GNN relation: $Q = I_{3L} + Y_W/2$, where Q is the non-chiral electric charge and I_{3L} and Y_W are respectively the third component of the "weak" isospin and "weak" hypercharge. The "weak" GNN relation (see §1.3c) is not to be confused with the "strong" GNN relation (see §1.2c): while Q is the same non-chiral electric charge in both relations, chiral I_{3L} is the third component of the "weak" isospin I_L (assigned to chiral quarks and leptons in the electroweak interaction) in contrast to the non-chiral third component I_3 of "strong" isospin I (assigned only to non-chiral quarks in the strong interaction); furthermore, Y_W is the "weak" hypercharge (assigned to chiral quarks and leptons in the electroweak interaction) in contrast to the "strong" hypercharge Y (assigned only to non-chiral quarks in the strong interaction). As we will see, the global "strong" GNN group is subsumed under the quark flavor group $SU(3)_F$ which is never gauged, whereas it is essentially the gauging of the global "weak" GNN group that produces the standard electroweak gauge group $SU(2)_L \times U(1)_Y$ (Y_W is replaced by Y).

But we are moving ahead of the transition year 1960 and must return briefly to the last couple of years at the end of the "Startup Period". By 1959, the V-A interaction — still without gauging — became the accepted theoretical framework for understanding all low-energy weak processes, and it was possible to turn to the resolution of more subtle questions in weak interaction physics, such as the relative strengths of $\Delta Y = 1$ and $\Delta Y = 0$ weak processes as well as the failure to observe the simplest of all leptonic weak decay processes $\mu \to e\gamma$. The first question — that of the relative strength of the $\Delta Y = 1$ and $\Delta Y = 0$ weak processes — did not really become acute until after 1960 (when reasonable "$\Delta Y = 1$" measurements were available) and was solved by the introduction of the Cabibbo angle [1.81] in 1963 (see §1.3c) during the "Heroic Period". The second question with regard to the absence of $\mu \to e\gamma$ decay came to a head before 1960, led to the two-neutrino hypothesis and the baryon-lepton symmetry principle for weak interactions, and is properly discussed within the context of the "Startup Period".

While the V-A theory only claimed to be an effective low energy theory, the V-A structure of the weak charged current-current theory immediately suggested the mediation of a charged intermediate vector boson (IVB) — with zero baryon charge and zero lepton charge — in the correct theory. However, it was precisely within the context of an IVB-mediated theory of weak interactions that the absence of the $\mu \to e\gamma$ decay process became a deep puzzle; this process appeared to satisfy conservation of lepton charge (and all other known conservation laws) and a straightforward calculation gave a much larger branching ratio than was found experimentally. The solution to the dilemma of $\mu \not\to e\gamma$ receiving the widest currency at the Ninth Rochester Conference in Kiev in 1959 was the idea that muon and electron neutrinos are distinct particles, i.e. $\nu_\mu \neq \nu_e$. A typical argument went as follows [1.82]:

> "The intermediate boson in weak interactions would be a nice thing to have, in order to explain that the weak interaction current is a charge exchange one...the negative experiment on the $\mu \to e + \gamma$ process is difficult to reconcile with the existence of the IBV meson but this difficulty is present only if there exists one neutrino. If there

are two neutrinos...there are no arguments against the existence of [IVB] mesons..."

More explicitly, the argument was that without a charged IVB, diagrams (a) of Fig. 1.3 give a null result; however, with a charged IVB and only one neutrino, the photon can also be emitted by a charged IVB [see diagram (b) of Fig. 1.3] and gives much too large a contribution. If $\nu_\mu \neq \nu_e$, diagram (b) is absent and $\mu \to e\gamma$ decay is forbidden. Hence, a simple solution to the dilemma is to assume $\nu_\mu \neq \nu_e$. As we know, ν_μ was found to be distinct from ν_e in 1962 [1.83] and the solution of the $\mu \neq e\gamma$ puzzle, in turn, raised interesting questions in neutrino physics (neutrino masses, neutrino oscillations, etc. — to be discussed in §6.4b).

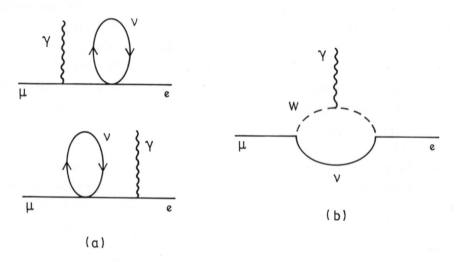

Fig. 1.3. Contrasting diagrams for $\mu \to e\gamma$ for one neutrino, without IVB (a) and with IVB (b).

The chief reason for mentioning the $\mu \to e\gamma$ puzzle as of 1959 — when the quark model was still unborn — is that the $\nu_\mu \neq \nu_e$ explanation provided the first test of a proposed baryon-lepton symmetry principle in weak interactions at that time [1.84]. Later, this baryon-lepton symmetry principle was translated into quark-lepton symmetry, which predicted the existence of the charm (c) quark (see below), to complete the second generation of quark doublets (s,c) and lepton doublets (μ^-, ν_μ) [once it was established that ν_μ is distinct from ν_e]. In the absence of quarks, the 1959 baryon-lepton symmetry principle was based on the Sakata model [1.61] and argued [1.84] that, if one postulates the invariance of all weak interactions under the simultaneous permutation:

$$p \leftrightarrow \nu, n \leftrightarrow e^-, \Lambda \leftrightarrow \mu^- \qquad (1.2)$$

then all observed weak processes "transform" into observed weak processes and, conversely. Furthermore, it was suggested as early as 1959 that one could introduce the concepts of "weak" isospin I_W and "weak" hypercharge Y_W [$\equiv B - L + S_W$ — where B and L are the usual baryon and lepton charges respectively and S_W is "weak" strangeness] so that the "strong" GNN relation $Q = I_3 + \frac{Y}{2}$ [see §1.2c] is converted into the "weak" GNN relation, thus:

$$Q = I_{3W} + Y_W/2; \qquad Y_W = B - L + S_W \qquad (1.3)$$

Equation (1.3) tells us that $S_W = 0$ for a "weak" isodoublet (of lefthanded baryons or leptons) so that $Y_W \equiv B - L$ for $I_L = \frac{1}{2}$, whereas $S_W = \pm 1$ for a "weak" isosinglet (of righthanded baryons or leptons) so that $Y_W = B - L \pm 1$ for $I_L = 0$ [where the plus sign is taken for "weak" isospin "up" righthanded quarks and leptons and the minus sign is used for "weak" isospin "down" righthanded quarks and leptons — with the proviso that there is no righthanded neutrino (see §6.2)].

The comparison of "weak" hypercharge $Y_W = B - L + S_W$ with "strong" hypercharge $Y = B + S$ is instructive: (1) B of the "strong" hypercharge is replaced by $B - L$ in "weak" hypercharge; and (2) S in "strong" hypercharge is replaced by S_W in "weak" hypercharge. The first point of comparison helps to explain why global B conservation in strong interactions gets translated into global $B - L$ conservation in weak interactions for the standard model of the electroweak interaction [e.g. we will see in §10.3a that while B charge conservation and L charge conservation are separately broken by the $SU(2)_L$ instantons, global $B - L$ charge conservation still holds]. The second point of comparison is equally interesting: it makes full use of the original concept of "displaced isomultiplets" [1.60] underlying the "strong" GNN relation where now chiral "weak" strangeness S_W, given by Eq. (1.3), vanishes for a (lefthanded) chiral weak isodoublet and is non-vanishing and $= \pm 1$, depending on whether the righthanded chiral (quark or lepton) isosinglet has "up" or "down" "weak" isospin.

Several other remarks should be made about the baryon-lepton symmetry implied by Eqs. (1.2) and (1.3): (1) the baryon-lepton permutation symmetry (1.2) immediately translates into an equivalent quark-lepton permutation symmetry: $u \to \nu_e, d \to e^-, s \to \mu^-$ because $p = (ud)u, n = (ud)d$, and $\Lambda = (ud)s$ in the simple "valence" quark picture of p, n and Λ; (2) since each of the three generations of quarks and leptons consists of a lefthanded quark "weak" isodoublet, two righthanded quark "weak" isosinglets plus a lefthanded lepton "weak" isodoublet and one righthanded charged lepton "weak" isosinglet, Eq. (1.3) predicts the phenomenologically correct values of the "weak" hypercharges for the quarks and leptons of each generation; and, finally, (3) when the two-neutrino hypothesis ($\nu_\mu \not\rightarrow \nu_e$) was put forward in 1959 [1.82] to explain the absence of $\mu \rightarrow e\gamma$ decay, the baryon-lepton symmetry principle was invoked [1.84] to point out that a second neutrino would imply the existence of a second type of proton p' (or, equivalently, a second type of u quark), for which there was then no experimental evidence. The situation changed dramatically when the second neutrino ν_μ was discovered in 1962 [1.83]; the 1959 baryon-lepton symmetry was then revived [1.85] and, as we will see in §1.3a, its translation into the quark-lepton symmetry principle in weak interactions, together with the GIM mechanism [1.86], led to the clinching argument for the c quark as the counterpart to ν_μ. We have thus come full circle: the initially-unexplained absence of $\mu \rightarrow e\gamma$ led to the anticipated existence of a second neutrino which, in turn, led to the expectation for a second quark [with the u quark quantum numbers (under the standard group)], namely, the c quark. The completion of the second quark-lepton generation of fermions prepared the way for the third quark-lepton generation and the recognition of the importance of quark-lepton symmetry in weak interactions.

§1.2e. Yang-Mills fields and spontaneous symmetry breaking

In the previous four sections (§1.2a–§1.2d), an attempt was made to identify the early contributions (both experimental and theoretical) — during the fifteen years following the end of World War II (i.e. the "Startup Period") — that impacted significantly on the new discipline of particle physics (whose "birthday" we set at ca. 1945). In §1.2a, we noted the rapid development of our understanding of quantum electrodynamics (QED) before QED became a partner in the electroweak interaction. In §1.2b–§1.2c, we traced the key formative ideas bearing on the theory of the strong interaction and tried to untangle the global internal (flavor) symmetry group that led later to the non-Abelian $SU(3)$ color gauge theory of the strong interaction (QCD). Finally, in §1.2d, we pointed out the good progress made — during the "Startup Period" — in dealing with the consequences of the V-A theory of the weak interaction at relatively low energies before it was joined with the electromagnetic interaction to become the relativistic quantum field theory known as quantum flavordynamics (QFD). The V-A theory — with its two-flavor lefthanded (chiral) currents — was a good start for non-Abelian QFD but the potential of using a spontaneously broken gauge theory for the electroweak interaction was still not appreciated.

Actually, by 1960, the seminal ideas of non-Abelian gauge symmetry and spontaneous symmetry breaking (SSB) of a global relativistic quantum field theory were in the "public domain" although the unifying idea of a spontaneously broken non-Abelian gauge symmetry — via the "Higgs" mechanism [1.87] — was not proposed until 1964. Nevertheless, we believe that our review of the "Startup Period" in particle physics would be greatly diminished if special mention is not made of the Yang-Mills work on non-Abelian $SU(2)$ gauge theory in 1954 and of the 1959–1960 ideas of Heisenberg and Nambu regarding SSB. We comment briefly on these two sets of contributions — partly to understand their early neglect — and return to discuss them in greater detail in Chapter 4 because they are, after all, essential ingredients of the "new look" in particle physics.

The 1954 paper by Yang and Mills [1.51] on the structure of the Lagrangian for non-Abelian $SU(2)$ (isospin) gauge fields met with initial hostility from Pauli [1.17], who worried about the equivocal status of the mass of the Yang-Mills gauge field. When it became clear that gauge invariance in the case of Yang-Mills fields requires massless gauge fields — just as in the case for the Abelian gauge field describing the photon in electromagnetism — it became dubious whether a non-Abelian gauge theory could be constructed for the short-range strong interaction. As we know, the difficulty of developing a successful theory of the strong interaction on the basis of massless Yang-Mills "color" fields (i.e. gluons) was finally overcome in the early 1970s and the great achievement of QCD is "previewed" in §1.3b and discussed at length in Chapter 5.

The breakthrough in the use of Yang-Mills gauge fields to formulate a theory of the weak interaction (strictly speaking, the electroweak interaction [1.88]) actually came before a similar approach led to QCD [1.89] because the V-A (two-flavor) charged current theory could only be mediated by a charged vector boson, thereby identifying a prime candidate for the Yang-Mills field in weak interactions. At first, the fact that the weak interaction is of even shorter range than the strong interaction militated against the idea of a viable Yang-Mills theory of the weak interaction. Moreover, the concept of "weak" isospin had not yet been invented (see §1.2d) so that it was unclear what global degrees of internal symmetry would have to be gauged in order to construct a non-Abelian gauge theory of the weak interaction. These obstacles were removed when the distinction between the SSB of gauge symmetries and that of global symmetries was understood in the mid-1960s and the path to QFD was opened when it was realized that the "Higgs" mechanism [1.87] generated the finite masses of the electroweak gauge fields that were needed (instead of the massless Nambu-Goldstone bosons of spontaneously broken global symmetries). A "preview" of the spontaneously broken non-Abelian gauge theory of the electroweak interaction (QFD) is given in §1.3c and the full story is recounted in Chapter 6.

Having emphasized the important role played by the concept of spontaneous symmetry breaking in the development of the standard electroweak theory, we can now comment briefly on the 1959–60 contributions of Nambu and Heisenberg to the evolution of this concept. In 1957, a major breakthrough occurred in condensed matter physics with the announcement of the Bardeen-Cooper-Schrieffer (BCS) theory of (low temperature) superconductivity [1.90]. Within a few years, the far-reaching implications of the BCS theory began to be appreciated, including the role of the "Cooper pairs" (of electrons) in providing a "quasi-Higgs" mechanism to spontaneously break the $U(1)$ electromagnetic gauge symmetry inside a superconductor. The BCS theory provided the long-awaited "microscopic" underpinning of the "macroscopic" Landau-Ginzburg equation [1.91] in superconductivity. Knowledge of this important BCS development and the close connection between "chirality invariance" and the V-A theory, inspired Nambu and Jona-Lasinio [1.92] to suggest that fermion mass generation in a chirally-invariant massless (nonlinear) four-fermion-interacting theory proceeds via an SSB mechanism analogous to that triggered by "Cooper pairs" in generating the "energy gap" in superconductivity [1.90]. At the 1960 Rochester Conference, Nambu [1.93] made a prescient statement:

> "It is an old and attractive idea that the mass of a particle is a self-energy due to interaction. According to the present analogy, it will come about because of some attractive correlation between massless bare particles, and will be determined in a self-consistent way rather than by simple perturbation. Since a free massless fermion conserves chirality, let us further assume that the interaction also preserves chirality invariance, just as the electron-phonon system preserves gauge invariance [in superconductivity]. Then if an observed fermion (quasi-particle) can have a finite mass, there should also exist collective excitations of fermion pairs. Such excitations will behave like bosons, of zero fermion number, so that they may be called

mesons. They will play the role of preserving the overall conservation of chirality, and from this we will be able to infer that they are
pseudoscalar mesons, like the pions found in nature...".

This 1960 statement by Nambu is noteworthy not only because he anticipated the Goldstone theorem (this is why Goldstone bosons have been renamed Nambu-Goldstone bosons or, for short, N-G bosons) but also because he made the first serious attempt — using the analogy with the BCS
theory drawn from condensed matter physics — to work out a dynamical
theory of fermion mass generation. In the process of doing this (with Jona-
Lasinio), Nambu discovered that the finite-mass quasi-particle fermion excitations are accompanied by finite-mass scalar "fermion-pair" excitations
(with a mass approximately twice the mass of the fermion), in addition to
the massless pseudoscalar "fermion-pair" (collective) excitations. Nambu's
dynamical chirality-symmetry-breaking approach to fermion mass generation not only uncovered the massless pseudoscalar bosons and massive scalar
excitations — with their many applications [e.g. nuclear physics [1.94]] —
but may shed light on the nature of the "Higgs" mechanism in electroweak
physics and the possible connection of the Higgs mass to the t quark mass
(see §6.4c).

Heisenberg [1.95] also invoked an analogy with condensed matter
physics, namely ferromagnetism, to argue for a possible role of the SSB
of global symmetries in particle physics. His argument may be summarized
with the statement that the SSB of a global symmetry occurs when the global
symmetry holds for the Lagrangian but not for the physical vacuum, which
must be degenerate. Heisenberg started with a massless fermion Lagrangian
of a single flavor (which he termed "Urmaterie"), whose mass is also supposed
to be dynamically generated by a (local) non-linear four-fermion interaction,
as in Nambu's case. However, Heisenberg's choice of Lorentz structure for
the "four-fermion interaction" was different from that of Nambu: he chose
an axial vector Lorentz structure to incorporate the largest "hidden" symmetry in his "one-flavor" Lagrangian, namely $SU(2)_I \times U(1)_B$, in order to
account for the isodoublet (two-flavor) nucleon; moreover, Heisenberg used
the "vertex" rather than the "bubble" diagram propagator for the fermion
mass. Unlike Nambu, Heisenberg was not willing to accept a phenomenological cutoff for his truncated non-perturbative expression for the nucleon
mass, and insisted on a relativistic cutoff that led to indefinite metric and
negative probability problems. Moreover, his attempt to account for strange
baryons led him to arbitrarily assign isospin to the degenerate vacuum in his

model and to use the scaling operation to define parity. Overall, it is fair
to say that, while Heisenberg's program to construct a full-blown theory of
elementary particles on the basis of a one-flavor quantum field was unsuccessful, his ideas about the SSB of a global symmetry, its connection with a
degenerate vacuum and the role of scaling in a theory with massless fermions
[1.95]) were very useful for later developments in particle theory.

§1.3. Heroic Period (1960–75): formulation of the standard model of strong and electroweak interactions

In §1.2–§1.2e, a brief review was given of a broad range of early contributions to the basic concepts of modern particle physics achieved during
the "Startup Period" (1945–60). While the increasingly higher energies of
particle accelerators made possible many of these early experimental contributions, cosmic ray experiments and relatively low energy but carefully
— targeted experiments (e.g. in beta decay) made important contributions
during this period as well. And, towards the end of the 1950s, the experimental contributions were matched by equally important theoretical contributions that, together, laid the conceptual foundations for the present-day
standard model. In this section, we recount the advances — during the
1960–75 "Heroic Period" in particle physics — that led to the definitive
formulation of the present-day "standard" non-Abelian gauge theory of the
strong and electroweak interactions. This remarkable progress is recorded
under four headings: (1) from quarks as "mathematical fictions" to quarks
as "physical entities"; (2) from global $SU(3)$ quark flavor to gauged $SU(3)$
color (QCD); (3) from the global V-A charged baryon-lepton weak current
theory to the gauged chiral quark-lepton flavor theory of the electroweak interaction (QFD); and, finally (4) summary of the properties of the standard
model of particle interactions.

§1.3a. From quarks as "mathematical fictions" to quarks as "physical entities"

We have already pointed out that, by 1959, a Japanese group had proposed an $SU(3)$ baryon "flavor" model of the observed hadronic states [1.62],
based on the (Sakata) baryon triplet (p, n, Λ) [1.61]; the Sakata idea was that
the lowest mass states of the observed mesonic and baryonic hadrons (other
than p, n, and Λ) would consist, respectively, of $B\bar{B}$ and $BB\bar{B}$ composites
(B is p, n or Λ) in zero orbital angular momentum states. This $SU(3)$ baryon
flavor model correctly predicted the $J = 0^-$ (pseudoscalar) meson octet but
could not account for the $J = \frac{1}{2}^+$ baryon octet (whose members include $p, n,$

and Λ). This defect of the model was soon traced to the assignment of integral B, Y and Q charges to the members of the fundamental three-fermion representation (p, n, Λ) of the $SU(3)$ flavor group instead of the fractional charges (in multiples of $\frac{1}{3}$) of B, Y and Q dictated by the mathematical requirements of an $SU(3)$ group. The mathematical point is that the commuting observables I_3 (the third component of isospin) and Y (hypercharge) of the rank 2 $SU(3)$ group are generators of this group and the matrices for these generators, corresponding to irreducible $SU(3)$ representations, must be traceless. For the fundamental (p, n, Λ) triplet, there is no problem on this score with I_3, but Tr Y (summation over the p, n, Λ triplet) does not vanish. On the other hand, still requiring that the observed baryons are three-fermion composites of the fundamental $SU(3)$ triplet and that this triplet consist of an isodoublet with zero strangeness and an isosinglet with strangeness -1 [analogous to the Sakata (p, n, Λ) triplet] — but assigning $B = \frac{1}{3}$ to each member of the $SU(3)$ triplet — the fractional values of Y and Q (derived from the "strong" GNN relation) of the $SU(3)$ triplet are those given in Table 1.1. The fractional charges (both Y and Q) in Table 1.1 define the new $J = \frac{1}{2}^+$ fermion triplet of the $SU(3)$ flavor group and were soon to characterize the u, d and s quarks belonging to the fundamental **3** representation of the global $SU(3)$ quark flavor group $[SU(3)_F]$ of the standard model (see §3.2c).

Table 1.1: Quantum numbers of the three light quarks.

quark	B	I	I_3	Y	Q
u	$\frac{1}{3}$	$\frac{1}{2}$	$+\frac{1}{2}$	$\frac{1}{3}$	$\frac{2}{3}$
d	$\frac{1}{3}$	$\frac{1}{2}$	$-\frac{1}{2}$	$\frac{1}{3}$	$-\frac{1}{3}$
s	$\frac{1}{3}$	0	0	$-\frac{2}{3}$	$-\frac{1}{3}$

Historically, the members of the fundamental fermion triplet of $SU(3)_F$ were not immediately identified with the three lightest (u, d, s) quarks; the quark hypothesis was not proposed until 1963 [1.96] and, during the early 1960s, the members of the fundamental fermion triplet of $SU(3)_F$ — denoted by (q_1, q_2, q_3) — were treated as "mathematical fictions", whose group-theoretic properties compelled the $q\bar{q}$ and qqq composite hadrons to follow the "eightfold way" [1.63]. Indeed, it was argued in 1961 [1.63] that the type of baryon-lepton symmetry noted in the 1959 baryon-lepton symmetry paper

[1.84], made provision for the type of fundamental massless fermion triplet needed to yield the mathematical underpinning of an $SU(3)$ flavor group describing the observed "eightfold" groupings of mesonic and baryonic composites. More importantly, the "eightfold way" predicted deviations in mass of the members of an $SU(3)$ octet [from a common $SU(3)$ octet mass], in accordance with a mathematically well-defined symmetry-breaking assumption, [i.e. that the symmetry-breaking is proportional to the Y generator], leading to the Gell-Mann-Okubo mass formula [1.97].

The success of the "eightfold way" and of the Gell-Mann-Okubo mass formula — not only for meson and baryon octets but also for a baryon decuplet, having the $\Delta(\frac{3}{2}, \frac{3}{2})$ resonances as four of its members — began to erode the notion that the fundamental fermion triplet of $SU(3)_F$ comprises a set of abstract "mathematical fictions" and to favor the view that these "carriers" of $SU(3)$ flavor symmetry are "physical entities" with measurable properties, i.e. quarks. From this altered point of view, it was legitimate to search for particles with fractional charges (in particular, fractional electric charges) but no solid evidence was found for these mysterious building blocks of hadrons in the "free state" [1.98]. The futile search for "free" fractionally charged quarks was not a serious impediment to the development of the quark flavor model of hadrons because the answers to key questions concerning hadronic quark composites did not depend initially on whether the quarks are "mathematical fictions" or "physical entities".

Even if the quarks are "mathematical fictions", the group-theoretic properties of suitable quark composites can have important experimental consequences with regard to the classification of particle masses and other hadronic properties. Thus, accepting that the $SU(3)_F$ properties of mesons and baryons are determined by respective combinations $q\bar{q}$ and qqq, one can predict the irreducible $SU(3)_F$ representations of these composite quark states and, by simple group theory, one can then predict the $SU(2)_I$ and $U(1)_Y$ quantum numbers corresponding to each irreducible representation of $SU(3)_F$. One finds: (1) since $q\bar{q}$ constitute the meson states, one can say that $\mathbf{3} \times \bar{\mathbf{3}} = \mathbf{1} + \mathbf{8}$ and hence that mesons should exist in singlet and octet representations of $SU(3)$ [the case of maximal mixing between the singlet and octet vector states — leading to a "pseudo-nonet" of vector mesons is discussed in §3.2c]; and (2) since baryons are qqq composites, one has $\mathbf{3} \times \mathbf{3} \times \mathbf{3} = \mathbf{1} + \mathbf{8} + \mathbf{8} + \mathbf{10}$ so that baryons can exist in $SU(3)_F$ decuplet representations — in addition to singlet and (two kinds of) octet representations. The decompositions of the $\mathbf{8}$ and $\mathbf{10}$ $SU(3)_F$ representations under

$SU(2)_I \times U(1)_Y$ — holding for both mesons and baryons — are purely mathematical statements, as follows:

$$8 \rightarrow (2,1) + (1,0) + (3,0) + (2,-1)$$
$$10 \rightarrow (4,1) + (3,0) + (2,-1) + (1,-2) \tag{1.4}$$

The masses and representations for the two observed lowest-lying $SU(3)_F$ meson and baryon multiplets and the correct predictions of the $SU(2)_I$ and $U(1)_Y$ quantum numbers, are shown in Table 1.2. Table 1.2, together with the successes of the Gell-Mann–Okubo mass formula [1.97] and other phenomenological predictions [1.99], placed the $SU(3)_F$ quark flavor model

Table 1.2: Isospin and hypercharge quantum numbers of the two lowest mass $SU(3)_F$ meson and baryon multiplets.

$J = 0^-$ meson octet $(B = 0)$			$J = 1^-$ meson quasi-nonet $(B = 0)$		
Particle	Mass (MeV)	$(2I+1)(Y)$	Particle	Mass (MeV)	$(2I+1)(Y)$
$\pi^0, \pi^\pm$	134.97 to 139.57	3(0)	$\rho^\pm, \rho^0$	768.1	3(0)
			ω	782.0	1(0)
K^+, K^0	493.65 to 497.67	2(1)	K^{*+}, K^{*0}	891.6 to 896.1	2(1)
$\overline{K}^0, K^-$	497.67 to 493.65	2(−1)	$\overline{K}^{*0}, K^{*-}$	896.1 to 891.6	2(−1)
η	547.45	1(0)	ϕ	1019.4	1(0)

$J = \frac{1}{2}^+$ baryon octet $(B = 1)$			$J = \frac{3}{2}^+$ baryon decouplet $(B = 1)$		
Particle	Mass (MeV)	$(2I+1)(Y)$	Particle	Mass (MeV)	$(2I+1)(Y)$
p,n	938.27 to 939.57	2(1)	$\Delta^{++}, \Delta^+, \Delta^0, \Delta^-$	1230 to 1234	4(1)
			$\Sigma^{*+}, \Sigma^{*0}, \Sigma^{*-}$	1382.8 to 1387.2	3(0)
Λ^0	1115.63	1(0)			
$\Sigma^+, \Sigma^0, \Sigma^-$	1189.37 to 1197.43	3(0)			
Ξ^0, Ξ^-	1314.9 to 1321.3	2(−1)	Ξ^{*0}, Ξ^{*-}	1531.8 to 1535.0	2(−1)
			Ω^-	1672.43	1(−2)

of hadrons on a sound footing, independent of the physical status of quarks, and encouraged further global group-theoretic enlargments.

As a global symmetry group, the $SU(3)$ flavor model of hadrons served very well as a "classification" group of physical quantities, and it became even more useful when it was enlarged into the mixed global (non-relativistic) $SU(6)$ quark flavor — spin group. The purpose of the $SU(6)$ model [1.100] was to understand the ordering of the low-lying $SU(3)_F$ mass multiplets in Table 1.2 [no one expects a simple explanation of "high-lying" (in mass) $SU(3)_F$ multiplets], i.e. why the lowest $J = 0^-$ (pseudoscalar) meson octet is followed by the $J = 1^-$ (vector) "pseudo-nonet", and the lowest $J = \frac{1}{2}^+$ baryon octet is followed by the $J = \frac{3}{2}^+$ baryon decuplet. The spins and parities of these four lowest-lying (in mass) $SU(3)_F$ hadronic multiplets follow immediately from the assumption that the respective $q\bar{q}$ and qqq (quark) composites are in zero orbital angular momentum states but with anti-parallel and parallel spins of the constituent quarks. The two different spin configurations of the $q\bar{q}$ and the qqq composites are natural enough, but it is difficult to understand why, for example, there is no low-lying $J = \frac{1}{2}^+$ baryon decuplet nor $J = \frac{3}{2}^+$ octet of baryons, since no coupling has been postulated between $SU(3)_F$ (unitary) spin and ordinary spin.

This is where the $SU(6)$ model enters the picture [1.100], wherein global $SU(3)$ flavor symmetry is mixed with global (non-relativistic) $SU(2)$ spin symmetry to form the $SU(6)$ group. The $SU(6)_{F-\text{SPIN}}$ group is taken to be the simple group that "unifies" $SU(3)_F$ and $SU(2)_{\text{SPIN}}$, where the fundamental **6** representation of $SU(6)_{F-\text{SPIN}}$ is hypothesized to consist of the six "spin-oriented" light quarks, namely $(u^\uparrow, u^\downarrow, d^\uparrow, d^\downarrow, s^\uparrow, s^\downarrow)$. In order to take full advantage of the power of the Pauli exclusion principle, we focus on the two lowest-lying baryonic three-quark composites; if we denote by q' the fundamental 6 representation of $SU(6)_{F-\text{SPIN}}$, accept vanishing orbital angular momentum for the constituent quarks, and allow for anti-parallel and parallel quark spins, the $q'q'q'$ composites must represent baryons with $J = \frac{1}{2}^+$ or $J = \frac{3}{2}^+$. In group-theoretic language, one or more of the irreducible $SU(6)_{F-\text{SPIN}}$ three-quark representations must decompose into the observed representations, under $SU(3)_F \times SU(2)_{\text{SPIN}}$, of the two lowest baryonic multiplets in Table 1.2. To see how this works, write out $q'q'q'$:

$$\mathbf{6} \times \mathbf{6} \times \mathbf{6} = \mathbf{20} + 2 \cdot \mathbf{70} + \mathbf{56} \tag{1.5a}$$

$$(A) \quad (M) \quad (S)$$

where A, M and S denote respectively the anti-symmetric, mixed and symmetric irreducible representations of $SU(6)$. If one now decomposes the A, M and S representations of $SU(6)_{F-\text{SPIN}}$ under $SU(3)_F \times SU(2)_{\text{SPIN}}$, one finds that only the symmetric 56 representation is relevant because its decomposition yields:

$$56 \rightarrow (\mathbf{8}, \mathbf{2}) + (\mathbf{10}, \mathbf{4}) \qquad (1.5\text{b})$$

where the decomposite $(\mathbf{8}, \mathbf{3})$ represents the $J = \frac{1}{2}^+$ octet and the decomposite $(\mathbf{10}, \mathbf{4})$ represents the $J = \frac{3}{2}^+$ decuplet; these representations correspond exactly to the two observed lowest-lying baryon representations in Table 1.2.

The perfect match of the predictions of the irreducible **56** representation of $SU(6)_{F-\text{SPIN}}$ to the quantum numbers of the two observed lowest baryon $SU(3)_F$ multiplets gave a great boost to the quark flavor model of hadrons even though it was recognized that the $SU(6)$ flavor-spin group could only be approximate and, at best, could only give a reasonable description of the lowest-mass $SU(3)_F$ multiplets (see §3.2d). Nevertheless, the credibility of this result was sufficiently great to raise the more probing question of why the composite three-quark (fermionic) baryons do not satisfy the Pauli exclusion principle since the wavefunction of the combined flavor-spin **56** representation is symmetric and the spatial zero orbital angular momentum part also contributes a symmetric term; consequently, the total wavefunction for the composite three-quark system is symmetric. This contradiction with the Pauli principle by the three-fermion (quark) composite is unacceptable and it is necessary to introduce some new degree of freedom for the quark (called the "color" degree of freedom) to restore anti-symmetry to the total three-quark wavefunction [1.101].

Within the framework of the $SU(6)_{F-\text{SPIN}}$ model, the simplest way to identify the new "color" degree of freedom appears to be to postulate a new simple internal symmetry group, say G_C (C stands for "color", which is still global), which, when adjoined to $SU(6)_{F-\text{SPIN}}$, i.e. the combined group $G_C \times SU(6)_{F-\text{SPIN}}$, yields a totally anti-symmetric wavefunction; moreover, G_C must be such that "color" is not detected (i.e. is in the singlet state) in the three-quark system. Mathematically, this is equivalent to two requirements: (1) the group G_C must contribute an anti-symmetric part to the wavefunction of the three-quark system in the new degree of freedom; and (2) the condition $R_C \times R_C \times R_C \supset \mathbf{1}$ — where R_C is the hypothesized fundamental representation of G_C [the choice of fundamental representation for R_C is the most economical and is supported by the other symmetry groups in particle physics] — must be satisfied. It is easy to prove that only three groups satisfy these two conditions, namely $SU(3)$ and the two exceptional Lie groups: G_2 and E_8. However, the G_2 and E_8 groups only possess real representations — in contradistinction to $SU(3)$ — and can be ruled out (for further details, see §3.2d). Thus, $SU(3)_C$ is the only global "color" group (with a complex representation) that is consistent with the Pauli principle and yields a totally anti-symmetric wave function for the composite three-quark baryon. The global $SU(3)$ "color" hypothesis is, of course, consistent with meson, not just baryon, phenomenology.

The necessity to introduce a new (global) degree of internal symmetry for the quarks in order to understand the mass and spin regularities of the lowest-lying hadron multiplets threw further doubt into the idea that the "colored and flavored" quarks might be merely "mathematical fictions". However, the decisive impetus to the acceptance of the fractionally charged quarks as physical particles — albeit "confined" — came with the systematic experiments first carried out on the "deep" inelastic scattering of leptons by hadrons (chiefly protons) in the multi-GeV region [1.102] ["deep" means large energy loss by the lepton and large four-momentum (squared) transfer to the hadron]. The "deep" inelastic scattering of leptons by hadrons is the analog of the famous Rutherford experiment on the scattering of α particles by atoms [1.9] — which established the existence of "point-like" scattering centers within atoms (i.e. atomic nuclei) at the "high" energies (several hundred KeV!) then available. In the case of "deep" inelastic lepton scattering by nucleons — if the nucleon is assumed to consist of point-like "partons" [1.103] — the inelastic lepton-nucleon scattering can be replaced by the sum of incoherent elastic scatterings of the leptons on the "partons" (really quarks) of the nucleon. The resulting cross section for, say, inelastic electron-nucleon scattering can then be expressed (by virtue of gauge invariance) in terms of two structure functions that are, in general, functions of ν and Q^2 [where ν is the energy loss by the electron, i.e. $\nu = E - E'$ (with E and E' the initial and final electron energies) and Q^2 is the four-momentum transfer squared to the nucleon]. The condition of "deep" inelastic lepton-nucleon scattering is satisfied when ν and Q^2 are both large and, in that case, it can be shown that the hypothesis of a point-like "parton" structure for the nucleon implies a scaling law, i.e. "Bjorken scaling" [1.104], for the two structure functions. In the "Bjorken scaling" régime, all form factors become functions only of the dimensionless ratio $Q^2/2M\nu$ (M is the nucleon mass), independent of Q^2.

The "Bjorken scaling" hypothesis was strongly supported by the "deep" inelastic electron-nucleon experiments — which established that the "partons" behave like Dirac $J = \frac{1}{2}$ point-like particles — and, when combined with the "deep" inelastic muon-nucleon and neutrino-nucleon experiments, confirmed that the "partons" inside the nucleon possess all the properties of quarks, e.g. the correct fractional charges (more fully discussed in §5.2c). Equally conclusive experimental evidence — before the end of the "Heroic Period" — that the "partons" are quarks came from the measurement of the R value in electron-positron annihilation in the deep inelastic region [1.105], where $R = \sigma(e^+e^- \rightarrow \text{hadrons})/\sigma(e^+e^- \rightarrow \mu^+\mu^-)$ is supposed to equal $\sum_i e_i^2$ [with e_i the electric charge of the ith "parton" (quark) whose (current) mass is below the energy threshold (see §5.2c)]. In deriving this expression for R — which measures directly the electric charges of the "confined" quarks — it is assumed that hadron production takes place via electron pair annihilation into virtual quark pairs, which quickly combine into the observed hadrons. Confirming evidence for the simple asymptotic expression for R came when the energies in the electron-positron collider beams exceeded the c quark production threshold: it was found that R increased — after the c quark threshold is passed — by the predicted amount $3e_c^2 = \frac{4}{3}$ ($e_c = \frac{2}{3}$), for the c quark (3 is the number of colors for each quark) [1.105].

Theoretical analysis of the evidence for "Bjorken scaling", given by the "deep" inelastic lepton-nucleon scattering experiments and the R value measurement, indicated that the quarks inside the nucleon are behaving — at short distances (high energies) — as if they are "free" point-like particles and that any correct dynamical theory of hadronic interactions must explain this important feature of the strong interaction, which is called "asymptotic freedom" (see §5.2a). The measurement of the spin $\frac{1}{2}$ and fractional charge (in units of $\frac{1}{3}$) of the "confined" quarks settled the question of whether quarks are "physical entities" or "mathematical fictions"; "confined" quarks are "physical entities" with well-defined measurable properties which do not include the possibility of detecting them as "free" particles. The stage was set for the construction of a truly dynamical theory of the strong interaction, as is explained in the next section.

§1.3b. From global $SU(3)$ quark flavor to gauged $SU(3)$ quark color (QCD)

There were at least four reasons why the time was ripe by the early 1970s for the construction of a dynamical and phenomenologically correct theory of strong interactions: (1) the $SU(3)$ quark flavor model of hadrons was highly successful, leading through its mixing with quark spin to the $SU(6)_{F-\text{SPIN}}$ model which — by virtue of its ability to give such an excellent description of the relations among hadron masses and the observed quantum numbers of the lowest-lying hadron multiplets — strongly suggested a new global (color) internal degree of freedom for each quark; (2) the clearcut evidence that quarks are much more than "mathematical fictions" and should be taken seriously as "physical constituents" of hadrons whose "confinement" is an important property to be explained by theory; (3) an example had been set several years earlier [1.88] — in 1967–68 — that a promising non-Abelian gauge theory of the electroweak interaction could be constructed (although the spectacular experimental triumphs were still in the future) once the proper global non-Abelian internal symmetries could be identified. The use of spontaneous symmetry breaking in the electroweak gauge theory created some confusion but 't Hooft's proof of renormalizability for non-Abelian gauge theories, without (as well as with) spontaneous symmetry breaking, was completed by 1971 [1.106]; and, finally (4) the confirmation of "Bjorken scaling" in the deep inelastic lepton-hadron scattering experiments was clearly inexplicable in an Abelian type of gauge theory (like QED) and the hope was entertained that the "asymptotic freedom" property could be reconciled with non-Abelian gauge theory. We propose to recount in this section how the creators of QCD took advantage of the favorable circumstances enumerated above but, before doing so, we briefly comment on the advances in S matrix theory that took place during the decade of the 1960s before the S matrix approach to hadronic physics began to lose ground to the increasingly successful gauged quark symmetry approach of QCD.

It is fair to say that dispersion theory — originating in the early 1950s in an attempt to perform semi-quantitative (quasi-dynamical) calculations of the pion-nucleon system [which were clearly outside the ken of perturbative quantum field theory modeled on Abelian QED] — continued to be tailored to more complex hadron systems as more and more mesonic and baryonic particles and resonances were uncovered during the subsequent decade. The promising developments during the 1960s in the domain of the quark flavor model of hadrons and the enlarged dimensionality of global internal symmetries applicable to multi-quark systems, did not immediately discourage continued interest in the refinement of dispersion-theoretic methods to deal directly with hadron systems. One such refinement — that of Regge theory [1.107] — was warmly received at the beginning of the 1960s

and offered hope for gaining some dynamical understanding of high energy
hadron-hadron scattering. Regge's original motivation was to prove the validity of the Mandelstam representation [1.55] in non-relativistic potential
scattering. He accomplished his aim by treating the partial wave scattering
amplitude as a well-behaved analytic function of complex ℓ (orbital angular
momentum), whose poles in the complex ℓ plane corresponded to the bound
states (on the real axis) and resonances (off the real axis) in the s channel
for the specified potential. With this approach, Regge could demonstrate
the existence of the Mandelstam representation for potential scattering (represented by a superposition of Yukawa potentials) with the help of a finite
number of subtractions in the t channel (see Fig. 1.1). The Regge representation was assumed to hold as well in the relativistic region and greatly
extended the Mandelstam domain of analyticity for the scattering amplitude.
The moving Regge poles were taken to define a "Regge trajectory" obeying
certain rules (e.g. the values of J given by the Regge poles must differ by two
units of angular momentum [1.108]) and, soon, a variety of "Regge trajectories" could be plausibly fitted to the observed meson and baryon particles
and resonances. Surprisingly enough, the "Regge trajectories" turned out to
be approximately linear in t over a considerable range and with roughly the
same slope of about 1 $(\text{GeV})^{-2}$, a hint of universality for the Regge approach.

Thus encouraged, the Mandelstam representation was combined with
Regge pole theory to study the conditions under which high energy scattering in the s channel is dominated by Regge poles in the crossed t channel.
This led to an intriguing result for the asymptotic form $(s \to \infty)$ for the
s channel two-particle scattering amplitude $A(s,t)$ at fixed t ("Regge behaviour"), to wit:

$$A(s,t) \simeq \underset{s \to \infty (t \text{ fixed})}{\sim} \beta(t) s^{\alpha(t)} \tag{1.6}$$

where $\alpha(t)$ (known as the "trajectory function") is the position of the Regge
pole which lies farthest to the right in the complex ℓ-plane (the so-called
"leading" pole) and $\beta(t)$ (called the "residue function") contains the remaining t-dependent factors in the Regge pole term. It should be noted that Eq.
(1.6) exhibits the same s-dependence as would result from the "exchange"
of a particle in the t channel with spin $\alpha(t)$. Since $\alpha(t)$ is well represented
by a straight line [i.e. $\alpha(t) \simeq \alpha(0) + t\alpha'(0)$], Eq. (1.6) implies that — in
the "leading" Regge pole approximation — the two-particle scattering cross

section for small $|t|$ becomes:

$$\frac{d\sigma}{d|t|} \sim \gamma(t) \; s^{2\alpha(0)-2} \; \exp[2t\alpha'(0)\ln s] \tag{1.7}$$

where $\gamma(t) \sim \beta^2(t)$.

Equation (1.7) possesses the interesting property that it predicts an approximately exponential form for the diffraction peak at high energies and a logarithmic shrinking with energy of the width of the diffraction peak. The first part of the prediction has been found to hold for all two-particle scattering processes (e.g. $pp, \pi^{\pm}p, \bar{p}p$, etc.) but the second part is not always confirmed. In any case, if all total cross sections tend to finite limits at high energies, then each elastic forward scattering amplitude (by virtue of the "optical theorem") must be dominated by a single Regge trajectory $\alpha_P(t)$ (the so-called Pomeranchuk trajectory or "pomeron") such that $\alpha_P(0) = 1$. Since the "pomeron" controls all forward elastic scattering amplitudes, it must possess the quantum numbers of the vacuum and this leads to the Pomeranchuk theorem [1.109], which states that the total cross sections for pp and $\bar{p}p$ scattering become equal as $s \rightarrow \infty$. Another interesting property of Regge poles — that served later as a paradigm for the duality hypothesis — is "factorization": the property that, in Regge pole exchange, the residue can be written as a product of two factors (as in the exchange of ordinary particles); these two factors can be interpreted as the couplings of the Regge pole to the initial and final states in the t channel and leads to relations among different processes in which the same Regge pole can be exchanged [e.g. $\sigma(\pi\pi)\sigma(NN) \sim \sigma^2(\pi N)$, where the σ's are the total cross sections at high energy].

Regge theory encountered a number of technical difficulties — such as the onset of "cuts" (generated by the simultaneous exchange of two or more Regge poles) at high energy; however, the key ideas enriched S matrix theory and it led to a number of finite energy sum rules [1.108] that tried to relate the integral over the energy of the contribution to the imaginary part of the scattering amplitude (of the relatively low-energy "resonances" in the s channel) to the same sum taken over the contributions of the Regge poles in the t channel. This led to the "duality hypothesis" [1.110], wherein it was proposed that one could write down (in the "narrow resonance approximation") the very symmetrical ansatz:

$$A(s,t) = \sum_n \frac{R_n(t)}{n - \alpha(s)} = \sum_n \frac{R_n(s)}{n - \alpha(t)} \tag{1.8}$$

where the R_n's are the residues at the Regge poles in the s and t channels respectively. Since $A(s,t)$ is a meromorphic function (i.e. it possesses no branch cuts or essential singularities), it can be constructed by means of the poles in the s or t channel — as indicated by Eq. (1.8).

By 1968, Veneziano [1.111] observed that the "duality hypothesis", expressed by Eq. (1.8), can be described in terms of a single Beta function $B(x,y)$ as follows:

$$A(s,t) = B[-\alpha(s), -\alpha(t)] \tag{1.9}$$

where the linear approximation to the "Regge trajectory" is assumed. Equation (1.9) then yields a simple explicit expression for the residues $R_n(t)$, in terms of the Gamma function, namely:

$$R_n(t) = \frac{\Gamma[\alpha(t) + n + 1]}{n! \, \Gamma[\alpha(t) + 1]} \tag{1.10}$$

where $R_n(t)$ is a polynomial of degree n in t and describes angular momenta $\ell \leq n$. Further work on the Veneziano scattering amplitude generalized the four-point amplitude to m-point amplitudes possessing factorization properties [1.112] and, in turn, these m-point amplitudes were demonstrated to describe the interactions of a relativistic string [1.113].

In sum, by the early 1970s, while Regge pole theory had extended the horizons of S matrix theory and had scored a significant number of phenomenological successes, it began to run into serious complications and ambiguities. The "dual resonance model" — which combined in an elegant fashion the key concepts of dispersion theory and Regge pole theory — excited fresh interest in the S matrix program but, again, the future of "duality" was quickly clouded by hidden "ghosts" and "tachyons" (faster-than-light particles) [1.114]. The fact that the generalization of Veneziano duality to m-point scattering amplitudes exhibited mathematical properties that bore a striking resemblance to relativistic strings which, in some models automatically predicted the presence of a massless "graviton" [1.114], played an important historical role in the evolution of the superstring theory of the 1980s. However, in the early 1970s, the intimation of a gravitational connection, achievable through some form of string theory, did not appear to be particularly helpful in pinning down the dynamical and phenomenologically correct theory of the strong interaction [although the "Regge pole" and "duality" refinements of S matrix theory did catalyze the supersymmetry, supergravity and superstring programs that ensued a decade later (see

§1.4c)]. The final blow to S matrix theory was delivered in the early 1970s by the discovery of "Bjorken scaling" and asymptotic freedom in the strong interaction, for which S matrix theory was completely unprepared. In contrast, the proponents of the quark flavor model of hadrons were aware of the need for global $SU(3)$ "color" [emerging from the $SU(6)$ model] and, furthermore, were in a good position to take advantage of the 1971 proof of the renormalizability of non-Abelian gauge theories [1.106]. The fully dynamical theory of the strong interaction was accomplished by following, despite initial skepticism, the QED paradigm of gauging a field-theoretic internal symmetry [1.89]. One might inquire — after witnessing the great success — by 1950 — of the dynamical Abelian gauge theory of electromagnetism (i.e. QED) — why it took so long to exploit the magical power of gauging quantum field theories to achieve a successful dynamical theory of the strong interaction. [One can ask the same question about the weak interaction — although QFD was formulated several years before QCD — and, as we will see in §1.3c, the answer is different.] As we know, already in 1954, Yang and Mills [1.51] had shown how to gauge a quantum field theory with a non-Abelian global $SU(2)$ internal symmetry (e.g. isospin). However, it was not immediately apparent that a gauged quantum field theory would work for the strong interaction for several reasons: (1) there was the problem of what new global internal symmetries — presumably non-Abelian — associated with the strong interaction could be gauged and progress on this score was only possible when quarks were identified as "physical entities" and did not serve merely as "mathematical fictions"; (2) it was known that the strong interaction is of short range ($\sim 10^{-13}$cm) — in contradistinction to the long-range Coulomb interaction in electromagnetism. Since the gauge invariance of non-Abelian gauge groups requires the gauge fields to be massless — just as for Abelian gauge groups — the question of major departures from the long-range QED-type predictions for the strong interaction had to be resolved; and, finally, (3) there was the problem of whether a non-Abelian quantum gauge theory is renormalizable, and this problem was only solved in 1971 [1.106]. Let me comment on these points in turn.

(1) The choice of the appropriate non-Abelian global internal symmetries to gauge in strong interactions only became apparent when the need for global $SU(3)$ "color" was established. It had been shown by 1967–68 [1.88] that the gauging of global non-Abelian flavor symmetries (in the case of QFD, chiral quark flavor together with chiral lepton flavor) had produced a promising non-Abelian gauge theory of the electroweak interaction, and

quark "color" looked like a likely candidate for gauging. The choice of $SU(3)$ color was highly felicitous: it was quickly demonstrated that "asymptotic freedom" is a unique property of non-Abelian gauge theories [1.89] and, hence that the "Bjorken scaling" observed in deep inelastic lepton-hadron scattering experiments [1.102] could be understood. The origin of "asymptotic freedom" in gauged $SU(3)$ color [1.89] stems from the fact that, in non-Abelian gauge theories (e.g. QCD), the gauge field quanta (e.g. gluons) carry non-vanishing (color) charges like the matter field quanta (e.g. quarks). [This is the essential difference from Abelian gauge theory (e.g. QED) where the gauge field quantum (e.g. photon) does not carry any charge (i.e. electric charge)]. The non-vanishing "color" charge of the spin 1 (bosonic) gluon produces the dominant contribution to the renormalized "running" gauge coupling constant and, despite a competing contribution from the fermionic quarks, results in a logarithmically decreasing value of the renormalized coupling constant as the distance between the quarks decreases (or as the energy increases). This onset of "asymptotic freedom" contrasts with gauged Abelian QED where the photon carries no charge and the fermions (charged leptons) are the sole contributors to the renormalized "running" gauge coupling constant (electric charge) so that the renormalized electric charge increases logarithmically with the inverse distance (or increasing energy). The renormalization group analysis that leads to these contrasting results between non-Abelian and Abelian gauge theories is given in §5.2a.

(2) The striking difference between non-Abelian and Abelian gauge theories — as exemplified by "asymptotic freedom" — helps to reconcile the observed short-range character ($\sim 10^{-13}$cm) of the strong interaction with the massless gauge fields (gluons) in QCD. This becomes clear when the renormalization group is applied to non-Abelian $SU(3)$ color (in §5.2a) and it is shown how the "running" strong gauge coupling constant starts with a low value at high energy (asymptotic freedom) and attains a maximum at the "nuclear" QCD scale Λ_{QCD} [$\Lambda_{QCD} \sim 150$ MeV (see §5.2a)]; beyond Λ_{QCD}, "confinement" takes over. In contrast to QCD, the QED "running" coupling constant increases monotonically with inverse energy until it "blows up" at the "Landau singularity" (see §4.5). In a word, seemingly dimensionless QCD (for massless quarks) acquires a finite scale as a consequence of the "asymptotic freedom" property of non-Abelian gauge groups, of the need for a finite renormalization reference scale (to avoid the infrared divergences present in any quantum field theory) and of a "confinement" mechanism. This results in a finite QCD scale Λ_{QCD}.

(3) The third major obstacle to the construction of a non-Abelian gauged quantum field theory for the strong interaction — the question of renormalizability — was, as we have said, solved in 1971 [1.106] and all impediments were removed from the novel idea that it is possible for a dynamically correct theory of the strong interaction (QCD) to emerge as the gauged, non-Abelian, $SU(3)_C$ counterpart to gauged, Abelian $U(1)_{\mathrm{EM}}$ QED.

It must be admitted that unbroken, non-Abelian and gauged QCD — with three generations of massive quark doublets and an octet of massless gluons — is a far cry from the original Yukawa pseudoscalar meson theory of nuclear forces, and yet it soon became apparent that QCD is capable of shedding light on the early discoveries in pion physics [apart from the close simulation by $\Lambda_{\mathrm{QCD}}^{-1}$ ($\Lambda_{\mathrm{QCD}} \simeq 150$ MeV) of the finite range $\sim 1/m_\pi$ (m_π is the pion mass) of the two-nucleon interaction]: the pseudoscalar nature of the pion, its small mass and its three charge states. Furthermore, other attractive features of QCD unfolded as detailed analysis of the $SU(3)_C$ gauge theory proceeded: (a) the (color) singlet character of the quark-antiquark condensates under the confining $SU(3)_C$ gauge group keeps $SU(3)_C$ unbroken so that asymptotic freedom is maintained; (b) the global chiral quark flavor group $SU(3)_L \times SU(3)_R$ follows from the QCD Lagrangian for massless fermions (i.e. the three light u, d and s quarks) as does the spontaneous breaking of $SU(3)_L \times SU(3)_R$ to the global quark flavor group $SU(3)_F$. Moreover, the identification of the global chiral quark flavor currents — associated with the $SU(3)_L \times SU(3)_R$ group — with the physical quark currents in weak and electromagnetic interactions is fully justified (and, consequently, the success of current algebra is understood); (c) the identification of the lowest-mass pseudoscalar octet ("pions") with the octet of N-G bosons resulting from the SSB of the global chiral quark flavor group $SU(3)_L \times SU(3)_R$ to $SU(3)_F$ also receives an explanation within the framework of QCD; and, finally, (d) at least a semi-quantitative — if not a fully rigorous — explanation for the absence of "free" quarks and gluons (i.e. "confinement") is found within the context of non-Abelian, non-chiral QCD (see §5.3–§5.3b). That is to say, there exist plausible arguments and some positive evidence from lattice gauge theory for color confinement in QCD, and the experiments on heavy quarkonia spectroscopy certainly support the long-distance behavior of the quark-antiquark (non-relativistic) potential expected of a "confining" theory [1.115]. All of these properties, and others, of QCD will be discussed in detail in Chapter 5.

As more sophisticated mathematical (e.g. topological) methods have been applied to the study of non-Abelian gauge fields, some new opportunities have arisen for QCD (e.g. the usefulness of the QCD anomaly) as well as new problems (e.g. the "strong CP" problem), related to the topologically non-trivial character of non-Abelian gauge solutions. These features of QCD will be discussed in §7.2b and §10.3c and no more will be said here. Instead, we turn, in the next section, to a "preview" of the progression of key steps that brought the global V-A charged baryon-lepton weak current theory to the gauged $SU(2)_L \times U(1)_Y$ chiral quark-lepton flavor theory of the electroweak interaction (QFD).

§1.3c. From the global V-A charged baryon-lepton weak current theory to the gauged chiral quark-lepton flavor theory of the electroweak interaction (QFD)

By 1960, all experiments were confirming the V-A Lorentz structure of the charged baryon and lepton currents participating in weak interactions, and these successes encouraged speculation that the universal V-A charged current-current interaction is mediated by charged IVB 's with masses much greater than that of the pion mass. Unfortunately, IVB calculations of "higher order" (in the Fermi coupling constant G_F) weak processes came to the same non-renormalizable grief as the localized V-A charged current-current theory [1.116]. [The point is that the IVB mass serves as an effective ultraviolet cutoff for lowest-order weak processes but leads to unacceptable divergences for higher-order weak processes [1.116].] A possible non-Abelian (Yang-Mills) gauge formulation for the universal weak interaction — combined with the (Abelian) electromagnetic interaction — became increasingly attractive during the 1960s when it was realized that the charged two-flavor chiral quark-lepton currents fitted naturally into an $SU(2)_L$ group [1.117] and when the significance of the sharing of "chirality invariance" by both the charged lefthanded (V-A) weak current and the neutral non-chiral (V) electromagnetic current, became clear [1.80]. There was, however, the big dilemma posed by the phenomenological need for a finite mass of the hypothesized $SU(2)_L$ gauge field (IVB). Fortunately, by 1964, the solution to the dilemma of a finite mass weak IVB was, in principle, available when the "Higgs" mechanism for spontaneous symmetry breaking (SSB) in gauged quantum field theories was worked out [1.87]. The 1961 Goldstone theorem [1.118] had demonstrated that the SSB of a global non-Abelian symmetry group is accompanied by massless N-G bosons and it was the extension of the

SSB concept to a gauged non-Abelian quantum field theory that yielded the important new result that the "would-be" N-G bosons generate finite masses for the initially massless gauge fields. The combination of gauging a global non-Abelian internal symmetry group plus the use of the SSB Higgs mechanism were, together, the key to the successful construction of a dynamical, phenomenologically correct theory of the electroweak interaction.

Let us expand somewhat on how the essential ingredients of the highly successful standard gauge theory of the electroweak interaction were finally brought together by the end of the 1960s. With a non-Abelian gauge field and SSB as crucial working concepts for the construction of a dynamical finite-range theory of the electroweak interaction, it was clearly necessary to identify the proper global non-Abelian internal symmetries to be gauged and spontaneously broken before the theory could achieve its definitive formulation. The process of identifying the smallest non-Abelian electroweak group that could be gauged was started in 1961 when Glashow [1.88] realized that the minimal algebra that makes provision for the weak $SU(2)_L$ current and the $U(1)$ electromagnetic current is the enlargement of $SU(2)_L$ to $SU(2)_L \times U(1)_{Y_W}$ [where Y_W is the chiral "weak" hypercharge, related to non-chiral electric charge Q through the "weak" GNN relation: $Q = I_{3L} + Y_W/2$ — see §1.2d]. This was a happy choice because the gauging of $SU(2)_L \times U(1)_{Y_W}$ provides just four gauge fields for the "unified" electroweak interaction: two charged gauge fields that couple to the two weak charged lefthanded (quark and lepton) currents (and acquire masses through the SSB mechanism) and two neutral gauge fields that couple — in suitable linear combinations — to the single non-chiral electromagnetic current and the single chiral neutral weak current [the first combination staying massless (photon) and the second combination acquiring mass through the SSB mechanism]. Glashow got as far as suggesting the correct electroweak gauge group — but without spontaneous symmetry breaking and hence without the possibility of renormalizing the theory (since intermediate charged vector bosons were present).

Without SSB, a theory based on $SU(2)_L \times U(1)_{Y_W}$ could not ascribe finite mass to the three weak bosons nor, equivalently, explain the extremely short range of the weak interaction. The QCD solution to the short-range character of the strong interaction is not available to explain the extremely short range of the weak interaction (much shorter than that of the strong interaction — by a factor of 10^3); however, with the Higgs SSB mechanism in hand after 1964, this important step could be taken in the construction of the electroweak part of the standard model. The electroweak program was completed by Weinberg and Salam [1.88] (in 1967–68) and the resulting theory of quantum flavordynamics (QFD) is based on the chiral gauge group $SU(2)_L \times U(1)_Y$ [the "weak" hypercharge Y_W is replaced by Y — as is the custom — but this Y is the chiral "weak" hypercharge], spontaneously broken by a "Higgs" doublet of complex scalars down to the unbroken non-chiral electromagnetic group $U(1)_{\text{EM}}$ (so that $Q = I_{3L} + Y/2$). The matter field associated with the $SU(2)_L \times U(1)_Y$ gauge group consists of Weyl (quark-lepton) fermions with a three-fold degeneracy of the quantum numbers (i.e. there are three quark-lepton generations) but with distinct (and very different) masses for the quarks and charged leptons of the three generations; it should be noted that all neutrino masses are taken to vanish in the standard model. Thus, for each generation there are five chiral quark-lepton representations: three chiral quark representations [a lefthanded "weak" isodoublet and two righthanded "weak" isosinglets] and two chiral lepton representations [one lefthanded "weak" isodoublet and one righthanded "weak" isosinglet — the righthanded neutrino is absent, thereby guaranteeing a massless (Dirac) neutrino].

QFD did not immediately gain acceptability because its renormalizability — that of a spontaneously broken non-Abelian theory — still had to be demonstrated. In particular, it was necessary to prove that the non-renormalizable quantum field theory of the massive charged vector boson is converted into a renormalizable quantum field theory when the charged, vector gauge boson acquires mass through the "Higgs" SSB mechanism. When this problem arose, the proof had not as yet been given for an unbroken non-Abelian gauge theory (with matter fermions) and both proofs were accomplished by 't Hooft in 1971 [1.106], thereby placing QFD on a firm foundation and justifying serious confrontation of QFD with a great variety of experimental tests (of which the first was the observation of weak neutral currents in 1973 [1.119], all of which have been passed thus far with flying colors. The detailed structure of QFD and the salient experimental tests will be discussed in Chapter 6 (although some mention has been made of the more recent precision tests in §1.4a).

Actually, the "renormalization" certification of the non-Abelian electroweak gauge theory in 1971 gave impetus to the final formulation of non-Abelian QCD, which was completed within a few years [1.89]. Indeed, by 1975 — the end of the "Heroic Period" — the modern dynamical gauge theory of the strong and electroweak interactions (i.e. the standard model) among three generations of leptons and confined (but "physical") quarks was in hand. The unremitting probing tests of the standard model after 1975 has led to occasional doubts about its resilience, but to everyone's surprise, the standard model has withstood the scrutiny of the most recent fifteen-year period (1975–90), that followed the "Heroic Period", and which we call the period of "Consolidation and Speculation".

This introductory chapter will conclude with a review (in §1.4–§1.4c) of the period of "Consolidation and Speculation" but before we comment on the most recent fifteen-year period of experimental and theoretical activity in particle physics, we add a final summary section on the established properties of the standard model to our account of the "Heroic Period". We make use of extensive tables and figures to record the salient features of the standard model, the interactions involved and the known fermion and boson quantum numbers, to further clarify the conceptual foundations of the standard model — which is the major purpose of this book — and to provide reference material for the detailed discussions that follow later in the book.

§1.3d. Summary of key properties of the standard model of particle interactions

The brief qualitative discussion of QED, QCD and QFD given thus far should serve as a quasi-historical prelude to the more mathematical and detailed treatment of the standard model that is undertaken in Chapters 4–7. As we have remarked, it seems worthwhile to supplement this qualitative discussion with some tables and figures that pertain to the standard model and provide a bird's-eye view of the similarities and differences among the component gauge groups of the standard model. We start by listing in Table 1.3 the properties of the gauge groups involved in the interaction part of the total Lagrangian for the standard model; these groups are $SU(3)_C$ (QCD), which is unbroken, and $SU(2)_L \times U(1)_Y$ (QFD), which breaks spontaneously to unbroken $U(1)_{\rm EM}$ (QED). In constructing Table 1.3, we distinguish among the QED, QCD, QFD $[U(1)_Y]$ and QFD $[SU(2)_L]$ groups and employ the categories of unbroken vs. (spontaneously) broken, Abelian vs. non-Abelian and non-chiral vs. chiral, to characterize the different gauge groups. Table 1.3 tells us that there is one gauge field apiece associated with $U(1)_{\rm EM}$ and $U(1)_Y$, three with $SU(2)_L$ and eight with $SU(3)_C$ [since the number of gauge fields present in an $SU(N)$ group is 1 for $U(1)$ but otherwise equal to the dimension $(N^2 - 1)$ of the adjoint representation of $SU(N)$ $(N \geq 2)$]; further, we learn that the unbroken $SU(3)_C$ group retains its eight massless

Table 1.3: Gauge interactions for QED, QCD and QFD.

QED: $U(1)_{EM}$ is an unbroken, Abelian, non-chiral gauge group

$$\mathcal{L}_I[U(1)_{EM}] = e \cdot j_\mu^Q(x) \cdot A^\mu(x)$$

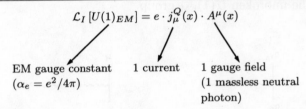

| EM gauge constant | 1 current | 1 gauge field |
| $(\alpha_e = e^2/4\pi)$ | | (1 massless neutral photon) |

QCD: $SU(3)_C$ is an unbroken, non-Abelian, non-chiral gauge group

$$\mathcal{L}_I(SU(3)_C) \sim g_s \cdot j_\mu^a(x) \cdot G_a^\mu \quad (a = 1, 2, \ldots 8)$$

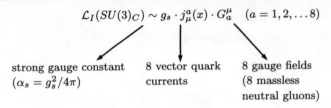

| strong gauge constant | 8 vector quark | 8 gauge fields |
| $(\alpha_s = g_s^2/4\pi)$ | currents | (8 massless neutral gluons) |

QFD $[U(1)_Y]$: $U(1)_Y$ is a broken, Abelian, chiral gauge group

$$\mathcal{L}[U(1)_Y] \sim g_Y \cdot j_\mu^Y(x) \cdot B(x) \quad (\text{Y is "weak" hypercharge})$$

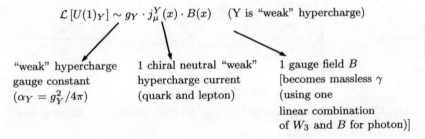

| "weak" hypercharge gauge constant $(\alpha_Y = g_Y^2/4\pi)$ | 1 chiral neutral "weak" hypercharge current (quark and lepton) | 1 gauge field B [becomes massless γ (using one linear combination of W_3 and B for photon)] |

QFD $[SU(2)_L]$: $SU(2)_L$ is a broken, non-Abelian, chiral gauge group

$$\mathcal{L}_I[SU(2)_L] \sim g_W \cdot j_{\mu L}^{(i)}(x) \cdot W_{(i)}^\mu(x) \quad (i = 1, 2, 3)$$

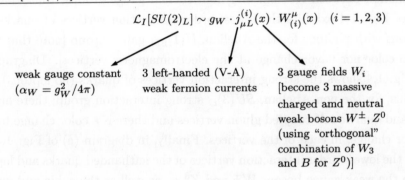

| weak gauge constant $(\alpha_W = g_W^2/4\pi)$ | 3 left-handed (V-A) weak fermion currents | 3 gauge fields W_i [become 3 massive charged amd neutral weak bosons $W^\pm, Z^0$ (using "orthogonal" combination of W_3 and B for Z^0)] |

gauge fields (gluons) while the four massless gauge fields associated with the spontaneously broken $SU(2)_L \times U(1)_Y$ group are converted into three massive (charged and neutral) weak bosons ($W^\pm$, Z^0) and one massless gauge boson (photon) of the unbroken $U(1)_{EM}$ group.

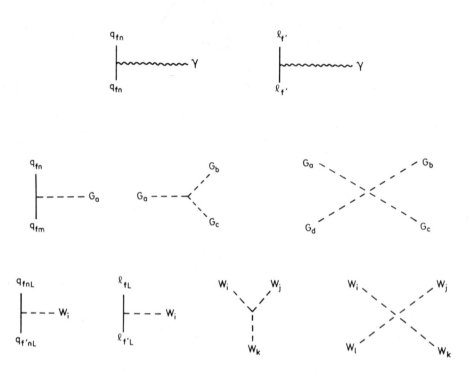

Fig. 1.4. Interaction vertices of $U(1)_{EM}$, $SU(3)_C$ and $SU(2)_L$ gauge groups.

We follow up Table 1.3 with Fig. 1.4 in order to pinpoint an important distinction between Abelian and non-Abelian gauge groups; thus, diagram (a) of Fig. 1.4 exhibits the lowest-order interaction vertices of quarks and leptons with photons for the Abelian, $U(1)_{EM}$ gauge group [note that there is no color nor flavor change at the electromagnetic vertices]. Diagram (b) of Fig. 1.4 shows the strong interaction vertices of quarks (no leptons) with gluons, for the non-Abelian, $SU(3)_C$ strong interaction group; there are cubic and quartic self-coupled gluon vertices and there is a color change but no flavor change, in each of the vertices. Finally, in diagram (c) of Fig. 1.4, we give the lowest-order interaction vertices of the lefthanded quarks and leptons with the weak gauge bosons $W^\pm$ and Z^0 — as well as the cubic and quartic self-coupled weak boson vertices — for the non-Abelian $SU(2)_L$ electroweak group [the $U(1)_Y$ part of the electroweak does not add any new features];

here the charged weak current vertices involve a fermion flavor change (but no color change) whereas the neutral weak current vertices necessarily involves no flavor change (and no color change). The self-coupled weak boson vertices obviously require "flavor conservation" (i.e. charge conservation) and no color change. The absence of self-interaction vertices among the gauge bosons for $U(1)_{\text{EM}}$ — compared to non-Abelian $SU(2)_L$ and $SU(3)_C$ — follows, of course, from the fact that the "Abelian" photon carries no (electric) charge whereas the "non-Abelian" gluons and weak bosons carry color charge and flavor charge in the QCD and QFD cases respectively. The physical consequences of this difference, as we have already remarked, are striking: e.g. there is "asymptotic freedom" in QCD but not in QED.

Table 1.4: Non-relativistic "potentials" of electromagnetic, weak and strong gravitational interactions.

	Fermion interaction	Boson mediator	Strength (α)	Range (R)	Non-relativistic "potential"
q, ℓ	QED: fermion V current-photon interaction (non-chiral)	one $J = 1^-$ massless photon (γ)	$\alpha_e = 1/137 \sim$ .007 $(Q^2 \sim m_e^2$, where Q^2 is four-momentum squared)	$R = \infty$	$V(r) = \frac{\alpha_e q_1 q_2}{r}$ (attractive: unlike sign of charges; repulsive: like sign of charges)
q, ℓ	$SU(2)_L$ part of QFD: $(V\text{-}A)$ fermion current-weak boson interaction (chiral)	three $J = 1^-$ massive weak bosons $(W^{\pm}, Z^0)$	$\alpha \simeq 4.3\alpha_e$ $\simeq .034$ $(Q^2 \sim M_W^2)$	$R \sim M_W^{-1} \sim$ 10^{-16} cm (electroweak scale $=$ $\Lambda_{\text{QFD}} \sim$ 250 GeV)	$V(r) \sim \delta(r) \cdot \frac{\alpha_W}{M_W^2}$ $\times \{(VV \text{ term})$ $+\underline{\sigma}_1 \cdot \underline{\sigma}_2 \ (AA \text{ term})$ $+\underline{\sigma} \cdot \underline{p}/m_f (VA$ term)$\}$ $(m_f$ is fermion mass, p the momentum)
q	QCD: V quark current-gluon interaction (non-chiral)	eight $J = 1^-$ massless gluons	$\alpha_s = 0.20$ $(Q^2 \sim 2$ GeV$^2)$	$R \sim 10^{-13}$ cm (strong interaction scale $=$ $\Lambda_{\text{QCD}} \sim$ 150 MeV)	$V = -\frac{4}{3}\frac{\alpha_s}{r} + kr$ $[k \sim 0.25$ (GeV)^2 for $q - \bar{q}$ system]
q, ℓ	Einsteinian gravity: symmetric tensor fermion current-tensor graviton interaction (non-chiral)	one $J = 2^+$ massless graviton	$\alpha_{\text{grav.}} =$ $K m_N^2 \sim 10^{-39}$ $(K$ is gravitational constant)	$R = \infty$	$V(r) = -\frac{m_1 m_2}{r}$ (only attractive)

We present one more table, Table 1.4, to show how the different mathematical properties of the three groups: $U(1)_{\text{EM}}$, $SU(2)_L$ and $SU(3)_C$ are reflected in the distinguishing features of the non-relativistic potentials that describe the static forces between two (slowly moving) fermions (quarks and/or leptons) governed by the three groups. Clearly, Table 1.4 — which is self-explanatory — can not be used, by and large, for serious calculations since most particle processes are relativistic although there are exceptions like heavy quarkonium. Table 1.4 is chiefly intended to firm up some qualitative distinctions among the four basic forces of nature (we have included gravity) that have emerged during our discussion of the "Startup" and "Heroic" periods in particle physics. We have four comments concerning Table 1.4:

(1) While QED and QCD are both governed by unbroken, non-chiral gauge groups, they differ in one essential respect: the single charge of Abelian QED commutes with itself but the eight charges of non-Abelian QCD do not, with the consequence that the "non-relativistic" potential in QED is Coulombic whereas the strong color potential between, say, a heavy quark and an antiquark, i.e. heavy quarkonium, is given approximately by [1.115]:

$$V(r)_{q\bar{q}} = -\frac{4}{3}\alpha_s/r + kr; \quad [k \simeq 0.25 \text{ GeV}^2] \qquad (1.11)$$

where the first term in Eq. (1.11) is the one-gluon "Coulombic" contribution, reflecting the "asymptotic freedom" character of the strong color force and $\alpha_s = g_s^2/4\pi$ (with g_s the strong gauge coupling constant). The coefficient $\frac{4}{3}$ in the first term of Eq. (1.11) arises from summing over the different color components corresponding to the color singlet nature of the heavy quarkonium system. The second term of Eq. (1.11) represents the non-perturbative "confining" long-distance contribution to the $q-\bar{q}$ potential and expresses the fact that the gluons in the non-Abelian color interaction enjoy self-interactions (of the 3-gluon and 4-gluon types — see Fig. 1.4); these gluonic self-interactions compress the "color-electric" flux lines between q and $\bar{q}$, thereby producing a term proportional to the distance r between q and $\bar{q}$ (see §5.3a). The proportionality constant k is fixed by matching the heavy quarkonium $J/\psi(c\bar{c})$ and $\Upsilon(b\bar{b})$ binding energies. It is surprising how well the simple form (1.11) of the $q-\bar{q}$ potential explains the details of heavy quarkonium spectroscopy (see §5.3b).

(2) The use of the concept of non-relativistic potential in comparing QED and QCDis straightforward: the binding energy of the $q-\bar{q}$ system is small compared to the quark masses and taking non-relativistic wave functions

to describe the heavy quarkonium states is justified and very useful. The concept of the non-relativistic potential is employed in a somewhat different sense in connection with weak interactions. In point of fact, all observable weak processes involve some "relativistic" particles; however, the very large masses of the three weak bosons ($W^{\pm}, Z^0$) compared to the fermion (quark and lepton) masses (the t quark mass may be an exception — see §6.4c) justifies the "non-relativistic" use of $\delta(r)$ [with $g^2/8M_W^2$ replaced by $G_F/\sqrt{2}$, where G_F is the Fermi constant] for weak processes. The last statement is equivalent to saying that use of the local four-fermion interaction is justified whenever $Q^2 << M_W^2$. The large weak boson masses — which result from SSB — clearly differentiate the $SU(2)_L \times U(1)_Y$ electroweak group from unbroken QED and QCD with their massless gauge fields. It is also responsible for the early misperception that the weak interaction is much "weaker" than the electromagnetic interaction [the (dimensionless) weak gauge coupling constant g is actually more than twice as large as the electromagnetic gauge coupling constant e (see Table 1.4)]; the point is that in beta decay (when $Q^2 << M_W^2$), g is effectively multiplied by (m_N/M_W), with m_N the nucleon mass) so that G_F is reduced by a factor of $(m_N/M_W)^2 \sim 10^4$, thus creating the impression that the weak interaction is intrinsically much weaker than the electromagnetic interaction.

(3) Another fundamental difference (apart from SSB) between the electromagnetic and strong interactions, on the one hand, and the weak interaction, on the other, is that the former two are parity-conserving (non-chiral) whereas the latter is maximally parity-violating (chiral); this difference is reflected in the expressions for the "non-relativistic" potential given in the last column of Table 1.4. It is seen that the QED and QCD potentials are purely scalar while the weak potential contains a pseudoscalar term. Specifically, because of the (V-A) coupling of the weak bosons to chiral (lefthanded) fermions, the weak potential contains two (parity-conserving) scalar terms: one from the V part — giving rise to the so-called Fermi selection rules in β decay — and the other from the A part — giving rise to the so-called Gamow-Teller selection rules in β decay; the cross $V \cdot A$ part is responsible for the parity-violating pseudoscalar term, which is a spin-momentum-dependent term. The chiral (V-A) structure is of great importance in pinning down the structure of the standard model through three chiral gauge anomaly-free constraints (see Chapter 7). The need for chiral fermions to describe the weak interaction also imposes conditions on the structure of grand unification groups (see §8.2) and even on "super-grand" unification groups [in

our terminology, a "super-grand" unification group is one that aims to unify gravity with the other three basic forces and is sometimes called a "theory of everything" (TOE) (see §1.4c)].

(4) As a final comment, we note that we have included the gravitational potential in Table 1.4 in order to emphasize three points: (a) the effective coupling constant of the gravitational potential at the nucleon mass scale is exceedingly small (of the order of 10^{-39}) and only becomes appreciable at the Planck scale, of the order of 10^{19} GeV. Even $\Lambda_{QFD} \sim 250$ GeV (the symmetry-breaking scale of the electroweak interaction) is miniscule compared to the Planck scale and explains why attempts to "unify" gravity with the other three interactions encourages the construction of models [e.g. Kaluza-Klein, superstring, etc.] that initially operate at the Planck scale and require compactification before "settling down" to the four space-time dimensions in which we live (see §1.4c); (b) the constancy of sign of the gravitational interaction (always attractive) is incompatible with a gauge (vector) field-theoretic origin and thereby underlines the difficulty of unifying the gravitational with the strong and electroweak interactions. There are, of course, other reasons for the intractability of "super-grand" unification: e.g. "general covariance" relates directly to the space-time coordinates, whereas "gauge invariance" involves the gauging of global internal symmetries, and there is no theory of "quantum gravity" in four space-time dimensions [1.120]; and (c) while the tiny ratio of the presently available particle energies to the Planck energy seems to imply a miniscule effect on particle physics, even in the multi-TeV region of the new accelerators, there is a "triangular-type" mixed gauge-gravitational anomaly, associated with the chiral "weak" hypercharge vertex and two graviton vertices of the triangle fermion loop, which is significant for the physics governed by the standard model; the relevance of the mixed anomaly to standard model physics is due to the fact that the mixed anomaly is a "triangular-type" anomaly and, as such, is not affected by the "short wavelength" "quantum gravity" problem in four dimensions (see §7.4).

We conclude this summary section on some basic properties of the standard model with two final tables. First, in Table 1.5, we list the quantum numbers of the quarks and leptons of one generation (these quantum numbers are identical for all three generations) under the standard group $SU(3)_C \times SU(2)_L \times U(1)_Y$, as well as under the unbroken $SU(3)_C \times U(1)_{\text{EM}}$ group, into which the standard group is broken spontaneously by the "Higgs" doublet. Table 1.5 is constructed as follows: (a) before the SSB of $SU(3)_C \times$

$SU(2)_L \times U(1)_Y$ gives rise to $SU(3)_C \times U(1)_{EM}$, the quark and lepton flavor representations are purely lefthanded (L) "weak" isospin doublets and righthanded (R) "weak" isospin singlets. Since the standard model assumes the absence of ν_R, no "weak" isospin "up" R state is listed for the leptons — in contrast to the quarks; (b) the $U(1)_Y$ part of the standard group is also chiral so that there are distinct values, Y_L and Y_R — corresponding to the different "weak" isospin L and R representations, respectively — for the quarks and leptons [fractional — in units of $\frac{1}{3}$ — for the quarks and integral for the leptons — the factor 3 arises because each baryon is a 3-quark composite; it has already been mentioned in §1.3c (and will be discussed at greater length in §8.4) that $Y_L = B - L$ whereas $Y_R = B - L \pm 1$, with $+1$ corresponding to the "weak" isospin "up" quark and lepton states and -1 to the "weak" isospin "down" quark and lepton states, as indicated in column 5 of the table; (c) since the "weak" isospin chiral doublet contains two L states, we have noted in column 3 that the chiral weight of this state is 2 — the same as the dimension of the fundamental representation; correspondingly, the chiral weight of each R singlet state is 1 and is so indicated in column 4. Furthermore, since $SU(2)_L$ is a non-Abelian group, singlet representations of this group are "inert" under this group (i.e. possess no weak interaction);

Table 1.5: Quantum numbers of one generation of quarks and leptons under $SU(3)_C \times SU(2)_L \times U(1)_Y$ (and under $U(1)_{EM}$).

Particle	Rep. [] of $SU(3)_C$ (color weight)	L [] of $SU(2)_L$ (chiral weight)	R [] of $SU(2)_L$ (chiral weight)	$Y_L = B - L$; $Y_R = B - L \pm 1$	Q
$\begin{bmatrix} q^+ \\ q^- \end{bmatrix}$	[3]	[2]	[1]	$\begin{bmatrix} \frac{1}{3} \\ \frac{1}{3} \end{bmatrix}_L$	$\begin{bmatrix} \frac{2}{3} \\ -\frac{1}{3} \end{bmatrix}$
	(color triplet)	(weak doublet)	(weak singlets)	$\left[\frac{4}{3}\right]_R^+, \left[-\frac{2}{3}\right]_R^-$	
	(q_1, q_2, q_3)	$\begin{bmatrix} q^+ \\ q^- \end{bmatrix}_L$	$[q_R^+], [q_R^-]$		
$\begin{bmatrix} \ell^+ \\ \ell^- \end{bmatrix}$	[1]	[2]	[1]	$\begin{bmatrix} -1 \\ -1 \end{bmatrix}_L$	$\begin{bmatrix} 0 \\ -1 \end{bmatrix}$
	(color singlet)	(weak doublet)	(weak singlet)	$[-]_R^+, [-2]_R^-$	
	(ℓ)	$\begin{bmatrix} \ell^+ \\ \ell^- \end{bmatrix}_L$	$\left[\text{no } \ell_R^+\right], [\ell_R^-]$		

(d) the fact that three colors must be assigned to each quark implies that the quark of each generation belongs to the fundamental **3** representation of $SU(3)_C$ ($\bar{q}$ belongs to the conjugate $\bar{\mathbf{3}}$ representation), as indicated in column 2, and that each chiral quark flavor [whether "weak" isospin "up" (q_i^+) or "weak" isospin "down" (q_i^-)] carries 3 colors, i.e. the "color weight" of each q_i^+ or q_i^- is 3. On the other hand, the absence of a strong (color) interaction for the leptons implies that each chiral lepton state (ℓ_i^+) or (ℓ_i^-) belongs to a singlet representation of $SU(3)_C$ and carries the "color weight" 1. If one uses the "chiral weight" and "color weight" assignments given in columns 2-4. simple counting tells us that the standard gauge group must deal with 15 chiral (Weyl) fermion states per generation, for a total of 45 chiral fermion states for the three generations.

Further comments concerning Table 1.5 are: (e) we list in Table 1.5 the quantum numbers of the fermions in each generation for $U(1)_{\text{EM}}$. The quantum number under $U(1)_{\text{EM}}$ is, of course, the non-chiral electric charge $Q = I_{3L} + Y/2$. The chiral quark and flavor states are converted — through the SSB of $SU(2)_L \times U(1)_Y$ — into the non-chiral fermion states with the usual quark and lepton flavor names [the names of the three generations of quarks and leptons are given in Table 1.6]. The electric charges of the quarks and leptons of each generation are listed in Table 1.5 so that the "weak" GNN relation: $Q = I_{3L} + Y/2$ is seen to be confirmed. It is to be noted that the electric charges and the "weak" hypercharges of the quarks and leptons are quantized, i.e. we have quantization of charges associated with the Abelian groups $U(1)_{\text{EM}}$ and $U(1)_Y$; this unexpected result must be explained, as is done in §7.6; (f) it should be pointed out that, whereas the three charged fermions of each generation acquire masses through the "Higgs" mechanism (using appropriate Yukawa coupling constants), the neutrinos of the three generations stay massless — despite the "Higgs" mechanism — because there is no ν_R to construct a finite Dirac mass. After the SSB of the standard group by the "Higgs" doublet, each generation therefore consists of two non-chiral quark flavors, one non-chiral lepton flavor and one chiral massless ν flavor — for a total of 12 fermion flavors in the three generations (assuming that the t quark is found); and, finally (g) the replicative character of the three generations of quarks and leptons (always assuming the existence of the as-yet-unobserved t quark) is highlighted in Table 1.5 since the quantum numbers — under the standard group — listed for one generation of quarks and leptons are identical for the other two generations [the actual number of generations is three — fixed by a precision measurement of the

Z^0 decay width into neutrino pairs (see §6.3b)]. The physical quantity that distinguishes — in a striking fashion — the twelve fermion flavors in the three generations from one another is the mass, which varies by more than a factor of 10^5 from the lightest first-generation charged lepton [e, having a mass of 0.5 MeV] to the heaviest third-generation quark [t — having a mass exceeding 92 GeV (see §6.4c)].

Table 1.6: Q, B, L and "current" masses of the three generations of quarks and leptons (G = generation).

Particle	Q	B	L	G1 (current mass)	G2 (current mass)	G3 (current mass)
q^+	$\frac{2}{3}$	$\frac{1}{3}$	—	$u(4-5$ MeV$)$	$c(1350-1500$ MeV$)$	$(t)(>92$ GeV$)$
q^-	$-\frac{1}{3}$	$\frac{1}{3}$	—	$d(7-9$ MeV$)$	$s(140-175$ MeV$)$	$b(4.95$ GeV$)$
ℓ^+	0	—	1	$\nu_e(<7.3$ eV$)$	$\nu_\mu(<0.27$ MeV$)$	$\nu_\tau(<35$ MeV$)$
ℓ^-	-1	—	1	$e^-(0.51$ MeV$)$	$\mu^-(106$ MeV$)$	$\tau(1.78$ GeV$)$

We have listed in a separate table (Table 1.6) the presently known ranges and limits on the masses for the three quark-lepton generations. Insofar as the leptons are concerned, the charged lepton masses are known quite accurately and are listed with their errors in Table 1.6, while only the upper limits on the neutrino masses are listed. Because the quarks are "confined", their "current" masses can only be deduced from the masses of "free" color-singlet multi-quark systems (e.g. mesons, baryons, quarkonium) and this introduces considerable error, which is expressed by giving ranges for the quark masses in Table 1.6. The ratio $m_t/m_e > \simeq 2 \times 10^5$ is not the only striking feature of the quark and lepton masses listed in Table 1.6; the large variation in average mass between the quarks and leptons of the same generation, between the "up" and "down" quarks and between the "up" and "down" leptons within the same generation, and the mass "inversion" pattern of the first generation quarks (with the "up" quark being lighter than the "down" quark) as compared to the other two generations are all unsolved problems as of now. We have also listed in Table 1.6 the values of Q, B and L [the lepton number L is really distinct for the three generations (i.e. $L_e \neq L_\mu \neq L_\tau$) but this distinction will not be made until §6.3d] for the three generations and note that $\Sigma\, Q$ (electric charge) and $\Sigma\,(B-L)$ vanish (when summed over all quarks and leptons of the first and second generations); the t quark [the "up" quark of the third generation] has not as yet been seen but, for a variety of reasons (to be discussed in §6.4c), it is expected that $\Sigma\, Q = 0$ and $\Sigma\,(B - L) = 0$ also hold for the third generation of quarks and leptons. We will see in later chapters (in particular §7.5, §8.4 and §10.3a) why the interrelated conditions $\Sigma Q = 0$ and $\Sigma(B - L) = 0$ for each generation of quarks and leptons (in a sense, these two conditions define a "generation of fermions") are not accidental but follow from non-trivial properties of the gauge groups that comprise the standard model. Indeed, these two conditions are only part of the general mystery of the "fermion generation problem", to which we return in Chapter 9.

§1.4. Period of consolidation and speculation (1975–90): precision tests of the standard model and theories beyond the standard model

In the previous sections of this introductory chapter, an attempt has been made to provide a historical-conceptual framework for the understanding of the highlights in the evolution of particle physics from its early beginnings to its fairly definitive formulation by 1975 as the standard model in particle physics. The first section — covering the long period from 1687 to 1945 — recounted the pinnacles of achievement in the broadest strokes: the universal law of gravitation (transformed into the general theory of relativity); Maxwell's equations; special relativity,quantum mechanics and the relativistic theory of the electron (that merged into relativistic quantum field theory); the nuclear model of the atom and the nucleonic model of the atomic nucleus; and, finally, the identification of the "first circle" of "elementary particles" [electron, neutrino, proton, neutron — and their antiparticles — augmented by the photon, graviton and meson] that laid the foundations for the "birth" of modern particle physics ca. 1945.

When the highlights of the "Startup Period" (1945–60) in modern particle physics were reviewed in §1.2–§1.2e, we called attention to a large number of experimental surprises — made possible by the new particle accelerators and the "old" cosmic rays with improved experimental techniques — such as the distinctly different π and μ mesons, the menagerie of strange particles, the $\theta - \tau$ puzzle and parity violation in weak interactions. Theory also made its seminal contributions to the maturation of particle physics: QED, the two-meson theory, strangeness (and other global internal flavor symmetries), Yang-Mills gauge fields, the renormalization group in gauge theory, the V-A theory of weak interactions and the concept of spontaneous symmetry breaking. Not surprisingly, semi-phenomenological paths (e.g. S matrix theory), were pursued to gain insight into the intractable strong interaction and much of this effort cleared the way for the fundamental theory of the strong interaction that finally emerged.

It is fair to say that the "Heroic Period" (1960–75) — discussed in §1.3–§1.3d — saw, with the help of further crucial experiments [such as the detection of the muon neutrino, the detection of the Ω^- hyperon, the discovery of scaling in deep inelastic lepton-hadron scattering, the detection of neutral weak currents, the discovery of the τ lepton, and the detection of the c and b quarks] — the bringing together, into the standard gauge theory of the strong and electroweak interactions of many of the disparate theoretical ideas put forward during the "Startup Period". The theoretical triumphs of the "Heroic Period" included the quark flavor model of hadrons, the use of the chiral quark flavor symmetry group (plus PCAC and soft pion techniques) to underpin current algebra calculations with the weak and electromagnetic quark currents, the $SU(6)$ model leading to global $SU(3)$ "color", "Bjorken scaling", the Wilson operator product expansion for quark currents, the gauging of $SU(2)_L \times U(1)_Y$ for the chiral weak currents and the non-chiral electromagnetic currents, the gauging of $SU(3)$ global "color" and the recognition of the "asymptotic freedom" property of unbroken non-Abelian gauge groups, the quantization and proof of renormalization of non-Abelian gauge groups (either unbroken or spontaneously broken) and, finally, the recognition of triangular (fermion) axial anomalies in gauge theories. The major accomplishments of the "Heroic Period" are reexamined in considerably more detail in Chapters 4–7 after some preparatory discussions in Chapters 2–3.

The above list of achievements — both experimental and theoretical — during the "Heroic Period" is truly impressive and it is quite understandable that the next fifteen-year period (1975–90) in modern particle physics was a period of "consolidation" [the initiation of more incisive experimental tests of the standard model and theoretical attempts to place the standard model on a more rigorous mathematical foundation] and "speculation" [consisting of a large number of theoretical excursions "beyond the standard model" and a vast-ranging experimental effort to verify some of the theoretical speculations]. In the concluding sections of this chapter (§1.4a–§1.4c), we summarize some of the more exciting episodes during the "Period of Consolidation and Speculation" in modern particle physics and, in the process, we indicate which of these developments will be discussed in later chapters of the book (Chapters 8–10). The summary account of the "Period of Consolidation and

Speculation" is covered in three sections: §1.4a. Experimental consolidation of the standard model (1975–90); §1.4b. Theoretical consolidation of the standard model (1975–90); and §1.4c. Theoretical speculations beyond the standard model (1975–90).

§1.4a. Experimental consolidation of the standard model (1975–90)

In considering some of the key experiments that have justified the name "standard model", we distinguish between the final unbroken version of the "standard group", namely $SU(3)_C \times U(1)_{EM}$, and the spontaneously broken "standard group": $SU(3)_C \times SU(2)_L \times U(1)_Y$. We do so to call attention to the precision tests during the past fifteen years that have given remarkable confirmation of the QED part of the unbroken standard model before discussing the more indirect tests of the QCD part of the unbroken standard model. On the other hand, it must be acknowledged that the precision tests of the spontaneously broken part of the standard model (QFD) during the past fifteen years have been the most spectacular, thanks to the high energies and intensities available in colliders. Our brief comments in this section bearing on "experimental consolidation" in QED, QCD and QFD during the 1975–90 period are intended only as previews of more detailed treatments in later chapters.

Table 1.7: Some precision tests of QED.

Process	Experimental value	Theoretical value
Lamb shift	1057851 kHz	1057853(13) kHz [for rms radius of proton = 0.805(11) fm]
Anomalous magnetic moment of electron $\left[\frac{1}{2}(g-2)\right]$	$1159652188.2(\pm 3.0) \times 10^{-12}$ (average over e^+ and e^-)	$1159652140(31) \times 10^{-12}$ [for $\alpha_e = 137.0359979(32)$]

The program of testing the QED predictions has been led by Kinoshita and co-workers [1.121] on the theoretical side and several groups on the experimental side. We list in Table 1.7 two of the impressive comparisons between theory and experiment that have been established, namely the Lamb shift and the anomalous magnetic moment of the electron — the same two experiments that were discussed at the first Shelter Island Conference back in 1947 (see §1.2a) and catalyzed the formulation of QED by 1950. We return to the precision-testing program of QED in §4.5a, where the comparison between theory and experiment for the anomalous $(g-2)$ magnetic moment of the muon is discussed in some detail.

When we turn to the experimental testing of unbroken non-Abelian QCD, a number of developments during the past fifteen-year period can be cited. Quite a few experiments have been dedicated to verifying the deviations from "Bjorken scaling" [1.122] — those due to the difference between "Bjorken scaling" and the "asymptotic freedom" property of QCD [1.123] (see §5.2c) as well as those due to perturbative QCD corrections [1.124] — and the agreement between theory and experiment is satisfactory. Other attempts have been made to establish the existence of the "color charged" gluon: strong indirect support comes from the observation of gluon jets in hadron-hadron collisions and the determination — from the distribution of energy with angle of a third jet with respect to the two "quark" jets — that the third jet is, indeed, gluon-initiated [1.125]. Direct evidence for the basic tenets of QCD would come from the observation of color singlet "glueball" condensates; some claims have been made for the detection of $J = 2$ "glueballs" in the 2 GeV region [1.126] although there is no consensus on the subject [1.127]. One problem is possible competition from $q\bar{q}$ systems with the same quantum numbers as "glueballs" and there is not sufficient theoretical guidance as of now [from lattice gauge theory (LGT) [1.128] or elsewhere] to pin down the search for "glueballs".

Another area in which much work has been done during the past fifteen years, to establish QCD as the correct dynamical theory of the strong interaction, bears on the color confinement problem in QCD. According to QCD, no free quarks nor free gluons should be observed and have not been [1.98]; however, abnormal conditions (such as high "temperatures") might be achieved in relativistic heavy ion collisions [1.129] and give rise to a "deconfined" quark-gluon plasma; this would be a spectacular experiment and argue strongly for QCD but it still remains to be done. Probably, the most convincing support, until now, for color confinement in QCD comes from the extensive studies of heavy quarkonium spectroscopy [1.130], wherein the energy levels, quantum numbers, etc. of the heavy quarkonium systems are fitted extremely well by the long-distance type of behavior (linear in r) that ensures quark confinement; the linear long-distance potential (see Table 1.4) is expected from qualitative arguments based on the "color-electric" flux tube between q and $\bar{q}$ (see §5.3a). The confinement question can be settled by means of lattice gauge theory — which has given it a great deal of attention in recent years — if one can prove (see §5.3a) that: (1) the "area law" is satisfied by the Wilson loop for the observed value of α_s; and (2) the continuum limit of the lattice can be reached without encountering a first-order "phase transition". The first condition has been proved in the strong coupling limit [1.131] and the second condition still remains to be established (see §5.3a). One can hardly claim great success — let alone precision tests — for the non-perturbative (low energy) predictions of non-Abelian QCD, and not because of the lack of sufficiently high energy accelerators!. More will be said about this frustrating situation in the next section when we discuss the topological properties of non-Abelian QCD.

On the other hand, precision tests of various aspects of the QFD part of the standard group, and the nature of its spontaneous breaking to the electromagnetic group, has proved to be a much more productive undertaking during the past fifteen years because, among other reasons, half of the (chiral) fermions (the leptons) are "unconfined" and so are the four gauge fields $(W^{\pm}, Z^0, \gamma)$. The much larger scale of interest in QFD, $\Lambda_{QFD} \simeq 250$ GeV — a factor of more than 1000 larger than the QCD scale $\simeq 150$ MeV — requires experiments with the highest available energies and, when they succeed, the results are gratifying. As a result of the comprehensive experimental program to test QFD, a number of major achievements was registered during the past fifteen years; these include: precision determinations of the two key parameters in the electroweak theory (i.e. $\sin^2 \theta_W$ and ρ); the detection of the three weak bosons $(W^{\pm}, Z^0)$ and precision measurements of their masses and decay widths; the determination that there are only three neutrino "species" (as long as the neutrinos possess masses < 40 GeV) and, *a fortiori*, precisely three quark-lepton generations; the establishment of lower limits on the t quark mass ($m_t > 92$ GeV) and the Higgs doublet mass ($M_H > 60$ GeV [1.132]); the discovery of large mixing in the neutral B_d system; almost complete determination of the CKM quark mixing matrix; and, finally, substantial reductions of the upper limits on the branching ratios of a variety of rare weak decays. [We do not give references for these many results because all of them are discussed in greater detail in Chapter 6 (where references are given).] As is well known, none of the results is in disagreement with the electroweak part of the standard model but it must be said that the 92 GeV lower limit on the t quark mass is quite unexpected and its measurement, together with that of M_H, is receiving the highest priority. [The unexpectedly large t quark mass has opened up the possibility that there is a common dynamical symmetry-breaking origin of both m_t and M_H and the renewed interest in this possibility will be discussed in §6.4c.] Despite valiant efforts, the measurement of the parameter ϵ'/ϵ in neutral kaon decay (to detect "direct" weak CP violation) is inconclusive, and there is no convincing evidence for finite neutrino masses nor for neutrino oscillations (topics also to be discussed in more detail in Chapter 6).

Overall, the standard model has withstood a large battery of probing experimental tests during the past fifteen years and it is interesting to judge (in the next section) how the feats of theoretical consolidation (*vis-à-vis* the standard model) fared during the same period.

§1.4b. Theoretical consolidation of the standard model (1975–90)

In reviewing the highlights of experimental consolidation of the standard model during the past fifteen years, we commented on those aspects of theoretical consolidation that complemented the experimental effort. In this section, we focus instead on those aspects of theoretical consolidation during this period that attempted to probe more deeply into the conceptual foundations of the standard model, without the expectation of experimental tests in the near future. It is again useful to record in turn our comments on "theoretical consolidation" of the standard model during the 1975–90 period under the headings of QED, QCD and QFD. We defer to the last section of this chapter (§1.4c) a brief review of the theoretical speculations during the past fifteen years that purport to move substantially beyond the standard model: grand unification, preon models and super-unification.

Insofar as further theoretical consolidation of QED is concerned, we have already mentioned the ambitious higher-order "radiative" calculations that have been carried out [1.121] and the excellent agreement obtained between theory and experiment. However, there is still intermittent interest in gaining a clearer understanding of the "Landau singularity" or "zero charge" problem [1.58] (see §4.5a) and some suspicions that some new physical concepts are needed when "the electromagnetic interaction becomes strong" [1.133], especially since the much-pursued Higgs sector in the electroweak gauge theory possesses a similar asymptotic behavior to that of QED but with a possibly much larger coupling constant. Just as questions were raised back in the 1950s about the "reality" of renormalized electric charge, so are questions now being raised about the "reality" of the Higgs sector in electroweak theory ("triviality question") [1.134]. Some authors have been moved by the surge of interest in topological conservation laws (see Chapter 10) during the past fifteen years to even suggest the possibility of soliton solutions in the "strongly-interacting" region of electromagnetism and the Higgs sector [1.135]. These highly tentative developments are noted under the rubric of the consolidation period in particle physics because they basically draw their motivation from an attempt to fathom the deeper meaning of those quantum fields in the standard model that are not asymptotically free.

Even greater theoretical efforts have been made during the period of consolidation to fully comprehend not only asymptotic freedom and confinement but also the topologically non-trivial features of non-Abelian QCD that are known to exist (e.g. instantons), and not merely hypothesized (e.g. 't Hooft-Polyakov magnetic monopoles). Curiously enough, Dirac's hypothetical magnetic monopole in Abelian QED (see §4.6) has been a useful tool in defining the conditions under which topologically non-trivial soliton solutions appear in non-Abelian gauge theory. The search for such solutions in classical Yang-Mills theory led to the discovery of instantons [1.136] with integral "Pontryagin number" n $(n = \pm 1, \pm 2, \ldots \pm \infty)$ which have a marked effect on the QCD vacuum [converting it into the so-called θ-vacuum (see §10.3)], with the unhappy consequence that the effective Yang-Mills Lagrangian acquires a potential parity-violating ($\not{P}$) and CP-violating ($\not{C}\not{P}$) term, contrary to experiment (see §10.3c). While, on the one hand, the θ-vacuum in QCD is welcome because it solves the so-called $U(1)$ problem [i.e. the absence of a ninth N-G boson — in addition to the pseudoscalar meson octet [1.137] (see §10.3b)], the additional $\not{P}$ and $\not{C}\not{P}$ term must be extremely small to be compatible with the known upper limit on the electric dipole moment of the neutron (see §10.3b). The "unnaturalness" of this miniscule term in the augmented QCD Lagrangian (due to instantons) has led to a herculean effort to modify the initial QCD Lagrangian so that the unwanted term disappears in a "natural" way. The Peccei-Quinn (PQ) hypothesis of an additional invariance of the QCD Lagrangian under an Abelian chiral group, called $U(1)_{PQ}$ [1.138], seemed very attractive at the beginning of the "Period of Consolidation and Speculation". However, since $U(1)_{PQ}$ can not be an exact symmetry, the low mass pseudoscalar particle (called axion) produced from the SSB of $U(1)_{PQ}$ — must be found. The search for the axion — "visible" and "invisible" (with further astrophysical and cosmological consequences) — has been extraordinarily intensive during the past fifteen years (see §10.3c), with disappointing results so far. It is unlikely that the Peccei-Quinn mechanism is capable of resolving the so-called "strong CP" problem in QCD. This is a serious matter and one can only hope that the "strong CP" problem will disappear through more careful treatment of the topologically non-trivial, non-perturbative QCD vacuum. Some comments regarding this possibility will be found in §10.3c.

Interest in exploring the topological implications of the new non-Abelian QCD theory of the strong interaction was pursued with considerable fervor in a totally different direction during the period of consolidation, namely in the reactivation of the old Skyrme model [1.139] in modern QCD "dress". It will be recalled that the Skyrme model of more than thirty years ago, pointed out that a topological baryonic soliton could emerge from a set of (purely) pionic fields subjected to a suitably chosen quartic interaction among them; while the existence of a soliton solution was demonstrated, it could not be established whether the topological solitons are fermions or bosons. The "new look" at the Skyrme model [1.140] starts with the non-linear σ model and works with the non-linear realization of the underlying $SU(3)_L \times SU(3)_R$ group for three flavors and the usual Skyrme type of quartic interaction among the eight N-G "pion" fields. The real innovation consists of adding a Wess-Zumino term [1.141] to explicitly incorporate triangular axial anomalies in the Skyrme action; one can then demonstrate that the resulting topological solitons (for $B = 1$) are fermions if the number of colors is odd (which is true for QCD) and that these solitons correspond to the phenomenologically correct low-lying baryonic representations. While the agreement between the detailed predictions of the "new" Skyrme model and experiment is only fair (approximately 30% agreement for various low-energy parameters), the "new" Skyrme model makes an intriguing attempt to simulate the non-perturbative (low-energy) features of QCD — via the topologically non-trivial properties of a non-Abelian field theory — and is discussed at considerable length in §10.5.

Our concluding comment in this section bears on one of the rather successful examples of the theoretical consolidation of the standard model during the past fifteen years, namely the attempt to probe more deeply into the role of chiral gauge anomalies in four dimensions in fixing the structure of the standard model. The first anomaly to be discovered — the perturbative, triangular axial (ABJ) anomaly [1.142] — was identified in connection with $\pi^0 \to 2\gamma$ decay where it plays a "harmless" and constructive role in explaining the observed decay width of $\pi^0 \to 2\gamma$. It was soon realized that the chiral (lefthanded) coupling of fermions to gauge fields in QFD introduces a triangular axial anomaly that is no longer "harmless" and that the "triangular" chiral gauge anomalies must be cancelled [1.143] in order to maintain the renormalizability of QFD. While there is no "harmful" triangular axial anomaly in QCD (because the coupling of the fermions to the gluon fields is "vector-like"), there is a "harmless" global triangular axial anomaly in QCD [resulting from the coupling of an axial vector quark current to one of the three vertices (in the fermion triangle) when the other two (vector) quark currents are coupled to the non-Abelian gluon fields]. Like the ABJ anomaly, the triangular axial anomaly in QCD has useful applications [e.g. in connection with the $U(1)$ problem in QCD and the "proton spin problem" (see §7.2b)]. A second "harmful" chiral gauge anomaly in four dimensions [having a topological (instanton) origin], namely the non-perturbative, global $SU(2)$ anomaly, was uncovered in the early 1980s [1.144], wherein it was shown that the absence of the global $SU(2)$ anomaly is required to maintain the mathematical self-consistency of the chiral gauge part of the standard model (i.e. QFD). Finally, a third perturbative, chiral gauge anomaly in four dimensions that can be troublesome for electroweak theory, namely the mixed gauge-gravitational anomaly, was identified in the mid-1980s [1.145] and it — like the perturbative triangular axial anomaly — has to be cancelled (by summing over the quarks and leptons of one generation) to guarantee renormalizability of the theory. It turns out that all three chiral gauge anomalies in four dimensions are absent when the known representations and quantum numbers of each quark-lepton generation (see Table 1.5) are introduced into the anomaly coefficients (see §7.5).

Since the two perturbative anomalies (the "triangular" and the "mixed" anomalies) possess long wavelength limits, and the existence of the non-perturbative global $SU(2)$ anomaly depends upon the topologically non-trivial gauge transformations associated with the $SU(2)_L$ part of the electroweak group, the three chiral gauge anomalies are, in a sense, reflecting the non-perturbative aspects of the standard model, and their absence in the standard model should be highly significant. Indeed, one can turn the question around, and inquire into the uniqueness of the representations and quantum numbers of the Weyl fermions in the standard model when one accepts an $SU(3) \times SU(2) \times U(1)$ group (with an arbitrary set of representations) as the starting point and imposes the three chiral gauge anomaly-free constraints. One finds a surprising degree of uniqueness emerging from the imposition of the three anomaly-free constraints and, furthermore, an unexpected quantization of the "weak" hypercharge, as well as electric charge (after the "Higgs" breaking of the electroweak group) at the standard model level [1.146]. This instructive example of theoretical consolidation of the standard model during the past fifteen years is fully considered in Chapter 7.

§1.4c. Theoretical speculations beyond the standard model (1975–90)

The great success of the standard model, and at the same time the recognition of its inadequacies, encouraged an assortment of theoretical speculations hoping to improve upon it or, more ambitiously, to go beyond it in the direction of a theory of everything (TOE) — which includes gravity. It is difficult to give a precise definition of TOE since it is an evolving concept depending on the state of knowledge at the time; as of now, a reasonable definition of TOE might be: a TOE is a theoretical synthesis that aims to understand the fundamental constituents of matter (including mass), to achieve a "super-unification" of the four basic forces of nature, and to arrive at a plausible description of the universe's development from its earliest ("big bang") stages to the present and into the future. This is a tall order and what distinguishes the past fifteen-year period from earlier ones is that the achievements of the standard model in particle physics, combined with important progress in astrophysics and cosmology, has created an intellectual environment that encourages TOE speculations.

The theoretical speculations beyond the standard model during the past fifteen years fall roughly into two categories: (A) going beyond the standard model in four-dimensional Minkowski space (with experimental tests where available); and (B) searching for a TOE. The "speculations" reviewed under category (A) involve model building under the rubric of the traditional precepts of relativistic quantum field theory (such as four space-time dimensions, point interactions, and no supersymmetry). With these stipulations, we offer comments under Category A on: (A1) unification models (partial and grand) and proton decay; (A2) technicolor models and the Higgs boson; (A3) composite preon models and the fermion generation problem; (A4) sphalerons and baryon charge violation in the electroweak theory and (A5) majorons and familons. The TOE speculations — grouped under category (B) — all attempt to unify the gravitational interaction with the other three interactions of the standard model and are compelled — at least at present — to surrender one or more of the traditional precepts of relativistic quantum field theory. TOE theoretical activity has been so overwhelming during the past fifteen years that we do not pretend that the topics selected for cursory review in the final section of this introductory historical chapter do justice to this vast output. With this disclaimer, we offer brief remarks concerning the TOE speculations in category (B) under the sub-headings:

(B1) supersymmetry and supergravity; (B2) Kaluza-Klein models and; (B3) superstring theories.

Category A: Speculations within the framework of traditional quantum field theory

(A1): partial and grand unification models. While the non-semisimple standard gauge group of the strong and electromagnetic interactions does not mix the quark and lepton representations, the requirements of chiral gauge anomaly-free constraints on the standard group are only fulfilled when the quantum numbers associated with both the quark and lepton representations are taken into account. These anomaly-related "linkages" between QCD and QFD [to be discussed in §8.1] are supplemented by several other "linkage" arguments and a series of plausibility arguments supporting unification of QCD and QFD — to be discussed in §8.2. Whatever the motivation, partial unification groups [e.g. the left-right-symmetric group $SU(2)_L \times SU(2)_R \times U(1)_{B-L}$ [1.147] and the Pati-Salam group [1.148] (see §8.4a)] as well as grand unification groups (GUTs) were studied extensively during the past fifteen years. The most attractive (lowest rank) $SU(5)$ GUT group [1.149] makes a unique prediction about proton decay (via the process $p \to \pi^0 + e^+$) and a world-wide effort to check this prediction was unsuccessful [1.150]. For a time, the $SU(5)$ GUT model acquired additional interest because $SU(5)$ GUT (implies topologically-favorable conditions (§10.2c) for the production of a 't Hooft-Polyakov-type of magnetic monopole[1.151] with a predicted mass of approximately M_U/α_e [where the unification mass $M_U \simeq 10^{15}$ GeV] $\sim 10^{17}$ GeV $\sim 10^{-5}$gm; the search for the 't Hooft-Polyakov monopole was taken up — albeit not as seriously as for proton decay — and was equally unsuccessful. The failure of $SU(5)$ GUT on several counts opened the door to further GUT speculation with regard to the two remaining candidate GUT groups: $SO(10)$ [1.152] and E_6 [1.153]. The higher rank GUT groups: $SO(10)$ and E_6, can easily be made compatible with the proton decay experiment but none of their other predictions (e.g. the existence of more than one neutral weak boson in the multi-100 GeV region) has thus far received experimental confirmation. The status of unification models beyond the standard model is discussed more fully in Chapter 8.

(A2): Technicolor models and the Higgs boson. A major unsolved problem of the standard model is the question of whether the "electroweak" Higgs boson is an "elementary" particle (with the same status, say, as the gauge fields) or is a quasi-Higgs boson (e.g. a fermion condensate along the lines of the Cooper pair of electrons in superconductivity [1.90]). The analogy with the BCS theory of superconductivity (see §4.4a) suggested very early that a reasonable choice for a quasi-Higgs boson in the electroweak theory is a $q\bar{q}$ condensate; however, until recently, the very small observed masses of the five known quarks (u, d, s, c, b) compared to the electroweak mass scale Λ_{QFD} (~ 250 GeV) greatly discouraged such an approach. This did not prevent the hypothesizing of a new class of fermions, called "technifermions", bound by a "technicolor" confining force that could give rise to a (quasi-) Higgs boson. Since, in the original technicolor model [1.154], no provision was made for an interaction between the technifermions and the ordinary fermions, the "technifermion-produced" Higgs bosons could generate finite mass for the weak gauge fields but not for quarks and leptons. Attempts to refine the technicolor theory to overcome this deficiency, in the form of an "Extended Technicolor theory" [1.155], ran into the problem of large flavor-changing neutral currents in QFD. While the technicolor model has not been abandoned — interesting papers are still being written on the subject ("walking technicolor" [1.156], etc.) — interest has shifted back to the possibility that the $t\bar{t}$ condensate is, after all, the quasi-Higgs boson required to break electroweak symmetry. Experiments of the past few years, placing a lower limit on the t quark mass of 92 GeV [1.157] has reopened the question of whether a quasi-Higgs boson can result from a dynamical symmetry-breaking mechanism involving the $t\bar{t}$ condensate; this possibility is discussed in §6.4c.

(A3): The fermion generation problem and preon models. Apart from the question of whether the electroweak Higgs boson is "elementary" or a fermionic composite, there is the interesting question of whether ordinary quarks and leptons are themselves "preon" composites. While the point-like structure of quarks and leptons has been established down to the 10^{-16} level (or, equivalently, up to 100 GeV [1.158]), the hypothesis of composite quarks and leptons has been stimulated by the observational fact that there are three quark-lepton generations — with precisely the same quantum numbers under the standard group $SU(3)_C \times SU(2)_L \times U(1)_Y$ — which differ only in mass (this is the "fermion generation problem"). A "natural" composite model approach to the "fermion generation problem" is to assume (in analogy to the quark flavor model of hadrons) that quarks and leptons are three-preon composites and to ascertain the conditions under which — employing standard quantum gauge field theory and the 't Hooft anomaly matching condition (between the "elementary" preons and the three-preon composites [1.159]) — the preon model can yield three fermionic generations with the known quantum numbers of quarks and leptons. The preon model was pursued vigorously during the early 1980s and, while the qualitative conclusions for some of the models were rather encouraging, quantitative calculations of the quark and lepton masses of the three generations did not fare too well [1.160]. Nevertheless, preon model building is a good example of trying to push beyond the standard model in fairly straightforward ways to solve the "fermion generation problem" and is discussed in some detail in Chapter 9.

(A4): Baryon charge violation and sphalerons. We have already mentioned the discovery, in 1975, of the (classical) instanton solutions in non-Abelian gauge theory [1.136] and have commented (in §1.4b) on the topologically non-trivial effects (the θ-vacuum) that the "color instantons" induce in QCD. It is evident that there should be "flavor-instanton" solutions associated with the non-Abelian $SU(2)_L$ part of the electroweak group and, indeed, the baryon and lepton charge-violating consequences of these "flavor instantons" were examined as early as 1976 [1.161], resulting in two conclusions (see §10.3a): (1) the B- and L-violating effects are miniscule, being of the order of $e^{-2\pi/\alpha_e} \sim e^{-400}$; and (2) the B- and L-violations just cancel each other so that there is no flavor-instanton-induced global $B - L$ violation. These conclusions seemed uninteresting and were put aside, until fairly recently, after repeated attempts failed to find a "particle physics" mechanism to support the plausible Sakharov-type explanation of the baryon-asymmetric universe (BAU — see §2.2d). Revival of interest in the possibility of suppressing the "flavor-instanton"-inhibiting effects *vis-à-vis* electroweak baryon violation has taken place because of the realization that the early instanton calculations had assumed zero temperature and that the extraordinarily high temperatures present during the early universe might create a more favorable situation for electroweak baryon violation. Such high temperature calculations have been carried out [1.162] and, indeed, it is found that there is well-nigh complete cancellation of "flavor instanton" suppression by the unstable (saddle point) topological solutions (called "sphalerons") to the Yang-Mills-Higgs equations. Put another way, in the electroweak case, there is a Higgs-induced "sphaleron" barrier of about 10 TeV and, consequently, there is a large exponential enhancement of baryon violation when the temperature $T \sim 10$ TeV. This result may provide a new handle for an explanation of BAU. A similar enhancement of baryon violation within the framework of the spontaneously broken electroweak group may also occur — according to some calculations [1.163] — in very high energy hadron-hadron collisions (at energies in excess of 10 TeV) provided the end products in the collision contain high multiplicities of weak bosons $(Z, W^{\pm})$ and Higgs particles. Further comments bearing on "sphalerons" and related phenomena will be found in §10.3a.

(A5): Majorons and familons. Needless to say, theoretical speculation during the past fifteen years — to push beyond the standard model within the framework of three basic forces (electromagnetic, strong and weak) and Minkowski space — did not stop with the efforts reviewed under A1–A4 above. It is impossible to comment on all such efforts and so this category A of §1.4c is concluded with some brief remarks concerning "majorons" and "familons". The "majaron" and "familon" are only two examples of N-G bosons that have been invented to solve real problems in the standard model. In principle, these N-G bosons are difficult to detect, because the long-range forces to which they give rise are spin-dependent; however, there are more phenomenologically attractive ways in which such N-G bosons could make their existence known, as will be illustrated in the cases of the "majoron" and the "familon". The real problem that has created interest in the "majoron" is the question of whether the three separate lepton charges L_e, L_μ and L_τ are absolutely conserved or whether there is an SSB mechanism that is responsible for the spontaneous breakdown of one or more of these lepton charges. We are in a curious situation with regard to the lepton charges of the three observed lepton generations; we know that for certain purposes, L_e, L_μ and L_τ can be treated on a par: for example, the leptonic "weak" hypercharges that ensure triangular anomaly cancellation for each generation do not distinguish among L_e, L_μ and L_τ (see §6.3d) nor is a distinction made among L_e, L_μ and L_τ in deducing global $B - L$ conservation in the presence of "flavor instantons" (see (A4) above)]. However, we know that, at the relatively low energies available to us, $L_e \neq L_\mu \neq L_\tau$ and that only the sum $(L_e + L_\mu + L_\tau)$ is conserved. The general idea of "majorans" is that the presumed spontaneous breakdown of some linear combination (other than the sum) of the three (generational) lepton charges can be achieved by coupling a weak isospin singlet (or triplet) scalar (Higgs) field to a lepton pair and postulating that the scalar field acquires a VEV (vacuum expectation value) which gives rise to the N-G "majoron" [1.164]. The existence of a "majoron" could give rise to inter-generational neutrino decay, a modification of neutrinoless double β decay, and so on; experiments that have been undertaken to test some of the predictions have not met with success [1.165].

The "familon" has a similar origin to that of the "majoron": the replication of three families (generations) of quarks and leptons with identical quantum numbers under the standard group can be ascribed to a global family group, say, $SU(3)_f$, which must be (spontaneously) broken to explain the quite different quark and lepton masses in each generation, thereby generating N-G bosons, which are called "familons" [1.166]. Experimental searches for "familons" [e.g. by looking for the decay $K \rightarrow \pi + f$) where f is a "familon"] have been no more successful than those for "majorons" and have merely succeeded in setting a lower limit on the "familon" SSB scale, namely $\Lambda_f > 10^{10}$ GeV [1.167]. Some further remarks concerning global (and gauge) family groups will be found in §9.1.

Category B: "Theory of Everything" (TOE) speculations

The theoretical speculations considered above under category A all accept the basic tenets of conventional quantum field theory in Minkowski space (or "Wick-rotated" Euclidean four-space): new particles (both fermions and bosons), new groups (both global and gauge) and new SSB mechanisms are permitted, but increasing the number (D) of space-time dimensions, or surrendering the point-like character of the fundamental constituents of matter, or deviating from the assumption that all generators describing global internal symmetries should obey commutation (not anti-commutation) relations, are not allowed. It was the last piece of conventional wisdom that was sacrificed by supersymmetry and supergravity, which is the first topic (B1) that is discussed under the rubric of category B, covering TOE speculations. $D = 4$ space-time dimensions and "point-like" (structureless) particles (quarks, leptons, gauge fields, graviton, etc.) were still accepted by the early versions of supersymmetry and supergravity. The consideration of relativistic quantum field theory in more than $D = 4$ space-time dimensions (namely 5), i.e. the original Kaluza-Klein theory is a pre-"modern particle physics" idea dating back to the 1920s but its reincarnation in $D = 11$ space-time dimensions and in combination with supersymmetry, only occurred during the past fifteen years and its prospects as a TOE are briefly reviewed in part (B2). Finally, the most ambitious of all TOEs — superstring theory — foresakes more key elements of the conventional wisdom in quantum field theories [by accepting supersymmetry, moving to $D = 10$ space-time dimensions and substituting string-like for point-like particles] and is treated briefly in part (B3).

(B1): Supersymmetry and supergravity. The idea of $D = 4$ supersymmetry was launched during the mid-1970s [1.168], shortly after the "spinning string-dual resonance" model [1.169] was developed and several years after the "no-go" theorem [1.170] (see §4.1) clarified the conditions under which it is possible to mix a global Lie internal symmetry group with a Poincaré group in order to achieve a closed finite-dimensional algebra. Resorting to the gauging of global non-Abelian internal symmetries (the path that led first to QFD and then to QCD as a response to the "no-go" theorem) was, in a sense, the more conservative approach (albeit remarkably successful) but the less traditional "supersymmetry" response to the "no-go" theorem stimulated the theoretical speculations that have occupied many in the particle theory community during the past fifteen years.

The basic idea of supersymmetry is to include generators (corresponding to global internal symmetries) that satisfy (Fermi-Dirac) anti-commutation relations (the "supercharges") so as to render "harmless" the "no-go" theorem; if one "supercharge" is introduced, one has $N = 1$ supersymmetry and $N \geq 2$ supersymmetry is a rather straightforward generalization of $N = 1$ supersymmetry. The fermionic "supercharges" Q_α can then be mixed in a non-trivial way with the Poincaré generators to produce what is called a "graded" supersymmetric algebra that closes upon itself. Further, from the anti-C.R.'s (commutation relations) of the Q_α's, it follows that the Q_α's commute with P^2 (the four-momentum squared) so that there is mass degeneracy of the fermions and bosons in the same supersymmetric multiplet. On the other hand, the Q_α's do not commute with W^2 [W^μ is the Pauli-Lubanski "spin-momentum" vector — see §2.1c] so that, in unbroken supersymmetry, the fermionic and bosonic members of the supersymmetric multiplet are always degenerate in mass but differ in spin — by a half integer for $N = 1$ supersymmetry or several half integers for $N \geq 2$ supersymmetry.

A brief review of supersymmetry (primarily $N = 1$ global supersymmetry) is given in §2.3a and, in this introductory historical section, we merely add some supplementary remarks. One reason for the rapid growth of interest in supersymmetry during the past fifteen years is that, while the dynamical non-Abelian gauge theory of the strong and electroweak interactions (QCD and QFD) rendered the "no-go" theorem moot during the early days of supersymmetry, it was soon realized [1.171] that supersymmetry is not incompatible with the $D = 4$ standard model but might enlarge its scope (e.g. to include gravity) and overcome some of its unpleasant features (e.g. the "gauge hierarchy"). It was clear at the outset that, within the framework of unbroken global supersymmetry, the number of "fermionic" degrees of freedom (i.e. "on-shell polarizations") must equal the number of bosonic "on-shell polarizations", and that a distinction has to be made between the "on-shell" and "off-shell" polarizations [e.g. the "on-shell" photon has 2 polarizations but the "off-shell" photon has 3]. Thus $N = 1$ global supersymmetry can bring into the same "on shell" (supersymmetric) representation one $J = \frac{1}{2}$ Weyl fermion and two $J = 0$ massless bosons, or one $J = \frac{1}{2}$ Weyl fermion and one $J = 1$ gauge field, or one $J = 2$ graviton and one $J = \frac{3}{2}$ massless fermion (called a "gravitino"); if the massless "on-shell" supersymmetric representation is to contain a mixture of bosonic and fermionic fields with spins from $J = 2$ to $J = 0$ — so that a place can be found for the presently recognized "elementary" particles: the $J = 2$ graviton, the $J = 1$ gauge fields, the $J = \frac{1}{2}$ quarks and leptons, and, provisionally, the $J = 0$ Higgs field — it is necessary to have $N \geq 4$ supersymmetry.

For a particular choice of $N \geq 4$, the distribution of multiplicities of the massless "on-shell" fields from $J = 0$ to $J = \frac{3}{2}$ is determined [there is always just one $J = 2$ graviton]. Thus, flexibility in the choice of N allows the bringing together — under one supersymmetric representation — the gravitational field with specified numbers of gauge, Weyl and massless scalar fields — at the price of accepting a fixed number (depending on the choice of N) of $J = \frac{3}{2}$ "gravitino" fields; this result opens up the possibility of moving on from global supersymmetry to local (gauged) supersymmetry [1.171] [without and with spontaneous symmetry breaking — thereby giving rise to "gauginos" and "Higgsinos"] and, even more ambitiously, to supergravity [1.172]. The basic strategy in extending global supersymmetry to local supersymmetry and supergravity is to insist that the supersymmetry transformations, which maintain the gauge and/or general covariance of the respective Yang-Mills and/or Einstein fields, satisfy the appropriate "on-shell" supersymmetry algebra. A very useful tool for carrying out this program and retaining manifest supersymmetry at each step of the calculations is the introduction of so-called "superspace" and the correlative concept of "superfield" [1.173].

We do not enter into the mathematical details of the many supersymmetric Yang-Mills and supergravity theories that were considered during the early 1980s. It was an exciting time for particle-theoretic speculation, especially when it was recognized that the "co-existence" of fermions and bosons in the same (supersymmetric) representation possesses the welcome property that the opposite sign contributions of fermions and bosons to radiative corrections (see §2.3a) go far towards alleviating the "gauge hierarchy" problem (see §8.4). Despite this promise for the supersymmetry program, a major deficiency of supersymmetry soon became apparent, namely the absence of supersymmetric partners of the same mass in nature. In view of the failure to detect supersymmetric partners of any type — after persistent efforts — it became necessary to consider the SSB of supersymmetry and to examine the consequences of a variety of SSB mechanisms. Plausible arguments were given that supersymmetric partners (assuming that supersymmetry is even an approximate symmetry) ought to possess masses that are not grossly different from each other [1.174]. An example of this type of argument concerns the relative masses of the gauge field and its superpartner, the "gaugino": it is known that the one-loop radiative corrections (due to the gauge field) to the mass of the Higgs boson is softened to a logarithmic divergence (if supersymmetry does its job — see §2.3a); it is also known that there is an upper bound to the Higgs boson mass of 1 TeV (see §6.3c). It follows that the "gaugino" mass ought not to exceed about 1 TeV (which is within range of the soon-to-be-constructed accelerators). More complicated arguments lead to the expectation that squarks and sleptons (the superpartners of the quarks and leptons respectively) should be observed with masses on the order of M_W (see §2.3a). While "gauginos", "squarks", "sleptons" and other supersymmetric partners like "Higginos", "photinos", and "gluinos" are not ruled out, it must be admitted that there is no phenomenological support, thus far, for supersymmetric partners of any kind (of known particles) in the 100 GeV energy region.

On the theoretical front, the extension of the supersymmetry principle to local supersymmetry and, through it, to supergravity for $D = 4$ (space-time dimensions) raised hopes for a more convergent quantum gravity theory as well as for general progress towards the unification of all four basic forces of nature (i.e. in the direction of a TOE). However, such hopes were soon tempered by the realization that any realistic supergravity theory requires a host of new particles (besides one graviton), a multiplicity of fermions and bosons with J's ranging from $\frac{3}{2}$ to 0, and, furthermore, that the supergravity theory must have at least four "supercharges" (i.e. $N \geq 4$). It became apparent that the elegance of the supersymmetry principle is undermined by the need for large numbers of unseen "exotic" fermions and bosons and the arbitrariness of additional parameters required for $D = 4$ space-time dimensions. With the old Kaluza-Klein (KK) theory [1.175] and the old "dual resonance" model [1.110] lurking in the background, the time seemed ripe to co-opt the attractive supersymmetry principle but to refocus the direction of TOE model building by moving beyond four dimensions. The two most sustained efforts to construct TOE's by increasing the number of space-time dimensions beyond the four of Minkowski space, have clearly taken their cues from the early KK theory and the "dual resonance — dual string" model, and we comment briefly on the Kaluza-Klein (KK) class of theories in part (B2) and the superstring (SS) class of theories in part (B3) of this section.

(B2): Kaluza-Klein models. We start with the KK theories because they have a much longer history (dating back to 1920) and ran into serious trouble just about the time that the SS theories came into their own — in the mid-1980s. The first KK theory was a classical five-dimensional ($D = 5$) field theory (four space dimensions — one of which is compactified — plus one time dimension) in which it was shown that one could employ the compactified space dimension to represent the gauged internal (charge) symmetry of electromagnetism; the $U(1)$ gauge symmetry arises from the fact that the compactification of the extra space dimension is done on a circle which possesses a $U(1)$ isometry [i.e. a set of coordinate transformations — the rotations of the circle — which leave the metric invariant]. Because of general covariance, the circle can be translated by a different amount at every point in Minkowski space and hence the $U(1)$ isometry is converted into a $U(1)$ gauge symmetry for a "Minkowski" observer; one has, so to speak, unification of gravitation and electromagnetism in four dimensions [1.175]. [This was the type of "unified theory" on which Einstein labored for several decades, without success [1.176], and which led Weyl to invent his version of the "gauge principle" [1.177]]. The $D = 5$ KK theory became archaic when the two additional basic forces of nature (i.e. the strong and weak interactions) were identified in the 1930s and the KK theory was only revived — in a larger number of dimensions ($D > 5$) after the non-Abelian gauge theories of the strong and electroweak interactions were in hand. It was soon realized [1.178] that if one wished to follow the traditional KK approach, i.e. identify the gauge fields as due to the isometry group of the metric tensor in the compactified dimensions and still account for the twelve gauge fields associated with the standard group $SU(3)_C \times SU(2)_L \times U(1)_Y$ [rather than the one gauge field of $U(1)_{\rm EM}$], it is necessary to choose $D = 11$.

The choice of a supersymmetric KK model in $D = 11$ dimensions (10 space plus 1 time) possesses several virtues: (1) it is possible to use the smallest number of "supercharges" (which simplifies the theory), i.e. $N = 1$ in $D = 11$ space-time dimensions; (2) there are enough additional "space" dimensions to compactify in order to generate a gauge group that contains the phenomenologically correct standard gauge group in $D = 4$ space-time dimensions; and (3) $N = 1$ supergravity in $D = 11$ space-time dimensions is uniquely translated into $N = 8$ supergravity in $D = 4$ space-time dimensions. As the $D = 11$ KK theory unfolded, a variety of problems began to surface, e.g. it is not finite to higher than the first order. However, the greatest setback occurred when it was realized that a $D = 11$ KK model has no place for chiral (Weyl) fermions [1.179], which is indispensable to the standard model. The point is that odd D does not allow for chiral fermions whereas even D permits chiral fermions; it is true that the chiral fermions introduce anomalies into the theory that must be cancelled but this is welcome because one can take advantage of the required cancellation to constrain the theory [as is done in the $D = 10$ superstring theory (see below)]. It has been shown recently that one can construct a $D = 11$ KK model that possesses many of the properties of the heterotic string model [see (B3)] by introducing the gauge fields *ad hoc* [1.180]; this ploy, however, surrenders a very attractive feature of the $D = 11$ KK theory [whereby the gauge fields arise "naturally" from the compactification of the additional space dimensions] and it does not fully dispose of the chiral fermion problem.

(B3): Superstring theories. While the $D = 11$ KK theory is, in certain ways, a fairly straightforward generalization (with supersymmetry added!) of the early $D = 5$ KK model, the modern version of superstring theory [1.181] is a far cry from the four-dimensional "spinning string" model of twenty years ago [1.169] that inspired it. The old "spinning string" model was supposed to be a theory of hadrons in four space-time dimensions but the appearance of scalar tachyons, massless vectors and massless tensors (gravitons) seemed to require $D = 10$ (rather than $D = 4$) space-time dimensions in order to maintain self-consistency. The "spinning string" model does possess a two-dimensional "world-sheet" supersymmetry [1.182] — an interesting property, whose generalization to four dimensions by Wess and Zumino [1.168] triggered the extensive supersymmetry program underway ever since — but that property did not, at the time, enhance interest in the "spinning string" model as a theory of hadrons. When the asymptotic freedom property of non-Abelian $SU(3)$ color was established in 1973 [1.89], the higher-dimensional "spinning string" model was cast aside as a precursor to a dynamical and phenomenologically correct theory of the strong interaction.

The revival of the ebbing fortunes of higher-dimensional superstring models as a possible path to a "Theory of Everything" is due to a courageous leap taken in 1974 by Scherk and Schwarz [1.183] who proposed to use the predicted presence of a $J = 2$ massless particle (graviton) in the spectrum of the "spinning string" model as the rationale for regarding string theory as having the potential to become a TOE instead of merely a theory of hadrons; the additional dimensions [e.g. the six additional (space) dimensions in the case of the $D = 10$ "spinning string" theory] could be rescaled — so the argument went — to the Planck mass 10^{19} GeV (i.e. 10^{-33}cm) rather than sit at the hadronic energy scale of 1 GeV (i.e. 10^{-14}cm). With the acceptance of the Planck scale for the extra "unphysical" dimensions, the focus shifted drastically from the three interactions (strong, electromagnetic and weak) to the gravitational interaction as a major actor in TOE — and modern superstring theory was born.

While the birth of modern superstring theory can be traced to the mid-1970s, it was not until Green and Schwarz [1.184] established the existence of anomaly-free (both gravitational and mixed gauge-gravitational) higher-dimensional superstring theories that modern superstring theory really took off. And it was not long before several classes of superstring theories were identified [1.185]: (1) Type I — where the superstrings can be open or closed. Without interactions, the mass spectrum of the open strings is $\alpha'm^2 = 0, 1, 2, \ldots$ (where α' is the Regge slope and is related to the string tension T by $\alpha' = 1/2\pi T$) and the massless states comprise a super-Yang-Mills multiplet. Correspondingly, the mass spectrum of the closed strings is $\alpha'm^2 = 0, 4, 8, \ldots$ and the massless states constitute an $N = 1$ supergravity multiplet. As far as the parameters of this type of theory is concerned — since the Regge slope α' is in units of L^2 (L is a length), the gauge coupling g is L^3 and Newton's constant κ is L^4 — one can understand the relation $\kappa\alpha' \simeq g^2$ in Type I superstring theory. The single "supercharge" ($N = 1$) is assumed to be simultaneously Majorana and Weyl (so that it possesses sixteen real components), and the gauge group — which keeps the theory anomaly-free and finite — is $SO(32)$; (2) the heterotic superstring contains closed, oriented strings so that string interactions preserve the orientation. The massless sector is identical with that of Type I superstring theory but another gauge group, $E_8 \times E_8'$ — in addition to $SO(32)$ — can keep the theory anomaly-free and finite. Another difference between the heterotic superstring and the Type I superstring is that the relation among the three parameters κ, g and α' is $\kappa^2 \simeq g^2\alpha'$ for the heterotic string, instead of $\kappa\alpha' \sim g^2$ as in the Type I theory; and, finally (3) Type II superstring theory (with two categories), which only contains oriented closed string states and only secondary Yang-Mills fields [in the spirit of the KK $D = 11$ theory where the Yang-Mills fields arise "naturally" from the compactification of the extra dimensions]. However, one category of Type II superstring theory predicts non-chiral supergravity and both categories are quadratically divergent at one loop; hence, Type II superstring theories are not interesting.

Of the various types of superstring models enumerated above, the most promising appears to be the heterotic string theory [1.185], for a variety of reasons, of which we mention a few: (1) the heterotic string is a rather unique combination of 26-dimensional "left" movers [i.e. 26-dimensional closed strings with a "left" orientation] and 10-dimensional "right" movers [i.e. 10-dimensional closed strings with a "right" orientation]; the difference — 16 — between the number of "critical" bosonic dimensions [i.e. the 26 space-time dimensions required to ensure freedom from anomalies for a bosonic string] and the number of "critical" fermionic dimensions [i.e. the number 10 of space-time dimensions required to guarantee freedom from anomalies for a fermionic string] is precisely the rank of the two gauge groups $SO(32)$ or $E_8 \times E_8'$ permitted for the heterotic string. The combination of 26-dimensional "left movers" and 10-dimensional "right movers" is possible because the resulting "heterotic" string theory is Lorentz invariant in the 10 shared dimensions since the "left movers" and "right movers" are separately Lorentz invariant; (2) while the $SO(32)$ gauge group is possible for a Type I superstring, the phenomenologically more interesting $E_8 \times E_8'$ gauge group in 10 dimensions is only possible for the heterotic string. Clearly, compactification to $D = 4$ must take place for any $D = 10$ superstring model and it has been argued that the most promising 6-dimensional manifold in which compatification can take place is a "Calabi-Yau" space, which is Ricci-flat (to satisfy the requirements of Einsteinian gravity) and possesses a so-called Kähler structure so that the holonomy group is basically $SU(3)$. If one accepts the "Calabi-Yau" space for compactification, the $SU(3)$ holonomy group can be embedded in one of the E_8 factors of the $E_8 \times E_8'$ gauge group so that the compactification from $D = 10$ to $D = 4$ "naturally" involves the breaking of E_8 to $E_6 \times SU(3)$. Such a breaking pattern for the original $E_8 \times E_8'$ group is intriguing because it opens the door to the use of the $D = 4$ E_6 gauge group (or one of its subgroups) as a grand unification (GUT) or a partial unification (PUT) group [see §8.5 for a discussion of "superstring-inspired" E_6 GUT]. Moreover, the choice of E_6 as a GUT model at the $D = 4$ level permits one to define the number of fermion generations in terms of the excess of the number of **27**-representations of E_6 over the number of $\overline{\mathbf{27}}$ representations and to relate this number to the Euler characteristic of the compactification manifold; and (3) although not unique in this respect, the heterotic string theory carries an unbroken $N = 1$ supersymmetry at the $D = 10$ level; this is the simplest type of supersymmetry (with only one "supercharge") and could be helpful in the search for the supersymmetry-breaking mechanism which must be present for phenomenological reasons. Finally, it should be noted that compactification converts the D-dimensional fermion-gauge vertices into $D = 4$ fermion-scalar vertices which is equivalent to providing a clue for the mass generation of fermions.

These positive features of the heterotic superstring theory: the identification of the E_6 gauge group (or one of its subgroups) as the unification group beyond the standard model, the potential topological understanding of the number of fermion generations through the Euler characteristic of the 6-dimensional manifold in which compactification takes place, the attribution of the origin of Yukawa couplings (and, *a fortiori*, the origin of mass) of the fermions to a well-defined property of the compactification process, etc. give credibility to this most ambitious TOE ever proposed and have sustained a major theoretical effort during the past decade to work out detailed consequences of the heterotic string theory. Unfortunately, the heterotic superstring theory must overcome major obstacles: (1) no principle has been uncovered to fix the ground state of the system; (2) the compactification of the additional "space" dimensions to bring the $D = 10$ superstring action down to a $D = 4$ effective action has proved to be a very difficult and equivocal procedure [e.g. there are at least ten thousand "Calabi-Yao" manifolds possible! [1.186] and there is no clear guidance as to how to select a unique one]; (3) the SSB mechanism that is supposed to break the $D = 10$ $E_8 \times E_8'$ gauge group down to the $D = 4$ E_6 gauge group (or one of its subgroups) is quite obscure [e.g. the "gluino" condensates that presumably perform the SSB job through the so-called "hidden" E_8' group is a bold hypothesis without clear justification]; (4) the minimum contact that one might expect between a TOE and the standard model at this time, is that a TOE should be able to predict the precise number (3) of quark-lepton generations; this is one aspect of the problem of reaching the intermediate electroweak scale in $D = 4$ spacetime dimensions in a theory which starts out at $D = 10$ with only massless and supermassive (of the order of Planck energy) states. Another aspect of this problem is to unravel the mechanism and scale of supersymmetry-breaking in the $D = 4$ world starting with a $N = 1$ supersymmetry in the $D = 10$ "Planck" world; (5) there is no overarching physical principle as yet — comparable to the Mach equivalence principle for Einsteinian gravity — that selects the beautiful mathematical edifice of the heterotic string theory as the likely candidate for TOE; and, finally (6) since the heterotic string theory claims to contain Einsteinian gravity, it should predict the small observed value of the "cosmological constant". However, there is no hint in the heterotic string theory as to why the "cosmological constant" essentially vanishes. These shortcomings are well-known to workers in superstring theory and are mentioned only to suggest that, despite the great success of the "Heroic Period" in constructing the standard model of the strong and electroweak interactions, the last fifteen years of the period of "consolidation and speculation" have brought us within sight of a potential TOE but can give us no assurance that the obstacles are surmountable.

References

[1.1] I. Newton, in *Opticks*, ed. I.B. Cohen, Dover Publications, New York (1952).

[1.2] I. Newton, *Philosophiae naturalis principia mathematica*, Streater, London (1687); Engl. transl.: *Mathematical Principles of Natural Philosophy*, Univ. of California Press, Berkeley (1934).

[1.3] A. Einstein, *Preuss. Akad. Wiss. Berlin, Sitzber* **788**, **799**, **844** (1915).

[1.4] M.B. Green, J.H. Schwarz and E. Witten, *Superstring Theory*, Cambridge Univ. Press, Cambridge (1987).

[1.5] J.C. Maxwell, *Collected papers*, Vol. 1, Dover Publications, New York (1952).

[1.6] A. Einstein, *Ann. der Phys.* **17** (1905) 639.

[1.7] M. Planck, *Ann. der Phys.* **4** (1901) 564.

[1.8] A. Einstein, *Ann. der Phys.* **17** (1905) 132; Engl. transl.: A.B. Arons and M.B. Pippard, *Am. J. Phys.* **33** (1965) 367.

[1.9] E. Rutherford, *Phil. Mag.* **21** (1911) 669.

[1.10] N. Bohr, *Phil. Mag.* **26** (1913) 1.

[1.11] L. de Broglie, *Comptes Rendus* **177** (1923) 507; *ibid.* **177** (1923) 548.

[1.12] W. Heisenberg, *Zeits. f. Phys.* **33** (1925) 879.

[1.13] E. Schrödinger, *Ann. der Phys.* **79** (1926) 361; Engl. transl. in *Collected papers on wave mechanics by E. Schrödinger*: transl. J. Shearer and W. Deans, Blackie, Glasgow (1928).

[1.14] P.A.M. Dirac, *Proc. Roy. Soc.* **A117** (1928) 610; *ibid.* **A118** (1928) 351.

[1.15] E. Fermi, *Rev. Mod. Phys.* **4** (1932) 131.

[1.16] J. Schwinger, *Phys. Rev.* **73** (1948) 416; *ibid.* **75** (1949) 898; S. Tomonaga, *Phys. Rev.* **74** (1948) 224; R.P. Feynman, *Phys. Rev.* **74** (1948) 939, 1430; F.J. Dyson, *Phys. Rev.* **75** (1949) 486.

[1.17] C.N. Yang, *Selected Papers 1945–1980 with Commentary*, W.H. Freeman, San Francisco (1983), p. 563.

[1.18] J.J. Thomson, *Phil. Mag.* **48** (1899) 547.

[1.19] J.D. Cockcroft and E.T.S. Walton, *Proc. Roy. Soc.* **A137** (1932) 229.

[1.20] E.O. Lawrence and M.S. Livingston, *Phys. Rev.* **37** (1931) 1707.

[1.21] C.D. Ellis and W.A. Wooster, *Proc. Roy. Soc.* **A117** (1927) 109.

[1.22] J. Chadwick, *Nature* **129** (1932) 312.

[1.23] W. Heisenberg, *Zeits. f. Phys.* **78** (1932) 156.

[1.24] H. Yukawa, *Proc. Phys. Math. Soc. (Japan)* **17** (1935) 48.

[1.25] *W. Pauli, Collected Scientific Papers*, ed. R. Kronig and V. Weisskopf, Vol. 2, p. 1313, Interscience, New York (1964).

[1.26] E. Fermi, *Nuovo Cim.* **11** (1934) 1; *ibid.*, *Zeits. f. Phys.* **88** (1934) 161.

[1.27] C.D. Anderson and S.H. Neddermeyer, *Phys. Rev.* **51** (1937) 894; *ibid.* **54** (1938) 88; J.C. Street and E.C. Stevenson, *Phys. Rev.* **52** (1937) 1003.

[1.28] C.M.C. Lattes, H. Muirhead, G.P.S. Occhialini and C.F. Powell, *Nature*, **159** (1947) 694.

[1.29] W.E. Lamb and R.C. Retherford, *Phys. Rev.* **72** (1947) 972.

[1.30] M. Conversi, E. Pancini and O. Piccioni, *Phys. Rev.* **71** (1947) 209.

[1.31] V.I. Veksler, *Dokl. Ak. Nauk* **43** (1944) 329; *ibid.* **44** (1944) 365; *J. of Phys. USSR* **9** (1945) 153; E.M. McMillan, *Phys. Rev.* **68** (1945) 143.

[1.32] R.E. Marshak and H.A. Bethe, *Phys. Rev.* **72** (1947) 506.

[1.33] B. Pontecorvo, *Phys. Rev.* **72** (1947) 246.

[1.34] S.S. Schweber, in *Shelter Island II*, ed. R. Jackiw, N.N. Khuri, S. Weinberg, and E. Witten, M.I.T. Press, Cambridge, Mass. (1985).

[1.35] *Proc. Tenth Intern. Conf. on High Energy Phys. at Rochester*, Studer, Geneva (1960).

[1.36] J.E. Nafe, E.B. Nelson and I.I. Rabi, *Phys. Rev.* **71** (1947) 914; D.E. Nagel, R.S. Julian and J.R. Zacharias, *Phys. Rev.* **72** (1947) 921.

[1.37] J.C. Ward, *Phys. Rev.* **78** (1950) 1824; Y. Takahashi, *Nuovo Cim.* **6** (1957) 370.

[1.38] Y. Aharonov and B. Bohm, *Phys. Rev.* **15** (1959) 485.

[1.39] P.A.M. Dirac, *Proc. Roy. Soc.* **A133** (1931) 60.

[1.40] E.C.G. Stueckelberg and A. Peterman, *Helv. Phys. Acta* **26** (1953) 499; M. Gell-Mann and F.E. Low, *Phys. Rev.* **95** (1954) 1300.

[1.41] L.D. Landau and Y. Ya. Pomeranchuk, *Doklady, Acad. Nauk*, USSR **102** (1955) 489.

[1.42] J.G. Wilson, *Proc. Roy. Soc.* **174** (1940) 73.

[1.43] S. Sakata and T. Inoue, *Prog. Theor. Phys.* **1** (1946) 143; Y. Tanikawa, *Prog. Theor. Phys.* **2** (1947) 220.

[1.44] S. Tomonaga and G. Araki, *Phys. Rev.* **58** (1940) 90.

[1.45] E. Fermi, E. Teller and V. Weisskopf, *Phys. Rev.* **71** (1947) 314.

[1.46] J.R. Oppenheimer, in *Shelter Island II* [1.34], p. 338.

[1.47] A. Pais, *Inward Bound*, Oxford University Press, New York (1986) p. 453.

[1.48] R.E. Marshak, *Meson Physics*, McGraw-Hill, New York (1952).

[1.49] H.L. Anderson, E. Fermi, E.A. Long and D.E. Nagle, *Phys. Rev.* **85** (1952) 936.

[1.50] W. Pauli, *Meson Theory of Nuclear Forces*, Interscience, New York (1946); G. Wentzel, *Rev. Mod. Phys.* **19** (1947) 1.

[1.51] C.N. Yang and R. Mills, *Phys. Rev.* **46** (1954) 191.

[1.52] M.L. Goldberger, *Phys. Rev.* **99** (1955) 975; M.L. Goldberger, H. Miyazawa and R. Oehme, *Phys. Rev.* **99** (1955) 986.

[1.53] G.F. Chew and F.E. Low, *Phys. Rev.* **101** (1956) 1570.

[1.54] *Dispersion Relations*, ed. G.R. Screaton, Oliver and Boyd, Edinburgh (1961); N.N. Bogoliubov and D.V. Shirkov, *Introduction to the Theory of Quantized Fields*, Wiley-Interscience, New York (1980).

[1.55] S. Mandelstam, *Phys. Rev.* **112** (1958) 1344; *ibid.* **115** (1959) 1741, 1752.

[1.56] G.F. Chew, *S-Matrix Theory of Strong Interactions*, W.A. Benjamin, Mass. (1961); R.E. Cutkowsky, *J. Math. Phys.* **1** (1960) 429.

[1.57] W. Heisenberg, *Z. Physik* **120** (1943) 513, 673.

[1.58] L.D. Landau, *Nucl. Phys.* **13** (1959) 181.

[1.59] M. Gell-Mann, "Particle Theory from S-Matrix to Quarks," in *Symmetries in Physics (1600-1980), Proc. of the First Intern. Meeting on the History of Scientific Ideas*, ed. M.G. Doncel, A. Hermann, L. Michel, and A. Pais, Bellaterra, Barcelona (1987), pp. 474-97.

[1.60] M. Gell-Mann, *Phys. Rev.* **92** (1953) 833; *Nuovo Cim. Suppl.* **4** (1956) 2848; T. Nakano and K. Nishijima, *Prog. Theor. Phys.* **10** (1953) 581; K. Nishijima, *Prog. Theor. Phys.* **12** (1954) 107; *ibid.* **13** (1955) 285.

[1.61] S. Sakata, *Prog. Theor. Phys.* **16** (1956) 686.

[1.62] S. Ogawa, *Prog. Theor. Phys.* **21** (1959), 209; M. Ikeda, S. Ogawa and Y. Ohnuki, *Prog. Theor. Phys.* **22** (1959) 715.

[1.63] M. Gell-Mann and Y. Ne'eman, *The Eightfold Way*, W.A. Benjamin (1964).

[1.64] G. Gamow and E. Teller, *Phys. Rev.* **49** (1936) 895.

[1.65] J. Tiomno and J.A. Wheeler, *Rev. Mod. Phys.* **21** (1949) 144; G. Puppi, *Nuovo Cim.* **5** (1948) 587; T.D. Lee, M. Rosenbluth and C.N. Yang, *Phys. Rev.* **75** (1949) 905.

[1.66] L. Michel, *Proc. Phys. Soc.* **A63** (1950) 154, 1371.

[1.67] M. Ruderman and R. Finkelstein, *Phys. Rev.* **76** (1959) 1458.

[1.68] B.M. Rustad and S.L. Ruby, *Phys. Rev.* **89** (1953) 880.

[1.69] S.S. Gershtein and Y.B. Zeldovich, *Soviet Phys. JETP* **2** (1956) 576.

[1.70] W.P. Alford and D.R. Hamilton, *Phys. Rev.* **95** (1954) 1351.

[1.71] *Proc. of the Sixth Annual Rochester Conf. on High Energy Nuclear Physics*, April 3–7, (1956), ed. J. Ballam, V.L. Fitch, T. Fulton, K. Huang, R.R. Rau, and S.B. Treiman, Interscience Publishers, New York (1956).

[1.72] T.D. Lee and C.N. Yang, *Phys. Rev.* **104** (1956) 254.

[1.73] C.S. Wu, E. Ambler, R.W. Hayward, D.D. Hoppes, and R.P. Hudson, *Phys. Rev.* **105** (1957) 1413.

[1.74] R. Garwin, L. Lederman and M. Weinrich, *Phys. Rev.* **105** (1957) 1415; J.I. Friedman and V.L. Telegdi, *Phys. Rev.* **105** (1957) 1681.

[1.75] L. Landau, *Nucl. Phys.* **3** (1957) 127; A. Salam, *Nuovo Cim.* **5** (1957) 299; T.D. Lee and C.N. Yang, *Phys. Rev.* **105** (1957) 1671.

[1.76] E.C.G. Sudarshan and R.E. Marshak, *Proc. Padua-Venice Conf. on Mesons and Recently Discovered Particles* (1957); *ibid.*, *Development of Weak Interaction Theory*, ed. P. Kabir, Gordon & Breach, New York (1963); R.P. Feynman and M. Gell-Mann, *Phys. Rev.* **109** (1958) 193.

[1.77] H.L. Anderson, T. Fuji, R.H. Miller, and L. Tau, *Phys. Rev. Lett.* **2** (1959) 53.

[1.78] E.C.G. Sudarshan and R.E. Marshak, *Phys. Rev.* **109** (1958) 860; I. Tamm, *Proc. of Ninth Intern. Conf. on High Energy Physics, Kiev*, Vol. 2, p. 412, Acad. of Sci., Moscow (1959)

[1.79] M. Goldhaber, L. Grodzins and A. Sunyar, *Phys. Rev.* **109** (1958) 1015.

[1.80] E.C.G. Sudarshan and R.E. Marshak, *Proc. Padua-Venice Conf. on Mesons and Recently Discovered Particles* (1957); R.E. Marshak,

Riazuddin and C.P. Ryan, *Theory of Weak Interactions in Particle Physics*, Wiley-Interscience, New York (1969).

[1.81] N. Cabibbo, *Phys. Rev. Lett.* **10** (1963) 531.

[1.82] B.M. Pontecorvo, *Proc. Ninth Annual Intern. Conf. on High Energy Physics, Kiev*, Vol. 2, Acad. of Sci., Moscow (1959); G. Feinberg, *Phys. Rev.* **110** (1958) 1482.

[1.83] G. Danby *et al.*, *Phys. Rev. Lett.* **9** (1962) 36.

[1.84] A. Gamba, R.E. Marshak and S. Okubo, *Proc. Nat. Acad. Sci. U.S.* **45** (1959) 881; R.E. Marshak, *Proc. Ninth Annual Intern. Conf. on High Energy Physics, Kiev*, Rapporteur talk on "Weak Interactions", Vol. 2, Acad. of Sci., Moscow (1959).

[1.85] J.D. Bjorken and S.L. Glashow, *Phys. Lett.* **11** (1964) 255.

[1.86] S.L. Glashow, J. Iliopoulos and L. Maiani, *Phys. Rev.* **D2** (1970) 1285.

[1.87] P.W. Higgs, *Phys. Rev. Lett.* **12** (1964) 132; F. Englert and R. Brout, *Phys. Rev. Lett.* **13** (1964) 321; G.S. Guralnik, C.K. Hagen and T.W.B. Kibble, *Phys. Rev. Lett.* **13** (1964) 585.

[1.88] S.L. Glashow, *Nucl. Phys.* **22** (1961) 579; S. Weinberg, *Phys. Rev. Lett.* **19** (1967) 1264; A. Salam in *Elementary Particle Theory*, ed. N. Swartholm, Almquist and Wissell, Stockholm (1968).

[1.89] D. Gross and F. Wilczek, *Phys. Rev. Lett.* **30** (1973) 1343; H.D. Politzer, *Phys. Rev. Lett.* **30** (1973) 1346.

[1.90] J. Bardeen, L.N. Cooper and J.R. Schrieffer, *Phys. Rev.* **108** (1957) 1175.

[1.91] V.L. Ginzburg and L.D. Landau, *J. Expl. Theoret. Phys. USSR* **20** (1950) 1064.

[1.92] Y. Nambu and G. Jona-Lasinio, *Phys. Rev.* **122** (1961) 345.

[1.93] Y. Nambu, *Proc. of Tenth Intern. Conf. on High Energy Physics, Rochester*, Studer, Geneva (1960).

[1.94] Y. Nambu, *Physics* **15D** (1985) 147.

[1.95] H. Dürr, W. Heisenberg, H. Mitter, S. Schlieder, and K. Yamazaki, *Zeits. Naturfor.* **14a** (1959) 441; W. Heisenberg, *Introduction to the Unified Field Theory of Elementary Particles*, Interscience, New York (1966).

[1.96] M. Gell-Mann, *Phys. Lett.* **8** (1964) 214; G. Zweig, CERN report No. 8182/TH401 (1964).

[1.97] M. Gell-Mann, Cal Tech Report CTSL-20 (1961); S. Okubo, *Prog. Theor. Phys.* **27** (1962) 949.

[1.98] G.S. LaRue, J.D. Phillips and W.M. Fairbank, *Phys. Rev. Lett.* **46** (1981) 967.

[1.99] Particle Data Group, *Phys. Rev.* **D45**, No. 11 (Part II), (1992).

[1.100] B. Sakita, *Phys. Rev. Lett.* **13** (1964) 643; F. Gürsey and L.A Radicati, *Phys. Rev. Lett.* **13** (1964) 173.

[1.101] O.W. Greenberg, *Phys. Rev. Lett.* **13** (1964) 598.

[1.102] E.D. Bloom *et al.*, *Phys. Rev. Lett.* **23** (1969) 930; M. Breidenbach *et al.*, *Phys. Rev. Lett.* **23** (1969) 935.

[1.103] R.P. Feynman, *Phys. Rev. Lett.* **23** (1969) 1415; *ibid.*, *Photon-Hadron Interactions*, W.A. Benjamin, Mass. (1973).

[1.104] J.D. Bjorken, *Proc. of 3rd Intern. Symposium on Lepton and Photon Interactions*, Stanford (1967); *ibid.*, *Phys. Rev.* **179** (1969) 1547; J.D. Bjorken and E.A. Paschos, *Phys. Rev.* **185** (1969) 1975.

[1.105] N. Cabibbo, G. Parisi and M. Testa, *Nuovo Cim. Lett.* **4** (1970) 35.

[1.106] G. 't Hooft, *Nucl. Phys.* **35** (1971) 167.

[1.107] T. Regge, *Nuovo Cim.* **14** (1959) 951; *ibid.* **18** (1960) 947.

[1.108] N.M. Queen and O. Violini, *Dispersion Theory in High Energy Physics*, John Wiley, New York (1974).

[1.109] I. Ya. Pomeranchuk, *Soviet Physics JETP* **7** (1958) 499.

[1.110] P.H. Frampton, *Dual Resonance Models*, W.A. Benjamin, Mass. (1974).

[1.111] G. Veneziano, *Nuovo Cim.* **57A** (1968) 190.

[1.112] K. Bardakci and H. Ruegg, *Phys. Rev.* **181** (1969) 1884; C.J. Goebel and B. Sakita, *Phys. Rev. Lett.* **22** (1969) 256.

[1.113] Y. Nambu, *Proc. Intern. Conf. on Symmetries and Quark Models*, Wayne State Univ. (1969); L. Susskind, *Nuovo Cim.* **69A** (1970) 457.

[1.114] J.H. Schwarz, *Physics in Higher Dimensions*, ed. T. Piran and S. Weinberg, World Scientific, Singapore (1986).

[1.115] E. Eichten, K. Gottfried, T. Kinoshita, K. Lane, and T. Yan, *Phys. Rev.* **D21** (1980) 203.

[1.116] B.L. Ioffe and E.P. Shabalin, *J. Nucl. Phys. (USSR)* **6** (1967) 328; R.N. Mohapatra, J. Subba Rao and R.E. Marshak, *Phys. Rev. Lett.* **20** (1968) 19.

[1.117] T.D. Lee and C.S. Wu, *Ann. Rev. Nucl. Sci.* **15** (1965) 381.

[1.118] J. Goldstone, *Nuovo Cim.* **19** (1961) 154; J. Goldstone, A. Salam and S. Weinberg, *Phys. Rev.* **127** (1962) 965.

[1.119] F.J. Hasert *et al.*, *Phys. Lett.* **46B** (1973) 138.

[1.120] R. Utiyama, *Phys. Rev.* **101** (1956) 1597; T.W.B. Kibble, *J. Math. Phys.* **2** (1961) 212.

[1.121] T. Kinoshita, *Quantum Electrodynamics*, World Scientific, Singapore (1991).

[1.122] L. Mo, *Proc. of Intern. Symp. on Lepton and Photon Interactions at High Energies*, Stanford (1975); H.L. Anderson *et al.*, *Phys. Rev. Lett.* **37** (1976) 4.

[1.123] K. Wilson, *Phys. Rev.* **D3** (1971) 1818.

[1.124] G. Altarelli and G. Parisi, *Nucl. Phys.* **B126** (1977) 298; *Perturbative Quantum Chromodynamics*, ed. D.W. Duke and J.F. Owens, AIP, New York (1981).

[1.125] M. Jacob, *Proc. 25th Intern. Conf. on High Energy Phys., Singapore*, World Scientific, Singapore (1991).

[1.126] S. Lindenbaum, *Comments Nucl. Part. Phys.* **13** (1984) 285.

[1.127] S. Lindenbaum, *Proc. of BNL Workshop on "Glueballs, Hybrids, and Exotic Hadrons"*, ed. S.U. Chung, AIP, New York (1989).

[1.128] P. van Baal and A.S. Kronfeld, *Nucl. Phys.* **B** *(Proc. Suppl.)* **9** (1989) 227.

[1.129] W.J. Willis, *Proc. 25th Intern. Conf. on High Energy Phys., Singapore*, World Scientific, Singapore (1991).

[1.130] *Heavy Quark Physics*, ed. P.S. Drell and D.L. Rubin, AIP, New York (1989).

[1.131] K. Wilson, *Phys. Rev.* **D14** (1974) 2455.

[1.132] F. Jegerlehner, *Proc. of XX Intern. Symposium on Multiparticle Dynamics*, Dortmund (1990).

[1.133] J.B. Kogut, E. Dagotto and A. Kocic, *Nucl. Phys.* **B317** (1989) 253.

[1.134] D. Callaway, *Phys. Rep.* **167** (1988) 241.

[1.135] J. Gipson and H.C. Tze, *Nucl. Phys.* **B183** (1981) 524.

[1.136] A.A. Belavin, A.M. Polyakov, A.S. Schwartz, and Yu.S. Tyupkin, *Phys. Lett.* **59B** (1975) 85.

[1.137] R. Peccei, *CP Violation*, ed. C. Jarlskog, World Scientific, Singapore (1989).

[1.138] R.D. Peccei and H.R. Quinn, *Phys. Rev. Lett.* **38** (1977) 1440; *ibid.*, *Phys. Rev.* **D16** (1977) 1791.

[1.139] T.H.R. Skyrme, *Proc. Roy. Soc.* **A260** (1961) 12; *Nucl. Phys.* **31** (1962) 556; L.D. Faddeev. *Lett. Math. Phys.* **1** (1976) 289.

[1.140] E. Witten, *Nucl. Phys.* **B223** (1983) 433.

[1.141] J. Wess and B. Zumino, *Phys. Lett.* **37B** (1971) 95.

[1.142] S.L. Adler, *Phys. Rev.* **177** (1969) 2426; J.S. Bell and R. Jackiw, *Nuovo Cim.* **60A** (1969) 47.

[1.143] C. Bouchiat, J. Iliopolous and Ph. Meyer, *Phys. Lett.* **38B** (1972) 519.

[1.144] E. Witten, *Phys. Lett.* **117B** (1982) 324.

[1.145] R. Delbourgo and A. Salam, *Phys. Lett.* **40B** (1972) 381; T. Eguchi and P. Freund, *Phys. Rev. Lett.* **37** (1976) 1251. L. Alvarez-Gaumé and E. Witten, *Nucl. Phys.* **B234** (1983) 269.

[1.146] C.Q. Geng and R.E. Marshak, *Phys. Rev.* **D39** (1989) 693; *ibid.* **D41** (1990) 717.

[1.147] G. Senjanović and R.N. Mohapatra, *Phys. Rev.* **D12** (1975) 1502; R.N. Mohapatra and R.E. Marshak, *Phys. Rev. Lett.* **44** (1980) 1316.

[1.148] J.C. Pati and A. Salam, *Phys. Rev.* **D10** (1974) 275; R.N. Mohapatra and J.C. Pati, *Phys. Rev.* **D11** (1975) 566, 2559.

[1.149] H. Georgi and S.L. Glashow, *Phys. Rev. Lett.* **32** (1974) 438.

[1.150] Y. Totsuka, W. Gajewski and J. Ernwein, *Last Workshop on Grand Unification*, ed. P. Frampton, World Scientific, Singapore (1989).

[1.151] G. t'Hooft, *Nucl. Phys.* **B79** (1974) 276; A.M. Polyakov, *JETP Lett.* **20** (1974) 194.

[1.152] H. Fritsch and P. Minkowski, *Ann. Phys.* **93** (1975) 193.

[1.153] F. Gürsey and P. Sikivie, *Phys. Rev. Lett.* **36** (1976) 775; *ibid.*, *Phys. Rev.* **D16** (1977) 816.

[1.154] L. Susskind, *Phys. Rev.* **D20** (1979) 2619; S. Weinberg, *Phys. Rev.* **D19** (1979) 1277.

[1.155] S. Dimopolous and L. Susskind, *Nucl. Phys.* **B155** (1979) 237; E. Eichten and K.D. Lane, *Phys. Lett.* **90B** (1980) 125.

[1.156] B. Holdom, *Phys. Rev.* **D24** (1981) 1441; T. Appelquist and L.C.R. Wijewardhana, *Phys. Rev.* **D35** (1987) 774; *ibid.* **D36** (1987) 568; K. Yamawaki, M. Bando and K. Matumoto, *Phys. Rev. Lett.* **56** (1986) 1335.

[1.157] C. Campagnari, *Proc. of 25th Intern. Conf. on High Energy Physics, Singapore*, p. 1186, World Scientific, Singapore (1991).

[1.158] H. Harari, *Phys. Rep.* **104** (1984) 159.

[1.159] G. 't Hooft, *Recent Developments in Gauge Theories*, Proc. of the NATO Advanced Study Institute, Cargése (1979), ed. G. 't Hooft, Plenum, New York (1980).

[1.160] R.D. Peccei, *Proc. of Intern. Symp. on Composite Models of Quarks and Leptons*, ed. H. Terazawa and M. Yasue, Tokyo (1985).

[1.161] G. 't Hooft, *Phys. Rev. Lett.* **37** (1976) 8; *ibid.*, *Phys. Rev.* **D14** (1976) 3432.

[1.162] F.R. Klinkhamer and N.S. Manton, *Phys. Rev.* **D30** (1984) 2212; V. Kuzmin, V. Rubakov and M. Shaposhnikov, *Phys. Lett.* **155B** (1985) 36.

[1.163] P. Arnold and L. McLerran, *Phys. Rev.* **D36** (1987) 581; A. Ringwald, *Nucl. Phys.* **B330** (1990) 1.

[1.164] Y. Chikashige, R.N. Mohapatra and R.D. Peccei, *Phys. Lett.* **98B** (1981) 265; G. Gelmini and M. Roncadelli, *Phys. Lett.* **99B** (1981) 411.

[1.165] V. DeLapparent, M. Geller and J. Huchra, *Ap. J.* **302** (1986) L1.

[1.166] F. Wilczek, *Phys. Rev. Lett.* **49** (1982) 1549.

[1.167] G. Gelmini, S. Nussinov and T. Yanagida, *Nucl. Phys.* **B219** (1983) 31.

[1.168] J. Wess and B. Zumino, *Nucl. Phys.* **B70** (1974) 139.

[1.169] P. Ramond, *Phys. Rev.* **D3** (1971) 2415; A. Neveu and J.H. Schwarz, *Nucl. Phys.* **31** (1971) 86.

[1.170] S. Coleman and J. Mandula, *Phys. Rev.* **159** (1967) 1251.

[1.171] S. Ferrara, D.Z. Freedman and P. van Nieuwenhuizen, *Phys. Rev.* **D13** (1976) 3214; S. Ferrara and B. Zumino, *Nucl. Phys.* **B79** (1974) 413.

[1.172] P. van Nieuwenhuizen, *Phys. Rep.* **68** (1981) 189.

[1.173] A. Salam and J. Strathdee, *Nucl. Phys.* **B76** (1974) 477; *ibid.*, *Phys. Lett.* **51B** (1974) 353.

[1.174] J. Ellis, *Phys. Rep.* **105** (1984) 121.

[1.175] T. Kaluza, Sitzungsber. Preuss. Akad. Wiss. Berlin, *Math: Phys.* **K1** (1921) 966; O. Klein, *Z. Physik* **37** (1926) 895; *ibid.*, *Arkiv. Mat. Astron. Fys.* B **34A** (1946).

[1.176] A. Einstein and W. Mayer, *Preuss. Akad.* (1931) 541, (1932) p. 130; A. Einstein and P. Bergmann, *Ann. Math.* **39** (1938) 683.

[1.177] H. Weyl, *Gesammelte Abhandlungen*, ed. K. Chandrasekharan, Vol. II, p. 29, Springer, Berlin (1968).

[1.178] M.J. Duff and C.N. Pope, *Supersymmetry and Supergravity*, ed. S. Ferrara, J.G. Taylor and P. van Nieuwenhuizen, World Scientific, Singapore (1983).

[1.179] E. Witten, *Nucl. Phys.* **B186** (1981) 412; *ibid. Shelter Island II*, MIT Press, Cambridge (1983).

[1.180] M.J. Duff, E.W. Nilsson and C.N. Pope, CERN preprint TH 4217/85 (1985).

[1.181] M.B. Green and J.H. Schwarz, *Nucl. Phys.* **B243** (1984) 536; D.J. Gross, J.A. Harvey, E. Martinec, and R. Rohm, *Nucl. Phys.* **B256** (1985) 257; *ibid.*, **B267** (1986) 75.

[1.182] J.L. Gervais and B. Sakita, *Phys. Rev.* **D4** (1971) 2291.

[1.183] J. Scherk and J.H. Schwarz, *Nucl. Phys.* **B81** (1974) 118.

[1.184] M.B. Green and J.H. Schwarz, *Phys. Lett.* **149B** (1984) 117; *Phys. Lett.* **151B** (1984) 21.

[1.185] P. Candelas, G. Horowitz, A. Strominger, and E. Witten, *Nucl. Phys.* **B258** (1985) 46; *String Theory in Four Dimensions*, ed. M. Dine, North-Holland, New York (1988).

[1.186] E. Calabi, *Algebraic Geometry and Topology: A Symposium in Honor of S. Leftschetz*, Princeton Univ. Press, Princeton (1957); S.T. Yau, *Proc. Nat. Acad. Sci.* **74** (1977) 1798.

Chapter 2

SPACE-TIME SYMMETRIES IN QUANTUM FIELD THEORY

§2.1. Continuous space-time symmetries

As we pointed out in Chapter 1, the decisive instrument of progress in achieving the dynamical, phenomenologically correct standard model of the strong and electroweak interactions has been gauged quantum field theory. While QED was recognized by 1950 as the unique Lorentz-invariant, gauge-invariant and renormalizable quantum field theory for Abelian gauge groups, it took more than two additional decades to establish that field theories of strong and electroweak interactions can be formulated with equal success by gauging appropriate global non-Abelian internal symmetries. In order to appreciate the breakthroughs that have occurred during the past twenty years in the development of the gauged non-Abelian quantum field theories that define QCD and QFD, we devote the next three chapters (2–4) to a review of the basic mathematical formalism of relativistic quantum field theory under three headings: (1) space-time symmetries in quantum field theory (Chapter 2); (2) global internal symmetries and their spontaneous breakdown (Chapter 3); and (3) gauge symmetry groups and their spontaneous breakdown (Chapter 4). With this mathematical background (and an assortment of physical examples) in hand, it will be possible to discuss in some detail (in Chapters 5 and 6 respectively) the physical content of QCD and QFD.

We start this chapter on space-time symmetries in quantum field theory with a discussion of the continuous space-time symmetries, to be followed by the three discrete symmetries (including the CPT theorem) — with particular reference to the Klein-Gordon, Proca, Maxwell, Dirac and Weyl fields —

and conclude with comments on some special topics such as supersymmetry and scale invariance in four dimensions. In starting with continuous space-time symmetries, our aim is to recapture some well-known properties of the Poincaré group within the context of quantum field theory. Most derivations given here are based on the complex (charged) scalar (Klein-Gordon) quantum field — which is the mathematically simplest quantum field theory that still contains a continuous global [$U(1)$] internal symmetry; a real (Hermitian) scalar quantum field theory is mathematically still simpler but the absence of "charge" renders it uninteresting for most purposes. We also record the results for the spin $\frac{1}{2}$ with mass (Dirac), spin $\frac{1}{2}$ without mass (Weyl), spin 1 with mass (Proca) and spin 1 without mass (Maxwell) fields so that we have before us the notation for the three lowest spin ($J = 0, \frac{1}{2}, 1$) fields (both massive and massless) that comprise the basic Lagrangian of the standard model. We first work out the special cases of invariance under space-time translations and Lorentz "boosts" and then comment on the properties of the full Poincaré group and its "Casimir" invariants (of mass and spin) for both massive and massless fields.

The derivation of the continuous space-time subgroups and the full Poincaré group for a charged scalar field (with mass) is well-known: let $\phi(\underline{x}, t)$ represent the (second) quantized complex scalar field $\phi(\underline{x}, t)$, where $\underline{x}$ denotes the three space coordinates and t the time. The covariant notation $x \equiv x^\mu (\mu = 0, 1, 2, 3) = (t, \underline{x})$ and $x_\mu = (t, -\underline{x})$ is used so that $x_\mu x^\mu = t^2 - \underline{x}^2$; the (Minkowski) metric is defined by $g_{00} = 1$, $g_{11} = g_{22} = g_{33} = -1$, $g_{\mu\nu} = 0 \, (\mu \neq \nu)$ so that $x_\mu = g_{\mu\nu} x^\nu$. The Lagrangian density is given by:

$$\mathcal{L}(x) = \partial_\mu \phi^\dagger(x) \partial^\mu \phi(x) - m^2 \phi^\dagger(x) \phi(x) \tag{2.1}$$

where $\phi^\dagger(x)$ is the Hermitian conjugate field to $\phi(x)$ and m is the mass of the scalar field quantum ($\hbar = c = 1$). [The basic formalism which follows is unaltered for a massless ($J = 0$) scalar field since there is only one "polarization"; this is no longer true for $J > 0$ and it is necessary to distinguish between $m \neq 0$ and $m = 0$ fields possessing $J > 0$.] The Lagrangian density (henceforth, we refer to it as the Lagrangian) leads to the same Klein-Gordon equation for $\phi(x)$ and $\phi^*(x)$ [2.1], namely:

$$(\partial_\mu \partial^\mu + m^2)\phi(x) = (\Box + m^2)\phi(x) = 0 \tag{2.2}$$

The next step is to write down the expressions for the canonically conjugate field momenta $\pi(x)$ and $\pi^\dagger(x)$ in terms of $\phi(x)$ and $\phi^\dagger(x)$:

$$\pi(x) = \partial_0 \phi^\dagger(x) = \partial^0 \phi^\dagger(x); \quad \pi^\dagger(x) = \partial^0 \phi(x) = \partial_0 \phi(x) \tag{2.3}$$

The equal-time commutation rules (C.R.) among the scalar ϕ's and π's are:

$$[\phi(\underline{x},t),\phi(\underline{y},t)] = 0; \quad [\pi(\underline{x},t),\pi(\underline{y},t)] = 0$$

$$[\pi(\underline{x},t),\phi(\underline{y},t)] = -i\delta(\underline{x}-\underline{y}) \tag{2.4}$$

From Eqs. (2.1) and (2.3), the Hamiltonian (density) $\mathcal{H}$ is derived:

$$\mathcal{H} = \pi(x)\pi^\dagger(x) + \nabla\phi(x)\cdot\nabla\phi^\dagger(x) + m^2\phi(x)\phi^\dagger(x) \tag{2.5}$$

Hence, the C.R. of the total Hamiltonian $H[\equiv \int d^3x\mathcal{H}]$ with the field operator $\phi(x)$ is:

$$[H,\phi(x)] = -i\partial^0\phi(x), \tag{2.6}$$

with a similar C.R. for H with $\phi^\dagger(x)$. The Heisenberg equation of motion for a function of the field operators, $F[\phi(x),...]$ is:

$$[H,F(x)] = -i\partial^0 F(x) \tag{2.7}$$

The total spatial momentum, $\underline{P}$, in the charged scalar field can be written as:

$$P^k = \int d^3x[\pi(x)\partial^k\phi(x) + \pi^\dagger(x)\partial^k\phi^\dagger(x)] \tag{2.8}$$

Since $H = P^0$, we have, using Eqs. (2.4) and (2.7):

$$[P^\mu, F(x)] = -i\partial^\mu F(x) \tag{2.9}$$

It is easy to show — using the C.R. — that $[P^\mu, P^\nu] = 0$ so that the total four-momentum is conserved. In addition to the conserved total energy and total spatial momentum in the quantized field — that derive their conservation property from the invariance of the field Lagrangian under space-time translations (see below) — there is another conserved quantity, the charge Q that derives its important status from the invariance of the Lagrangian (including interaction) under a global constant phase transformation (see below). Invariance of the field Lagrangian under a global constant phase transformation leads to a conserved vector current, i.e. $j^\mu = i[\phi^\dagger(x)\partial^\mu\phi(x) - \phi(x)\partial^\mu\phi^\dagger(x)]$ that satisfies $\partial_\mu j^\mu = 0$; the continuity equation $\partial_\mu j^\mu = 0$ implies that there exists a conserved charge: $Q = \int d^3x\, j^0(x)$ [with $j^0 = i(\phi^\dagger\pi^\dagger - \phi\pi)$].

For particle physics — where particles are created and destroyed — it is convenient to work in "Fourier transform space" (i.e. occupation number space, better known as Fock space) so that one gets Fourier expansions of $\phi(x)$ and $\phi^\dagger(x)$ in terms of the creation and destruction operators:

$$\phi(t,\underline{x}) = \sum_k \frac{1}{\sqrt{2\omega_k}}\{a_k e^{i(\underline{k}\cdot\underline{x}-\omega_k t)} + b_k^\dagger e^{-i(\underline{k}\cdot\underline{x}-\omega_k t)}\} \tag{2.10a}$$

$$\phi^\dagger(t, \underline{x}) = \sum_k \frac{1}{\sqrt{2\omega_k}}\{a_k^\dagger e^{-i(\underline{k}\cdot\underline{x}-\omega_k t)} + b_k e^{i(\underline{k}\cdot\underline{x}-\omega_k t)}\} \qquad (2.10b)$$

where $\omega_k = \sqrt{\underline{k}^2 + m^2}$, $a_k^\dagger$ and a_k are the creation and destruction operators (corresponding to particles with spatial momentum $\underline{k}$), and $b_k^\dagger$ and b_k refer to the antiparticles. From the C.R. among the field operators $\phi(x)$ and $\pi(x)$ [and their Hermitian conjugates — see Eq. (2.4)], it is possible to derive the (Bose-Einstein) C.R. among the creation and destruction operators $a_k^\dagger$ and a_k, as follows:

$$[a_k, a_{k'}] = 0; \quad [a_k^\dagger, a_{k'}^\dagger] = 0; \quad [a_k, a_{k'}^\dagger] = \delta_{kk'} \qquad (2.11)$$

with similar C.R. for $b_k^\dagger$, b_k and all other pairs of operators commuting. The total energy E, total spatial momentum P, and total charge Q of the quanta (particles) in the complex scalar field $\phi(x)$ can be expressed in terms of the creation and destruction operators, thus:

$$E = \sum_k (a_k^\dagger a_k + b_k^\dagger b_k)\omega_k = \sum_k (n_k + \bar{n}_k)\omega_k$$

$$\underline{P} = \sum_k (a_k^\dagger a_k + b_k^\dagger b_k)\underline{k} = \sum_k (n_k + \bar{n}_k)\underline{k} \qquad (2.12a)$$

$$Q = q\sum_k (a^\dagger a_k - b_k^\dagger b_k) = q\sum_k (n_k - \bar{n}_k) \quad \text{(q is the unit of charge)} \qquad (2.12b)$$

where the operator combinations $a_k^\dagger a_k$ and $b_k^\dagger b_k$ represent the numbers n_k and $\bar{n}_k$ of particles and antiparticles with momentum $\underline{k}$ respectively. Equation (2.12b) tells us that the charge operator Q possesses the interesting property that it takes on opposite signs for particles and antiparticles so that it vanishes for a real (Hermitian) field, i.e. when $\phi^\dagger(x) = \phi(x)$.

The opposite sign of the charge Q operator for particle and antiparticle also shows up (even when there are interacting fields) in the C.R. of Q with $\phi(x)$ and $\phi^\dagger(x)$, i.e.

$$[Q, \phi(x)] = -\phi(x); \quad [Q, \phi^\dagger(x)] = \phi^\dagger(x) \qquad (2.13)$$

That is to say, $\phi(x)$ is the charge-lowering operator (by one unit of "charge" q — if q is positive) whereas $\phi^\dagger(x)$ is the charge-raising operator (by one unit of q). It is easy to understand the important equation (2.13) if one recalls — from Eq. (2.10a) — that $\phi(x)$ contains the destruction operator for the particle, a_k, and the creation operator for the antiparticle, $b_k^\dagger$, whereas $\phi^\dagger(x)$ contains the obverse of the particle and antiparticle operators [i.e. $a_k \rightarrow$

$b_k, a_k^\dagger \rightarrow b_k^\dagger$]. Equation (2.13) is useful in discussing the global internal symmetry of (electric) charge (see §3.1).

With the formalism for the quantized complex scalar field in hand, one can readily derive the subgroups and the full Poincaré group if one recalls the transformations by means of which the generators of the unitary symmetry group in ("second" quantized) Fock space are determined by the space-time symmetry operation on the "first" quantized field [2.2]. The argument is straightforward: the unitary symmetry group is fixed by the condition that the unitary transformation $U(\Lambda)$ operating on the "second" quantized field yields the same probability density — in Fock space — for a physical parameter, when the Lagrangian of the "first" quantized field is invariant under the specified symmetry transformation Λ on the space-time coordinates. More explicitly, if Ψ_A and $\Psi_{A'}$ are the initial and final physical states belonging to the same Hilbert vector (Fock) space, the unitary transformation $U(\Lambda)$, resulting from $x' = \Lambda x$, satisfies the relation: $\Psi_{A'} = U(\Lambda)\Psi_A$ with $U^\dagger U = UU^\dagger = 1$. Hence, if the result of a measurement on the scalar field $\phi(x)$ is to remain unchanged by the symmetry operation $x' = \Lambda x$, one must obtain the same value of the physical quantity independently of whether the measurement is made on the physical state $\Psi_{A'}$ or Ψ_A, i.e. $< A'|\phi(x')|A' > = < A|\phi(x)|A >$; this last equality requires:

$$< U(\Lambda)A|\phi(\Lambda x)|U(\Lambda)A > = < A|\phi(x)|A > \qquad (2.14)$$

from which one derives the crucial formula:

$$\phi(\Lambda x) = U(\Lambda)\phi(x)U^{-1}(\Lambda) \qquad (2.15)$$

Equation (2.15) allows one to determine $U(\Lambda)$ once Λ is specified and we now apply Eq. (2.15) to: (a) space-time translations and (b) space-time rotations, leading to (c) the general Poincaré transformation.

§2.1a. Invariance under space-time translations

The presumed homogeneity of space-time implies the invariance of quantum field theory under a space-time translation, i.e. under $\Lambda(a) : x_\mu \rightarrow x_\mu + a_\mu$, where a_0 corresponds to the time translation and the three a_i's ($i = 1, 2, 3$) represent the three-dimensional space translation. To exploit Eq. (2.15), one writes $U(\Lambda) = e^{i\eta}$ (η is a Hermitian operator since U is unitary) so that Eq. (2.15) becomes:

$$\phi(x + a) = e^{i\eta}\phi(x)e^{-i\eta} \qquad (2.16)$$

To find η, we assume that a_μ is infinitesimal so that Eq. (2.16) reduces to:

$$\phi(x) + a_\mu \partial^\mu \phi + \ldots = \phi(x) + i[\eta, \phi(x)] + \ldots \qquad (2.17a)$$

from which, using the relation $\partial^\mu \phi(x) = i[P^\mu, \phi(x)]$ [see Eq. (2.9)], one obtains:

$$\eta = a_\mu P^\mu \quad \text{and} \quad U(\Lambda) = e^{i a_\mu P^\mu} \qquad (2.17b)$$

The generators, P^μ, of the unitary transformation group $U(\Lambda)$ obviously satisfy the C.R. $[P^\mu, P^\nu] = 0$, and, hence satisfy an Abelian algebra [the group defined by Eq. (2.17b) is $T(3,1)$] so that the values of the constants of motion, P^μ, are not quantized; lack of quantization always holds for $T(3,1)$ and is generally true of Abelian groups although a gauged Abelian group coupled to chiral fermions turns out to be an exception (see §7.5). If $J > 0$ is the spin of the quantum field, the generators of the space-time translation group will be the same as for the $J = 0$ case and hold for all the components of the higher spin field [in general, there will be $(2J + 1)$ "polarizations" except when the mass m of the field vanishes, in which case there are just two "polarizations" — see below]; on the other hand, as will be seen in §2.1b, the generators of the space-time "rotation" group depend on J.

§2.1b. Invariance under space-time rotations

We now turn to space-time rotations, beginning with the $J = 0$ quantum field. Space-time rotations include spatial rotations and Lorentz "boosts" and assumed isotropy of four-dimensional space-time (Minkowski space) results in the invariance of the quantum field theory under the general space-time rotation, giving rise to the Lorentz group. To prove this result for the $J = 0$ "scalar" field, we make use of Eq. (2.15) where Λ is simply the general four-dimensional orthogonal (Minkowski) transformation defined by: $x' = \Lambda x$, or $x'^\mu = \Lambda^\mu_\nu x^\nu$, where:

$$g_{\mu\nu} \Lambda^\mu_\alpha \Lambda^\nu_\beta = g_{\alpha\beta} \qquad (2.18)$$

with $g_{\mu\nu}$ the previously defined Minkowski metric. We now consider the Lorentz transformation in infinitesimal form so that:

$$\Lambda^\mu_\nu = \delta^\mu_\nu + \alpha^\mu_\nu. \qquad (2.19)$$

Using Eq. (2.18), Eq. (2.19) implies that $\alpha_{\mu\nu}$ is an antisymmetric 4×4 matrix, i.e. $\alpha_{\mu\nu} = -\alpha_{\nu\mu}$. Again, writing $U(\Lambda) = e^{i\eta}$, one gets:

$$\phi(x^\mu + \alpha^\mu_\nu x^\nu) = \phi(x) + \alpha^\mu_\nu x^\nu \partial_\mu \phi + \ldots = \phi(x) + i[\eta, \phi(x)] + \ldots \qquad (2.20)$$

and hence:

$$i[\eta, \phi(x)] = \frac{1}{2}\alpha^{\mu\nu}(x_\nu\partial_\mu - x_\mu\partial_\nu)\phi(x) + ... \qquad (2.21)$$

If one writes η in the form: $\eta = \frac{1}{2}\alpha^{\mu\nu}M_{\mu\nu}$, Eq. (2.21) becomes:

$$[M_{\mu\nu}, \phi(x)] = -\frac{1}{i}(x_\mu\partial_\nu - x_\nu\partial_\mu)\phi(x) \qquad (2.22)$$

Because of the anti-symmetry of the 4×4 matrix $M_{\mu\nu}$, the number of generators corresponding to the (four-dimensional) Lorentz group is six: three generators M_{ij} $(i, j = 1, 2, 3)$, related to the three components of the orbital angular momentum in the quantum field (see below), and the other three M_{0i} $(i = 1, 2\,3)$ corresponding to the "Lorentz boost" generators. The three orbital angular momenta, denoted by J_i, are expressed in terms of the "space-space" generators, M_{ij} by $J_i = -\frac{1}{2}\epsilon_{ijk}M_{jk}$ and satisfy the non-commutative algebra of the $O(3)$ group:

$$[J_i, J_j] = i\epsilon_{ijk}J_k \quad (i = 1, 2, 3) \qquad (2.23)$$

The fact that the J_i's do not commute implies that the group $O(3)$ is non-Abelian; moreover, the eigenvalues are quantized as integers. The three "Lorentz-boost" generators M_{0i} are usually denoted by K_i, and if one goes to Euclidean four-space ($x_4 = it$), one obtains the four-dimensional orthogonal group $O(4)$, which is locally isomorphic to $SU(2) \times SU(2)$, with one $SU(2)$ defined by the three generators J_i and the other $SU(2)$ by the three generators K_i. We note that the Higgs boson and the pion are prime examples (in the standard model) of charged (and neutral) scalar quantum fields whose Lorentz generators satisfy Eq. (2.22); since the pion is a pseudoscalar particle, its intrinsic negative parity is not decided by the Lorentz transformation but is taken into account by the separate discrete transformation called space inversion (see §2.2a).

Equation (2.23) also represents the algebra of the non-Abelian $SU(2)$ group — which is therefore isomorphic with $O(3)$ and is its "covering" group with the property that its spinor representations [absent for the $SO(3)$ group] allow half-integral values of the angular momentum in addition to integral ones. From the definition of J_i in terms of M_{jk}, it is evident that in the absence of "Lorentz boosts", the Lorentz group becomes the rotation group defined by Eq. (2.23) where the J_i represent only the orbital angular momenta associated with the scalar field $\phi(x)$. The situation changes when the quantum field possesses a non-vanishing intrinsic spin and then M_{jk} becomes more complicated — see below — and the (non-relativistic) spatial rotation is composed of a spin in addition to an orbital contribution. If the intrinsic spin is half-integral, the total J must take on half-integral values — as well as integral values — and the group must be global $SU(2)$ (allowing for spinor as well as tensor representations).

In working out the Lorentz group for $J = \frac{1}{2}$ and $J = 1$ fields, we treat the bosonic $J = 1$ field before the $J = \frac{1}{2}$ fermionic field because the former is a straightforward generalization of the $J = 0$ case: from one to two "polarizations" of the quantum field for the massless $J = 1$ Maxwell field and a further increase to three "polarizations" for the massive $J = 1$ Proca field [2.3]. On the other hand, the change in Lorentz group in going from the bosonic $J = 0$ (Klein-Gordon) field to the fermionic $J = \frac{1}{2}$ (Dirac or Weyl) field requires a careful taking into account of the effect of a Lorentz transformation on the "first" quantized Dirac (or Weyl) field which, in turn, requires a change from C.R. to anti-C.R. for the "second"-quantized creation and destruction operators in the fermionic case. Turning to the $J = 1$ Proca field, the basic strategy in dealing with a massive charged vector field is to represent it by a complex four-vector field $\phi^\mu(x)$ (x is the space-time four-vector) and to impose the "Lorentz condition" $\partial_\mu \phi^\mu = 0$ so that $\phi^0(x)$ (the "scalar" polarization) can be discarded and the Klein-Gordon equations emerge for the other three components $\phi^i (i = 1, 2, 3)$. The expressions for $\phi^i(x)$ [and $\phi^{i\dagger}(x)$] can then be written in the form [see Eqs. (2.10)]:

$$\phi^i(x) = \frac{1}{(2\pi)^{\frac{3}{2}}} \sum_{\lambda=1}^3 \int \frac{d^3k}{\sqrt{2\omega_k}} e^i(k,\lambda)[a_k(\lambda)e^{-ikx} + b_k^\dagger(\lambda)e^{ikx}]$$

$$\phi^{i\dagger}(x) = \frac{1}{(2\pi)^{\frac{3}{2}}} \sum_{\lambda=1}^3 \int \frac{d^3k}{\sqrt{2\omega_k}} e^i(k,\lambda)[b_k(\lambda)e^{-ikx} + a_k^\dagger(\lambda)e^{ikx}]$$

(2.24)

where $a_k^i(\lambda)$ and $a_k^{i\dagger}(\lambda)$ are the destruction and creation operators of a vector particle with "polarization" λ [and, correspondingly, for the $b_k^i(\lambda)$ and $b_k^{i\dagger}(\lambda)$ for the antiparticle]; the $e^i(k,\lambda)$ are c-number functions which are chosen so that:

$$[a_k(\lambda), a_{k'}(\lambda')] = [a_k^\dagger(\lambda), a_{k'}^\dagger(\lambda')] = 0; \quad [a_k(\lambda), a_{k'}^\dagger(\lambda')] = \delta_{\lambda\lambda'}\delta_{kk'} \quad (2.25)$$

with similar C.R. for $b_k^\dagger(\lambda), b_k(\lambda)$, and all other pairs of operators commuting. The condition satisfied by the $e^i(k,\lambda)$'s is:

$$\sum_{\lambda=1}^3 e^i(k,\lambda)e^j(k,\lambda) = \delta_{ij} + \frac{k^i k^j}{m^2} \quad (2.26)$$

If the momentum $\underline{k}$ is taken along the z axis — so that $\underline{k} = (0, 0, k)$, then Eq. (2.26) gives the solution:

$$e^i(k, 2) = \delta_{i2}; \quad e^i(k, 3) = 0 \quad (i = 1, 2); \quad e^i(k, 1) = \delta_{i1};$$

$$e^3(k, 3) = 1 + \frac{k^2}{m(\omega_k + m)} = \frac{\omega_k}{m} \qquad (2.26a)$$

Equation (2.26a) records the expressions for the three "polarizations" of a massive vector field — two transverse and one longitudinal — and one can now proceed to derive the generators of the Lorentz group. One simply generalizes Eq. (2.15) for the massive scalar field to the massive vector field, namely:

$$U(\Lambda)\phi^\mu(x)U^{-1}(\Lambda) = (\Lambda^{-1})^\mu_\nu \phi^\nu(\Lambda x) \qquad (2.27)$$

The use of $(\Lambda^{-1})^\mu_\nu$ in Eq. (2.27) occurs because one wishes to maintain the vectorial transformation property (Λ^μ_ν) between the expectation values over the transformed $\Psi_{A'}$ and initial Ψ_A states, i.e. $< A'|\phi^\mu(\Lambda x)|A' > = \Lambda^\mu_\nu < A|\phi^\nu(x)|A >$. One next applies the infinitesimal method to obtain the $M_{\mu\nu}$ generators of the Lorentz group and, as might be expected from the fact that the intrinsic spin of the field is 1 (instead of 0), one obtains the result:

$$[M_{\mu\nu}, \phi_\sigma(x)] = i(x_\mu \partial_\nu - x_\nu \partial_\mu)\phi_\sigma(x) + i[g_{\mu\sigma}\phi_\nu(x) - g_{\nu\sigma}\phi_\mu(x)] \qquad (2.28)$$

where the first and second terms on the R.H.S. of Eq. (2.28) (for μ and $\nu \neq 4$) correspond to the three components of orbital angular momentum and intrinsic spin of the vector field respectively. More precisely:

$$[M_{ij}, \phi_\sigma(x)] = \epsilon_{ijk}\{L_k \phi_\sigma(x) + (S_k)_{\sigma\rho}\phi_\rho(x)\} \qquad (2.29)$$

where:

$$L_k = -i(x^i \partial_j - x^j \partial_i); \quad (S_k)_{\rho\sigma} = -i\{\delta_{\rho i}\delta_{\sigma j} - \delta_{\rho j}\delta_{\sigma i}\} \qquad (2.29a)$$

Equation (2.29a) for L_k obviously represents the orbital angular momentum and it is easily checked that S_k satisfies the same algebra as L_k, namely:

$$[S_i, S_j] = i\epsilon_{ijk}S_k \qquad (i = 1, 2, 3) \qquad (2.30)$$

with:

$$(S^2)_{ij} = 2\delta_{ij} \qquad (2.30a)$$

Equation (2.30) defines the algebra of the group $O(3)$ — [not $SU(2)$] — since only integral quantum numbers are possible and Eq. (2.30a) confirms that the intrinsic spin of the Proca field is, indeed, 1.

From the form of Eq. (2.29), it follows directly that the statements about angular momentum for the charged vector (Proca) field remain valid for neutral massive vector (Proca) fields. The transition from a neutral massive vector field to a neutral massless vector field (i.e. the Maxwell field) is less straightforward since the Maxwell field only possesses two (transverse) polarizations rather than the three "polarizations" (two transverse plus one longitudinal) of the massive vector field. Equations (2.26) and (2.26a) already indicate the qualitative difference between the two transverse polarizations and the one longitudinal "polarization" of a Proca field. We do not enter into the derivation of the further reduction — from three to two — in the number of "polarizations" for a massless vector field, except to point out that, while the elimination of one of the four polarizations (i.e. the "scalar" polarization) is accomplished for the Proca field by means of the Lorentz condition, the reduction from three to two "polarizations" is achieved for a Maxwell field by utilizing the condition of gauge invariance; gauge invariance permits the use of the (arbitrary) gauge function to achieve a cancellation of both the longitudinal and "scalar" polarizations, thereby leaving just the two transverse polarizations for a Maxwell field. [The actual proof of the reduction from three to two polarizations for a Maxwell field is non-trivial because the imposition of gauge invariance during the course of "second quantization" of the Maxwell field brings in the "indefinite metric" that yields "negative norm" states (with "negative probabilities"); however, it can be shown [2.4] that the use of gauge invariance eliminates the physical effects of the "indefinite metric".] The result of cancelling the longitudinal polarization for a Maxwell field is that $e^3(k, 3)$ [in Eq. (2.26a)] is totally absent and only $e^i(k, 1) = 1$ and $e^2(k, 2) = 1$ remain; effectively, one ends up with the two transverse components of the Maxwell field and one can write:

$$A^r(x) = \frac{1}{(2\pi)^{\frac{3}{2}}} \int \frac{d^3k}{\sqrt{2k}} [a_k^r e^{-ikx} + \text{h.c.}] \quad (r = 1, 2) \qquad (2.31)$$

where r is the polarization index. Comparing Eq. (2.24) with Eq. (2.31) makes it clear that the b_k (and $b_k^\dagger$) operators are absent in the Maxwell case (the photon is its own antiparticle) and that the Poincaré generators for the Maxwell field are given by Eq. (2.29a).

We next move on to a derivation of the Poincaré generators for the fermionic Dirac field; now the replacement of the C.R. for the bosonic ($J = 0$ and $J = 1$) fields by the fermionic $J = \frac{1}{2}$ anti-C.R. yields the "spinorial" counterpart of the second term on the R.H.S. of Eq. (2.28) for the Poincaré generators $M_{\mu\nu}$. The Lagrangian for the Dirac field is:

$$\mathcal{L}(x) = \bar{\psi}(x)(i\gamma^\mu \partial_\mu - m)\psi(x) \tag{2.32}$$

where $\psi(x)$ is the four-component (single-column) Dirac wavefunction and $\bar{\psi}(x) = \psi^\dagger \gamma^0$ [$\psi^\dagger(x)$ is the Hermitian conjugate of $\psi(x)$] with the Dirac operators γ^μ satisfying the anti-C.R.:

$$[\gamma^\mu, \gamma^\nu]_+ = 2g^{\mu\nu}. \tag{2.33}$$

Since we have chosen γ^0 Hermitian, $\gamma^i (i = 1, 2, 3)$ anti-Hermitian and $\gamma_5 = -i\gamma^0\gamma^1\gamma^2\gamma^3$ (see §1.2d), the Dirac equations for $\psi(x)$ and $\bar{\psi}(x)$ become:

$$(i\gamma^\mu\partial_\mu - m)\psi(x) = 0; \qquad i\partial_\mu\bar{\psi}(x)\gamma^\mu + m\bar{\psi}(x) = 0 \tag{2.34}$$

The current $j^\mu(x) = \bar{\psi}(x)\gamma^\mu\psi(x)$ has a simple form in terms of $\bar{\psi}$ and ψ and since $\partial_\mu j^\mu(x) = 0$, we have for the conserved charge Q:

$$Q = \int d^3x j^0(x) = \int d^3x \{\bar{\psi}(x)\gamma^0\psi(x)\} = \int d^3x \psi^\dagger(x)\psi(x)$$

$$= \int d^3x \sum_\rho |\psi_\rho(x)|^2 \tag{2.35}$$

where the summation over ρ (1, ...4) is over the four components of the one-column Dirac wavefunction.

It would seem from Eq. (2.35) that the total charge Q is positive definite and one wonders whether the Dirac theory makes provision for the negative charge associated with the antiparticles of the theory (if the particles possess positive charge). The answer is in the affirmative when one takes into account "second quantization": the "second-quantized" Dirac theory predicts the same reversal of charge sign between particle and antiparticle as does the Klein-Gordon scalar theory (or Proca theory) since anti-C.R. are satisfied by the $J = \frac{1}{2}$ Dirac spinors in contrast to the C.R. satisfied by the $J = 0$ and $J = 1$ fields. Another important difference emerges between the fermionic $J = \frac{1}{2}$ case and the bosonic $J = 0$ and $J = 1$ cases; for the boson fields, the second-order Klein-Gordon and Proca equations do not depend on the sign of the energy whereas the first-order Dirac equation depends on the sign of the energy (since $E = \pm\sqrt{p^2 + m^2}$); in the "second-quantized" version of the Dirac field, the positivity of the total energy for a Dirac field is guaranteed through the use of independent positive and negative frequency solutions and the imposition of anti-C.R. on the creation and destruction operators associated with these independent solutions. [In the mid-1930s, Pauli was very much aware of these differences between the "second-quantized" Dirac theory and the Pauli-Weisskopf theory (i.e. the "second-quantized" version of the Klein-Gordon equation) and, for a time, referred to the Pauli-Weisskopf theory as the "anti-Dirac" theory! [2.5]]

If we Fourier transform Eq. (2.34) and denote the positive and negative frequency solutions by $u(p)$ and $v(p)$ respectively, the four-component $\psi(x)$ can be expressed in terms of $u(p)$ and $v(p)$ satisfying the equations:

$$(\gamma^\mu p_\mu - m)u(p) = 0; \qquad (\gamma^\mu p_\mu + m)v(p) = 0 \qquad (2.36)$$

Each of the Eqs. (2.36) has two solutions, which we denote by $u^r(p)$ and $v^r(p)$ $(r = 1, 2)$, where r corresponds to the two spin directions; for processes involving chiral fermions, it is more convenient to let r represent the helicity and to take account of the fact that:

$$\mathcal{H}\, u^r(p) = r u^r(p); \qquad \mathcal{H}\, v^r(p) = -r v^r(p) \qquad (2.37)$$

where $\mathcal{H} = \underline{\sigma} \cdot \underline{p}/|\underline{p}|$ (see below). Using the above notation, the "first-quantized" Dirac field $\psi(x)$ can now be "second-quantized" by expanding it [and $\bar{\psi}(x)$] in terms of the creation and destruction operators associated with the u and v solutions, namely:

$$\psi(x) = \frac{1}{(2\pi)^{\frac{3}{2}}} \sum_{r=1}^{2} \int d^3p\, [a^r(p)u^r(p)e^{-ipx} + b^{r\dagger}(p)v^r(p)e^{ipx}]$$

$$\bar{\psi}(x) = \frac{1}{(2\pi)^{\frac{3}{2}}} \sum_{r=1}^{2} \int d^3p\, [a^{r\dagger}(p)\bar{u}^r(p)e^{ipx} + b^r(p)\bar{v}^r(p)e^{-ipx}]$$

$$(2.38)$$

If one proceeds to derive an expression for the total energy — after introducing the appropriate canonically conjugate field momenta — one obtains a result that is not positive definite, namely:

$$E = \sum_{r=1}^{2} \int d^3p\, [E_p a^{r\dagger}(p)a^r(p) - E_p b^r(p)b^{r\dagger}(p)] \qquad (2.39)$$

Equation (2.39) is a consequence of the spinor character of the Dirac field and can be rectified if one imposes anti-C.R. among the creation and destruction operators associated with the particles and antiparticles. This is achieved by postulating the following equal-time anti-C.R. for $\psi(x)$ and $\psi^\dagger(x)$:

$$[\psi_\alpha(\underline{x}, t), \psi_\beta(\underline{y}, t)]_+ = [\psi_\alpha^\dagger(\underline{x}, t), \psi_\beta^\dagger(\underline{y}, t)]_+ = 0$$

$$[\psi_\alpha(\underline{x}, t), \psi_\beta^\dagger(\underline{y}, t)]_+ = \delta_{\alpha\beta}\delta(\underline{x} - \underline{y}) \qquad (2.40)$$

which leads to the anti-C.R. for the creation and destruction operators, namely:

$$[a^r(p), a^{s\dagger}(p')]_+ = [b^r(p), b^{s\dagger}(p')]_+ = \delta_{rs}\delta(p - p') \qquad (2.41)$$

with all other anti-C.R. vanishing. The use of Eq. (2.41) — allowing for a physically irrelevant infinite constant — then leads to the physically correct positive-definite expression for the total energy [in contrast to Eq. (2.39)]:

$$E = \sum_{r=1}^{2} \int d^3p \, [E_p a^{r\dagger}(p) a^r(p) + E_p b^{r\dagger}(p) b^r(p)] \qquad (2.42)$$

With the anti-C.R. (2.41), not only does the total energy in the field (and the total momentum) acquire a physically meaningful form (in terms of the occupation numbers $n^r(p) = a^{r\dagger}(p) a^r(p)$ and $\bar{n}^r(p) = b^{r\dagger}(p) b^r(p)$) but also so does the total charge. The total charge Q is given by:

$$Q = q \sum_{r=1}^{2} \int d^3p \, [a^{r\dagger}(p) a^r(p) - b^{r\dagger}(p) b^r(p)] = \sum_{r=1}^{2} \sum_{p} \{n^r(p) - \bar{n}^r(p)\} \quad (2.43)$$

As anticipated, when we impose the anti-C.R. on the creation and destruction operators for particles and antiparticles in the $J = \frac{1}{2}$ fermionic case (in contrast to the C.R. for the boson case), the behavior of the charge for both cases is identical and Q changes sign (but not magnitude) in going from particle to antiparticle.

With the basic "first-quantized" and "second-quantized" Dirac field formalism before us, we can derive the generators for the Lorentz group corresponding to the Dirac field. In the case of the scalar field, the form of the (scalar) field operator is preserved when a Lorentz transformation is performed on the space-time coordinates, i.e. $\phi'(x') = \phi(x)$ when $x' = \Lambda x$. This is no longer true for the Dirac field, where the transformation $x' = \Lambda x$ must change the form of the spinor field in the new Lorentz frame so that $\psi'(x')$ satisfies the same Dirac equation as $\psi(x)$, namely:

$$(i\gamma^\mu \partial'_\mu - m)\psi'(x') = 0 \qquad (2.44)$$

If we write $\psi'(x') = S(\Lambda)\psi(x)$, it can be shown that the Dirac equation retains its Lorentz-invariant form when:

$$S(\Lambda) = e^{-(i/2)\alpha_{\mu\nu}\Sigma^{\mu\nu}}; \qquad \Sigma^{\mu\nu} = -\frac{1}{4i}[\gamma^\mu, \gamma^\nu] \qquad (2.45)$$

With $S(\Lambda)$ known, it is easy to show that the Dirac field analog of Eq. (2.15) is:

$$U(\Lambda)\psi(x)U^{-1}(\Lambda) = S^{-1}(\Lambda)\psi(\Lambda x) \qquad (2.46)$$

Applying the infinitesimal method to Eq. (2.46) — by letting $U(\Lambda) = 1 + i/2\,\alpha^{\mu\nu}M_{\mu\nu}$, $S^{-1}(\Lambda) = 1 + i/2\,\alpha^{\mu\nu}\Sigma_{\mu\nu}$ and $\psi(\Lambda x) = \psi(x^\mu + \alpha^\mu_\nu x^\nu) \simeq \psi(x^\mu) +$

$\alpha_\nu^\mu x^\nu \partial_\mu \psi(x^\mu)$ — one obtains the relation:

$$[M_{\mu\nu}, \psi(x)] = \{\frac{1}{i}(-x_\mu\partial_\nu + x_\nu\partial_\mu) + \Sigma_{\mu\nu}\}\psi(x) \qquad (2.47)$$

Comparing Eq. (2.47) with Eq. (2.22) (for the scalar field), it is clear that the intrinsic spin of the Dirac field is responsible for the additional term, $\Sigma_{\mu\nu}$, in the generators for the Lorentz group; to verify this, we consider only spatial rotations so that $\mu, \nu \neq 0$ and we then get [from Eq. (2.45)] $\Sigma_{ij} = \frac{1}{2}\epsilon_{ijk}\sigma_k$, which is precisely the spin of the Dirac field. Comparison of Eqs. (2.47) and (2.28) confirms the translation of the "tensorial" spin contribution to the Lorentz generators of the $J = 1$ (Proca) field into the "spinorial" spin contribution to these generators for the $J = \frac{1}{2}$ (Dirac) field.

In deriving the Lorentz generators for the Dirac field, we have assumed a finite mass for the Dirac particle. This assumption works well for all the quarks and charged leptons of the three fermion generations but must be modified if the fermion is massless, as appears to be the case for the neutrinos of the three fermion generations. Moreover, in the standard model, the quarks and charged leptons are assumed to be massless chiral fermions — like neutrinos — prior to the onset of the Higgs SSB mechanism which generates distinct finite (Dirac) masses for all the quarks and charged leptons by means of Yukawa couplings with the Higgs boson. Consequently, it is necessary to discuss the replacement of the finite-mass Dirac equation by the massless Weyl equation which — in contrast to the 4-component parity-conserving Dirac equation — is the 2-component maximal parity-violating equation. It is easy to see how setting $m = 0$ in the Dirac equation (2.34) gives rise to two Weyl equations of opposite helicity. One first observes that the Dirac equation for $m = 0$ is invariant under the global γ_5 (chirality) transformation: $\psi \rightarrow e^{i\alpha\gamma_5}\psi$ (α is constant) or in the infinitesimal form: $\psi \rightarrow i\alpha\gamma_5\psi$. It follows that the chiral-projected fields $\psi_\pm(x) = \frac{1}{2}(1\pm\gamma_5)\psi(x)$ also satisfy the Dirac equation and one then obtains the two uncoupled (2-component) Weyl equations for the chiral-projected fields $\psi_\pm(x)$:

$$(\frac{\partial}{\partial t} + \underline{\sigma} \cdot \underline{\nabla})\,\psi_\pm(x) = 0 \qquad (2.48)$$

where $\underline{\sigma}$ is the 2-component spin and the $\psi_\pm$ are 2-component Weyl spinors, with $\psi_+(x)$ representing a Weyl particle with positive chirality (and a Weyl antiparticle with negative chirality), whereas $\psi_-(x)$ represents a Weyl particle of negative chirality (and, correspondingly, a Weyl antiparticle with positive chirality).

To exhibit more clearly the physical differences between the equations describing the massive and massless $J = \frac{1}{2}$ particles, it is convenient to work in momentum space; in momentum space, Eq. (2.48) becomes:

$$(\mp |\underline{p}| + \underline{\sigma} \cdot \underline{p})\psi_{\pm}(p) = 0 \quad \text{or} \quad \mathcal{H}\psi_{\pm}(p) = \pm\psi_{\pm}(p) \qquad (2.49)$$

which yields the helicity $\mathcal{H} = \underline{\sigma} \cdot \underline{p}/|\underline{p}| = +1(-1)$ for the positive (negative) chirality Weyl particle (antiparticle) and $-1(+1)$ for the negative (positive) chirality Weyl particle (antiparticle) [see Eq. (2.37)]. The fact that a Weyl particle may possess a righthanded helicity ($\mathcal{H}_R = +1$) or a lefthanded helicity ($\mathcal{H}_L = -1$) emphasizes the importance of having an independent means of deciding whether the lefthanded or righthanded state corresponds to the particle; in the case of the neutrino, it is the sign of the lepton charge $(+1)$ that decides that ν_L is the particle. Insofar as the Lorentz generators are concerned, one can expand the Weyl spinors $\psi_+(x)$ and $\psi_-(x)$ in terms of creation and destruction operators $[a_+(p)$ and $b_+^\dagger(p)$ for $\psi_+(x)$ and $b_-(p)$ and $a_-^\dagger(p)$ for $\psi_-(x)]$ and proceed to calculate the Weyl analog of $\Sigma_{\mu\nu}$ in Eq. (2.47); it is evident that $\Sigma_{\mu\nu}$ becomes the 2×2 $\underline{\sigma}$ analog of $\Sigma_{\mu\nu}$ in Eq. (2.45). The basic structure of the Lorentz generators remains the same in going from the massive Dirac field to the massless Weyl field as long as one replaces $\Sigma_{\mu\nu}$ by $\sigma_{\mu\nu}$; the big change comes in connection with the discrete parity and charge conjugation operations, which yield chirality (γ_5) invariance and CP invariance of the Weyl equation but not of the Dirac equation (because of the finite mass).

This completes our review of the basic formalism of quantum field theory for the $J = 0$ and 1 bosonic fields and the $J = \frac{1}{2}$ fermionic field, all of which play a role in the standard model: the massless $J = 0$ fields perform as N-G bosons, the massive $J = 0$ fields as Higgs bosons, the $J = \frac{1}{2}$ fields serve as the "matter" fields [either as massless neutrinos or as massive charged leptons and quarks], and the $J = 1$ fields play the role of gauge (force) fields [as massless photons and gluons in QED and QCD respectively, or when the gauge symmetry is spontaneously broken — as in QFD — they play the role of massive weak bosons]. Beyond the standard model, the graviton of Einsteinian gravity is described by a massless $J = 2$ (tensor) equation and a "bestiary" of speculative particles — such as the $J = \frac{3}{2}$ [2.6] "gravitino" of supergravity (see §1.4c) — has been added to the list. We do not write down the explicit expressions for higher spin ($J > 1$) bosonic and fermionic fields but we do note that the group properties of the ten ($M^{\mu\nu} + P^{\mu}$) continuous $(3+1)$-space-time generators — six for the $(3+1)$ "rotations" in Minkowski space leading to the $M^{\mu\nu}$ generators of $SO(3,1)$, and four for the $(3+1)$-translations leading to the P^μ generators of $T(3,1)$ — hold for quantum fields of arbitrary spin (integral or half-integral). In the next section, we discuss some general properties of the Poincaré group which is defined by these ten generators.

§2.1c. Poincaré group

In this section, we define the general Poincaré transformation, show that it gives rise to a group (the Poincaré group) with the ten $(M^{\mu\nu} + P^\mu)$ generators, and explain the relation between mass and the number of spin "polarizations" of a quantum field with arbitrary mass and arbitrary spin. The general Poincaré transformation is defined as the inhomogeneous Lorentz transformation:

$$x'^\mu = \Lambda^\mu_\nu x^\nu + a^\mu \tag{2.50}$$

satisfying the conditions: $\det(\Lambda) = 1$ and $\Lambda^0_0 > 1$. The condition $\det(\Lambda) = +1$ excludes the discrete transformation of space inversion and the further condition $\Lambda^0_0 \geq 1$ excludes the discrete transformation of time inversion; when the two conditions are obeyed, the Poincaré transformation is said to be "proper" and "orthochronous" respectively. These two conditions are invoked in deriving the expressions for the Poincaré generators because $\det(\Lambda) = -1$ changes the parity and $\Lambda^0_0 \leq -1$ reverses the time and it is not possible to change the parity nor reverse the time by means of the "infinitesimal method" that is used to derive the generators of the Poincaré group.

With the understanding that we are dealing with a proper, orthochronous Poincaré transformation, it is easy to show that the inhomogeneous Lorentz transformations defined by Eq. (2.50) constitute a group (the Poincaré group): i.e. two successive Poincaré transformations represent a Poincaré transformation, an inverse to Eq. (2.50) exists, etc. The unitary transformation in the Hilbert space of state vectors corresponding to the Poincaré transformation may be written in the form $U(a, \Lambda)$ and it is obvious that $U(a, \Lambda) = U(a)U(\Lambda)$ where $U(a) = U(a, 1)$ and $U(\Lambda) = U(0, \Lambda)$. The fact that $U(a, \Lambda) = U(a)U(\Lambda)$ immediately tells us that the Poincaré group has ten generators: P^μ (the four generators of the space-time translation group) and $M^{\mu\nu}$ (the six generators of the Lorentz group). It is straightforward to write down the commutators of the ten generators P^μ and $M^{\mu\nu}$; we have already noted that the four "energy-momentum" generators commute among themselves, i.e. $[P^\mu, P^\nu] = 0$. The other C.R. are more complicated because we are dealing with a semi-direct product group; they are:

$$[M^{\mu\nu}, P^\sigma] = -i(P^\mu g^{\nu\sigma} - P^\nu g^{\mu\sigma})$$

$$[M^{\mu\nu}, M^{\rho\sigma}] = -i(M^{\nu\sigma} g^{\mu\rho} - M^{\mu\sigma} g^{\nu\rho} + + M^{\rho\nu} g^{\mu\sigma} - M^{\rho\mu} g^{\nu\sigma}). \qquad (2.51)$$

The group-theoretic properties of the $M_{\mu\nu}$ are more transparent when we use the "spatial rotation" generators $J_i = -\frac{1}{2}\epsilon_{ijk}M^{jk}$ $(i = 1, 2, 3)$ and the "Lorentz boost" generators $K_i = M_{i0}$ $(i = 1, 2, 3)$; the C.R. of Eq. (2.51) then transform into:

$$[J_i, P_k] = i\epsilon_{ikl}P^l; \quad [J_i, P_0] = 0; \quad [K_i, P_k] = iP_0 g_{ik}; \quad [K_i, P_0] = -iP_i;$$

$$[J_m, J_n] = i\epsilon_{mnk}J_k; \quad [J_m, K_n] = i\epsilon_{mnk}K_k; \quad [K_m, K_n] = -i\epsilon_{mnk}J_k \qquad (2.52)$$

The first set of commutators among the J's close on the $SU(2)$ algebra but the K_i's do not; the reason is that the metric is the Minkowski metric and the Lorentz group is $O(3, 1)$. If we change to the metric of Euclidean four-space (i.e. $x^4 = ix^0 = it$), the redefined "Lorentz-boost" generators also have a plus sign in their C.R. and the Lorentz group is $O(4)$ [which is locally isomorphic to $SU(2) \times SU(2)$]. It is possible to define two non-Hermitian sets of generators $N_i = \frac{1}{2}(J_i + iK_i)$ $(i = 1, 2, 3)$ and $N_i^\dagger = \frac{1}{2}(J_i - iK_i)$ $(i = 1, 2, 3)$ so that both the N_i's and $N_i^\dagger$'s close on the Lie algebra of $SU(2) \times SU(2)$, to wit:

$$[N_i, N_j] = i\epsilon_{ijk}N_k; \quad [N_i^\dagger, N_j^\dagger] = i\epsilon_{ijk}N_k^\dagger; \quad [N_i, N_j^\dagger] = 0 \qquad (2.53)$$

Since both N_i and $N_i^\dagger$ close on the $SU(2)$ algebra, the eigenvalues are, in both cases, integral or half-integral; the algebra defined by Eq. (2.53) is called the Cartan algebra [2.7]. The Cartan algebra is especially useful for designating the representations of Weyl spinors; for example, $(\frac{1}{2}, 0)$ is the representation of a right-handed Weyl spinor and $(0, \frac{1}{2})$ that of a lefthanded Weyl spinor [see Eq. (2.37)]. Of course, these two basic representations of the Cartan algebra can be employed to construct more complicated fields, e.g. $[(\frac{1}{2}, 0) + (0, \frac{1}{2})]$ is the "Cartan" representation for a Dirac field and $(\frac{1}{2}, \frac{1}{2})$ that of a vector field. We will have occasion to use this notation in §9.2 when we attempt to construct three-preon Weyl spinor composites to represent Weyl quarks and leptons.

The next useful step is to find the two second-order Casimir invariants that commute with all ten generators of the Poincaré group; the eigenvalues of these Casimir invariants can be then employed to characterize a quantized field theory with arbitrary mass and arbitrary spin. The two second-order Casimir invariants are $P_\mu P^\mu$ — whose eigenvalues are obviously m^2 (m is the mass) — and $W_\sigma W^\sigma$, where W_σ is the so-called Pauli-Lubanski (mixed) operator $W_\sigma = -\frac{1}{2}\epsilon_{\mu\nu\rho\sigma}M^{\mu\nu}P^\rho$. The W_σ satisfy the following C.R. with the $M_{\mu\nu}$ and P_μ:

$$[W_\sigma, P^\mu] = 0; \quad [M_{\mu\nu}, W_\sigma] = -i(W_\mu g_{\nu\sigma} - W_\nu g_{\mu\sigma});$$
$$[W_\lambda, W_\sigma] = i\epsilon_{\lambda\sigma\alpha\beta}W^\alpha P^\beta \tag{2.54}$$

With the help of Eq. (2.54), it can be shown that $W_\sigma W^\sigma$ commutes with all ten generators $M_{\mu\nu}$ and P_μ and possesses the eigenvalues $-m^2 s(s+1)$ (s is the spin); the number of spin states is, of course, $(2s+1)$ [one can also speak of $(2s+1)$ "polarizations" (the "polarization" or "helicity" representation is more useful for massless fields — see below)]. This is the general result for finite-mass quantum fields that are invariant under the Poincaré transformation, and tells us that one can assign arbitrary values of mass and spin to a quantum field insofar as the Poincaré transformation is concerned; the previous discussion of the $J = 0, \frac{1}{2}$ and 1 massive quantum fields is, of course, consistent with this general result.

For massless fields — where it is not possible to choose a "rest frame" — the general conclusions from an analysis of the Poincaré group are somewhat different, namely: (1) the number of spin states, or, equivalently, the number of polarization or helicity states, is reduced to two, for arbitrary "quantized" (non-vanishing) spins (i.e. half-integral or integral $s > 0$). It is easy to show that for $m = 0$ [since both $P_\mu P^\mu = 0$ and $W_\mu W^\mu = 0$ in addition to $P_\mu W^\mu = 0$], the four-vectors W_μ and P_μ are proportional. Hence, we may write $W_\mu = \pm h P_\mu$ (h is a positive constant) so that there are two (transverse) "polarizations" or (longitudinal) helicities for arbitrary $s > 0$. For $s = \frac{1}{2}$ (Weyl case), this result is in agreement with the two values of the helicity $\mathcal{H}$ deduced from the Weyl equation (2.49); (2) the number of possible "polarizations" or helicities is two, provided that one does not insist on $\det \Lambda = +1$, and allows for spatial inversion where $\det \Lambda = -1$. Otherwise, only one "polarization" or helicity state is possible for arbitrary non-vanishing "quantized" spin $s > 0$ since the helicity $\mathcal{H} = \underline{J} \cdot |p|/|p|$ changes sign under spatial inversion; (3) for massless fields, the spin need not belong to the compact Lie group $SU(2)$; indeed, Bargmann and Wigner [2.8] have shown that the spin behaves like the generator of an Abelian $U(1)$ group so that it can take on a continuum of real values. [The Bargmann-Wigner statement applies to $(3+1)$-space-time dimensions; for fewer dimensions (e.g. $2+1$-space-time dimensions), fractional spin does not require massless fermions [2.9].]

As far as phenomenology is concerned, massless fields do exist: the bosonic photon, gluon and graviton fields and, as far as we know, the fermionic neutrino field. The massless photon is the gauge field for electromagnetism and, since parity is conserved, the two (transverse) polarizations of the photon are precisely what one finds. The $SU(3)$ color gauge group, describing the strong interaction, possesses eight massless gluons (see Table 1.3) and each gluon — like the massless photon — must have two "polarizations" because parity is conserved in the strong interaction. In the case of the graviton (see Table 1.4), Einsteinian gravity tells us that it is massless and should possess two "polarizations" since parity is conserved in the gravitational interaction; apparently, "self-dual" gravitons can be assigned definite "helicities" (or "orientations") and whether, under certain circumstances, there is preferential coupling for one of the two "helicities" (as for neutrinos), remains to be seen [2.10]. When we come to massless fermions, the neutrino is the most likely candidate: maximal parity violation is observed in the weak interaction and only the lefthanded helicity state (one "polarization") enters into the standard electroweak theory (see §6.2). It should be pointed out that the "suppressed polarizations" are restored in the theory when the symmetry principle responsible for the "suppression" is broken spontaneously and this statement applies to both bosons and fermions in the standard electroweak gauge theory; thus, the masses of the weak gauge fields are generated by means of the Higgs mechanism which restores the "longitudinal" polarization of the weak gauge fields by breaking the gauge invariance of the electroweak group. For fermions in the electroweak theory, the Higgs mechanism is also available to give mass to the quarks and leptons by breaking the chirality invariance of the electroweak group. Another option for fermion mass generation and restoring "suppressed" degrees of polarization — other than the Higgs mechanism in the electroweak theory — is the possibility of dynamical generation of fermion mass within the framework of the electroweak theory — see §6.4c.

§2.2. Discrete symmetries

In the previous section, we have considered continuous space-time symmetries: space-time translations, the Lorentz transformation (including spatial rotations) and the Poincaré transformation (i.e. the proper, orthochronous inhomogeneous Lorentz transformation) and we have derived the unitary transformations on the ("second-quantized") field operators acting in the infinite-dimensional Hilbert space of the many-particle (Fock) states, as determined by the particular continuous space-time transformation on the "singly-quantized" $J = 0, \frac{1}{2}$ or 1 field. The unitary transformations on the "second-quantized fields" represent well-known symmetry groups: translation group $[T(3,1)]$, Lorentz group $[SO(3,1)]$, and Poincaré group $[SO(4,2)]$, and reflect the invariance of the ("single-quantized") field Lagrangian under the specified continuous space-time transformation. Thus far, there have been no known phenomenological departures from the conservation laws that follow from the continuous space-time symmetries. However, since the Poincaré transformation refers to the proper orthochronous inhomogeneous Lorentz group, the discrete symmetries of space inversion (P) and time inversion (T) must be considered separately. In addition, there is the important discrete symmetry of charge conjugation (C) which must be treated on a par with P and T and enters into the crucial CPT theorem to be discussed in §2.2d.

While Poincaré invariance appears to be an absolute conservation law for all quantum field theories, some of the spectacular surprises in particle physics in recent decades have occurred in connection with the breakdown of the two discrete space-time symmetries and the discrete symmetry of charge conjugation. Maximal parity (P) violation and a small departure from time-reversal invariance (T) — deduced from CP violation and the CPT theorem (see §2.2d) — are observed in weak interactions (albeit not in the electromagnetic and strong interactions) and we propose, in this section, to examine the status of the discrete space and time symmetries as well as that of charge conjugation. We consider in turn: space inversion in §2.2a; charge conjugation in §2.2b; time inversion in §2.2c; and, finally, the CPT theorem in §2.2d. We treat charge conjugation prior to time inversion for technical reasons: charge conjugation — like parity inversion — involves the customary unitary transformations on the field operators whereas time inversion requires the introduction of "anti-unitary" transformations. The CPT theorem plays an important role in quantum field theory because the conservation law associated with it should be as absolute as Lorentz invariance and microcausality for quantum fields. In fact, as remarked earlier, the breakdown of time reversal invariance is deduced from the breakdown of CP invariance and the assumption of strict CPT invariance.

§2.2a. Space inversion

We first derive the unitary transformations induced by the space inversion operation on the "second-quantized" bosonic $J = 0$ and $J = 1$ quantum fields (limiting ourselves to the Klein-Gordon and Maxwell fields) and on the fermionic $J = \frac{1}{2}$ fields (covering both the Dirac and Weyl fields). We then consider some striking physical examples of parity conservation and parity violation. We start by first defining the meaning of space inversion for each field in "first-quantized" form and then applying Eq. (2.14) for the relation between the unitary transformation on the "second-quantized" field operator and the effect of space inversion on the "first-quantized" field. The space inversion operation is simply defined by the transformation:

$$\underline{x} \to \underline{x}' = -\underline{x}; \quad t \to t' = t \tag{2.55}$$

For the "first-quantized" (complex) scalar field, the P transformation is:

$$P: \quad \phi'(\underline{x}',t') = \eta_P \phi(-\underline{x},t) \quad (|\eta_P| = 1) \tag{2.56}$$

where η_P is, in general, a complex number of modulus unity since the C.R. must remain invariant under space inversion and a second application of P restores the initial state. In the special case of Hermitian fields, it follows that $\eta_P^2 = +1$ so that $\eta_p = \pm 1$. For the (neutral) Maxwell field (describing the photon), we get:

$$P: \quad A_i'(\underline{x}',t') = -A_i(-\underline{x},t) \quad (i = 1,2,3); \quad A_0'(\underline{x}',t') = A_0(-\underline{x},t) \tag{2.57}$$

where $A_\mu (\mu = 0,1,2,3)$ is the electromagnetic four-potential that transforms under space inversion in precisely the same fashion as the space-time four-vector $(t,\underline{x})$, i.e. $\eta_p = -1$ for the three-vector potential $\underline{A}$ and $\eta_p = +1$ for the time-component A_0 (scalar potential). To complete the roster, we list the P transformation for the "first-quantized" Dirac field (the P transformation can not be applied to the Weyl field unless the particle is transformed into an antiparticle — see below):

$$\psi'(x',t') = \eta_P \gamma_0 \psi(-x,t); \quad |\eta_P| = 1 \tag{2.58}$$

We next examine the properties of the unitary transformation, U_P, which must be applied to each "second-quantized" field (Klein-Gordon, Maxwell and Dirac) resulting from the above P transformation for each "first-quantized" field. From Eq. (2.56), it follows for the Klein-Gordon field that:

$$U_P \, \phi(\underline{x},t) \, U_P^{-1} = \eta_P \, \phi(-\underline{x},t); \quad U_P \, \phi^\dagger(\underline{x},t) U_P^{-1} = \eta_P^* \, \phi^\dagger(-\underline{x},t) \tag{2.59}$$

where the ϕ's and $\phi^\dagger$'s are now the "second-quantized" field amplitudes. Using Eqs. (2.10) and (2.59), one readily finds that (space inversion changes

$\underline{k} \to -\underline{k})$ [2.11]:

$$U_P a_k U_P^{-1} = \eta_P \, a_{-k}; \qquad U_P b_k^\dagger U_P^{-1} = \eta_P \, b_{-k}^\dagger$$
$$U_P a_k^\dagger U_P^{-1} = \eta_P^* \, a_{-k}^\dagger; \qquad U_P b_k U_P^{-1} = \eta_P^* \, b_{-k} \qquad (2.60)$$

In Eq. (2.60), we have translated Eq. (2.59) into conditions on the particle and antiparticle creation and destruction operators in the Fourier expansions of $\phi(x)$ and $\phi^\dagger(x)$; this approach is convenient for the three discrete symmetries whereas the total field operators $\phi(x)$ and $\phi^\dagger(x)$ are more suitable for the continuous space-time symmetries. The usefulness of Eq. (2.60) can be exhibited by letting the creation operators $a_k^\dagger$ and $b_k^\dagger$ operate on the vacuum (which is assigned an intrinsic positive parity, i.e. $U_P|0> = |0>$):

$$U_P \, a_k^\dagger |0> = \eta_P^* \, a_{-k}^\dagger |0>; \qquad U_P \, b_k^\dagger |0> = \eta_P \, b_{-k}^\dagger |0>$$
$$U_P \, a_k^\dagger b_{-k}^\dagger |0> = |\eta_P|^2 \, a_{-k}^\dagger b_k^\dagger |0> \qquad (2.61)$$

Consequently, the intrinsic parity of a Klein-Gordon particle-antiparticle system is always positive since $|\eta_P|^2 = 1$; as we will see, this is in contradistinction to the negative intrinsic parity of a Dirac particle-antiparticle system. For a single Klein-Gordon particle, such as the neutral pion (which is its own antiparticle), the a_k and b_k operators are identical and it follows from Eq. (2.60) that $\eta_P^2 = 1$; it turns out experimentally that $\eta_P = -1$ for π^0. The unitary transformation U_P for the neutral "second-quantized" Maxwell fields follows readily from the U_P for the Klein-Gordon field when due account is taken of the fact that the Maxwell field is neutral (so that the photon is its own antiparticle) and there are two (transverse) polarizations for the Maxwell field instead of the one for the Klein-Gordon field. It is easily seen from Eqs. (2.57) and (2.31) that $\eta_P = -1$ for the photon.

The derivation of U_P for the Dirac field $\psi(x)$, is more complicated; with the help of Eq. (2.58), we get:

$$U_P \, \psi_\alpha(\underline{x}, t) \, U_P^{-1} = \eta_P \, (\gamma_0)_{\alpha\beta} \psi_\beta(-\underline{x}, t) \qquad (2.62)$$

which leads to the following defining equations for the unitary operator U_P:

$$U_P a_k^r U_P^{-1} = \eta_P \, a_{-k}^r; \qquad U_P b_k^{r\dagger} U_P^{-1} = -\eta_P \, b_{-k}^{r\dagger};$$
$$U_P a_k^{r\dagger} U_P^{-1} = \eta_P^* \, a_{-k}^{r\dagger}; \qquad U_P b_k^r U_P^{-1} = -\eta_P^* \, b_{-k}^r \qquad (2.63)$$

Under the U_P operation, one-particle and one-antiparticle states behave as:

$$U_P \, a_k^{r\dagger} |0> = \eta_P^* \, a_{-k}^{r\dagger} |0>; \qquad U_P \, b_k^{r\dagger} |0> = -\eta_P \, b_{-k}^{r\dagger} |0> \qquad (2.64)$$

From Eq. (2.64) it follows that the Dirac particle-antiparticle state, $a_k^{r\dagger} b_{-k}^{r\dagger} |0>$ (which defines positronium when the Dirac particle is the electron) behaves under the unitary operator U_P, thus:

$$U_P \, a_k^{r\dagger} b_{-k}^{r\dagger} |0> = -|\eta_P|^2 \, a_{-k}^{r\dagger} b_k^{r\dagger} |0 > \qquad (2.65)$$

Since $|\eta_P|^2 = 1$, Eq. (2.65) implies that the intrinsic parity of the (fermionic) Dirac particle-antiparticle system is negative, in contrast to a spin 0 or a spin 1 (boson) particle-antiparticle system. This result has many applications in particle physics; for example, it explains why the lowest $^1S_0$ state of positronium emits two photons with the two transverse polarizations perpendicular to each other (a property of a pseudoscalar particle) and it also explains, for example, the pseudoscalar character of the lowest-lying meson octet in the quark flavor model of hadrons.

In the above discussion of the effects of the space inversion transformation on the $J = 0, \frac{1}{2}$ and 1 quantum fields, we have assumed that the phase factor η_P associated with U_P satisfies the condition $|\eta_P| = 1$ so that double application of the space inversion operator recaptures the initial state. This last statement of parity conservation holds for free fields. If there is interaction between two quantum fields, the total field Lagrangian only conserves parity if it is invariant under space inversion of all the interacting fields. To decide whether the interaction part of the total Lagrangian is parity-conserving, it is useful to know the "parity transforms" of the Lorentz bilinear covariants for the Klein-Gordon, Maxwell and Dirac fields (the "parity transform"' of the Weyl field does not exist); they are given in Table 2.1. Table 2.1 shows that the space inversion transformation, not unexpectedly, changes the sign of the space-like parts of the vector and tensor bilinear covariants but not the sign for the time-like parts of these covariants. All interacting Lagrangians representing observed strong and electromagnetic interactions appear to be invariant under the space inversion operation, i.e. parity-conserving; the weak interaction is the exception where the charged V-A weak current — which is defined by two bilinear covariants (V and A) for each charged quark and lepton current — is not invariant under space inversion and, indeed, is maximally parity-violating.

The bilinear "parity transforms" given in Table 2.1 are useful in a variety of ways. For example, the parity-conserving electromagnetic decay $\pi^0 \to 2\gamma$ they single out the term $\underline{E} \cdot \underline{H}$ (where $\underline{E}$ and $\underline{H}$ are the electric and magnetic fields associated with the two photons) because π^0 is pseudoscalar. Table 2.1 has also played a useful role in the derivation of the V-A Lorentz structure of all charged weak quark and lepton currents on the basis of the principle of "chirality invariance" (see §1.2d). The V-A Lorentz structure obviously guarantees maximal parity violation and is equivalent to the statement that only lefthanded chiral fermions — no matter what the mass — enter into the charged weak current. When the same principle of "chirality invariance" is applied to a neutral current (weak or electromagnetic), chiral fermions are not mandated because the "chirality invariance" of a neutral current is maintained if the neutral current is an arbitrary linear combination of V and A interactions; hence the neutral electromagnetic V current is chirality invariant as is the combination of V-A neutral weak current and V electromagnetic neutral current that enters into the standard model; in fact, it is this result that allows the unification of the chiral-invariant (parity-violating) weak interaction group with the chiral-invariant (parity-conserving) electromagnetic group into a single chiral electroweak group (see §1.2d and §6.2).

Table 2.1: "Parity transforms" of bilinear covariants for $J = 0$, $1/2$ and 1 fields.

Klein-Gordon field

J	$U_P J U_P^{-1}$
$\varphi^\dagger \varphi$ (scalar)	$+\varphi^\dagger \varphi$
$\left. \begin{array}{l} \varphi^\dagger \frac{\partial \varphi}{\partial x_i} - \varphi \frac{\partial \varphi^\dagger}{\partial x_i} \\ \varphi^\dagger \frac{\partial \varphi}{\partial t} - \varphi \frac{\partial \varphi^\dagger}{\partial t} \end{array} \right\}$ (vector)	$\left\{ \begin{array}{l} -\left(\varphi^\dagger \frac{\partial \varphi}{\partial x_i} - \varphi \frac{\partial \varphi^\dagger}{\partial x_i}\right) \\ +\left(\varphi^\dagger \frac{\partial \varphi}{\partial t} - \varphi \frac{\partial \varphi^\dagger}{\partial t}\right) \end{array} \right.$

Maxwell field

J	$U_P J U_P^{-1}$
$\underline{E} \cdot \underline{H}$ (pseudoscalar)	$-\underline{E} \cdot \underline{H}$
$\underline{E}^2 - \underline{H}^2$ (scalar)	$+(\underline{E}^2 - \underline{H}^2)$

Dirac field

J	$U_P J U_P^{-1}$
$\bar{\psi}\psi$ (scalar) (S)	$+\bar{\psi}\psi$
$\left. \begin{array}{l} \bar{\psi}\gamma_i\psi \\ \bar{\psi}\gamma_0\psi \end{array} \right\}$ (vector) (V)	$\left\{ \begin{array}{l} -\bar{\psi}\gamma_i\psi \\ +\bar{\psi}\gamma_0\psi \end{array} \right.$
$\left. \begin{array}{l} \bar{\psi}\gamma_i\gamma_j\psi \\ \bar{\psi}\gamma_0\gamma_i\psi \end{array} \right\}$ (tensor) (T)	$\left\{ \begin{array}{l} +\bar{\psi}\gamma_i\gamma_j\psi \\ -\bar{\psi}\gamma_0\gamma_i\psi \end{array} \right.$
$\left. \begin{array}{l} \bar{\psi}\gamma_5\gamma_i\psi \\ \bar{\psi}\gamma_5\gamma_0\psi \end{array} \right\}$ (axial vector) (A)	$\left\{ \begin{array}{l} +\bar{\psi}\gamma_5\gamma_i\psi \\ -\bar{\psi}\gamma_5\gamma_0\psi \end{array} \right.$
$\bar{\psi}\gamma_5\psi$ (pseudoscalar) (P)	$-\bar{\psi}\gamma_5\psi$

In hindsight, it is surprising how well the principle of "chirality invariance" — which is modelled on the two-component (massless) "neutrino paradigm" (see §1.2d) — works for the observed finite-mass quarks and charged leptons. A possible explanation may reside in the fact that the quark and lepton masses are supposed to result from the Higgs SSB mechanism with a scale Λ_{QFD} much larger than the quark and lepton masses (except possibly for the t quark); one might then expect that corrections to "chirality invariance" would be of the order of $(m_q^2, \ m_\ell^2)/M_W^2$ (see [2.12]). If the departure from "chirality invariance" is of this order, one can understand the failure to observe any evidence for righthanded currents involving the charged leptons and five of the six quarks; It would appear that an admixture of righthanded currents in weak processes involving the massive t quark — when it is finally detected — may become observable.

§2.2b. Charge conjugation

We have seen that one of the surprises in modern particle physics has been the observation of maximal parity violation in the weak interactions and that the incorporation of this breakdown of the discrete symmetry P has strongly influenced the formulation of the standard model (see Chapter 6). The maximal breakdown of a second discrete symmetry, that of charge conjugation invariance C, has gone hand in hand with the maximal breakdown of P invariance, and can be treated rather quickly because it involves a unitary transformation on the "second quantized" fields — as does P — and, furthermore, the unitary operator for C can be defined directly in terms of the creation and destruction operators for the particle and antiparticle, without considering the space-time coordinates at all. The idea of charge conjugation is motivated by the observation that the charge associated with a global internal symmetry of an Abelian (or non-Abelian) group changes sign under the (charge conjugation) operation: particle ↔ antiparticle, and one can inquire whether the field Lagrangian is invariant under this transformation. We have noted that charge conjugation plays a "dynamical" role in dispersion theory (through "crossing" — see §1.2b) and a fundamental role in the CPT theorem (see §2.2d); we will see that the presence or absence of C invariance is decisive in connection with other key questions in modern particle physics, e.g. the "strong CP" problem in QCD and the origin of the baryon asymmetry of the universe (BAU) following the "big bang", apart from the deeply puzzling phenomenon of weak CP violation (see §6.3f).

To consider charge conjugation, we again start with a non-Hermitian Klein-Gordon field. In analogy to space inversion, the "charge conjugation" unitary transformation, U_C, must satisfy the following conditions [2.11]:

$$U_C a_k U_C^{-1} = \eta_C\, b_k; \quad U_C a_k^\dagger U_C^{-1} = \eta_C^*\, b_k^\dagger; \quad U_C b_k U_C^{-1} = \eta_C\, a_k;$$
$$U_C b_k^\dagger U_C^{-1} = \eta_C^*\, a_k^\dagger \tag{2.66}$$

where, again, $|\eta_C| = 1$ since a double application of the charge conjugation operation yields the identity. However, the phase factor η_C for the non-Hermitian Klein-Gordon field is non-measurable and only becomes measurable for a Hermitian field, i.e. when $b_k = a_k$ (and $b_k^\dagger = a_k^\dagger$); evidently, the four equations of Eq. (2.66) then become two equations:

$$U_C a_k U_C^{-1} = \eta_C\, a_k; \quad U_C a_k^\dagger U_C^{-1} = \eta_C\, a_k^\dagger \tag{2.67}$$

where now $\eta_C^2 = 1$ or $\eta_C = \pm 1$. In other words, for a Hermitian Klein-Gordon field, η_C is measurable and becomes the multiplicative quantum number "charge conjugation parity" (even or odd). As was the case for space inversion, the charge conjugation operation for the quantized Maxwell field can be handled in precisely the same way as the Hermitian Klein-Gordon field. One obtains an equation analogous to Eq. (2.67) for the two (transverse polarization) creation and destruction operators and, from an examination of the structure of the photon interaction with charged particles, one concludes that the photon has a definite "charge conjugation parity" (i.e. odd — see below).

The C operation for the $J = \frac{1}{2}$ Dirac field is only slightly more complicated than for a bosonic $J = 0$ or 1 field; one gets:

$$U_C a_k^r U_C^{-1} = \eta_C\, b_k^r; \qquad U_C a_k^{r\dagger} U_C^{-1} = \eta_C^*\, b_k^{r\dagger};$$
$$U_C b_k^{r\dagger} U_C^{-1} = \eta_C\, a_k^{r\dagger}; \qquad U_C b_k^r U_C^{-1} = \eta_C^*\, a_k^r \tag{2.68}$$

where $|\eta_C|^2 = 1$. The phase η_C is complex just as for the non-Hermitian Klein-Gordon field and it is, again, not possible to define a measurable "charge conjugation parity" for the (non-Hermitian) Dirac field. The C operation for the non-Hermitian Weyl field by itself (no more than the P transformation by itself) — does not exist because the helicity must change sign; however, the CP operation does exist because the P operation changes the sign of the helicity. [It is for this reason that the T operation exists for a Weyl field (since CPT invariance holds) — see below.]

It is possible and useful to define bilinear "charge conjugation parities" for the non-Hermitian Klein-Gordon and Dirac fields as well as of the Hermitian

Table 2.2: "Charge conjugation parities" of bilinear covariants for $J = 0$, $1/2$ and 1 fields.

Klein-Gordon field

J	η_C
$\varphi^\dagger \varphi$ (scalar)	$+1$
$(\varphi^\dagger \frac{\partial \varphi}{\partial x_\mu} - \frac{\partial \varphi^\dagger}{\partial x_\mu} \varphi)$ (vector)	-1

Maxwell field

J	η_C
$\underline{E} \cdot \underline{H}$ (pseudoscalar)	$+1$
$\underline{E}^2 - \underline{H}^2$ (scalar)	$+1$

Dirac field

J		η_C
$\bar{\psi}\psi$	(scalar)	$+1$
$\bar{\psi}\gamma_\mu\psi$	(vector)	-1
$i\bar{\psi}\gamma_\mu\gamma_\nu\psi$	(tensor)	-1
$i\bar{\psi}\gamma_5\gamma_\mu\psi$	(axial vector)	$+1$
$i\bar{\psi}\gamma_5\psi$	(pseudoscalar)	$+1$

Maxwell field; Table 2.2 lists the "charge conjugation parities" of the bilinear covariants of the three fields. The usefulness of Table 2.2 is seen in four examples: (1) the "charge conjugation parity" of the photon must be odd because it is described by the four-vector A_μ which, by virtue of Lorentz invariance, must be coupled to a vector fermion or boson (electromagnetic) current. Examination of Table 2.2 lists $\eta_C = -1$ for the neutral vector current (associated with a charged fermion or boson) and, hence, the photon must have $\eta_C = -1$ to ensure the known charge conjugation invariance of the electromagnetic interaction; (2) it is evident that $\eta_C = +1$ for π^0 if $\pi^0 \to 2\gamma$ decay is to be consistent with the charge conjugation invariance of the electromagnetic interaction; (3) one can apply a "generalized" Pauli principle to the positronium state which states that the wavefunction of a positronium state should be totally antisymmetric in space, spin, and the exchange of charge. More explicitly, the Pauli principle for a positronium state requires that $(-1)^L \times (-1)^{S+1} \times \eta_C = -1$ where the $(-1)^{S+1}$ factor makes provision for the intrinsic negative parity of the positronium $(e^- - e^+)$ system; it follows that the relation $\eta_C = (-1)^{L+S}$ must be satisfied. Thus, if we consider the emission of photons by a specified positronium state, the number n_γ of photons emitted must be consistent with the relation $\eta_C = (-1)^{L+S} = (-1)^{n_\gamma}$, a condition that is always satisfied. It should be noted that a similar selection rule emerges when G conjugation [where $U_G = e^{i\frac{\pi}{2}\tau_2}U_C$, with U_G and U_C the unitary operators corresponding to G and C conjugation respectively and τ_2 the second isospin generator] is applied to the number of pions emitted from a nucleonium system; and, finally, (4) the η_C's of the V and A fermion currents in Table 2.2 are opposite in sign and this explains why there is (maximal) breakdown of charge conjugation invariance for the V-A charged weak current interaction. Soon after the V-A charged weak current interaction was formulated and was receiving strong confirmation, it was proposed [2.13] that, while P and C may be maximally violated, the product CP is conserved. The Weyl neutrino was offered as an example of maximal violation of P and C while maintaining CP conservation. If CP is conserved, then, in accordance with the CPT theorem, time reversal invariance must hold. The observation of weak CP violation (see §6.3f) indicates that the time inversion operation (T) is weakly violated and so we give "equal" treatment in the next section to T. [It is perhaps significant that the "neutrino paradigm" that is invoked to motivate "chirality invariance" of the weak and electromagnetic currents (see §2.2a) can also help to explain why "combined" CP invariance (and, *a fortiori*, maximal C non-invariance) is so closely obeyed in weak processes; perhaps the deviation from "combined" CP invariance, i.e. the weak CP violation observed in the quark sector, is correlated with a possible departure from "chirality invariance" and both are due to the finiteness of the quark masses.

§2.2c. Time inversion

As indicated above, "weak" T violation is deduced from the observed small violation of CP invariance in weak interactions and the CPT theorem. Even though T violation is a small effect in weak interactions, the special features of the time inversion operation [e.g. the need for anti-unitary in place of unitary transformations, as for P and C] and the possible importance of T non-invariance or, equivalently, CP non-invariance, for major questions like the BAU (see §2.2d) and the "strong CP problem" in QCD (see §10.3c) necessitate treatment of the T operation.

The novel property of the time inversion operation [i.e. $\underline{x}' \to \underline{x}$; $t' \to -t$] stems from the fact that the T operation requires interchange of initial and final states (i.e. "bras" and "kets") and hence an anti-unitary operator. This need for an anti-unitary transformation can already be seen from the C.R. for one generalized coordinate $q(t)$ and its canonically conjugate momentum $p(t)$ in ordinary quantum mechanics, namely:

$$[q(t), p(t)] = i \qquad (2.69)$$

When time inversion is applied to Eq. (2.69), i.e. $t' \to -t$ (while $\underline{x}$ is unchanged), we have $q(-t) \to q(t)$ and $p(-t) \to -p(t)$ so that the invariance of (2.59) is not preserved. This deficiency in ordinary quantum mechanics is remedied by inverting the order of the operators q and p as part of the time inversion operation. This suggests that, whereas the time inversion operation T cannot be represented by a unitary transformation, it can be represented by the product of a unitary transformation U_T and a transformation K that takes the transpose of the product of operators; that is, we can write:

$$T = U_T K \qquad (2.70)$$

where K changes initial into final states (or "bras" into "kets") and conversely.

It is now possible to derive the unitary transformation U_T corresponding to the time inversion operation for "second-quantized" fields. We first write down the S_T transformations for the first-quantized Klein-Gordon, Maxwell, Dirac and Weyl fields; these transformations are [2.11]:

$$S_T(\text{Klein-Gordon}): \quad \phi(\underline{x}, t) \to \eta_T \, \phi^\dagger(\underline{x}, -t) \qquad (2.71)$$

$$S_T(\text{Maxwell}): \quad A(\underline{x}, t) \to -A(\underline{x}, -t); \quad A_0(\underline{x}, t) \to +A_0(\underline{x}, -t) \qquad (2.72)$$

$$S_T(\text{Dirac}): \quad \psi(\underline{x}, t) \to i\eta_T \, C^\dagger \bar{\psi}^T(\underline{x}, -t) \qquad (2.73)$$

$$S_T(\text{Weyl}): \quad \psi(\underline{x}, t) \to i\eta_T \, \sigma_2 \psi^{\dagger T}(\underline{x}, -t) \qquad (2.74)$$

where $|\eta_T| = 1$. In Eqs. (2.73) and (2.74), ψ^T is the transpose of ψ and C is a 4×4 matrix that satisfies the relations:

$$C\gamma^\mu C^{-1} = (\gamma^\mu)^\dagger \qquad (2.75)$$

whose explicit form depends on the particular representation of the γ matrices. We note that the Weyl spinor ψ in (2.74) is a two-component spinor in contrast to the Dirac spinor ψ in Eq. (2.73).

We next define the anti-unitary transformation T on the "second-quantized" fields by taking the transposes of all the operators on the R.H.S.

of Eqs. (2.71)–(2.74). If we use the expansion of the Klein-Gordon field $\phi(\underline{x}, t)$ and $\phi^\dagger(\underline{x}, t)$ in terms of the creation and destruction operators [see Eqs. (2.10)], we obtain for the unitary part U_T of T [see Eq. (2.70)]:

$$U_T a_k U_T^{-1} = \eta_T \, a^\dagger_{-k}; \quad U_T a_k^\dagger U_T^{-1} = \eta_T^* \, a_{-k}$$

$$U_T b_k U_T^{-1} = \eta_T^* \, b^\dagger_{-k}; \quad U_T b_k^\dagger U_T^{-1} = \eta_T \, b_{-k} \tag{2.76}$$

Further, we have for the K part of T:

$$K|0> = <0| \tag{2.77}$$

Combining the expressions (2.76) and (2.77), we find that the T operation for the single particle and antiparticle states gives:

$$T a_k^\dagger |0> = \eta_T <0|a_{-k}; \qquad T b_k^\dagger |0> = \eta_T <0|b_{-k} \tag{2.78}$$

From Eq. (2.78), we see that T has the effect of transforming a one-particle (-antiparticle) initial state with momentum $\underline{k}$ into a one-particle (-antiparticle) final state with momentum $-\underline{k}$. In Dirac terminology, we can say that the time inversion operation changes the sign of the momentum of the particle or antiparticle and transforms a "ket" into a "bra" and *vice versa*.

Since there are some nuances in connection with the time inversion operation for the Maxwell field, we record the key formulas; the transformation T for the "second-quantized" Maxwell field is defined by:

$$T\underline{A}(\underline{x}, t)T^{-1} = -\underline{A}(\underline{x}, -t); \qquad TA_0(\underline{x}, t)T^{-1} = +A_0(\underline{x}, -t) \tag{2.79}$$

Expanding $\underline{A}(\underline{x}, t)$ in terms of the creation and destruction operators [see Eq. (2.31)], we get ($\eta_T = -1$):

$$U_T a_k^r U_T^{-1} = -a^{r'}_{-k}; \qquad U_A a_k^{r\dagger} U_T^{-1} = -a^{r'\dagger}_{-k} \tag{2.80}$$

It is seen from Eq. (2.80) that the polarization index r changes into a new index r', which is related to r so as to retain the helicity of the photon; this follows from the definition of the helicity (i.e. $\mathcal{H} = \underline{J} \cdot \underline{p}/|\underline{p}|$) and the fact that the T transformation inverts the direction of the spin as well as the momentum. Since the electric field intensity $\underline{E}$ is the time derivative of $\underline{A}$ and the magnetic field $\underline{H}$ is curl $\underline{A}$, we can immediately write down the behavior of the "second-quantized" $\underline{E}$ and $\underline{H}$ field amplitudes under the transformation T, namely:

$$T\underline{E}T^{-1} = +\underline{E} \; ; \qquad T\underline{H}T^{-1} = -\underline{H} \tag{2.81}$$

It would appear from Eq. (2.81) that $\eta_T = \pm 1$ are the respective eigenvalues of the "time parity" of the electric and magnetic parts of the Hermitian

Maxwell field; however, one must remember that the time inversion operation includes the anti-unitary transposition operator K in addition to the unitary operator U_T and anti-unitary operators do not have eigenvalues. One thing is sure: "time parity" can not be defined for a relativistic quantum field.

Finally, we derive U_T operating on the creation and destruction operators of the Dirac and Weyl fields. Starting with the transformation given in Eq. (2.73) for the "first-quantized" Dirac field, the T transformation operating on the "second-quantized" Dirac spinor $\psi(x)$ (taking account of the transposition of operators) becomes:

$$T\psi_\alpha(\underline{x}, t)T^{-1} = -i\eta_T \, (\gamma_0)_{\alpha\beta}(C_{\gamma\beta})^*\psi_\gamma^\dagger(\underline{x}, -t)$$
$$= i\eta_T \, (C^{-1}\bar{\psi}^T(\underline{x}, -t))_\alpha \qquad (2.82)$$

where C is the matrix defined by Eq. (2.75). If we now expand ψ in terms of the creation and destruction operators, we obtain the following conditions on U_T (the helicity index $r = \pm$):

$$U_T a_k^r U_T^{-1} = -r \, \eta_T \, a_{-k}^{\dagger r}; \quad U_T a_k^{r\dagger} U_T^{-1} = -r \, \eta_T^* \, a_{-k}^r$$
$$U_T b_k^r U_T^{-1} = +r \, \eta_T^* \, b_{-k}^{\dagger r}; \quad U_T b_k^{r\dagger} U_T^{-1} = +r \, \eta_T \, b_{-k}^r \qquad (2.83)$$

As before: $K|0>=< 0|$ and, if we apply T as defined by Eq. (2.83) to the one-particle and one-antiparticle states, we obtain:

$$Ta_k^{r\dagger}|0>= -r \, \eta_T^* < 0|a_{-k}^r; \qquad Tb_k^{r\dagger}|0>= +r \, \eta_T < 0|b_{-k}^r \qquad (2.84)$$

Equation (2.84) tells us that the time inversion operation changes an initial Dirac particle (or antiparticle) state with momentum $\underline{k}$ and helicity index $r(=\pm)$ into a final particle (or antiparticle) state with momentum $-\underline{k}$ and the same helicity index r [the conservation of helicity in the T transformation is the reason why the polarization index had to change for the Maxwell field (see Eq. (2.80))].

The expressions for the U_T and T operations for the Weyl field can be taken over from the corresponding operations — given by Eqs. (2.83) and (2.84) — for the Dirac field. We have for the counterpart of Eq. (2.84) (since there is only one index, which we take as $r = +1$):

$$Ta_k^\dagger|0>= -\eta_T^* < 0|a_{-k}; \qquad Tb_K^\dagger|0>= \eta_T < 0|b_{-k} \qquad (2.85)$$

In §2.2a–§2.2b, we pointed out that the separate P and C operations do not exist for the Weyl field but that the combined CP operation — which is equivalent to the T operation — does exist and Eq. (2.85) is the explicit expression for the T operation for the massless neutrino. We give in Table 2.3 the bilinear "time inversion transforms" for the Klein-Gordon, Maxwell, Dirac and Weyl fields (the Weyl field is now included — in contrast to its absence in Tables 2.1 and 2.2).

Table 2.3: "Time inversion transforms" of bilinear covariants for $J = 0$, $1/2$ and 1 fields.

Klein-Gordon field

J	TJT^{-1}
$\varphi^\dagger \varphi$ (scalar)	$+\varphi^\dagger \varphi$
$\left.\begin{array}{l} \varphi^\dagger \frac{\partial \varphi}{\partial x_i} - \varphi \frac{\partial \varphi^\dagger}{\partial x_i} \\ \varphi^\dagger \frac{\partial \varphi}{\partial t} - \varphi \frac{\partial \varphi^\dagger}{\partial t} \end{array}\right\}$ (vector)	$\left\{\begin{array}{l} -(\varphi^\dagger \frac{\partial \varphi}{\partial x_i} - \varphi \frac{\partial \varphi^\dagger}{\partial x_i}) \\ +(\varphi^\dagger \frac{\partial \varphi}{\partial t} - \varphi \frac{\partial \varphi^\dagger}{\partial t}) \end{array}\right.$

Maxwell field

J	TJT^{-1}
$\underline{E} \cdot \underline{H}$ (pseudoscalar)	$-\underline{E} \cdot \underline{H}$
$\underline{E}^2 - \underline{H}^2$ (scalar)	$+(\underline{E}^2 - \underline{H}^2)$

Dirac field

J	TJT^{-1}
$\bar{\psi}\psi$ (scalar)	$+\bar{\psi}\psi$
$\left.\begin{array}{l} \bar{\psi}\gamma_i\psi \\ \bar{\psi}\gamma_0\psi \end{array}\right\}$ (vector)	$\left\{\begin{array}{l} -\bar{\psi}\gamma_i\psi \\ +\bar{\psi}\gamma_0\psi \end{array}\right.$
$\left.\begin{array}{l} \bar{\psi}\gamma_i\gamma_j\psi \\ \bar{\psi}\gamma_0\gamma_i\psi \end{array}\right\}$ (tensor)	$\left\{\begin{array}{l} -\bar{\psi}\gamma_i\gamma_j\psi \\ +\bar{\psi}\gamma_0\gamma_i\psi \end{array}\right.$
$\left.\begin{array}{l} \bar{\psi}\gamma_5\gamma_i\psi \\ \bar{\psi}\gamma_5\gamma_0\psi \end{array}\right\}$ (axial vector)	$\left\{\begin{array}{l} -\bar{\psi}\gamma_5\gamma_i\psi \\ +\bar{\psi}\gamma_5\gamma_0\psi \end{array}\right.$
$\bar{\psi}\gamma_5\psi$ (pseudoscalar)	$-\bar{\psi}\gamma_5\psi$

Weyl field

J	TJT^{-1}
$\left.\begin{array}{l} \psi^\dagger \sigma_i \psi \\ \psi^\dagger \psi \end{array}\right\}$ (four vector)	$\left\{\begin{array}{l} -\psi^\dagger \sigma_i \psi \\ +\psi^\dagger \psi \end{array}\right.$

We conclude this section on the discrete transformation of time inversion by commenting on the outcome of a substantial number of experiments to directly test time inversion (reversal) invariance in the strong, electromagnetic and weak interactions. In sum, direct experiments on time inversion invariance have thus far yielded completely negative results [2.14]. The most probing of these experiments — attempts to detect an electric dipole moment for the neutron — have continued with increasingly sensitive apparatus for over three decades [2.15], and have given a very small upper limit on the neutron's electric dipole moment. In essence, an electric dipole moment of the neutron can only arise from an electromagnetic interaction term in the total Lagrangian of the form (see §6.3f):

$$d_n \; \bar{\psi}_n \gamma_5 \sigma_{\mu\nu} F^{\mu\nu} \psi_n \simeq d_n \; \bar{\psi}_n \underline{\sigma} \cdot \underline{E} \psi_n \tag{2.86}$$

where d_n is the electric dipole moment. From Tables 2.1 and 2.3, it follows that Eq. (2.86) changes sign under both P and T operations. The measured upper limit on d_n is now $d_e \overset{<}{\sim} 1.2 \times 10^{-25} e$ cm [2.15]. [The standard model prediction of d_n is discussed in §6.3f and the role of d_n in the "strong CP" problem in QCD in §10.3c.] Many other experiments to detect the direct breakdown of time inversion invariance have also been fruitless [2.16] but indirect evidence for a small amount of time reversal non-invariance in weak interactions has been found; these experiments pertain to the weak decays of neutral K mesons and strictly speaking, measure a small breakdown of CP invariance; only when the CPT theorem is invoked, can one deduce the same degree of breakdown of time reversal invariance. The ramifications of weak CP violation will be discussed in some detail in §6.3f. In the next section (§2.2d), we comment briefly on the CPT theorem [2.17] in order to understand the above statement as well as the general restriction placed on the relation between spin and statistics.

§2.2d. CPT theorem and the relation between spin and statistics

In §2.1–§2.1c, we discussed the role of the continuous space-time symmetries in quantum field theory and explained how the invariance of the Lagrangian under the Poincaré transformation leads to the existence of the Poincaré group with its arbitrary mass and spin representations. We pointed out that the invariance under the Poincaré group of all local quantum field theories appears to be absolute. In §2.2a–§2.2c, we examined the role of the two discrete space-time symmetries (space inversion and time inversion) in quantum field theory, as well as that of the third discrete symmetry (charge conjugation), having the character of an internal symmetry. Gross departures from the conservation laws associated with space inversion and charge conjugation invariance (at least for the weak interaction) occur, and it was noted that a small departure from time reversal invariance (again, in weak interactions) follows from the CPT theorem and the observation of weak CP violation. The last statement depends on the correctness of a very important theorem in quantum field theory which asserts [2.17] that, if a local quantum field theory is Lorentz invariant and satisfies the spin-statistics requirement [i.e. boson fields obey C.R. and fermion fields anti-C.R.], then it is invariant under the product of the three discrete operations C, P and T, i.e. CPT, taken in any order. It is important that the CPT theorem holds even when one or more of the three discrete symmetries (C, P and T) breaks down. The CPT theorem is really quite remarkable because it relates the invariance of a quantum field theory — under continuous space-time (Lorentz) transformations and canonical commutation relations — to invariance under the combined interchange of particles with antiparticles, left with right, and past with future. Because, in actual fact, all three discrete symmetries — at least in weak interactions — separately break down, it is incumbent on us to sketch a proof of the CPT theorem and to comment on its power [e.g. CPT invariance guarantees the equality of particle and antiparticle mass as well as the equality of total decay widths] and its limitations [e.g. the partial decay widths for particle and antiparticle need not be equal — the caveat that underpins the BAU hypothesis (see below)].

The starting point of the CPT theorem is the observation that the combined CPT operation (sometimes called "strong reflection" S) corresponds to the following transformations on the Klein-Gordon, Maxwell, Dirac and Weyl fields [2.11]:

$$\text{Klein-Gordon:} \quad S\phi(\underline{x}, t)S^{-1} = \eta_S \; \phi^\dagger(-\underline{x}, -t)$$

$$\text{Maxwell:} \quad SA_\mu(\underline{x}, t)S^{-1} = -\eta_S \; A_\mu(-\underline{x}, -t)$$

$$\text{Dirac:} \quad S\psi(\underline{x}, t)S^{-1} = \eta_S \; (i\gamma_5)\psi^\dagger(-\underline{x}, -t)$$

$$\text{Weyl:} \quad S\psi(\underline{x}, t)S^{-1} = \eta_S \; \psi^\dagger(-\underline{x}, -t) \tag{2.87}$$

where η_S is the complex "strong reflection" phase. These transformations can be deduced from §2.2a–§2.2c if we note that the S transformation can be accomplished by a sequence of two rotations through 180^0 in two perpendicular planes: say, a rotation through 180^0 in the 1-2 plane followed by another rotation through 180^0 in the 3-0 plane. The operator representing this sequence of rotations is:

$$e^{i\pi J_{12}} e^{i\pi J_{30}} \tag{2.88}$$

where J_{12} and J_{30} are the "angular momentum" operators that are respectively the generators of space rotations "about" the third axis and of Lorentz transformations "along" the third axis. It can be checked that these are the transformation laws defined by Eq. (2.87). Hence, the CPT theorem is proved for the Klein-Gordon, Maxwell, Dirac and Weyl fields.

It is possible to prove the CPT theorem for quantum fields of arbitrary spin J by generalizing the operation of "strong reflection" defined by Eq. (2.87) [2.18]. Thus, for an integral spin (bosonic) field — where the field can always be represented by a Jth rank symmetric tensor function $B_{\mu_1...\mu_j}$ — we have:

$$SB_{\mu_1...\mu_j}(\underline{x},t)S^{-1} = \eta_S \, (-1)^J B^\dagger_{\mu_1...\mu_j}(-\underline{x},-t) \qquad (2.89)$$

On the other hand, for a general half-integral spin J (fermion) field $F_{\mu_1...\mu_{j-\frac{1}{2}}\alpha}$ — which can be represented by the direct product of a $(J-\frac{1}{2})^{th}$-rank symmetric tensor $B_{\mu_1...\mu_{j-\frac{1}{2}}}$ times a Dirac spinor ψ_α, i.e. $F_{\mu_1...\mu_{j-\frac{1}{2}}\alpha} = B_{\mu_1...\mu_{j-\frac{1}{2}}}\psi_\alpha$ [2.18] — we have:

$$SF_{\mu_1...\mu_{j-\frac{1}{2}}\alpha}(\underline{x},t)S^{-1} = (-1)^{J-\frac{1}{2}}(i\gamma_5)_{\alpha\beta}F^\dagger_{\mu_1...\mu_{j-\frac{1}{2}}\beta}(-\underline{x},-t) \qquad (2.90)$$

With the transformation laws (2.89) and (2.90), it is possible to show that a local, normal-ordered, n^{th}-rank Lorentz tensor function of bosons and fermions transforms as:

$$S: T_{\mu_1...\mu}(\underline{x},t): S^{-1} = (-1)^n : T^\dagger_{\mu_1...\mu_n}(-\underline{x},-t): \qquad (2.91)$$

[We note that the definition of the normal-ordered (Wick) product [2.19] of two operators A and B is $: AB : \; = \pm : BA :$, where the plus sign applies to integral spin fields obeying C.R. and the minus sign to half–integral spin fields obeying anti-C.R.]. More generally [2.18]:

$$: A_1(x_1)A_2(x_2)...A_n(x_n) :\equiv \delta_p A_1(x_{p_1})A_2(x_{p_2})...A_{p_n}(x_{p_n}) \qquad (2.92)$$

where $\delta_p = +1$ if the order of fermion operators on the L.H.S. of Eq. (2.92) is an even permutation of that on the R.H.S.; otherwise $\delta_p = -1$. The L.H.S. is arranged so that annihilation operators always appear to the right of the creation operators and $\{p_1,...,p_n\}$ is a permutation of $\{1,...,n\}$.

Equation (2.91) is now used to prove the invariance of the action under the "strong reflection" operation. Assuming that the Lagrangian density can be written in terms of a sum of normal-ordered products of ∂_μ, boson and fermion fields [equivalent to the assumptions of Lorentz invariance, validity of the spin-statistics theorem, and the existence of a vacuum state], in which all fields are taken at the same space-time point (assumption of locality), the transformation property of $\mathcal{L}$ under S is:

$$SL(\underline{x}, t)S^{-1} = \mathcal{L}^\dagger(-\underline{x}, -t) = \mathcal{L}(-\underline{x}, -t) \qquad (2.93)$$

since $\mathcal{L}$ is Hermitian. But the action requires the integration of $\mathcal{L}$ over all x and t and it is therefore invariant under "strong reflection". Consequently, if the vacuum is CPT-invariant, so will be the dynamics. As a corollary, since $\mathcal{H}^\dagger = \mathcal{H}$, the Hamiltonian density transforms as:

$$S\mathcal{H}(\underline{x}, 0)S^{-1} = \mathcal{H}(-\underline{x}, 0) \qquad (2.94)$$

and hence the Hamiltonian is invariant under "strong reflection", i.e.:

$$SHS^{-1} = H \qquad (2.95)$$

With the CPT theorem in hand, it is possible to deduce several interesting consequences for relativistic quantum field theory [2.18]. We focus on three predictions: (1) mass equality between particle and antiparticle; (2) opposite sign of charge between particle and antiparticle; and (3) equality of total decay widths of particle and antiparticle. With regard to (1), the equality of mass between particle and antiparticle is built into the definition of charge conjugation, i.e. the mass of a free particle should be that of the corresponding antiparticle. However, when account is taken of interactions, mass equality is not so obvious. Nevertheless, it does follow from CPT invariance even when C invariance breaks down. Thus, consider a particle p at rest with the z-component of angular momentum m; since the discrete operations C, P and T can be applied in any order, we have:

$$TPC|p>_m = TPe^{i\alpha}|\bar{p}>_m = Te^{i\alpha}|\bar{p}>_m = e^{i\alpha}|\bar{p}>_{-m} \qquad (2.96)$$

where $e^{i\alpha}$ denotes the (complex) phase factor under C and $\bar{p}$ is the antiparticle. We next write:

$$\begin{aligned}(\text{mass})_p &= <p|H|p>_m = <p|S^{-1}S|H|S^{-1}S|p>_m \\ &= <\bar{p}|H|\bar{p}>_{-m} = (\text{mass})_{\bar{p}}\end{aligned} \qquad (2.97)$$

where we have used Eq. (2.96) and the S invariance of H [see Eq. (2.95)]. The most accurate test of the mass equality between particle and antiparticle has come from studies of the $K^0 - \bar{K}^0$ system, where the CPT theorem is confirmed to an accuracy of 7×10^{-15}.

(2) A second consequence of the CPT theorem is that the charge of a particle must be the negative of the charge of the antiparticle. Using Eq.

(2.91) and the fact that $j_\mu(x)$ is a four-vector, we have:

$$S j_\mu(x) S^{-1} = -j_\mu^\dagger(-x) \qquad (2.98)$$

But $j_\mu(x) = (\underline{j}(x), i\rho(x))$ so that $S\rho(x)S^{-1} = -\rho(-x)$; hence:

$$Q_p = <p| \int d^3x \rho(x) |p>_m = <p|S^{-1}S \int d^3x \rho(x) S^{-1} S |p>_m$$

$$= <\bar{p}| \int d^3x [-\rho(-x)]|\bar{p}>_{-m} = -Q_{\bar{p}}. \qquad (2.99)$$

This prediction of the CPT theorem also holds to an extremely high accuracy (see §3.1). It should be noted that the same prediction (2.99) follows, for example, from the assumption of the invariance of the ("second-quantized") field Lagrangian under the global Abelian internal (charge) symmetry transformation $\phi \to e^{i\alpha}\phi$ (α is a constant) (see §3.1).

(3) The CPT predictions of equal masses and opposite charges for particle and antiparticle are absolute predictions, i.e. they hold to all orders of the particle interactions (strong, electromagnetic and weak, as well as gravitational). The CPT prediction concerning the relation between the total decay widths for particle and antiparticle is more qualified: equality has been proved to hold to all orders of the strong interaction but only to first-order in the weak interaction. We summarize the somewhat complicated proof by first defining the decay widths for particle and antiparticle, namely:

$$\Gamma_a = 2\pi \sum_b \delta(E_b - E_a)| < b_{\text{free}}|G(\infty,0)H_I|a >_m |^2 \qquad (2.100a)$$

and:

$$\Gamma_{\bar{a}} = 2\pi \sum_{\bar{b}} \delta(E_{\bar{b}} - E_{\bar{a}})| < \bar{b}_{\text{free}}|G(\infty,0)H_I|\bar{a} >_m |^2 \qquad (2.100b)$$

where $|a>_m$ and $|\bar{a}>_m$ are eigenstates of the strong interaction Hamiltonian H_s, b_{free} and $\bar{b}_{\text{free}}$ are the final (free) continuum states and $G(t,t_0)$ (with t_0 the initial time and t the time of interest) is the Green's function in the interaction representation for H_I (weak or electromagnetic interaction), which is only considered to first order. In our notation, the "scattering matrix" (which is denoted by $\bar{S}$ to avoid confusion with the "strong reflection" transformation S) is $\bar{S} = G(\infty, -\infty)$; if we now use the property of the Green's function $G(t,t_0)$ under "strong reflection", i.e. $SG(t,t_0)S^{-1} = G(-t,-t_0)$, we have the relation $S\bar{S}S^{-1} = \bar{S}^\dagger$, and can rewrite Eq. (2.100a) (dropping the subscript m, since the lifetime is independent of the z component of angular momentum) in the form ($E_b = E_{\bar{b}}, E_a = E_{\bar{a}}$):

$$\Gamma_a = 2\pi \sum_{\bar{b}} \delta(E_{\bar{b}} - E_{\bar{a}})| < \bar{b}_{\text{free}}|\bar{S}^\dagger G(\infty,0)H_I|\bar{a} > |^2$$

$$= 2\pi \sum_{\bar{b}} \delta(E_{\bar{b}} - E_{\bar{a}})| \sum_{\bar{b}'} < \bar{b}_{\text{free}}|S^\dagger|\bar{b}'_{\text{free}} > \cdot < \bar{b}'_{\text{free}}|G(\infty,0)H_I|\bar{a} > |^2 \tag{2.101}$$

In Eq. (2.101), a full set of intermediate states has been inserted; using the unitarity of $\bar{S}$, one finally arrives at Eq. (2.100b), which completes the proof that $\Gamma_a = \Gamma_{\bar{a}}$ to all orders of H_S and to lowest order in H_I. It is important to note that the proof of $\Gamma_a = \Gamma_{\bar{a}}$ depends on the summation over all states of equal energy b and $\bar{b}$ in Eqs. (2.100a) and (2.100b) respectively, and on the fact that $E_b = E_{\bar{b}}$; if this summation is not taken — so that only a partial decay width is computed, the equality of the partial decay widths for particle and antiparticle is not required by the CPT theorem. Measurements of the total decay widths have been made for muon, pion and kaon with accuracies in the neighborhood of 0.05% [2.18] (not very great accuracy) but no violation of the CPT theorem has been observed. It is the acceptance of the CPT theorem and the weak CP violation observed in the K^0 system that justifies the earlier conclusion that time reversal invariance does not hold absolutely for all known particle interactions.

The loophole allowed by the CPT theorem for unequal partial decay widths for particle and antiparticle is confirmed by interference studies of the long-lived and short-lived neutral kaons K_L and K_S (which are linear superpositions of K^0 and its antiparticle state $\bar{K}^0$) into the well-defined channels $\pi^\pm + \ell^\mp + \bar{\nu}_\ell$ (or ν_ℓ); these measurements [2.20] yield deviations from unity of several tenths of 1% for the ratio of the partial decay widths for particle and antiparticle. This is a very satisfactory result and it highlights the loophole in the CPT theorem — emphasized by Okubo [2.21] — that permits unequal partial decays width for particle and antiparticle. The best-known exploitation of this loophole in the CPT theorem bears on the attempt to understand the present baryon asymmetry of the universe (BAU). In his 1967 paper — building on Okubo's work demonstrating the possibility of unequal partial decay widths for particle and antiparticle — Sakharov [2.22] proposed to explain BAU by adding two additional hypotheses: (1) baryon charge is not conserved during the initial stages of the "big bang"; and (2) the CP-violating decay is a non-equilibrium process. By putting together these three hypotheses, Sakharov defined the major direction taken during the past two decades for attempts to solve the BAU problem. We do not intend to give a full treatment of the BAU problem in this book — a definitive theory does not exist as yet — but we will discuss it further in connection with the possibility of electroweak-induced baryon violation (see §10.3a).

§2.3. Special topics in space-time symmetries

Relativistic local quantum field theory, with its built-in microcausality and absolute invariance under the Poincaré group for $D = 4 = (3 + 1)$ space-time dimensions, has provided the framework within which the highly successful standard gauge theory of the strong and electroweak interactions has been formulated. The consequences of Poincaré invariance and its corollary, CPT invariance, have been confirmed without exception and without the slightest deviation. It is clear that the Poincaré and CPT invariances are major building blocks of the modern theory of particle interactions (at least for three of the four basic forces in nature). However, equally important for the construction of the standard model has been the introduction of global internal symmetry groups into relativistic quantum field theory and the use of the gauge principle to achieve a tractable and Lorentz-invariant, dynamical theory of the particle interactions. The idea of extending the Abelian (QED) gauge principle to the global non-Abelian internal symmetry groups of the strong and weak interactions only came after it was recognized that any attempt to mix the Poincaré group with a global internal symmetry (compact Lie) group — in order to obtain a dynamical theory of the strong and weak interactions [2.23] — is doomed to failure by virtue of the "no-go" theorem [2.24] (see §4.1)). The "no-go" theorem also focused people's attention on supersymmetry as a way to circumvent the strictures on combining global internal symmetries with the Poincaré group.

The idea of supersymmetry in four dimensions — first put forward in 1974 — has been extensively explored ever since, first leading to a class of "graded" supersymmetric algebras, then to gauged supersymmetric field theory and supergravity, and finally, to superstrings and all its ramifications (see §1.4c). These are important developments and may lead to a major breakthroughs in modern particle theory; however, this has not happened as yet and so, for the purposes of this book, we content ourselves with a brief introduction (in §2.3a) to the basic concepts of supersymmetry — chiefly $N = 1$ global supersymmetry — in order to appreciate the clever way in which supersymmetry bypasses the "no-go" theorem and opens the door to possible improvements on present-day quantum field theory and the unification of the gravitational interaction with the other three basic interactions.

Another topic pertaining to space-time symmetries in quantum field theories — which merits special treatment in this chapter — is the question of the scale transformation and its relation to the conformal group in four dimensions; again, recent work in connection with the conformal group (not necessarily in four space-time dimensions) may provide new insights into the construction of candidate TOE's. The scale transformation and a few related topics are discussed in the last section (§2.3b) of this chapter [2.25].

§2.3a. Poincaré group and supersymmetry

It was shown in §2.2d that the important connection between the Poincaré group and the CPT theorem is dependent on the use of the spin-statistics relation, i.e. boson field operators must satisfy C.R. whereas fermion field operators must obey anti-C.R. No exception to the spin-statistics relationship had ever been found in the four-dimensional world of quantum field theory and — during the early 1960s — efforts were made to develop dynamical theories of the strong and weak interactions by mixing global internal symmetries with the Poincaré group. As will be explained in §3.2d, the mixing of the $SU(3)$ quark flavor group with the (non-relativistic) spin subgroup of the Poincaré group is partially successful. However, the "no-go" theorem [2.24] made it clear that the full Poincaré group could not be mixed with a global internal symmetry (compact Lie) group. In particular, the "no-go" theorem showed (see §4.1) that any internal symmetry group of the S matrix, whose generators obey closed C.R., must commute with the Poincaré generators. Hence, within the context of a Lie group of internal symmetries, only a direct product of the Poincaré and internal symmetry group is possible. Moreover, since the second-order Casimir invariants of the Poincaré group, $P_\mu P^\mu$ and $W_\mu W^\mu$ (see §2.1c), commute with the internal symmetry generators, all members of an irreducible multiplet of the internal symmetry group must possess the same mass and spin (unless the symmetry is spontaneously broken). The weak link in the argument is that the "no-go" theorem only applies to the mixing of the Poincaré group with a Lie group of internal symmetries, i.e. for symmetries that satisfy (bosonic) C.R. It is possible to arrange a non-trivial mixing of Poincaré and internal symmetry groups if one is willing to modify the Lie algebra so as to include generators applying to internal symmetries that satisfy anti-C.R.. This possibility leads to a class of "graded" supersymmetric algebras and was the beginning of the supersymmetry program.

The starting point for the derivation of the supersymmetry algebra is the observation that a "graded" $\mathcal{Z}_2$ structure allows for "even charges" and "odd charges" (that satisfy C.R. and anti-C.R. relations respectively); schematically, we may write:

$$[\text{even, even}] = \text{even}; \quad [\text{odd, odd}]_+ = \text{even}; \quad [\text{odd, even}] = \text{odd} \quad (2.102)$$

Before discussing the simplest "graded" supersymmetric algebra (i.e. the $N = 1$ supersymmetric algebra with one "odd" charge, i.e "supercharge"), we recall a simple example [2.26] that exhibits the novel features of a quantum field theory into which a single "supercharge" is introduced. In this example, one postulates the existence of two pairs of creation and annihilation operators, $(a, a^\dagger)$ and $(b, b^\dagger)$, with the first pair obeying (bosonic) C.R. and the second obeying (fermionic) anti-C.R., thus:

$$[a, a^\dagger] = [b, b^\dagger]_+ = 1. \quad (2.103)$$

Defining the Hamiltonian in the usual way — with ω the energy per field quantum, one gets:

$$H = \omega_a a^\dagger a + \omega_b b^\dagger b. \quad (2.104)$$

Introducing the fermionic ("supercharge") operator:

$$Q' = b^\dagger a + a^\dagger b \quad (2.105)$$

one finds that Q' takes bosonic states (represented by $a^\dagger|0>$) into fermionic states (represented by $|b^\dagger|0>$) since:

$$[Q', a^\dagger] = b^\dagger; \quad [Q', b^\dagger]_+ = a^\dagger. \quad (2.106)$$

Moreover, the commutator of Q' with the Hamiltonian yields:

$$[Q', H] = (\omega_a - \omega_b)Q'. \quad (2.107)$$

so that H is supersymmetric only for equal energies (masses) for bosons and fermions. Finally, the anti-commutator for the supercharges Q' and $Q'^\dagger$ gives:

$$[Q', Q'^\dagger]_+ = \frac{2}{\omega}H. \quad (2.108)$$

It is evident from Eqs. (2.106) and (2.108) that a (fermionic) "supercharge" can couple the fermionic and bosonic degrees of freedom in a quantum mechanical system (provided there is mass degeneracy and equal numbers of them) and that the anti-commutator of the "supercharge" closes upon the Hamiltonian. These basic results are duplicated in $N = 1$ supersymmetric algebra, to which we now turn.

In considering any supersymmetric algebra, we must take cognizance of the fact that the "no-go" theorem still applies to the ten generators of the Poincaré group and the "even" generators T^a of the global internal symmetry group, thus;

$$[P_\mu, T^a] = [M_{\mu\nu}, T^a] = 0 \tag{2.109}$$

where the T^a's [or, equally well, the "even" charges Q^a — see Eq. (2.102)] satisfy the Lie algebra condition:

$$[Q^a, Q^b] = i f_{abc} Q^c \tag{2.110}$$

with f_{abc} the (usual) structure constants. The simplest "graded" (Lie) supersymmetric algebra ($N = 1$ supersymmetry algebra) consists of the "even" charges Q^a [$a = 1, 2, ...(n_{even}^2 - 1)$, where n_{even} is the dimension of the global internal symmetry group $SU(n_{even})$] and one ("odd") "supercharge" Q'. The Q^a's transform as scalars under the Lorentz group but Q' transforms as a $J = \frac{1}{2}$ "spinor" under the Lorentz group. Since it can be shown [2.26] that 2-component "supercharges" are acceptable as generators of "graded" Lie algebras of symmetries of the S matrix, it is convenient to represent Q' in terms of a chiral (say, lefthanded) Weyl spinor $Q'_\alpha (\alpha = 1, 2)$, thus:

$$Q' = \begin{pmatrix} Q_\alpha \\ i\sigma_2 Q_\alpha^* \end{pmatrix}, \tag{2.111}$$

It is well known that Q_α and $i\sigma_2 Q_\alpha^*$ belong to the $(\frac{1}{2}, 0)$ and $(0, \frac{1}{2})$ representations of the Lorentz group and that the $SO(3, 1)$ Lorentz group is locally isomorphic to the $SL(2, C)$ group consisting of arbitrary complex 2×2 matrices with unit determinant. Hence, if Q_α belongs to the fundamental representation of the $SL(2, C)$ group, $\bar{Q}_\alpha$ belongs to the conjugate representation; taking advantage of the fact that $i\sigma_2 = \epsilon^{\alpha\beta}$, Q' can be written in the equivalent Majorana spinor form [2.27]:

$$Q' = \begin{pmatrix} Q_\alpha \\ \bar{Q}^\alpha \end{pmatrix} \tag{2.112}$$

where the overbar denotes the complex two-dimensional representation of $SL(2, C)$.

With the notation for the "odd" (Majorana) "supercharge" given by Eq. (2.112), one can readily work out the anti-C.R. among the spinor components of Q', as well as the C.R. between them and the "even" charges Q and the generators of the Poincaré group, $M_{\mu\nu}$ and P_μ. First, consider the anti-commutators $[Q_\alpha, \bar{Q}_\beta]_+$ and $[Q_\alpha, Q_\beta]_+$; the first anti-commutator transforms as $(\frac{1}{2}, \frac{1}{2})$ under the Lorentz transformation and, since P_μ is the only generator of the Poincaré algebra in such a representation, one gets:

$$[Q_\alpha, \bar{Q}_\beta]_+ = 2\,(\sigma^\mu)_{\alpha\beta} P_\mu \qquad (2.113)$$

The factor 2 in Eq. (2.113) respects the normalization convention and the positive sign follows from the requirement that the energy P_0 is a semi-positive definite operator, namely:

$$\sum_{\alpha=1}^{2} [Q_\alpha, \bar{Q}_\alpha]_+ = 2\,Tr[\sigma^\mu P_\mu] = 4P_0 \qquad (2.114)$$

To evaluate $[Q_\alpha, Q_\beta]_+$, one notes that the anti-commutator of Q_α and Q_β must be a linear combination of the Poincaré generators in the $(0,1)$ and $(1,0)$ representations of the Lorentz group; however, this anti-commutator must commute with P_μ and since this is impossible, we have:

$$[Q_\alpha, Q_\beta]_+ = 0 \qquad (2.115)$$

Insofar as the commutator between the "even" charges Q_a $[a = 1, 2...(n_{\text{even}}^2 - 1)]$ with Q_α and $\bar{Q}_\alpha$ is concerned, only a single global $U(1)$ internal symmetry, known as R symmetry, remains and the C.R. [see last of Eqs. (2.102)] are:

$$[Q_\alpha, R] = Q_\alpha ; \qquad [\bar{Q}_\alpha, R] = \bar{Q}_\alpha \qquad (2.116)$$

It should be noted that $Q_\alpha \to \bar{Q}^\alpha$ under the parity operation so that $R \to -R$, i.e. the $U(1)$ R symmetry group is chiral.

Next, we consider the C.R. of Q_α and $\bar{Q}_\alpha$ with the Poincaré generators $M_{\mu\nu}$ and P_μ. Using the "graded" Jacobi identities and the known C.R. $[P_\mu, P_\nu] = 0$, it can be shown that:

$$[Q_\alpha, P_\mu] = [\bar{Q}_\alpha, P_\mu] = 0 \qquad (2.117)$$

The C.R. of Q_α and $\bar{Q}_\alpha$ with $M_{\mu\nu}$ do not vanish and are:

$$[M_{\mu\nu}, Q_\alpha] = \frac{1}{2}(\sigma_{\mu\nu})_\alpha^\beta Q_\beta; \qquad [M_{\mu\nu}, \bar{Q}_\alpha] = -\frac{1}{2}(\bar{\sigma}_{\mu\nu})_\alpha^\beta \bar{Q}_\beta \qquad (2.118)$$

From Eqs. (2.113)–(2.118), we see that, for the $N = 1$ supersymmetric algebra, the single "supercharge" Q' and the generators of the Poincaré group close upon themselves. The consequence is that for the $N = 1$ supersymmetry case, one can work with a direct product of the $N = 1$ supersymmetry and the Abelian or non-Abelian internal symmetry group so that the Poincaré group serves as the space-time symmetry. The situation becomes more complicated when one moves on to $N \geq 2$ supersymmetry [i.e. two or more "supercharges"]; for $N \geq 2$, if the "supercharges" $Q_{\alpha i}$

$(i = 1, 2, ...N)$ belong to some representation T_j of the internal symmetry group [of "even" charges, $SU(n_{\text{even}})$], one has

$$[Q_{\alpha i}, T_j] = f_{ijk} Q_{\alpha k} \qquad (2.119)$$

where f_{ijk} are the structure constants of the group $SU(n_{\text{even}})$. In comparison, for $N = 1$ supersymmetry, one has only to deal with the Abelian R charge satisfying the C.R. (2.116). The C.R. for the $N \geq 2$ "supercharges" with the generators of the Poincaré group are even more complicated (requiring, for example, the introduction of so-called central charges [2.28]) and we will not elaborate further on $N \geq 2$ global supersymmetry. Instead, we mention several other theoretical properties of $N = 1$ global supersymmetry and conclude with some comments regarding the phenomenological status of supersymmetry.

First, it is worth mentioning that the C.R. and anti-C.R. defining the $N = 1$ supersymmetry algebra can be used to derive statements about the supersymmetric vacuum and the conditions for the SSB of supersymmetry. Thus, if there exists a supersymmetrically-invariant state (vacuum) such that:

$$Q_\alpha |0> = 0 \quad \text{and} \quad \bar{Q}_\alpha |0> = 0 \qquad (2.120)$$

then it follows from Eq. (2.114) that $E_{vac} = 0$. Since Eq. (2.114) tells us that the spectrum of H is semi-positive definite, $E_{vac} = 0$ implies that the supersymmetrically-invariant ground state is at the absolute minimum of the potential so that we have a "true" vacuum. This result differs from the non-supersymmetric case where a symmetric state can exist without being the ground state. Hence, if $Q_\alpha |0> \neq 0$ and/or $\bar{Q}_\alpha |0> \neq 0$, a supersymmetrically-invariant ground state does not exist and we must have $E_{vac} > 0$. That is to say, $E_{vac} = 0$ requires — and is required by — unbroken supersymmetry whereas $E_{vac} > 0$ requires — and is required by — -spontaneous breaking of supersymmetry. These conclusions are a consequence of the fact that the anti-commutator of Q_α and $\bar{Q}_\alpha$ given in Eq. (2.114) involves the Hamiltonian, whereas the bosonic charge commutator does not. The special property of the supersymmetric vacuum can be exploited to derive the analog of the Goldstone theorem, i.e. the "super-Goldstone" theorem. Thus, consider the state obtained by acting on the vacuum with Q_α or $\bar{Q}_\alpha$:

$$|\Psi> = Q_\alpha |0> \quad (\text{or } |\Psi> = \bar{Q}_\alpha |0>) \qquad (2.121)$$

SSB is defined by the condition $|\Psi> \neq 0$ and, since $|\Psi>$ is a state with odd fermion number and $[Q_\alpha, H] = 0$, it must be degenerate with the vacuum.

But the existence of such a state, i.e. a state of a single massless fermion of momentum p (in the limit $p \to 0$) is precisely the Goldstone "fermion" or so-called "Goldstino".

Another important prediction of $N = 1$ supersymmetry algebra (which obtains as well for $N \geq 2$ supersymmetry algebra) follows from the observation that $[Q_\alpha, P^2] = 0$ and $[Q_\alpha, W^2] \neq 0$ (where W^μ is the Pauli-Lubanski vector — see §2.1c). Since P^2 and W^2 are the two second-order Casimir operators of the Poincaré group, it follows that while supersymmetry multiplets are always degenerate in mass, they must contain different spins. And since the "supercharge" is a spin $\frac{1}{2}$ object, fermions must coexist in the same supersymmetry (mass degeneracy) multiplet as bosons. This is equivalent to the statement that, for $N = 1$ supersymmetry, the action of the $J = \frac{1}{2}$ "supercharge" Q' on a state of spin J results in a state of spin $J \pm \frac{1}{2}$, thereby mixing fermions and bosons (i.e. "matter" and "forces"). The fermion-boson mass degeneracy is a striking consequence of supersymmetry since the members of the composite quark flavor multiplets are all bosons (mesons) or fermions (baryons) with the same spin (and parity) and, even the mixed $SU(6)$ $[SU(3)_F$ — spin] model (see §3.2d) only couples (bosonic) meson $SU(3)_F$ multiplets of different integral spins (0 and 1) or (fermionic) baryon $SU(3)_F$ multiplets of different half-integral spins ($\frac{1}{2}$ and $\frac{3}{2}$), but not bosons and fermions. Since no fermion-boson mass degeneracy has been observed in nature, supersymmetry must be broken either explicitly or spontaneously if it is to be physically relevant. In order to test the supersymmetry principle and its possible breaking, it is therefore necessary to obtain an estimate of the mass splitting between the fermion and boson multiplets.

To obtain an estimate of the mass difference between a normal particle (i.e. one that enters into the standard gauge theory of strong and electroweak interactions) and its supersymmetric partner, a well-defined model is required. Since $N = 1$ global supersymmetry is itself not unique and there are many supersymmetry-breaking models, we necessarily obtain a rough estimate — within the context of $N = 1$ global supersymmetry — between two supersymmetric partners differing by $\Delta J = \frac{1}{2}$. The example we choose is the elusive but, in certain respects, the most favorable case of the Higgs boson in the standard model. Since the value of the Higgs mass is unknown — a credible theoretical upper limit of about 1 TeV has been derived (see §6.3c) — we sketch the argument why the mass of the supersymmetric partner of the weak gauge boson, i.e. the $J = \frac{1}{2}$ "weak gaugino" should approximate the upper limit of the Higgs mass. The argument exploits the fact that the one-loop radiative corrections to the mass of a scalar boson — due to bosons or fermions — give rise to quadratic divergences of opposite sign so that the quadratic divergence can be softened to a logarithmic divergence by suitable choice of masses and coupling constants of the mediating bosons and fermions. While we present the argument in fairly general terms, it is useful to exhibit in Fig. 2.1 the contributions of the weak gauge boson and the "weak gaugino" to the one-loop radiative corrections to the Higgs mass. The one-loop radiative corrections to the Higgs mass from two supersymmetric partners [the subscript B denotes the boson (read "weak" gauge boson in Fig. 2.1a) and F the fermion (read the "gaugino" in Fig. 2.1b)] can typically be written in the form [2.29]:

$$\Delta M_H^2 = C_B \alpha_B \int_{M_H^2}^{\Lambda^2} \frac{d^4 k}{(k^2 - m_B^2)} - C_F \alpha_F \int_{M_H^2}^{\Lambda^2} + \frac{d^4 k}{(k^2 - m_F^2)} + L(m_B^2, m_F^2)$$

$$(2.122)$$

$$\simeq (C_B \alpha_B - C_F \alpha_F)\, \Lambda^2 + C\alpha(m_B^2 - m_F^2)\, \ln\left(\frac{\Lambda^2}{M_H^2}\right) \qquad (2.123)$$

where (C_B, C_F), (g_B, g_F) and (m_B, m_F) are the respective "internal symmetry" coefficients, coupling constants and masses of the boson and fermion (supersymmetric) partners; Λ is the effective cutoff scale (arising from whatever "new physics" exists beyond the standard model) and $L(m_B^2, m_F^2)$ is the residual logarithmically divergent correction. In going from Eq. (2.122) to Eq. (2.123), we have assumed that Λ is large compared to M_H, m_B and m_F, and we have allowed $m_B \neq m_F$.

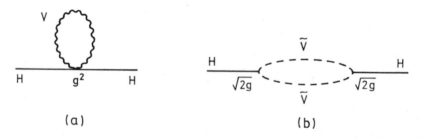

Fig. 2.1. Gauge supermultiplet one-loop radiative corrections to Higgs mass; V is the weak gauge boson ($J = 1$) and $\tilde{V}$ is the "gaugino" ($J = \frac{1}{2}$).

We can draw several conclusions from Eq. (2.123): (1) if Λ is large compared to m_B and m_F — as we have assumed — the coefficient of the quadratically divergent term in Eq. (2.123), namely $(C_B \alpha_B - C_F \alpha_F)$ vanishes because we can use the unbroken supersymmetric limit. Whether the separate equalities $C_B = C_F$ and $\alpha_B = \alpha_F$ hold, depends on the model; for example, in the Wess-Zumino model [2.29] — where two real scalar fields and one four-component Majorana spinor are used, the separate equalities hold. However, if the two supersymmetric partners coupled to the scalar Higgs field are a $J = 1$ bosonic field and a $J = \frac{1}{2}$ fermionic field — as in the case in Fig. 2.1 — $C_B \neq C_F$ and $\alpha_B \neq \alpha_F$ but $C_B \alpha_B = C_F \alpha_F$ so that the quadratic divergence is still cancelled; (2) the second term on the R.H.S. of Eq. (2.123) allows for the breaking of the $N = 1$ global supersymmetry by permitting $m_B \neq m_F$ although we have set $C_B \alpha_B = C_F \alpha_F = C\alpha$ (the relation between coupling constants that holds for unbroken supersymmetry). This enables us to make an important qualitative point, namely that the radiative corrections to the Higgs mass is of the order of a coupling constant times the difference between the mass squared of the two supersymmetric partners (which we denote by ΔM^2); (3) if we come closer to phenomenology and consider the one-loop contribution to the Higgs mass by the weak gauge boson field and its supersymmetric "weak gaugino" partner (as in Fig. 2.1), we deduce (using the theoretical upper limit on the Higgs mass and our knowledge of the weak gauge coupling constant) that the mass difference between the weak gauge boson and the "weak gaugino" fermion is also of the order of a TeV; and, finally, (4) noting that the ΔM^2 between two supersymmetric partners contributing to the Higgs mass varies inversely as the coupling constant, the same type of argument, applied to the Higgs coupling to quarks and leptons in the standard model, tells us that the ΔM^2 between quark and "squark" and/or lepton and "slepton" superpartners should be substantially larger than a TeV (since the Yukawa coupling of the Higgs to quarks and leptons is substantially less than that between the Higgs and the weak gauge bosons). Unfortunately, experimental searches for supersymmetric partners of any kind to the known particles of the standard model have thus far proved fruitless but the critical TeV energy region has not been reached and there may be some surprises.

We have reviewed in this section the elements of global supersymmetry with $N = 1$ "supercharge". During the two decades that followed the discovery of global $N = 1$ supersymmetry, much work was done on field theories with $N \geq 2$ global supersymmetry, spontaneously broken global supersymmetry, local supersymmetry and supergravity [2.30], also on supersymmetric field theories in four dimensions with an infinite number of higher spins [2.31] and on superstring theory in a large number of dimensions (> 4)

[1.4c]. This vast program has made repeated efforts to make phenomenological contact with the four-dimensional world of the standard model but, thus
far, no phenomenological support has been forthcoming. Phenomenological
success necessarily involves physics "beyond the standard model" and may
revive the hope that supersymmetry — particularly in its later versions —
opens the door to a "super-grand" unification of all four basic forces of nature (TOE). However, apart from our earlier brief discussion — in §1.4c —
of supersymmetry, supergravity and superstrings within a quasi-historical
context, we leave to others the recounting of these speculative forays into
TOE [1.185] and turn, instead, to a final section on scale invariance and the
conformal group in four-dimensional space-time. Again, we merely introduce
the subject [2.25] and leave to others the chronicling of the latest developments in $D = 2$ conformal field theory and related subjects as a prelude to
the construction of a TOE from a new direction [2.32].

§2.3b. Scale invariance and the conformal group in four
space-time dimensions

We will see in §4.5 that the Abelian gauge theory of the electromagnetic
interaction (QED) with massive fermions (charged leptons), and the non-
Abelian gauge theory of the strong interaction (QCD) with massive fermions
(quarks) are renormalizable. One might think that when the fermions are
taken to be massless (the gauge fields and the coupling constants already are
dimensionless), and the resulting Lagrangians for QED and QCD become
fully dimensionless, that the two theories are invariant under scale transformations (in space-time) of their respective fields. This is true for the
classical fields of the two theories but the renormalization requirements after
quantization introduce finite reference points [to avoid infrared singularities
(see §4.2)] and consequently the "full" scale invariance in QED and QCD
is no longer maintained. Because it is the renormalization process that destroys the "full" scale invariance of the quantum field theory, one might
inquire whether some sort of "residual" scale invariance survives through
the renormalization group equations — which can be derived (see §5.2a)
assuming the "scale invariance" of the renormalization reference point —
when the energies (and four-momentum transfers) are large compared to
the fermion masses. We will explain in some detail in §5.2a and §5.2c the
sense in which the onset of "Bjorken scaling", in connection with deep inelastic lepton-hadron scattering, is a manifestation of the "scale invariance"
discussed in this section under the rubric of space-time symmetries. In §5.2b,

we will see how the "residual" scale invariance reflecting the existence of the renormalization group shows up as "anomalous dimensions" in the "asymptotic freedom" region of QCD. To provide some background for these applications, we review in this section the properties of scale transformations on quantum fields in four-dimensional ($D = 4$) space-time as well as the relationship between scale invariance and the conformal group (for $D = 4$).

A scale transformation (or a dilatation) on the $(3 + 1)$ space-time coordinates is defined by: $x' = e^{\alpha}x$ (α a real number). If the dilatation acts linearly on quantum fields and the representation is fully reducible, the field transforms as [2.25]:

$$\phi'(x') \to e^{-\alpha d}\phi(x) \qquad (2.124)$$

The infinitesimal transformation is then:

$$\delta\phi(x) = -\delta\alpha(d + x^{\mu}\partial_{\mu})\phi(x) \qquad (2.125)$$

The quantity d in Eqs. (2.124) and (2.125), is called the "canonical" dimension of the field (see §4.5) and possesses the value $d = 1$ for boson fields and $d = \frac{3}{2}$ for fermion fields. These choices of d maintain the invariance under dilatation of the equal time C.R. for the boson fields and anti-C.R. for the fermion fields. It is worth pointing out how the invariance of the C.R. for a scalar field under dilatation yields $d = 1$; thus, the C.R. for the transformed scalar field is given by:

$$[\phi'(\underline{x}',t'), \dot{\phi}'(\underline{y}',t')] = [e^{-\alpha d}\phi(\underline{x},t), e^{-\alpha-\alpha d}\dot{\phi}(\underline{y},t)]$$
$$= e^{-\alpha(2d+1)}[\phi(\underline{x},t), \dot{\phi}(\underline{y},t)] = ie^{-\alpha(2d+1)}\delta^3(\underline{x} - \underline{y}) = i\delta^3(\underline{x}' - \underline{y}') \qquad (2.126)$$

for $d = 1$. The values of d for the boson and fermion fields coincide with the "canonical" dimensions of these fields within the framework of the usual dimensional analysis. However, the scale dimensions d that arise in connection with "scale invariance" should not be confused with the "dimensions" of ordinary dimensional analysis [2.32]. In the latter case, not only are the dynamical variables of the physical theory (the quantum fields in our case) scaled but so are the "non-dimensional" numerical parameters (e.g. masses if they are present). To put it another way, the transformations of dimensional analysis convert one physical theory into another (e.g. with different masses) and maintain the existing symmetries. On the other hand, scale transformations do not affect numerical parameters (e.g. mass) and scale invariance is actually broken by non-vanishing mass. At this stage, we can say that scale invariance for classical field theory is an exact symmetry if the coupling constants and masses all vanish and d is chosen as 1 for boson fields and $\frac{3}{2}$ for fermion fields; however, when the classical field theory is quantized and renormalized (renormalizability is always assumed), scale invariance is broken and the "canonical" dimensions of the fields are modified in a manner determined by the renormalization group requirements of the quantum field theory in question.

Before discussing some ramifications of broken scale invariance, let us consider further the implications of exact scale invariance for a classical field theory with massless fields and dimensionless coupling constants. In general, it is possible to define a dilatation current s_μ, in terms of the energy-momentum tensor $T_{\mu\nu}$ associated with the field, namely:

$$s^\mu = x_\nu T^{\mu\nu} \tag{2.127}$$

Hence, if there is exact scale invariance, the current s^μ is conserved and is equivalent to the condition that the trace of the energy-momentum tensor vanishes, i.e.

$$\partial^\mu s_\mu = T^\mu_\mu = 0 \tag{2.128}$$

It should be pointed out that the construction of the energy-momentum tensor $T_{\mu\nu}$, that enters into Eqs. (2.127) and (2.128), is only possible if the following condition holds [2.33]:

$$\frac{\partial \mathcal{L}}{\partial(\partial_\nu \phi)} \cdot [g^{\mu\nu}d - \Sigma^{\mu\nu}]\phi = \partial_\nu \theta^{\mu\nu} \tag{2.129}$$

where $\mathcal{L}$ is the Lagrangian, ϕ the field, $\Sigma_{\mu\nu}$ the spin matrix [see Eq. (2.47)] and $\theta^{\mu\nu}$ is some tensor function of the fields and their derivatives. It can be shown that all renormalizable field theories (which includes all the interactions of the standard model) satisfy Eq. (2.129). A typical energy-momentum tensor $T_{\mu\nu}$ — taken from QCD — that can be used in Eqs. (2.127) and (2.128), is the following:

$$T_{\mu\nu} = -G^a_{\mu\alpha}G^a_{\nu\alpha} + \frac{1}{4}g_{\mu\nu}G^a_{\alpha\beta}G^a_{\alpha\beta} + \frac{i}{4}\sum_q \{\bar{q}(\gamma_\mu D_\nu + \gamma_\nu D_\mu)q$$

$$- \bar{q}(\gamma_\mu \overleftarrow{D}_\mu + \gamma_\nu \overleftarrow{D}_\mu)q\} \tag{2.130}$$

Different choices of $T^{\mu\nu}$ are possible which differ in terms that do not affect the total four-momentum nor the Lorentz generators but do fit the requirements of the scale current.

With a suitable definition of the energy-momentum tensor $T_{\mu\nu}$, it is possible to define four additional conserved tensor currents as follows ($\lambda = 0, 1, 2, 3$):

$$K^{\lambda\mu} = x^2 T^{\lambda\mu} - 2x^\lambda x_\rho T^{\rho\mu}; \quad \partial_\mu K^{\lambda\mu} = 2x_\mu T^{\lambda\mu} - 2x_\rho T^{\rho\lambda} - 2x^\lambda T^\rho_\rho = -2x^\lambda T^\rho_\rho.$$
$$(2.131)$$

It is seen from Eq. (2.131) that the divergence of the tensor current $K^{\lambda\mu}$ vanishes when $T^\rho_\rho = 0$ (i.e. is traceless). Since the scale current s_μ is also divergenceless when the energy-momentum tensor is traceless [see Eq. (2.128)], it follows that when an energy-momentum tensor $T_{\mu\nu}$ can be defined to satisfy Eqs. (2.127) and (2.128), scale invariance implies conformal invariance (the $K^{\lambda\nu}$ transformations are the "special" conformal transformations in four dimensions — see below) and conversely. It should be emphasized that the crucial equation (2.128) — which relates the divergence of the scale (or dilatation) current to the trace of the energy-momentum tensor is a classical equation and acquires an anomaly when it is "quantized" and when the one-loop (and higher-order loop) corrections are calculated. The trace anomaly (or "dilaton") pinpoints the reason why the prediction of the conformal group in four dimensions (which contains the dilatation generator as one of its fifteen generators) — that the mass spectrum in nature should be continuous (or that all the masses vanish) (see below) — is a "classical" and not a "quantum mechanical" prediction. The trace anomaly (which is $\sim G_{\mu\nu} G^{\mu\nu}$ in QCD [2.34]) possesses both physical and mathematical analogies to the triangular axial QCD anomaly (which is $\sim G_{\mu\nu} \tilde{G}^{\mu\nu}$ in QCD) that is discussed in §7.2a as a "harmless" but useful anomaly in quantum field theory.

It is interesting to inquire into the meaning of the $K^{\mu\nu}$ tensor currents. It is possible to show that the four conserved tensor currents, given by Eq. (2.131), arise from inversions applied to the four space-time translations, with which are associated the four "special" conformal generators (of the fifteen generators) of the so-called conformal group. To explain these last statements, we recall that a Minkowski space inversion is defined by

$$I: \quad x^\mu \to -x^\mu/x^2. \tag{2.132}$$

If the inversion transformation is applied to the right and left of the four-parameter group of space-time translations: $x^\mu \to x^\mu + a^\mu$, we get the "special" conformal transformations:

$$x^\mu \to \frac{x^\mu - a^\mu x^2}{1 - 2a_\mu x^\mu + a^2 x^2}. \tag{2.133}$$

Just as the infinitesimal scale transformation on the field [given by Eq. (2.125)] followed from the scale transformation on the space-time coordinates [i.e. $x^\mu \to e^\alpha x^\mu$ (α is a real number)], so the infinitesimal "special" conformal transformation on the field follows from Eq. (2.133) with the result [see Eqs. (2.131) and (2.129)]:

$$\delta^\alpha \phi(x) = (2x^\alpha x^\nu - g^{\alpha\nu} x^2) \partial_\nu \phi(x) + 2x_\nu (g^{\nu\alpha} d - \Sigma^{\nu\alpha}) \phi(x) \qquad (2.133a)$$

Equation (2.133a) anticipates the interplay of the "special" conformal transformation with the other transformations of the conformal group (see below).

Following up on the last remark, we add to the "special" conformal transformations (2.133) the Poincaré transformation: $x^\mu \to \Lambda^\mu_\nu x^\nu + a^\mu$ and the dilatation $x^\mu \to e^\alpha x^\mu$; we then have the ensemble of transformations that define the conformal group with its fifteen generators: P_μ, $M_{\mu\nu}$, D and K_μ. The four P_μ generators and six $M_{\mu\nu}$ generators comprise the ten generators of the Poincaré group, D is the one generator of the dilatation group [see Eq. (2.125)] and the K_μ [$K_\mu = \int K_{o\mu} dx$] are the four generators corresponding to the "special" conformal transformations (2.133a), yielding a total of fifteen generators. We can now write down the C.R. among the fifteen generators and exhibit their closure to define the "conformal algebra" in four space-time dimensions:

$$[M_{\mu\nu}, M_{\rho\sigma}] = -i(g_{\mu\rho} M_{\nu\sigma} - g_{\nu\rho} M_{\mu\sigma} + g_{\mu\sigma} M_{\rho\nu} - g_{\nu\sigma} M_{\rho\mu}) \qquad (2.134a)$$

$$[M_{\mu\nu}, P_\rho] = i(g_{\nu\rho} P_\mu - g_{\mu\rho} P_\nu); \quad [P_\mu, P_\nu] = 0 \qquad (2.134b)$$

$$[M_{\mu\nu}, K_\rho] = -i(g_{\mu\rho} K_\nu - g_{\rho\nu} K_\mu) \qquad (2.134c)$$

$$[P_\mu, K_\nu] = 2i(g_{\mu\nu} D - M_{\mu\nu}) \qquad (2.134d)$$

$$[M_{\mu\nu}, D] = 0; \quad [P_\mu, D] = iP_\mu \qquad (2.134e)$$

$$[K_\mu, K_\nu] = 0; \quad [D, K_\mu] = iK_\mu \qquad (2.134f)$$

Equations (2.134a) and (2.134b), of course, constitute the Poincaré algebra whereas the remaining C.R. of K_μ and D with the ten Poincaré generators and with each other show clearly that the conformal algebra closes upon itself. In mathematical language, the full conformal group in Minkowski space is an extension of the Poincaré group $SO(3,1)$ into a higher-dimensional homogeneous orthogonal group $SO(4,2)$, which represents the "connected" part of the smallest group of transformations in Minkowski space containing both the Poincaré group and inversion.

The conformal group (with its fifteen generators) in Minkowski space is quite elegant and certainly can be applied to classical field theories with dimensionless constants and massless fermions. However, we are interested in quantum field theories with dimensionless coupling constants and massless fermions and, unfortunately, as we have stated, quantization breaks conformal invariance and gives rise to an anomaly, just as chirality invariance with massless coupling constants and massless fermions is broken by the triangular axial vector anomaly. The simplest way to establish that the conformal group does not hold in the real four-dimensional world is to draw out the consequence of the second C.R. of Eq. (2.134e); thus, using the C.R., $[P_\mu, D] = iP_\mu$, it follows that:

$$[P^2, D] = P^\mu[P_\mu, D] + [P^\mu, D]P_\mu = 2iP^2 \tag{2.135}$$

Successive application of Eq. (2.135) yields:

$$e^{i\beta D}P^2 e^{-i\beta D} = e^{2\beta}P^2 \tag{2.136}$$

where β is some constant; but $P^2 = m^2$ and hence Eq. (2.136) implies that the mass spectrum in nature is either continuous or that all the masses vanish. This is unacceptable and hence there must be an anomaly in a phenomenologically correct quantum field theory; this is the previously mentioned trace anomaly or "dilaton" and impacts importantly on the naive Ward-Takahashi identities associated with the initial classical field theory. The expression for the dilaton is simply the non-vanishing, quantized and renormalized form for T^μ_μ for the field theory in question. The point we wish to emphasize here is that, at least for Yang-Mills theories, classical scale invariance in the field theory is perpetuated after quantization in what we have called the "residual scale-invariant" behavior in the asymptotic region (energy large compared to the particle masses), where it displays a certain universal character with operators being assigned "anomalous" dimensions (see §5.2b). Since the significance of the "anomalous dimensions" — as a measure of the quantum breakdown of classical scale invariance — has to be explained within the framework of the "quantum mechanical" renormalization group equation, we postpone this discussion to §5.2a.

References

[2.1] O. Klein, *Zeits. f. Phys.* **37** (1926) 895; W. Gordon, *Zeits. f. Phys.* **40** (1926) 117.

[2.2] S. Gasiorowicz, *Elementary Particle Physics*, John Wiley, New York (1966).

[2.3] A. Proca, *J. Phys. Rad. Radium* **7** (1936) 347.

[2.4] K. Bleuler, *Helv. Phys. Acta.* **23** (1950) 567; S. Gupta, *Proc. Phys. Soc. (London)* **A63** (1950) 681; *ibid.* **A64** (1951) 850.

[2.5] W. Pauli and V. Weisskopf, *Helv. Phys. Act.* **7** (1934) 709; W. Pauli, Lecture notes, Institute for Advanced Study, Princeton (1936).

[2.6] W. Rarita and J. Schwinger, *Phys. Rev.* **60** (1940) 61.

[2.7] E. Cartan, *Leçons sur la Théorie des Spineurs*, Tome I et II, Hermann et Cie, Paris (1938).

[2.8] V. Bargmann and E.P. Wigner, *Proc. Nat. Acad. Sci. U.S.* **34** (1948) 211.

[2.9] *Physics and Mathematics of Anyons*, ed. S.S. Chern, C. W. Chu and C.S. Ting, World Scientific, Singapore (1991).

[2.10] C. Soo, *Classical and Quantum Gravity with Ashtekar Variables*, Ph.D. thesis, Virginia Tech (1992), preprint # VPI-IHEP-92/11.

[2.11] R.E. Marshak and E.C.G. Sudarshan, *Introduction to Elementary Particle Physics*, John Wiley, New York (1961).

[2.12] R.E. Marshak, *Proc. of Virginia Tech Workshop*, A.I.P. Publication, No. 72 (1981), p. 665.

[2.13] L.D. Landau, *Nucl. Phys.* **3** (1957) 127.

[2.14] R.G. Sachs, *The Physics of Time Reversal*, Univ. of Chicago Press, Chicago (1987).

[2.15] N.F. Ramsey, *Phys. Rep.* **43** (1978) 409; *ibid.*, *Ann. Rev. Nucl. Part. Sci.* **32** (1982) 211; I.S. Altarev, *Phys. Lett.* **B136** (1984) 327.

[2.16] E.D. Cummins and P.H. Buchsbaum, *Weak Interactions of Leptons and Quarks*, Cambridge Univ. Press (1983).

[2.17] G. Luders, *Kongl. Dansk. Medd. Fys.* **28** (1954) No. 5; W. Pauli, *Niels Bohr and the Development of Physics*, ed. W. Pauli, L. Rosenfeld and V. Weisskopf, McGraw-Hill, New York (1955).

[2.18] T.D. Lee, *Particle Physics and Introduction to Field Theory*, Harwood, New York (1981).

[2.19] G.C. Wick, *Phys. Rev.* **80** (1950) 268.

[2.20] J. Bennet *et al.*, *Phys. Rev. Lett.* **19** (1967) 993; D. Dorfan *et al.*, *Phys. Rev. Lett.* **19** (1963) 987.

[2.21] S. Okubo, Univ. of Rochester preprint NYO-2099 (1957); A.D. Sakharov, *Memoirs*, A. A. Knopf, New York (1990), p. 252.

[2.22] A.D. Sakharov, *Soviet Physics JETP Lett.* **5** (1967) 24.

[2.23] A. Salam, *The Second Coral Gables Conference*, ed. B. Kursunoglu, A. Perlmutter and I. Sakmar, W.H. Freeman, San Francisco (1965).

[2.24] L. O'Raifeartaigh, *Phys. Rev.* **139B** (1965) 1052; S. Coleman and J.E. Mandela, *Phys. Rev.* **159** (1967) 1251; Y.A. Gelfand and E.S. Likhtman, *JETP Lett.* **13** (1972) 323; W.D. McGlinn, *Phys. Rev. Lett.* **12** (164) 467.

[2.25] S. Coleman, *Aspects of Symmetry*, Cambridge Univ. Press (1989).

[2.26] R.N. Mohapatra, *Unification and Supersymmetry*, Springer (1986).

[2.27] P. West, *Introduction to Supersymmetry and Supergravity*, World Scientific, Singapore (1986).

[2.28] A. Salam and J. Strathdee, *Nucl. Phys.* **B98** (1975) 293; P. Fayet, *Nucl. Phys.* **B113** (1976) 135; M.F. Sohnius, *Nucl. Phys.* **B138** (1978) 109.

[2.29] J. Wess and B. Zumino, *Phys. Lett.* **49B** (1974) 52.

[2.30] *Introduction to Supersymmetry and Supergravity*, ed. M. Jacob, North-Holland/World Scientific, Singapore (1986).

[2.31] E.S. Fradkin and V. Ya. Linetsky, *Phys. Lett.* **253** (1991) 107.

[2.32] A. Belavin, A.M. Polyakov and A.B. Zamolodchikov, *Nucl. Phys.* **B241** (1984) 333.

[2.33] C.G. Callan, S. Coleman and R. Jackiw, *Annals of Phys.* **59** (1970) 42.

[2.34] J. Pasupathy, *Proc. of Intern. Seminar on Direct Nuclear Reactions*, India (1991).

Chapter 3

GLOBAL INTERNAL SYMMETRIES AND THEIR SPONTANEOUS BREAKDOWN

§3.1. Global Abelian internal symmetries

In §2.1–2.2, we discussed space-time symmetries — both continuous and discrete — for Klein-Gordon, Proca, Maxwell, Dirac and Weyl quantum fields, all of which enter into the standard model. In §2.1, we covered the continuous space-time symmetries embodied in the Poincaré transformation and explained how the invariance of the Lagrangian under these global space-time symmetries gives rise to the absolute conservation laws of energy, momentum and angular momentum as well as Lorentz invariance. In §2.2, we went beyond the proper, orthochronous Poincaré transformation, to consider the two discrete space-time transformations: space inversion (P) and time inversion (T), which do not lead to absolute conservation laws for all particle interactions and, indeed, yield maximal P and C violation [and a small T violation — assuming CPT invariance] for the weak interactions. The third discrete symmetry considered in §2.2 — charge conjugation (C) — is, to all intents and purposes, a global internal symmetry that is linked to P and T through the CPT theorem, which mandates CPT invariance for a local quantum field theory, obeying Lorentz invariance and the usual spin-statistics relation; it is the CPT theorem that permits the conclusion of a small T violation in the weak interactions from the observed CP violation (see §6.3f). Finally, in §2.3, we discussed some special topics connected with space-time symmetries, of which the more relevant for the present chapter on global internal symmetry groups is §2.3a, where we drew attention to the "no-go" theorem [2.24], which forbids meaningful mixing of Lie groups of global internal

146

symmetries with the Poincaré group. In a sense, the gauging of well-chosen
Lie groups of global internal symmetries — leading to the breakthroughs of
the standard model — is an extremely clever way to circumvent the "no-go"
theorem and achieve a dynamical quantum field theory (see §4.1).

In this chapter, we review the progression of steps that led — prior to the
final step of gauging — to successive enlargements of the global quark and
lepton internal (flavor) symmetry groups (chiefly non-Abelian) that played
such a prominent role — in the late 1960s and early 1970s — in the construction of the highly successful standard model. When global internal
symmetries of quantum field theories are considered in their "pure" form
(i.e. unmixed with space-time symmetries), they give rise — not surprisingly — to mass degeneracies or, when explicitly broken, to relations among
particle masses of the same spin and parity. The idea of mixing global internal symmetry with space inversion to deduce mass relations among hadrons
(mesons) of the same spin but of opposite parity [3.1] or of mixing global internal symmetry with spatial rotation (non-relativistic spin) to predict mass
relations among hadrons (baryons) with different spin (but the same parity)
[i.e. the $SU(6)$ model (see §3.2d)] helped to clarify the difficulties of mixing
global internal symmetry with discrete space-time operations or the Poincaré
group. These attempts to construct quasi-dynamical quantum field theories
were soon undermined by the powerful "no-go" theorems [2.24] (see §4.1)
and, before long, were overtaken by the much more successful concept of
gauging the global non-Abelian internal symmetries of color (for quarks)
and chiral flavor (for both quarks and leptons) to overcome the final hurdles
in the formulation of truly dynamical theories of the strong and electroweak
interactions. However, the decade and a half of hard work (from the mid-
1950s to the end of the 1960s), probing ever more deeply into the nature
of the global internal symmetries governing the strong and, to a lesser extent, the weak interaction, was by no means a failure and its achievements
(e.g. current algebra) must be recounted as a prelude to an understanding of
later developments. The purpose of this chapter — in line with the title of
the book — is to try to communicate some sense of the accomplishments of
the global internal symmetry years in laying the conceptual foundations for
the present era in particle physics.

In this first section of Chapter 3, we begin the story with the simplest
global internal symmetries (of the Abelian variety) that were around since
the early days of quantum field theory (in the 1930s) and comment on some
physical examples thereof. This will be followed in §3.2–§3.2d with a review

of the global non-Abelian internal symmetries, primarily in the strong interaction sector, as they grew in scope from $SU(2)$ isospin to the quark flavor
model of hadrons and "global color". We next consider (in §3.3) global chiral quark flavor symmetries and their quasi-dynamical implementation — by
means of current algebra — in both the strong and weak interactions. In
§3.4, we continue with the spontaneous breaking of global internal symmetries and the Goldstone theorem, and, then, in §3.5 — §3.5c, discuss several
applications of current algebra, used in conjunction with PCAC, including
the sigma (σ) model of pions and nucleons. Finally, in §3.5d, we discuss
an example of the usefulness of current algebra in deriving high energy sum
rules in the early years.

Resuming with global Abelian internal symmetries, we have already invoked the "Noether" theorem for the single conserved charge Q of an Abelian
Klein-Gordon field, and have mentioned the important property of Q changing sign in going from particle to antiparticle. Our task now is to derive
the "Noether" theorem more explicitly for the Abelian Klein-Gordon field
so that the generalization of the "Noether" theorem to non-Abelian global
internal symmetries is transparent. One way to prove that the conservation of Q follows from the invariance of the theory under the constant phase
transformation $\phi(x) \rightarrow e^{-i\alpha}\phi(x)$ (α is a constant) is to find the unitary
transformation $U(Q)$ that operates on the "second-quantized" Klein-Gordon
field to reproduce the conserved charge and the associated "additive" conservation law. The alternative Lagrangian method — equivalent to the first —
proves directly that the assumed invariance of the field Lagrangian under a
global (constant) phase transformation on the quantum field leads to an "additive" conservation law for charge. Both methods are summarized because
the comparison is instructive and both are capable of easy generalization,
depending on the purpose.

The first (Hamiltonian) method starts with Eq. (2.15):

$$U(\Lambda)\phi(x)U^{-1}(\Lambda) = \phi(\Lambda x) \tag{3.1}$$

and a corresponding equation for $\phi^\dagger(x)$. Making use of the "charge-lowering"
and "charge-raising" relations [see Eq. (2.13)]:

$$[Q, \phi(x)] = -\phi(x); \quad [Q, \phi^\dagger(x)] = \phi^\dagger(x) \tag{3.2}$$

we deduce the defining equation for the unitary transformation $U(Q)$ in terms of $\phi(x)$, to wit:

$$e^{i\alpha Q}\phi(x)e^{-i\alpha Q} = \phi(x) + i\alpha[Q,\phi(x)] + \frac{1}{2}(i\alpha)^2[Q,[Q,\phi(x)]] + \cdots$$

$$= \phi(x)[1 - i\alpha + \frac{(i\alpha)^2}{2!} - \frac{(1\alpha)^3}{3!} + \cdots] = e^{-i\alpha}\phi(x) \quad (3.3)$$

The corresponding equation for $\phi^\dagger(x)$ is:

$$e^{i\alpha Q}\phi^\dagger(x)e^{-i\alpha Q} = e^{i\alpha}\phi^\dagger(x) \quad (3.4)$$

Hence:

$$U(Q) = e^{i\alpha Q} \quad (3.5)$$

From Eqs. (3.3) and (3.4), charge conservation follows since:

$$e^{i\alpha Q}He^{-i\alpha Q} = H \quad (3.6)$$

where H is the Hamiltonian.

In the Lagrangian formulation, we consider the invariance of $\mathcal{L}(\phi,\phi^\dagger;\partial_\mu\phi,\partial_\mu\phi^\dagger)$ under the global phase transformations:

$$\phi(x) \to e^{-i\alpha}\phi(x); \quad \phi^\dagger(x) \to e^{i\alpha}\phi^\dagger(x) \quad (3.7)$$

In this proof, we do not assume the particular "Klein-Gordon" form for $\mathcal{L}$ — nor whether the quantum field is free or interacting — but work with $\mathcal{L}$ as a function of the indicated ϕ, $\phi^\dagger$ and their first derivatives. With a Lagrangian of this form, the equations of motion can be written:

$$\frac{\partial}{\partial x_\mu}\frac{\partial\mathcal{L}}{\delta(\frac{\partial\phi}{\partial x^\mu})} = \frac{\delta\mathcal{L}}{\delta\phi} \quad ; \quad \frac{\partial}{\partial x_\mu}\frac{\delta\mathcal{L}}{\delta(\frac{\partial\phi^\dagger}{\partial x^\mu})} = \frac{\delta\mathcal{L}}{\delta\phi^\dagger} \quad (3.8)$$

The general four-vector current j_μ is:

$$j_\mu = -i\left(\frac{\delta\mathcal{L}}{\delta\left(\frac{\partial\phi}{\partial x_\mu}\right)}\phi - \frac{\delta\mathcal{L}}{\delta\left(\frac{\partial\phi^\dagger}{\partial x_\mu}\right)}\phi^\dagger\right) \quad (3.9)$$

and, using Eq. (3.8), we obtain:

$$\sum_\mu\frac{\partial j_\mu}{\partial x_\mu} = -i\{[\frac{\partial\mathcal{L}}{\partial\phi}\phi + \frac{\partial\mathcal{L}}{\partial\left(\frac{\partial\phi}{\partial x_\mu}\right)}\left(\frac{\partial\phi}{\partial x_\mu}\right)] - [\frac{\partial\mathcal{L}}{\partial\phi^\dagger}\phi^\dagger + \frac{\partial\mathcal{L}}{\partial\left(\frac{\partial\phi^\dagger}{\partial x_\mu}\right)}\left(\frac{\partial\phi^\dagger}{\partial x_\mu}\right)]\}$$

$$(3.10)$$

It is evident that the R.H.S. of Eq. (3.10) represents the variation $\delta\mathcal{L}$ under the infinitesimal global phase transformations corresponding to (3.7), i.e. $\phi \to (1 - i\alpha)\phi$ and $\phi^\dagger \to (1 + i\alpha)\phi^\dagger$. But the invariance of $\mathcal{L}$ under the transformations (3.7) implies $\delta\mathcal{L} = 0$ for arbitrary $\delta\alpha$ and consequently $\partial_\mu j^\mu = 0$. This completes the proof because we can now define the conserved charge $Q = \int d^3x \, j_0$ and hence the choice of $U(Q) = e^{i\alpha Q}$ in Eq. (3.5) is validated. It should be noted that $U(Q) = e^{i\alpha Q}$ defines an (Abelian) $U(1)$ group with the single generator Q and that Q obeys an "additive" conservation law (see §2.1). We discuss below the phenomenological status of the additive conservation laws for the Abelian Q_e, B and L charges.

The procedure for demonstrating the existence of an "additive" charge conservation law as a consequence of the invariance of the field Lagrangian under a global $U(1)$ internal symmetry group, can easily be extended to higher ($J > 0$) spin fields. We have already written down expressions for Q for $J = \frac{1}{2}$ and $J = 1$ quantum fields (see §2.2b) — possessing an Abelian $[U(1)]$ global internal symmetry, and it is straightforward to do the same for $J > 1$ quantum fields as long as commutation relations (C.R.) are employed for even J fields and anti-C.R. for odd J fields. It should be noted that, whereas for bosonic fields, the occupation numbers can take on all non-negative integral values, they can take on only the values 0 or 1 in the case of fermionic fields. In sum, for an Abelian internal symmetry of a (free or interacting) quantum field, the charge conservation law is separately additive for each "polarization", no matter what the spin happens to be. We will see that for non-Abelian internal symmetries, the associated conservation laws are more complicated [e.g. for $SU(2)$, one has $[Q_i, Q_j] = i\epsilon_{ijk}Q_k$ $(i = 1, 2, 3)$ — which defines a "vectorial" ("spin") conservation law] because the "non-Abelian" Q's do not commute with each other.

In the remainder of this section, we briefly examine the status of several known global Abelian conservation laws in particle physics. As particle physics has matured, quite a number of evanescent global Abelian internal symmetries have been proposed; however, we limit our discussion to Q_e (Q_e refers to electric charge), B and L, which continue to play major roles for the quarks and leptons of the standard model. The Q_e and B charges are repetitive for the three quark-lepton generations and their values were listed in Table 1.6 for one generation. Strictly speaking, the L charge is different for each generation, i.e. $L_e \neq L_\mu \neq L_\tau$ (since $\nu_e \neq \nu_\mu \neq \nu_\tau$); the situation is further confused by the possibility that the three neutrinos are finite-mass Majorana particles so that the L charge is undefined (see §6.4b). Nevertheless, we have assumed $L_e = L_\mu = L_\tau = L$ (listed as L in Table 1.6) because statements made about L in our discussion of the standard model (e.g. the special role of global $B - L$ symmetry) are "generation-independent" and it is obvious when the difference matters.

One striking feature of the two global charges Q_e and B is the remarkable degree to which they are conserved and it is worth noting the observed upper limits on such departures. In the case of Q_e conservation, one searches for processes like $e^- \rightarrow \nu + \gamma$. The possibility of the electron decaying by means of this reaction, and thereby violating charge conservation, was explored thirty years ago by looking for the formation of K electron "holes" resulting from the postulated decay of atomic electrons [3.2]. A lower limit of 10^{19} years was placed on the "partial" lifetime for the decay $e^- \rightarrow \nu + \gamma$. One can write down an explicit Q_e symmetry-breaking "effective" interaction, to produce the decay $e^- \rightarrow \nu + \gamma$, of the form:

$$\frac{f_Q}{m_e} \bar{\nu}_L \sigma_{\alpha\beta} e_R F^{\alpha\beta} \tag{3.11}$$

where e_R is the "righthanded" part of the electron wavefunction and ν_L is the lefthanded neutrino wavefunction, $F^{\alpha\beta}$ is the electromagnetic field, $\sigma_{\alpha\beta}$ is the spin, m_e is the electron mass, and f_Q is a dimensionless constant; one obtains an upper limit $f_Q^2 \lesssim 10^{-47}$ from the 10^{19} year lower limit on the lifetime for the indirectly sought $e^- \rightarrow \nu + \gamma$ decay (from the absence of γ rays in a NaI scintillation counter [3.2]). [The choice of the coefficient f_Q/m_e in Eq. (3.11) is somewhat arbitrary and a more "natural" choice is $e m_e/\Lambda^2$ (Λ is an arbitrary mass scale — see §9.3) and, with that choice, $f_Q^2 \leq 10^{-47}$ translates into $\Lambda \geq 10^8$ GeV.] A recent experiment searching for deviations from Q_e charge conservation has raised the lower limit on Λ to 10^{10} GeV [3.3].

The upper limit on the effective coupling constant f_B^2, associated with the departure from B charge conservation, is substantially lower than that on Q charge conservation, thanks to the extensive set of experiments several years ago to test the prediction of $SU(5)$ grand unification theory. According to $SU(5)$ GUT (see §8.3), the decay $p \rightarrow e^+ + \pi^0$ should have a partial lifetime τ_p of the order of 10^{30-31} years. The most accurate experiment was carried out by the IMB group [1.150] — using Cerenkov detectors in a large pool of water located in a deep mine — and, after several years of data collection, the lower limit on τ_p was pushed up to 8×10^{32} years. To obtain an estimate of the upper limit on the departure from B charge conservation, one writes the "effective" baryon charge-violating interaction in the form:

$$f_B \bar{\psi}_e \gamma_5 \psi_P \phi_{\pi^0} \tag{3.12}$$

where ψ_P is the proton wave function, ψ_e is the electron wavefunction, ϕ_{π^0} is the π^0 wave function and f_B is a dimensionless constant. Again, the choice of the constant in front of Eq. (3.12) is somewhat arbitrary but, in any case, $\tau_p > 8 \times 10^{32}$ year translates into $f_B^2 \leq 10^{-62}$ [using the form given in Eq. (3.12)], a value that immediately tells us that, phenomenologically, the conservation law for baryon charge is more "absolute" than the conservation law for electric charge. This is a very interesting situation because, while the standard group makes no provision for the breakdown of Q_e conservation, it does leave room for B violation. It is true that the color gauge part of the standard model, $SU(3)_C$, implies global B conservation [see §5.1]; however, the electroweak part of the standard model, $SU(2)_L \times U(1)_Y$, predicts an instanton-induced breakdown of B conservation. This predicted B violation at the electroweak level is miniscule (see §10.3a) but there is a possibility that the "sphaleron" process compensates for the instanton-induced suppression of baryon charge violation at very high temperatures and energies and this possibility is discussed in §10.3a.

One concluding remark should be made about global $U(1)$ internal symmetries before we move on to global non-Abelian internal symmetry groups. We note that the values of Q_e given in Table 1.6 are "quantized" despite the fact that the global $U(1)$ group allows a continuous range of charges — in contrast to the quantized values predicted by global non-Abelian groups. For example, the three angular momentum generators J_i ($i = 1, 2, 3$) belong to the global non-Abelian $SU(2)$ group and the total angular momentum J (related to the second-order Casimir operator) can only take on integral or half-integral values. But, in general, no such "quantization" effect is required for the Abelian $U(1)$ group. The reasons for the "quantization" of the Abelian electric charges of the quarks and leptons of the standard model are rather subtle, devolving upon the chiral gauge anomaly-free constraints required to ensure the renormalizability and self-consistency of the standard model, and will be discussed in §7.5 and §7.6. It should be noted that these arguments for the quantization of the electric charge can not be used to explain the observed baryon charge and lepton lepton charge quantization because electric charge is a gauge group generator in contrast to the global status of the baryon and lepton charge generators. The quantization of baryon and lepton charge appears to have a different origin; a possible explanation for baryon charge may be that it is a topological quantity contained in the QCD-certified "skyrmion model" of the baryon (see §10.5b) [a similar topological origin of lepton charge is much more speculative (see §6.3d)].

§3.2. Global non-Abelian internal symmetry groups

We now turn our attention to global non-Abelian internal symmetry groups. Since it is difficult to cope with groups having infinite-dimensional representations, it has been a long-standing hope that any identified internal symmetry in particle physics can be described by finite-dimensional representations, i.e. by compact Lie groups, and thus far, this hope has been well fulfilled, certainly at the level of the standard model. It is quite possible that non-compact Lie groups will be needed to achieve a TOE [2.31] (see §1.4c) but, until now, all four-dimensional excursions "beyond the standard model" (e.g. grand unification, preon models, etc.) have taken the conservative path of working within the framework of compact Lie groups. In any case, we focus the next few sections on the compact Lie internal symmetry groups since they suffice to describe the strong and electroweak interactions of the standard model to be discussed in Chapters 4–7. For the same reason, we limit our considerations in this chapter to the classical Lie groups: the special unitary groups $SU(N)$, the orthogonal groups $O(N)$, and the symplectic groups $Sp(N)$. The five exceptional compact Lie groups: G_2, F_4, E_6, E_7 and E_8, will come up for discussion in §3.2d of this chapter, when we consider the uniqueness of the $SU(3)$ color group $[SU(3)_C]$, and in later chapters (Chapters 8–10), when we consider a number of attempts to "go beyond the standard model".

Since, in actual fact, the standard model itself is formulated primarily in terms of $SU(N)$ groups, we begin our discussion in this section with some basic mathematical facts about the general global non-Abelian $SU(N)$ group $(N > 1)$ and utilize this formalism to trace the fruitful development of particle physics as larger [up to $SU(6)$] and more varied (chiral as well as non-chiral, spontaneously broken as well as unbroken) non-Abelian global internal symmetry groups finally led to the recognition of the non-Abelian global internal symmetry groups suitable for gauging in order to achieve the highly successful standard model. The formal definition of $SU(N)$ is that $SU(N)$ is the group of $N \times N$ unitary matrices with unit determinant, i.e. the group element U satisfies the conditions: $U^\dagger U = UU^\dagger = 1$ and $\det U = 1$. Since any unitary matrix U can be written in terms of a Hermitian matrix H: $U = e^{iH}$, it follows from the condition $\det U = 1$ that $Tr\, H = 0$. With $Tr\, H = 0$, it is easily seen that there are $(N^2 - 1)$ traceless Hermitian $N \times N$ matrices $T^a [a = 1, 2, ... (N^2 - 1)]$ in terms of which the general group element

U of $SU(N)$ can be written:

$$U = \exp\{i \sum_{a=1}^{N^2-1} \alpha_a T^a\} \tag{3.13}$$

where the α_a's are the $(N^2 - 1)$ real phase constants (i.e. the group parameters) and the T^a's are the $(N^2 - 1)$ group generators represented by the traceless Hermitian matrices. It can be shown that only $(N - 1)$ of the $(N^2 - 1)$ generators can be simultaneously diagonalized (thereby commuting with each other) which define the rank of the group $SU(N)$. [For example, the $SU(2)$ isospin group is of rank 1 since only τ_3 can be diagonalized; the $U(1)$ group is an exception with regard to rank in that it also possesses one generator, like $SU(2)$.] The $(N^2 - 1)$ generators satisfy the commutation relations:

$$[T_a, T_b] = if_{abc}T_c \tag{3.14}$$

where f_{abc} are the structure constants, which can be taken as real and completely antisymmetric with respect to a, b, and c. The smallest non-singlet irreducible representation is called the fundamental representation of $SU(N)$ and has dimensionality N. The so-called adjoint representation of $SU(N)$ — with dimensionality $(N^2 - 1)$ — is always real, even when the fundamental representation of $SU(N)$ $(N \geq 3)$ is complex; the matrix elements of the adjoint representation can be written in terms of the structure constants:

$$(T_a)_{bc} = -if_{abc}. \tag{3.14a}$$

Some further properties of the class of $SU(N)$ groups are worth enumerating: if N denotes the fundamental representation of the $SU(N)$ group, this N representation can be real, pseudo-real or complex. If N is the fundamental representation of the particles, the conjugate representation $\overline{N}$ is the fundamental representation of the antiparticles so that $\overline{N} = N$ if the fundamental representation is real or pseudo-real whereas $\overline{N} \neq N$ if the fundamental representation is complex. With these definitions, we can say that the fundamental **2** representation of $SU(2)$ is pseudo-real, whereas all fundamental representations of $SU(N)$ $(N \geq 3)$ are complex. The importance of the fundamental representation stems from the fact that all irreducible representations of an $SU(N)$ group (whether complex or real) can be constructed out of the fundamental representation. An $SU(N)$ group is called "vector-like" (or non-chiral) if N and $\overline{N}$ chiral representations are present so that the helicities of the fermions in the N representation are cancelled out by the same helicities of the antifermions in the $\overline{N}$ representation; this is the case

for the $SU(3)$ color group where the quark helicities in the color 3 representation are matched by an equal number of antiquark helicities in the color
$\bar{3}$ representation. The $SU(N)$ group is called chiral if the representations
with one helicity are not matched by the conjugate representations with the
opposite helicity; thus, the $SU(N)_L$ part of the standard electroweak group
is purely chiral because the quarks or leptons in the flavor 2 representation
of $SU(2)_L$ are all lefthanded and there are no righthanded antiquarks or
antileptons in the flavor 2 representation of $SU(2)_L$.

An interesting property of the class of $SU(N)$ groups whose fundamental
representations are complex (i.e. $N \geq 3$) is the existence of a non-trivial
centrum (i.e. other than the identity) which is the set of elements of $SU(N)$
which commutes with all other group elements; the centrum is necessarily
an Abelian group, consisting of the N roots of unity, and is denoted by the
discrete group Z_N. One important consequence of a non-trivial centrum
for an $SU(N)$ group ($N \geq 3$) is that the members of the fundamental representation must carry fractional charges (in units of $1/N$). The $SU(N)$
representations (other than the fundamental) possess fractional or integral
charges depending on whether the so-called N-ality (for $N = 3$, this is the
triality) is or is not a multiple of N. This is why for $SU(3)_F$ — nature's
choice for the quark flavor constituents of baryons — three-quark composites
are sufficient to guarantee integral charge for the baryons. This also explains
why $U(3)_F$ — which is isomorphic to $SU(3)_F \times U(1)/Z_3$ — only admits
representations with zero triality and, *a fortiori*, integral charges [3.4].

A crucial property of a global non-Abelian internal symmetry group is
that there is more than one generator — in contrast to the Abelian group —
and hence more than one current and associated charge. Since we continue to
focus on the class of $SU(N)$ groups, we can say that the number of conserved
charges Q^a increases from $a = 1$ [for the Abelian $U(1)$ group] to $a = (N^2 - 1)$
for the non-Abelian $SU(N)$ ($N \geq 2$) group]. Assuming that we have an
ensemble of N fields, denoted by $\phi = \{\phi_i\}$ ($i = 1, 2, \ldots N$) belonging to the
fundamental N representation of $SU(N)$, the global $SU(N)$ transformation
on ϕ can be written in the form:

$$\phi(x) \rightarrow e^{i\alpha_a T^a} \phi(x) \qquad (3.15)$$

where α_a [$a = 1, 2, \ldots(N^2 - 1)$] are the $(N^2 - 1)$ constant phases and the T^a
generators constitute the set of $N \times N$ matrices satisfying the Lie algebra of
the $SU(N)$ group [see Eq. (3.14)]. To derive expressions for the $(N^2 - 1)$
charges, it is convenient to use the Lagrangian approach (rather than the

Hamiltonian) to obtain the generalization of Eq. (3.9). The infinitesimal variation in the Lagrangian — due to infinitesimal changes in the α_a's — now yields:

$$\delta\mathcal{L} = \sum_i \{\frac{\delta\mathcal{L}}{\delta\phi_i}\delta\phi_i + \frac{\delta\mathcal{L}}{\delta(\partial_\mu\phi_i)}\delta(\partial_\mu\phi_i)\} \tag{3.16}$$

subject to the infinitesimal version of Eq. (3.15):

$$\delta\phi_i(x) = i\alpha_a T_{ij}^a \phi_j(x) \tag{3.16a}$$

Using the Lagrangian equations of motion:

$$\frac{\delta\mathcal{L}}{\delta\phi_i} = \partial_\mu \frac{\delta\mathcal{L}}{\delta(\partial_\mu\phi_i)} \tag{3.16b}$$

we get:

$$\delta\mathcal{L} = i\alpha_a \sum_i \partial^\mu j_\mu^{ai} \tag{3.17}$$

where:

$$j_\mu^{ai} = \frac{\delta\mathcal{L}}{\delta(\partial^\mu\phi_i)} T_{ij}^a \phi_j \tag{3.17a}$$

The assumption of the invariance of the Lagrangian under the transformation (3.15) implies $\delta\mathcal{L} = 0$ for arbitrary $\delta\alpha_a = 0$ (for all a) and this leads to a divergenceless equation for the $(N^2 - 1)$ currents j_μ^a, i.e. $\partial^\mu j_\mu^a = 0$ $[a = 1, 2, ...(N^2 - 1)]$ where $j_\mu^a = \sum_i j^{ai}$. Consequently, the $(N^2 - 1)$ conserved charges Q^a become:

$$Q^a = \int d^3x \, j_0^a(x) \qquad [a = 1, 2, ...(N^2 - 1)] \tag{3.18}$$

We emphasize that the Q^a's are non-commuting charges corresponding to the global non-Abelian internal symmetry group $SU(N)$, satisfy the Lie algebra (3.14) for $SU(N)$, and can be considered the generators of the non-Abelian group $SU(N)$.

The fact that the $(N^2 - 1)$ Q^a's are conserved charges implies, of course, that all the Q^a's commute with the Hamiltonian for the field, i.e. $[H, Q^a] = 0$. It follows that any element $U = e^{i\alpha_a Q^a}$ $[a = 1, 2, ...(N^2 - 1)]$ also commutes with H. But, by definition, U connects states that form an irreducible representation (basis) of the $SU(N)$ group, i.e.

$$U|A\rangle = |B\rangle. \tag{3.19}$$

where $|A\rangle$ and $|B\rangle$ belong to the same irreducible representation of the group. Using an obvious notation for the energy of the state, it is evident that:

$$E_A = \langle A|H_0|A\rangle = \langle B|H_0|B\rangle = E_B. \tag{3.20}$$

Hence, the invariance of H under $SU(N)$ yields the degeneracy of the energy (or mass) eigenstates contained in the irreducible representations of $SU(N)$. These mass degeneracies [one mass multiplet corresponding to each irreducible representation of $SU(N)$] are, however, only present when the ground state of the field is invariant under the global internal symmetry transformation, i.e. $U|0\,>=\,|0\,>$. When the ground state (vacuum) is not invariant under the global symmetry group U — but the Lagrangian (or Hamiltonian) for the field stays invariant — the régime of spontaneous symmetry breaking (SSB) prevails. The consequences of the SSB of global internal symmetries are discussed in §3.4.

We now turn to the physical applications of global non-Abelian internal symmetry $SU(N)$ groups in the strong interaction domain. The $SU(N)$ groups: $SU(2)_I$, $SU(2)_I \times U(1)_Y$ and $SU(3)_F$ global internal symmetry groups and the mixed global $SU(6)$ quark flavor-spin group played prominent roles in the conceptual evolution of QCD. We review in §3.2a–3.2d the rationale for these four groups, starting with the $SU(2)$ isospin group proposed during the early 1950s.

§3.2a. $SU(2)$ isospin group and the pion-nucleon interaction

In particle physics, the $SU(2)$ isospin group became the first global internal symmetry group that made excellent predictions [e.g. "charge independence" of the pion-nucleon and two-nucleon interactions (see §1.2b)]. Global $SU(2)$ is also mathematically the simplest non-Abelian group to consider because its representations are real or pseudo-real, in addition to having the smallest number of dimensions. Since global $SU(2)$ is the group of 2×2 unitary complex matrices with unit determinant, an element of $SU(2)$ can be written as:

$$U = \exp\{i \sum_{a=1}^{3} \alpha_a \tau_a\} \qquad (a = 1, 2, 3) \tag{3.21}$$

where the $\alpha_a's$ are real parameters and the τ_a's are the three (isospin) generators belonging to the (adjoint) **3** representation of $SU(2)$. Since the τ_a's are traceless 2×2 Hermitian matrices, they can be given the Pauli form:

$$\tau_1 = \begin{pmatrix} 0 & 1 \\ 1 & 0 \end{pmatrix}, \tau_2 = \begin{pmatrix} 0 & -i \\ i & 0 \end{pmatrix}, \tau_3 = \begin{pmatrix} 1 & 0 \\ 0 & -1 \end{pmatrix} \tag{3.22}$$

satisfying the C.R.:

$$[\tau_a, \tau_b] = i\epsilon_{abc}\tau_c \tag{3.23}$$

where ϵ_{abc} is the totally antisymmetric Levi-Civita symbol (with $\epsilon_{123} = 1$); the ϵ_{abc} are clearly the structure constants of the $SU(2)$ group. The Lie algebra of $SU(2)$ is governed by Eq. (3.23) and all representations of the generators must satisfy this set of C.R. The $SU(2)$ group is clearly of rank 1 because only one of the three generators (τ_3) can be diagonalized.

The $SU(2)_I$ (isospin) group was originally invented to explain the "charge independence" of the two-nucleon interaction [3.5]. An internal symmetry — isospin — was postulated because the two electric charge (flavor) states of the nucleon N (p and n) differ so little in mass, and the "two-flavor" nucleon N was placed in the fundamental 2 representation of $SU(2)_I$ ($I = \frac{1}{2}$). If we write $\Psi_N = \binom{\psi_p}{\psi_n}$, where ψ_p and ψ_n are the separate Dirac spinors for p and n respectively, then the three isospin "charges" of the nucleon are (we now have three charges instead of the one charge of the Abelian group):

$$I_a^{(N)} = \frac{1}{2} \int d^3x \ \Psi_N^\dagger \tau_a \Psi_N \qquad (a = 1, 2, 3) \qquad (3.24)$$

The discovery of the strongly interacting pion with its three electric charge states, again with small mass separations, led to the assignment of the 3 representation of $SU(2)_I$ ($I = 1$) to the pion; if we write the complex (charged) pion field in terms of real Hermitian (Klein-Gordon) fields, thus: $\phi = 1/\sqrt{2}(\phi_1 + i\phi_2)$ [so that $\phi^\dagger = 1/\sqrt{2}(\phi_1 - i\phi_2)$], and the neutral real (Hermitian) pion field as ϕ_3, we can denote the isotriplet pion field (π^+, π^-, π^0) by $\Phi(x) = [\phi_1(x), \phi_2(x), \phi_3(x)]$ and its canonically conjugate isotriplet field momentum by $\Pi(x) = [\pi_1(x), \pi_2(x), \pi_3(x)]$. The three isospin "charges" of the pion then become:

$$I_a^\pi = \int d^3x \ \Phi^\dagger T_a \Pi(x) \qquad (a = 1, 2, 3) \qquad (3.25)$$

where the T_a's are the three isospin generators in the adjoint representation. Using the special "adjoint" property of T_a, i.e. $(T_a)_{bc} = -i\epsilon_{abc}$, [see Eq. (3.14a)], Eq. (3.25) reduces to the familiar "iso-orbital angular momentum" form:

$$I_a^{(\pi)} = i\epsilon_{abc} \int d^3x \ \pi_b(x)\phi_c(x) \qquad (a = 1, 2, 3) \qquad (3.25a)$$

The next step is to define the total isospin I in terms of the second-order Casimir operator for $SU(2)_I$ and one obtains the correct $I(I+1)$ eigenvalues for the total isospins of N and π respectively. The nucleon and pion states can then be specified by the eigenvalues I and I_3 with the (electric) charge $Q = I_3 + B/2$ (where B is the baryon charge). Evidently, there are $(2I_3 + 1)$ electric charge states for a given I but the physical properties of the pion-nucleon system (and, *a fortiori*, the two-nucleon system) do not depend on I_3

("charge independence") if the pion-nucleon interaction is isospin invariant, as is assumed to be the case for the strong interaction.

The hypothesis of global $SU(2)_I$ isospin invariance of the pion-nucleon interaction gives rise to many predictions: the ratios of the pion-nucleon elastic scattering processes for different values of I_3 for π and N, the isospin scalar ("charge-independent") structure of the two-nucleon interaction, the cross-section ratios for pion-nucleus and nucleon-nucleus reactions, etc. These predictions were tested for a great variety of strong interaction processes during the first half of the 1950s and the usefulness of global $SU(2)$ isospin as a "classification" (non-dynamical) group for the pion-nucleon and nucleon-nucleon systems was demonstrated without exception. By 1954, the Yang-Mills paper [1.51] showed how to gauge global non-Abelian internal symmetry groups [in particular, the $SU(2)$ group] in order to achieve a dynamical theory of the strong interactions. However, the difficulty of using the gauged $SU(2)$ group to explain the strong interactions (e.g. the short-range character of the two-nucleon interaction) seemed insurmountable at that time (see §1.2e), and attention shifted to the search for larger global non-Abelian internal symmetry groups governing the strong interactions or in the direction of dispersion theory and its many ramifications (see §1.2b). [QED was regarded as much more suitable for the gauging of a global internal symmetry (i.e. electric charge) in view of its long-range Coulomb potential and weak coupling; it did not appear, in the mid-1950s, that the phenomenologically impressive dynamical consequences of gauged Abelian QED could easily be translated into a comparable success for the strong interaction.] Fortunately, the "enlargement of global internal symmetry" — approach to the strong interaction ultimately did pay off when the quark flavor model of hadrons led to global "color" after a series of intermediate steps.

§3.2b. $SU(2)_I \times U(1)_Y$ group and strange particles

The success of isospin invariance soon encouraged the enlargement of the $SU(2)_I$ group — when the strange particles were discovered — to what we choose to call the "strong" Gell-Mann-Nakano-Nishijima (GNN) group, i.e. the global non-Abelian internal symmetry group $SU(2)_I \times U(1)_Y$ [1.60] [we discussed the distinction between the "strong" and "weak" GNN groups in §1.2d and come back to it in §6.2]; the hypercharge Y (rather than the strangeness S) Abelian quantum number is used because it allows a unified group-theoretic treatment of mesons and baryons (see below). The observation of "associated" production of strange mesons and baryons, together with the particular patterns observed, led to the introduction of the additional Abelian $U(1)_Y$ symmetry group, which operated in a "product" relationship to $SU(2)_I$, i.e. as the factor group $SU(2)_I \times U(1)_Y$, with the electric charge Q of any meson or baryon related to I_3 and Y through the "strong" GNN relation: $Q = I_3 + Y/2$. All strange particle experiments were consistent with the conservation of isospin and hypercharge under the "strong" GNN group. Information concerning masses, spins and parities of strange mesons and baryons accumulated rapidly during the late 1950s and presently it was possible to discern some major groupings in terms of these three parameters. We have already listed in Table 1.2 the values of the masses, spins and parities of the two lightest families of mesons and baryons, as well as the quantum numbers of the particles under the groups $SU(2)_I$ and $U(1)_Y$. The important point to note is that the $J = 0^-$ and $J = 1/2^+$ families are both families of eight particles (octets) with identical I and Y quantum numbers under $SU(2)_I \times U(1)_Y$ and happen to be the lightest families of mesons and baryons respectively. These two octet families, as well as the other two families listed in Table 1.2, point to the approximate validity of a larger global flavor group $SU(3)_F$ that contains the "strong" GNN group as a subgroup and, as we will see, constitute the basic empirical data that first suggest and then test the quark flavor model of hadrons.

§3.2c. $SU(3)_F$ group and quark flavor model of hadrons

With hindsight, the determination that the enlargement of the "strong" GNN group should be $SU(3)_F$ came in two steps. The first useful step was taken by Sakata in 1955 [1.61] when the baryon triplet (p, n, Λ)–a combination of $(I = 1/2, Y = 1)$ and $(I = 0, Y = 0)$ representations under $SU(2)_I \times U(1)_Y$ — was used to construct all known hadrons as composites. [The precursor of the Sakata model was the Fermi-Yang [3.6] model in which the newly-discovered pions were regarded as two-fermion $(N - \overline{N})$ composites.] The (relatively) small mass difference between N and Λ then made it attractive to neglect this mass difference [just as the $(n - p)$ mass difference is neglected to construct the two-flavor $SU(2)_I$ group] and to hypothesize a three-flavor $SU(3)_F$ group with the baryon triplet $\mathcal{B} = (p, n, \Lambda)$ as its fundamental $\mathbf{3}$ representation [1.62]. With $\mathcal{B}$ as the fundamental representation of $SU(3)_F$, simple group-theoretic manipulation shows that the Sakata model correctly predicts the meson octet [since $\mathbf{3} \times \overline{\mathbf{3}} = \mathbf{1} + \mathbf{8}$], whereas it fails dismally for the baryon octet since $\mathbf{3} \times \mathbf{3} \times \overline{\mathbf{3}} = \mathbf{3} + \mathbf{3} + \overline{\mathbf{6}} + \mathbf{15}$ [baryon charge conservation requires one of the $\mathbf{3}$ representations to be the conjugate representation $\bar{3}$] and does not contain an irreducible $SU(3)$ octet representation among its composites.

The imperatives of the "eightfold way" for baryons [1.63] soon brought Gell-Mann [1.96] and Zweig [1.96] to the realization that the Sakata triplet (p, n, Λ) had to be replaced by a new fundamental triplet of fermionic constituent (called quarks) with the same isospin and hypercharge quantum numbers as the Sakata triplet but allowing for fractional baryon charge of $\frac{1}{3}$ — so that the three-quark (baryonic) composite $(qqq) = \mathbf{3} \times \mathbf{3} \times \mathbf{3} = \mathbf{1} + \mathbf{8} + \mathbf{8} + \mathbf{10}$ contains a baryon octet as one of its decomposites. Because the "strong" GNN group is a subgroup of $SU(3)_F$, the "strong" GNN relation $Q = I_3 + Y/2$ still holds and the fractional baryon charge translates into fractional hypercharges and fractional electric charges. It was natural to think of the fundamental quark triplet as consisting of the three lightest quarks (u, d, s) with the values of Y and Q listed in Table 1.1. In more mathematical language, the essential reason for the fractional charges, Y and Q, of the quark triplet is that the $SU(3)$ group possess a non-trivial centrum, Z_3, so that the members of the fundamental $\mathbf{3}$ representation (i.e. u, d, s quarks) have fractional charges in multiples of $\frac{1}{3}$. As was pointed out in §3.2, the presence of the Z_3 centrum in the $SU(3)$ group gives rise to a new quantum number — the "triality" t, defined by the condition $t = 0, 1, 2$ mod 3 — which can be assigned to each irreducible representation of $SU(3)$. It is easy to work out the t value for each representation, starting with $t = 1$ for the fundamental $\mathbf{3}$ representation, $t = 2$ for the $\mathbf{6}$ representation, and so on for the higher representations with non-vanishing t; non-vanishing t implies fractional charges for each member of the representation. On the other hand, beginning with the trivial singlet $(\mathbf{1})$ representation with $t = 0$, the next $t = 0$ representation is $\mathbf{8}$ [whose $t = 3$ (mod 3)= 0], and so on, for the higher $(\mathbf{10}, \mathbf{27}$, etc.) $t = 0$ representations; all $t = 0$ representations carry integral charges. In a word, the $SU(3)$ group, while not permitting the assignment of integral charges to its fundamental representation (as was done in the Sakata model), does allow integral charges for three-quark composites belonging to the $t = 0$ representations.

Thus was the global $SU(3)_F$ quark flavor model of hadrons born, wherein the meson composites were assumed to consist of $q\bar{q}$ (the same group-theoretic structure as the $\mathcal{B}\bar{\mathcal{B}}$ of the Sakata model) and the baryonic composites of qqq (very different from the $\mathcal{B}\mathcal{B}\bar{\mathcal{B}}$ group-theoretic structure of the Sakata model). The first formulation of the quark flavor model of hadrons was couched in cautious terms — with the unseen fractionally charged quarks being thought of as "mathematical fictions" and not as physical particles. It is interesting to quote from a historical lecture of Gell-Mann in 1984 [1.59] on his perception of the role of quarks in hadronic physics at the time the quark model was invented:

> "it occurred to me that if the bootstrap approach were correct, then any fundamental hadrons would have to be unobservable, incapable of coming out of the baryons and mesons to be seen individually, and that if they were unobservable, they might as well have fractional charge...I referred to these trapped fundamental entities as "mathematical quarks" — by which I always meant that they were permanently stuck inside the baryons and mesons..."

It took a decade before the "mathematical" view was superceded by the "physical" view and this only happened after the experiments on the deep inelastic scattering of leptons by hadrons established the physical reality of the fractionally charged quarks (see §1.3a and §5.2c). Mathematical fiction or physical entity, the predictions of the $SU(3)$ quark flavor model were increasingly impressive and soon became the decisive instrument for uncovering the new global "color" degree of freedom of the quarks. This fascinating next step in the genesis of QCD was a major triumph for the usefulness of global non-Abelian internal symmetries and is discussed in §3.2d.

Before continuing our review of the role played by non-Abelian global internal symmetries in the conceptual evolution of the quark flavor model of hadrons to "global color" [and, finally, to gauged $SU(3)$ color], we add some mathematical details concerning the pivotal $SU(3)$ group. The eight generators T_a ($a = 1, 2, ...8$) satisfy the C.R. (3.14) where the structure constants are listed in Table 3.1. For purposes of reference, we also record the explicit 3×3 matrices that represent the eight generators of $SU(3)$ [with the usual notation $\lambda_a/2 = T_a$ ($a = 1, 2, ...8$)], namely:

$$\lambda_1 = \begin{pmatrix} 0 & 1 & 0 \\ 1 & 0 & 0 \\ 0 & 0 & 0 \end{pmatrix} ; \quad \lambda_2 = \begin{pmatrix} 0 & -i & 0 \\ i & 0 & 0 \\ 0 & 0 & 0 \end{pmatrix} ; \quad \lambda_3 = \begin{pmatrix} 1 & 0 & 0 \\ 0 & -1 & 0 \\ 0 & 0 & 0 \end{pmatrix}$$

$$\lambda_4 = \begin{pmatrix} 0 & 0 & 1 \\ 0 & 0 & 0 \\ 1 & 0 & 0 \end{pmatrix} ; \quad \lambda_5 = \begin{pmatrix} 0 & 0 & -i \\ 0 & 0 & 0 \\ i & 0 & 0 \end{pmatrix} ; \quad \lambda_6 = \begin{pmatrix} 0 & 0 & 0 \\ 0 & 0 & 1 \\ 0 & 1 & 0 \end{pmatrix}$$

$$\lambda_7 = \begin{pmatrix} 0 & 0 & 0 \\ 0 & 0 & -i \\ 0 & i & 0 \end{pmatrix} ; \quad \lambda_8 = \begin{pmatrix} \frac{1}{\sqrt{3}} & 0 & 0 \\ 0 & \frac{1}{\sqrt{3}} & 0 \\ 0 & 0 & -\frac{2}{\sqrt{3}} \end{pmatrix} \qquad (3.26)$$

We note that the two generators λ_3 and λ_8 are diagonal and are proportional to I_3 and Y respectively; this is consistent with the expectation that the rank of $SU(3)$ is 2. The eight generators $\lambda_a (a = 1, 2, ...8)$ have the normalization Tr $(\lambda_a \lambda_b) = 2\,\delta_{ab}$ and satisfy the C.R. $[\lambda_a, \lambda_b] = 2i f_{abc} \lambda_c$, as they must.

Table 3.1: Antisymmetric structure constants f_{abc} of $SU(3)$.

a	1	1	1	2	2	3	3	4	6
b	2	4	5	4	5	4	6	5	7
c	3	7	6	6	7	5	7	8	8
f_{abc}	1	$\frac{1}{2}$	$-\frac{1}{2}$	$\frac{1}{2}$	$\frac{1}{2}$	$\frac{1}{2}$	$-\frac{1}{2}$	$\sqrt{3}/2$	$\sqrt{3}/2$

With the necessary formalism for the $SU(3)$ group in hand and the assignment of quark quantum numbers as listed in Table 1.5, it is easy to show that all mesons and baryons listed in Table 1.2 can be constructed out of $q\bar{q}$ "valence" pairs and qqq "valence" triplets respectively, to generate the two lightest $SU(3)_F$ mass multiplets of mesons and baryons. Insofar as the spins and parities of these four low-lying (in mass) hadron multiplets are concerned, it is seen from Table 1.2 that, on the assumption of zero orbital angular momentum for both the $q\bar{q}$ and qqq systems, the lowest mass $J = 0^-$ $SU(3)_F$ meson octet corresponds to $q\bar{q}$ with antiparallel spins (the negative parity of the mesons result from the intrinsic negative parity of $q\bar{q}$ — see §2.2a), whereas the next lowest $J = 1^-$ meson "nonet" corresponds to $q\bar{q}$ with parallel spins. Similarly, the lowest mass $J = 1/2^+$ $SU(3)_F$ baryon octet corresponds to qqq having two parallel and one anti-parallel spin [the mixed (M) symmetry state of the $SU(2)_{\text{SPIN}}$ group] whereas the next lowest $J = \frac{3}{2}^+$ baryon decuplet has all three spins parallel [the completely symmetric (S) state of the $SU(2)_{\text{SPIN}}$ group]. Thus, Table 1.2 confirms the expectation that the spins of the two lowest-lying mass $SU(3)_F$ multiplets are $J = 0^-$ (pseudoscalar) and $J = 1^-$ (vector) for the mesons and $J = 1/2^+$ and $J = 3/2^+$ for the baryons.

It is clear why the particle groupings listed in Table 1.2 represent particle multiplets with the same values of spin since there is no mixing of the Poincaré group with $SU(3)_F$. The mixing of the space inversion-operation and a global internal symmetry group like $SU(3)_F$ is possible [3.1] but it is not realized in nature [3.7] and provides the chief argument for the SSB of the global chiral quark flavor group $SU(3)_L \times SU(3)_R$ (see §5.1). What is not so clear is the justification for particle groupings within $SU(3)_F$ which

— if $SU(3)_F$ is-an exact symmetry — predicts mass degeneracy within each irreducible representation of $SU(3)_F$. It is true that the sub-groupings of particles with identical quantum numbers under $SU(2)_I \times U(1)_Y$ [a subgroup of $SU(3)_F$] are clearcut because the variations in mass within the subgroup are only a few percent at most. However, the variation in mass within the presumed $SU(3)$ multiplets is substantial — anywhere from 30%–50% for the baryon multiplets to more than a factor 3 for the $J = 0^-$ meson octet (see Table 1.2) — and some compelling reasons should be forthcoming to justify the assignments to the $SU(3)_F$ multiplets. The reasons are twofold: the first is that if one ascribes the breakdown $SU(3)_F \rightarrow SU(2)_I \times U(1)_Y$ to an explicit ("hard") symmetry-breaking term proportional to the eighth $SU(3)$ generator λ_8 [which is essentially the hypercharge (in analogy to the explicit symmetry-breaking of the $SU(2)_I$ group by electromagnetism via the third isospin generator τ_3], the decomposition of the octet and decuplet representations of $SU(3)_F$ yields:

$$8 = (3,0) + (2,1) + (2,-1) + (1,0); \quad 10 = (4,1) + (3,0) + (2,-1) + (1,-2)$$
$$(3.27)$$

where (m,n) are the quantum numbers under $SU(2) \times U(1)$ required by group theory. But these are precisely the $SU(2)_I$ and $U(1)_Y$ quantum numbers belonging to the $SU(3)_F$ multiplets listed in Table 1.2. One caveat should be mentioned: one of the mass groupings in Table 1.2, that of the vector mesons, contains nine members: the $SU(3)$ vector octet (consisting of three ρ's, four K^*'s and one ω) plus one $SU(3)_F$ vector singlet (consisting of ϕ). It will be seen below that the Gell-Mann-Okubo relation [1.97] is not satisfied by the vector meson octet but a closer study [3.8] reveals "maximal" mixing between the ω and ϕ mesons so that it is possible to regard the combination of octet and singlet as an $SU(3)_F$ "nonet".

The second test of the $SU(3)_F$ assignments follows from the first and is more exacting, namely the "hard" breaking of $SU(3)_F$ by λ_8 leads to a well-defined mass relation among the members of each irreducible $SU(3)_F$ multiplet, i.e. the Gell-Mann-Okubo formula [1.97]; it should be noted that for the baryons, the mass relation is among the m's of the baryons [to simulate the mass term in the Lagrangian for fermions], whereas the Gell-Mann-Okubo formula is expected to hold among the m^2's of the mesons [to simulate the m^2 mass term in the Lagrangian for bosons]. We can therefore write:

$$m(I,Y) = k_0 + k_1 Y + k_2[I(I+1) - \frac{Y^2}{4}] \qquad (3.28)$$

where $m(I, Y)$ is the mass of a baryonic hadron with I and Y quantum numbers and $m(I, Y)$ is to be replaced by $m^2(I, Y)$ for a mesonic hadron with I and Y quantum numbers; the quantities k_0, k_1 and k_2 are three arbitrary constants. For an (irreducible) $SU(3)_F$ octet representation, the Gell-Mann-Okubo relation is:

$$\frac{m(\frac{1}{2}, 1) + m(\frac{1}{2}, -1)}{2} = \frac{3m(0, 0) + m(1, 0)}{4} \qquad \text{(for baryons)}$$

$$m(I, Y) \to m^2(I, Y) \qquad \text{(for mesons)} \qquad (3.29a)$$

For an (irreducible) $SU(3)_F$ decuplet (10) representation, where the further relation $I = Y/2 + 1$ holds, Eq. (3.28) reduces to the "equal spacing" formula for baryons:

$$m(0, -2) - m(\frac{1}{2}, -1) = m(\frac{1}{2}, -1) - m(1, 0) = m(1, 0) - m(\frac{3}{2}, 1) \quad (3.29b)$$

These predictions are all in excellent agreement with the experimental mass values listed in Table 1.2 for the $J = \frac{1}{2}^+$ baryon octet and $J = \frac{3}{2}^+$ baryon decuplet, as well as for the pseudoscalar meson octet. Insofar as the vector meson "nonet" is concerned, the Gell-Mann-Okubo prediction for the m^2's works well if the "maximal mixing angle" (i.e. $\sin\theta = \sqrt{\frac{2}{3}}$) between the ω and ϕ mesons is utilized; in quark language, the quark-antiquark composites comprising ω and ϕ are $(u\bar{u} + d\bar{d})/\sqrt{2}$ and $s\bar{s}$ respectively. The mixing angle required to understand the mass relation among the members of the vector meson "nonet" is corroborated by the observed decay products and decay widths of the ω and ϕ mesons [3.8].

The evidence that we have cited for the applicability of the global non-Abelian $SU(3)_F$ quark flavor symmetry group to a description of the multi-quark hadron composites was quite convincing in the mid-1960s and the search for even larger global non-Abelian internal symmetry groups to describe the strong interactions received new impetus. Progress was achieved on two fronts: in one direction, an attempt was made to understand why the $SU(3)_F$ quark flavor model predicts $J = \frac{1}{2}^+$ octet and $J = \frac{3}{2}^+$ decuplet states for the two low-lying baryon multiplets and not the obverse (i.e. $J = \frac{1}{2}^+$ decuplet and $J = \frac{3}{2}^+$ octet of baryons); in the other direction, an effort was made to exploit the approximate global chiral quark flavor symmetry $SU(3)_L \times SU(3)_R$ — associated with the three lightest (u, d, s) quarks — for the purpose of making quasi-dynamical calculations in the realm of weak and strong interactions on the basis of the C.R. postulated to hold for the chiral quark currents (i.e. current algebra). We propose to consider the first achievement — which led to the non-relativistic mixed global

$SU(6)$ quark flavor-spin group — in the next section (§3.2d), and to discuss accomplishments of the current algebra program in subsequent sections.

§3.2d. Mixed global $SU(6)$ quark flavor-spin group and "global color"

The global $SU(6)$ quark flavor-spin model was invented to explain why the two lowest-lying baryon mass groupings consist of a $J = \frac{1}{2}^+$ octet and a $J = \frac{3}{2}^+$ decuplet and not the obverse, nor some other combination of spin and $SU(3)_F$ quantum numbers. It is true that the "no-go" theorem (see §4.1) forbids a true mixing of the Poincaré group with a global internal symmetry group but it was hypothesized that combining the purely (static) "spatial rotation" $SU(2)$ subgroup of the Poincaré group — denoted by $SU(2)_{\text{SPIN}}$ — with global $SU(3)_F$ into a simple (mixed) $SU(6)$ quark flavor-spin group might yield more restrictive and correct predictions for the decomposite $[SU(3)_F, SU(2)_{\text{SPIN}}]$ representations of the two lowest meson and two lowest baryon groupings in mass and spin [1.100]. The idea was to place the three lightest (u, d, s) quark flavors with "up" and "down" spins [i.e. $(u^\uparrow, u^\downarrow, d^\uparrow, d^\downarrow, s^\uparrow, s^\downarrow)$] in the fundamental 6 representation and to inquire into the decomposition into $SU(6)$ irreducible representations of the three-quark (qqq) composite (baryon) representations $6 \times 6 \times 6$; one finds [see Eqs. (1.5)] that the purely symmetric irreducible 56 representation of $SU(6)$ decomposes — under the subgroup $SU(3)_F \times SU(2)_{\text{SPIN}}$ — into precisely the observed $(8, 2)$ $(J = \frac{1}{2}^+$ octet) plus $(10, 4)$ $(J = \frac{3}{2}^+$ decuplet) lowest-lying (three-quark composite) baryon representations. It is interesting to note that the two lowest-lying $[SU(3)_F, SU(2)_{\text{SPIN}}]$ representations for the mesons also receive a plausible explanation on the basis of the $SU(6)$ model; i.e. the two irreducible representations of $SU(6)$ resulting from $6 \times \bar{6}$ are 1 and 35; 35, in turn, decomposes into: $(8, 1) + (9, 3)$, i.e. a $J = 0^-$ octet plus a $J = 1^-$ "nonet" under $[SU(3)_F, SU(2)_{\text{SPIN}}]$. This agrees with Table 1.2 for the lowest-mass meson representations if the strong $\omega - \phi$ mixing is taken into account.

The successful application of the mixed global $SU(6)$ quark flavor-spin group to the two lowest-lying meson and baryon multiplets, can be understood. One is dealing here with $q\bar{q}$ and qqq composite hadrons with zero orbital angular momentum; when the orbital angular momentum within the quark composites increases, one expects the masses to increase and the static $SU(6)$ model to become a poor approximation. Indeed, one can show that the commutators of the generators corresponding to the total angular momentum mixed with $SU(3)_F$ do not close so that one must confront the complicated mathematics of an infinite-dimensional Lie algebra and a non-local interaction [3.9].

Sticking to zero orbital angular momentum, it is gratifying that the symmetric 56 representation of the combined quark flavor-spin $SU(6)$ group neatly explains the uniqueness of the observed lowest-lying $[SU(3)_F,$ $SU(2)_{\mathrm{SPIN}}]$ baryonic multiplets. But now a new puzzle — that of consistency with the Pauli principle for fermionic baryons — must be solved. This dilemma results from the fact that the total wavefunction — for space, spin and flavor — for the (qqq) system with zero orbital angular momentum (out of which the two lowest-lying fermionic baryons are constructed) is symmetric [the zero (orbital angular momentum) space contribution is symmetric, as is the spin-flavor contribution from the 56 representation of $SU(6)$], contrary to the Pauli principle. A good way to save the Pauli principle — supported by prior successes in particle physics [e.g. when isospin was added (to space and spin) as a new degree of freedom for the nucleon] — is to add a new internal degree of freedom to quarks, which we call "global color"; as usual, this new degree of freedom is introduced via a global internal symmetry group G_C (where C stands for "color") which, when adjoined to $SU(6)$, i.e. $G_C \times SU(6)$, must give rise to a totally anti-symmetric wavefunction for the composite three-quark baryonic system. Moreover, since "global color" appears to be a "hidden" symmetry, it is necessary to assume that, whatever be the irreducible representation R_C of G_C corresponding to a quark, the representation for the three-quark baryonic system must contain a "color" singlet.

It is instructive to exhibit the extent to which the global group G_C is fixed by the two requirements of agreement with the Pauli principle and undetectability of "color" in the observed hadrons. Since the 56 representation of $SU(6)$ contributes a symmetric part to the total wavefunction for the three-quark system, the Pauli principle is obeyed if G_C contributes an anti-symmetric part. We add two technical conditions on G_C that are eminently plausible: (1) G_C is a compact Lie group; and (2) R_C is a fundamental representation of G_C, a property that seems to hold for all internal symmetry groups of fundamental constituents in nature. The requirement of a "color" neutral three-quark system requires that $R_C \times R_C \times R_C \supset 1$ and there are only five simple groups G_C that satisfy this condition; they are listed in Table 3.2, together with the symmetries these groups contribute to the "color" singlet wavefunction of the qqq system. From Table 3.2, it follows that the

E_6 and F_4 groups are eliminated because they give symmetric contributions to the "color" singlet wavefunction of the three-quark system. The remaining groups: $SU(3)$, G_2 and E_8, are candidates for the "global color" group G_C, with $R_C[SU(3)]$ being a complex (fundamental) representation and the R_C's of G_2 and E_8 being real (fundamental) representations. However, it is possible to rule out the "real" G_2 and E_8 groups because they are "centrum-free" (see §3.2) with the consequence that there exists for these two groups a generalized G conjugation invariance [3.10] [the generalized G conjugation — denoted by $\bar{\mathcal{G}}$ — is defined in terms of the usual G conjugation [3.11] — denoted by $\mathcal{G}$ — by means of $\bar{\mathcal{G}} = \Theta^{-1}\mathcal{G}\Theta$, where Θ is an inner automorphism of the "color" group G_C and is called by Okubo "color G parity" [3.10]], that leads to a violation of the OZI (quark-line) rule [3.12] (see §5.4c). This leaves the $SU(3)$ group with its (non-trivial) Z_3 centrum as the only group satisfying the two technical conditions above and agreeing with the OZI rule despite invariance under $\bar{\mathcal{G}}$.

Table 3.2: Lie groups that give $R_c \times R_c \times R_c \supset 1$.

Group	R_C (fundamental representation)	G_C is real or complex	Symmetry for singlet wave function of qqq system
$SU(3)$	3	complex	A
E_6	27	complex	S
G_2	7	real	A
F_4	26	real	S
E_8	248	real	A

The arguments given for the uniqueness of the "global color" $SU(3)$ group include one "first principle" argument (concordance with the Pauli principle for the fermionic three-quark baryonic composites) and several highly plausible arguments of simplicity and economy; but the deduction of $SU(3)_C$ finally depends on the empirical fact that the OZI rule is strongly supported by experiment [3.13]. Probably, the most direct phenomenological argument for $SU(3)_C$ comes from the asymptotic freedom property of the gauged (not global) $SU(3)$ color group and the measurement of the R value $[R = \sigma(e + \bar{e} \to \text{hadrons}) / \sigma(e + \bar{e} \to \mu + \bar{\mu})]$ in the deep inelastic region (see §5.2c). Since the gauging of the $SU(3)$ color group for quarks yields the dynamically correct non-Abelian quantum field theory for the strong interaction, it is to be hoped that some day the uniqueness of $SU(3)_C$ can be derived solely from a set of "first principles".

The success of the static $SU(6)$ model in predicting the correct $SU(3)_F$, spin content and other properties of the lowest mass $J = \frac{1}{2}^+$ and $J = \frac{3}{2}^+$ baryonic $SU(3)_F$ multiplets encouraged attempts to "relativize" the $SU(6)$ model. However, all such efforts to mix the Poincaré group and global internal symmetries in a compact Lie group were soon shown to be impossible on the basis of the "no-go" theorem (see §4.1). The other success of the $SU(6)$ model — that of making manifest the need for another internal degree of freedom for the quark (i.e. "color") was more consequential in that it identified the non-Abelian global internal symmetry whose gauging led to the QCD theory of the strong interaction [1.89]. In fact, we could move on directly from the global $SU(3)$ color stage of the "strong interaction " story to the final steps in the construction of the gauged $SU(3)$ color group of the standard model. But such a course would overlook the important contributions (during the 1960s) of current algebra and related developments in providing the critical tools for quickly assessing the merits of the newly-created gauge theories of the strong (and electroweak) interactions in the early 1970s. One might say that current algebra — like S matrix theory — tried to probe the dynamics of the strong interaction by correcting for the inadequacies of the quantum field-theoretic approach of weakly-coupled QED but — unlike S matrix theory — it drew its conceptual strength from the quark flavor model of hadrons and its associated symmetries rather than from dispersion theory and the analytic properties of the S matrix. It is for this reason that current algebra had a greater impact on the development of the standard model than did S matrix theory (it is always possible that a successful post-standard model will negate this quasi-historical remark!).

Gell-Mann's original paper [3.14] on $SU(3) \times SU(3)$ current algebra (initially an $SU(3) \times SU(3)$ algebra of charges) drew heavily, not only on the quark model, but also the universal $(V - A)$ theory of the weak interaction and hence launched the current algebra program on a sound phenomenological foundation when the vector and axial vector currents (or charges) were ascribed to the three light quarks. The basic idea of writing down equal-time C.R. among the eight vector and eight axial-vector currents that close upon themselves, guarantees the persistence of this structure (i.e. equal-time C.R.) even when the $SU(3) \times SU(3)$ symmetry of the Lagrangian is lost — as long as the symmetry-breaking is reasonable (i.e. no derivatives, etc.) and the canonical field variables are unchanged. Both $SU(3) \times SU(3)$ and

$SU(2) \times SU(2)$ current algebra soon revealed further desirable properties that permitted easy incorporation of highly useful concepts such as spontaneous symmetry breaking (SSB), the Goldstone theorem, partially conserved axial vector current (PCAC) and soft pion theorems. Overall, the current algebra program produced a rich body of theoretical ideas and experimental tests that could be utilized to scrutinize the new gauge theories of QFD and QCD when they were proposed in the late 1960s and early 1970s. We do not propose to review this extensive program in great detail but we do sketch in the next few sections the key developments in the current algebra program and complete the review with examples of the low energy theorems and high energy sum rules that evolved during the "current algebra" period.

§3.3. Global chiral quark flavor symmetry group and current algebra

As we have already remarked, the genesis of current algebra can be traced to two seminal developments in the late 1950s and early 1960s, namely: (1) the universal V-A current-current theory of the weak interaction; and (2) the quark flavor model of the strongly-interacting hadrons. The chirality-invariance derivation of the V-A theory was originally applied to the lepton and hadron charged currents and immediately determined the Lorentz structure as maximally parity-violating, with equal contributions from V and A currents in the absence of the strong interaction radiative corrections. Within this V-A framework, the early formulation of the conserved vector current (CVC) [3.15] and PCAC {[3.16], [3.17]} hypotheses placed the V and A currents on an approximately equal footing in the hadron sector and called attention to a possible role for a higher chiral symmetry for the hadron currents. The possibility of an underlying chiral symmetry in the hadron sector was exhibited about 1960 in several models [3.18] that assigned zero "bare" mass to the hadronic constituents of the theory [the kinetic energy term in the Lagrangian is manifestly chiral invariant] and regarded the dynamical breaking of the chiral symmetry as the source of the finite mass hadronic composites.

Shortly thereafter, the quark flavor model of hadrons entered the picture and the weak hadron currents were quickly reinterpreted as weak quark currents (resembling the weak lepton currents) and it was realized that the kinetic energy of the Lagrangian for n flavors of massless quarks possesses an $SU(n)_L \times SU(n)_R$ chiral symmetry. The crucial steps in the formulation of current algebra [3.14] occurred when it was recognized that: (1) equal-time

C.R. hold among the V and A quark currents even when the chiral symmetry of the Lagrangian is broken (see above); (2) the V and A quark currents are simply sums and differences of lefthanded (L) and righthanded (R) chiral quark currents which comprise the chiral symmetry group of the Lagrangian for massless quarks and that, hence, the three-flavor $SU(3)_L \times SU(3)_R$ and two-flavor $SU(2)_L \times SU(2)_R$ chiral groups are good starting points for quasi-dynamical calculations; and, finally (3) the L and R chiral quark currents and, *a fortiori*, the V and A quark currents that enter into current algebra, can be identified with the vector and axial vector currents that enter into the universal V-A theory among quarks and leptons. Since reasonable symmetry-breaking interactions among the quarks do not modify the equal-time C.R. among the L and R currents (or V and A) currents, we derive the equal-time C.R. from the free massless quark model and then comment upon possible corrections.

In the free massless quark model, the Lagrangian $\mathcal{L}_0$ for n flavors is simply:

$$\mathcal{L}_0 = i \sum_{j=1}^{n} \bar{q}_j \gamma^\mu \partial_\mu q_j \tag{3.30}$$

where $q_j(x)(j = 1, 2 \ldots n)$ (we use n for flavors and N for colors) belong to the fundamental representation [denoted by $q(x)$] of $SU(n)$. If we take the negative chiral (L helicity) and positive chiral (R helicity) projections of $q(x)$, thus:

$$q_{L,R}(x) = \frac{(1 \mp \gamma_5)}{2} q(x) \tag{3.31}$$

$\mathcal{L}_0$ becomes:

$$\mathcal{L}_0 = i\bar{q}_L \gamma^\mu \partial_\mu q_L + i\bar{q}_R \gamma^\mu \partial_\mu q_R \tag{3.32}$$

It is obvious from Eq. (3.32) that $\mathcal{L}_0$ is invariant under separate infinitesimal L and R constant phase transformations $[a = 1, 2, \ldots (n^2 - 1)]$:

$$q_L \to q_L + i\alpha_L^a T_L^a \, q_L; \qquad q_R \to q_R + i\alpha_R^a T_R^a \, q_R \tag{3.33}$$

where the T^a's are the generators of $SU(n)$. Equations (3.33) give rise to the L and R currents:

$$j_{\mu_L}^a(x) = \bar{q}_L(x)\gamma_\mu T^a q_L(x); \quad j_{\mu_R}^a(x) = \bar{q}_R(x)\gamma_\mu T^a q_R(x) \tag{3.34}$$

with the corresponding charges defined by:

$$Q_L^a(t) = \int d^3x j_{0L}^a(x); \qquad Q_R^a(t) = \int d^3x j_{0R}^a(x) \tag{3.34a}$$

If one makes use of the equal-time anti-C.R. among the components of the Dirac fields, namely:

$$[q_{\alpha i}(\underline{x}, t), q^\dagger_{\beta j}(\underline{y}, t)]_+ = \delta_{ij}\delta_{\alpha\beta}\ \delta^3(\underline{x} - \underline{y}) \tag{3.35}$$

(where i, j are the flavor indices and α, β the Dirac indices), one can derive the equal-time C.R. among the L and R charges, i.e.

$$[Q^a_L(t), Q^b_L(t)] = if^{abc}\ Q^c_L(t); \qquad [Q^a_R(t), Q^b_R(t)] = if^{abc}\ Q^c_R(t);$$
$$[Q^a_L(t), Q^b_R(t)] = 0 \tag{3.36}$$

Clearly, the Q^a_L's generate the $SU(n)_L$ algebra while the Q^a_R's generate the $SU(n)_R$ algebra so that we are dealing with the chiral $SU(n)_L \times SU(n)_R$ algebra.

For physical applications of the $SU(n)_L \times SU(n)_R$ group, as previously noted, it is more convenient to work with the V and A quark currents; for this purpose, we use Eq. (3.34) to define the vector and axial vector currents, thus:

$$V^a_\mu(x) = \frac{1}{2}(j^a_{\mu L} + j^a_{\mu R}); \qquad A^a_\mu(x) = \frac{1}{2}(j^a_{\mu L} - j^a_{\mu R}) \tag{3.37}$$

so that the corresponding charges are:

$$Q^a = \frac{1}{2}(Q^a_L + Q^a_R) = \int d^3x\ V^a_0(x); \qquad Q^{5a} = \frac{1}{2}(Q^a_L - Q^a_R) = \int d^3x\ A^a_0(x) \tag{3.37a}$$

Hence, from Eq. (3.36), the "chiral" algebra of the L and R charges is converted into the algebra of V and A charges:

$$[Q^a(t), Q^b(t)] = if^{abc}Q^c(t); \qquad [Q^a(t), Q^{5b}(t)] = if^{abc}Q^{5c}(t);$$
$$[Q^{5a}(t), Q^{5b}(t)] = if^{abc}Q^c(t) \tag{3.38}$$

It is evident from Eq. (3.38) that the "chiral" algebra of the physical charges "closes" on itself — a *sine qua non* for useful applications.

It can easily be shown that one can extend the charge algebra shown in Eq. (3.38) to the equal-time commutators of charges and currents; one can derive:

$$[Q^a(t), V^b_\mu(x, t)] = if^{abc}\ V^c_\mu(x, t) \tag{3.39}$$

with similar generalizations for the other two charge algebra relations of Eq. (3.38). One can even write down equal-time current-current C.R. for the zero (time) components of the V^a and A^a currents, by simply substituting $V^a_0(\underline{x}, t)$ for $Q^a(t)$ and $A^a_0(\underline{x}, t)$ for $Q^{5a}(t)$ in the three equations of Eq. (3.38); e.g. the first equation of (3.38) becomes:

$$[V^a_0(\underline{x}, t), V^b_0(\underline{y}, t)] = if^{abc}\ V^c_0(\underline{x}, t)\delta^3(\underline{x} - \underline{y}). \tag{3.40}$$

However, if one tries to derive the equal-time C.R. of two currents, one of which is temporal, i.e. $j_0^a = V_0^a(\underline{x}, t)$ or $A_0^a(\underline{x}, t)$, and the other spatial, i.e. $j_i^a(\underline{x}, t) = V_i^a(\underline{x}, t)$ or $A_i^a(\underline{x}, t)$ $(i = 1, 2, 3)$, the theory becomes ill-defined because of the presence of the so-called Schwinger terms [3.19], thus:

$$[j_0^a(\underline{x}, t), j_i^b(\underline{y}, t)] = i f^{abc} j_i^c(\underline{x}, t) \delta^3(\underline{x} - \underline{y}) + S_{ij}^{ab}(\underline{x}) \frac{\partial}{\partial y_j} \delta^3(\underline{x} - \underline{y}) \quad (3.41)$$

where $S_{ij}^{ab}(\underline{x})$ is a c- or q-operator depending on $j_i^a(\underline{x})$. One way to avoid the Schwinger term is to integrate both sides of Eq. (3.41) over all space — so that the Schwinger term vanishes — but then one is not working with equal-time (temporal current)-(spatial current) C.R. but with equal-time charge-temporal current C.R. For "safe" applications of current algebra — and most applications belong to this category — the "Schwinger-troubled" equal-time C.R. are avoided. In other words, the applications of current algebra to particle physics devolve upon the charge-charge C.R. [Eq. (3.38)], the charge-temporal current C.R. [Eq. (3.39)] and the temporal current-temporal current C.R., given by Eq. (3.40); the term "current algebra" is used indiscriminately for calculations based on any one of the three sets of C.R. given by Eq. (3.38)–(3.40).

Before we turn to a partial review of the numerous applications of current algebra relevant to the standard model, we comment on the mathematical and/or physical significance of the three C.R. comprising current algebra. We use the charge-charge C.R., given by Eq. (3.38), to make these comments. Thus, if set $n = 3$ in Eq. (3.38), the first C.R. among the three "scalar" charges $[Q^a \ (a = 1, 2, 3)]$ is merely an expression of $SU(3)_F$ group symmetry, the second C.R. in Eq. (3.38) — among the three "scalar" and three "pseudoscalar" charges $[Q^{5a} \ (a = 1, 2, 3)]$ — follows from the fact that the "pseudoscalar" charges contained in the $SU(3)_L \times SU(3)_R$ symmetry group belong to an octet of the $SU(3)_F$ group. It is the third set of equal-time C.R. in Eq. (3.38) — among the "pseudoscalar" charges themselves — which closes the $SU(3)_L \times SU(3)_R$ algebra and contains the "quasi-dynamical" content of the current algebra approach by normalizing the "pseudoscalar" charge (and hence the axial vector current).

A second comment reiterates the reason why current algebra became such a useful calculational tool for dealing with quark currents. The point is that equal-time C.R. (for charges and/or appropriate currents) hold even in the presence of interactions that induce the breaking of the $SU(n)_L \times SU(n)_R$ group $(n = 2 \text{ or } 3)$. More precisely, while the L and R charges or, equivalently, the "scalar" and "pseudoscalar" charges are no longer constant in time, the equal-time C.R. continue to hold as long as the field momentum canonically conjugate to the field is unmodified by the symmetry-breaking interaction. That is to say, when the field Lagrangian is not invariant under the global internal symmetry group, one can still define the charge $Q^a(x)$:

$$Q^a(t) = \int d^3x \; j_0^a(x) = -i \int \frac{\delta \mathcal{L}}{\delta(\partial^0 q_i)} T_{ij}^a q_j \; d^3x \qquad (3.42)$$

and if $\mathcal{L}$ — in the presence of the symmetry-breaking term — is still only a function of the field components and the canonically conjugate field momenta, the equal-time C.R. among the charges and/or currents, are unchanged. The charges Q^a $(a = 1, 2, ...n^2 - 1)$ are no longer constants of motion but the algebra of charges (and/or currents) holds at each instant of time and this result justifies the many quasi-dynamical calculations based on the algebra of charges (and/or currents).

The final comment that helps to explain the great success of current algebra is that once the chiral quark flavor currents of $SU(n)_L \times SU(n)_R$ $(n = 2$ or $3)$ are identified with the vector and axial vector quark currents in the weak and electromagnetic interactions, one can make use of the phenomenologically-inspired (see §6.1) simplifying properties of the divergences of vector and axial vector quark currents, i.e. CVC and PCAC. It was the exploitation of the CVC and PCAC hypotheses within the framework of the algebra of vector and axial vector quark currents that made possible many current algebra calculations of weak and electromagnetic processes that would otherwise have been intractable. CVC simplifies current algebra calculations by allowing the divergence of the vector quark current to be replaced by zero, whereas PCAC helps out by permitting the divergence of the axial vector quark current (as an operator) to be replaced by the pion field operator. Both CVC and PCAC are especially good approximations when used in conjunction with the $SU(2)_L \times SU(2)_R$ chiral quark flavor group, with $SU(2)_{L+R} = SU(2)_V$ [$\equiv SU(2)_I$] leading to CVC and $SU(2)_{L-R} = SU(2)_A$ leading to PCAC.

While the enlargement of $SU(2)_L \times SU(2)_R$ to $SU(3)_L \times SU(3)_R$ introduces only small errors into the CVC hypothesis when applied to three flavors (thanks to the Ademollo-Gatto theorem [3.20]), larger errors are introduced into the PCAC hypothesis in going from two to three chiral flavors and it is necessary to understand the underlying physics of the PCAC hypothesis. The fact is that the PCAC hypothesis ultimately derives its validity (within the framework of QCD) from the SSB of the global chiral quark group $SU(3)_L \times SU(3)_R$ to $SU(3)_F$, generating in the process the octet of N-G bosons that are identified with the pseudoscalar "pion" octet; in a sense, the finite-mass (quasi-N-G) "pions" serve as "Higgs" bosons to break the global "axial vector" quark current symmetry and the "pion" masses measure the amount of spontaneous symmetry breaking. This explains the larger anticipated error in the application of PCAC to the three-flavor global chiral quark group as compared to the two-flavor case. Consequently, to clarify the physical meaning of the PCAC hypothesis, we digress — in the next section (§3.4) — to examine the spontaneous breaking of global internal symmetries and the resulting Goldstone theorem before concluding this chapter with a sampling of current algebra applications to particle physics.

§3.4. Spontaneous breaking of global internal symmetries and the Goldstone theorem

We have already pointed out that in a quantum field theory, the invariance of the field Hamiltonian under a global internal symmetry group G gives rise to mass degeneracies for those eigenstates belonging to irreducible representations of G, provided the ground (vacuum) state maintains its invariance under the symmetry transformation. We showed that if U is an element of the group G under which the Hamiltonian H is invariant [i.e. $(UHU^{-1} = H)$], then U connects the states A and B belonging to an irreducible representation of G, i.e. $U|A > = |B >$ and $E_A = E_B$ provided $U|0 > = |0 >$ [see Eqs. (3.19)–(3.20)]. Thus, the SSB of a quantum field theory can only take place if the vacuum state is not invariant under the global internal symmetry group U, i.e. $U|0 > \neq |0 >$. But since the group element U is $e^{i\alpha_a Q_a}$, the condition $U|0 > \neq |0 >$ implies that at least one charge Q_a does not annihilate the vacuum, i.e.

$$Q_a|0 > \neq 0. \qquad (3.43)$$

Equation (3.43) can be stated in an equivalent fashion in terms of the vacuum expectation value (VEV) of ϕ, i.e. $< 0|\phi(0)|0 > \neq 0$ (see below). In other words, we have SSB when the Hamiltonian is invariant under a global continuous internal symmetry but the vacuum is broken by one or more generators of this continuous symmetry. For example, in QCD, the global chiral quark flavor group $SU(3)_L \times SU(3)_R$ is spontaneously broken down — by means of QCD quark condensates as delineated in §5.1 — to the vector quark flavor group $SU(3)_F$ and the eight broken axial charges Q_A^a ($a = 1, 2, ...8$) do not annihilate the vacuum.

We must now prove the Goldstone theorem [1.118], i.e. that the SSB of a global continuous symmetry is accompanied by the creation of massless spinless excitations (called "Goldstone bosons"), the number of such excitations being equal to the number of broken generators. [In view of Nambu's earlier hypothesis (see §1.2e) — that the breaking of "chirality invariance" is accompanied by a massless pseudoscalar particle [1.93] — the "Goldstone boson" has been renamed the "Nambu-Goldstone boson" (N-G boson for short); however, we continue to refer to the "Goldstone theorem".] To prove the Goldstone theorem, we first assume that the Lagrangian is invariant under a global continuous (Abelian) $U(1)$ internal symmetry $G = e^{i\alpha Q}$, so that by Noether's theorem, the j_μ current associated with the symmetry group is conserved, i.e. $\partial^\mu j_\mu = 0$. Current conservation implies: $\int d^3x [\partial^\mu j_\mu^a(\underline{x}, t), \phi(0)] = 0$ and hence (for a large enough spatial surface):

$$\frac{d}{dt}[Q(t), \phi(0)] = 0 \tag{3.44}$$

But, if $U(1)$ is spontaneously broken, $< 0|\phi(0)|0 > \neq 0$ and, since $[Q(t), \phi(0)] = -\phi(0)$ [see Eq. (2.13)], it follows that:

$$< 0|[Q(t), \phi(0)]|0 > = \eta \neq 0 \tag{3.45}$$

where η is a constant. Inserting a complete set of intermediate states $|n >$ and using the translation operator (of the Poincaré group), Eq. (3.45) becomes:

$$\sum_n (2\pi)^3 \delta^3(\underline{p}_n)\{< 0|j_0(0)n >< n|\phi(0)|0 > e^{-iE_n t}$$
$$- < 0|\phi(0)|n >< n|j_0(0)|0 > e^{iE_n t}\} = \eta \tag{3.46}$$

Since η is a constant, there can only be cancellation of the positive and negative frequency terms of Eq. (3.46) when $E_n = 0$ for $\underline{p}_n = 0$; consequently, $\eta \neq 0$ is only possible if a state with $E_n = 0$, $\underline{p}_n = 0$ exists, i.e. a massless spinless state (called the N-G excitation or boson) must be present. It is evident that there is only one N-G boson when the global $U(1)$ group is spontaneously broken because there is only one broken generator (charge); from Eq. (3.46), we can say that the N-G excitation n satisfies two conditions:

$$< n|\phi(0)|0 > \neq 0; \qquad < 0|j_0(0)|n > \neq 0 \tag{3.47}$$

One can readily generalize the above derivation of the Goldstone theorem to global non-Abelian internal symmetry groups — in order to establish the relation between the number of N-G bosons and the number of broken generators (charges). However, we prefer to re-derive the Goldstone theorem for a global Abelian group within the framework of the Lagrangian formulation before generalizing to the non-Abelian case. We first prove that, for a charged (complex) scalar field which contains a global $U(1)$ symmetry, the non-invariance of the vacuum state allows us to define a new "vacuum" (i.e. state of lowest energy) that explicitly generates a massless, spinless (N-G boson) particle, in addition to a massive neutral scalar particle that remains after SSB. We write down the Lagrangian for a complex scalar field $\phi(x)$ in terms of two real (Hermitian) fields $\sigma(x)$ and $\pi(x)$ [where $\phi = \frac{1}{\sqrt{2}}(\sigma + i\pi)$], namely:

$$\mathcal{L} = (\partial^\mu \phi^* \partial_\mu \phi) - \mu^2 |\phi|^2 - \lambda |\phi|^4 = \frac{1}{2}\partial^\mu \sigma \partial_\mu \sigma + \frac{1}{2}\partial^\mu \pi \partial_\mu \pi - V(\sigma^2 + \pi^2) \quad (3.48)$$

where:

$$V(\sigma^2 + \pi^2) = \frac{\mu^2}{2}(\sigma^2 + \pi^2) + \frac{\lambda}{4}(\sigma^2 + \pi^2)^2 \quad (3.49)$$

with μ and λ constants. The choice of $V(\sigma^2 + \pi^2)$ as a quartic polynomial (in σ and π) is dictated by the fact that the corresponding quantum field theory is renormalizable. It is evident that Eq. (3.48) is invariant under the global phase transformation $\phi(x) \rightarrow e^{i\alpha}\phi(x)$ (α is constant) so that we are dealing with a global $U(1)$ symmetry group.

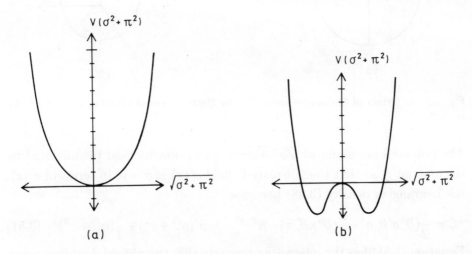

Fig. 3.1. The Higgs potential $V(\phi)$ for a charged scalar field: a) for $\mu^2 > 0$; b) for $\mu^2 < 0$.

Up to this point, we have not specified the signs of μ^2 and λ in the "Higgs" potential (3.49). The case $\lambda < 0$ can be discarded because there is no lower bound on the energy. With $\lambda > 0$, we can consider two possibilities: (1) $\mu^2 > 0$, in which case the "Higgs" potential V is shown in Fig. 3.1a. The minimum of the potential is at $\phi = 0$ so that the vacuum state of the theory has ϕ vanishing everywhere. There is no SSB, the vacuum continues to be (trivially) invariant under the global $U(1)$ phase transformation $e^{i\alpha Q}$ and the mass of the scalar quantum is μ; and (2) $\mu^2 < 0$, in which case the potential V is shown in Fig. 3.1b and $V(\sigma^2 + \pi^2)$ exhibits two minima at $(\sigma^2 + \pi^2) = \pm v$ $(v = \sqrt{\frac{-\mu^2}{\lambda}})$. One must choose one of the two minima (usually v) to define the new vacuum. It is evident that $\sigma^2 + \pi^2 = v^2$ defines a circle of radius v around the origin (see Fig. 3.2a). Clearly, we can pick any point on this circle as the vacuum state so that we have an infinite number of possible vacua; we choose the vacuum on the positive real axis so that its coordinates are:

$$< 0|\sigma|0 >= v \; ; \qquad < 0|\pi|0 >= 0 \qquad (3.50)$$

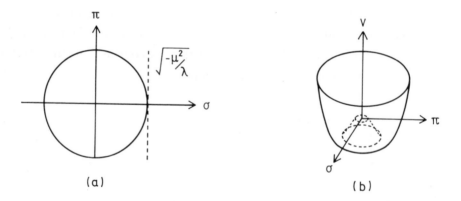

Fig. 3.2. a) Origin of Goldstone boson; b) the Higgs potential function $V(\sigma^2 + \pi^2)$ for $\mu^2 < 0$.

The potential minimum at $\sqrt{\sigma^2 + \pi^2} = v$ can now be used to shift $\sigma(x)$ by the amount v, so that the "displaced" field $\sigma'(x)$ is $\sigma - v$; in terms of $\sigma'(x)$, the Lagrangian density (3.48) becomes:

$$\mathcal{L} = \frac{1}{2}(\partial^\mu \sigma' \partial_\mu \sigma' + \frac{1}{2}(\partial^\mu \pi \partial_\mu \pi) - \mu^2 \sigma'^2 - \lambda v \sigma'(\sigma'^2 + \pi^2) + \frac{\lambda}{4}(\sigma'^2 + \pi^2)^2. \quad (3.51)$$

Equation (3.51) has the interesting property that the σ' field describes a particle of mass $\sqrt{2}|\mu|$, while the π field is massless and becomes the N-G boson resulting from the SSB of the global $U(1)$ symmetry of the Lagrangian. The masslessness of the N-G π field is due to the fact that it describes excitations tangential to the circle in Fig. 3.2a and thus encounters no "resistance" from the potential. It is instructive to give a "three-dimensional" view of the massless excitation characterizing the N-G state by means of the "Mexican hat" diagram (see Fig. 3.2b). We note that the new vacuum is not invariant under the $U(1)$ group and that the quantum field theory with a single complex scalar field has been converted by SSB into a quantum field theory with a single real field of mass $\sqrt{2}|\mu|$ plus a single N-G boson.

The above treatment of the SSB of a single complex field ϕ — composed of two real fields — can readily be extended to $n > 2$ real scalar fields, thereby permitting an easy generalization to global non-Abelian $SO(n)$ groups. The generalization is easy because the $U(1)$ group is locally isomorphic to the $SO(2)$ group, as becomes evident when we rewrite the constant phase transformation $\phi'(x) = e^{-i\alpha}\phi(x)$ (α is a constant) — which defines the $U(1)$ group — in the form [since $\phi(x) = \frac{1}{\sqrt{2}}(\sigma(x) + i\pi(x))$]:

$$\begin{pmatrix} \sigma'(x) \\ \pi'(x) \end{pmatrix} = \begin{pmatrix} \cos\alpha & \sin\alpha \\ -\sin\alpha & \cos\alpha \end{pmatrix} \begin{pmatrix} \sigma(x) \\ \pi(x) \end{pmatrix} \tag{3.52}$$

Equation (3.52) clearly implies an $SO(2)$ global symmetry and so the isomorphism of $U(1)$ and $SO(2)$ is established. If we change the notation slightly and write for the two real scalar fields $\sigma(x) = \phi_1(x)$ and $\pi(x) = \phi_2(x)$, the SSB conditions (3.51) for $U(1)$ become:

$$< 0|\phi_1|0 >= v; \qquad < 0|\phi_2|0 >= 0 \tag{3.52a}$$

In the language of Eq. (3.52a), the SSB of $SO(2)$ occurs via a non-vanishing VEV of the first real field ϕ_1. The translation to n real scalar fields $\phi_i(x)$ ($i = 1, 2, \ldots n$) is now completely straightforward; the Lagrangian becomes:

$$\mathcal{L} = -\frac{1}{2}\sum_{i=1}^{n}(\partial_\mu\phi_i)^2 - \frac{\mu^2}{2}\sum_{i}^{n}(\phi_i^2) - \frac{\lambda}{4}(\sum_{i=1}^{n}\phi_i^2)^2 \tag{3.53}$$

Equation (3.53) is invariant under the global n-dimensional orthogonal group $SO(n)$. For $\mu^2 < 0$, the minimum of the Higgs potential is at $\sum_{i=1}^{n}\phi_i^2 = v^2 = -\mu^2/\lambda$, i.e. the minimum of $V(\phi)$ occurs on the n-dimensional sphere of radius v in the n-dimensional space defined by the fields ϕ_i ($i = 1, 2, \ldots n$). Because of the $SO(n)$ invariance of the Lagrangian (3.53), we may choose, as in the $SO(2)$ case, $< 0|\phi_1|0 >= v$ and $< 0|\phi_j|0 >= 0$ ($j = 2, 3, \ldots n$) and, evidently, the vacuum is still invariant under the group $SO(n-1)$. With the non-zero VEV for ϕ_1, there are still $(n-1)$ linearly independent directions to leave this point and still stay on the n-dimensional sphere defined by

$$\sum_{i=1}^{n} \phi_i^2 = v^2.$$ Consequently, under the specified SSB condition, $(n-1)$ N-G bosons are generated together with one massive scalar field of mass $\sqrt{2|\mu|}$.

The above result can be couched in more general terms: since the original group $SO(n)$ possesses $\frac{1}{2}n\,(n-1)$ generators and the residual group $SO(n-1)$ has $\frac{1}{2}(n-1)(n-2)$ unbroken generators (which do not annihilate the vacuum), it follows that the difference is $(n-1)$ spontaneously broken generators, precisely the number of N-G bosons created in the process. The final result is even more general and applies to any global non-Abelian group G that is broken down spontaneously to the subgroup H: the number of massless scalar states (N-G bosons) that arise in the particle spectrum is equal to $(n_G - n_H)$, where n_G is the dimension of the adjoint representation of the group G and, correspondingly, for n_H. It is this result that is invoked in connection with the SSB breakdown of the global chiral quark flavor symmetry group $SU(3)_L \times SU(3)_R$ to $SU(3)_{L+R} = SU(3)_F$ when it is stated that 8 N-G bosons ("pions") are generated since $n_G = 16$ and $n_H = 8$; or, to put it another way, the axial vector group $SU(3)_{L-R}$ with 8 generators is completely broken which manifestly generates 8 N-G "pions". As a matter of general terminology, we say that when the Hamiltonian and the physical vacuum are both invariant under a global internal symmetry group G, the symmetry is realized in the Wigner-Weyl mode; on the other hand, when the Hamiltonian is invariant under a global internal symmetry group G and the physical vacuum is only invariant under the subgroup group H (which results from the SSB of G), the symmetry is realized in the Nambu-Goldstone mode, with $(n_G - n_H)$ N-G bosons.

From the above discussion, it is evident that the Goldstone theorem has a wide range of applicability and provides the signature for the common phenomenon of spontaneous symmetry breaking (SSB) in a system with global internal symmetries and many degrees of freedom, whether it is a relativistic quantum field theory or a non-relativistic many-body system. In modern particle physics, we deal with the SSB of relativistic quantum field theories possessing global internal symmetries and/or gauge symmetries and the Goldstone theorem enters the picture in its "pristine" form (in connection with the SSB of global symmetries) or in its modified form (as the Higgs mechanism in connection with the SSB of gauge symmetries — see §4.4b). We promised a justification of the PCAC hypothesis on the basis of the Goldstone theorem and this promise is fulfilled in the next section (§3.5), followed (in §3.5a–§3.5c) by a number of applications of current algebra plus

PCAC that led to very useful low energy theorems during the 1960s and are still relevant for the standard model. Insofar as the role of the Goldstone theorem in non-relativistic many-body systems is concerned, we need only mention the Abelian $[U(1)]$ "phonons" of finite crystalline solids, the non-Abelian $[O(3)]$ "magnons" of ferromagnetism, and so on.

§3.5. Some applications of current algebra to particle physics

The global chiral quark symmetry for both $n = 2$ and $n = 3$ flavors is clearly realized in the Nambu-Goldstone mode because there is no evidence for the parity degeneracy in particle masses that is expected if the global chiral quark symmetry were realized in the Wigner-Weyl model. The SSB of the global chiral quark group generates (n^2-1) N-G bosons and the essence of the PCAC hypothesis is to relate directly the divergence of the (n^2-1) axial vector quark currents to the appropriate $(n^2 - 1)$ N-G pseudoscalar boson fields — as operator equations. We derive the PCAC operator equation for $SU(2)_L \times SU(2)_R$ and $SU(3)_L \times SU(3)_R$ but we motivate these derivations by starting with the earlier formalism developed for the SSB of the global $U(1)$ group for the complex scalar field $\phi(x)$. With the notation $\phi(x) = \frac{1}{\sqrt{2}}(\sigma(x) + i\pi(x))$, the expression for the current is:

$$j_\mu(x) = [(\partial_\mu \pi)\sigma - (\partial_\mu \sigma)\pi] \tag{3.54}$$

with the associated conserved charge:

$$Q = \int d^3x \, j_0(x) = \int d^3x [(\partial_0 \pi)\sigma - (\partial_0 \sigma)\pi]. \tag{3.55}$$

Using the canonical equal-time C.R. in the new notation:

$$[\partial_0 \pi(\underline{x}, t), \pi(\underline{y}, t)] = -i\delta^3(\underline{x} - \underline{y}); \quad [\partial_0 \sigma(\underline{x}, t), \sigma(\underline{y}, t)] = -i\delta^3(\underline{x} - \underline{y}) \tag{3.56}$$

one finds:

$$[Q, \pi(0)] = -i\sigma(0); \quad [Q, \sigma(0)] = i\pi(0) \tag{3.57}$$

But, in deriving the Goldstone theorem, it was shown that a N-G state $|n>$ exists satisfying the two conditions of Eq. (3.50) which translate, in the present notation $[|n> = |\pi>, < 0|\sigma|0> = v]$ into:

$$< 0|\pi(0)|\pi(p) > = 1 \quad ; \quad < 0|j_0(0)|\pi(p) > = ivp_0 \tag{3.58}$$

where the first equation becomes a normalization condition. With the help of Lorentz covariance, the second Eq. (3.58) becomes:

$$< 0|j_\mu(0)|\pi(p) > = ivp_\mu. \tag{3.59}$$

and the divergence of j_μ yields:

$$< 0|\partial^\mu j_\mu(0)|\pi(p) >= v m_\pi^2 \tag{3.60}$$

Equation (3.60) tells us that $j_\mu(0)$ is the axial vector current [which we denote by $A_\mu(0)$] and that v is the pion decay constant ($v = f_\pi = 93$ MeV); further, since $v \neq 0$, A_μ is only conserved if $m_\pi = 0$. However, Eq. (3.60) is not quite the PCAC hypothesis because it only provides for the evaluation of a matrix element; the PCAC hypothesis is stronger because it is the operator equation patterned after Eq. (3.60).

The "operator" formulation of the PCAC hypothesis can readily be derived starting with the non-Abelian $SU(2)$ generalization of Eq. (3.60). There are now three axial vector currents $A_\mu^a(0)$ and three pions fields $\pi^a(p)$ ($a = 1, 2, 3$) and Eq. (3.60) becomes:

$$< 0|\partial^\mu A_\mu^a(0)|\pi^b(p) >= f_\pi m_\pi^2 \delta^{ab} \tag{3.61}$$

But we can make use of the fact that the pion field operator, $\phi^a(0)$, has the normalization [see Eq. (3.58)]:

$$< 0|\phi^a(0)|\pi^b(p) >= \delta^{ab}. \tag{3.62}$$

Inserting Eq. (3.62) into Eq. (3.61), we get the form appropriate for the PCAC hypothesis, namely:

$$< 0|\partial^\mu A_\mu^a(0)|\pi^b(p) >= f_\pi m_\pi^2 < 0|\phi^a(0)|\pi^b(p) > \tag{3.63}$$

The PCAC hypothesis simply postulates that Eq. (3.63) is not only an equality between matrix elements but is actually a field operator equation, thus:

$$\partial^\mu A_\mu^a = f_\pi m_\pi^2 \, \phi^a \qquad (a = 1, 2, 3) \tag{3.64}$$

Equation (3.64) constitutes the PCAC hypothesis for the axial vector currents associated with the $SU(2)_L \times SU(2)_R$ global chiral quark flavor group; the generalization to the $SU(3)_L \times SU(3)_R$ group — which is not expected to hold to the same accuracy — is clear:

$$\partial^\mu A_\mu^a = f_a m_a^2 \, \phi^a \qquad (a = 1, 2, \ldots, 8) \tag{3.65}$$

where there are now eight axial vector currents and allowance is made (on the R.H.S.) for different decay constants f_a [sometimes, we use the term "pion" to designate all eight members of the pseudoscalar meson octet]. Equation (3.64) or Eq. (3.65) can be applied to the real world of the low-lying (in mass) pseudoscalar mesons [pion triplet or "pion" octet] and the low-lying $J = \frac{1}{2}^+$ baryon octet. Clearly, if m_π or m_a vanishes, we are dealing strictly with

N-G bosons whereas the observed finite values of m_π and m_a — presumably due to the "Higgs" mechanism in the QFD part of the standard model — indicate the departures from the Goldstone theorem [in QCD, the divergence of the axial vector quark current is non-vanishing — even when the quarks are massless — because of the triangular axial anomaly (see §7.2a)].

With PCAC in hand, we can turn to a sampling of the physical applications of current algebra, worked out during the 1960s [3.21] that continue to be relevant for the present-day standard model. The applications of current algebra fall into two major categories: (A) Low energy theorems, based on the use of PCAC and soft pion techniques; and (B) High energy sum rules, which take advantage of the Lagrangian-independent charge and/or current commutation relations. Under category (A), we consider: (1) the Goldberger-Treiman relation, which is a significant triumph for PCAC and stands as the archetypal phenomenologically-confirmed relation between the strong and the weak interactions; (2) the linear sigma (σ) model, which starts out as a chiral Lagrangian for two fermion flavors (isospin doublet), three pseudoscalar flavors and one scalar flavor, and makes provision for the SSB of the chiral symmetry by the non-vanishing VEV of the scalar field. The linear sigma model lends itself to generalization into the non-linear sigma model which becomes a very useful probe of the low-energy behavior of QCD (see §10.5a); and (3) the computation of the "current" masses of the three light quarks, which are needed for a number of unresolved problems in the standard model (e.g. "strong CP problem" in QCD — see §10.3c). The three examples under (A) demonstrate the power of the current algebra method — with PCAC and soft pion techniques — to make quasi-dynamical predictions at low energy and are treated in §3.5a–3.5c.

Under category (B) of current algebra applications, we sketch as an illustration the derivation in §3.5d of the Adler high energy sum rule for neutrino scattering, which tests the local current commutation relations for arbitrary four-momentum transfer squared. There are quite a number of other interesting high energy sum rules, some of which test "Bjorken scaling" in deep inelastic lepton-hadron processes [e.g. the Bjorken sum rule for the proton spin structure function which has acquired new interest recently because of the "proton spin crisis" (see §5.2c and §5.4a)]. Apart from the "positive" predictions of the current algebra method, it has also been very useful in calling attention to its own limitations [e.g. its prediction — in the soft pion limit — of vanishing width for $\pi^0 \to 2\gamma$ decay and its incorrect prediction of the η' meson mass (see §7.2b)]; when these calculations were scrutinized more carefully, they led to a recognition of the importance of the triangular axial anomaly [3.22] in QED [to explain $\pi^0 \rightarrow 2\gamma$ decay (see §7.2a)] and in QCD [to explain the size of $m_{\eta'}$ (see §7.2b)]. We consider in the next four sections (§3.5a–§3.5d) "positive" applications of current algebra and defer until §7.2a–§7.2b the treatment of the triangular axial anomalies which correct for current algebra limitations.

§3.5a. Goldberger-Treiman relation

The Goldberger-Treiman relation [3.23] — by predicting the correct relation between the "weak" axial vector coupling constant and the "strong" pion-nucleon coupling constant — was an early success for PCAC. It is simple to derive the Goldberger-Treiman relation from Eq. (3.64), which is the statement of the PCAC hypothesis; writing $A_\mu^+ = (A_\mu^1 + iA_\mu^2)/\sqrt{2}$ [on the L.H.S. of Eq. (3.64)] and $\phi^+ = (\phi_1 + i\phi_2)/\sqrt{2}$ [on the R.H.S.], the matrix element of A_μ^+ between neutron and proton states becomes [3.24]:

$$< p(k')|A_\mu^+|n(k) >= \frac{\bar{u}_p(k')}{\sqrt{2}}[\gamma_\mu\gamma_5 g_A(q^2) + iq_\mu\gamma_5 h_A(q^2)]u_n(k) \qquad (3.66)$$

where $q = k - k'$ is the four-momentum transfer between n and p, and $g_A(q^2)$ and $h_A(q^2)$ are the respective "axial vector" and "pseudoscalar" form factors. [The "axial vector" and "pseudoscalar" form factors in Eq. (3.66) are associated with the "first-class" hypercharge-conserving axial vector current A_μ^+, which satisfies the condition $\mathcal{G}A_\mu^+\mathcal{G}^{-1} = -A_\mu^+$ with the G conjugation $\mathcal{G} = Ce^{i\pi I_2}$ (C is the charge conjugation operator and I_2 is the second component of the isospin); a "second-class" axial vector current satisfies the condition $\mathcal{G}A_\mu^+\mathcal{G}^{-1} = +A_\mu^+$ and carries the form factor corresponding to $\gamma_5\sigma_{\mu\nu}q_\nu$ [3.24].]

Taking the four-divergence of both sides of Eq. (3.66) (and using the Dirac equation) yields:

$$< p(k')|\partial^\mu A_\mu^+|n(k) >= \frac{i}{\sqrt{2}}\bar{u}_p(k')\gamma_5 u_n(k)[2m_N g_A(q^2) + q^2 h_A(q^2)] \quad (3.67)$$

where m_N is the nucleon mass. The use of Eq. (3.64) (i.e. PCAC) gives [3.24]:

$$< p(k')|\partial^\mu A_\mu^+|n(k) > = f_\pi m_\pi^2 < p(k')|\phi_\pi^+|n(k) >$$
$$= \frac{\sqrt{2}f_\pi m_\pi^2}{-q^2 + m_\pi^2} \, g_{\pi NN}(q^2)i\bar{u}_p(k')\gamma_5 u_n(k)$$
$$(3.68)$$

where use has been made (on the R.H.S.) of the vertex function $g_{\pi NN}(q^2)$ [which is related to the physical pion-nucleon coupling constant by $g_{\pi NN} =$

$2g_{\pi NN}(m_\pi^2)]$. It follows from Eqs. (3.67) and (3.68) that:

$$\frac{2f_\pi m_\pi^2}{-q^2 + m_\pi^2}g_{\pi NN}(q^2) = 2m_N g_A(q^2) + q^2 h_A(q^2) \tag{3.69}$$

and, consequently, setting $q^2 = 0$:

$$f_\pi g_{\pi NN}(0) = m_N g_A(0). \tag{3.70}$$

Equation (3.70) is the Goldberger-Treiman relation and implicitly assumes a slow variation from $g_{\pi NN}(m_\pi^2)$ to $g_{\pi NN}(0)$ so that the replacement of $m_\pi^2 \to 0$ (soft pion limit) is justified; empirically, a recent value of $g_{\pi NN}$ (m_π^2) is about 10% lower than the accepted value [3.25] and the Goldberger-Treiman relation is then satisfied to a few percent.

There is an equivalent derivation of the Goldberger-Treiman relation that focuses on the chiral quark flavor group $SU(2)_L \times SU(2)_R$ in the "Goldstone" limit (i.e. $m_\pi \to 0$) [3.26]. In this limit, $\partial^\mu A_\mu^+ = 0$ so that the R.H.S. of Eq. (3.67) becomes:

$$2m_N g_A(q^2) + q^2 h_A(q^2) = 0 \tag{3.71}$$

In the $m_\pi \to 0$ limit, the pole term of $h_A(q^2)$ is:

$$\lim_{q^2 \to 0} h_A(q^2) = \frac{-2f_\pi g_{\pi NN}(0)}{q^2} \tag{3.72}$$

which, when combined with Eq. (3.71), again yields the Goldberger-Treiman relation (3.70). This second method of derivation of the Goldberger-Treiman relation exhibits more directly the relation between the PCAC hypothesis and the SSB of the global chiral quark flavor symmetry group $SU(2)_L \times SU(2)_R$ or, with an obvious extension, $SU(3)_L \times SU(3)_R$.

§3.5b. Linear sigma (σ) model of pions and nucleons

It is clear from the second method of derivation above of the Goldberger-Treiman relation that the pion is the N-G boson resulting from the SSB of the global chiral quark flavor group $SU(2)_L \times SU(2)_R$ and that PCAC supplies the correction to the divergence of the axial vector current due to the non-zero mass of the pion. The continuing interest that attaches to the original linear σ model stems from the fact that it shows explicitly how the initial global chiral symmetry $SU(2) \times SU(2)$ is spontaneously broken by the isosinglet scalar σ boson and, in the process of chiral symmetry-breaking, generates an isotriplet of pseudoscalar N-G bosons (pions), and finite masses for the σ meson and the nucleon. It is straightforward to modify the linear σ model in a variety of ways to probe low-energy, non-perturbative aspects of the strong interaction (e.g. to convert the linear σ model into the non-linear

σ model and, in turn, to modify the non-linear σ model into topologically distinct two-flavor and three-flavor Skyrme models — see §10.5). For these reasons, we review the linear σ model of pions and nucleons at this point as an especially interesting low energy application of current algebra.

The linear σ model [3.27] took as its starting point the following fields: an isotriplet of pseudoscalar fields (pions): $\underline{\pi} = (\pi_1, \pi_2, \pi_3)$, an isosinglet scalar field (σ) and an isodoublet (two flavors) of massless Dirac fields (nucleons): $N = (p, n)$. The masses of the three π's and σ were taken to be equal so that the initial global symmetry is $O(4)$ but, instead of considering the SSB of $O(4)$ (with its 6 generators) to $O(3)$ (with its three generators) in the traditional fashion, the σ model took advantage of the local isomorphism between $O(4)$ and $SU(2)_L \times SU(2)_R$ to examine the consequences of the SSB of $SU(2)_L \times SU(2)_R$ (with its six generators) to $SU(2)_{L+R}$ (with its three generators) in the presence of massless N. The presence of the fermionic nucleon is not essential for this analysis and, indeed, the non-linear σ model without initial fermions, has been used to generate the nucleon as a topological soliton out of interacting N-G pion fields (see §10.5a and §10.5b).

The linear σ model starts with the following Lagrangian for the $\underline{\pi}, \sigma$ and N fields:

$$\mathcal{L} = \frac{1}{2}[(\partial_\mu \sigma)^2 + (\partial_\mu \underline{\pi})^2] + \bar{N}i\gamma^\mu \partial_\mu N + g\bar{N}(\sigma + i\underline{\tau} \cdot \underline{\pi}\gamma_5)N - V(\sigma^2 + \pi^2) \quad (3.73)$$

with the potential $V(\sigma^2 + \pi^2)$ having the usual "Higgs" quartic form [see Eq. (3.49)]. It can readily be shown that the Lagrangian (3.73) is invariant under the pair of $SU(2)$ "vector" and "axial vector" transformations (which we designate as I and II respectively), thus:

$$\text{I:} \quad \sigma' = \sigma; \quad \underline{\pi}' = \underline{\pi} + \underline{\alpha} \times \underline{\pi}; \quad N' = N + i\underline{\alpha} \cdot \frac{\underline{\tau}}{2}N \quad (3.74a)$$

$$\text{II:} \quad \sigma' = \sigma + \underline{\beta} \cdot \underline{\pi}; \quad \underline{\pi}' = \underline{\pi} - \underline{\beta}\sigma; \quad N' = N + i\underline{\beta} \cdot \frac{\underline{\tau}}{2}\gamma_5 N \quad (3.74b)$$

The conserved current associated with transformation I is:

$$V_\mu^a = \bar{N}\gamma_\mu \frac{\tau^a}{2}N + \epsilon^{abc}\pi^b \partial_\mu \pi^c \quad (a = 1, 2, 3) \quad (3.75)$$

with charges:

$$Q^a = \int d^3x \, V_0^a(x) \quad (a = 1, 2, 3) \quad (3.75a)$$

The conserved current associated with transformation II is:

$$A_\mu^a = \bar{N}\gamma_\mu \gamma_5 \frac{\tau^a}{2}N + (\partial_\mu \sigma)\pi^a - (\partial_\mu \pi^a)\sigma \quad (a = 1, 2, 3) \quad (3.76)$$

with charges:

$$Q^{5a} = \int d^3x A_0^a(x). \tag{3.76a}$$

The important result is that the charges Q^a and Q^{5a} generate the $SU(2)_L \times SU(2)_R$ algebra:

$$[Q^a, Q^b] = i\epsilon^{abc}Q^c; \quad [Q^a, Q^{5b}] = i\epsilon^{abc}Q^{5c}; \quad [Q^{5a}, Q^{5b}] = i\epsilon^{abc}Q^c$$

$$(a = 1, 2, 3) \tag{3.77}$$

The SSB of the underlying global chiral $SU(2)_L \times SU(2)_R$ group in the σ model is handled in the usual fashion (see §3.4); the "Higgs" potential $V(\sigma^2 + \underline{\pi}^2)$ has a minimum at:

$$\sigma^2 + \underline{\pi}^2 = v^2 \quad \text{with } v = (-\mu^2/\lambda)^{\frac{1}{2}}. \tag{3.78}$$

and the choice $< 0|\sigma|0 >= v$ (so that $\sigma' = \sigma - v$) produces an equation for the purely scalar field part of Eq. (3.73) identical with Eq. (3.51) and for the fermion part of Eq. (3.73) the following:

$$\mathcal{L}_F = -gv\bar{N}N \tag{3.78a}$$

Clearly, the fermion (nucleon) has acquired a mass $m_N = gv$. Since the purely scalar part of Eq. (3.73) is identical with Eq. (3.51), we know that the following two relations hold:

$$[Q^{5a}, \pi^b] = -i\sigma\delta^{ab}; \quad < 0|A_\mu^a(0)|\pi^a >\neq 0 \quad (a = 1, 2, 3) \tag{3.79}$$

where the axial charges Q^{5a} and axial vector currents A_μ^a ($a = 1, 2, 3$) are given by Eqs. (3.76a) and (3.76) respectively.

Four comments can be made about the results in the linear σ model: (1) Eq. (3.79) implies that the global chiral flavor $SU(2)_L \times SU(2)_R$ group is spontaneously broken down to the $SU(2)_{L+R}$ group with the three charges Q^a ($a = 1, 2, 3$) serving as its three generators, whereas the three axial charges Q^{5a} ($a = 1, 2, 3$) do not annihilate the vacuum (i.e. $Q^{5a}|0 >\neq 0$) and are consequently broken; (2) as a consequence of the SSB of the $SU(2)_L \times SU(2)_R$ group, the isodoublet nucleon acquires a mass ($m_N = gv$) where g is the meson-nucleon coupling constant and v [given by Eq. (3.78)] is the "displacement" of the vacuum; (3) the isotriplet $\underline{\pi}$ emerges as the N-G boson triplet (i.e. with mass $m_\pi = 0$) whereas the isosinglet scalar σ boson acquires mass $m_\sigma = \sqrt{2}|\mu|$; and, finally (4) from Eq. (3.78), it follows that m_σ varies as $\lambda^{\frac{1}{2}}$ (for constant v) and can be made arbitrarily large by increasing the coupling constant λ in the Higgs potential given by Eq. (3.73). With the further dropping of the N term [in Eq. (3.73)], this leads to the non-linear

σ model, expressed solely in terms of "unitarized" massless interacting N-G (pion) fields; by adding a suitable quartic term in the "pion" fields, one arrives at the Skyrme model (see §7.5).

§3.5c. Current algebra calculation of the light quark masses

Another "low-energy" current algebra calculation of long standing that has implications for modern particle physics is the calculation (with extensive use of PCAC) of the "current" masses of the three light quarks. We now know that three generations of quark and lepton doublets — six quark flavors (assuming that the t quark is found, which is likely) and six lepton flavors — suffice to define the physics of the standard model so that it is important to pin down the mass values of these twelve "elementary" particles that serve as the basic constituents of matter (at least up to 100 GeV). More precisely, we wish to know the values of the masses that should be inserted into the Lagrangian of the standard model in order to perform calculations of strong and electroweak processes. The value of m in the Lagrangian is called the "current" mass, which is identical with the "free" mass of each of the six (color-singlet) leptons in the standard model; the directly measured values of the charged lepton masses and the measured upper limits on the neutrino masses of the three generations are listed in Table 1.6. The situation is drastically different for the quarks, which are "confined" (color-triplet) particles and one must make a distinction between the "current" and the "constituent" quark mass (i.e. the "effective" quark mass within a color-singlet hadron); this distinction is important for the three lightest (u, d, s) quarks but becomes increasingly unimportant for the c, b, and t quark masses. Thus, when $m_q \gg 150$ MeV (the QCD scale, Λ_{QCD}) — which is true for the c, b and t quarks — the "current" quark masses can be deduced, to a fairly good accuracy, from the ground state masses of the heavy quarkonia $(q\bar{q})$ composites $[J/\psi = (c\bar{c})$ and $\Upsilon = (b\bar{b})$ have been seen but "toponium" $(t\bar{t})$ must still be found], where the binding energy corrections (of order Λ_{QCD}) are quite small.

Another method has to be found to compute the masses of the "confined" (u, d, s) quarks in terms of "free" hadron mass measurements, and it has turned out — at least until now — that the most reliable method is based on current algebra [plus PCAC plus soft pion techniques], supplemented by the quark flavor model to specify the chiral-symmetry-breaking structure that generates finite masses for the three lightest quarks. It should be emphasized that the standard model is of no help in this regard since it attributes all quark and charged lepton masses to the SSB of the electroweak gauge group $SU(2)_L \times U(1)_Y$ by a Higgs doublet, with arbitrary Yukawa coupling constants. [The assignment of null masses to the neutrinos of the three generations is even more arbitrary since massive neutrinos could be Majorana particles (see §6.4b).] Our focus, then, is on the three lightest (u,d and s) quark "current" masses — in contradistinction to the "constituent" quark masses derived as phenomenological parameters from the "free" masses of the quark composites known as mesons and baryons. It should not be surprising that the "constituent" mass values of the u, d and s quarks are larger than the derived "current" masses by an amount comparable to Λ_{QCD}.

We proceed to show how current algebra plus PCAC plus soft pion techniques permit a "quasi-dynamical" calculation of the u, d and s quark "current" masses in terms of the measured masses of the pseudoscalar meson octet (consisting of the π, K and η mesons). The starting point is the PCAC Eq. (3.65) for the eight axial vector currents associated with the global chiral quark flavor group $SU(3)_L \times SU(3)_R$; one takes the matrix element of Eq. (3.65) between vacuum and the pseudoscalar meson state $P_b(k)$ to obtain [see Eq. (3.65)]:

$$< 0|\partial^\mu A_\mu^a|P_b(k) >= \delta_{ab} m_a^2 f_a. \qquad (3.80)$$

By using the reduction formula [3.26] and PCAC for the pseudoscalar meson state $|P_b(k)>$, Eq. (3.80) is transformed into (T is the time-ordered product):

$$\delta_{ab} m_a^2 f_a = \frac{i(m_b^2 - k^2)}{f_b m_b^2} \int d^4x\, e^{-ik \cdot x} < 0|T(\partial^\mu A_\mu^a(0)\partial^\nu A_\nu^b(x))|0 >$$

$$= \frac{i(m_b^2 - k^2)}{f_b m_b^2} \{ ik_\nu \int d^4x < 0|T(\partial^\mu A_\mu^a(0) A_\nu^b(x))|0 > e^{-ik \cdot x}$$

$$- \int d^4x\, e^{-ik \cdot x} < 0|\delta(x_0)[A_0^b(x), \partial^\mu A_\mu^a(0)|0 >\}. \qquad (3.81)$$

In the soft pion limit, Eq. (3.81) becomes:

$$\delta_{ab} m_a^2 f_a^2 = i \int d^4x < 0|\delta(x_0)[A_0^b(x), \partial^\mu A_\mu^a(0)]|0 > \qquad (3.82)$$

which, by taking account of the fact that $Q^{5a} = \int d^3x\, A_0^a(x)$ and $\partial^0 A_0^b = i[\mathcal{H}, A_0^b]$ ($\mathcal{H}$ is the Hamiltonian density), can be rewritten as the so-called $\bar{\sigma}$ term:

$$\bar{\sigma}^{ab} =< 0|[Q^{5a}, [Q^{5b}, \mathcal{H}(0)]]|0 > . \qquad (3.83)$$

We see from Eqs. (3.82) and (3.83) that the mass of each member of the pseudoscalar meson octet, m_a ($a = 1, 2, ...8$), is related to a double commutator of the axial charge with the total Hamiltonian (density), denoted by $\bar{\sigma}^{ab}$.

From the structure of Eq. (3.83), it follows that the $\bar{\sigma}$ term vanishes if $\mathcal{H}$ retains its initial global chiral symmetry $SU(3)_L \times SU(3)_R$. The calculation of the $\bar{\sigma}$ term therefore depends upon some hypothesis with regard to the tensorial behavior of the chiral symmetry-breaking terms.

As we have already pointed out, QCD tells us (see §5.1) that the color-induced quark condensates are responsible for the SSB of $SU(3)_L \times SU(3)_R$ to $SU(3)_F$; while we do not have precise knowledge of this symmetry-breaking mechanism, we do know that the finite and unequal masses of the three light quarks must break the global chiral symmetry $SU(3)_L \times SU(3)_R$ and it is plausible to guess the tensorial behavior of the chiral symmetry-breaking from the quark mass term in the Hamiltonian. We also know that the global chiral quark symmetry for two quark flavors, $SU(2)_L \times SU(2)_R$, is a much better approximation than the three-flavor $SU(3)_L \times SU(3)_R$, because of the significant mass difference between the s quark and the u and d quarks. The breakings of $SU(3)_L \times SU(3)_R$ [to $SU(2)_L \times SU(3)_R$], then of $SU(2)_L \times SU(2)_R$, [to $SU(2)_I$] and, finally, of $SU(2)_I$ [to $U(1)_{EM}$] can be taken into account by first allowing for a common zero mass of the u and d quarks and a finite mass of the s quark, then allowing for a common finite mass for the u and d quarks, and, finally, assuming that the u and d quarks differ in mass. With this physics in mind, we write down the Hamiltonian for the three light quarks in the form:

$$\mathcal{H} = \mathcal{H}_0 + \mathcal{H}_1 + \mathcal{H}_2 + \mathcal{H}_3 \qquad (3.84)$$

where $\mathcal{H}_0$ is the Hamiltonian when $m_u = m_d = m_s = 0$, $\mathcal{H}_1 = m_s \bar{s}s$ with $m_u = m_d = 0$, and $\mathcal{H}_2 = m_u \bar{u}u + m_d \bar{d}d$ with $m_u = m_d \neq 0$. The last term on the R.H.S. of Eq. (3.84), $\mathcal{H}_3$, is responsible for the breaking of $SU(2)_I$ to $U(1)_{EM}$ and receives contributions from the quark mass term when $m_u \neq m_d$ as well as from the electromagnetic current; $\mathcal{H}_3$ produces the smallest perturbation on the quark masses and is considered separately after the affects of $\mathcal{H}_1$ and $\mathcal{H}_2$ are evaluated.

From the form of $\mathcal{H}$ in Eq. (3.84), it is obvious that $\mathcal{H}_0$ does not contribute to the $\bar{\sigma}$ term in Eq. (3.83) while $\mathcal{H}_1 + \mathcal{H}_2 + \mathcal{H}_3$ do. Dropping $\mathcal{H}_3$ until later, we calculate the $\bar{\sigma}$ term due to $\mathcal{H}_1 + \mathcal{H}_2$ by examining the tensorial chiral-symmetry-breaking behavior of $\mathcal{H}_1 + \mathcal{H}_2$ under the group $SU(3)_L \times SU(3)_R$. Since $\mathcal{H}_1 + \mathcal{H}_2 = m_u \bar{u}u + m_d \bar{d}d + m_s \bar{s}s$, and q_L and q_R have the respective transformation properties $(3, 1)$ and $(1, 3)$ [under the $SU(3)_L \times SU(3)_R$ group], $\mathcal{H}_1 + \mathcal{H}_2$ possesses the structure $\bar{q}_L q_R + \bar{q}_R q_L$. This quark structure implies that $(\mathcal{H}_1 + \mathcal{H}_2)$ transforms as $[(\bar{3}, 3) + (3, \bar{3})]$

($\bar{3}$ is the conjugate representation to 3) under $SU(3)_L \times SU(3)_R$ and, if $m_u = m_d = m_s$, $SU(3)_L \times SU(3)_R$ breaks down to $SU(3)_{L+R}[\equiv SU(3)_F]$.

To perform the actual calculation of the $\bar{\sigma}_{ab}$'s [in Eq. (3.83)], it is convenient to rewrite the $SU(3)_L \times SU(3)_R$ — breaking term, $(\mathcal{H}_1 + \mathcal{H}_2)$, in terms of three scalar densities: u_0, u_3 and u_8, thus [3.28]:

$$\mathcal{H}_1 + \mathcal{H}_2 = c_0 u_0 + c_3 u_3 + c_8 u_8 \qquad (3.85)$$

where the scalar densities are given in terms of the quark fields and the $SU(3)_F$ transformation matrices λ_a $(a = 1, 2, ...8)$ [see Eq. (3.26) — a ninth scalar density u_0 is added since the group is actually $U(3)$, not $SU(3)$]:

$$u_a = \bar{q}\lambda_a q \qquad (a = 0, 1, 2, \ldots, 8) \qquad (3.86)$$

where $\lambda_0 = \sqrt{(2/3)}I$ (I is the identity matrix). Because λ_0, λ_3 and λ_8 are all diagonal matrices, u_0, u_3 and u_8 assume the familiar quark forms:

$$u_0 = \sqrt{\frac{2}{3}}(\bar{u}u + d\bar{d} + s\bar{s}); \quad u_3 = (u\bar{u} - d\bar{d}); \quad u_8 = \sqrt{\frac{1}{\sqrt{3}}}(u\bar{u} + \bar{d}d - 2\bar{s}s) \quad (3.87)$$

and hence:

$$c_0 = \frac{1}{\sqrt{6}}(m_u + m_d + m_s); \quad c_3 = \frac{1}{3}(m_u - m_d); \quad c_8 = \frac{1}{\sqrt{3}}(\frac{m_u + m_d}{2} - m_s)$$
$$(3.88)$$

By working out the equal-time commutators of the Q^{5a}'s with the scalar densities u_a and the corresponding set of pseudoscalar densities:

$$v_a = -i\bar{q}\lambda_a \gamma_5 q \qquad (a = 0, 1, 2 \ldots 8) \qquad (3.89)$$

one obtains — with the help of the canonical anti-C.R. for the quark fields $(\alpha, \beta$ are the Dirac and spin indices): $\{q_\alpha^\dagger(\underline{x}, t), q_\beta(\underline{y}, t)\} = \delta_{\alpha\beta}\,\delta^3(\underline{x} - \underline{y})$ — the equations [3.26]:

$$\delta(x_0)[Q_a^5(x_0), u_j(\underline{x}, x_0)] = -id_{ajk}v_k(0)\,\delta^4(x)$$

$$\delta(x_0)[Q_a^5(x_0), v_j(\underline{x}, x_0)] = id_{ajk}u_k(0)\,\delta^4(x). \quad (a = 1, ...8) \quad (j, k = 0, 1, ...8)$$
$$(3.90)$$

$$\delta(x_0)[Q_a(x_0), u_j(\underline{x}, x_0)] = if_{ajk}u_k(0)\,\delta^4(x)$$

$$\delta(x_0)[Q_a(x_0), v_j(\underline{x}, x_0)] = if_{ajk}v_k(0)\,\delta^4(x).$$

where: $d_{abc}(d_{0ab} = \sqrt{\frac{2}{3}}\delta_{ab})$ and f_{abc} $(f_{ab0} = 0)$ are the usual $SU(3)$ symmetric and antisymmetric structure constants.

We are now in a position to relate the double commutator $\bar{\sigma}_{ab}$ to the observed pseudoscalar octet masses $(m_\pi, m_K$ and $m_\eta)$ [see Eqs. (3.82) and

(3.83)], and we get:

$$f_\pi^2 m_\pi^2 = \frac{(m_u + m_d)}{2} < 0|\bar{u}u + \bar{d}d|0 >$$

$$f_K^2 m_K^2 = \frac{(m_u + m_s)}{2} < 0|\bar{u}u + \bar{s}s|0 >$$

$$f_\eta^2 m_\eta^2 = \frac{(m_u + m_d)}{6} < 0|\bar{u}u + \bar{d}d|0 > +\frac{4m_s}{3} < 0|s\bar{s}|0 >$$

$$(3.91)$$

If we take the vacuum to be $SU(3)_F$-symmetric — an approximation that is reasonable since there is no SSB of $SU(3)_F$ symmetry — we can equate the VEV's of the three "quark condensates" and set $f_\eta = f_k = f_\pi$; Eq. (3.91) then predicts two relations:

$$4m_K^2 = 3m_\eta^2 + m_\pi^2; \qquad \frac{m_u + m_d}{2m_s} = \frac{m_\pi^2}{2m_K^2 - m_\pi^2} \simeq \frac{1}{25}. \qquad (3.92)$$

Equation (3.92) recaptures the Gell-Mann-Okubo mass formula for the octet of pseudoscalar mesons [see Eq. (3.28)] — which is encouraging — and predicts a large ratio ($\simeq 25$) of the strange quark mass to the average of the u and d quark masses. Another encouraging aspect of this calculation is that Eq. (3.92) implies $c_8/c_0 = -1.25$ [see Eq. (3.88)], quite close to the value $-\sqrt{2}$ that is consonant with the $SU(2)_L \times SU(2)_R$ symmetric value.

The above very plausible results suggest an extension of the current algebra approach based on Eq. (3.83) to calculate the ratio m_u/m_d by taking into account not only the $\mathcal{H}_3$ contribution in Eq. (3.84) (due to $m_u \neq m_d$) but also to incorporate the electromagnetic current into the calculation by adding the term $\mathcal{H}_\gamma$ to Eq. (3.84), where:

$$\mathcal{H}_\gamma = e^2 \int d^4x \, T(j^\mu(x) j^\nu(0)) \, D_{\mu\nu}(x) \qquad (3.93)$$

with $j^\mu(x)$ the electromagnetic current operator and $D_{\mu\nu}(x)$ the photon propagator. Using both terms $\mathcal{H}_3$ and $\mathcal{H}_\gamma$, one obtains for the m_d/m_u ratio:

$$\frac{m_d}{m_u} = \frac{m^2(K^+) - m^2(K^0) - m^2(\pi^+)}{m^2(K^0) - m^2(K^+) + m^2(\pi^+) - 2m^2(\pi^0)} \simeq 1.8 \qquad (3.94)$$

The absolute values of the m_u and m_d current masses are finally pinned down by computing the $(n - p)$ mass difference [3.29] which, together with Eq. (3.92), yields:

$$m_u \simeq 4 \text{ MeV}; \qquad m_d \simeq 7 \text{ MeV} \qquad (3.95)$$

from which it follows that $m_s \simeq 130$ MeV. The separate values of m_u and m_d are not too accurate [3.30] but the general consensus is that $m_d > m_u$

and that $m_u > 0$. The inequality $m_d > m_u$, i.e. m (weak isospin "down" quark) $> m$ (weak isospin "up" quark) for the first generation of quark doublets [see Table 1.6] is important because it is the reverse of the situation for the second and third generations of quark doublets, where m ("up" quark) $> m$ ("down" quark), i.e. $m_c > m_s$ and $m_t > m_b$; this "mass reversal" phenomenon has not as yet been explained within the framework of the standard model. The second conclusion, that $m_u \neq 0$, is also critical for the standard model because of the "strong CP problem" in QCD (see §10.3c); the existence of topologically non-trivial gauge transformations in non-Abelian QCD requires the "θ-vacuum" which, in turn, generates a "strong CP violation" term in the QCD Lagrangian unless the θ parameter is zero. If the current algebra derivation of the $m_u \neq 0$ result is sufficiently inaccurate to permit $m_u = 0$, the "strong CP problem" would be solved immediately (see §10.3c); however, if the $m_u \neq 0$ result survives, then the "strong CP problem" in QCD is still without definitive solution. For this reason alone, the complicated sequence of steps in the current algebra derivation of the $m_u \neq 0$ result merits careful reexamination.

This completes our discussion of the current algebra determination of the three light quark masses (u, d and s) and of our brief review of several of the early triumphs in the application of current algebra (plus PCAC plus soft pion techniques) to particle physics that are still relevant for the present-day standard model in the low-energy domain. Before moving on in the next chapter to the actual gauging of global internal symmetry groups, we briefly review (in §3.5d) one high energy sum rule that is derived from current algebra (without PCAC).

§3.5d. Adler sum rule for inelastic neutrino scattering on nucleons

The Adler sum rule for inelastic neutrino scattering on nucleons [3.31] is interesting in several respects: (1) while the sum rule does not focus on deep inelastic scattering (since the sum is taken over all inelastic energy transfers), it does single out the $W_2(Q^2, \nu)$ structure function that is so important in deep inelastic "Bjorken scaling" considerations (see §5.2c). Moreover, it takes advantage of the cancellation of unknown strong interaction effects by considering the difference between the neutrino and anti-neutrino W_2 structure functions; (2) the sum rule has the important feature that the Q^2 dependence is "integrated away" and the dependence of $W_2(Q^2, \nu)$ is necessarily testing local commutation relations for $Q^2 \neq 0$; and (3) the sum rule is derived from current commutators involving hypercharge-changing as well as hypercharge-conserving currents, both vector and axial vector. At the end, the sum rule simply requires knowledge of the I_3 and Y quantum numbers of the nucleon in question. It must be emphasized that the Adler sum rule for neutrino scattering on a nucleon was derived before the deep inelastic lepton-nucleon scatterings had been performed and "Bjorken scaling" established and also before the parton model became the instrument of analysis of these experiments. However, it is interesting to see how Adler could nevertheless obtain a high energy sum rule from current algebra.

Adler derived his sum rule for inelastic neutrino scattering on nucleons [3.31] before quarks were invented and so he made do with "old-fashioned" nucleon currents. The process treated by Adler is shown in Fig. 3.3 where the neutrino ν is scattered off the nucleon target N and produces a charged lepton ℓ^- and some hadron state X_n (n particles with total momentum p_n), i.e.

$$\nu(k) + N(p) \rightarrow \ell^-(k') + X(p_n) \tag{3.96}$$

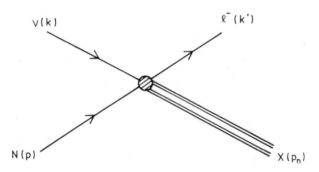

Fig. 3.3. Neutrino scattering on a nucleon.

We introduce the same variables as in the section (§5.2c) on deep inelastic lepton-hadron scattering, namely: $q = k - k'$, $\nu = p \cdot q/M$ (M is the nucleon mass) and hence, neglecting the lepton mass, $q^2 = -4EE' \sin^2 \theta/2$ (θ is the angle between $\underline{k}$ and $\underline{k}'$) and $\nu = (E - E')$ is the energy loss of the neutrino (E and E' are the initial neutrino energies respectively). Using the above notation, the key matrix element $T_{\mu\nu}$ acquires the form:

$$T_{\mu\nu} = i \int dx e^{-iq \cdot x} < p|\theta(x_0)[j_\nu^\dagger(x), j_\mu(0)]|p > . \tag{3.97}$$

where j_μ is a nucleon current and it is understood that $T_{\mu\nu}$ is averaged over the spin of the nucleon state $|p>$. Equation (3.97) can be converted into:

$$iq_\nu T_{\nu\mu} = i \int dx e^{iq\cdot x} < p|\theta(x_0)[\frac{\partial j_\nu(x)}{\partial x_\nu}, j_\mu(0)]|p >$$

$$+ i \int dx e^{-iq\cdot x} \delta(x_0) < p|[j_0(x), j_\mu(0)]|p > . \qquad (3.98)$$

Since the matrix element $T_{\mu\nu}$ is a tensor that depends only on the momenta p and q, the first integral on the R.H.S. of Eq. (3.98) can be written in the form:

$$\int dx\ e^{iq\cdot x} < p|\theta(x_0)[\frac{\partial j_\nu^\dagger(x)}{\partial x_\nu}, j_\mu(0)]|p >= Dp_\mu + Eq_\mu, \qquad (3.99)$$

where the functions D and E are scalar functions of q^2 and ν. The second integral on the R.H.S. of Eq. (3.98) is an equal-time commutator term and can be written in the form:

$$\int dx\ e^{iq\cdot x} \delta(x_0) < p|[j_0^\dagger(x), j_\mu(0)]|p >= Cp_\mu, \qquad (3.100)$$

where C is a constant and depends only on the quantum numbers of the symmetries assumed for the lepton and hadron currents. Finally, since $T_{\mu\nu}$ only depends on the momenta p and q, its most general decomposition is:

$$T_{\mu\nu} = Ap_\mu p_\nu + B_1\delta_{\mu\nu} + B_2 p_\mu q_\nu + B_3 q_\mu p_\nu + B_4 q_\mu q_\nu + B_5\epsilon_{\mu\nu\alpha\beta}q_\alpha p_\beta, \qquad (3.101)$$

where A, B_i ($i = 1, 2, ..5$) are functions of q^2 and ν; the B_5 term is parity-violating while all the others are parity-conserving.

The next step is to combine Eqs. (3.99) and (3.100) with Eq. (3.101); one finds that the coefficient of p_μ must satisfy the condition:

$$(q \cdot p)A + q^2 B_2 = D + C. \qquad (3.102)$$

At this point, Adler makes a major assumption to derive his sum rule, namely that the functions A, B_2 and D of q^2 and ν all satisfy unsubtracted dispersion relations in the variable ν for fixed q^2, namely:

$$A(\nu, q^2) = \frac{1}{\pi} \int_{-\infty}^{\infty} d\nu' \frac{\text{Im } A(\nu', q^2)}{\nu' - \nu} \qquad (3.103)$$

and similar equations for B_2 and D (for neutrino scattering, we are dealing with space-like q); the point now is to recognize that Im A is related to the $W_2(q^2, \nu)$ structure function that appears in the deep inelastic scattering problem (see §5.2c), thus:

$$\text{Im } A = \frac{\pi W_2}{M^2} \qquad (3.103a)$$

It can now be shown that in the limit $\nu \to \infty$, the functions B_2 and D do not contribute to Eq. (3.102) but only A does; hence:

$$\lim_{\nu \to \infty} (q \cdot p)A = \frac{M}{\pi} \int_{-\infty}^{\infty} d\nu' \, \text{Im} \, A(\nu', q^2) = C. \qquad (3.104)$$

Further, it is known that:

$$\text{Im} \, A(-\nu, q^2) = -\text{Im} \, \tilde{A}(\nu, q^2) \qquad (3.105)$$

where $\tilde{A}$ denotes A with the charge-raising and charge-lowering currents interchanged. Equation (3.104) can then be written in the form:

$$\frac{1}{\pi} \int_0^{\infty} d\nu [\text{Im} \, A(\nu, q^2) - \text{Im} \, \tilde{A}(\nu, q^2)] = \frac{C}{M}. \qquad (3.106)$$

The conversion of Eq. (3.106) into the more familiar Adler sum rule for neutrino scattering in terms of the $W_2(q^2, \nu)$ structure functions for ν and $\bar{\nu}$, i.e. $W_2^{\nu}(q^2, \nu)$ and $W_2^{\bar{\nu}}(q^2, \nu)$, is made possible by the identifications: $\text{Im} \, A(q^2, \nu) = \pi W_2^{\nu}(\nu, q^2)/M^2$ and $\text{Im} \, \bar{A}(\nu, q^2) = \pi W_2^{\bar{\nu}}(q^2, \nu)/M^2$. Consequently, the final expression for the Adler sum rule for neutrino and antineutrino scattering by a nucleon, using Eq. (3.106) and the normalization $< p|V_{\mu}^3|p > = \frac{p_{\mu}}{M} I_3$ and $< p|V_{\mu}^8|p > = \frac{p_{\mu}\sqrt{3Y}}{2M}$, becomes:

$$\int_0^{\infty} d\nu \{W_2^{(\bar{\nu})}(q^2, \nu) - W_2^{(\nu)}(q^2, \nu)\} = 4I_3 \cos^2 \theta_C + (3Y + 2I_3) \sin^2 \theta_C$$

$$= 2 + 2\sin^2 \theta_C \quad \text{for proton target}$$

$$= -2\cos^2 \theta_C + \sin^2 \theta_C \quad \text{for neutron target} \qquad (3.107)$$

The high energy Adler sum rule for inelastic neutrino and antineutrino scattering on nucleons was a precursor of the numerous high energy sum rules that were developed in connection with experiments on the deep inelastic scattering of leptons by nucleons; its verification would be an important success for "old-fashioned" current algebra.

References

[3.1] R.E. Marshak, N. Mukunda and S. Okubo, *Phys. Rev.* **137** (1965) 698.

[3.2] G. Feinberg and M. Goldhaber, *Proc. Nat. Acad. Sci.* **45** (1959) 1301.

[3.3] F.T. Avignone *et al.*, *Phys. Rev.* **D34** (1986) 97.

[3.4] S. Okubo, C. Ryan and R.E. Marshak, *Nuovo Cim.* **34** (1964) 759.

[3.5] B. Cassen and E.U. Condon, *Phys. Rev.* **50** (1936) 846.

[3.6] E. Fermi and C.N. Yang, *Phys. Rev.* **76** (1949) 12.

[3.7] R.E. Marshak, S. Okubo and J.H. Wojtaszek, *Phys. Rev. Lett.* **15** (1965) 463.

[3.8] J.J. Sakurai, *Phys. Rev. Lett.* **9** (1962) 472; S. Okubo, *Phys. Lett.* **5** (1963) 165.

[3.9] S. Okubo and R.E. Marshak, *Phys. Rev. Lett.* **13** (1964) 818.

[3.10] S. Okubo, *Phys. Rev.* **D16** (1977) 3535.

[3.11] T.D. Lee and C.N. Yang, *Nuovo Cim.* **10** (1956) 3.

[3.12] S. Okubo, *Phys. Lett.* **5** (1963) 163; G. Zweig, CERN report 84/19 TH412 (1964); J. Iizuka, *Prog. Theor. Phys.* **145** (1966) 1156.

[3.13] M.K. Gaillard, B.W. Lee and J. Rosner, *Rev. Mod. Phys.* **47** (1975) 277.

[3.14] M. Gell-Mann, *Phys. Rev.* **125** (1962) 1067; *ibid.*, *Physics* **1** (1964) 63.

[3.15] R.P. Feynman and M. Gell-Mann, *Phys. Rev.* **109** (1958) 193; S. Gershtein and Ya. Zeldovich, *Soviet Physics* **2** (1956) 576.

[3.16] Y. Nambu, *Phys. Rev. Lett.* **4** (1960) 380.

[3.17] Chou, Kuang-chao, *Soviet Physics JETP* **12** (1961) 492.

[3.18] Y. Nambu and Jona-Lasinio, *Phys. Rev.* **122** (1961) 345; R.E. Marshak and S. Okubo, *Nuovo Cim.* **19** (1961) 1226.

[3.19] J. Schwinger, *Phys. Rev. Lett.* **3** (1959) 296.

[3.20] M. Ademollo and R. Gatto, *Phys. Rev. Lett.* **13** (1964) 264.

[3.21] S.L. Adler and R.F. Dashen, *Current Algebras*, W.A. Benjamin, Mass. (1968).

[3.22] R. Jackiw, *Current Algebra and Anomalies*, ed. S. Treiman, R. Jackiw, B. Zumino, and E. Witten, Princeton Univ. Press (1985).

[3.23] M.L. Goldberger and S.B. Treiman, *Phys. Rev.* **110** (1958) 1178;

[3.24] R.E. Marshak, Riazuddin and C.P. Ryan, *Theory of Weak Interactions in Particle Physics*, Wiley-Interscience, New York (1969).

[3.25] R. Arndt, Z. Li, L.D. Roper, and R. Workman, *Phys. Rev.* **D44** (1991) 289.

[3.26] T.P. Cheng and L.F. Li, *Gauge Theory of Elementary Particle Physics*, Clarendon Press, 1984.

[3.27] M. Gell-Mann and M. Lévy, *Nuovo Cim.* **16** (1960) 705.

[3.28] M. Gell-Mann, R.J. Oakes and B. Renner, *Phys. Rev.* **175** (1968) 2195; S. Glashow and S. Weinberg, *Phys. Rev. Lett.* **20** (1968) 224.

[3.29] A. Gasser and H. Leutwyler, *Nucl. Phys.* **B94** (1975) 269.

[3.30] S. Weinberg, *Festschrift for I.I. Rabi*, ed. L. Motz, New York Acad. Sci., New York (1977).

[3.31] S.L. Adler, *Phys. Rev.* **143** (1966) 1144; S. Treiman, *Current Algebra and Anomalies*, ed. S. Treiman, R. Jackiw, B. Zumino, and E. Witten, Princeton Univ. Press (1985).

Chapter 4

GAUGE SYMMETRY GROUPS AND THEIR
SPONTANEOUS BREAKDOWN

§4.1. From global symmetries to dynamics in quantum field theory

In the previous chapter, we considered the global internal symmetries (both Abelian and non-Abelian) of the quantum field for both strong and weak interactions, and discussed the conservation laws, mass relations and selection rules that flowed from either absolute or approximate invariance of the field Lagrangian under the postulated global internal symmetry. We also treated in some detail the spontaneous breaking of global internal symmetries (and the Goldstone theorem), with particular attention given to the global chiral quark flavor symmetry (for two and three flavors) and its implementation in a variety of current algebra applications with and without PCAC. It was realized during the height of this "global symmetry" activity that one was ignorant of the dynamical content of the theories of the strong and weak interactions and that one was making do with the quasi-dynamical current algebra methods that give rise to low energy theorems and high energy sum rules.

The success of current algebra with PCAC (and the use of soft pion techniques) and the recognition that it is essentially an algebraic formulation of chiral symmetry, led to the notion that chiral Lagrangians could be constructed, in which the lowest-order perturbation calculations (without any closed loops) could replicate the current algebra predictions (as well as additional ones) in strong interactions at low energies [4.1]. The σ model [3.27] starts with a chiral $[SU(2)_L \times SU(2)_R]$ Lagrangian and shows explicitly — in the linear realization — how the chiral symmetry is broken by a non-vanishing VEV of the σ field and yields a triplet of N-G bosons. If one introduces a symmetry-breaking term proportional to the pion mass (finite but small), it can be shown that PCAC is automatically built into the chiral Lagrangian. When the σ field in this Lagrangian is expressed in terms of the π field, one obtains the non-linear realization of the Lagrangian and one is in a position to calculate a variety of new processes (e.g. the $\pi - \pi$ scattering length, multiple pion production, etc.) [4.2] by means of first-order perturbation theory. The chiral Lagrangian method was an improvement on current algebra, was a boon to nuclear physics and, indeed, it has inspired the modern version of the Skyrme model (see §10.5) — but it was still far from being a complete dynamical theory of the strong interaction.

Other attempts were made during the late 1960s to move towards a more complete dynamical theory of the strong interaction by mixing global internal symmetries with space-time symmetries but they were not really successful; e.g. the attempt to mix the parity operation with the global $SU(3)_F$ quark flavor group [3.1] brought the prediction — contrary to experiment — that there should be mass degeneracy (with respect to parity) for the $SU(3)_F$ multiplets. The absence of parity doublets among hadrons provided the rationale for the concept of spontaneous breaking of the global chiral quark flavor group that has survived in QCD (see §5.1). A more ambitious attempt to mix global internal symmetry groups with the "spin part" of the Poincaré group–along the lines of the static $SU(6)$ group (see §3.2d) — had good success at low energies. Treated as a non-relativistic group, $SU(6)$ does explain why the $J = \frac{1}{2}^+$ octet and $J = \frac{3}{2}^+$ decuplet are the lowest-mass baryon representations, an indispensable step in firming up the need for a new degree of freedom (global "color") for the quarks; however, it was shown (see §3.2d) — since the $SU(2)_{\text{SPIN}}$ part of $SU(6)$ neglects the orbital angular momentum — that the $SU(6)$ approach runs into trouble with relativistic invariance. Indeed, any attempt to enlarge $SU(6)$ and maintain Lorentz invariance falters because the generators of the larger group do not close upon themselves, thereby requiring infinite-dimensional representations characteristic of non-compact Lie groups [3.9].

Nevertheless, the partial success of the static $SU(6)$ model did encourage further work seeking to "relativize" the $SU(6)$ model and, somehow, mix the Poincaré group with the $SU(3)_F$ internal symmetry group [this search led to groups like $SU(6)_W$ and $\widetilde{SU}(12)$ [2.23]]. It was soon demonstrated — in particular, by O'Raifeartaigh [2.24] and Coleman and Mandula [1.170] in the mid-1960s — that all such efforts to mix the Poincaré group and global internal symmetries in a compact Lie group are doomed to failure by so-called "no-go" theorems. [It is possible to circumvent the "no-go" theorems by postulating Grassmann-type internal symmetry generators (obeying anti-C.R.), in addition to the usual internal symmetry generators (obeying C.R.) and the Poincaré generators; however, this requires "new physics", namely the introduction of "supercharges" and, thus far, supersymmetry, in its various "incarnations", has not received any phenomenological support (see §2.3a).] A simple example can be given to exhibit the difficulty of mixing, in a non-trivial way, the generator of a global internal symmetry and a generator of the Poincaré group. Thus, let us assume that the commutator of τ^+ [the isospin-raising operator, i.e. $\tau^+|n> = |p>$ (n and p are neutron and proton states] with $P^2(= P_\mu P^\mu)$ [P_μ is the translation generator in Minkowski space] is non-vanishing:

$$[\tau^+, P^2]|n> \neq 0 > \tag{4.1}$$

Equation (4.1) immediately leads to: $m_n^2 \neq m_p^2$, which contradicts the basic assumption that p and n belong to the same isodoublet with equal masses.

O'Raifeartaigh's "no-go" theorem [2.24] is more sophisticated than the above example and depends on a powerful lemma in the theory of finite-order Lie algebras. The lemma states that every Lie algebra G of finite order can be written as a semi-direct product of a semi-simple subalgebra E and an invariant solvable subalgebra S. Using this lemma, and letting M and P stand for the homogeneous and translation parts, respectively, of the Poincaré group, O'Raifeartaigh shows that there are only four possible options: (1) $S = P$ and $G = L \oplus T$ [L is the inhomogeneous Lorentz group (i.e. the Poincaré group) and T the global internal symmetry group] — an uninteresting option; (2) S is Abelian but larger than and containing P. Physically, this option introduces an invariant translational algebra of more than four dimensions; (3) S is non-Abelian and contains P. Physically, this option implies that, except for trivial Abelian representations, Hermitian conjugation cannot be defined for finite-dimensional representations; (4) $P \cap S = 0$ (i.e. P "intersects" S so that no element of P lies completely in S). Physically, this option means that the parameters corresponding to P_μ have a "non-compact" range so that multiplets can not be defined. Clearly, all four options are "no-go".

The O'Raifeartaigh "no-go" theorem went quite far in discouraging further attempts to mix the Poincaré group and global internal symmetries but it limited itself to Lie groups of finite order and focused its attention on single-particle spectra and their classification. The Coleman-Mandula "no-go" theorem [1.170] went even further in proving the impossibility of combining (in a non-trivial way) the Poincaré group and global internal symmetries by extending the theorem to infinite-parameter groups (not just compact Lie groups) and to the "dynamical" S matrix itself. By making a number of plausible assumptions: Lorentz invariance, particle finiteness (i.e. for any finite mass M, there is only a finite number of particle states with mass below M), "weak" elastic analyticity, and, finally, that the S matrix is not the identity, Coleman and Mandula were able to demolish all hope of a non-trivial mixing of space-time and global internal symmetries into a mathematically tractable and dynamically interesting combined group.

By the late 1960s, the stage was set for the big jump from global internal symmetries to gauge symmetries in the quantum field theories of the strong and weak interactions. On the negative side — in strong interactions — the afore-mentioned "no-go" theorems ruled out a "dynamical unification" of the Poincaré group with the known global internal symmetries. Moreover, the S matrix — Regge pole — duality hypothesis program in the strong interactions was running into snags and blind alleys (see §1.2b). In the case of weak interactions, the inability to calculate higher-order processes within the framework of the V-A theory was becoming frustrating. On the positive side, the quark flavor model of hadrons was in good shape [the best days were still to come with the deep inelastic lepton-hadron scattering experiments establishing quarks as "physical entities" rather than mere "mathematical fictions" and, above all asymptotic freedom]; the mixed $SU(6)$ quark flavor-spin model had given powerful indications of a new quark degree of freedom (i.e. global "color"), and the successes of current algebra with PCAC [giving respectability to the spontaneous breaking of global internal symmetries in particle physics] were multiplying. The time was ripe to rediscover the Yang-Mills 1954 paper [1.51] — as well as the papers about the same time on the renormalization group [1.40] — and to realize that the application of the gauge principle to non-Abelian global internal symmetries might bring success. With hindsight, the magical power of gauging a quantum field theory — to convert "symmetry into dynamics" [1.17] should not have come as such a great surprise. After all, the highly successful gauged non-chiral, Abelian quantum field theory of electromagnetism (QED) had been in existence since ca. 1950. And yet, it took until the early 1970s before the essential ingredients of QED were brought together with some new ideas to convert the global, non-chiral, non-Abelian "color" symmetry of the strong interaction into quantum chromodynamics (QCD), and to transform the global, non-Abelian quark-lepton chiral flavor symmetry of the weak and electromagnetic interactions into quantum flavordynamics (QFD).

The three seminal ideas that made possible the successful formulation of QCD and QFD as gauge theories encompassed the following: (1) instead of gauging one global internal symmetry — that of electric charge to yield a non-chiral (vector-like) $U(1)$ gauge group — as is done in QED, the global non-Abelian $SU(3)$ "color" group was gauged in a non-chiral fashion to yield the $SU(3)$ color gauge theory of the strong interaction (QCD). As for the weak interaction, the chiral quark and lepton quark flavors of each generation were gauged as lefthanded doublets and righthanded singlets under the non-Abelian chiral [lefthanded (L)] group $SU(2)_L \times U(1)_Y$ (Y is the "weak hypercharge") that — in conjunction with spontaneous symmetry breaking (SSB) — could accommodate the chirality-invariant (maximally parity-violating) weak and (parity-conserving) electromagnetic interactions; (2) the second seminal idea was the realization that the short-range character of the strong interaction among quarks — in apparent contradiction to the masslessness of the gluon gauge fields — could be explained by exploiting the asymptotic freedom and confinement properties of unbroken non-Abelian $SU(3)$ color, in combination with the finite scale introduced through the renormalization of QCD. In the case of the extremely short-range character of the weak interaction, the Higgs SSB mechanism to generate massive (weak) gauge bosons was utilized. The SSB of chiral $SU(2)_L \times U(1)_Y$ group down to the non-chiral $U(1)_{EM}$ group provided the link between the weak and electromagnetic interactions; and, finally (3) it would be necessary to extend the regularization and renormalizability methods that had proved to be so successful for unbroken Abelian gauge theory (i.e. electromagnetism) to non-Abelian (Yang-Mills) gauge theories, both unbroken and spontaneously broken. The actual proof of the renormalizability of unbroken and broken Yang-Mills theories did not come until 1971 [1.106]. These three seminal ideas were key ingredients of the conceptual foundation of the fully dynamical, renormalizable and surprisingly predictive gauge theory of the strong and electroweak interactions that is called the standard model.

The definitive formulation of the standard model was accomplished, during what we call the Heroic Period (1960–75), through a series of incremental steps that drew their inspiration from the three seminal ideas mentioned above, as well as from some major experimental breakthroughs. In the next four chapters, we give a rather detailed account of the conceptual evolution and "watershed" experiments that produced the remarkably successful standard model. In this first of the four chapters, we go back to Dirac's early derivation of the gauge principle and trace the development of this concept from its Abelian realization to its application in non-Abelian quantum field theories. In §4.1–§4.2a, we summarize the highlights of the Abelian $[U(1)]$ gauge theory of electromagnetism that are especially relevant for meeting the challenges of gauging non-Abelian global internal symmetries. We not only work out the basic mathematical formalism for a $U(1)$ gauge theory but also discuss Dirac's early derivation of the relation between the "non-integrable phase" associated with a quantum mechanical wave function and the Abelian gauge field of electromagnetism. Dirac's "non-integrable phase" is responsible for the Aharonov-Bohm effect encodes the quantum mechanical difference between the gauge field and the (antisymmetric) second-rank field tensor of electromagnetism; this difference becomes even more significant for Yang-Mills fields. In §4.3, we extend the gauge principle to global non-Abelian internal symmetry groups, both in Dirac's "non-integrable phase" language and in the conventional Lagrangian formulation.

We continue, in §4.4, with a review of the SSB Higgs mechanism for gauge groups (both Abelian and non-Abelian) and anticipate the result that SSB — despite its generation of finite masses for charged vector (gauge) fields — does not alter the renormalizability of either Abelian or Yang-Mills gauge theories. We also present some examples of spontaneously broken gauge groups (both Abelian and non-Abelian) in particle and condensed matter physics. Gauged quantum field theories are not very interesting if they lack renormalizability and, hence, in §4.5 and in §4.5b, we sketch the regularization and renormalization properties of both Abelian and Yang-Mills gauge theories in order to provide sufficient preparation for the discussion of the renormalization group to be given in §5.2a. Before moving on to the discussion of regularization and renormalization in the more complicated Yang-Mills gauge theories, we insert a brief section (§4.5a) to rebuttress the claimed renormalization triumphs of QED — ever since ca. 1950 — by citing several recent precision tests of QED; discussion of the experimental tests of renormalized non-Abelian QCD and QFD is left to Chapters 5 and 6 respectively. We conclude the present chapter with a final section (§4.6) on the hypothetical Dirac magnetic monopole, explain its connection to electric charge quantization through the "Dirac quantization condition", and note the topological ramifications of the Dirac monopole and its "kinship" to the 't Hooft-Polyakov monopole of Yang-Mills theories (see §10.2c) and to the "Wess-Zumino" term in the Skyrme model (see §10.5c).

§4.2. Abelian gauge theory of electromagnetism (QED)

The beginnings of QED date back, of course, to the Maxwell equations but Dirac's relativistic theory of the electron was needed to move quantum electrodynamics forward — in the hands of Heisenberg, Pauli and Dirac himself — during the 1930s (see §1.1). During this stage, good first-order relativistic calculations could be carried out — Möller scattering, Bhabha scattering, etc. — but the role of gauge invariance in electromagnetic theory was not fully appreciated until the ultraviolet divergences inherent in the higher-order processes (e.g. the Lamb shift and the anomalous magnetic moment of the electron) further propelled the development of the renormalization techniques that came to define QED [1.34]. It is true that as early as 1918 (a decade before quantum mechanics was born) Weyl — in his attempt to unify Einsteinian gravity and electromagnetism — had introduced the concept of "measure-invariance" or "calibration invariance" [1.177] (which was the "non-integrable phase factor" without the i [see Eq. (4.7)]. Einstein strongly objected to Weyl's "non-integrable scale factor" (i.e. "measure invariance") since it would imply that a "clock's measure of time depends on its history" and that would mean the end of physical law [4.3].

With the advent of quantum mechanics, it became clear that the well-known operator $(p_\mu - eA_\mu)$ should be replaced by $i(\partial_\mu + ieA_\mu)$ so that Weyl's original "non-integrable scale factor" $\exp(\int_C A_\mu dx_\mu)$ should be replaced by the "non-integrable phase factor" $\exp(-ie\int_C A_\mu dx_\mu)$ [4.4]. Weyl's initial "measure invariance" became "phase invariance" and was translated into English as "gauge invariance" [4.5]. It was Weyl's second formulation of the "non-integrable phase factor" and its implied gauge principle [4.6] that Dirac — in his 1931 paper [1.39] — established as fully compatible with the general principles of quantum mechanics and, indeed, a consequence of them. Before we discuss Dirac's "derivation" of the "non-integrable phase factor" and the implied gauge principle, we supply the standard Lagrangian formulation of gauge invariance in QED so that the formalism for discussing the "Aharonov-Bohm-type" consequences of the "non-integrable phase" in QED is available to us [there are some subtle distinctions between the "non-integrable phase" — denoted by Δ — and the "non-integrable phase factor", $e^{i\Delta}$ — which is discussed later]. With the knowledge of the gauge formalism for QED and Dirac's "non-integrable phase" approach to the gauge principle for Abelian groups, we can continue with Yang-Mills fields and treat their gauging both from the "non-integrable phase" and Lagrangian vantage points.

Gauging in QED within the Lagrangian framework consists of changing the global Abelian $U(1)$ group of electric charge, $e^{-i\alpha Q}$ (α is a constant), into $e^{-i\alpha(x)Q}$ [$\alpha(x)$ is a function of the space-time coordinates]. This simple modification has drastic consequences for a gauge (vector) field in the presence of matter (e.g. a charged lepton field). As is well-known, the Lagrangian $\mathcal{L}$ for the electromagnetic field in the presence of electrons is invariant under the global symmetry $\psi \to e^{-i\alpha Q}\psi$ (ψ is the Dirac spinor). However, if one wishes to maintain the invariance of $\mathcal{L}$ under the $U(1)$ gauge transformation: $\psi \to e^{-i\alpha(x)Q}\psi$, this invariance can only be achieved if ∂_μ is replaced by the covariant derivative $D_\mu = \partial_\mu + ieA_\mu$ in $\mathcal{L}$. This follows from the fact that the covariant derivative D_μ has the property:

$$D_\mu\psi \to e^{-i\alpha(x)Q}D_\mu\psi \tag{4.2}$$

i.e. $D_\mu\psi$ transforms under the gauge transformation as ψ does under the (global) constant phase transformation. Moreover, a vector gauge field (four-vector potential for electromagnetism) A_μ must be introduced that transforms under $\alpha(x)$ as follows:

$$A_\mu(x) \to A_\mu(x) + \frac{1}{e}\partial_\mu\alpha(x) \tag{4.3}$$

The invariance of the matter Lagrangian in the presence of the photon field A_μ is now guaranteed; but if we wish A_μ to have dynamical significance, i.e. to propagate, we must add a gauge-invariant "kinetic energy" term of the form $F_{\mu\nu}F^{\mu\nu}$ to $\mathcal{L}$ (where the antisymmetric second-rank field tensor, $F_{\mu\nu} = \partial_\mu A_\nu - \partial_\nu A_\mu$, is the electromagnetic field $\underline{E}$ and $\underline{H}$).

The requirement of gauge invariance for the $J = \frac{1}{2}$ electron field ψ, plus the requirements of Lorentz invariance, renormalizability and invariance under the discrete symmetries of space inversion and time inversion lead to the unique Lagrangian for the $J = \frac{1}{2}$ electron field in the presence of a (vector) gauge field (γ^μ is the Dirac operator and m the electron mass), namely:

$$\mathcal{L} = \bar{\psi}(i\gamma^\mu D_\mu - m)\psi - \frac{1}{4}F_{\mu\nu}F^{\mu\nu} \tag{4.4}$$

It should be noted here that the choice of $F_{\mu\nu}F^{\mu\nu}$ as the "kinetic energy" or (free-field) term is dictated by considerations of renormalizability ($F_{\mu\nu}F^{\mu\nu}$ is a dimension 4 term — see §4.5) and invariance under the discrete symmetries of space inversion and time inversion. The renormalizability argument actually permits the term $F_{\mu\nu}\tilde{F}^{\mu\nu}$ ($\tilde{F}^{\mu\nu} = \frac{1}{2}\epsilon^{\mu\nu\rho\sigma}F_{\rho\sigma}$ is the dual tensor to $F_{\mu\nu}$)

— which has the same dimension as $F_{\mu\nu}F^{\mu\nu}$; however, $F_{\mu\nu}\tilde{F}^{\mu\nu}$ is not invariant under space inversion nor time inversion (this term is the counterpart to the "strong" CP violation term in QCD — see §10.3c). By accepting the phenomenological evidence concerning parity conservation and time reversal invariance in electromagnetism, only the term $F_{\mu\nu}F^{\mu\nu}$ remains.

Before we discuss the implications of the basic Eq. (4.4) for the QED Lagrangian, we digress to record the comparable gauged Lagrangian for a complex scalar field with its global $U(1)$ charge symmetry; this will be useful later (e.g. §6.2) in considerations concerning the Higgs particle and clarifies a number of differences between boson and fermion fields. It will be recalled that the Lagrangian of a complex scalar field is [see Eq. (2.1)]:

$$\mathcal{L} = \partial_\mu \phi^\dagger \partial^\mu \phi - \mu^2 \phi^\dagger \phi - \lambda(\phi^\dagger \phi)^2 \qquad (4.5)$$

$\mathcal{L}$ is invariant under the global transformation $\phi(x) \rightarrow e^{-i\alpha}\phi(x)$, and yields the conserved charge $Q = -i \int d^3x (\phi\pi - \pi^\dagger \phi^\dagger)$. Equation (4.5) is gauged — following the procedure in the case of the Dirac field — by letting α become a function of the space-time coordinates [i.e. by replacing the constant α by $\alpha(x)$ and hence replacing $\partial_\mu \phi$ by $D_\mu \phi$]; we thus arrive at the result that the total gauged Lagrangian for a complex scalar field in the presence of electromagnetism is:

$$\mathcal{L} = (D_\mu \phi)^\dagger (D^\mu \phi) - \mu^2 \phi^\dagger \phi - \lambda(\phi^\dagger \phi)^2 - \frac{1}{4}F_{\mu\nu}F^{\mu\nu} \qquad (4.6)$$

The generalization of Eq. (4.6) to a non-Abelian scalar field [e.g. when ϕ belongs to the $SU(2)$ group as in the electroweak case] will be given in §4.3.

Returning now to Eq. (4.4) representing QED, we note that elaborate calculations employing renormalization techniques have been carried out on the basis of Eq. (4.4). We do not give full details of such calculations (since many excellent treatments are available [4.7]) although we summarize the renormalization formalism for gauge groups (both Abelian and non-Abelian) in §4.5 in preparation for the renormalization group-theoretic derivation of asymptotic freedom in QCD (in §5.2a). Instead, we comment on the key dynamical consequences of imposing gauge invariance on the global $U(1)$ symmetry of electric charge for the Dirac field, and then examine more carefully the physical meaning of the gauge principle in its application to electromagnetism by looking at Dirac's non-"integrable phase" (and the closely-related Aharonov-Bohm effect).

What are the consequences of imposing gauge invariance on a global Abelian charge group, which have motivated the use of QED as a paradigm

for the non-Abelian gauge theories of QCD and QFD? We list the consequences that are especially relevant for distinguishing between Abelian and
non-Abelian gauge theories:

(1) The most important consequence of gauging the global $U(1)$ (electric)
charge group is that the requirement of gauge invariance can only be met
through the introduction of a vector (gauge) field that induces an interaction
term with the matter field of a well-defined "Yukawa" ("minimal coupling")
form, i.e. $e \cdot j_\mu A^\mu$, where $j_\mu = \bar{\psi}\gamma_\mu\psi$ is the vector current associated with
the Dirac matter field and A^μ is the (photonic) gauge field. It is this "mini-
mal" interaction term that is responsible for the dynamical content of QED.
While the small value of the renormalized fine structure constant α_e at low
energies ($\alpha_e = \frac{e^2}{4\pi} \sim \frac{1}{137}$) justifies the use of perturbation theory in the calculation of "lowest-order" processes, the QED predictions of "higher-order"
processes diverge and can only be defined when renormalization methods are
applied. The tremendous success of "renormalized" QED in a wide range of
applications [1.121] prompted a major effort to prove that suitably modified
renormalization techniques also worked for Yang-Mills theories (both unbroken and spontaneously broken), a feat which was accomplished in the early
1970s [1.106]; (2) a non-vanishing mass for the gauge field (photon in Abelian
QED) requires a term of the form $m_\gamma^2 A_\mu A^\mu$ (m_γ is photon mass) in Eq. (4.4);
such a term is not gauge invariant and must therefore be absent, i.e. the gauge
field must be massless. A zero mass photon ($m_\gamma = 0$) is not only supported
by experiment to very high accuracy ($m_\gamma < 3 \times 10^{-54\pm2} g$ [4.8]) but it is also
the key to the renormalization program in QED (see §4.5a). In addition,
the long-range Coulomb potential follows from the massless gauge field feature of QED since the range of a gauge boson-mediated interaction (in the
non-relativistic limit) varies inversely as the mass of the gauge field (see Table 1.4). It can also be shown that the gauging of global non-Abelian internal
symmetry groups requires massless gauge fields [4.9]. However, such a requirement runs counter to the known short-range character of the strong and
weak interactions and the twin challenges were precisely to construct phenomenologically correct gauge theories of the strong and weak interactions
despite this constraint; and, finally, (3) the single gauge boson in Abelian
gauge theory (photon in QED) possesses no (electric) charge. This has the
consequence that a pure Abelian gauge Lagrangian (i.e. when the matter
field is absent) carries no interaction and is thus a "free field" theory. The
situation is very different for a non-Abelian gauge group, where the gauge
field quanta carry non-vanishing charges so that a pure non-Abelian gauge

field Lagrangian carries interaction. Indeed, it is precisely the gauge field charges present in a Yang-Mills theory (see Fig. 1.4) that are responsible for the strikingly different properties of a non-Abelian gauge group (e.g. asymptotic freedom and confinement) (see §5.2a and §5.3a) vis-a-vis an Abelian gauge group.

§4.2a. Dirac's non-integral phase and the Aharonov-Bohm effect

It is clear from the previous section that the gauging of a global internal symmetry of a quantum field theory has well-defined and far-reaching dynamical consequences and that the physics is significantly different for non-Abelian as compared to Abelian gauge theories. The power of the gauging operation has proved to be so impressive that it behooves us to step back and try to pinpoint the deeper physical meaning of "gauging a global internal symmetry". Within the context of Abelian gauge theory, Dirac long ago [1.39] grappled with the problem of justifying Weyl's "gauge principle" [4.6] in terms of the basic principles of quantum mechanics and was led to the concept of "non-integrable phase" of a wavefunction that only makes physical sense if a gauge field is present; it is the "non-integrable phase", defined in terms of the electromagnetic gauge field, that provides the rationale for the Aharonov-Bohm effect [1.38]. Dirac's 1931 paper [1.39] actually consists of two parts: the first part explains why the "non-integrable phase" and an electromagnetic gauge field (the "gauge principle") are necessary consequences of quantum mechanics and the correct physical interpretation of a wavefunction phase; the second part examines the conditions under which the existence of magnetic monopoles is compatible with electromagnetism and quantum mechanics. We discuss here Dirac's argument for the "non-integrable phase" (and its relation to the Aharonov-Bohm effect) and defer until §4.6 our comments on Dirac's magnetic monopole and its topological ramifications.

In arriving at the concept of "non-integrable phase", Dirac first pointed out that, since any (quantum mechanical) wave function $\phi(x)$ (whether relativistic or non-relativistic) is a complex function of the space-time coordinates x, it can be written in the form $\phi(x) = A(x)e^{i\gamma(x)}$ [where $A(x)$ and $\gamma(x)$ are real functions of x]. Normalization of ϕ requires that $\phi(x)$ is determined by the wave equation except for a constant of modulus unity; hence, $\gamma(x)$ is undetermined for a given x and only the difference between values of $\gamma(x)$ at two different points is physically meaningful. Based on this observation, Dirac then postulates that the difference in γ is only defined for two neighboring space-time points so that a definite phase difference in the wave function only emerges between points at a finite distance relative to some curve joining them. Consequently, different curves (taken between the same two points) can yield different phase differences so that the total change in phase around a closed curve C need not vanish. This total change in phase around C is called the "non-integrable phase" Δ — because it is path-dependent — and $e^{i\Delta}$ is the "non-integrable phase factor".

The crucial next step is to find a general expression for Δ. Dirac accomplishes this by pursuing his argument in a situation that requires knowledge of a transition amplitude between two quantum mechanical states (described by two different wavefunctions with different phases) instead of knowledge of the probability density (described by one quantum mechanical state). Consideration of the transition amplitude persuades Dirac that the integral which defines the transition probability can only possess a definite modulus if there is a definite phase difference (in the transition probability amplitude) between any two points, whether neighboring or not. This condition can only be met if the change in phase of the transition probability amplitude, taken around a closed curve, vanishes. If this is so, the change in phases of the two contributing wavefunctions (to the transition probability amplitude) around a closed curve must be equal and opposite in sign so that the change in phase of any wavefunction around any closed curve must be identical for all wavefunctions.

The last result is decisive: if Δ is to be independent of the wavefunction and to represent the total change in phase around a finite closed space-time curve C, the integrand of Δ can not itself be a function with a definite value of the phase difference — call it $\beta(x)$ — at each point of C but must be represented by the derivatives of the phase difference, i.e. $\alpha_\mu = \partial_\mu \beta(x)$, where α_μ is a function that, in general, does not satisfy the conditions of integrability (i.e. $\partial_y \alpha_x \neq \partial_x \alpha_y$, etc.). The "non-integrable phase" Δ now becomes $\Delta = \oint_C \alpha_\mu dx^\mu$, which can be shown to be consistent with the principle of superposition of quantum mechanical wavefunctions. Further, it is clear that the same value of Δ results for different wavefunctions if the β's themselves differ by the gradient of a scalar wavefunction $\phi(x)$. Having shown that the "non-integrable phase" Δ is independent of the wavefunction and that the $\beta(x)$ which defines it is arbitrary to the extent of the gradient of a scalar wavefunction, Dirac argues that Δ must be a quantity determined by a force field and that, in the case of electromagnetism $\partial_\mu \beta(x) = \alpha_\mu(x) = eA_\mu$, where A_μ is the electromagnetic four-potential; it follows that A_μ is arbitrary to the extent of $\frac{1}{e}\partial_\mu\beta(x)$, which is the "gauge principle" [see Eq. (4.3)].

In sum, Dirac's analysis of the physical meaning of quantum mechanical phase leads him to the conclusion that the "non-integrable phase" $\Delta = e\oint_C A_\mu dx^\mu$ is independent of the wavefunction and depends only on the line integral of the gauge field around a closed curve C which, by Stoke's theorem, can be written:

$$\Delta = e\oint_C \underline{A}\cdot d\underline{l} = e\int_S \text{curl}\,\underline{A}\cdot d\underline{\sigma} = e\Phi \qquad (4.7)$$

where Φ is the magnetic flux through the surface S bounded by the closed curve C. Equation (4.7) is manifestly invariant under the gauge transformation $\underline{A}(\underline{x}) \to \underline{A}(\underline{x}) + \frac{1}{e}\nabla\beta(\underline{x})$ and tells us that the condition for the integrability of $\oint \underline{A}\cdot d\underline{l}$, namely $B_k = \partial_i A_j - \partial_j A_i = 0$ $(k = 1, 2, 3)$ need not, in general, be satisfied.

Dirac's elucidation of the connection between the "non-integrable phase" Δ and the (Abelian) gauge theory of electromagnetism goes far towards explaining the physical meaning of quantum mechanical phase and in spelling out the rationale for a direct measurement of Δ to test the gauge principle in electromagnetism; this test is known as the Aharonov-Bohm effect. Actually, it was not until 1959 that the physical consequences of Dirac's "non-integrable phase" Δ in electromagnetism were spelled out by Aharonov and Bohm [1.38]. Aharonov and Bohm argued that, despite the fact that the vector potential (gauge field), A_μ, in electromagnetism is unspecified to the extent of the "gauged phase" transformation $\partial_\mu\beta(x)$ and produces no observable effects in classical electromagnetism [only the gauge-invariant second-rank tensor field $F_{\mu\nu}$ enters classical electromagnetism through the "Lorentz force"], A_μ should still produce observable (gauge-invariant) phase-shift (interference) effects in quantum mechanics under suitable conditions. That is, the "non-integrable phase", Δ, given by Eq. (4.7), does not vanish even when $F_{\mu\nu} = 0$ outside a closed curve, as long as $\oint A_\mu dx^\mu \neq 0$ on the closed curve that encloses the magnetic flux. The experiment proposed by Aharonov and Bohm to test the reality of Δ was the two-slit electron interferometer experiment first used by Davisson and Germer [4.10] to establish the wave nature of electrons.

In view of the importance of the "non-integrable phase" (Aharonov-Bohm) effect in clarifying the physics of the gauge principle in quantum field theories, we digress to explain how the two-slit electron interferometer experiment is utilized to detect the "Aharonov-Bohm" effect. The "Aharonov-Bohm" experiment has the following essential features: consider a source of electrons at P (see Fig. 4.1) moving through a field-free ($B = 0$) region, of which a small part (that is inaccessible to the electrons) contains a homogeneous magnetic field $B \neq 0$ (pointing in a vertical direction out of the paper). The magnetic field B, presumed to be produced by a long solenoid of small cross-section, is restricted to a sufficiently small area so that the two electron beams (1) and (2) can stay within the $B = 0$ region, and then meet in the "interference region" P'. Let us now assume that the current in the solenoid has been switched off so that $B = 0$ inside the small circle shown as "solenoid" in Fig. 4.1; the interference pattern (as shown by the solid curve on the R.H.S. of Fig. 4.1) is centered symmetrically (at $y = 0$). The wave functions representing the two electron beams are denoted by $A_1 e^{i\gamma_1}$ and $A_2 e^{i\gamma_2}$, where the phase difference $\delta = \gamma_1 - \gamma_2$ determines the interference pattern. If the distance between the slits and detector screen is L (see Fig. 4.1), and the difference in the path lengths for beams (1) and (2) is a, we get $\delta = a/\bar{\lambda}$ (with $\bar{\lambda} = \lambda/2\pi$ and λ the wavelength of the diffraction pattern). Further, if $y << L$, then $a = yd/L$ and hence:

$$\delta(B = 0) = \frac{y}{L} \cdot \frac{d}{\lambda} \qquad (4.8)$$

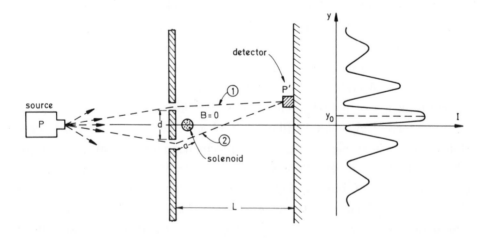

Fig. 4.1. Two-slit electron interferometer test of the Aharonov-Bohm effect.

From Eq. (4.8), it follows that there is maximum (constructive) interference when $\delta = 0$ (i.e. $y = 0$) and minimum (destructive) interference when $\delta = \pi$. This is the explanation for the statement that when $B = 0$ (solenoid turned off), the maximum of the interference pattern is centered at $y = 0$. The situation changes when $B \neq 0$ and the "non-integrable phase" effect (i.e. the "Aharonov-Bohm" effect) takes hold.

To find $\delta(B \neq 0)$, i.e. the phase shift δ when the magnetic field does not vanish, one must take into account the changes in the phases of the wavefunctions representing the two electron beams (1) and (2), intraversing the different paths from P to P', namely:

$$\gamma_1 \to \gamma_1 + e \int_P^{P'} \underline{A} \cdot d\underline{l} \qquad \text{(over path (1))}$$

$$\gamma_2 \to \gamma_2 + e \int_P^{P'} \underline{A} \cdot d\underline{l} \qquad \text{(over path (2))} \qquad (4.9)$$

Consequently, the difference in phase shifts for the two beams in the presence of a finite B is given by:

$$\delta(B \neq 0) = \delta(B = 0) + e[\int_P^{P'} \underline{A} \cdot d\underline{\ell} - \int_P^{P'} \underline{A} \cdot d\underline{\ell}] \qquad (4.10)$$
$$\text{(path (1))} \quad \text{(path (2))}$$

or:

$$\delta(B \neq 0) - \delta(B = 0) = e \oint_C \underline{A} \cdot d\underline{\ell} = \Delta = e\Phi \qquad (4.11)$$

where the closed path C is $PP'P$ (see Fig. 4.1) and the incremental phase difference $[\delta(B \neq 0) - \delta(B = 0)]$ — due to the solenoidal field $B \neq 0$ — is precisely the "non-integrable phase" $\Delta = e\Phi$ of Eq. (4.7). Since the diffraction maximum of the interference pattern arising from the two electron beams is centered at $y = 0$ when $B = 0$, Eq. (4.11) tells us that the diffraction maximum of the interference pattern when $B \neq 0$ is located at $y_0 = \Delta(L\bar{\lambda}/d)$. Since Δ is a gauge-invariant quantity, the shift in the diffraction maximum (i.e. y_0) is measurable; indeed, the existence and magnitude of the "Aharonov-Bohm" effect in the two-slit experiment has been confirmed by means of a "magnetic whiskers" experiment [4.11].

Three comments are worth making about the Aharonov-Bohm effect to highlight the relationship to Dirac's "non-integrable phase" analysis: (1) both electron beams (1) and (2) in the two-slit experiment (see Fig. 4.1) stay within the classical field-free region [so that there are no classical electromagnetic (Lorentz) forces operating on the electron beams] and, while the quantum mechanical phase shifts generated in the two wavefunctions (in going from P to P') by the vector potential A_μ are not separately measurable, the difference between the phase shifts is measurable because it reduces to a dependence on the closed path C, which is a gauge-invariant quantity. Hence, when the integral over the closed path C encircles a finite magnetic field, the resulting finite "non-integral phase" Δ has physical consequences in quantum mechanics even though the magnetic field never acts "classically" on the electron beams; (2) by means of Stoke's theorem, the closed path integral over the gauge-dependent vector potential is transformed into a surface integral (one dimension higher) over the gauge-invariant magnetic field, which implies that $[\delta(B \neq 0) - \delta(B = 0)]$ is a physically measurable quantity. This will be of interest later in connection with the topologically non-trivial Chern-Simons terms of QCD (see §10.3c and §10.5c); and (3) in QED, we are dealing with an Abelian gauge group with only one (self-commuting) charge and hence the infinitesimal local phase change in going from one space-time point to a neighboring one is parameterized simply by $eA_\mu(x)$ and the $F_{\mu\nu}$ fields depend linearly on A_μ. The situation is more complicated for Yang-Mills fields (required in QCD and QFD) — where the charges are non-commuting — and leads to a non-linear dependence of the antisymmetric second-rank tensor fields on the gauge fields (see §4.3). With the mathematical properties and physical significance of the gauge principle in Abelian QED spelled out, we extend our discussion in the next section (§4.3) to non-Abelian gauge groups.

§4.3. Non-Abelian gauge groups

In §3.2, we discussed the non-Abelian global internal symmetry groups [we continue to focus on the $SU(N)$ groups], and wrote the non-Abelian unitary global transformation in the form: $\exp(-i\alpha_a T^a)$ $[a = 1, 2, ...(N^2-1)]$ with the α_a's constant phases and the T^a's generators. Mathematically, the gauging of a non-Abelian global internal symmetry group consists of allowing the α_a's to be functions of space-time, in complete analogy to the Abelian $U(1)$ case considered in §4.2 of this chapter. Before working out the far-reaching consequences of this simple gauging operation, we exploit the "non-integral phase" approach to derive an expression for the non-Abelian anti-symmetric second-rank tensor field in terms of the gauge fields. In the "non-integrable phase" approach, one starts out with $e^{i\Delta_{AB}}$ where:

$$\Delta_{AB} = g \; P \int_A^B A_\mu(x) dx^\mu \qquad (4.12)$$

In Eq. (4.12), A_μ is no longer a single gauge field but a sum over the (N^2-1) gauge fields $A_\mu^a(x)$ [i.e. $A_\mu = A_\mu^a(x)T^a$], g is the gauge coupling constant, and the line integral in Eq. (4.12) is taken over an arbitrary curve from the space-time point A to B [P reminds us that the ordering of the A_μ's must be taken into account because of their non-commutativity at different space-time points].

The calculation of $F_{\mu\nu}$ (the anti-symmetric tensor field in the non-Abelian case) is fairly straightforward if one calculates the "non-integrable phase" around a closed path defined by an infinitesimal parallelogram, with sides dx^μ and dx'^ν [so that the area of the infinitesimal parallelogram is $d\sigma^{\mu\nu} = dx^\mu dx'^\nu$], and allows for the non-commutativity of A_μ^a and A_μ^b at different space-time points; one gets [4.12]:

$$\exp(i\Delta) \simeq 1 + ig\, F_{\mu\nu} d\sigma^{\mu\nu} \tag{4.13}$$

where:

$$F_{\mu\nu} = F_{\mu\nu}^a T^a \quad \text{with} \quad F_{\mu\nu}^a = \partial_\mu A_\nu^a - \partial_\nu A_\mu^a + g f_{abc} A_\mu^b A_\nu^c \tag{4.14}$$

and f_{abc} are the $SU(N)$ structure constants. The quadratic term in the A_μ^a's vanishes for the Abelian group [so that $F_{\mu\nu}$ reduces to the gauge-invariant fields ($\underline{E}, \underline{H}$) of electromagnetism] but this is not the case for the non-Abelian $F_{\mu\nu}^a$'s and reflects the fact that Yang-Mills fields carry charge. While each $F_{\mu\nu}^a$ is not gauge-invariant — only covariant — Tr $(F_{\mu\nu}^a T^a)$ is gauge invariant because the quadratic term in the A_μ^a's gives zero contribution by virtue of the anti-symmetry of the structure constants f_{abc} in Eq. (4.14). By considering non-Abelian processes (e.g. in QCD), which depend on Tr $(F_{\mu\nu}^a T^a)$, one can deduce from the "non-integrable phase factor" possible Aharonov-Bohm-type experiments in QCD [4.13]. However, none of these experiments appears feasible, and so we revert to the traditional approach in developing the mathematical formalism for non-Abelian gauge theories.

In proceeding with the customary derivation of the non-Abelian gauge group formalism, we denote the generators of the $SU(N)$ gauge group by Λ^a [$a = 1, 2, \ldots (N^2 - 1)$], where the Λ^a's are $N \times N$ traceless Hermitian matrices that satisfy the commutation relations:

$$[\Lambda^a, \Lambda^b] = i f^{abc} \Lambda^a \tag{4.15}$$

If the fermion matter field ψ belongs to some representation of $SU(N)$, with representation matrices T^a, then $\psi' = U\psi$ where $U = e^{-i\alpha_a T^a}$ [$a = 1, 2, \ldots (N^2 - 1)$], where the T^a's satisfy the same algebra (4.15) as the Λ^a's; of course, $T^a \equiv \Lambda^a$ when ψ belongs to the fundamental representation of $SU(N)$. Thus far, we have not specified whether the α_a's are constants or functions of space-time; if they are constants [i.e. $SU(N)$ is global], the free

Lagrangian for the fermion matter field is, as before:

$$\mathcal{L}_0 = \bar{\psi}(x)(i\gamma^\mu \partial_\mu - m)\psi(x) \qquad (4.16)$$

However, if the α_a's are functions of space-time, i.e. $SU(N)$ is a gauge group, Eq. (4.16) is no longer invariant under the unitary transformation $\psi \to U\psi$.

To construct a gauge-invariant Lagrangian for Yang-Mills fields, we form the covariant derivative, with the help of the infinitesimal form of U, to obtain:

$$D_\mu \psi = (\partial_\mu + igT^a \cdot A_\mu^a)\psi \quad \text{(summation over } a) \qquad (4.17)$$

where the $A_\mu^a(x)$ $[a = 1, 2, \ldots (N^2 - 1)]$ are the gauge fields and g is the gauge coupling constant. We next demand, as in the Abelian case, that $D_\mu \psi$ possesses the same transformation property as ψ itself, i.e.

$$D_\mu \psi' \to U D_\mu \psi. \qquad (4.18)$$

Equation (4.18) implies that the following relation holds between $A_\mu^{a'}$ and A_μ^a:

$$T_a \cdot A_\mu^{a'} = U T_a \cdot A_\mu^a U^{-1} - \frac{i}{g}[\partial_\mu U]U^{-1} \qquad (4.19)$$

Equation (4.19) defines the exact transformation law (called the "large" gauge transformation in contrast to the infinitesimal or "local") for a Yang-Mills field and is useful when non-trivial topological effects in Yang-Mills theories are considered (e.g. instantons — see §10.4a); however, for most purposes, the infinitesimal form of U, i.e. $U \simeq 1 - i\alpha_a T^a$, suffices and Eq. (4.19) yields:

$$A_\mu^{a'} = A_\mu^a - \frac{1}{g}\partial_\mu \alpha^a(x) + f^{abc}\alpha^b(x)A_\mu^c(x) \qquad (4.20)$$

The second term on the R.H.S. of Eq. (4.20) is an obvious generalization of the Abelian case but the third term on the R.H.S. of Eq. (4.20) — having the transformation property of the adjoint representation under $SU(N)$ — is new, and tells us that the A_μ^a's must carry charges (in contrast to the neutral photon of the Abelian electromagnetic gauge field).

We next deduce the "kinetic energy" term associated with the $SU(N)$ gauge field [i.e. the counterpart of $-\frac{1}{2}F_{\mu\nu}F^{\mu\nu}$ of the $U(1)$ gauge group]. This is most easily done by, again, modeling the Abelian case and defining the antisymmetric tensor $F_{\mu\nu}^a$ in terms of the covariant derivative, thus:

$$(D_\mu D_\nu - D_\nu D_\mu)\psi \equiv ig\left(T^a F_{\mu\nu}^a\right)\psi \quad [a = 1, 2, \ldots(N^2 - 1)] \qquad (4.21)$$

From Eq. (4.21), one gets:

$$T^a \cdot F_{\mu\nu}^a = \partial_\mu(T^a A^a) - \partial_\nu(T^a A_\mu^a) - ig[(T^b A_\mu^b), (T^c A_\nu^c)] \qquad (4.22)$$

The "infinitesimal" form of Eq. (4.22) translates into:

$$F_{\mu\nu}^a = \partial_\mu A_\nu^a - \partial_\nu A_\mu^a + g f^{abc} A_\mu^b A_\nu^c \qquad (4.23)$$

Equation (4.23) is identical with the non-Abelian tensor field required by the "non-integrable phase" approach, as it should be [see Eq. (4.14)]. Eq. (4.23) implies that $F_{\mu\nu}^a$ is not gauge-invariant — unlike $F_{\mu\nu}$ in the Abelian case — but transforms under the adjoint representation of $SU(N)$. Since the "kinetic energy" of the non-Abelian field is of the form Tr $F_{\mu\nu}^a F^{a\mu\nu}$ (summation over a), it follows that this term will not disturb the gauge invariance of the total Lagrangian for a spin $\frac{1}{2}$ matter field in the presence of a non-Abelian $SU(N)$ gauge group, which is given by [D_μ is defined by Eq. (4.17)]:

$$\mathcal{L}_{\text{total}} = \bar{\psi}(i\gamma^\mu D_\mu - m)\psi - \frac{1}{4}\text{Tr } F_{\mu\nu}^a F^{a\mu\nu} \qquad (4.24)$$

The most striking aspect of the non-Abelian Lagrangian defined by Eq. (4.24) is that the pure gauge part (second term on the R.H.S.) contains self-coupling terms that are trilinear and quadrilinear in the gauge fields A_μ^a, namely (see Fig. 1.4):

$$-g f^{abc} \partial_\mu A_\nu^a A^{b\mu} A^{c\nu} - \frac{g^2}{4} f^{abc} f^{ade} A_\mu^b A_\nu^c A^{d\mu} A^{e\nu} \qquad (4.25)$$

These pure gauge terms give the distinctive stamp to non-Abelian gauge theory vis-a-vis Abelian gauge theory. It is these non-linear couplings of the Yang-Mills fields which, when carried over to gauged $SU(3)$ color, are responsible for the novel asymptotic freedom and confinement properties of the present-day gauge theory of the strong interaction. The important asymptotic freedom property of Yang-Mills theory is revealed when one studies the short-distance behavior of non-Abelian compared to Abelian gauge groups; this comparison is carried out with the help of the renormalization group and the results are summarized in §5.2a.

Several remarks should be made in concluding this section: (1) the pure gauge field term $-\frac{1}{4}$ Tr $F_{\mu\nu}^a F^{a\mu\nu}$ in the total Lagrangian is a "dimension 4" term that satisfies all the discrete symmetries. However, as has been mentioned earlier, in connection with Abelian gauge groups, it is possible to include a gauge-invariant, renormalizable, "dimension 4" term of the form: $F_{\mu\nu}^a \tilde{F}^{a\mu\nu}$ (where $\tilde{F}^{a\mu\nu} = \frac{1}{2}\epsilon^{\mu\nu\rho\sigma} F_{\rho\rho}^a$ is the dual anti-symmetric second-rank tensor). At first, it was argued that this term should be omitted because it could be written as the divergence of a four-current so that it would contribute only a surface term to the action and therefore could be neglected. However, it has been shown that there are so-called (classical) "instanton" solutions with non-trivial topological properties that do not fall off fast enough at infinity to permit neglect of the surface terms (see §10.3). These "instanton" effects are responsible for the triangular QCD anomaly and the "strong CP problem" in QCD and are discussed respectively in §7.2b and in §10.3c; (2) the generalization to non-Abelian Lie groups other than the $SU(N)$ $(N > 1)$ groups is straightforward. One simply has to make a suitable choice of generators, structure constants and fermion (matter) representations. We will have occasion in later chapters (Chapters 8–10) to consider compact non-Abelian Lie groups other than the class of $SU(N)$ $(N > 1)$ groups but the standard model only makes use of the $SU(2)$ and $SU(3)$ non-Abelian gauge groups and hence our $SU(N)$ discussion suffices for Chapters 5–6; (3) the number of gauge fields is always equal to the dimension of the adjoint representation of the non-Abelian group, i.e. $(N^2 - 1)$ in the case of $SU(N)$, the same as the number of generators; (4) like the Abelian gauge field, the non-Abelian gauge fields are massless. At first, this creates a problem for the finite-range strong interaction — with its massless gluons — but we explain its resolution in §5.2a in terms of the asymptotic freedom and confinement properties of non-Abelian $SU(3)$ color, working in combination with the finite renormalization scale; and, finally (5) initially, the massless gauge fields of QFD also create a problem for the short-range weak interaction but the resolution of this problem in QFD comes from invoking the Higgs SSB mechanism in gauge theories. For this purpose, we review in the next section the salient features of SSB in gauged quantum field theories (for both Abelian and non-Abelian groups), with particular reference to the fate of the "would-be" N-G bosons resulting from the operation of the Higgs mechanism.

§4.4. Spontaneous symmetry breaking (SSB) of gauge groups

The importance of the concept of spontaneous symmetry breaking (SSB) of a gauge symmetry group in modern particle physics is highlighted by the crucial role the SSB Higgs mechanism plays in the standard gauge theory of the electroweak interaction. Actually, we have already made reference to the determining role that the quasi-Higgs mechanism of "Cooper pairs" of electrons plays in achieving a phenomenologically correct theory of low temperature superconductivity. The role of SSB in particle and condensed matter physics is now unquestioned — with numerous physical examples — for both global and gauge, Abelian as well as non-Abelian, symmetry groups. We discussed (in §3.4) the SSB of global Abelian and non-Abelian groups, and found — thanks to the Goldstone theorem — that SSB gives rise to massless scalar (N-G) states in non-relativistic many-particle systems and to N-G bosons in quantum field theory, and both condensed matter and particle physics have been enriched by the recognition of the existence of N-G excitations (whether states or bosons!). In a similar fashion, a major breakthrough in modern particle theory occurred with the realization that, for gauge symmetry groups, the "would-be" N-G bosons disappear from the physical spectrum of particle states and instead supply longitudinal modes to the initially massless (transverse) gauge fields so that they acquire mass. This striking SSB effect due to the Higgs mechanism can be activated by a "fermion condensate" (serving as a "quasi-Higgs particle") or result directly from the intervention of an "elementary" Higgs particle. In §4.4a, we discuss the SSB of $U(1)$ gauge theory and apply the result to a well-known example taken from low temperature superconductivity. We then consider, in §4.4b, the SSB of non-Abelian gauge symmetry groups in preparation for the application of these results to the spontaneously broken gauge theory of the electroweak interaction (QFD) (in Chapter 6) and to other applications (see Chapters 7–10).

§4.4a. SSB of Abelian gauge groups and application to low temperature superconductivity

We first work out the SSB Higgs mechanism for a $U(1)$ gauge group because the formalism, as expected, is simpler than for non-Abelian symmetry groups. Again, the starting point is the complex scalar field, whose gauge-invariant Lagrangian is:

$$\mathcal{L} = (D_\mu \phi)^\dagger (D_\mu \phi) - V(\phi) - \frac{1}{4} F_{\mu\nu} F^{\mu\nu} \tag{4.26}$$

where:

$$D_\mu \phi = (\partial_\mu - ig A_\mu)\phi; \quad F_{\mu\nu} = \partial_\mu A_\nu - \partial_\nu A_\mu; \quad V(\phi) = \mu^2 \phi^\dagger \phi + \lambda (\phi^\dagger \phi)^2 \tag{4.27}$$

As in the global $U(1)$ case, for $\mu^2 > 0$, the vacuum stays "normal"; but, for $\mu^2 < 0$, a new vacuum state is generated, i.e.

$$< 0|\phi|0 > = \frac{v}{\sqrt{2}} \quad \text{with} \quad v = \sqrt{\frac{-\mu^2}{2\lambda}} \tag{4.28}$$

Since we are now dealing with a gauged $U(1)$ group, it is more convenient to parametrize $\phi(x)$ so that advantage can be taken of the arbitrariness in the gauge field, thus:

$$\phi(x) = \frac{1}{\sqrt{2}}\{v + \rho(x)\}e^{iG(x)/v}, \tag{4.29}$$

With this parametrization, the covariant derivative, $D_\mu \phi$, becomes:

$$D_\mu \phi = \frac{1}{\sqrt{2}}e^{iG/v}[\partial_\mu \rho - ig(\rho + v)B_\mu], \tag{4.30}$$

where A_μ has been replaced by $B_\mu = A_\mu + \frac{1}{g}\partial_\mu \alpha(x)$ $[\alpha(x) = -\frac{G(x)}{v}]$, i.e.:

$$B_\mu = A_\mu - \frac{1}{gv}\partial_\mu G. \tag{4.31}$$

Hence, the Lagrangian can be rewritten in the form:

$$\mathcal{L} = -\frac{1}{4}B_{\mu\nu}B^{\mu\nu} + \frac{1}{2}M_B^2 B_\mu B^\mu + \frac{1}{2}(\partial_\mu \rho)^2 + \mu^2 \rho^2$$

$$+ \frac{1}{2}g^2(\rho^2 + 2\rho v)B_\mu B^\mu - \lambda v \rho^3 - \lambda \rho^4 \tag{4.32}$$

It is evident from Eq. (4.32) that the massless vector field A_μ and the massless "would-be" Goldstone field G have been replaced by a new massive vector field B_μ (with $M_B = gv$). The mass of the "residual" neutral Higgs particle — as specified by the ρ field — is $\sqrt{2}|\mu|$. The reason that the "would-be" Goldstone boson is "eaten up" in the case of $U(1)$ gauge symmetry [but not for $U(1)$ global symmetry] is that the freedom which exists to choose the gauge field A_μ — up to a local phase function $\alpha(x)$ [see Eq. (4.3)] — is exploited through the choice: $\alpha(x) = -G/v$ — to eliminate one of the two real scalar fields, $G(x)$ [the other being $\rho(x)$], with which we started [see Eq. (4.29)]. In the final Lagrangian of Eq. (4.32), the "would-be" Goldstone boson $G(x)$ is manifestly gone ("eaten up") and has been converted into the longitudinal mode of the vector (gauge) field. We therefore end up with a massive vector boson $B_\mu(x)$ (three degrees of freedom) plus a massive scalar (neutral) boson $\rho(x)$ (one degree of freedom), a total of four. This should equal the number of degrees of freedom with which we started and, indeed, this is so because initially the massless vector (gauge) boson A_μ possesses two degrees of freedom and the complex (charged) scalar boson $\phi(x)$ also has two degrees of freedom, a total of four.

The importance of the Higgs mechanism in particle physics is that theories involving massive vector bosons are, in general, non-renormalizable; however, if the mass of the vector boson is generated by the SSB of a gauge group, it can be shown that renormalizability continues to hold just as for massless (gauge) bosons. This can be seen roughly as follows: for the massless QED gauge field (photon), the photon propagator is given by:

$$\Delta_{\mu\nu}(k) = \frac{-ig_{\mu\nu}}{k^2 + i\epsilon} \xrightarrow[k\to\infty]{} O(k^{-2}).$$ (4.33)

Equation (4.33), with its asymptotic $1/k^2$ behavior, implies the renormaliz-ability of QED (see §4.5). In contrast, the propagator in momentum space for a vector field with finite mass M_V is:

$$D_{\mu\nu}(k) = \frac{-i(g_{\mu\nu} - k_\mu k_\nu/M_V^2)}{k^2 - M_V^2 + i\epsilon},$$ (4.34)

where the asymptotic behavior of $D_{\mu\nu}(k)$ is:

$$D_{\mu\nu}(k) \xrightarrow[k\to\infty]{} O(1).$$ (4.35)

which implies the failure of renormalizability for the quantum field theory (see §4.5). However, for a gauge field, the propagator is initially given by Eq. (4.33) and, if the finite mass of the gauge field is generated by an SSB mech-anism, the short-distance structure (high-energy behavior) of the theory — which is responsible for the non-renormalizability — is not affected by the symmetry-breaking [see §4.5]. We note that for Abelian gauge theory, SSB is really not necessary in order to achieve renormalizability. Here, a mass term for the gauge field can be introduced by hand, without spoiling renormaliz-ability, provided the neutral gauge field is coupled to a conserved current. This is not true any longer for a Yang-Mills theory where the spontaneous generation of gauge boson masses is the only way to guarantee renormaliz-ability.

As indicated, the question of the renormalizability of a spontaneously broken non-Abelian gauge theory will be further discussed in §4.5 and we conclude this section with some remarks concerning an impressive physical application of the SSB of an Abelian gauge symmetry, namely to low tem-perature superconductivity. This example is particularly illuminating since the "Cooper pairs" of low temperature superconductivity provide a concrete realization of a quasi-Higgs SSB mechanism operating within the framework of the $U(1)$ gauge theory of electromagnetism. A "Cooper pair" is a pair of electrons with opposite momenta and spins near the top of the Fermi surface of a metal, which acquire a "binding" energy due to the lattice phonon-mediated attractive interaction between the pair of electrons [see §6.4c)]. What Cooper showed [4.14] was that even the weak phonon-mediated at-tractive interaction between the two electrons is sufficient to create a ground state of bound Cooper pairs which is more stable than the ground state of the normal metal. Thereby, an energy gap Δ (associated with the Cooper pairs) is created which is a measure of the "condensation energy" between the normal and the superconductive state; that is, a "phase transition" takes place between the normal and superconductive state below a critical temperature (T_c) — depending on the metal — when a sufficient number of "Cooper pairs" has condensed into the ground state. The gap energy Δ in the superconductor possesses the characteristic (essential singularity) form of a "phase transition", namely $\Delta \sim 2W_D \; e^{-\frac{1}{\rho V}}$ where W_D is the Debye energy, ρ is the "Cooper pair" density per unit energy at the Fermi surface and V is the interaction.

The achievement of the BCS theory [1.90], which followed the Cooper pair paper by one year — was to write down a mathematically ingenious and physically cogent wavefunction (the BCS wavefunction) for the super-conductive ground state and to launch an amazing number of correct detailed predictions concerning the properties of low temperature superconductors. We do not intend to present in this section the essential mathematics of the BCS theory (this will be done to some extent in §6.4c in connection with the Nambu approach to fermion mass generation in particle physics) nor to describe the numerous theoretical and experimental successes of the BCS theory. We propose instead to exhibit the intimate relationship between the Ginzburg-Landau expression for the microscopic free energy F [1.91] for the superconductive state in the presence of a magnetic field and the spontaneously broken Abelian gauge theory of the complex scalar field defined by Eqs. (4.26)–(4.28).

Seven years before the BCS theory, Ginzburg and Landau introduced a complex wavefunction ϕ (we use the notation ϕ because it ultimately becomes the quasi-Higgs scalar field function for the "Cooper pairs") as an "order parameter" for the superconducting electrons such that the density of superconducting electrons (as defined by the London equations [4.4]) is given by $n_s = |\phi(x)|^2$, where $\phi(x)$ satisfies the Ginzburg-Landau equation for the free energy F_s (F_n is the free energy for the normal metal, F_s for the superconductive state) in the form [4.15]:

$$F_s = F_n + \int d^3x [\frac{1}{2m^*}|(\underline{\nabla} - ie^*\underline{A})\phi|^2 + a|\phi|^2 + \frac{b}{2}|\phi|^4 + \frac{H^2}{8\pi}] \qquad (4.36)$$

In Eq. (4.36), m^* and e^* are respectively mass and charge parameters [the BCS theory tells us that $m^* = 2m$ (the mass of the Cooper pair) and $e^* = 2e$ (the charge of the Cooper pair)], $\underline{A}$ is the externally applied vector potential (curl $\underline{A}$ is the magnetic field $\underline{H}$) and a and b are phenomenological constants in what we would now call the "Higgs potential". [The Ginzburg-Landau equation (4.36) was cleverly constructed before a number of experiments (particularly by Pippard [4.16]) indicated the presence of "non-local" effects in superconductivity, necessitating at least a quartic term in the free energy (4.36) (i.e. the term with the coefficient b).] When the Ginzburg-Landau Eq. (4.36) is compared with Eq. (4.26), it is evident that $b \simeq \lambda$ (and is positive to make physical sense for the minimum free energy) and $a \simeq -\mu^2$ is positive or negative (depending on whether or not the "phase transition" between the normal and superconductive state takes place); it turns out that $a(T)$ vanishes as $(T - T_c)$ (T_c is the critical temperature at which the "phase transition" takes place) so that in the superconductive state ($T < T_c$), $a(T)$ is negative and SSB occurs [see Eq. (4.28)]. In particle physics language, $|\phi|^2 = -\frac{a}{b} = v^2$.

At this point, the spontaneously broken Abelian gauge theory of low temperature superconductivity has an advantage over its analog in particle physics, QFD, in that two key parameters in the Ginzburg-Landau theory (the "coherence" length and the "penetration depth") — whose evaluation is given by the BCS theory and its subsequent modifications — are essentially the inverses of the quasi-Higgs mass of the "Cooper pair" and the "heavy photon" mass respectively; this situation contrasts with QFD where λ is unknown and the Higgs mass has not yet been measured. In the language of superconductivity, the "coherence" length is (far from T_c) the smallest size wave packet describing the "Cooper pair" and the "penetration depth" is the distance within the surface of the superconductor over which an applied magnetic field suffers an exponential decrease (the "Meissner effect"). In terms of the quantities defined by the Ginzburg-Landau equation, the "coherence" length λ and the "penetration depth" d can be written in the form:

$$\lambda = \frac{1}{\sqrt{2m^*|a(T)|}}; \quad d = \sqrt{\frac{m^*b}{4\pi e^* a(T)}} \tag{4.37}$$

From the vantage point of particle physics, the inverse of λ can be identified with the "quasi-Higgs mass" of the "Cooper pair" [and is known because $a(T)$ is known] although no new insight in obtained by using particle physics terminology. On the other hand, the inverse of d is clearly of the form gv [see Eq. (4.37)] and has been identified with the (transverse) magnetic modes of the "heavy photon" and, in the Coulomb gauge, with the plasmon mass [4.17] [i.e. the "would-be" N-G boson resulting from the SSB of the electromagnetic field is converted — through the interaction of the "longitudinal" component of the "heavy photon" with the density fluctuations of the plasma inside the superconductor — into the plasmon massive]. While it is prudent not to push analogies between condensed matter and particle physics too far, the fact remains that the Ginzburg-Landau equation (4.36) is equivalent to the Lagrangian of an Abelian gauged charged scalar field and that both undergo SSB. This perceived relationship is certainly a tribute to our present-day understanding of physical phenomena since the same SSB Higgs (or quasi-Higgs) mechanism are invoked in condensed matter and particle physics despite the vast disparity in energy scale [fractions of eVs in superconductivity in contrast to hundreds of GeVs in particle physics — a factor of 10^{12} or more].

There are other interesting properties of low temperature superconductivity that have a conceptual affinity to particle physics. It is well-known that the superconductive phase may have different macroscopic attributes, depending on the material, and may belong to one of two classes [4.18]: Type I superconductors (for which $\lambda >> d$), which possess the property that, for a given critical temperature T_c, there is only one transition magnetic field H_c, above which the full (diamagnetic) superconductive phase converts to the normal phase and Type II superconductors (for which $\lambda << d$), where there are two transition magnetic fields — a lower field H_{c1}, describing the onset of magnetic flux tubes (in the form of "vortices") and a (second) higher field H_{c2}, indicating when the penetration is complete and the diamagnetism of the superconductor disappears (see Fig. 4.2). Actually, there is a sharp transition between Type I and Type II superconductivity defined by the condition $\kappa = \frac{d}{\lambda} = 1/\sqrt{2}$ with $\kappa < 1/\sqrt{2}$ yielding "well-behaved" Type I superconductivity with a "first-order" breakdown of superconductivity at H_c, and $\kappa > 1/\sqrt{2}$ leading to Type II superconductivity with a continuous increase in flux penetration from H_{c1} to H_{c2}; H_{c2} is larger than H_c (it turns out that $H_{c2} = \sqrt{2}\kappa H_c$) (see Fig. 4.2) and makes possible the high-field superconducting magnets that are so essential for high energy accelerators. [The transition condition $\kappa = 1/\sqrt{2}$ on the ratio of the "penetration depth" to the "coherence length" can be translated — in particle physics parlance — into the condition $\frac{1}{\kappa} = \sqrt{2} = \frac{"M_H"}{"M_W"}$, where "$M_H$" is the Higgs or the quasi-Higgs mass and "M_W" is the mass of the gauge field after SSB]; at this stage, this translation is a mere curiosity because the particle physics language of "M_H" and "M_W" — rather than λ and d — is hardly necessary to make progress in superconductivity theory and particle physics is hardly in a position to attach deep physical significance, i.e. "phase transition" status (as between Type I and Type II superconductivity), to a particular ratio of the Higgs mass to the heavy gauge field mass, since the Higgs particle is still undetected.

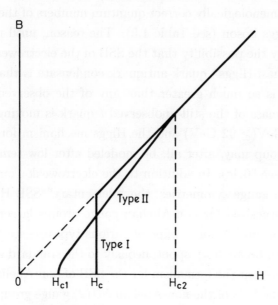

Fig. 4.2. Comparison of flux penetration behavior of Type I and Type II superconductors with the same thermodynamic critical field H_c; $H_{c2} = \sqrt{2}\kappa H_c$.

On the other hand, what adds further "particle physics" interest to Type II superconductivity — and constitutes another good example of the deeper conceptual affinity between particle and condensed matter physics — is the fact that the magnetic flux tubes in Type II superconductivity are quantized — because of their non-trivial topology — and obey an "additive" topological conservation law. The quantization property of magnetic flux tubes in Type II superconductivity is derived in §10.2b as part of our rather detailed discussion of topological conservation laws in particle and condensed matter physics. We note that we return to the mathematical formalism of the BCS theory in §6.4c when we discuss the Nambu hypothesis that the Higgs boson mass in the electroweak theory is generated by means of the same type of mechanism that is responsible for creating the energy gap Δ in low temperature superconductivity.

§4.4b. SSB of non-Abelian gauge groups

The standard gauge theory of the electroweak interaction depends crucially on the SSB of the non-Abelian gauge group $SU(2)_L$ [the full electroweak group is $SU(2)_L \times U(1)_Y$ — see §6.2] by an "elementary" Higgs doublet. To all intents and purposes, the quantum numbers of the Higgs boson under the electroweak group, $SU(2)_L \times U(1)_Y$, are put in "by hand" although it should be noted that the "quark-condensate" $< 0|\bar{q}_L q_R|0 >$ does give the phenomenologically correct quantum numbers of the postulated "elementary" Higgs boson (see Table 1.5). The reason, until recently, for not taking seriously the possibility that the SSB of the electroweak group is triggered by a "quasi-Higgs" quark-antiquark condensate is that the SSB scale ($\simeq 250$ GeV) is so much greater than any of the observed quark masses. However, the mass of the still-unobserved t quark is moving upward to the range of 100 GeV(≥ 92 GeV) and the Higgs mechanism for the SSB of the electroweak group may, after all, be modeled after low temperature superconductivity (see §6.4c). In addition to the electroweak example of the SSB of non-Abelian gauge symmetries, the "elementary" SSB Higgs mechanism has been hypothesized for non-Abelian gauge groups larger than the standard group (e.g. unification groups of various types — see Chapter 8) that are presumed to break down spontaneously to the standard group. For these reasons, we work out the formalism for the SSB of non-Abelian gauge groups starting with the SSB of the non-Abelian $SU(2)$ gauge group and then moving on to $SU(3)$ and the general gauge group G.

Just as in the Abelian case, we develop the formalism for the SSB of a non-Abelian SU(2) gauge theory by starting with the gauged Lagrangian for the Higgs scalar field. We consider a doublet of complex scalar fields $\phi = \binom{\phi_1}{\phi_2}$ interacting with the three gauge fields $A_\mu^a(x)$ ($a = 1, 2, 3$) in the adjoint representation of SU(2); the Lagrangian is:

$$\mathcal{L} = (D_\mu \phi)^\dagger (D_\mu \phi) - V(\phi) - \frac{1}{4} F_{\mu\nu}^a F^{a\mu\nu} \qquad (4.38)$$

where:

$$D_\mu \phi = (\partial_\mu - ig \frac{\tau^a}{2} \cdot A_\mu^a)\phi; \quad F_{\mu\nu}^a(x) = \partial_\mu A_\nu^a - \partial_\nu A_\mu^a + g\epsilon^{abc} A_\mu^b A_\nu^c \qquad (4.38a)$$

and:

$$V(\phi) = \mu^2(\phi^\dagger \phi) + \lambda(\phi^\dagger \phi)^2 \qquad (\lambda > 0). \qquad (4.39)$$

As usual, for $\mu^2 < 0$, we have SSB and $v = \sqrt{\frac{-\mu^2}{2\lambda}}$ and we write in analogy to Eq. (4.29):

$$\phi(x) = \tilde{\phi}(x) \, e^{i\tau^a/2 \cdot G^a(x)/v} \qquad (4.40)$$

where:

$$\bar{\phi}(x) = \frac{1}{\sqrt{2}} \begin{pmatrix} 0 \\ v + \rho(x) \end{pmatrix} \qquad (4.40a)$$

with $\rho(x)$ and $G^a(x)$ $(a = 1, 2, 3)$ real functions. The covariant derivative, operating on $\tilde{\phi}(x)$, becomes:

$$\tilde{D}_\mu \tilde{\phi} = \frac{1}{\sqrt{2}} \left(\partial_\mu - ig \frac{\tau^a}{2} \cdot B^a_\mu \right) \tilde{\phi}(x); \quad B^a_\mu = A^a_\mu - \frac{1}{gv} \tau^a \cdot \partial_\mu G^a(x) \quad (4.41)$$

and the Lagrangian is:

$$\mathcal{L} = (D_\mu \tilde{\phi})^\dagger (D_\mu \tilde{\phi}) - \frac{\mu^2}{2} (v + \rho)^2 - \frac{\lambda}{4} (v + \rho)^4 - \frac{1}{4} G^a_{\mu\nu} G^{a\mu} \quad (4.42)$$

where:

$$G^a_{\mu\nu}(x) = \partial_\mu B^a_\nu - \partial_\nu B^a_\mu + g\epsilon^{abc} B^b_\mu B^c_\nu \quad (4.42a)$$

The first term on the R.H.S. of Eq. (4.42) contains the mass contribution:

$$\frac{g^2}{8} (0, v)(\tau^a \cdot B^a_\mu)(\tau^a \cdot B^{a\mu}) \binom{0}{v} = \frac{1}{2} M_B^2 \, B^a_\mu B^{a\mu} \quad (a = 1, 2, 3) \quad (4.43)$$

where $M_B = gv/2$. Furthermore, it is seen from Eq. (4.42) that only one real field, ρ, of the original four real scalar fields (the two complex ϕ_1 and ϕ_2 fields), remains in the Lagrangian with mass $\sqrt{2}|\mu|$. Thus, the $SU(2)$ gauge symmetry is completely broken and the three "would-be" N-G fields, G^a_μ, are "eaten up" to provide the longitudinal components of the three massive gauge fields B^a_μ $(a = 1, 2, 3)$ of the $SU(2)$ group. Again, the number of "degrees of freedom" is "conserved", at 10: initially, there are three massless gauge fields (contributing two "polarizations" apiece) and two complex scalar fields (contributing two "polarizations" apiece) and, finally, there are three massive gauge bosons (contributing three "polarizations" apiece) and one massive real scalar field (contributing one "polarization"), a total of ten "degrees of freedom" before and after.

It is important to realize that the group representation content of the Higgs scalar field determines the pattern of symmetry breaking and, indeed, if we choose an $SU(2)$ triplet representation of real scalar fields (instead of two complex doublets), the $SU(2)$ gauge group breaks down to a $U(1)$ gauge group, instead of breaking down completely. For the triplet representation of real scalar fields $\phi^a (a = 1, 2, 3)$, the Lagrangian has the form:

$$\mathcal{L} = -\frac{1}{4} F^a_{\mu\nu} F^{a\mu\nu} + \frac{1}{2} (\partial^\mu \phi^a - g\epsilon^{abc} A^b_\mu \phi^c)(\partial_\mu \phi^a - g\epsilon^{ade} A^d_\mu \phi^e)$$

$$- \frac{1}{2} \mu^2 (\phi^a \phi^a) - \frac{1}{4} \lambda (\phi^a \phi^a)^2 \quad (4.44)$$

If $\mu^2 < 0$, SSB takes place and we can choose:

$$< 0|\phi|0 > = (0 \quad 0 \quad v). \quad (4.45)$$

Inserting Eq. (4.45) into the second term of Eq. (4.44), the mass term of the gauge bosons is given by:

$$\frac{1}{2}g^2\epsilon^{abc}A_\mu^b <0|\phi^c|0> \epsilon^{ade}A_\mu^d <0|\phi^e|0> . \tag{4.46}$$

Using $<0|\phi^a|0> = v\delta^{a3}$, we obtain from Eq. (4.46):

$$\frac{1}{2}g^2v^2[(A_\mu^1)^2 + (A_\mu^2)^2]. \tag{4.47}$$

The gauge bosons coupled to the broken generators T^1 and T^2 have acquired equal masses $M_G = gv$. No mass is generated for the third gauge field, A_μ^3, which is coupled to the conserved generator T^3.

As before, the number of "degrees of freedom" is "conserved", at 9: initially there are three gauge fields (with two "polarizations" apiece) and three real scalar fields (with one "polarization" apiece) and, finally, there are two massive vector gauge fields with (with three "polarizations" apiece), one massless gauge field (with two "polarizations"), and one real scalar field (with one "polarization") for a total of 9 before and after. An instructive way to understand this result is to note that the $SU(2)$ group is isomorphic to the $SO(3)$ group and that the triplet of real Higgs fields with only one of the fields having a vacuum expectation value $v \neq 0$ breaks the $SO(3)$ group down to the $SO(2)$ group which, itself, is isomorphic to the $U(1)$ group. The latter approach was used to discuss the SSB of non-Abelian global symmetry groups in order to motivate the general result that the number of N-G bosons, generated in the SSB process, is equal to $(n_G - n_H)$ where $n_{G,H}$ is the dimension of the adjoint representation of G, H when the group G spontaneously breaks down to the subgroup H (see §3.4). The number of "would-be" N-G bosons — which convert into massive gauge bosons — of course, satisfies the present result $[n_G(SO(3)) - n_H(SO(2))] = 2$ since the dimension of the adjoint representation of $SO(n)$ is $\frac{n(n-1)}{2}$. In any case, it is clear that the pattern of symmetry-breaking — which decides how many "would-be" N-G bosons are "eaten up" to form massive gauge particles — depends on the transformation property of the Higgs scalar field under the gauge group that is being spontaneously broken.

In principle, the transformation property of the Higgs scalar boson should be decided by theory but, until now, for the electroweak group, it has been decided by experiment. As will be seen in §6.2, the SSB of the electroweak group $SU(2)_L \times U(1)_Y$ is achieved by means of an "elementary" Higgs doublet (of complex scalar fields) with splendid results. It should be noted, however, that when a complex Higgs doublet completely breaks down an $SU(2)$ gauge group, equal masses are acquired by all three gauge fields [see Eq. (4.43)]; however, the same complex scalar (Higgs) doublet applied to the SSB of a non-semi-simple gauge group like $SU(2) \times U(1)$ [the electroweak group $SU(2)_L \times U(1)_Y$ possesses this structure and is broken down to the $U(1)_{EM}$ gauge group by the complex Higgs doublet] yields (for the electroweak group) one massless gauge field (photon) plus two gauge particles (charged weak bosons $W^\pm$) of equal mass and a fourth gauge particle (neutral weak boson Z^0) with different (and larger) mass. The experimental support for the doublet choice of Higgs representation in electroweak theory is discussed in §6.3a. It is, in principle, possible that more than one Higgs representation is responsible for the SSB of the electroweak group — the triplet representation is usually taken as the second candidate because its Yukawa coupling with the lefthanded neutrino can give rise to a finite Majorana mass — but the increasing accuracy in the measurement of the so-called ρ parameter (see §6.3a) places a very low upper limit on the possible admixture of the triplet representation. As we will see (in §6.3a), the SSB of $SU(2)_L \times U(1)_Y$ by more than one Higgs doublet representation (with $|Y| = 1$) can not be ruled out.

Since, as already mentioned, larger gauge groups and their SSB may be of interest, we discuss one more explicit example — an $SU(3)$ gauge group in the presence of an $SU(3)$ triplet of complex Higgs fields — and conclude with brief comments concerning the general case of a non-Abelian gauge group subjected to the SSB Higgs mechanism. For the $SU(3)$ case, the gauge fields form an $SU(3)$ octet, which we denote by B_μ^a ($a = 1, 2, ...8$), and the Higgs potential can be written:

$$V(\phi) = \mu^2 \phi^\dagger \phi + \lambda(\phi^\dagger \phi)^2 \qquad (4.48)$$

where $\phi = (\phi_1, \phi_2, \phi_3)$. For $\mu^2 < 0$, ϕ develops a non-vanishing VEV and, if we choose the simple form: $\phi_0 \equiv < 0|\phi|0 >= v(0, 0, 1)$, the $SU(3)$ gauge group breaks down spontaneously to $SU(2)$. Of the eight gauge fields associated with $SU(3)$, one then expects that five acquire mass and the remaining three — the gauge fields associated with unbroken $SU(2)$ — remain massless. This result is confirmed by writing down the gauge boson mass matrix that follows from the above choice of ϕ_0; we get:

$$(\phi_0^\dagger \lambda^a \lambda^b \phi_0) B_\mu^a B^{b\mu} \qquad (4.49)$$

where the λ^a's are the generators of $SU(3)$ (see §3.2c). Using the definition of ϕ_0 and the properties of the λ^a's, one finds from Eq. (4.49) that B^1, B^2, B^3

remain massless, B^4, B^5, B^6, B^7 acquire a common mass $M_G = gv$ and B^8 acquires a mass $2/\sqrt{3}M_G$. This statement follows from the fact that we started out with a triplet of complex scalar fields (i.e. six real fields), of which five are "eaten" (to serve as the longitudinal components of the five massive bosons) and one massive scalar boson survives.

With the last example in mind, we can quickly derive the general result for the SSB of a gauge symmetry group G, breaking down to the gauge group H; consider the general Lagrangian:

$$\mathcal{L} = \frac{1}{2}[(\partial_\mu \phi_i + ig T_{ij}^a A_\mu^a \phi_j)][(\partial^\mu \phi_i - ig T_{ik}^a A^{a\mu} \phi_k)] - V(\phi_i) - \frac{1}{4} F_{\mu\nu}^a F^{a\mu\nu} \quad (4.50)$$

where ϕ_i is a set of real fields belonging to a (possibly reducible) representation of G with n generators, thus:

$$\phi_i(x) \to \phi_i'(x) = \phi_i(x) + i\epsilon^a(x) T_{ij}^a \phi_j(x) \qquad (a = 1, 2, ..., n). \quad (4.51)$$

Since $V(\phi_i)$ is invariant under an arbitrary transformation belonging to G, we can write:

$$0 = \delta V = \frac{\partial V}{\partial \phi_i} \delta \phi_i = \epsilon^a \frac{\partial V}{\partial \phi_i} T_{ij}^a \phi_j \quad \text{so that} \quad \frac{\partial V}{\partial \phi_i} T_{ij}^a \phi_j = 0 \quad (a = 1, \dots, n)$$

$$(4.52)$$

Further, since we know that V is a minimum at $\phi_i = v_i$, we find, after differentiating Eq. (4.52), that:

$$\left. \frac{\partial^2 V}{\partial \phi_i \partial \phi_k} \right|_{\phi_i = v_i} T_{ij}^a v_j = 0. \quad (4.52a)$$

By expanding $V(\phi_i)$ around the minimum $V(v_i)$, it is easily seen that the mass matrix is:

$$(M^2)_{ik} = \left. \frac{\partial^2 V}{\partial \phi_i \partial \phi_k} \right|_{\phi_i = v_i}. \quad (4.53)$$

If the unbroken subgroup of G is H, with m generators, it follows that:

$$T_{ij}^b v_j = 0 \quad \text{for} \quad b = 1, 2 \dots m \quad (4.54)$$

(since these m generators leave the vacuum invariant). On the other hand, the remaining generators do not leave the vacuum invariant and one gets:

$$T_{ij}^c v_j \neq 0 \quad \text{for} \quad c = m + 1, \dots, n. \quad (4.55)$$

If the T^a's are linearly independent, these equations imply that M^2 has $n - m$ zero eigenvalues and hence $(n - m)$ N-G bosons, which in the gauge symmetry case, translate into $(n - m)$ "would-be" Goldstone bosons that produce $(n-m)$ massive gauge fields corresponding to the broken generators. The diagonalization of the mass matrix given by Eq. (4.53) gives rise to the mass values of the $(n-m)$ gauge bosons.

§4.5. Regularization and renormalizability of gauge theories

While quantum field theories are strongly constrained by the conditions of Lorentz invariance and gauge invariance, the condition of renormalizability not only adds further constraints (e.g. contributing to the uniqueness of Abelian QED) but it also ensures dynamical content and phenomenological predictability for any quantum field theory. Without renormalizability, one is limited to lowest-order perturbation calculations and can not explain quantitatively the Lamb shift (and other higher-order effects) in Abelian QED (see below). For unbroken non-Abelian gauge theories (e.g. QCD), renormalization theory (plus the renormalization group) have been the instrument for identifying the crucial property of asymptotic freedom in QCD that has led to a highly successful gauge theory of the strong interaction. Finally, in QFD, the SSB of the gauge group $SU(2)_L \times U(1)_Y$ raises the question as to whether the renormalizability of the electroweak gauge theory with massless gauge fields is destroyed by the finite-mass charged weak bosons that are generated by the Higgs SSB mechanism. Consequently, it is incumbent upon us to review the essential features of regularization and renormalization, first for Abelian gauge theory and then for non-Abelian gauge theory; this program is carried out in this and the next two sections (§4.5–§4.5b).

The need to understand the Lamb shift in atomic physics gave the great impulse to the renormalization program for QED at the Shelter Island Conference in 1947, where the subject was under intense scrutiny [1.34]. Bethe [4.19] published the first paper after the Shelter Island Conference, exploiting the renormalization ideas of charge and mass renormalization in a nonrelativistic calculation; Bethe's success in coming to within 5% of the measured Lamb shift [because he neglected the "relativistic" effect of vacuum polarization] gave encouragement to the papers of Schwinger and Feynman that shortly followed. Schwinger completed the first relativistic renormalized calculation of the Lamb shift [1.16] on the basis of Green's function techniques and judicious use of gauge invariance. [Tomanaga's independent work [1.16] on renormalizable Abelian gauge theories was close to Schwinger's approach]. Feynman's calculation was formulated in terms of the path integral method, got interpreted in terms of "Feynman diagrams", [1.16] and achieved the same correct prediction for the Lamb shift with considerably less labor

[4.20]. Dyson's contribution [1.16] consisted of demonstrating the mathematical equivalence of the Schwinger-Tomonoga and Feynman approaches. In view of the greater computational simplicity of the Feynman method, we employ Feynman diagram techniques to present our brief summary of regularization and renormalizability in Abelian and non-Abelian gauge theories.

For the usual pedagogical reasons, it is convenient to start our review of mass, wavefunction and coupling constant renormalization in quantum field theory within the framework of the scalar ϕ^4 theory [3.26]. We then employ the "power-counting method" to fix the "superficial" degree of divergence for various types of interactions (involving $J = 0, \frac{1}{2}$ and 1 fields) and explain why certain earlier relativistic quantum field theories had to be rejected because of their intrinsic non-renormalizability. We continue in §4.5a to translate the power-counting and renormalization considerations with regard to the ϕ^4 theory into Abelian QED and note the impressive precision tests of renormalized QED. We conclude our discussion of regularization and renormalization in gauge theories in §4.5b by reporting on the status of the renormalization program for non-Abelian gauge theories without and with SSB.

We take the ϕ^4 theory for the model calculation even though it is not a gauge theory because its mathematical simplicity exhibits clearly how renormalization shuffles the higher-order divergences (in the propagator and vertex of the ϕ^4 theory) into the "bare" mass, "bare" (self-) coupling constant ("charge") and "bare" wavefunction, so as to yield the finite mass, charge and wavefunction renormalization constants. [These results for the renormalization constants (really functions of the cutoff parameter) of the scalar ϕ^4 theory can be transformed in a plausible fashion into the electron self-energy, electric charge and vacuum polarization renormalization constants in QED (see §4.5a).] We begin with the unrenormalized Lagrangian density for the ϕ^4 theory, namely:

$$\mathcal{L}_0 = \frac{1}{2}[(\partial_\mu \phi_0)^2 - \mu_0^2 \phi_0^2] - \frac{\lambda_0}{4!} \phi_0^4 \qquad (4.56)$$

where the unrenormalized propagator and vertex are exhibited in Fig. 4.3. If we recall that the one-particle-irreducible (1PI) Green's functions, denoted by $\Gamma^n(p_i)$ $(i = 1, 2...n)$ receive contributions only from the 1PI Feynman diagrams (i.e. the diagrams that cannot be disconnected by cutting an internal line), we can write for the n-point Green's function (with n external lines):

$$G_0^{(n)}(x_1, ...x_n) = <0|T(\phi_0(x_1)...\phi_0(x_n))|0> \qquad (4.57)$$

The two-point Green's function (in momentum space) is the propagator $\Delta(p)$ so that we get:

$$i\Delta(p) = \int d^4x \, e^{-ip\cdot x} < 0|T(\phi_0(x)\phi_0(0))|0 > \qquad (4.58)$$

where Fig. 4.4 displays the decomposition of the propagator in terms of the 1PI self-energy parts, denoted by $\Sigma(p^2)$. Figure 4.4 tells us that the propagator becomes:

$$
\begin{aligned}
i\Delta(p) &= \frac{i}{p^2 - \mu_0^2 + i\epsilon} + \frac{i}{p^2 - \mu_0^2 + i\epsilon}(-i\Sigma(p^2))\frac{i}{p^2 - \mu_0^2 + i\epsilon} + \dots \\
&= \frac{i}{p^2 - \mu_0^2 + i\epsilon}\left[\frac{1}{1 + i\Sigma(p^2)\frac{i}{p^2-\mu_0^2+i\epsilon}}\right] = \frac{i}{p^2 - \mu_0^2 - \Sigma(p^2) + i\epsilon}.
\end{aligned} \qquad (4.59)
$$

Fig. 4.3. Unrenormalized propagator and vertex in the ϕ^4 theory.

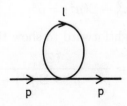

Fig. 4.4. The propagator as a sum of 1PI self-energy insertions in the ϕ^4 theory.

It is evident from Eq. (4.59) that $\Delta(p)$ is finite if the proper self-energy $\Sigma(p^2)$ is finite. Hence, the first task is to "finitize" $\Sigma(p^2)$ and, we will see that in the process, it is necessary to introduce the concepts of mass and wavefunction renormalization for the scalar ϕ^4 theory. The procedure is to first calculate the 1PI contribution to the self-energy $\Sigma(p^2)$ taking account of the one-loop contributions as shown in Fig. 4.5.

Fig. 4.5. One-loop mass corrections in the ϕ^4 theory.

The contribution to the self-energy graph — which is $\Sigma(p^2)$ — becomes:

$$-i\Sigma(p^2) = -\frac{i\lambda_0}{2} \int \frac{d^4l}{(2\pi)^4} \frac{i}{l^2 - \mu_0^2 + i\epsilon}. \tag{4.60}$$

In calculating $\Sigma(p^2)$ — which governs the degree of divergence of the propagator function — it is evident that Eq. (4.60) is quadratically divergent, the same degree of divergence which holds for multi-loop contributions to the self-energy $\Sigma(p^2)$ [see below for the general discussion of power counting and renormalizability]. The one-loop $\Sigma(p^2)$ possesses the special property of being independent of the external momentum p in the ϕ^4 theory [this is no longer true of two-loop and higher-loop contributions to $\Sigma(p^2)$]. Overlooking this point, it is customary to expand $\Sigma(p^2)$ about some arbitrary finite value μ^2, thus:

$$\Sigma(p^2) = \Sigma(\mu^2) + (p^2 - \mu^2)\Sigma'(\mu^2) + \tilde{\Sigma}(p^2) \tag{4.61}$$

where $\Sigma(\mu^2)$ retains the quadratic divergence but $\Sigma'(\mu^2)$ is logarithmically divergent and the remaining term $\tilde{\Sigma}(p^2)$ is finite. Allowing for the dependence of $\Sigma(p^2)$ on p^2, Eq. (4.59) is converted into:

$$i\Delta(p) = \frac{i}{p^2 - \mu_0^2 - \Sigma(\mu^2) - (p^2 - \mu^2)\Sigma'(\mu^2) - \tilde{\Sigma}(p^2) + i\epsilon}. \tag{4.62}$$

It should be noted that the choice of a finite μ^2 for the reference point of renormalization enables us to avoid the infrared divergences that confront all quantum field theories [4.21]. However, since the scaling law is obeyed by the renormalization group equation (see §5.2a), the arbitrariness in the choice of μ^2 does not affect the final "running" values of the renormalization constant. Moreover, as we will see (in §5.2a), a finite value of μ^2 is crucial for the finite scale Λ_{QCD} in non-Abelian QCD when the renormalization group analysis is carried out.

With this understanding of the role of finite μ^2 in renormalizing the ϕ^4 theory, one can define the physical mass as the pole of the propagator satisfying the following simple equation:

$$\mu_0^2 + \Sigma(\mu^2) = \mu^2. \tag{4.63}$$

From Eq. (4.63), it is straightforward to show that (to order λ_0^2):

$$i\Delta(p^2) = \frac{iZ_\phi}{p^2 - \mu^2 - \tilde{\Sigma}(p^2) + i\epsilon} \tag{4.64}$$

where:

$$Z_\phi = [1 - \Sigma'(\mu^2)]^{-1} = 1 + \Sigma'(\mu^2) + 0(\lambda_0^2). \tag{4.65}$$

The quantity Z_ϕ is a multiplicative factor and can be removed by rescaling the field operator ϕ_0, thus:

$$\phi = Z_\phi^{-\frac{1}{2}} \phi_0 \tag{4.66}$$

Hence, Z_ϕ is the wavefunction renormalization constant and relates the renormalized propagator function to the unrenormalized one, namely:

$$i\Delta_R(p^2) = \int d^4 x e^{-ip \cdot x} < 0|T(\phi(x)\phi(0))|0 >$$
$$= Z_\phi^{-1} \int d^4 x e^{-ip \cdot x} < 0|T(\phi_0(x)\phi_0(0))|0 >= iZ_\phi^{-1}\Delta_U(p^2) \tag{4.67}$$

With the wavefunction renormalization, we can relate the renormalized n-point Green's function to the n-point unrenormalized Green's function (in momentum space) through the equation:

$$G_R^{(n)}(p_1...p_n) = Z_\phi^{-n/2} \, G_U^{(n)}(p_1...p_n) \tag{4.68}$$

Since we are working with the 1PI Green's functions $\Gamma^n(p_i)$, and the connected n-point Green's functions $G^n(p_i)$ contain a propagator for each of the n external lines, the relation between $\Gamma_R^n(p_i)$ and $\Gamma_U^n(p_i)$ becomes [using Eqs. (4.67) and (4.68)]:

$$\Gamma_R^{(n)}(p_1...p_n) = Z_\phi^{n/2} \, \Gamma_U^{(n)}(p_1...p_n). \tag{4.69}$$

With the relation between the renormalized 1PI n-point Green's function $\Gamma_R^n(p_1,...p_4)$ and unrenormalized $\Gamma_U^n(p_1,...p_4)$ given by Eq. (4.69), we can proceed with the derivation of the vertex renormalization constant in the ϕ^4 theory for $n = 4$. The three diagrams of Fig. 4.6 are the one-loop 1PI contributions to the vertex corrections and the expressions representing them

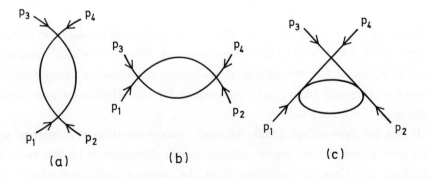

Fig. 4.6. One-loop vertex corrections in the ϕ^4 theory.

can be written in the form:

$$\Gamma_a = \Gamma(p^2) = \Gamma(s) = \frac{(-i\lambda_0)^2}{2} \int \frac{d^4l}{(2\pi)^4} \frac{i}{(l-p)^2 - \mu_0^2 + i\epsilon} \frac{i}{l^2 - \mu_0^2 + i\epsilon}$$

(4.70)

where $\Gamma_b = \Gamma(t)$, $\Gamma_c = \Gamma(u)$ with: $s = p^2 = (p_1 + p_2)^2$, $t = (p_1 + p_3)^2$, $u = (p_1 + p_4)^2$ (s, t and u are the Mandelstam variables — see §1.2b). Since $\Gamma_U^4(s,t,u)$ is a function of all three kinematical variables s, t and u, an arbitrary choice of these variables must be made to define the physical coupling constant; the symmetric point $s_0 = t_0 = u_0 = 4\mu^2/3$ for particles on the "mass-shell" $p_i^2 = \mu^2$ [μ is the (renormalized) physical mass] is usually taken and one has:

$$\Gamma_R^{(4)}(s_0, t_0, u_0) = -i\lambda \tag{4.71}$$

where λ is the renormalized coupling constant. The unrenormalized $\Gamma_U^4(s,t,u)$ — given by Eq. (4.70) — is written as:

$$\Gamma_U^{(4)}(s,t,u) = -i\lambda_0 + 3\Gamma(s_0) + \bar{\Gamma}(s) + \bar{\Gamma}(t) + \bar{\Gamma}(u) \tag{4.72}$$

where $\Gamma(s_0)$ is logarithmically divergent and $\bar{\Gamma}(s) = \Gamma(s) - \Gamma(s_0)$ is finite [and correspondingly for $\bar{\Gamma}(t)$ and $\bar{\Gamma}(u)$]. The vertex renormalization constant Z_λ can now be defined, namely:

$$-iZ_\lambda^{-1}\lambda_0 = -i\lambda_0 + 3\Gamma(s_0). \tag{4.73}$$

so that, at the symmetric point $s_0 = t_0 = u_0 = 4\mu^2/3$, we have the simple relation:

$$(\Gamma^4)_U(s_0, t_0, u_0) = -iZ_\lambda^{-1}\lambda_0 \tag{4.73a}$$

Also, it follows from Eq. (4.69) that:

$$\Gamma_R^{(4)}(s,t,u) = Z_\phi^2 \, \Gamma_U^{(4)}(s,t,u) \tag{4.74}$$

and, hence, the renormalized λ is related to the unrenormalized λ_0 by:

$$\lambda = Z_\phi^2 Z_\lambda^{-1} \, \lambda_0. \tag{4.75}$$

Equation (4.75) is a key equation since it expresses the relation between the renormalized "charge" to the unrenormalized "charge", which determines the β function of the renormalized group equation and, consequently, the absence or presence of asymptotic freedom for the Abelian or non-Abelian gauge group in question.

It can be shown that when "charge" renormalization is combined with mass and wavefunction renormalizations, all divergences in the two- and four-point 1PI Green's functions — in the one-loop approximation — are removed. These three types of renormalization in the ϕ^4 theory take care of any divergences that arise in the higher-point Green's functions. In sum, all 1PI Green's functions become finite when the "bare" (unrenormalized) quantities ϕ_0, μ_0 and λ_0 are expressed in terms of the renormalized quantities ϕ, μ and λ, thus:

$$\phi = Z_\phi^{-1/2}\phi_0; \qquad \mu^2 = \mu_0^2 + \Sigma(\mu^2); \qquad \lambda = Z_\lambda^{-1}Z_\phi^2\lambda_0 \qquad (4.76)$$

The finite renormalized n-point 1PI Green's functions (expressed in terms of the renormalized μ and λ) can be written in terms of the unrenormalized n-point 1PI Green's functions [expressed in terms of the unrenormalized μ_0 and λ_0] as follows:

$$\Gamma_R^{(n)}(p_1, ..., p_n; \lambda, \mu) = Z_\phi^{n/2} \, \Gamma_U^{(n)}(p_1, ..., p_n; \lambda_0, \mu_0, \Lambda). \qquad (4.77)$$

where Λ is the cutoff scale needed to define the divergent integrals contained in $Z_\phi, \Sigma(\mu^2)$ and Z_λ.

Equation (4.77) is the starting point for the renormalization group analysis — leading to the concept of the "running" renormalized charge — that will be discussed in §5.2a. However, the "running" coupling constant and related results require a knowledge of the behavior of the various renormalization constants as functions of the cutoff scale Λ and so we indicate how these expressions are derived. [We sketch the derivation of the Λ dependence of the renormalization constants so that we can deal directly with the renormalization group — for both QED and QCD — in §5.2a.] Two popular methods have been suggested for calculating the renormalization constants as functions of the cutoff scale; they are the "covariant" method [4.22] and the "dimensional regularization" method [4.23]. While the "dimensional regularization" method is more elegant (and suffices for the renormalization group discussion in §5.2a), we choose the more transparent "covariant" Pauli-Villars regularization method, which consists of modifying the divergent expression for a renormalization constant by means of a covariant convergence factor. In the Pauli-Villars method, the propagator in the ϕ^4 theory is written as:

$$\frac{1}{l^2 - \mu^2 + i\epsilon} \rightarrow \frac{1}{l^2 - \mu^2 + i\epsilon} + \sum_i \frac{a_i}{l^2 - \Lambda_i^2 + i\epsilon} \qquad (4.78)$$

where $\Lambda_i^2 >> \mu^2$, and the a_i's are chosen so that the propagator becomes finite for $\Lambda \rightarrow \infty$. For example, the propagator given by Eq. (4.78) can be "regularized" by choosing a_1 and a_2 as follows:

$$a_1 = \frac{\mu^2 - \Lambda_2^2}{\Lambda_2^2 - \Lambda_1^2}; \qquad a_2 = \frac{\Lambda_1^2 - \mu^2}{\Lambda_2^2 - \Lambda_1^2} \qquad (4.78a)$$

with the result that the two-point 1PI Green's function becomes:

$$-i\Sigma(p^2) = \frac{-i\lambda}{32\pi^2} [\Lambda^2 - \mu^2 \ln\frac{\Lambda^2}{\mu^2}]. \tag{4.79}$$

From the earlier definitions of Z_ϕ and $\Sigma(\mu^2)$ it follows, in the one-loop approximation, that $Z_\phi = 1$ and $\Sigma(\mu^2) = \lambda \Lambda^2/32\pi^2$.

A similar treatment of the four-point 1PI Green's function, given by Eq. (4.70), requires just one $a_i (a = -1)$ in the Pauli-Villars regularization scheme and yields (in the one-loop approximation) for large Λ^2:

$$\Gamma(0) \sim \frac{i\lambda^2}{32\pi^2} \ln\frac{\Lambda^2}{\mu^2}. \tag{4.80}$$

Using Eq. (4.73) and the fact that $\Gamma(s_0) = \Gamma(0)$ to within a (finite) constant, we get for the Λ-dependent charge (vertex) renormalization constant in the ϕ^4 theory (to order λ_0^2):

$$Z_\lambda^{-1} = 1 + \frac{3i\Gamma(0)}{\lambda} = 1 - \frac{3\lambda}{32\pi^2} \ln\frac{\Lambda^2}{\mu^2}. \tag{4.81}$$

In sum, the Pauli-Villars regularization method enables us to derive the following Λ-dependent values of the renormalization constants (in the one-loop approximation) in the ϕ^4 theory (Z_ϕ happens to be constant in the ϕ^4 theory — this is not generally true):

$$Z_\phi = 1; \qquad Z_\lambda = 1 + \frac{3\lambda}{32\pi^2} \ln\frac{\Lambda^2}{\mu^2}; \qquad \Sigma(\mu^2) = \frac{\lambda}{32\pi^2}\Lambda^2 \tag{4.82}$$

The results in Eq. (4.82) serve as a paradigm of the corresponding expressions for the renormalization constants in the Abelian gauge theory of QED and, to some extent, for non-Abelian gauge theories such as QCD.

Before summarizing the renormalization results for Abelian QED and non-Abelian QCD, in preparation for the renormalization group discussion in §5.2a, we again use the ϕ^4 quantum field theory as a model to define the simple power-counting method for predicting — without laborious calculation — the renormalizability of interacting spin $\frac{1}{2}$ and spin 0 quantum field theories. We then discuss the extension of the power-counting method to vector fields as well. This will enable us to understand why earlier candidate quantum field theories of the strong and weak interactions [e.g. the pion-nucleon interaction with axial vector coupling and the V-A four-fermion theory in weak interactions] could not pass the test of renormalizability whereas QED, QCD and QFD do. To explain the power-counting method, we recall that, in the ϕ^4 theory, the two-point 1PI Green's function is quadratically

divergent and the four-point 1PI Green's function is logarithmically divergent. To recapture these two results, the concept of the "superficial" degree
of divergence $\bar{D}$ is introduced [3.26]. Thus, consider $\Gamma^2(p_i)$, described by Fig.
4.5: Fig. 4.5 has $B = 2$ external lines, $b = 1$ internal line, and $n = 1$ vertex.
Similarly, $\Gamma^4(p_i)$ — described by Fig. 4.6 — has $B = 4$, $b = 2$ and $n = 2$.
Clearly, the formula:

$$4n = 2b + B \qquad (4.83)$$

relates the number of internal and external lines to the number of vertices.
To fix $\bar{D}$ for each $\Gamma^n(p_i)\,(i = 1, ...n)$, one must relate the number of loop
momenta (L) to B and b, keeping in mind that, because of overall momentum
conservation, only $(n - 1)$ vertices must be counted; hence:

$$L = b - (n - 1) \qquad (4.84)$$

In order to compute $\bar{D}$, we must assign a weight of 4 to each loop
(since each loop integration contributes four powers of loop momenta in
the numerator of the integral for the Feynman diagram) and a weight of -2
for each internal line (since the propagator contributes two powers of loop
momenta in the denominator); we get:

$$\bar{D} = 4L - 2b \qquad (4.85)$$

Equation (4.85) holds for the ϕ^4 theory since the vertices do not contribute
any momentum factors; appropriate account has to be taken of the weight
factors for L and b in a theory wherein the vertices contribute momentum
factors (see below). Using Eq. (4.85) and the relation (4.83), the "superficial" degree of divergence $\bar{D}$ is simply $4 - B$ for the ϕ^4 theory. The last
succinct result tells us that, in the ϕ^4 theory, $\bar{D} = 2$ (signifying a quadratic
divergence) for $\Gamma^2(p_1, p_2)$ and $\bar{D} = 0$ (signifying a logarithmic divergence)
for $\Gamma^4(p_1, \ldots p_4)$ — in agreement with previous statements. Power counting
of the higher-order loop approximations confirm the expectation that the
renormalization of the one-loop approximation maintains renormalization to
all orders of perturbation theory. It still has not been proved that the "renormalization sum" to all orders of perturbation theory is convergent; however,
we do know that the renormalization program has been vindicated in QED
to at least four orders in the fine structure constant α_e (see §4.5a) and the
general consensus is that renormalized QED is completely reliable up to huge
energies (see §1.2a and §4.5a).

The generalization of the power-counting method to determine $\bar{D}$ for any
combination of spin $\frac{1}{2}$ and spin 0 quantum field theories is straightforward

and we summarize the results. The inclusion of vector fields is more subtle — depending on whether or not the vector field is massless (i.e. a gauge field) or massive and/or neutral or charged — and requires a separate discussion. Limiting ourselves to an interaction Lagrangian consisting only of spin 0 (boson) fields and spin $\frac{1}{2}$ (fermion) fields, the following quantities can be defined [3.26]:

n = number of vertices; b' = number of scalar (boson) lines in vertex;

f' = number of Dirac (fermion) lines in vertex;

d = number of derivatives in vertex;

B = number of external scalar lines; F = number of external Dirac lines;

b = number of internal scalar lines; f = number of internal Dirac lines

$$(4.86)$$

The analog of Eq. (4.83) becomes:

$$nb' = B + 2b \quad ; \quad nf' = F + 2f \tag{4.87}$$

The analog of Eq. (4.84) is:

$$L = b + f - n + 1 \tag{4.88}$$

Using Eqs. (4.87) and (4.88), the "superficial" degree of divergence $\bar{D}$ can be written in the form:

$$\bar{D} = 4 - B - \frac{3}{2}F + n\delta \tag{4.89}$$

where:

$$\delta = b' + \frac{3}{2}f' + d - 4 \tag{4.89a}$$

The relative factor of $\frac{3}{2}$ before the numbers of fermion lines (F and f) as compared to the numbers of boson lines (B and b) arises from the simple fact that the mass term in the Dirac Lagrangian is $-m\bar{\psi}\psi$ in contrast to the $-m^2\phi^2$ term in the scalar Lagrangian. Equation (4.89) yields, of course, the same values of $\bar{D}$ for the $\Gamma^{(2)}$ and $\Gamma^{(4)}$ of the ϕ^4 model since $f' = F = f = d = 0$ and $b' = 4$.

Once the interaction Lagrangian density for any combination of spin 0 and spin $\frac{1}{2}$ fields is given, Eq. (4.89) for the "superficial" degree of divergence $\bar{D}$ can immediately decide — from a scrutiny of the 1PI graphs — whether the theory is basically renormalizable or not. Thus, in the early 1950s, when attempts were made to formulate a theory of the pion-nucleon interaction along the lines of the "Yukawa-coupled" QED, the two "candidate" interaction Lagrangians considered were: $g(\bar{\psi}\gamma_5\psi)\phi$ (pseudoscalar coupling) and $f(\bar{\psi}\gamma_\mu\gamma_5\psi)\partial_\mu\phi$ (axial vector coupling) where ϕ and ψ are the pion and nucleon fields respectively. These two pion-nucleon interaction Lagrangians involve two fermion fields and one boson field [unlike the QED interaction — to be discussed separately below — because the boson field is vector and not scalar] and it is easily shown that, for the pseudoscalar "pion-nucleon" interaction, the "superficial" degree of divergence for the (pion) vacuum polarization contribution is quadratic ($\bar{D} = 2$), for the (nucleon) self-energy contribution is linear ($\bar{D} = 1$) and, finally, for the coupling constant ("charge") is logarithmic ($\bar{D} = 0$). The pseudoscalar (pion-nucleon) interaction is therefore renormalizable but, unfortunately, the dimensionless coupling constant $g^2/4\pi \simeq 15$ is so large [compared to $\alpha_e = e^2/4\pi = 1/137$ for QED] that hope quickly faded (in the early 1950s) that a perturbation-theoretic approach to a renormalizable pion-nucleon interaction would work. In the axial vector case, the derivative before the pion field ϕ increases the values of $\bar{D}$ by two — to $\bar{D} = 4, 3, 2$ respectively for the "vacuum polarization", self-energy and coupling constant contributions to the "superficial" degree of divergence; the fact that $\bar{D} = 2$ for the coupling constant renormalization dooms the axial vector "pion-nucleon" interaction to non-renormalizability and, hence, to failure. Curiously enough, the small value of the pseudovec-tor coupling constant ($f^2/4\pi \simeq 0.08$) and some empirical successes of the non-relativistic version of the axial vector form of the "pion-nucleon" inter-action kept the meson field theory of the strong interaction alive in several different "incarnations" (e.g. Pauli-Wentzel's strong coupling meson theory [1.50], the Chew-Low model [1.53] and, most recently, the Skyrme model [1.139]). However, a truly relativistic renormalizable quantum field theory of the strong interaction has only been realized with fermionic quark fields and non-Abelian bosonic (gluon) gauge fields.

Before turning to a brief account of the power-counting method in con-nection with vector fields, we show how simple use of Eq. (4.89) rules out renormalizability for the four-fermion interaction as a relativistic quantum field theory. For the simplest four-fermion interaction Lagrangian: $g(\bar{\psi}\psi)^2$, we can use the 1PI graphs given by Figs. 4.5 and 4.6 of the ϕ^4 theory, where the fermion field ψ replaces the scalar field ϕ; the vast difference between the scalar ϕ^4 theory and the four-fermion theory emerges immediately because $\delta = 2$ for the four-fermion interaction in contrast to $\delta = 0$ for the four-(scalar) boson interaction (due to the simple fact that the $J = \frac{1}{2}$ fermion possesses the canonical dimension $\frac{3}{2}$ whereas the $J = 0$ boson has the canon-ical dimension 1). The analogs of Figs. 4.5 and 4.6 (with the scalar lines replaced by fermion lines) yield $\bar{D} = 9$ and $\bar{D} = 6$ — in contrast to the $\bar{D} = 2$ and $\bar{D} = 0$ for the scalar case — for the degrees of "superficial" divergence of Γ^2 and Γ^4 respectively. Of course, this result for the four-fermion interaction is equivalent to the statement that the four-fermion interaction possesses a dimension 2 coupling constant (because the fermion dimension is $\frac{3}{2}$), in contrast to the dimensionless coupling constant of the ϕ^4 theory (because the boson dimension is 1); nevertheless, it is striking to see how bad the "superficial" degree of divergence really is for a relativistic quantum field theory based on the four-fermion interaction. Fortunately, the admittedly non-relativistic use of the V-A four-fermion theory of the weak interaction (see Table 1.4) is highly successful into the several of GeV region because of the large masses (~ 100 GeV) of the weak gauge bosons. Just as for the strong interaction, a truly relativistic renormalizable quantum field theory of the weak interaction has required the gauging of non-Abelian global internal symmetries.

The power-counting method for deciding the renormalizability of a quantum field theory can be extended to vector fields if some care is exercised. If the vector field is a massless (gauge) field — as in QED — power counting is the same as for the scalar field. This can be verified in Fig. 4.7 where the three diagrams — Fig. 4.7a, Fig. 4.7b and Fig. 4.7c — exhibit respectively the one-loop corrections to the (photon) vacuum polarization, the electron (or any charged lepton) self-energy, and the electric charge (vertex) renormalization [4.24]. [We will presently write out the expressions for the three renormalization QED constants: Z_3, Z_2 and Z_1 — for future use in §5.2a — that are associated with the three respective diagrams of Fig. 4.7; but, for the present, we limit ourselves to power-counting.] Hence, if the vector field is a (massless) gauge field, the quadratically divergent vacuum polarization term (see Fig. 4.7a) and the linearly divergent electron self-energy term (see Fig. 4.7b) are no more troublesome than the corresponding terms in the pseudoscalar pion-nucleon interaction because the degree of divergence of the coupling constant ("charge") term for both QED and the pseudoscalar pion-nucleon interaction settles down to a logarithmic divergence (which persists in the higher-loop approximations), which is compatible with renormalization. Actually, the "superficial" degree of divergence of the first two QED one-loop diagrams in Fig. 4.6 is reduced to that of logarithmic for both diagrams [because of gauge invariance in the case of Fig. 4.7a and because of a simple cancellation of the linearly divergent contribution to the electron self-energy in the case of Fig. 4.7b (see below)].

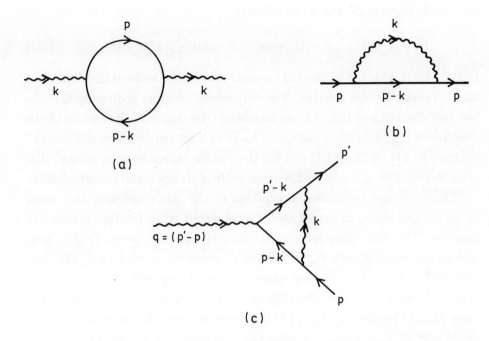

Fig. 4.7. One-loop radiative corrections in QED: (a) vacuum polarization; (b) electron self-energy; (c) vertex diagram.

It is important to note that the conclusion drawn from power counting as regards the equivalence of a scalar boson field and a massless vector field in quantum field theories is supported by the observation that the massless photon propagator has the asymptotic form [see Eq. (4.33)]:

$$\Delta_{\mu\nu}(k) = \frac{-ig_{\mu\nu}}{k^2 + i\epsilon} \xrightarrow[k\to\infty]{} O(k^{-2}). \qquad (4.90)$$

The situation changes drastically for a quantum field theory with massive vector fields since the asymptotic behavior of the propagator for such a field [see Eq. (4.34)] is:

$$D_{\mu\nu}(k) = \frac{-i(g_{\mu\nu} - k_\mu k_\nu/M_V^2)}{k^2 - M_V^2 + i\epsilon} \xrightarrow[k\to\infty]{} O(1) \qquad (4.91)$$

where M_V is the mass of the vector boson. The higher "superficial" degree of divergence of the propagator for the massive vector field (by $+2$) [as given by Eq. (4.91)] — compared to the massless vector field [as given by Eq. (4.90)] — is equivalent to assigning the weight 2 (rather than 1) to a vector line in a quantum field theory with massive vector fields so that $\bar{D}$ and δ of

Eq. (4.87) become $\bar{D}'$ and δ', as follows:

$$\bar{D}' = 4 - B - \frac{3}{2}F - 2V + n\delta; \quad \delta' = b + \frac{3}{2}f + 2v + d. \tag{4.92}$$

In Eq. (4.92), V is the number of external vector lines and v is the number of internal vector lines in a vertex. The "superficial" degrees of divergence of the last two diagrams of Fig. 4.7 are increased (the degree of divergence of the diagram in Fig. 4.7a stays the same) to $\bar{D}' = 3$ for the "fermion self-energy" diagram in Fig. 4.7b and $\bar{D} = 2$ for the "vector boson-fermion vertex" diagram in Fig. 4.7c. Clearly, such a quantum field theory is non-renormalizable.

There are two interesting exceptions to the last conclusion that must be mentioned since, in actual fact, the observed weak (vector) bosons are massive. The first exception occurs when the massive vector field is neutral and is coupled to a conserved vector current. In that case, the non-renormalizability of an Abelian quantum field theory with a massive vector field is changed to renormalizability because the massive vector field propagator $D_{\mu\nu}(k)$ [as given by Eq. (4.91)] appears between the conserved currents $J^\mu(k)$ and $J^\nu(k)$, thereby removing the troublesome $k_\mu k_\nu$ part of $D_{\mu\nu}$. The Abelian quantum field theory with massive vector field becomes equivalent — insofar as renormalizability is concerned — to the massless vector case (the statement is not true for non-Abelian fields — see §4.5b). A second exception to the conclusion that quantum field theories with massive vector fields are non-renormalizable occurs when the vector boson masses have an SSB origin. With the SSB mechanism operative, the initial Lagrangian consists of massless (gauge) fields and is renormalizable; the SSB Higgs mechanism is associated with the alteration of the physical vacuum of the theory — resulting from a non-vanishing VEV of one or more Higgs fields — and does not affect the asymptotic behavior of the n-point 1PI Green's functions that determine renormalizability. It is this loophole that is exploited in the formulation of the spontaneously broken, non-Abelian gauge theory of the electroweak interaction (see §6.2). In the light of the above remarks, we are in a good position to pick up our discussion of renormalization in the unbroken Abelian gauge theory of electromagnetism and to cite several examples of its remarkable predictive power.

§4.5a. Renormalization and precision tests in QED

The unique Lorentz-invariant, gauge-invariant and renormalizable theory of the electromagnetic interaction that is QED [1.16] has been subjected to increasingly searching experimental tests during the past forty years and its capability to pass all of them — at least until now [1.121] — is a great triumph for relativistic quantum field theory. In this section, we translate the renormalization methods and results developed in the previous section for the simple (non-gauge-invariant) scalar ϕ^4 theory to gauge-invariant Abelian QED. The gauge invariance property of QED is very helpful in reducing the "superficial" degree of divergence of the "one-loop" corrections to QED (e.g. the gauge invariance of QED converts the quadratic divergence of the vacuum polarization term to a logarithmic divergence — which already characterizes the "one-loop" corrections to the fermion self energy and coupling constant terms) so that the proof of renormalization for QED becomes more tractable. By translating the results of the ϕ^4 model into QED language, we keep in mind the need to provide sufficient detail concerning the cutoff dependence of the QED renormalization constants so that a comparison can be made (in §5.2a) — by means of renormalization group analysis — with the cutoff behavior of the renormalization constants associated with non-Abelian gauge groups (in particular QCD).

The scalar ϕ^4 theory can serve as a model for QED because it shows how the need for regularization and renormalization arises when one tries to "finitize" the one-loop radiative corrections. However, it must be admitted that the ϕ^4 theory is a self-interacting scalar theory (through the quartic term) and without gauge invariance, whereas QED involves a Dirac field interacting with a vector (gauge) field. Consequently, there are more basic (1PI) diagrams entering into the one-loop radiative corrections in QED than in the ϕ^4 theory; on the other hand, as we will see below, the gauge invariance in QED moderates the "superficial" degree of divergence in QED — compared to the ϕ^4 theory — and is computationally more tractable because of the vector Ward-Takahashi identities. The three basic (1PI) "one-loop" diagrams in QED — corresponding respectively to the vacuum polarization, the fermion self energy and the vertex — are those shown in Fig. 4.7; Figs. 4.7a, 4.7b and 4.6c are to be compared with the two basic (1PI) diagrams in the ϕ^4 theory (see Figs. 4.5 and 4.6). Translated into the language of QED (for electrons and photons), we can say that: (1) Fig. 4.7a is the "vacuum polarization" diagram $\Pi_{\mu\nu}(k)$ for the photon with the renormalized photon propagator being an infinite sum of the repetitive vacuum polarization diagrams, which requires renormalization; (2) Fig. 4.7b is the electron self-energy diagram and contributes in a similar way to the renormalized electron propagator; and (3) Fig. 4.7c is the photon-electron vertex diagram whose renormalization is called the (electric) charge renormalization [4.24].

We next write down the divergent integrals describing the three diagrams of Fig. 4.7 and point out how gauge invariance in QED lowers the "superficial" degree of divergence of the vacuum polarization to a logarithmic divergence and how the naive linear divergence of the electron propagator resulting from the electron self-energy diagram is also converted to a logarithmic divergence; in addition, the electron mass resulting from the electron self-energy diagram possesses a logarithmic divergence (a result noted by Weisskopf [4.25] in the 1930s!). Just as in the ϕ^4 theory, the divergence of the electron-photon vertex function is intrinsically logarithmic. The three QED diagrams of Figs. 4.7a–4.7c yield respectively:

$$\Pi_{\mu\nu}(k) = -(-ie)^2 \int \frac{d^4p}{(2\pi)^4} \, \text{Tr}(\frac{1}{\not{p} - m + i\epsilon}\gamma_\mu \frac{1}{\not{p} - \not{k} - m + i\epsilon}\gamma_\nu), \quad (4.93)$$

$$-i\Sigma(p) = (-ie)^2 \int \frac{d^4k}{(2\pi)^4 i}\frac{1}{k^2}\gamma_\mu \frac{1}{(\not{p} - \not{k} - m + i\epsilon)}\gamma^\mu, \quad (4.94)$$

$$-ie\Gamma_\mu(p',p) = (-ie)^3 \int \frac{d^4k}{(2\pi)^4 i}\frac{1}{k^2}\gamma_\nu \frac{1}{\not{p}' - \not{k} - m + i\epsilon}\gamma_\mu \frac{1}{(\not{p}' - \not{k} - m + i\epsilon)}\gamma^\nu \quad (4.95)$$

If one now takes into account the gauge-invariant property of QED, the "normal" quadratic degree of divergence ($\bar{D} = 2$) is reduced to $\bar{D} = 0$ (i.e. logarithmic divergence); this result immediately follows when one imposes gauge invariance on the vacuum polarization Green's functions $\Pi_{\mu\nu}(x)$ in momentum space [see Eq. (4.93)]:

$$k^\mu \Pi_{\mu\nu}(k) = 0 \quad (4.96)$$

Equation (4.96) enables $\Pi_{\mu\nu}$ to be written in the form:

$$\Pi_{\mu\nu}(k) = (k^2 g_{\mu\nu} - k_\mu k_\nu)\Pi(k^2). \quad (4.97)$$

where $\Pi(k^2)$ is a logarithmically divergent expression. Further, it becomes evident that the linearly divergent self-energy Green's function $\Sigma(p)$ in Eq. (4.94) is converted into a logarithmically divergent expression (because its linear dependence on the internal momentum is integrated over momentum space) so that the divergences of all three QED diagrams are logarithmic. It is true that one can manage with quadratically and linearly divergent diagrams — as in the ϕ^4 theory — as long as the higher-loop approximations settle down to $\bar{D} = 0$ (logarithmic divergence) but the immediate reduction of the superficial degrees of divergence of the first-order radiative corrections simplifies the calculations in QED.

In preparation for the renormalization group discussion of the two unbroken gauge theories (QED and QCD) in §5.2, we write out the explicit one-loop expressions for the three renormalization constants (vacuum polarization, electron propagator and vertex) that result from the renormalization method proceeding along the lines sketched earlier for the ϕ^4 theory. Every expression is written in terms of the "bare" charge, the reference renormalization point μ and the cutoff scale Λ (really Λ/μ since μ is arbitrary). Using the standard notation for the three Z's: Z_3 for the vacuum polarization renormalization constant, Z_2 for the electron propagator renormalization constant, and Z_1 for the charge (vertex) renormalization constant, we have [4.24]:

$$Z_3 = 1 - \frac{\alpha_e}{3\pi}\ln\frac{\Lambda^2}{\mu^2}; \quad Z_2 = 1 - \frac{\alpha_e}{4\pi}\ln\frac{\Lambda^2}{\mu^2}; \quad Z_1 = 1 - \frac{\alpha_e}{4\pi}\ln\frac{\Lambda^2}{\mu^2} \qquad (4.98)$$

where, in each expression, we have replaced the unrenormalized coupling constant e_0 by the renormalized coupling constant e because we are working in the "one-loop" approximation. It turns out that Z_2 and Z_1 are identical by virtue of one of the Ward-Takahashi (vector) identities, i.e.

$$\Gamma_\mu(p,p) = -\frac{\partial\Sigma(p)}{\partial p^\mu} \qquad (4.98a)$$

The equality of Z_2 and Z_1 simplifies the relation between the renormalized charge e to the unrenormalized charge e_0 through the equation:

$$e = Z_1^{-1}Z_2 Z_1^{\frac{1}{2}} e_0 = Z_3^{\frac{1}{2}} e_0 \quad (\text{since } Z_1 = Z_2) \qquad (4.99)$$

Equation (4.99) will be useful when we compare the β functions — which depend on the gauge coupling constants — for QED and QCD in §5.2a. We note that the "physical" renormalized electron mass is given by: $\frac{3\alpha_e}{4\pi} m(\ln\frac{\Lambda^2}{m^2} + \frac{1}{2})$.

Before leaving the subject of renormalized QED — to briefly review the renormalization situation in Yang-Mills theory — we extend remarks already made in §1.4a concerning precision tests of QED. In Table 1.7 of §1.4a, we compared the experimental and theoretical values of the Lamb shift and of the anomalous magnetic moment of the electron, the two earliest experiments that catalyzed the successful renormalization program in quantum electrodynamics. The excellent agreement exhibited in Table 1.7 is subject to some small caveats: a slight uncertainty in the R.M.S. radius of the proton in connection with the Lamb shift and a small uncertainty in the fine structure constant in making the comparison for $(g-2)$ of the electron. Over the years, further confirmatory comparisons between QED theory and experiment have been made for the hyperfine structure of hydrogen, the fine structure of helium, the hyperfine structure of muonium and the fine and hyperfine structure of positronium, among others [1.121]. Here we only comment on the comparison between QED theory and experiment for the $(g-2)$ value of the muon because the theoretical calculation (already to fourth order in α_e), involves not only the photon and electron fields, but also the more massive charged lepton fields (μ and τ), virtual hadrons (representing the strong interaction), and the virtual W, Z and Higgs particles of the electroweak interaction. The theoretical value of a_μ $[a_\mu = \frac{1}{2}(g-2)_\mu]$ can be written [4.26]:

$$a_\mu(\text{theor}) = a_\mu(\text{QED}) + a_\mu(\text{strong}) + a_\mu(\text{weak}) \qquad (4.100)$$

where:

$$a_\mu(\text{QED}) = 1\,165\,846951\,(44)\,(28) \times 10^{-12} \qquad (4.100a)$$

[the first error is calculational, the second is due to uncertainty in α_e]

$$a_\mu(\text{hadrons}) = 7027\,(175) \times 10^{-11} \qquad (4.100b)$$

$$a_\mu(\text{weak}) = 195(10) \times 10^{-11} \qquad (4.100c)$$

Putting these numbers together, one gets:

$$a_\mu(\text{theor}) = 116\,591\,917\,(176) \times 10^{-11} \qquad (4.100d)$$

The QED contribution to a_μ — as given by Eq. (4.100a) — is based on calculations to fourth order in α_e (to which 469 Feynman diagrams contribute! [4.27]) and an estimate of the contribution in fifth order of α_e; the accuracy of this impressive calculation (with supercomputers) now exceeds that which is obtainable in the calculation of the hadronic contribution to a_μ and is even better than the small weak interaction contribution to a_μ. It should be noted that the size of the hadronic contribution to a_μ is much larger than the corresponding contribution to a_e (see §1.2a) because the hadronic contribution arises from the second-order QED vertex diagram which favors the internal momenta of the muon by a substantial margin [by $(\frac{m_\mu}{m_e})^2$ [4.28]] over those associated with the electron (with its smaller mass). The theoretical calculation of the hadronic contribution to a_μ is strongly dependent on measurements of the R value (see §5.2a) in the neighborhood of 1 GeV, whose evaluation, in turn, depends on a proper treatment of the low-lying hadronic resonances such as the ρ meson.

The most accurate experimental value of a_μ is: $a_\mu(\text{expt}) = 11\,659\,230$ $(85) \times 10^{-10}$ (7.2 ppm). The agreement between $a_\mu(\text{expt})$ and $a_\mu(\text{theor})$ is highly satisfactory and informs us that despite the uncertainty of the hadron contribution to a_μ, the charged lepton of the second generation (μ) is a close copy of the first generation charged lepton (e) [i.e. μ is a "heavy electron"]; furthermore, the agreement is sufficiently good to place a lower limit on the scale of any possible muon structure of $(800 \text{ GeV})^{-1}$, i.e. the muon is "point-like" down to approximately 10^{-17}cm. More accurate measurements of $(g-2)_\mu$ are in the planning stage [4.29] in order to learn more about the strong and electroweak interactions through their virtual contributions to the $(g-2)_\mu$ of the muon. A high precision measurement of a fundamental quantity like $(g-2)_\mu$ may even reveal discrepant features beyond the present standard model that are inaccessible to the present generation of high energy accelerators. Before trying to move "beyond the standard model" in QED at low energy or, in QCD and QFD at high energy — we must review the renormalization situation for non-Abelian gauge theories which comprise the standard model, either without SSB as in QCD or with SSB as in QFD. This is the subject of an admittedly sketchy treatment in the next section (§4.5b) of how matters stand with regard to renormalization in Yang-Mills theories, prefatory to renormalization group considerations in §5.2a.

§4.5b. Renormalization in non-Abelian gauge theories

It was not immediately apparent that the application of the simple power counting argument to non-Abelian gauge theories (with fermions) — which indicates renormalizability — would survive more careful examination, particularly when SSB takes place. The reasons for this uncertainty stemmed from the fact that even in the absence of fermions, the pure Yang-Mills Lagrangian contained triple and quartic self-coupling terms among the gauge fields (with gauge coupling constant appreciably larger than α_e) in contrast to Abelian QED where the interaction term requires a coupling between the gauge field (photon) and charged fermions. Indeed, the pure gauge field contribution to the Yang-Mills Lagrangian provides the dominant contribution to the β function in the renormalization group equation and is responsible for the unexpected asymptotic freedom property of QCD. For these reasons, we start our brief review of renormalization in non-Abelian gauge theories with the pure Yang-Mills theory and, after calculating the renormalization constants associated with the pure theory, we merely cite the corrections to the dominant term due to interactions of fermion and scalar fields with the gauge fields.

It is not surprising that the renormalization techniques at work for Abelian gauge theory (e.g. QED) must be significantly modified in deal-

ing with non-Abelian gauge theories. It is true that one can work out —
for non-Abelian theories — the gauge-fixing condition within the framework
of the path-integral formulation and the use of the axial gauge. However,
it is more convenient to perform Feynman-type calculations in the covariant
gauge for non-Abelian gauge theories and this requires the identification of
an "effective" Lagrangian which takes "non-Abelian" gauge-fixing fully into
account. In Abelian gauge theory, the elimination of the unwanted degrees
of freedom associated with the covariant gauge can be achieved by imposing constraints on the gauge fields already introduced into the Lagrangian
("Abelian" gauge-fixing), as is done in the Gupta-Bleuer formulation [2.4].
However, in Yang-Mills theory, the elimination of the non-dynamical degrees
of freedom requires the introduction of additional fields (i.e. the Fadeev-
Popov (FP)) "ghost" fields [4.30] (in addition to the "Abelian"-type gauge-
fixing term).

   It can be shown that the "effective" Lagrangian for a "pure" Yang-Mills
theory, suitable for the performance of Feynman-type calculations in the
covariant gauge, is:

$$\mathcal{L}_{eff} = \mathcal{L}_{YM} + \mathcal{L}_{gf} + \mathcal{L}_{FP} \tag{4.101}$$

where:

$$\mathcal{L}_{YM} = -\frac{1}{4}F^a_{\mu\nu}F^{a\mu\nu}; \quad \mathcal{L}_{gf} = -\frac{1}{2\zeta}(\partial^\mu A^a_\mu)^2;$$

$$\mathcal{L}_{FP} = ic^\dagger_a \partial^\mu[\delta_{ab}\partial_\mu - g\epsilon_{abc}A^c_\mu]c_b \tag{4.101a}$$

In Eq. (4.101a), the first term is the "pure" Yang-Mills Lagrangian, the second term is the "Abelian"-type gauge-fixing contribution to the Lagrangian
with the "gauge-fixing" parameter ζ [where $\zeta = 1$ gives the "'t Hooft-
Feynman" gauge and $\zeta = 0$ the "Landau" gauge] and the third term contains
the "FP ghost" field (i.e. c_a is the "FP ghost" field). With Eqs. (4.101)–
(4.101a) as starting point, one can proceed with the usual renormalization
calculations of the "one-loop" radiative corrections to the vertex function
(needed for the β function) in the covariant gauge; for this purpose, we
record the Feynman rules for the gauge and "FP ghost" field propagators
[4.24]:

   (i) Gauge field propagator:

$$i\Delta^{ab}_{\mu\nu}(k) = -i\delta_{ab}\left[\frac{g^{\mu\nu}}{k^2 + i\epsilon} - (1-\zeta)\frac{k_\mu k_\nu}{(k^2 + i\epsilon)^2}\right] \tag{4.102}$$

(ii) "FP ghost" field propagator:

$$i\Delta^{ab}(k) = i\delta_{ab}\,\frac{1}{k^2 + i\epsilon}.\qquad(4.103)$$

We note that the "ghost field" is a massless complex scalar field that only propagates in closed loops and obeys Fermi-Dirac statistics (i.e. anti-C.R.).

In working out the renormalization constants for non-Abelian gauge theory (e.g. QCD), provision must not only be made for the FP "ghost" fields but also for the triple and quartic gauge field self-energy graphs shown in Figs. 4.8a–4.8b and 4.9a–4.9b. The three-point and four-point vertices are:

$$\Gamma^{abc}_{\mu\nu\lambda} = g\epsilon^{abc}[g_{\mu\nu}(k_1 - k_2)_\lambda + g_{\nu\lambda}(k_2 - k_3)_\mu + g_{\mu\lambda}(k_3 - k_1)_\nu]$$
$$(k_1 + k_2 + k_3 = 0).\qquad(4.104a)$$

$$\Gamma^{abcd}_{\mu\nu\lambda\rho} = -g^2 \left[\epsilon^{abe}\epsilon^{cde}(g_{\mu\lambda}g_{\nu\rho} - g_{\nu\lambda}g_{\mu\rho} + \epsilon^{ace}\epsilon^{bde}(g_{\mu\nu}g_{\lambda\rho} - g_{\lambda\nu}g_{\mu\rho})\right.$$
$$\left.+ \epsilon^{ade}\epsilon^{cbe}(g_{\mu\lambda}g_{\rho\nu} - g_{\rho\lambda}g_{\mu\nu})\right]\quad (k_1 + k_2 + k_3 + k_4 = 0)$$
$$(4.104b)$$

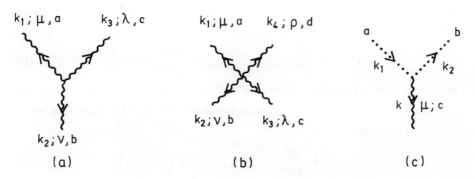

Fig. 4.8. Yang-Mills and FP "ghost"-Yang-Mills vertices: (a) triple Yang-Mills vertex; (b) quartic Yang-Mills vertex; (c) FP "ghost"-Yang-Mills vertex (dotted lines are the FP "ghost" fields).

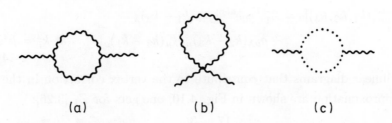

Fig. 4.9. Gauge field self-energy graphs (the dotted loop is that of the FP "ghost").

In addition to the purely gauge field vertices exhibited in Eqs. (4.104a) and (4.104b), one must take into account the coupling of two "FP ghost" fields to a gauge field [with polarization vector $\epsilon^\mu(k_1 + k_2)$] [see Eq. (4.101a) and Fig. 4.8c], thus:

$$\Gamma_\mu^{abc} = ig\epsilon^{abc}k_{2\mu} \quad (k + k_2 - k_1 = 0) \tag{4.105}$$

With the Feynman rules and the 1PI Green's functions in hand, one can derive the gauge field wavefunction renormalization constant (Z_3) and the gauge field charge (vertex) renormalization constant (Z_1), for a pure Yang-Mills theory. Amusingly enough, the "pure" Yang-Mills theory is close to the scalar ϕ^4 theory since there are only two renormalization constants for the "pure" Yang-Mills theory as for the ϕ^4 theory; however, the Yang-Mills theory is a gauge theory — unlike the ϕ^4 theory — and its non-Abelian character requires the introduction of the FP "ghost" fields. In any case, the renormalization constant Z_3 for the Yang-Mills theory can be defined in terms of the unrenormalized transverse gauge field propagator at the arbitrary reference point $k^2 = \mu^2 > 0$) [as in the case of Abelian QED, the renormalization reference point $\mu^2 = 0$ must be rejected in order to avoid infrared singularities (see §4.5)]:

$$[\Delta_{\mu\nu}^{ab}(k)]_0^{\text{Tr}}|_{k^2=\mu^2} = Z_3 \, (g_{\mu\nu} + \frac{k_\mu k_\nu}{\mu^2})\delta^{ab}/\mu^2. \tag{4.106}$$

The gauge field self-energy diagrams that contribute to Z_3 in the one-loop approximation are shown in Fig. 4.9; one finds for the wavefunction renormalization constant Z_3:

$$Z_3 = 1 + \frac{g_0^2}{32\pi^2}(\frac{13}{3} - \zeta) \, C_2(V) \, \ln(\Lambda^2/\mu^2) \tag{4.107}$$

where g_0 is the unrenormalized gauge coupling constant and $C_2(V)$ is defined in terms of the group structure constants (see below). The vertex renormalization constant Z_1 can be defined in terms of the unrenormalized vector three-point vertex written, thus:

$$[\Gamma_{\mu\nu\lambda}^{abc}(k_1, k_2, k_3)]_0 = Z_1^{-1}g_0\epsilon^{abc}[g_{\mu\nu}(k_1 - k_2)_\lambda$$
$$+ g_{\nu\lambda}(k_2 - k_3)_\mu g_{\lambda\mu}(k_3 - k_1)_\nu] \qquad \text{at } k_i^2 = \mu^2. \tag{4.108}$$

The trilinear diagrams that contribute to the vertex correction in the one-loop approximation are shown in Fig. 4.10; one gets for Z_1 [3.26]:

$$Z_1 = 1 + \frac{g_0^2}{32\pi^2}(\frac{17}{6} - \frac{3\zeta}{2}) \, C_2(V) \, \ln(\Lambda/\mu)^2 \tag{4.109}$$

The quantity $C_2(V)$ in Eq. (4.109) [and in Eq. (4.107)] is the second-order Casimir defined in terms of the structure constants of the $SU(3)$ group, thus:

$$C_2(V)\delta^{ab} = \epsilon^{acd}\epsilon^{bcd}. \tag{4.110}$$

Equation (4.110) can also be written as:

$$C_2(V)\delta^{ab} = \mathrm{Tr}\{T^a(V)T^b(V)\}. \tag{4.111}$$

Thus, $C_2(V)$ is simply the sum of the squared (symmetry) charges of the gauge particles and is 3 for the $SU(3)$ group.

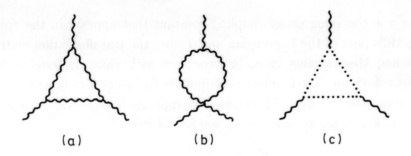

$$(a) \qquad\qquad (b) \qquad\qquad (c)$$

Fig. 4.10. Trilinear gauge field-boson vertex corrections.

In contrast to QED, where the relation between the renormalized charge e and the unrenormalized charge e_0 [see Eq. (4.99)] involves three renormalization constants: Z_3, Z_2, and Z_1 — with $Z_1 = Z_2$ (due to gauge invariance), the relation between the renormalized Yang-Mills gauge coupling constant g and its "bare' value g_0 involves the two renormalization constants Z_3 and Z_1 and both are needed to achieve independence of the gauge parameter ζ; one has:

$$g = Z_3^{\frac{3}{2}} Z_1^{-1} g_0 \tag{4.112}$$

Equation (4.112) is properly compared with the third equation of (4.76) for the relation between the renormalized and unrenormalized scalar coupling constants in the ϕ^4 theory; the difference between the exponent $\frac{3}{2}$ of Z_3 in Eq. (4.112) and 2 for Z_ϕ in Eq. (4.76) is due to the differing definitions in the two cases. It will be seen (in §5.2a) that it is precisely the combination $Z_3^{\frac{3}{2}} Z_1^{-1}$ (which is independent of the gauge parameter ζ) that enters into the definition of the renormalization group β function, which decides whether or not the gauge theory (Abelian or non-Abelian) possesses the crucial asymptotic freedom property.

Until now, we have focused on the "pure" Yang-Mills theory since it is the part of the total Lagrangian (even when fermions are included) that mandates the introduction of the novel features of non-Abelian gauge theories (e.g. the "ghost" fields) in the renormalization procedure. In the realistic QCD theory, fermions (i.e. quarks) must be included. However, the inclusion of fermions in the total Lagrangian does not introduce a new coupling constant and it is only necessary to add to the "effective" Lagrangian $\mathcal{L}_{eff}$, given by Eq. (4.101), the term:

$$\mathcal{L}_F = \bar{\psi}(i\partial_\mu - m - gT^a A^a_\mu)\psi, \tag{4.113}$$

where g is the same gauge coupling constant that appears in the "pure" Yang-Mills part of the Lagrangian and T^a are the transformation matrices of the non-Abelian gauge group in accordance with which the fermion Dirac spinors transform. The fermions contribute to the gauge wavefunction renormalization constant Z_3 — in the one-loop approximation — via the diagram shown in Fig. 4.8a; we have for Z_3 [see Eq. (4.98)]:

$$Z_3^F = -\frac{g^2}{16\pi^2}\frac{4}{3}C_2(F)\ln(\Lambda^2/\mu^2) \tag{4.114}$$

where $C_2(F)$ is the second order Casimir for the fermions (quarks). The addition of scalar fields (when there is SSB) to the Yang-Mills Lagrangian can be treated in essentially the same fashion as the inclusion of the fermion fields and one obtains for Z_3^S the following:

$$Z_3^S = 1 - \frac{g^2}{16\pi^2}\cdot\frac{1}{3}C_2(S)\ln(\Lambda/\mu)^2 \tag{4.115}$$

where $C_2(S)$ is the second order Casimir for the scalar fields. For Dirac fermions and complex scalars in the fundamental representation of $SU(N)$, we get: $C_2(F) = C_2(S) = \frac{1}{2}$; for Weyl fields or real scalars, both $C_2(F)$ and $C_2(S) = \frac{1}{4}$.

With the various renormalization constants known for both Abelian and non-Abelian gauge groups, with or without fermions and/or scalars, it is possible to write down the expressions for the β function in §5.2a as part of the renormalization group analysis of non-Abelian gauge theories. The comparison of the β functions of Abelian QED and non-Abelian QCD will exhibit in a striking way why the "pure" Yang-Mills gauge part of the non-Abelian "effective" Lagrangian is so important for the presence of asymptotic freedom in non-Abelian QCD.

§4.6. Dirac's magnetic monopole and electric charge quantization

We have completed our introductory review of the mathematical formalism and renormalization results for non-Abelian gauge theories in preparation for their application to QCD and QFD, the two parts of the standard model of the strong and electroweak interactions. Before starting our discussion of QCD in Chapter 5, we pause to consider the earliest treatment of a topological result in Abelian gauge theory — without phenomenological support thus far but with interesting paradigmatic consequences for non-Abelian gauge theories — namely Dirac's magnetic monopole. Dirac's hypothesis of the magnetic monopole was made in the same 1931 paper [1.39] in which he examined the connection between the physical meaning of the phase factor of a wavefunction and the gauge principle in quantum mechanics. This led him to the concept of "non-integrable phase" expressed in terms of the line integral of the gauge field around a closed space-time curve C, which is independent of the wavefunction and depends only on C. In the same 1931 paper, Dirac explored further the consequences of having singularities (i.e. "nodal lines") in the (complex) quantum mechanical wavefunction and inferred that magnetic monopoles are needed to deal with "nodal singularities". We give, in this section, a résumé of Dirac's early argument for the existence of magnetic monopoles within the framework of "QED" and, then, discuss some other approaches to Dirac's magnetic monopole to provide the language and the topological paradigm for our later treatment of the 't Hooft-Polyakov magnetic monopole in non-Abelian gauge theories (in §10.2c) and the "Wess-Zumino" term in the Skyrme model (in §10.5c).

We were reminded by Dirac in 1931 (see §4.2) that a quantum mechanical wavefunction ψ can be expressed as $\psi(x) = \psi'(x)e^{i\beta(x)}$ with $\psi'(x)$ an ordinary wavefunction (with a definite phase at each point) and β a real function of x. Pursuing a sequence of arguments that takes careful account of what is, and what is not, observable in quantum mechanics, Dirac arrives at the conclusion that the change in the derivative of the phase $[\partial_\mu \beta]$ around a closed curve, C, i.e. $\Delta = \oint_C \partial_\mu \beta \cdot d\ell^\mu$, must be the same for all wavefunctions; this result, in turn, implies that a field of force A_μ is present that is related to Δ by:

$$\Delta = \oint_C A_\mu d\ell^\mu \tag{4.116}$$

The quantity Δ is called the "non-integrable phase" and, in general, need not vanish, giving rise to the Aharonov-Bohm effect (see §4.2). Continuing this line of argument in his 1931 paper [1.39], Dirac realized that the independence of the "non-integrable phase" of a wavefunction, $\Delta = \int_C A_\mu \cdot d\ell^\mu$, only holds in the absence of "nodal singularities" in the wavefunction. If there are "nodal singularities", Δ does depend on the wavefunction and it is this dependence of Δ on a wavefunction with "nodal singularities" that was carefully analyzed by Dirac. Dirac argues that a singular situation arises when the quantum mechanical wavefunction vanishes since its phase then is physically meaningless; furthermore, since the wavefunction is complex, the points at which the wavefunction vanishes must lie along a "nodal line". Different nodal lines can then be characterized by different values of an integer n. Thus, for a curve enclosing one nodal line, the total change in the "non-integrable phase" Δ is given by [henceforth, we regard A_μ as the electromagnetic three-potential $\underline{A}$ and insert the charge e before the $\int$ defining Δ without "nodal singularities"]:

$$\Delta = 2\pi n + e \oint_C \underline{A} \cdot d\underline{\ell} = 2\pi n + e \oint_S \underline{B} \cdot d\underline{\sigma} \qquad (4.117)$$

where n is associated with the "nodal line" under consideration. If the curve encloses more than one nodal line, with different n_i ($i = 1, 2, ...m$), Δ becomes:

$$\Delta = 2\pi \sum_{i=1}^{m} n_i + e \oint_S \underline{B} \cdot d\underline{\sigma} \qquad (4.118)$$

where the summation is over the points in S with nodal lines through them and the proper sign is given to n_i. The R.H.S. of Eq. (4.118) consists of two terms, the second being the same for all wavefunctions, and the first depending on the particular wavefunction (according to the number of nodal lines it possesses).

Dirac next assumes S is a closed surface and points out that Δ must vanish for all wavefunctions so that $\sum_{i=1}^{m} n_i$ (which is summed for all "nodal lines" crossing the closed surface S) must be identical for all wavefunctions and equal to $-e/2\pi \int_S \underline{B} \cdot d\underline{\sigma}$. Clearly, the only contributions to $\sum_{i=1}^{m} n_i$ must come from "nodal lines" that have endpoints inside the closed surface S since a "nodal line" passing through the closed surface must cross the surface at least twice and contribute equal and opposite amounts to $\sum_{i=1}^{m} n_i$. Further, since the resultant sum is identical for all wavefunctions and holds for any closed surface, it follows that the endpoints of "nodal lines" must be identical for all wavefunctions and, consequently, that these endpoints are singular points of the electromagnetic field. It is these "nodal singularities" of the electromagnetic field that Dirac proposes to relate to magnetic monopoles.

While Dirac's magnetic monopole has never been found, his novel quantization condition— with its topological overtones (see §10.1) — has led to numerous studies of its significance in electromagnetism if it did exist; we mention a few of them: (1) what is the relation between the magnetic monopole and the dual electromagnetic field?; (2) is the "Dirac string", which turns up when the Maxwell equations are solved in the presence of a magnetic monopole, an artifact and, if so, can it be eliminated by means of the gauge principle? and, finally (3) what is the relation between the Dirac magnetic monopole and the topological (solitonic) magnetic monopoles arising within the framework of non-Abelian gauge theories and their spontaneous symmetry breaking (e.g. the 't Hooft-Polyakov monopole")? We discuss the first two aspects of Dirac's magnetic monopole concept in this section and relegate our comments on the non-Abelian topological ramifications of the Dirac monopole to §10.2b.

Dirac's second paper on the magnetic monopole in 1948 [4.31] reexamined the properties of the Schrödinger wavefunctions for an electron in the field of a magnetic monopole and clearly identified the string of singularities (i.e. the "Dirac string") — his earlier "nodal singularities" — that are encountered in the solution of this problem. To recapture Dirac's derivation — and, in the process, answer question (1) above — we recall that the Maxwell equations in covariant form are given by:

$$\partial_\nu F^{\mu\nu} = -j^\mu; \quad \partial_\nu \tilde{F}^{\mu\nu} = 0 \tag{4.119}$$

where:

$$j^\mu = (\rho, \underline{j}), \quad F^{0i} = E^i, \quad F^{ij} = -\epsilon^{ijk} B^k \tag{4.120}$$

and the dual field tensor is:

$$\tilde{F}^{\mu\nu} = \frac{1}{2} \epsilon^{\mu\nu\rho\sigma} F_{\rho\sigma}. \tag{4.121}$$

In the special case that $j_\mu = 0$ (vacuum), the Maxwell equations are symmetric under the "duality" transformation:

$$F^{\mu\nu} \to \tilde{F}^{\mu\nu}, \tilde{F}^{\mu\nu} \to -F^{\mu\nu} \tag{4.122}$$

or, equivalently, the interchange of $\underline{E} \to \underline{B}, \underline{B} \to -\underline{E}$. When $j_\mu \neq 0$, this "duality" symmetry is lost but can be restored if a magnetic current $k^\mu = (\sigma, \underline{k})$ is added to the R.H.S. of Eq. (4.120), provided the interchange

$F^{\mu\nu} \to \tilde{F}^{\mu\nu}, \tilde{F}^{\mu\nu} \to -F^{\mu\nu}$ is supplemented by the interchange:

$$j^\mu \to k^\mu, \; k^\mu \to -j^\mu. \tag{4.123}$$

Clearly, the introduction of the "dual" magnetic current k^μ (to the electric current j^μ) requires the existence of magnetically charged particles (i.e. the magnetic monopoles) in addition to the electrically charged particles of classical electromagnetism. It is important to note that the second Eq. (4.120) is replaced by:

$$\partial_\nu \tilde{F}^{\mu\nu} = -k^\mu \tag{4.124}$$

so that k^μ is the source of the dual electromagnetic field tensor $\tilde{F}^{\mu\nu}$, just as j^μ is the source of the (regular) electromagnetic field tensor $F^{\mu\nu}$; as will be seen later, the relation between $\tilde{F}^{\mu\nu}$ and k^μ helps to explain the topological character of Dirac's quantization condition between the electric and magnetic charges (see §10.2c).

We are now in a position to exhibit the relationship between the "Dirac string" and the magnetic monopole solution, and to derive Dirac's quantization condition from the non-observability of the "Dirac string". We know that, in the absence of the magnetic monopole (i.e. $k^\mu = 0$), we have:

$$\underline{B} = \nabla \times \underline{A}; \quad \underline{E} = -\nabla\phi - \frac{\partial \underline{A}}{\partial t} \tag{4.125}$$

and that Schrodinger's equation for a (non-relativistic) particle moving in the electromagnetic field has the form:

$$[\frac{1}{2m}(\underline{p} - e\underline{A})^2 + e\phi]\psi = i\frac{\partial\psi}{\partial t}, \tag{4.126}$$

An important property of Eq. (4.126) is its invariance under the gauge transformation:

$$\underline{A}(\underline{x}) \to \underline{A}(\underline{x}) + \frac{1}{e}\nabla\alpha(\underline{x}); \quad \psi(\underline{x}) \to e^{i\alpha(\underline{x})}\psi(\underline{x}) \tag{4.127}$$

On the other hand, in the presence of a magnetic monopole, the gauge field (vector potential) can not exist everywhere because $\tilde{F}^{\mu\nu}$ no longer vanishes but satisfies Eq. (4.124). To see that the "Dirac string" is a convenient device for taking account of the effect of the magnetic monopole, we first point out that "Gauss' law" for the magnetic charge requires $g = \int_S \underline{B} \cdot d\underline{\sigma}$ and hence $B \neq \text{curl } \underline{A}$ everywhere. The trick is to recognize that $\underline{A}$ can be defined in such a way that $\underline{B} = \text{curl } \underline{A}$ everywhere except on a line joining the origin to infinity (i.e. the "Dirac string").

The "Dirac string" can be replaced by an infinitely long and thin solenoid placed along the negative z axis with its positive pole (of strength g) at the

origin. The magnetic field due to the solenoid can now be split into two terms:
the first term due to the positive quasi-magnetic monopole of strength g plus
the second term due to the singular magnetic flux along the solenoid, namely
[3.26]:

$$\underline{B}_{\text{sol}} = \frac{g}{4\pi r^2}\hat{r} + g\ \theta(-z)\delta(x)\delta(y)\ \hat{z} \tag{4.128}$$

Since the magnetic field $\underline{B}_{sol}$ is source-free, it can be written as $\underline{B}_{sol} = \nabla \times \underline{A}$
so that the first term (the monopole field) becomes:

$$\underline{B} = \frac{g^2}{4\pi r^2}\hat{r} = \nabla \times \underline{A} - g\ \theta(-z)\delta(x)\delta(y)\ \hat{z} \tag{4.129}$$

Figure 4.11 shows how the magnetic field of the monopole is related to the
magnetic field due to the solenoid and the magnetic field due to the "Dirac
string". To continue the derivation, we know the vector potential due to the
solenoid and hence, using polar coordinates, $\underline{A}$ can be written in the form:

$$\underline{A} = \frac{g}{4\pi r}(\frac{1 - \cos\theta}{\sin\theta})\hat{\phi} \tag{4.130}$$

In sum, the magnetic monopole field can be represented by the vector potential $\underline{A}$ [given by Eq. (4.130)] together with the contribution from the "Dirac
string" (see Fig. 4.11).

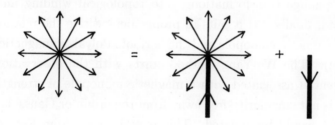

Fig. 4.11. Field of the Dirac magnetic monopole and the "Dirac string".

Apart from the question of the existence of the Dirac magnetic monopole,
there is the question of the theoretical status of the "Dirac string", i.e. does
it have a gauge-invariant status? The answer is in the negative and it can
be shown that the "Dirac string" can be moved around by a suitable gauge
transformation and consequently that it would not be observable. The connection between the non-observability of the "Dirac string" and the gauge
principle can be demonstrated directly [3.26] but we prefer to exhibit this
connection by means of the Wu-Yang method [4.32] (see below). Before doing
so, we show in a simple way that the "Dirac string" is unphysical by proving
that it does not contribute to the Aharonov-Bohm effect. The condition for

the absence of the Aharonov-Bohm effect (see §4.2) is that the two electron beams in the two-slit experiment moving on opposite sides of the "Dirac string" do not experience any detectable phase shift in their interference pattern when they strike the detector screen (see Fig. 4.1). The condition for no change in the interference pattern between the two electron beams is that the "non-integrable phase" $\Delta = e \int_C \underline{A} \cdot \underline{dl} = e \int_S \underline{B} \cdot \underline{d\sigma} = 2\pi n$ (n an integer); but if we insert the value of the flux due to the "Dirac string", in correct units, namely $4\pi g$, we get $eg = \frac{n}{2}$ (the Dirac quantization condition) and there is no Aharonov-Bohm effect.

We now consider the Wu-Yang interpretation of the "Dirac string" [4.32] and the consequences thereof. The Wu-Yang paper eliminates Dirac's string of singularities in the presence of a magnetic monopole by partitioning the space outside the origin of the magnetic monopole into two overlapping regions — in each of which the vector potential due to the monopole is singularity-free — but which are related by a "transition" gauge transformation (see below). The Wu-Yang approach not only eliminates the "Dirac string" but, in the process, does three things: (1) it provides a derivation of Dirac's quantization condition in a geometric context; (2) it exhibits the close relationship between a "global" (large) gauge transformation — in contrast to a "local" gauge transformation — to topological winding numbers (see §10.4a); and, finally, (3) it puts in proper perspective the relation between the Dirac magnetic monopole and the 't Hooft-Polyakov magnetic monopole (see §10.2c). The Wu-Yang method starts with the observation that the vector potential associated with a magnetic monopole of strength $g \neq 0$ at the origin is not singularity-free away from the origin or Gauss' law for magnetic charge would be violated. This is easily seen from Fig. 4.12a since $\oint_C \underline{A} \cdot \underline{dl} = \int_S \underline{B} \cdot \underline{d\sigma} = 4\pi g$ must vanish if $\underline{A}$ does not possess a singularity outside the origin.

To deal with the string of singularities in the presence of the magnetic monopole, one divides the space of Fig. 4.12a into two regions (called "patches" in more recent terminology [2.25]), as shown schematically in Fig. 4.12b. We give in Eqs. (4.131) and (4.132) the precise definitions of the overlapping regions R_a and R_b, as well as the expressions for the vector potentials in the two regions, whose curls define the magnetic field of the monopole:

$$R_a: \quad 0 \leq \theta < \frac{\pi}{2} + \delta, \quad 0 < r, \quad 0 \leq \phi < 2\pi$$

$$R_b: \quad \frac{\pi}{2} - \delta < \theta \leq \pi, \quad 0 < r, \quad 0 \leq \phi < 2\pi \qquad (4.131)$$

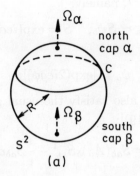

Fig. 4.12a. A sphere of radius R with a magnetic monopole at its center; curve C divides sphere into two caps α and β.

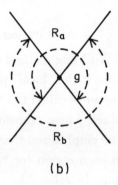

Fig. 4.12b. Division of space outside monopole g into overlapping regions: R_a and R_b.

$$(A_0)_a = (A_r)_a = (A_\theta)_a = 0; \quad (A_\phi)_a = \frac{g}{r\sin\theta}(1-\cos\theta)$$
$$(A_0)_b = (A_r)_b = (A_\theta)_b = 0; \quad (A_\phi)_b = -\frac{g}{r\sin\theta}(1+\cos\theta)$$

$$(4.132)$$

where r, θ and ϕ are the usual spherical coordinates and $\delta > 0$. The next step in the derivation is the important one: since the two sets of vector potentials share the same curl (i.e. magnetic field), their difference must be curlless and hence representable by the gradient of some function $\alpha(x)$, thus:

$$(A_\mu)_a - (A_\mu)_b = \partial_\mu\alpha \qquad \text{where } \alpha = 2g\phi \qquad (4.133)$$

When the two vector potentials $(A_\mu)_a$ and $(A_\mu)_b$ are inserted into the Schrodinger equation (4.126) giving rise to ψ_a and ψ_b as their respective wavefunctions, it is easy to show that ψ_a and ψ_b are related by the "local" gauge transformation $e^{ie\alpha(x)}$, namely:

$$\psi_a = S\psi_b, \quad S = \exp(ie\alpha),$$

(4.134)

or:

$$\psi_a = [\exp(2ieg\phi)]\psi_b$$

(4.135)

The two vector potentials also satisfy the usual gauge transformation relationship, thus [see Eq. (4.19)]:

$$(A_\mu)_a = S_{ab}(A_\mu)_b S_{ab}^{-1} - \frac{i}{e}S_{ab}\partial_\mu(S_{ab}^{-1})$$

(4.136)

or:

$$(A_\phi)_a = (A_\phi)_b + \frac{2g}{r\sin\theta}$$

(4.137)

Since both ψ_a and ψ_b are single-valued around the equator, we must have the condition $S_{ab}(\phi = 2\pi) = S_{ab}(\theta = 0)$; but this is precisely Dirac's quantization condition

$$eg = \frac{n}{2} \qquad (n \text{ is integer}).$$

(4.138)

We have thus seen how the Wu-Yang method eliminates the Dirac string of singularities and substitutes for it the "local" phase factor transformation $S = e^{ie\alpha(x)}$ $[\alpha(x) = 2g\phi]$ — sometimes called the "transition" function [4.32] — which must be a single-valued gauge transformation between the two overlapping regions. It should be emphasized that one is free to perform the usual gauge transformations in each region (or "patch" [2.25]) but the two potentials and wavefunctions must be related by the "transition" function in the overlap region. Dirac's quantization condition (4.138) is then an immediate corollary. Moreover, the "transition" function around the closed curve C is — from a topological viewpoint — a continuous map from a circle onto the gauge group $U(1)_{EM}$. All such maps can be classified by the homotopic group Π_1, thus: $\Pi_1[U(1)] = \mathcal{Z}$ (see §10.4c) which, again, informs us of Dirac's quantization condition $S = e^{in\phi}$. Thus, the "transition" function around a closed curve provides a map in which n is simply the number of times the circle is mapped into $U(1)$ group parameter space. The homotopic classification of magnetic monopoles of the Dirac (Abelian) type [by $\Pi_1(U(1)) = \mathcal{Z}$] can be extended to appropriate problems in non-Abelian gauge theories and, we will discuss in Chapter 10 (§10.2c and §10.5c) on "topological conservation laws", the fruitful topological insights that are conveyed by Dirac's hypothetical magnetic monopole.

We conclude this section by pointing out that Dirac was greatly taken by his quantization condition between electric charge and magnetic charge. In his second paper on the theory of magnetic monopoles [4.31], he argued with considerable conviction that: "the mere existence of one pole of strength g would require all electric charges to be quantized in units of $\frac{1}{2g}$ and, similarly, the existence of one charge would require all poles to be quantized. The quantization of electricity is one of the most fundamental and striking features of atomic physics, and there seems to be no explanation for it apart from the theory of poles. This provides some grounds for believing in the existence of these poles." Dirac's argument for electric charge quantization (ECQ) is not to be taken lightly because it is the first "topological conservation law" uncovered in Abelian gauge theory. Since Dirac's two papers on the magnetic monopole were published, other potential physical examples of "topological conservation laws" have been identified in non-Abelian gauge theories (e.g. instantons, 't Hooft-Polyakov magnetic monopoles, etc.) which have revealed the rich topological structure of QCD (see Chapter 10). Moreover, with regard to Dirac's ECQ argument, it is likely that — within the framework of the standard model — ECQ is well explained by the chiral gauge anomaly-free constraints which, to some extent, contain topological information (see §7.5 and §7.6).

References

[4.1] J. Gasser and H. Leutwyler, *Annals of Phys.* **158** (1984) 142; *ibid.*, *Nucl. Phys.* **250** (1985) 465.

[4.2] S. Weinberg, *Phys. Rev. Lett.* **17** (1966) 616; *ibid.*, *Phys. Rev.* **166** (1968) 1568.

[4.3] Discussion between Weyl, Pauli and Einstein at Ban Nauheim, *Phys. Z.* **21** (1920) 649.

[4.4] F. and H. London, *Proc. Roy. Soc. (London)* **A149** (1935) 71.

[4.5] C.N. Yang, *Hermann Weyl Centenary Lectures*, ed. K. Chandrasekharan, Spring, Berlin (1986).

[4.6] H. Weyl, *Z. Physik* **56** (1929) 330.

[4.7] R.P. Feynman, *Quantum Electrodynamics: a Lecture Note and Reprint Volume*, Benjamin (1962).

[4.8] A. Barnes and J.D. Scargle, *Phys. Rev. Lett.* **35** (1975) 1117.

[4.9] C.N. Yang, *Selected Papers 1945-1980 with Commentary*, W.H. Freeman, San Francisco (1983), p. 20.

[4.10] C.J. Davisson and L.H. Germer, *Phys. Rev.* **30** (1927) 705.

[4.11] R.G. Chambers, *Phys. Rev. Lett.* **5** (1960) 3.

[4.12] J.L. Rosner, *Proc. of 1987 Theor. Advanced Study Institute*, World Scientific, Singapore.

[4.13] T.T. Wu and C.N. Yang, *Phys. Rev.* **D12** (1975) 3845.

[4.14] L.N. Cooper, *Phys.* **104** (1956) 1189.

[4.15] V.L. Ginzburg and L.D. Landau, *Zh. Eksperim i Teor. Fiz.* **20** (1950) 1054.

[4.16] A.B. Pippard, *Proc. Roy. Soc. (London)* **A216** (1953) 547.

[4.17] P.W. Anderson, *Phys. Rev.* **130** (1963) 439. 1364.

[4.18] A. Abrikosov, *Soviet Phys. JETP* **5** (1957) 1174.

[4.19] H.A. Bethe, *Phys. Rev.* **72** (1947) 339.

[4.20] J.B. French and V.F. Weisskopf, *Phys. Rev.* **75** (1949) 1240; R.P. Feynman, *Phys. Rev.* **76** (1949) 769.

[4.21] F. Bloch and A. Nordsieck, *Phys. Rev.* **52** (1937) 54.

[4.22] W. Pauli and F. Villars, *Rev. Mod. Phys.* **21** (1949) 434.

[4.23] G. 't Hooft and M. Veltman, *Nucl. Phys.* **B44** (1972) 189.

[4.24] C. Izykson and J-B. Zuber, *Introduction to Quantum Field Theory*, McGraw-Hill, New York (1980).

[4.25] V.F. Weisskopf, *Zeit. f. Phys.* **89** (1934) 27; *ibid.* **90** (1934) 817.

[4.26] T. Kinoshita, *Proc. of Workshop on Future of Muon Physics*, Heidelberg (1991) (to be published in *Zeit. f. Phys. C*).

[4.27] H. Kawai, T. Kinoshita and Y. Okamoto, *Phys. Lett.* **B260** (1991) 193.

[4.28] C. Bouchiat and L. Michel, *J. Phys. Radium* **22** (1961) 121.

[4.29] J. Bailey *et al.* (CERN-Mainz-Darebury Collaboration), *Nucl. Phys.* **B150** (1979) 1.

[4.30] L.D. Faddeev and V.N. Popov, *Phys. Lett.* **25B** (1967) 29.

[4.31] P.A.M. Dirac, *Phys. Rev.* **74** (1948) 817.

[4.32] C.N. Yang, *Phys. Rev. Lett.* **33** (1974) 445; T.T. Wu and C.N. Yang, *Phys. Rev.* **D12** (1975) 3845; C.N. Yang, Festschrift for R.E. Marshak: *Five Decades of Weak Interactions*, N.Y. Academy of Sci. Annals (1977).

Chapter 5

GAUGE THEORY OF THE
STRONG INTERACTION (QCD)

§5.1. Massless quark limit in gauged $SU(3)$ color

In Chapter 3, we described the progression of steps that led from the simplest global non-Abelian internal symmetry (isospin) group $[SU(2)_I]$ to the "strong" GNN group $SU(2)_I \times U(1)_Y$ and then on to the larger global $SU(3)_F$ quark flavor group for hadrons. We then showed how the behavior of the low-lying (in mass) baryon multiplets under the static $SU(6)$ quark flavor-spin group led to a new global "color" internal symmetry group, $SU(3)_C$, in order to satisfy the Pauli principle. Finally, we spelled out the important role played by the global chiral quark flavor symmetry group $SU(3)_L \times SU(3)_R$ [and its subgroup $SU(2)_L \times SU(2)_R$] in underpinning the current algebra calculations of the quark current contributions to weak and electromagnetic interactions. The introduction of the concept of spontaneous symmetry breaking (SSB) to break $SU(3)_L \times SU(3)_R$ down to $SU(3)_F$ [and $SU(2)_L \times SU(2)_R$ to $SU(2)_I$], with the concomitant generation of N-G "pions", together with the realization that PCAC defines an operator relation between the divergence of the axial vector quark current and the "pion" field, further enriched the scope of current algebra calculations during the 1960s. The successes of $SU(2)_L \times SU(2)_R$ current algebra led to the construction of a chiral Lagrangian that depended on isotopic-spin-conserving functions of a covariant pion derivative, plus other fields and their covariant derivatives [4.2]. This chiral Lagrangian — which bore some resemblance to the σ model [3.27] and the Skyrme model [1.139] — was used in the late 1960s and early 1970s to calculate strong interaction processes at low energies [5.1] and

received several reformulations after QCD was established [as chiral perturbation theory with the three light quarks [5.2] and, most recently as a chiral
invariant effective Lagrangian for the heavy quarks [5.3]]. However, neither
current algebra nor the chiral Lagrangian approach could provide a truly
dynamical foundation for a theory of the strong interaction that would hold
at all energies. Indeed, as discussed in Chapter 4, the "no-go" theorems of
the mid-1960s — announcing the mathematical impossibility of mixing the
Poincaré group with global internal symmetries into a compact Lie group
— signaled the need to pursue a completely different path — that of gauging global non-Abelian internal symmetries — in order to achieve a fully
dynamical theory of the strong interaction.

It is the purpose of this chapter to review the basic tenets of the standard
(QCD) theory of the strong interaction among quarks based on the gauging
of the unbroken, non-Abelian, non-chiral $SU(3)$ color group. Since the key to
QCD is the gauging of a non-Abelian global internal symmetry group — in
contradistinction to the gauging of a global Abelian internal symmetry group
which led to QED — we reviewed in Chapter 4 the mathematical formalism
and physical significance of Yang-Mills groups in preparation for this chapter
on the gauge theory of the strong interaction (QCD) and the next chapter
(Chapter 6) on the gauge theory of the electroweak interaction (QFD). [We
discuss QCD before QFD because it resembles QED in its unbroken and
non-chiral aspects and its non-Abelian differentiation from Abelian QED
introduces us to the novel properties of asymptotic freedom and confinement
characteristic of Yang-Mills theories. Spontaneously broken and chiral QFD
is, in certain ways, more complicated than QCD although, historically, its
theoretical formulation preceded that of QCD by several years (see §1.3b–
§1.3c).]

We start this QCD chapter with an examination of the special properties
of gauged $SU(3)$ color for three massless quark flavors in order to exhibit the
role of the (color) quark-condensates in maintaining $SU(3)_C$ as an unbroken
(confining) gauge symmetry, while at the same time spontaneously breaking the "hidden" global chiral quark three-flavor group $SU(3)_L \times SU(3)_R$ to
the global non-chiral ("diagonal") quark flavor group $SU(3)_{L+R} = SU(3)_F$
(see §3.3). Actually, for three massless quarks, the "hidden" global chiral
quark symmetry is $U(3)_L \times U(3)_R$ with the extra $U(1)_L$ and $U(1)_R$ — in the
combinations $U(1)_{L+R}$ (vector) group and $U(1)_{L-R}$ (axial vector) group —
possessing very different properties when subjected to the quark-antiquark
condensates present in the theory. The variety of "hidden" symmetries flow-

ing from the massless quark limit of $SU(3)$ color is discussed in the remainder of this section, to be followed in §5.2–5.2c with a theoretical and experimental review of the status of asymptotic freedom in QCD: the renormalization group derivation of asymptotic freedom is given in §5.2a, the relation between "Bjorken scaling" and asymptotic freedom — as defined by Wilson's operator product expansion — is contained in §5.2b, and §5.2c concludes with a report on the experimental tests of "Bjorken scaling" and asymptotic freedom. While the property of asymptotic freedom in non-Abelian QCD is fully confirmed on both theoretical and experimental grounds, the situation with regard to color confinement in QCD is less decisive but still quite positive on the basis of theoretical arguments (lattice gauge theory, etc.) and experimental evidence from studies of heavy quarkonium; the discussion of quark confinement is given in §5.3–§5.3b. In the final sections of this chapter (§5.4–§5.4e), several special topics are treated (including one on the theoretical and experimental evidence for confined gluons and/or "glueballs"), either because they further elucidate the physical content of QCD or raise questions about the non-perturbative aspects of QCD that must still be resolved.

In searching for the "hidden" symmetries of the QCD Lagrangian when some of the quark flavors are taken to be massless, we begin by writing down the QCD Lagrangian for three quark flavors with distinct masses:

$$\mathcal{L}_{\text{QCD}} = -\frac{1}{4}\text{Tr}\, G_{\mu\nu}G^{\mu\nu} + \sum_{k=1}^{3} \bar{q}_k(i\slashed{D} - m_k)q_k \tag{5.1}$$

where:

$$G_{\mu\nu} = \partial_\mu A_\nu - \partial_\nu A_\mu + ig[A_\mu, A_\nu]; \quad \slashed{D} = \gamma_\mu D_\mu$$

$$D_\mu q_k = (\partial_\mu - igA_\mu)q_k; \quad A_\mu = \sum_{a=1}^{8} A_\mu^a \lambda^a/2 \quad (a = 1, \dots 8) \tag{5.1a}$$

with the q_k's being the quark fields ($k = 1, 2, 3$ is the flavor index) and the m_q's are arbitrary. The eight gauge fields A_μ^a ($a = 1, 2, \dots 8$) are the gluons and $\lambda_a/2$ ($a = 1, 2, \dots 8$) are the $SU(3)_C$ generators defined by Eq. (3.26). Note that the gauge coupling constant is the same for all quark flavors [i.e. the strong interaction among quarks is flavor-independent — see Fig. 1.4] so that the strangeness, charm, bottom (and, presumably, top) quantum numbers are all conserved by the strong interaction. The Lagrangian in Eq. (5.1) also conserves parity because no preference is given to the coupling of lefthanded chiral quarks *vis-à-vis* righthanded chiral quarks (in contrast to the weak interaction); the QCD Lagrangian is said to be "vector-like" (i.e. non-chiral).

The purely gluonic term in Eq. (5.1) contains the trilinear and quadrilinear self-couplings of the gluon field (see Fig. 1.4) and is responsible for the properties of asymptotic freedom and confinement in QCD, as is seen below.

The Lagrangian of QCD in Eq. (5.1), possesses an explicit dependence on the mass m_k of each quark flavor. Since the masses of the u, d and s quarks are small (compared to the QCD scale $\Lambda_{\rm QCD}$ — see §5.2a), we examine the consequences of setting $m_k = 0$ ($k = 1, 2, 3$) in Eq. (5.1) by identifying the "hidden" global symmetries; we get:

$$\mathcal{L}_0 = -\frac{1}{4} \, {\rm Tr} \, G_{\mu\nu} G^{\mu\nu} + \sum_{k=1}^{3} \bar{q}_k (i \not{D}_\mu) q_k \qquad (5.2)$$

The "truncated" QCD Lagrangian of Eq. (5.2) — like the "free" massless quark model [when the gauge field is absent (see §3.3)] — is invariant under the global chiral unitary transformations (L and R, as usual, stand for left- and righthanded chiral fermions respectively):

$$q_{L,R} \to e^{i\alpha^a_{L,R}(\frac{\lambda^a}{2})} q_{L,R} \qquad (5.3)$$

where $q_{L,R} = (q_{1L,1R}, q_{2L,2R}, q_{3L,3R})$. The local chiral transformations for the three light quarks become:

$$q_{jL} \to q_{jL} + i\alpha^a_L (\lambda^a_{jk}/2) q_{kL}; \quad q_{jR} \to q_{jR} + i\alpha^a_R (\lambda^a_{jk}/2) q_{kR} \quad (j = 1, 2, 3) \qquad (5.3a)$$

Invariance of $\mathcal{L}_0$ under the local chiral quark flavor transformations (5.3a) implies that the theory is invariant under the "hidden" global chiral quark flavor group $SU(3)_L \times SU(3)_R$ [the global chiral quark flavor group is called "hidden" because it only arises when the quark masses vanish — "hidden" chiral symmetry groups were first identified in connection with four-fermion interaction models [5.4]]. There is an additional "hidden" chiral quark flavor symmetry $U(1)_L \times U(1)_R$ in the Lagrangian of Eq. (5.2), as can easily be checked by applying the global chiral transformations:

$$q_{L,R} \to e^{i\alpha^a_{L,R}} q_{L,R} \qquad (5.3b)$$

Thus, the "hidden" chiral symmetry group in Eq. (5.2) is really $U(3)_L \times U(3)_R$ not $SU(3)_L \times SU(3)_R$.

Returning to the QCD Lagrangian (5.2), one now expects that the quark-gluon color gauge interaction produces "quark-antiquark condensates" that activate a Higgs-type mechanism to spontaneously break the global chiral groups $SU(3)_L \times SU(3)_R$ and $U(1)_L \times U(1)_R$ down to their respective "diagonal" global vector subgroups [5.5] $SU(3)_{L+R}$ [$= SU(3)_F$] and

$U(1)_{L+R}$ $[= U(1)_B]$ which are realized in the normal (Wigner-Weyl) mode (see §3.4). The $SU(3)_{L+R}$ group is really $SU(3)_F$ [the non-chiral (vector) quark flavor group] and $U(1)_{L+R}$ is similarly $U(1)_B$ (B is the baryon charge or quark charge) that gives rise to baryon charge conservation for the hadrons. [It is possible to prove rigorously that global internal symmetries (like isospin or "unitary spin" or baryon charge) can not be spontaneously broken in "vector-like" gauge theories like QCD provided the "strong CP" parameter $\theta = 0$ (see §10.3) [5.6].] The proof depends on the reality and positivity of the fermion determinant resulting from the integration of the path integral for the action over the Dirac spinors [these conditions are similar to those that guarantee the absence of a global $SU(2)$ anomaly for an even number of Weyl doublets (see §7.3)]. On the other hand, $SU(3)_{L-R}$ — which is the axial vector $SU(3)_A$ subgroup of $SU(3)_L \times SU(3)_R$ — is completely broken down (spontaneously), giving rise to an octet of (pseudoscalar) N-G bosons that can be identified with the observed "pion" octet (when account is taken of the finite quark masses due to the Higgs SSB mechanism in QFD). The SSB of the $U(1)_{L-R}$ [i.e. $U(1)_A$] subgroup of the $U(1)_L \times U(1)_R$ group [down to $\mathcal{Z}_6$ (for $n = 3$ flavors)] is achieved by a mechanism different from that of "quark condensates", namely by the color instanton solutions of the $SU(3)_C$ Yang-Mills equations (see §10.3).

The above statements are confirmed when we write down (in Table 5.1) the representations and quantum numbers of the quarks, anti-quarks and "quark-antiquark condensates" (for n flavors of massless quarks) under the relevant groups. Thus, Table 5.1 enables us to make the following clarifying comments concerning the effects of the "quark-antiquark condensates" on the global chiral quark flavor groups involved in QCD with $n = 3$ massless flavors: (1) the confining $SU(3)_C$ gauge group produces a "quark anti-quark condensate" that is a color singlet, which does not break $SU(3)_C$. This follows from the fact that $q\bar{q} = \mathbf{3} \times \bar{\mathbf{3}}$ [under $SU(3)_C$] decomposes into the $\mathbf{1} + \mathbf{8}$ irreducible representations of $SU(3)_C$; since the second-order Casimir gives attraction for the $\mathbf{1}$, but repulsion for the $\mathbf{8}$ representation (see §9.2), the $q\bar{q}$ condensate does not break $SU(3)_C$; (2) the global chiral flavor quantum numbers of the $q\bar{q}$ condensate are $\{(\mathbf{3}, \bar{\mathbf{3}}) + (\bar{\mathbf{3}}, \mathbf{3})\}$ [since $< 0|q\bar{q}|0 > = < 0|\bar{q}_L q_R + \bar{q}_R q_L|0 >$]; this set of "symmetrical" quantum numbers for the $q\bar{q}$ condensate serves to break down the $SU(3)_L \times SU(3)_R$ chiral group to the vector group $SU(3)$ [5.5], at the same time breaking down completely the axial vector group $SU(3)$ and generating the octet of (pseudoscalar) N-G bosons. This last statement is crucial for understanding the successes of the entire program of current algebra (together with PCAC and soft pion theorems) and, while a completely rigorous proof has not been given for $N_C = 3$ colors, such a proof does exist for the large N_C limit [5.7]. Based on other successful deductions drawn from large N_C limit calculations (e.g. the OZI rule [3.12]), the "quark anti-quark condensate" mechanism is generally accepted for the SSB of $SU(3)_L \times SU(3)_R$ down to $SU(3)_F$.

Table 5.1: Quantum numbers of q, $\bar{q}$ and $\langle 0|q\bar{q}|0 \rangle$ in QCD with $n = 2, 3$ massless flavors.

	$SU(3)_C$	$SU(n)_L$	$SU(n)_R$	$U(1)_L$	$U(1)_R$	$U(1)_{L+R}$	$U(1)_{L-R}$
q	3	n	1	1	0	1	1
$\bar{q}$	$\bar{3}$	1	$\bar{n}$	0	-1	-1	1
$\langle 0|q\bar{q}|0 \rangle$	1	n	$\bar{n}$	1	-1	0	2

There are two other clarifying remarks about Table 5.1 that are worth making: (3) Table 5.1 shows that the $U(1)_{L+R}$ charge for the $q\bar{q}$ condensate is zero so that the $q\bar{q}$ condensate in QCD does not break $U(1)_{L+R}$, i.e. there is baryon charge conservation as previously stated. While it is true that vector-like QCD, by itself, predicts baryon charge conservation for the quarks (and, *a fortiori*, for the hadrons), the adjoining of the electroweak chiral gauge group $SU(2)_L \times U(1)_Y$ for quarks (as for leptons) — to comprise the standard gauge group $SU(3)_C \times SU(2)_L \times U(1)_Y$ of the strong and electroweak interactions — introduces "flavor instantons" into the picture to produce a miniscule amount of baryon charge violation (as well as lepton charge violation — see §10.3a). Because of "quark-lepton flavor doublet universality" (see §1.2d), the baryon and lepton charge violations — due to the "flavor instantons" — are equal and opposite in sign so that the full standard gauge group predicts global $B - L$ charge conservation even in the presence of "flavor instantons". It will be seen in §10.3a that a new type of topological soliton ("sphaleron") can arise for a spontaneously broken non-Abelian gauge symmetry group — an unstable, saddle-point solution of the classical Yang-Mills-Higgs equations — which may overcome the strongly suppressive "flavor instanton" effects on baryon charge conservation at very high temperatures — presumably present in the early universe — and in extremely high energy collisions; and, finally (4) the SSB of $U(1)_{L-R}$ down to $\mathcal{Z}_6$ by the "color instantons" eliminates the need for a ninth (pseudoscalar) N-G boson [the other eight being the octet of (pseudoscalar) N-G bosons resulting from the SSB of $SU(3)_L \times SU(3)_R$ down to $SU(3)_A$] and helps to explain the "hard" character of the "harmless" QCD anomaly (see §7.2a) and to solve the so-called "$U(1)$ problem in QCD" (see §10.3b).

Putting our comments together, we draw two important conclusions: (1) the confining gauge-induced "quark anti-quark condensates" are responsible for the SSB of the "hidden" global chiral quark flavor symmetry group $SU(3)_L \times SU(3)_R$ for the three light quarks and hence justifies the identification of the vector and axial vector quark currents with the physical currents in electromagnetic and weak interactions. A rationale is thereby provided for a significant linkage between the strong and weak interactions (see §8.1) and for the use of PCAC and soft pion theorems in conjunction with current algebra; (2) since the "quark anti-quark condensates" leave the $SU(3)_C$ group unbroken, the predicted properties of asymptotic freedom and confinement are expected to hold in the QCD theory of the strong interaction. In succeeding sections, we focus on the ramifications of these two key properties of Yang-Mills theories — asymptotic freedom and confinement [1.89] — that distinguish so sharply between unbroken non-Abelian QCD and unbroken Abelian QED.

§5.2. Asymptotic freedom in the unbroken $SU(3)$ color group (QCD)

A major turning point in the conceptual evolution of the theory of the strong interaction occurred in the late 1960s when the deep inelastic lepton-hadron experiments were undertaken [5.8] and soon showed substantial agreement with Bjorken's scaling hypothesis [1.104] and Feynman's parton model of hadrons [1.103]. This success — which depended upon the decreasing strength of the quark-quark interaction as the energy increased — triggered a search for a theory of the strong interaction that would also incorporate the knowledge that each quark possesses three (global) color degrees of freedom. The $SU(3)$ color gauge group became the prime candidate when it was discovered that non-Abelian gauge theories — and only such theories — possess the property of asymptotic freedom [1.89].

In preparation for §5.2a — which deals with the renormalization group derivation of asymptotic freedom in QCD, we reviewed in §4.5 the renormalization formalism for both Abelian and non-Abelian gauge theories; without renormalizability, neither QED nor QCD can be considered a dynamical quantum field theory with verifiable phenomenological predictions. We brought the discussion in §4.5 to the point where the relations between the unrenormalized and renormalized 1PI Green's functions in the one-loop approximation were written down for both Abelian and non-Abelian gauge groups. These relations can then be employed to derive the "Abelian" and "non-Abelian" β functions defined by the renormalization group — whose zeros play a critical role in deciding whether there is a stable "infrared" or stable "ultraviolet" fixed point at the "origin" (see below) and thereby whether asymptotic freedom is absent or present. Earlier renormalization group analysis of QED had provided a simple derivation of the "Landau singularity" phenomenon which could be understood on the basis of the existence of a stable "infrared" fixed point at the "origin" (for QED). [This result for QED does not rule out the possible existence of stable "ultraviolet" fixed points away from the "origin" for QED; apart from the technical difficulty of finding an "ultraviolet" fixed point away from the "origin", the obscure physical consequences of such a determination have not encouraged its pursuit (see, however, [5.9]).] In actual fact, Kadanov and Wilson [5.10] exploited the stable fixed point concept away from the "origin" in statistical mechanics where the behavior of the β function in the neighborhood of a stable fixed point (where β vanishes) can be related to the so-called "critical exponent" associated with a specified phase transition. In contrast to QED, the application of renormalization group analysis to non-Abelian gauge theories reveals the existence of a stable "ultraviolet" fixed point at the "origin" — and hence asymptotic freedom — for all such theories. Having demonstrated that QCD is asymptotically free — in §5.2a — we next review in §5.2b how, and under what conditions, the short-distance behavior of QCD translates — by means of Wilson's operator production expansion [5.11] — into asymptotic freedom and the closely-related "Bjorken scaling" hypothesis for the structure functions governing the deep inelastic scattering of leptons by hadrons. Finally, in §5.2c, we survey the evidence gleaned from experiments on the deep inelastic scattering of leptons by hadrons in favor of "Bjorken scaling" [1.104] and the manner in which the observed (logarithmic) deviations from "Bjorken scaling" support the asymptotic freedom property of non-Abelian QCD. We also comment on some high energy sum rules derived for the "deep inelastic" structure functions on the basis of the quark-parton model.

§5.2a. Renormalization group derivation of asymptotic freedom in QCD

In §4.5 will be found the expressions for the renormalized constants (in terms of the unrenormalized constants) — in the one-loop approximation — for Abelian and non-Abelian gauge theories. We can now derive the renormalization group equation for the renormalized n-point 1PI Green's functions that, in turn, leads to the identification of the β function of the renormalized coupling constant, $\beta(g)$, whose asymptotic behavior (with energy) determines the nature of the stable fixed point at the "origin" ("infrared" or "ultraviolet") in the quantum field theory under consideration. If we denote the unrenormalized 1 PI Green's function, with n_B external boson lines and n_F external fermion lines, by $\Gamma^U_{n_B,n_F}(p_i, g_0, \Lambda)$ [where p_i denotes the external momenta, g_0 is the unrenormalized coupling constant and Λ is the cutoff momentum], then the renormalized Green's function $\Gamma^R_{n_B,n_F}(p_i, g, \mu)$ [where g is the renormalized coupling constant and $\mu \neq 0$ is the reference point of renormalization], can be written as follows:

$$\Gamma^R(p_i, g, \mu) \equiv \lim_{\Lambda \to \infty} (Z_B)^{n_B}(Z_F)^{n_F}\Gamma^U_{n_B,n_F}(p_i, g_0, \Lambda) \tag{5.4}$$

where the dimensionless renormalization constants Z_B and Z_F are functions of the unrenormalized coupling constant g_0 and the dimensionless factor Λ/μ. It is possible to give a succinct derivation of the renormalization group equation on the basis of Eq. (5.4) by observing that the unrenormalized Green's function $\Gamma^U_{n_B,n_F}$ is independent of the arbitrary scale μ; thus, differentiating both sides of Eq. (5.4) with respect to μ, and remembering that the renormalization parameters Z_B and Z_F must be functions of Λ/μ, we obtain:

$$[\mu\frac{\partial}{\partial\mu} + \mu\frac{\partial g}{\partial\mu}\frac{\partial}{\partial g}]\,\Gamma^R_{n_B,n_F}(p_i, g, \mu)$$
$$= \lim_{\Lambda\to\infty}[\frac{n_B}{Z_B}\mu\frac{\partial Z_B}{\partial\mu} + \frac{n_F}{Z_F}\mu\frac{\partial Z_F}{\partial\mu}]Z_B^{n_B}Z_F^{n_F}\,\Gamma^U_{n_B,n_F}(p_i, g_0, \Lambda) \tag{5.5}$$

or:

$$[\mu\frac{\partial}{\partial\mu} + \beta(g)\frac{\partial}{\partial g} - n_B\gamma_B(g) - n_F\gamma_F(g)]\,\Gamma^R_{n_B,n_F}(p_i, g, \mu) = 0 \tag{5.6}$$

where:

$$\beta(g) \equiv \mu\frac{\partial g}{\partial\mu}; \quad \gamma_B(g) \equiv \frac{\mu}{Z_B}\frac{\partial Z_B}{\partial\mu}; \quad \gamma_F(g) \equiv \frac{\mu}{Z_F}\frac{\partial Z_F}{\partial\mu} \tag{5.6a}$$

Equation (5.6) is known as the homogeneous Callan-Symanzik equation [5.12].

An alternative, and perhaps more perspicuous, way to derive the renormalization group equation (5.6) is to argue that the finite n-point 1 PI renormalized Green's functions should not depend on the choice of the renormalization reference scale μ. Hence, one expects that a change to a new renormalization reference scale μ' is accompanied by compensating changes in all renormalized quantities so that the n-point 1PI renormalized Green's functions are unchanged. The requirement that the renormalized Green's functions at different reference scales μ and μ' are related, leads to the equation:

$$\Gamma^{R'}(p_i, g', \mu') = Z(R', R)\ \Gamma^R(p_i, g, \mu) \qquad (5.7)$$

where $Z(R', R) = Z(R')/Z(R)$ with $Z(R) = [Z_B(R)]^{n_B} \cdot [Z_F(R)]^{n_F}$ and, similarly, for $Z(R')$ [see Eq. (5.4)]. It is evident from Eq. (5.7) that $Z(R', R)$ is a multiplicative constant and is finite [despite the divergences of $Z(R)$ and $Z(R')$] because the two renormalized Green's functions entering into Eq. (5.7) are both finite. The quantity Z can be thought of as a function of g and the ratio μ'/μ, i.e. $Z = Z(g, \mu'/\mu)$, and the same can be said for g', i.e. $g' = g'(g, \mu'/\mu)$. The statement that Eq. (5.7) is invariant under differential (rescaling) changes in μ yields the renormalization group equation (5.6). It is clear from this second method of derivation of Eq. (5.6) that $\beta(g)$ represents the differential transformation of g and the $\gamma(g)$'s represent the differential rescalings of the respective operators Z_B and Z_F (weighted by the numbers of external lines for bosons and fermions respectively), corresponding to the differential change in μ. It also becomes clear why Eq. (5.6) is called the renormalization group equation because the operation of going from one renormalization scheme R (with renormalization reference scale μ) to another renormalization scheme R' (with renormalization reference scale μ') is a multiplicative transformation from R to R' so that the set of all such transformations can — with some qualification (there is no "inverse" operation) — be said to constitute the renormalization group.

Since the renormalization group equation (5.6) tells us what happens for fixed couplings and momenta, when there is a change in μ, it is possible to derive — by means of dimensional analysis — the dependence of the 1PI renormalized Green's function, for fixed coupling g and fixed μ, on the rescaled momenta (i.e. when $p_i \rightarrow \sigma p_i$ with σ the scaling constant). If d is the "canonical" dimension of Γ (see §4.5), Γ scales like σ^d and the rescaled Green's function (we drop the superscript R) satisfies the equation:

$$\Gamma(\sigma p_i, g, \mu) = \sigma^d \Gamma(p_i, \bar{g}, \mu) \exp\{-\int_0^t \gamma(\bar{g}) dt'\} \qquad (5.8)$$

where $\gamma(g) = n_B \gamma_B(g) + n_F \gamma_F(g)$, $t = \ln \sigma$ and $\bar{g}(g, t)$, the "running" coupling constant, satisfies the differential equation:

$$\frac{d\bar{g}}{dt} = \beta(\bar{g}) \tag{5.9}$$

with the boundary condition:

$$\bar{g}(g, 0) = g(g_0, \Lambda/\mu) = g \tag{5.9a}$$

Actually, Eq. (5.9) is equivalent to another equation which manifestly possesses the form of the renormalization group equation (5.6), namely:

$$\{\frac{\partial}{\partial t} - \beta(\bar{g})\frac{\partial}{\partial g}\}\bar{g}(g, t) = 0 \tag{5.10}$$

subject to the same boundary condition (5.9a).

Equation (5.10) [or Eq. (5.9)] is the crucial equation that decides whether or not there is a so-called stable fixed point at the "origin" [i.e. whether $\beta(\bar{g}) = 0$ for $\bar{g} = 0$] of either the "infrared" or "ultraviolet" type — with implications for the absence or presence of asymptotic freedom in the gauge theory under consideration. The distinction between an "infrared" and an "ultraviolet" stable fixed point depends on whether the derivative $\beta'(\bar{g}) > 0$ or $\beta'(\bar{g}) < 0$ at $\bar{g} = 0$ (the "origin") respectively. In both kinds of gauge theories (Abelian and non-Abelian), there is a stable fixed point at the "origin" but it turns out that, while $\beta'(0) > 0$ (stable "infrared" fixed point) for an Abelian gauge theory, the reverse is true for a non-Abelian gauge theory, i.e. $\beta'(0) < 0$ (stable "ultraviolet" fixed point). To draw conclusions from the nature of the stable fixed point, we must clarify the distinction between "infrared" and "ultraviolet" stable fixed points, and then prove that $\beta'(0) > 0$ for Abelian QED and $\beta'(0) < 0$ for non-Abelian QCD. For this purpose, let us consider the hypothetical curve for $\beta(g)$, shown in Fig. 5.1, which has zeros of $\beta(g)$ for $g = g_0 = 0, g_1, g_2$ etc.; these zeros of $\beta(g)$ correspond to the zeros of $\beta(\bar{g})$ [by virtue of Eq. (5.9a)] and are called "fixed points" because, as we will see, $\bar{g}(g, t)$ is "driven" to $g_0, g_1, g_2, \ldots$ as $t \to \infty$ [in which case, the "fixed point" is a stable "ultraviolet" fixed point] or as $t \to 0$ [in which case the "fixed point" is a stable "infrared" fixed point]. Thus, in Fig. 5.1 as drawn, $\beta(g) > 0$ for $0 < g < g_1$ so that, from Eq. (5.9), $\bar{g}(g, t)$ increases with increasing t and hence is "driven" to g_1 as t approaches infinity. On the other hand, in Fig. 5.1, $\beta(g) < 0$ for $g_1 < g < g_2$ so that the flow of $\bar{g}(g, t)$ is reversed and $\bar{g}(g, t)$ is "driven" back to g_1 as t approaches infinity. Hence, g_1 is a stable "ultraviolet" fixed point. The same argument leads to the conclusion that g_3 is also a stable "ultraviolet" fixed point. It is easy to see that g_0 and g_2 are stable "infrared" fixed points in Fig. 5.1, because $\bar{g}(g,t)$ is "driven" to these two zeros of $\beta(g)$ when $t \to 0$. Evidently, if the curve in Fig. 5.1 is inverted about the x axis, g_0 and g_2 become the stable "ultraviolet" fixed points whereas g_1 and g_2 become the stable "infrared" fixed points.

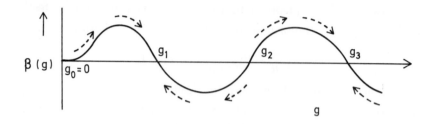

Fig. 5.1. Schematic representation of $\beta(g)$, with zeroes $g_0 = 0, g_1, g_2, \ldots$

In general, it is extremely difficult to establish the existence of stable fixed points of a quantum field theory of $\beta(g)$ for finite values of g. The mathematical situation evidently becomes more tractable in trying to decide whether there is a stable fixed point at the "origin", i.e. when $g = 0$; the physics also becomes more clearcut. [Although, as previously mentioned, Wilson has built his theory of "critical exponents" in statistical mechanics on the basis of stable fixed points away from the "origin" [5.10]]. Thus, a stable "infrared" fixed point of a quantum field theory at the "origin" implies that there is almost complete screening at low energies (or large distances) since $\beta(\bar{g})$ is "driven" to $g = 0$ when $t \to 0$; this is the case for Abelian QED which predicts a stable "infrared" fixed point at the "origin" for the β function of the renormalization group equation (5.6). The reverse is true for non-Abelian QCD which predicts a stable "ultraviolet" fixed point at the "origin" since $\beta(\bar{g})$ is "driven" to $g = 0$ as $t \to \infty$; this translates into a vanishing interaction at high energies (or, equivalently, short distances), i.e. asymptotic freedom.

Our remaining task is to demonstrate that Abelian QED really predicts the stable "infrared" fixed point at the "origin" for the β function whereas non-Abelian QCD predicts a stable "ultraviolet" fixed point at the "origin". For this purpose, we employ the expressions for the QED and QCD renormalization constants (in the one-loop approximation) given in §4.5 to work out the QED and QCD β functions with the help of Eq. (5.6a). We summarize the results for QED where the renormalized electric charge is given by [see Eq. (4.99)]:

$$e = Z_2^{-1} Z_1 Z_3^{\frac{1}{2}} e_0 = Z_3^{\frac{1}{2}} e_0 \tag{5.11}$$

with e_0 the "bare" coupling constant and Z_3, Z_2 and Z_1 the vacuum polarization, electron propagator and vertex (charge) renormalization constants respectively; because of the Ward identity $Z_1 = Z_2$, α_e depends only on Z_3. The one-loop contribution to Z_3 turns out to be [see Eq. (4.98)]:

$$Z_3 = 1 - \frac{e_0^2}{12\pi^2} \ln (\Lambda/\mu)^2 \tag{5.12}$$

and, consequently, to lowest-order, the β function for QED is:

$$\beta_e = -e \frac{\partial \left[\ln Z_3^{1/2}(e_0, \Lambda/\mu) \right]}{\partial \ln \Lambda} = \frac{+e^3}{12\pi^2} + O(e^5) \tag{5.13}$$

The positive sign of the leading term in Eq. (5.13) implies $\beta_e'(0) > 0$ so that QED possesses a stable "infrared" fixed point at the "origin" and there is no asymptotic freedom in QED, as is well-known.

It is interesting to solve Eq. (5.13) for the "running" coupling constant $\alpha_e(t)$; using the value of β_e given in Eq. (5.13), one finds:

$$\alpha_e(t) = \frac{\alpha_e(0)}{1 - \frac{\alpha_e t}{3\pi}} \tag{5.14}$$

Equation (5.14) is only valid for small coupling and leads to the "Landau singularity" at $t \sim 7 \times 10^3$, which is an enormous energy, and has, obviously, never caused concern for QED predictions. It should be remarked that Landau arrived at this "singularity" during the 1950s [1.58] by summing the leading logarithmic divergences contributing to the renormalized charge by means of what he called the method of "graphs" (see §1.2b). While Landau's method was an improvement over the straightforward perturbation-theoretic approach, its complications are avoided through the use of the renormalization group. However the "Landau singularity" is derived, it is worth recalling that Landau and his collaborators [1.58] argued from Eq. (5.14) that the only way to avoid the singularity at $t = 3\pi/\alpha_e$ is to assume that the unrenormalized charge $\alpha_e(0)$ vanishes. At the time, this assertion was termed the problem of the "Moscow zero" [5.13] in the USSR but was not taken too seriously in the West [1.59] (since the weak interaction enters the picture much before the "Landau singularity" is reached). The problem of the "Moscow zero" has been revived during the past few years as the "triviality problem" in the Higgs sector of the electroweak gauge theory (QFD) (see §6.3c).

We come now to non-Abelian gauge theory where we combine the results for the wavefunction and coupling constant renormalization constants

278 *Conceptual Foundations of Modern Particle Physics*

corresponding to the pure Yang-Mills theory [see Eqs. (4.107) and (4.109)] with the contribution from the fermion fields [see Eq. (4.114)] to derive the "non-Abelian" β function of the renormalized group equation. We then particularize the total β function to the $SU(3)$ color case (QCD). The renormalized coupling constant for a pure non-Abelian gauge field can be written in the form (see §4.5b):

$$g = Z_3^{\frac{3}{2}} Z_1^{-1} g_0 \tag{5.15}$$

where g_0 is the "bare" non-Abelian gauge coupling constant and Z_3 and Z_1 are the Yang-Mills wavefunction renormalization and vertex renormalization constants respectively. The expressions for Z_3 and Z_1 become [see Eqs. (4.107) and (4.109)]:

$$Z_3 = 1 + \frac{g^2}{32\pi^2}(\frac{13}{3} - \zeta)C_2(V) \ln (\Lambda/\mu)^2$$

$$Z_1 = 1 + \frac{g^2}{32\pi^2}(\frac{17}{6} - \frac{3\zeta}{2})C_2(V) \ln (\Lambda/\mu)^2 \tag{5.16}$$

where ζ is the gauge-fixing parameter. From Eq. (5.16), one obtains the β function in one-loop approximation (denoted by β_0):

$$\beta_0 = -g_s\frac{\partial}{\partial \ln \Lambda}[\frac{3}{2}\ln Z_3 - \ln Z_1] = -\frac{g_s^3}{16\pi^2} \cdot \frac{11}{3}C_2(V) \tag{5.17}$$

It is important to note that ζ has disappeared from the final expression for the lowest-order β function in Eq. (5.17) so that the β function is independent of the gauge-fixing parameter, as it should be. Recalling that $C_2(V) = N_C$ ($N_C = 3$ for QCD), it follows that there is a stable "ultraviolet" fixed point at the "origin" for a pure Yang-Mills theory and, hence, a pure Yang-Mills theory is asymptotically free.

Since realistic QCD contains fermions (three generations of quark doublets), the "anti-asymptotic freedom" effect of the fermions — as in QED — must be taken into account. We already know the dependence of the β function — in one-loop approximation — on the number of fermions, namely [see Eq. (4.114)]:

$$\beta_0 = \frac{g_s^3}{16\pi^2}[-\frac{11}{3}C_2(V) + \frac{4}{3}C_2(F)] \tag{5.18}$$

where $C_2(F) = \frac{1}{2}$ for Dirac fermions but is reduced by a factor of $\frac{1}{2}$ for Weyl fields. For the specific case of QCD — Eq. (5.18) becomes:

$$\beta_0(g_s) = g_s^3/16\pi^2[-11 + \frac{2}{3}n_f] \tag{5.19}$$

where n_f is the number of quark flavors. Equation (5.19) implies that $\beta_0(g_0) < 0$ as long as $n_f < 17$ and certainly for $n_f = 6$ (the QCD number when the t quark is included).

An alternative semi-classical derivation [5.14] of asymptotic freedom in QCD seizes upon the dominant role played by the self-interacting color-charged gluons — through their trilinear and quadrilinear self-couplings (see Fig. 1.4) — in converting the QCD vacuum into a quasi-color-paramagnetic medium despite the quasi-color-diamagnetic effects of the fermionic quarks on the QCD vacuum. The color-polarizable QCD vacuum is assigned a color-magnetic succeptability χ_C [and hence a "color-magnetic permeability" $\mu_C = 1 + 4\pi\chi_C$] and a "color-dielectric" constant ϵ_C. A calculation is first made of χ_C by computing the vacuum energy of a massless charged vector field in the presence of an external magnetic field — to simulate the gluon field — with the Landau method [5.15]; the Landau solution for ordinary electromagnetism is then transcribed into the "color-electric" and "color-magnetic" fields of QCD. One finds that $\chi_C > 0$ for the bosonic self-interacting gluon field and, *a fortiori*, $\mu_C > 1$; i.e. the pure Yang-Mills field acts as a "color-paramagnetic" medium. When the Yukawa-type quark-gluon interaction is taken into account, its contribution to the QCD vacuum energy is of opposite sign, i.e. "color-diamagnetic", and of lesser strength than the purely gluonic contribution; indeed, it turns out that χ_C possesses the same dependence on the number of quark flavors (n_f), namely $\chi_C \sim (11 - \frac{2}{3}n_f)$ as does the β function in the renormalization group analysis of QCD [see Eq. (5.19)]. Clearly, for $n_f = 6$, the net χ_C is positive and μ_C exceeds unity. Since the condition $\epsilon_C\mu_C = 1$ holds for any relativistic quantum field theory vacuum, it follows that $\epsilon_C < 1$. Hence, the QCD vacuum is "anti-screening" in character, in contrast to the "screening" property of the QED vacuum (where $\epsilon > 1$). The finding that the QCD vacuum "anti-screens" is equivalent to the conclusion from renormalization group analysis that QCD is asymptotically free.

With the asymptotic freedom of QCD established, it is useful to recast the "running" coupling constant $\bar{g}_s(g_s, t)$ into a form that exhibits the effective range of the strong quark-quark interaction. We continue to use the one-loop approximation for $\beta(g_s)$, which we write $\beta_0(g_s) = -bg_s^3$ $[b = 1/16\pi^2(11 - \frac{2}{3}n_f)]$; we return to Eq. (5.9), which becomes:

$$\frac{d\bar{g}_s}{dt} = -b\bar{g}_s^3 \qquad (5.20)$$

whose solution is:

$$\bar{g}_s^2(t) = \frac{g_s^2}{1 + 2bg_s^2 t} \tag{5.21}$$

If we let $t = \frac{1}{2}\ln(Q^2/\mu^2)$ [Q^2 is the four-momentum transfer squared of interest and μ is the reference scale for renormalization], we can rewrite Eq. (5.21a) as:

$$\alpha_s(Q^2) = \frac{\alpha_s(\mu^2)}{1 + 4\pi b\alpha_s(\mu^2)\ln(Q^2/\mu^2)} \tag{5.22}$$

where $\alpha_s(Q^2) = \bar{g}_s^2(t)/4\pi$ and $\alpha_s(\mu^2) = g_s^2/4\pi$. Strictly speaking, the quantity $\alpha_s(Q^2)$ is the "running coupling constant" and clearly decreases as Q^2 becomes larger than μ^2 because b is positive (for three generations of quark doublets). It is possible to further rewrite Eq. (5.22) in terms of a single parameter Λ_{QCD}, defined by:

$$\ln \Lambda_{\text{QCD}}^2 = \ln \mu^2 - \frac{1}{\alpha_s(\mu^2)4\pi b} \tag{5.23}$$

to obtain:

$$\alpha_s(Q^2) = \frac{4\pi}{(11 - \frac{2}{3}n_f)\ln(Q^2/\Lambda_{\text{QCD}}^2)} \tag{5.24}$$

Since — unlike Eq. (5.22) — the "running" coupling constant $\alpha_s(Q^2)$ does not depend on μ and is a function of the single parameter Λ_{QCD}, Λ_{QCD} becomes the QCD scale, effectively defining the energy at which the "running" coupling constant attains its maximum. Λ_{QCD} must be deduced from experiment and depends to a slight extent on the particular renormalization procedure; in the so-called modified minimal subtraction scheme, Λ_{QCD} turns out to be 160^{+47}_{-37} MeV above the charmonium threshold when $n_f = 4$ (see Eq. (5.24)) [5.16]. [The value of Λ_{QCD} changes when it is evaluated for a different value of n_f [5.17]; we will use the rounded value $\Lambda_{\text{QCD}} \simeq 150$ MeV throughout the book.] This value of Λ_{QCD} is of the order of a pion mass and we can now understand why the two-nucleon interaction, which possesses a finite range $\sim \Lambda_{\text{QCD}}^{-1}$, happens to be of the order of the pion Compton wavelength (m_π^{-1}); thus is explained some of the successes of Yukawa's meson theory of nuclear forces a half century ago!

Having expressed satisfaction with the finite (short-range) interaction predicted by QCD, one might inquire how a finite value of Λ_{QCD} emerges since there is no canonical dimension in the QCD Lagrangian (5.2): the coupling constant is dimensionless, the gluons are massless and the massless quarks suffice for asymptotic freedom. The answer is that QCD must be renormalized and, in the process, a finite scale (i.e. μ) is set (to avoid infrared divergences [4.21]) which, in combination with the asymptotic freedom and confinement properties of non-Abelian QCD, generates an effective, finite, phenomenologically correct QCD scale (Λ_{QCD}). The situation is very different for QED with its enormously large QED scale defined by the "Landau singularity" (see above) [the charge renormalization in QED is carried out with the help of the Thomson formula (for the scattering of light by electrons at low energy)]. To all intents and purposes, the range of the electromagnetic interaction is infinite, i.e. we have the "long-range" Coulombic interaction between two electrically charged particles in QED.

As a final comment, we note that because of the asymptotic freedom property of QCD, the problem of making perturbation-theoretic predictions in QCD is turned around (as compared to QED), and must be confined to the highest possible energies (and momentum transfer squared). A more accurate statement is that this "ultraviolet" asymptotic limit must lie in the deep "Euclidean" region where all particles are far from their mass shells, which is frequently not true for strong interaction processes. However, there is a class of physical processes where the "external particles" are far from their mass shells, and the asymptotic freedom predictions of QCD can be tested. This class of processes comprises the deep inelastic scattering of leptons by hadrons, and we present a summary of the striking experimental evidence in support of asymptotic freedom in the strong interaction in §5.2c. The results of the deep inelastic lepton-hadron scattering experiments are usually presented in terms of verifying "Bjorken scaling", which is then equated to support for asymptotic freedom in QCD. However, there are some subtle differences between "Bjorken scaling" and asymptotic freedom and we devote the next section (§5.2b) to a brief discussion of the differences between them and the experimental consequences of this distinction. This discussion requires having recourse to Wilson's "operator product expansion" (OPE) method [5.11] in combination with the renormalization group equation for QCD and some of these details are sketched in §5.2b. The final section bearing on asymptotic freedom (§5.2c) contains a rather detailed discussion of the favorable experimental situation with regard to both "Bjorken scaling" and asymptotic freedom.

§5.2b. Breakdown of scale invariance, operator product expansion and Bjorken scaling

In §2.3b, we showed that since the particle masses of the real world neither vanish nor form a continuous spectrum, scale invariance must break down for any quantum field theory purporting to describe the real world. Nevertheless, one might hope that, for energies much larger than the observed particle masses, a "residual" scale invariance would remain and simplify the description of certain classes of experiments at increasingly higher particle energies. Indeed, "Bjorken scaling" and asymptotic freedom are both manifestations of this "residual" scale invariance for large space-like momenta and, in this section, we explain how the renormalization group equation provides the framework — in conjunction with Wilson's "operator product expansion" (OPE) — for distinguishing between "Bjorken scaling" and asymptotic freedom in terms of the "anomalous dimensions" defined by the renormalization group equation (see below). We also explain why small (logarithmic) departures from "Bjorken scaling" (for which there is experimental evidence — see §5.2c) are expected from the calculation of the "short-distance" behavior of deep inelastic lepton-hadron scattering (which gives rise to asymptotic freedom) — in contradistinction to the calculation of the light-cone behavior of "free" quark currents (that leads to "Bjorken scaling").

To proceed, we rewrite the renormalization group equation (5.6), i.e. the homogeneous Callan-Symanzik equation (without mass insertions) [5.12] after rescaling ($p_i \to \sigma p_i$):

$$[\sigma \frac{\partial}{\partial \sigma} - \beta(g) \frac{\partial}{\partial g} + n\gamma(g) + (n-4)]\Gamma^n(\sigma p_i, g) = 0 \qquad (5.25)$$

where $\Gamma^n(\sigma p_i, g)$ denotes the 1PI renormalized Green's function with n external lines (we do not distinguish between external boson and fermion lines), and $(4-n)$ is its "canonical" dimension. If we separate out the "canonical" dimensional dependence of the renormalized Green's function $\Gamma^{(n)}(\sigma p_i, g)$, we can write it in the form:

$$\Gamma^{(n)}(\sigma p_i; g) = \sigma^{4-n} z(\sigma)^{-n} \Gamma^{(n)}[p_i; \bar{g}(\sigma)] \qquad (5.26)$$

where:

$$\sigma \frac{d}{d\sigma} \bar{g}(\sigma) = \beta[\bar{g}(\sigma)]; \qquad \bar{g}(1) = g \qquad (5.26a)$$

and:

$$z(\sigma) = \int_g^{\bar{g}(\sigma)} \frac{dg'}{\beta(g')} \gamma(g'); \qquad z(1) = 1 \qquad (5.26b)$$

The behavior of $z(\sigma)$, defined by Eq. (5.26b), determines whether there is an "anomalous dimension" or not; if $z(\sigma)$ develops a power dependence σ^d (as $\sigma \to \infty$), the anomalous dimension for $\Gamma^{(n)}(\sigma p_i; g)$ is $(4-n-nd)$ instead of

$(4-n)$ and becomes a measure of the "residual" scale invariance maintained through the mechanism of the renormalization group equation.

From Eq. (5.26b), it is evident that the γ function plays the key role in fixing the "anomalous dimension". In particular, an interesting question is whether, as $\sigma \to \infty$, $z(\sigma) \underset{\sigma \to \infty}{\sim} \gamma_\infty$ [where $\gamma_\infty = \gamma(g_\infty)$] so that we can write:

$$\Gamma^{(n)}(\sigma p_i, g) \underset{\sigma \to \infty}{\sim} \sigma^{4-n(1+\gamma_\infty)} \Gamma^{(n)}[p_i, g_\infty] \tag{5.27}$$

i.e. whether the effective "anomalous dimension" d_{eff} becomes simply:

$$d_{eff} = 1 + \gamma_\infty; \quad \gamma_\infty = \gamma[g_\infty] \tag{5.28}$$

The answer to this question is in the affirmative if $\bar{g}(\infty)$ is associated with a stable "ultraviolet" fixed point — which need not necessarily be at the "origin". Unfortunately, it is, in general, very difficult to calculate the zeros of the β function (either of the "infrared" or "ultraviolet" kind) away from the "origin" and so this result is not very useful in quantum field theory [unless fast computers are available and one is dealing with phase transitions in many-body systems susceptible to statistical mechanics treatment (e.g. critical phenomenon, polymer theory, etc. [5.18])]. However, it is easy to ascertain whether there is a stable "ultraviolet" fixed point at the "origin" and then one obviously obtains $\gamma_\infty = 0$ and the "canonical" behavior of the Green's function in the asymptotic region is recaptured.

This last statement is correct as long as we inquire into the power dependence of σ and essentially justify the Bjorken scaling hypothesis; however, the derivation of Eq. (5.27) neglects leading logarithmic terms of the form $(\ln \sigma)^k$ (k is a constant) which, when summed in perturbation theory, can produce significant logarithmic deviations from "Bjorken scaling" and more accurately define the asymptotic freedom property in QCD. The point is that the Bjorken scaling hypothesis was proposed before the advent of QCD and its first derivation by Bjorken was given in terms of current algebra [1.104], was then justified by the parton model and was finally given a more formal status through a study of the light-cone behavior of the deep inelastic structure functions when the hadron currents are described by "free" quark fields. This conceptual evolution of "Bjorken scaling" took place before the QCD β and γ functions were shown to be responsible for the novel properties of asymptotic freedom and "anomalous dimensions" which encompass "Bjorken scaling". We propose, in this section, to sketch the "light-cone plus free quark" derivation of "Bjorken scaling" (the "parton model" derivation is presented in the experimental section 5.2c on deep inelastic lepton-hadron processes) and then to review the more rigorous "QCD renormalization group" derivation of asymptotic freedom and the resulting logarithmic corrections to "Bjorken scaling".

The "light-cone plus free quark" confirmation of "Bjorken scaling" in the deep inelastic region for lepton-hadron scattering [for the purpose of this derivation, we assume that the lepton is an electron and the hadron is a proton] is based on examining the light-cone behavior of the quark hadronic tensor $W_{\mu\nu}$ (which decomposes into the structure functions and is fully derived in §5.2c), namely:

$$W_{\mu\nu}(p,q) = \frac{1}{4M} \sum_{\sigma} \int \frac{d^4x}{2\pi} e^{iq \cdot x} < p, \sigma [J_\mu(x), J_\nu(0)] p, \sigma > \qquad (5.29)$$

where $W_{\mu\nu}$ is a second-rank Lorentz tensor depending on the four-momentum p_μ and spin σ of the initial proton and q_μ, the four-momentum transfer to the electron; the summation over σ implies that the electron and proton are unpolarized. For deep inelastic $e - p$ scattering, $W_{\mu\nu}(p,q)$ can be expressed in terms of two structure functions $W_1(p,q)$ and $W_2(p,q)$ [3.26], thus:

$$W_{\mu\nu}(p,q) = -W_1 \left(g_{\mu\nu} - \frac{q_\mu q_\nu}{q^2} \right) + \frac{W_2}{M^2} \left(p_\mu - \frac{p \cdot q}{q^2} q_\mu \right) \left(p_\nu - \frac{p \cdot q}{q^2} q_\nu \right)$$
$$(5.30)$$

We now show that light-cone variables precisely reflect the conditions for deep inelastic lepton-hadron scattering $[-q^2, \nu \to \infty$ (ν is the energy loss of the electron after its collision with the proton) with $-q^2/2M\nu$ fixed] and that, when the electromagnetic currents J_μ in the expression for the quark hadronic tensor $W_{\mu\nu}$ [see Eq. (5.29)] are taken to be "free-field" currents, the two structure functions in Eq. (5.30) are correctly expressed in terms of the "Bjorken scaling" variable $x = -q^2/2M\nu$. First, we expand $q \cdot x$ in the expression for $W_{\mu\nu}$ [see Eq. (5.29)] as follows:

$$q \cdot x = \frac{(q_0 + q_3)}{\sqrt{2}} \frac{(x_0 - x_3)}{\sqrt{2}} + \frac{(q_0 - q_3)}{\sqrt{2}} \frac{(x_0 + x_3)}{\sqrt{2}} - \vec{q}_T \cdot \vec{x}_T \qquad (5.31)$$

where:

$$\vec{q}_T = (q_1, q_2); \qquad \vec{x}_T = (x_1, x_2) \qquad (5.31a)$$

In the rest frame of the target (nucleon), we have:

$$p_\mu = (M, 0, 0, 0); \quad q_\mu = (\nu, 0, 0, \sqrt{\nu^2 - q^2}) \qquad (5.32)$$

Since the deep inelastic limit is defined by: $-q^2, \nu \to \infty$ ($-\frac{q^2}{2M\nu}$ fixed), the light-cone variables can be approximated by:

$$q_0 + q_3 \sim 2\nu; \quad q_0 - q_3 \sim \frac{q^2}{2\nu} \qquad (5.33)$$

As usual, the dominant contributions to the $W_{\mu\nu}$'s come from regions with less rapid oscillations, i.e. $q \cdot x = O(1)$, or:

$$x_0 - x_3 \sim O(\frac{1}{\nu}); \qquad x_0 + x_3 \sim O(-\frac{\nu}{q^2}) \qquad (5.33a)$$

Hence:

$$x_0^2 - x_3^2 \sim O(\frac{1}{-q^2}); \qquad x^2 = x_0^2 - x_3^2 - x_T^2 \leq x_0^2 - x_3^2 \sim O(\frac{1}{-q^2}) \quad (5.33b)$$

It follows from Eq. (5.33b) that $x^2 \to 0$ as $-q^2 \to \infty$. Thus, light-cone behavior occurs when the conditions of deep inelastic lepton-hadron scattering are satisfied.

We must now show that "Bjorken scaling" ensues when the quark currents (in the hadronic tensor $W_{\mu\nu}$) are taken to be "free-field" currents; we write:

$$J^\mu(x) = \bar{\psi}(x)\gamma^\mu Q\psi(x) \qquad (5.34)$$

with ψ the "free" fermion field and Q the charge matrix. Since, in the light-cone limit, we have:

$$[\psi(x), \bar{\psi}(y)]_+ \simeq \frac{1}{2\pi}\not{\partial}_x\, \epsilon(x^0 - y^0)\, \delta[(x - y)^2] \qquad (5.34a)$$

the current commutator in $W_{\mu\nu}$ becomes:

$$[J_\mu(x), J_\nu(y)] \underset{(x-y)^2 \to 0}{\simeq} [\bar{\psi}(x)Q^2\gamma_\mu\gamma_\alpha\gamma_\nu\psi(y) - \bar{\psi}(y)Q^2\gamma_\nu\gamma_\alpha\gamma_\mu\psi(x)]$$

$$\times \partial^\alpha \frac{\epsilon(x^0 - y^0)}{2\pi}\delta[(x - y)^2] \qquad (5.35)$$

Since it is convenient to express $W_{\mu\nu}$ in terms of the commutator of $j_\mu(x)$ and $j_\nu(-x)$, we expand the product $\bar{\psi}(x)\,\psi(-x)$ in a Taylor's series, thus:

$$\bar{\psi}(x)\psi(-x) = \sum \frac{(-1)^n}{n!} x^{\mu_1} \cdots x^{\mu_n} \bar{\psi}(0)\overleftrightarrow{\partial}_{\mu_1}, \dots \overleftrightarrow{\partial}_{\mu_n}\psi(0) \qquad (5.35a)$$

Using Eqs. (5.35) and (5.35a) and the identity:

$$\frac{1}{2}(\gamma_\mu\gamma_\alpha\gamma_\nu + \gamma_\nu\gamma_\alpha\gamma_\mu) = (g_{\mu\alpha}g_{\nu\beta} + g_{\mu\beta}g_{\nu\alpha} - g_{\mu\nu}g_{\alpha\beta})\gamma^\beta \qquad (5.35b)$$

we can write:

$$\frac{1}{2}\{[J_\mu(x), J_\nu(-x)] + [J_\nu(x), J_\mu(-x)]\} = \sum_{n \text{ odd}} x_{\mu_1} \dots x_{\mu_n} \frac{1}{n!}$$

$$\times \mathcal{O}^{\beta\mu_1 \cdots \mu_n}(g_{\mu\alpha}g_{\nu\beta} + g_{\mu\beta}g_{\nu\alpha} - g_{\mu\nu}g_{\alpha\beta}) \times \frac{1}{8\pi}\partial^\alpha \epsilon(x^0)\delta(x^2)$$

$$(5.36)$$

where:

$$\mathcal{O}^{\beta\mu_1 \cdots \mu_n} \equiv i\bar{\psi}(0)Q^2\gamma^\beta\overleftrightarrow{\partial}^{\mu_1}\dots\overleftrightarrow{\partial}^{\mu_n}\psi(0) \qquad (5.36a)$$

Finally, we can evaluate the hadronic tensor $W_{\mu\nu}$ if we express it in the form:

$$W_{\mu\nu} = \frac{1}{2\pi} \int d^4 y e^{iq \cdot y} \frac{1}{2} \sum < p|[J_\mu(\frac{y}{2}), J_\nu(-\frac{y}{2})]|p > \qquad (5.37)$$

where the matrix elements of Eq. (5.36a) become:

$$\frac{1}{2} \sum < p|\frac{1}{2^n} 0^{\beta \mu_1 \cdots \mu_n}|p > = a_{n+1}(p^\beta p^{\mu_1} \ldots p^{\mu_n} + \text{trace terms}) \qquad (5.37a)$$

The summation in Eqs. (5.37) and (5.37a) is over nucleon polarization. Further, in Eq. (5.37a), the trace terms (containing contractions of two indices) do not contribute to $W_{\mu\nu}$ and the (constant) a_{n+1} coefficients are the moments of the distribution function, denoted by $f(x)$, and defined by:

$$\sum_{n \text{ odd}} \frac{(y \cdot p)^n}{n!} a_{n+1} = \frac{1}{2i} \int dx\, e^{ix(y \cdot p)} \frac{f(x)}{x} \qquad (5.38)$$

Recognizing that we are interested in the deep inelastic limit [where $p^2 = M^2 << p \cdot q$], we find that, in the deep inelastic limit, $W_{\mu\nu}$ becomes:

$$W_{\mu\nu} = \frac{i}{4\pi} \int d^4 y e^{iq \cdot y} \int dx\, e^{ix(y \cdot p)} \frac{f(x)}{x}$$

$$\times [p_\mu(q + xp)_\nu + p_\nu(q + xp)_\mu - g_{\mu\nu}p \cdot q] \times \frac{\epsilon(y^0)}{\pi} \delta(y^2)$$

$$= \frac{f(x)}{2M}(\frac{q_\mu q_\nu}{q^2} - g_{\mu\nu}) + x\frac{f(x)}{\nu M^2}(p_\mu - \frac{\nu}{q^2}q_\mu)(p_\nu - \frac{\nu}{q^2}q_\nu) \qquad (5.39)$$

where $x = -\frac{q^2}{2M\nu}$ is now the Bjorken variable. If we compare Eq. (5.39) with Eq. (5.30), we confirm "Bjorken scaling", i.e.

$$2MW_1 = f(x); \qquad \nu W_2 = xf(x) \qquad (5.40)$$

In sum, "Bjorken scaling" results from the assumption of the canonical "free-field" light-cone structure of the commutator of the electromagnetic currents in the hadronic tensor describing deep inelastic electron-proton scattering.

We have seen that "Bjorken scaling" occurs when we apply the well-defined conditions for "deep" inelastic lepton-hadron scattering to non-interacting quark currents in the quark hadronic tensor. As we have already stated, the assumptions underlying the "parton model" — to be defined in detail in §5.2c — which include no interaction among the quarks comprising the hadrons when the "deep" inelastic region is considered, also confirm the "Bjorken scaling" hypothesis. *Prima facie*, the "non-interacting quark" assumption in the "deep" inelastic region is bold but unnatural because we are accustomed to strong interactions among the multi-quark composites comprising hadrons. This paradoxical situation was cleared up with the discovery of the properties of asymptotic freedom and confinement in non-Abelian QCD. The question then arises as to whether asymptotic freedom is identical with "Bjorken scaling" or whether there are identifiable and measurable differences. The answer is that asymptotic freedom in QCD, strictly speaking, requires logarithmic corrections to "Bjorken scaling" and that these corrections lead to experimental deviations from "Bjorken scaling" that have actually been observed. It should be pointed out that our derivation of the "asymptotic freedom" corrections to "Bjorken scaling" is primarily pedagogical in intent — to clarify the relationship between the renormalization group and "residual" scale invariance in QCD; the precise quantitative corrections to "Bjorken scaling" under a variety of conditions are contained in the work of Altarelli and Parisi [5.19], which takes account of the "q^2 evolution" of the naive quark-parton distribution and simulates OPE to leading order.

The QCD renormalization group calculation of asymptotic freedom is best accomplished by invoking the "operator production expansion" (OPE) method devised by Wilson [5.11] to study the interplay between the scale invariance of a quantum field theory (see §2.3b) and the Callan-Symanzik renormalization group equation [5.12]. The OPE method starts by expressing the short-distance singular behavior of the product of two currents $j_A(x)$ and $j_B(y)$ in the form:

$$j_A(x)j_B(y) \xrightarrow[(x-y)_\mu \to 0]{} \sum_i c_i(x-y)\mathcal{O}_i(\frac{x+y}{2}) \qquad (5.41)$$

where the c_i's are the "Wilson coefficients" that are singular c-number functions (taking care of the singularities of the operator product), satisfying the renormalization group (Callan-Symanzik) equation, and the $\mathcal{O}_i$'s are local non-singular field operators with quantum numbers matching those of $j_A(x)$ $j_B(y)$ and whose "non-perturbative" matrix elements are evaluated by means of standard QCD methods (e.g. dispersion theory, sum rules, etc.). We note that the higher the dimension of $\mathcal{O}_i$, the less singular are the c_i coefficients; this signifies that the dominant operators at short distance have the smallest dimensions (in units of mass). Specific examples of $\mathcal{O}_i$'s and their dimensions will be given when the OPE method is discussed in §5.4d in connection with QCD sum rules.

Insofar as the Wilson coefficients c_i are concerned, one expects their short-distance behavior to exhibit the form:

$$c_i(x) \underset{x \to 0}{\sim} x^{d_{\mathcal{O}_i} - d_A - d_B} \left[\text{mod } \ln |x|\right] \qquad (5.42)$$

where d_A, d_B and the $d_{\mathcal{O}_i}$'s are the "canonical" dimensions of j_A, j_B and the $\mathcal{O}_i$'s respectively and mod $\ln|x|$ allows for the expected logarithmic corrections. Because the stable "ultraviolet" fixed point of QCD under consideration occurs at the "origin", the "anomalous dimensions" are shifted to the $\ln|x|$ term [the "anomalous dimensions" would displace the "canonical" dimensions in the neighborhood of a stable "ultraviolet" fixed point away from the "origin"], which we now show. Since the c_i coefficients satisfy the renormalization group equation, we have:

$$[\mu\frac{\partial}{\partial\mu} + \beta(g)\frac{\partial}{\partial g} + d_A + \gamma_A(g) + d_B + \gamma_B(g) - d_{\mathcal{O}_i} - \gamma_{\mathcal{O}_i}(g)]c_i(x) = 0 \quad (5.43)$$

leading — when there is a stable "ultraviolet" stable fixed point for some finite g [see Eq. (5.27)] — to the asymptotic behavior of $c_i(x)$ ($|x| \to 0$):

$$c_i(x) \underset{x \to 0}{\sim} |x|^{d_{\mathcal{O}_i} + \gamma_{\mathcal{O}_i}(g_\infty) - d_A - \gamma_A(g_\infty) - d_B - \gamma_B(g_\infty)} \quad (5.44)$$

where the $\gamma(g_\infty)$'s are the anomalous dimensions corresponding to the A and B currents and the Wilson operator $\mathcal{O}_i$ [see Eq. (5.28)]. But for an asymptotically free gauge theory such as QCD, the "ultraviolet" stable fixed point is at the "origin" so that we can use Eq. (5.26b) to calculate the "anomalous" dimensions. In the asymptotic freedom limit and in the "one-loop" approximations for $\beta(g_s)$ [i.e. $\beta(g_s) = -bg_s^3$] and $\gamma(g_s)$ [i.e. $\gamma(g_s) = cg_s^2$] we get $z(\sigma) \sim (2b\, g_s^2(\ln\sigma)^{c/2b}$ [see Eq. (5.26b)] and the Wilson c_i coefficients assume the form:

$$c_i(x) \sim |x|^{d_{\mathcal{O}_i} - d_A - d_B} \left(\ln\frac{1}{|x|}\right)^{(c_A + c_B - c_{\mathcal{O}_i})/2b} \quad (5.44a)$$

where c_A, c_B and $c_{\mathcal{O}_i}$ are the "one-loop" coefficients in the γ functions and b is the common coefficient of the β function of the QCD renormalization group. Equation (5.44a) is the final result and exhibits the logarithmic departure from ("canonical") "Bjorken scaling" predicted for the asymptotic freedom region in QCD. The logarithmic departure from "Bjorken scaling" translates into a $\ln Q^2$ dependence of the W_1 and W_2 structure functions, i.e. a logarithmic deviation from pure "Bjorken scaling" (i.e. exclusive dependence on $x = Q^2/2M\nu$) of the form $[\ln(Q^2)]^{-c_{\theta_i}/2b}$. The experimental evidence for "Bjorken scaling" and the naive quark-parton model, as well as for the asymptotic freedom (or, equivalently, the Altarelli-Parisi) corrections are treated in the next section; §5.2c also reviews the experimental status of some of the deep inelastic lepton-hadron scattering sum rules.

§5.2c. Experimental evidence for Bjorken scaling, the quark-parton model and asymptotic freedom

If non-Abelian QCD is the correct gauge theory of the strong interaction, the unique property of this type of theory, asymptotic freedom (and its limiting version of "Bjorken scaling" in the quark-parton model) should have well-defined experimental consequences that confirm the theory. Probing tests of "Bjorken scaling", the "quark-parton" model and asymptotic freedom in QCD have come from experiments on the deep inelastic scattering of leptons (e, ν, μ) by hadrons and measurements of the R value [with R defined by $\sigma(e^+e^- \to \text{hadrons})/\sigma(e^+e^- \to \mu^+\mu^-)$] and have been highly successful. In this section, we review the results of these key experiments and spell out some of the high energy sum rules that have been tested for unpolarized deep inelastic lepton-hadron experiments and mention several sum rules — to be discussed more fully in §5.4a — that are being put to the test with regard to the deep inelastic scattering of polarized leptons by polarized protons. We start by giving a brief derivation of the cross section for the scattering of electrons by nucleons. The novel feature of this class of experiments is that the initial and final four-momenta of the electrons are measured without paying attention to the hadronic end-products. By limiting the measurements to large energy losses for the electrons, as well as large transfers of space-like momenta (squared) from the electrons to the target nucleons, the resulting "deep" inelastic lepton-nucleon scattering cross sections can, to a first approximation, be compared to the "Bjorken scaling" predictions [1.104]. The derivation of the deep inelastic cross section proceeds as follows: the kinematic variables in the inelastic lepton-hadron scattering reaction are defined with the help of Fig. 5.2 [5.20]:

$$\ell(k) + N(p) \to \ell(k') + X(p_n) \tag{5.45}$$

The notation refers to the laboratory frame, where X is the hadronic final state with total (four-) momentum p_n and the other momenta are given by: $p = (M, 0)$, $k = (E, \underline{k})$ and $k' = (E', \underline{k}')$, with M the initial hadron (nucleon) mass. The four-momentum transfer, $q = (k - k')$, becomes (setting $m_\ell = 0$):

$$q^2 = (k - k')^2 = -4EE' \sin^2 \frac{\theta}{2} \le 0; \qquad Q^2 = -q^2 \tag{5.46}$$

where θ is the scattering angle of the lepton (see Fig. 5.2), Q^2 is always positive and $\nu = \frac{p \cdot q}{M} = (E - E')$ is the energy loss of the lepton.

The inelastic scattering amplitude, corresponding to Fig. 5.2, takes on two different forms depending on whether the electromagnetic or weak

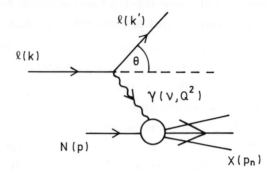

Fig. 5.2. Inelastic scattering of a lepton by a nucleon.

interaction is responsible for the inelastic scattering; the electromagnetic-generated inelastic scattering amplitude — which we denote by T_n^ℓ — can be written:

$$T_n^\ell = e^2 \bar{u}(k', \lambda') \gamma^\mu u(k, \lambda) \frac{1}{q^2} < n|J_\mu^{em}(0)|p, \sigma > \qquad (5.47)$$

where the u's are the charged lepton spinors (with λ the polarization) and J_μ^{em} is the hadronic electromagnetic current between the initial state of the nucleon (with polarization σ) and the final hadronic state n. The terms in front of $1/q^2$ in Eq. (5.47) represent, of course, the charged lepton electromagnetic current, which has no form factors — in contrast to the hadronic electromagnetic current — because the leptons are regarded as points [thus far, the experimental evidence tells us that the electron and muon are point-like down to 10^{-16} and 10^{-17} cm (or up to 10^2 and 10^3 GeV) respectively (see §4.5a). If neutrinos, rather than charged leptons, are involved in the inelastic scattering by hadrons, the inelastic scattering amplitude, which is now denoted by T_n^ν, can be written [$\gamma_\mp^\mu = \frac{\gamma^\mu}{2}(1 \mp \gamma_5)$]:

$$T_n^\nu = \frac{G_F}{\sqrt{2}M} \bar{u}(k', \lambda') \gamma_\mp^\mu u(k, \lambda) \frac{1}{q^2 + M^2} < n|J_\mu^W(0)|p, \sigma > \qquad (5.48)$$

where e^2 [in Eq. (5.47)] is replaced by $G_F/\sqrt{2}M^2$ [$M = M_W$ if the neutrino is transformed into a charged lepton whereas $M = M_Z$ if the final lepton is a neutrino], q^2 is replaced by $(q^2 + M^2)$, γ^μ is replaced by $\gamma_\mp^\mu/2 = \gamma^\mu(1 \mp \gamma_5)$ [the minus sign corresponds to an initial (lefthanded) neutrino and the plus sign to an initial (righthanded) neutrino], and the hadronic electromagnetic current J_μ^{em} is replaced by the hadronic weak current J_μ^W.

We proceed to calculate the unpolarized differential scattering cross section $d\sigma_n$ in terms of the general T_n (which can be T_n^ℓ or T_n^ν) and only later distinguish between T_n^ℓ and T_n^ν (further distinguishing between ν and $\bar{\nu}$ inelastic scattering by writing T_n^ν or $T_n^{\bar{\nu}}$); we have:

$$d\sigma_n = \frac{1}{|v|} \frac{1}{2M} \frac{1}{2E} \frac{d^3k'}{(2\pi)^3 2k_0'} \prod_{i=1}^n [\frac{d^3p_i}{(2\pi)^3 2p_{i0}}]$$

$$\times \frac{1}{4} \sum_{\sigma\lambda\lambda'} |T_n|^2 (2\pi)^4 \delta^4(p + k - k' - p_n) \qquad (5.49)$$

where $\underline{v}$ is the relative velocity between the initial lepton and nucleon. Since the final hadronic states are not observed, it is necessary to sum over all final hadronic states to obtain the "inclusive" cross section for lepton-proton scattering:

$$\frac{d^2\sigma}{d\Omega dE'} = f(q^2) \left(\frac{E'}{E}\right) L^{\mu\nu} W_{\mu\nu} \qquad (5.50)$$

where $f(q^2) = \dfrac{2e^4}{q^4}$ for $T_n = T_n^\ell$ and $f(q^2) = \dfrac{G_F^2 M^2}{(q^2 + M_W^2)^2}$ for $T_n = T_n^\nu$ or $T_n^{\bar{\nu}}$. The quantities $L_{\mu\nu}$ and $W_{\mu\nu}$ are the leptonic and hadronic tensors respectively, and take different forms depending on whether the initial lepton is charged (to which we refer subsequently as the electron) or neutral (we distinguish between ν and $\bar{\nu}$ because their helicities are different). We have for the purely kinematical lepton tensor $L_{\mu\nu}$, the following three expressions:

$$L_{\mu\nu}^\ell = kk_\mu k_\nu' + k_\mu' k_\nu - k \cdot k' g_{\mu\nu}; \qquad L_{\mu\nu}^\nu = L_{\mu\nu}^\ell + i\epsilon_{\mu\nu\alpha\beta} k^\alpha k'^\beta;$$

$$L_{\mu\nu}^{\bar{\nu}} = L_{\mu\nu}^\ell - i\epsilon_{\mu\nu\alpha\beta} k^\alpha k'^\beta \qquad (5.51)$$

where the additional terms in Eq. (5.51) for $L_{\mu\nu}^\nu$ and $L_{\mu\nu}^{\bar{\nu}}$ are due to the lefthanded and righthanded helicities of ν and $\bar{\nu}$ respectively; the term $\pm i\epsilon_{\mu\nu\alpha\beta} k^\alpha k'^\beta$ must be added to $L_{\mu\nu}^\ell$ in Eq. (5.42) if one is performing a polarization experiment with the charged leptons.

The parity violation associated with ν (or $\bar{\nu}$ or longitudinally polarized ℓ) inelastic scattering by hadrons must also be taken into account in writing down the general covariant decomposition of the quark hadronic tensor $W_{\mu\nu}$, which can be expressed in terms of the current commutator of J_μ and J_ν ($J_\mu = J_\mu^{em}$ or J_μ^W):

$$W_{\mu\nu}(p, q) = \frac{1}{4M} \sum_\sigma \int \frac{d^4x}{2\pi} e^{iq \cdot x} <p, \sigma \, [J_\mu(x), J_\nu(0)] \, p, \sigma> \qquad (5.52)$$

In general, $W_{\mu\nu}$ is a second-rank Lorentz tensor depending on the four-momenta p_μ and q_μ, and it is easy to show that the covariant decomposition of $W_{\mu\nu}$ contains six terms (called structure functions) if gauge invariance and parity conservation are not assumed; we have [see §3.5d]:

$$W_{\mu\nu} = -W_1 g_{\mu\nu} + \frac{W_2}{M^2} p_\mu p_\nu - \frac{i}{2} \frac{W_3}{M^2} \epsilon_{\mu\nu\alpha\beta} p^\alpha q^\beta$$

$$+ \frac{W_4}{M^2} q_\mu q_\nu + \frac{W_5}{M^2} (p_\mu q_\nu + q_\mu p_\nu) + \frac{W_6}{M^2} (p_\nu q_\mu - p_\mu q_\nu)$$

(5.53)

If gauge invariance holds — as is always the case — it follows that $q^\mu W_{\mu\nu} = 0$ and when this condition is applied to Eq. (5.53), the W_4 and W_5 structure functions can be expressed in terms of the W_1 and W_2 structure functions, thus:

$$W_4 = W_2 (p \cdot q/q^2)^2 - W_1 M^2/q^2; \qquad W_5 - W_6 = -W_2(p \cdot q/q^2) \quad (5.54)$$

If parity conservation holds — which is true for the electromagnetic interaction— the parity-violating W_3 structure function must be deleted; on the
other hand, if there is parity violation, the W_3 structure function must be
included. Finally, if the initial nucleon is unpolarized — as will be the case
unless stated otherwise — -$W_{\mu\nu}$ must be symmetric in μ and ν so that the
W_6 structure function is absent. Putting the above conditions together, one
obtains the following general expression for $W_{\mu\nu}$:

$$W_{\mu\nu} = -W_1(Q^2,\nu)(g_{\mu\nu} - \frac{q_\mu q_\nu}{q^2}) + \frac{W_2(Q^2,\nu)}{M^2}[(p_\mu - \frac{p \cdot q}{q^2} q_\mu)(p_\nu - \frac{p \cdot q}{q^2} q_\nu)]$$

$$\frac{-i}{2M^2} W_3(Q^2,\nu)\epsilon_{\mu\nu\alpha\beta} p^\alpha q^\beta$$

(5.55)

where W_1, W_2 and W_3 are Lorentz-invariant structure functions of the target
nucleon expressed in terms of Q^2 and ν.

With the expressions for $L_{\mu\nu}$ and $W_{\mu\nu}$ before us, we first consider the
deep inelastic (unpolarized) electron-nucleon scattering experiments; in that
case, W_3 can be dropped so that $W_{\mu\nu}$ becomes:

$$W_{\mu\nu}(p,q) = -W_1 \left(g_{\mu\nu} - \frac{q_\mu q_\nu}{q^2} \right) + \frac{W_2}{M^2} \left(p_\mu - \frac{p \cdot q}{q^2} q_\mu \right) \left(p_\nu - \frac{p \cdot q}{q^2} q_\nu \right)$$

(5.56)

Equation (5.56) leads to the standard differential "inclusive" cross section
for eN scattering in terms of directly-measured quantities:

$$\frac{d^2\sigma}{d\Omega dE'} = \frac{\alpha_e^2}{4E^2 \sin^4 \frac{\theta}{2}} [2W_1(Q^2,\nu) \sin^2 \frac{\theta}{2} + W_2(Q^2,\nu) \cos^2 \frac{\theta}{2}] \qquad (5.57)$$

Equation (5.57) is a key equation that enables one — by direct comparison
with experiment — to establish that the nucleon is not a point-like particle
(like the electron or the muon) nor a "diffuse" structure with a "pion cloud"
surrounding a core. Rather, the deep inelastic lepton-nucleon scattering ex-

periments demonstrate that the nucleon consists of three point-like "quarks" [or at least, point-like particles with all the properties — spin, electric charge, etc. — that are assigned to quarks] that behave as free particles at very close distances, but are otherwise confined (see §5.3). We use Eq. (5.57) to rule out the (single) point-like and "diffuse" possibilities and then — with the help of the quark-parton model [1.103] — show why experiment favors a quark-like structure for the nucleon. First, if the nucleon (with mass M) were a point-like particle like the muon, say, we would have $p^2 = p'^2 = M^2$ and $p \cdot q = M\nu$ so that $\nu = Q^2/2M$ [see Eq. (5.46)]; furthermore, $W_1(Q^2, \nu)$ and $W_2(Q^2, \nu)$ would reduce respectively to $Q^2/4M^2$ and 1. Simply put, this would correspond to the elastic scattering of two point-like particles (e.g. electron-muon scattering) and Eq. (5.57) would simplify to:

$$\frac{d^2\sigma}{d\Omega dE'} = \frac{4\alpha_e^2(E')^2}{Q^4}[\cos^2\frac{\theta}{2} + \frac{Q^2}{2M^2}\sin^2\frac{\theta}{2}]\delta(\nu - \frac{Q^2}{2M}) \qquad (5.58)$$

where $E' = E - Q^2/2M$. It is trivial in this case to integrate Eq. (5.58) over dE' to obtain the familiar formula:

$$\frac{d\sigma}{d\Omega} = \left(\frac{d\sigma}{d\Omega}\right)_{\text{Mott}} \frac{E'}{E}(1 + \frac{Q^2}{2M^2}\tan^2\frac{\theta}{2}) \qquad (5.59)$$

The "Mott" differential cross section [for a singly-charged point-like target assumed to have infinite mass] is $\alpha_e^2 \cos^2\frac{\theta}{2}/4E^2\sin^4\frac{\theta}{2}$; it is easy to establish experimentally that Eq. (5.59) does not describe the elastic scattering of an electron by a nucleon.

The second possibility — that the electron is scattered elastically by a nucleon with a "diffuse" structure — can be computed by representing the "diffuse" structure by the electromagnetic form factor of the nucleon. In that case, Eq. (5.58) can be written (assuming $M << E, E'$) in the form:

$$\frac{d^2\sigma}{d\Omega dE'} = \frac{4\alpha_e^2(E')^2}{Q^4}\delta(\nu - \frac{Q^2}{2M})$$
$$\times \{[\frac{G_E^2(Q^2) + (Q^2/4M^2)G_M^2(Q^2)}{1 + Q^2/4M^2}]\cos^2\frac{\theta}{2} + \frac{G_M^2(Q^2)}{2M^2}\sin^2\frac{\theta}{2}\} \qquad (5.60)$$

where $G_E(Q^2)$ and $G_M(Q^2)$ are the electric and magnetic form factors of the nucleon. From measurements carried out on the elastic scattering of electrons by protons (and, to a lesser extent, by neutrons), a good approximation to the cross sections for elastic electron-nucleon scattering is obtained by representing G_E and G_M by the "dipole" form factor [5.20], to wit:

$$\frac{G_E(Q^2)}{G_E(0)} = \frac{G_M(Q^2)}{G_M(0)} = (1 + \frac{Q^2}{0.7\,\text{GeV}^2})^{-2} \qquad (5.61)$$

Clearly, the finite-size ("diffuse") nucleon causes the elastic scattering cross section for electron scattering to decrease rapidly as Q^2 increases — in contrast to the behavior of a point-like nucleon [as given by Eq. (5.58)]. Thus, as long as one insists on doing an elastic electron scattering experiment on nucleons (so that the only hadron in the final state is the nucleon), there is an extremely rapid dropoff of cross section [see Eq. (5.60)] for $Q^2 \gg 0.7$ GeV^2.

The situation changes dramatically if, instead of measuring the elastic lepton-nucleon scattering, one measures the scattering cross section of electrons by nucleons in the "deep" inelastic region (large Q^2 and ν with $\frac{Q^2}{2M\nu}$ held constant) where Eq. (5.57) can be applied to the scattering of leptons by the "quark-parton" constituents of the nucleon, taking proper account of the "quark-parton" momentum distribution inside the nucleon. The crucial observation in connection with the first deep inelastic electron-proton scattering experiment was that the experiment effectively measured the forward Compton scattering of virtual photons (associated with the incident electrons) by the hypothesized "quark-parton" constituents of the proton; since the virtual photons carried large controllable Q^2 and the electron energy loss was also measured, the Bjorken scaling hypothesis could be tested for $W_1(Q^2, \nu)$ and $W_2(Q^2, \nu)$ in Eq. (5.57). The Bjorken scaling hypothesis stated [1.104] that, for large values of Q^2 and ν, the combinations $2MW_1(Q^2, \nu)$ and $\nu W_2(Q^2, \nu)$ are functions only of the dimensionless ratio $x = Q^2/2M\nu$, instead of being functions of the two variables Q^2 and ν.

Without the Bjorken scaling hypothesis applied to a "quark-parton" structure of the nucleon, one expects a rapidly decreasing cross section for inelastic lepton-nucleon scattering with increasing Q^2 at fixed x. But with "Bjorken scaling" plus the "quark-parton model" of the nucleon, it can be shown (see below) that the decrease of the inelastic cross section with Q^2 is much more moderate at constant x. This type of behavior reminds one of the old Rutherford experiments on the large-angle scattering of α particles by atoms (see §1.1), where the surprisingly large cross section led to the conclusion that the positive charge of the target atom is concentrated in a "point-like" nucleus. The analogy suggests that the anomalously large cross section for deep inelastic lepton-hadron scattering has its origin in the presence of "point-like" constituents inside the nucleon [these "point-like" constituents were first called "partons" [1.103] and then shown to possess all the properties of quarks]. Our task in the remainder of this section is to provide convincing evidence for the primacy of the "Bjorken scaling" hypothesis [the (logarithmic) deviations that are found serve as further evidence for asymptotic freedom] and the validity of the "quark-parton model" of the nucleon. Apart from the crucial evidence provided for a Yang-Mills type of theory for the strong interaction, the second outcome of the deep inelastic lepton-hadron scattering experiments, i.e. that the quarks are "physical entities" and not merely "mathematical fictions" (see §1.3a), was equally significant for the development of the standard model. In anticipation of the physical identification of "partons" with "quarks", we henceforth refer to the point-like constituents of nucleons as quarks (although we continue to refer to the "quark-parton model").

We have stated that the original (or naive) "quark-parton model" leads to "Bjorken scaling" in deep inelastic lepton-hadron scattering, but before we confront the experimental data with the "Bjorken scaling" predictions, we demonstrate the connection between the "quark-parton model" and "Bjorken scaling". We have already proved one connection between the light-cone behavior of free quark currents in the quark hadronic tensor and "Bjorken scaling" but it is instructive to give a second proof — couched in "quark-parton model" language — because this is the language that is most useful for extracting the physical meaning of the comparison between experiment and theory. We note three basic assumptions of the "quark-parton model": (1) the (inclusive) inelastic scattering is due to an incoherent superposition of elastic scatterings by the weakly-bound (quasi-free) constituents (quarks) with an unspecified momentum distribution; (2) the final state quarks recombine over a sufficiently long time scale into the observed final-state hadrons so that the final state quark interactions can be ignored and, finally, (3) to a good approximation, the nucleon is presumed to consist of three "valence" quarks — to each of which is assigned a momentum distribution in terms of the fraction of the original nucleon momentum that it carries. It is possible to take into account the "sea" of quark pairs by assigning to each quark and antiquark (of the quark pair) its own momentum distribution. With these three basic assumptions, we proceed to show that the two structure functions $2MW_1(Q^2, \nu)$ and $\nu W_2(Q^2, \nu)$ in Eq. (5.57) are functions only of the dimensionless ratio $x = Q^2/2M\nu$ (i.e. we have "Bjorken scaling") and that, in addition, there is a simple relation between the two structure functions ("Callan-Gross" [5.21] relation for a $J = \frac{1}{2}$ particle). Assume then, that each quark carries a fraction ξ of the original nucleon momentum p ($0 \leq \xi \leq 1$)

— see Fig. 5.3 — and neglect quark momentum transfers to the nucleon momentum; we can write the "quark-parton" contribution — denoted by $W_{\mu\nu}(\xi)$ — to the hadronic tensor $W_{\mu\nu}$, given by Eq. (5.52), in the form:

$$W_{\mu\nu}(\xi) = \frac{1}{4\xi M} \sum_{\text{spin}} \frac{d^3 p'}{(2\pi)^3 2p'_0} \times < \xi p, \sigma | J_\mu^{em}(0) | p', \sigma' >$$

$$< p', \sigma' | J_\nu^{em}(0) | \xi p, \sigma > (2\pi)^3 \delta^4(p' - \xi p - q)$$

$$= \frac{1}{4\xi M} \sum_{\text{spin}} \bar{u}(\xi p)\gamma_\mu u(p')\bar{u}(p')\gamma_\nu u(\xi p)\delta(p'_0 - \xi p_0 - q_0)/2p'_0 \tag{5.62}$$

where $W_{\mu\nu}(p, q) = \int_0^1 d\xi f(\xi) W_{\mu\nu}(\xi)$ with $f(\xi)$ the "quark-parton" distribution function inside the nucleon.

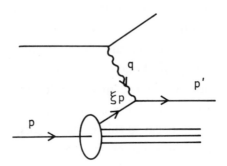

Fig. 5.3. Inelastic lepton-nucleon scattering treated as incoherent elastic scattering from quarks.

The simple final form of the "quark-parton" tensor in Eq. (5.62) is due to the hypothesis that the electrons are scattered elastically by the "point-like" quarks [assumption (1) above]. The factor ξ in the denominator of Eq. (5.62) takes care of the change in relative flux from p to ξp. The δ function in Eq. (5.62) can be rewritten as:

$$\delta(p'_0 - \xi p_0 - q_0)/2p'_0 = \theta(\xi p_0 + q_0)\frac{\delta(\xi - x)}{2M\nu} \tag{5.63}$$

The spin sum over the quark spinor wavefunctions in Eq. (5.62) is also straightforward to compute, with the result (neglecting quark mass):

$$W_{\mu\nu}(\xi)(=)\frac{1}{2} \sum_{\text{spin}} \bar{u}(\xi p)\gamma_\mu u(\xi p + q)\bar{u}(\xi p + q)\gamma_\nu u(\xi p)$$

$$= 4M^2\xi^2(p_\mu p_\nu/M^2) - 2M\nu\xi g_{\mu\nu} + \dots \tag{5.64}$$

Combining Eqs. (5.63) and (5.64), one finds for the two dominant terms of $W_{\mu\nu}(\xi)$:

$$W_{\mu\nu}(\xi) = \delta(\xi - x)(\frac{\xi}{\nu}\frac{p_\mu p_\nu}{M^2} - \frac{1}{2M}g_{\mu\nu} + \dots) \qquad (5.65)$$

and hence:

$$W_{\mu\nu} = \int_0^1 f(\xi)W_{\mu\nu}(\xi)d\xi = \frac{xf(x)}{\nu}\frac{p_\mu p_\nu}{M^2} - \frac{f(x)}{2M}g_{\mu\nu} + \dots \qquad (5.66)$$

Equation (5.66) yields the required result for the two structure functions W_1 and W_2:

$$2MW_1(Q^2,\nu) = f(x); \qquad \nu W_2(Q^2,\nu) = xf(x) \qquad (5.67)$$

It is important to note: (a) it is the δ function in Eq. (5.65), expressing the mass-shell condition on the final quark, that leads to the dependence of W_1 and W_2 only on the dimensionless variable $x = Q^2/2M\nu$; (b) there is the simple Callan-Gross relation between the two structure functions W_1 and W_2 [i.e. $2MW_1 = x(\nu W_1)$], resulting from the assumption that the quark is a spin $\frac{1}{2}$ particle; and (c) the scaling function $f(x)$ is truly a measure of the momentum distribution of the "quark-parton" in the target nucleon. We have now proved that "Bjorken scaling" in deep inelastic lepton-hadron scattering is a consequence of both the naive "quark-parton model" and the light-cone behavior for free quark currents, which is not surprising since in both cases the quarks within the nucleon are assumed to be non-interacting. The more accurate QCD renormalization method for studying the deep inelastic domain in lepton-hadron scattering pays attention to the quark interactions and, understandably, predicts corrections to "Bjorken scaling"; the only reason that the QCD corrections to "Bjorken scaling" are logarithmic in character — and not larger and more complicated — is that the stable "ultraviolet" fixed point in QCD is at the "origin".

With this theoretical background, we are in a position to subject Eq. (5.67) to a variety of experimental tests. An early experimental test consisted of a series of measurements on the deep inelastic scattering of electrons by protons over a wide range of values of Q^2 and ν and then plotting the structure function $\nu W_2(Q^2,\nu)$ as a function of the dimensionless ratio $\frac{1}{x} = \frac{2M\nu}{Q^2}$. The resulting curve is shown in Fig. 5.4 [5.20], where many measurements were made for widely varying values of Q^2 and ν corresponding to the same value of x; it is seen that several measurements were made for each value of x. The statistical weight of the data is less good for the smaller values of x (although the deviations can be explained — see below) but it is clear that over most of the range of x, the agreement is excellent. This is striking confirmation of the Bjorken scaling hypothesis for deep inelastic electron-proton scattering. In addition, the large observed inelastic scattering cross sections — much larger than the elastic ones — is strongly supportive of the "quark-parton hypothesis", postulating a quasi-free behavior of the quark constituents of the proton in their large Q^2 collisions with the leptons.

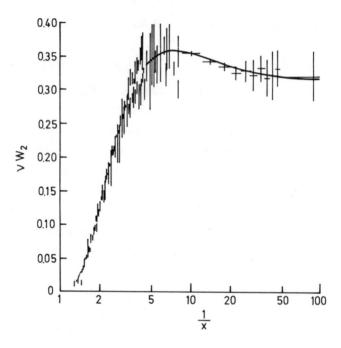

Fig. 5.4. Dependence of structure function νW_2 on $\left(\frac{1}{x}\right)$ for deep inelastic $e-p$ scattering.

Many other experiments have been performed on the deep inelastic scattering of charged leptons by hadrons [e.g. on deuteron targets, to probe the validity of the "quark-parton model" for the neutron], with muons [to check whether there are any differences in the deep inelastic cross sections for different generations of charged leptons], with neutrinos [to test whether "Bjorken scaling" holds as well for deep inelastic neutrino-hadron scattering where the hadronic current $J_W^\mu(x)$ (interacting with the weak boson field) replaces the hadron current $J_{EM}^\mu(x)$ (interacting with the electromagnetic field), etc.)] All experiments give good agreement with "Bjorken scaling" and the "quark-parton" hypotheses, subject to the following caveats: (1) the quark "sea" effect (i.e. the contribution of virtual quark-anti-quark pairs) must be taken into account; it is expected to alter the small x values of the structure functions (see below). The deep inelastic neutrino-hadron experiments provide confirmation of the "sea" effect; (2) there are logarithmic deviations from the "Bjorken scaling" predictions — due to the asymptotic property of QCD — and these are also discussed below; and (3) the "quark-parton model" may be too crude an approximation and it may be necessary to deal more carefully with the "quark fusion" into hadrons in the final stage of the scattering process [5.22]; for much larger values of $Q^2 [\leq 10^5 \ (\text{GeV})^2]$ — which are available on the HERA machine — it may also be necessary to consider deviations from the point-like structure of quarks and leptons (see §9.3).

We speak to some of the above caveats presently but first we must present in somewhat greater detail the strong evidence for "Bjorken scaling" and the "quark-parton model". We have shown that both the Bjorken scaling hypothesis predicts that the two structure functions $W_1(Q^2, \nu)$ and $W_2(Q^2, \nu)$ can be written in terms of $f(x)$ as follows [see Eq. (5.67)]:

$$2MW_1(Q^2, \nu) = F_1(x) = f(x); \quad \nu W_2(Q^2, \nu) = F_2(x) = xf(x) \qquad (5.68)$$

When there are m quark flavors with electric charges e_i $(i = 1, 2, \ldots m)$ and momentum distribution functions $f_i(x)$ $(i = 1, 2, \ldots m)$, and we allow for the "quark sea", we have $f(x) = \Sigma_i e_i^2 f_i(x)$ $(i = 1, 2, \ldots m)$ and Eq. (5.68) becomes:

$$2MW_1(Q^2, \nu) = \sum_{i=1}^{m} e_i^2 f_i(x); \quad \nu W_2(Q^2, \nu) = \sum_{i=1}^{m} e_i^2 x f_i(x) \qquad (5.69)$$

Equation (5.69) is the "master" equation, incorporating "Bjorken" scaling and the "quark-parton model" and yields a variety of high energy sum rules that can be checked directly with experiment and, in the process, establish that the "quark-partons" of theory are truly the quarks of QCD, i.e. "physical entities" (albeit confined) and not "mathematical fictions" (see §1.3a).

One of the first sum rules to be derived — for $\nu W_2(Q^2, \nu)$ — follows directly from Eq. (5.69):

$$\int_0^1 \nu W_2(Q^2, \nu) \frac{dx}{x} = \sum_{i=1}^{m} e_i^2 \int_0^1 f_i(x) dx \quad (i = 1, \ldots, m) \qquad (5.70)$$

It should be noted that there is no factor 3 due to color in Eq. (5.70) because only the color singlet quark currents contribute (the situation is different for the R value — see below). Clearly, the value of the R.H.S. of Eq. (5.70) depends on the assumed quark composition of the hadron; if only the "valence" quarks are considered for the nucleon, then the proton consists of uud and the neutron of udd and the R.H.S. of Eq. (5.70) is 1 and $\frac{2}{3}$ respectively if the u and d "quark-parton" distributions are taken to be the same. However, it is expected that the quark "sea" contributes to the R.H.S. of Eq. (5.70), especially for small x (see below), but this effect can be compensated for — very crudely — by making measurements for both proton and neutron targets, since the quark "sea" is presumed to be isotopically neutral (i.e. an isotopic singlet). Figure 5.4 exhibits the singular behavior of νW_2 for small x; if one makes measurements of the νW_2 structure function for both proton and neutron, one obtains the much better-behaved difference curve given in Fig. 5.5, to be compared with the naive "valence" sum rule:

$$\int_0^1 (\nu W_2^{ep} - \nu W_2^{en}) \frac{dx}{x} = \frac{1}{3} \tag{5.70a}$$

The most recent experimental determination of the L.H.S. of Eq. (5.70a) is 0.240 ± 0.016 [5.23], to be compared with the theoretical value of 0.33 — a 25% discrepancy; considering the omission of quark "sea" effects and the approximations entering into the "quark-parton model", the agreement is fair but not impressive and calls for improvements in the underlying assumptions.

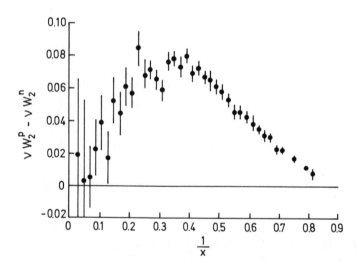

Fig. 5.5. Plot of difference in νW_2 for the $e - p$ and $e - n$ reactions as a function of $(\frac{1}{x})$.

One way to improve the theoretical prediction is to fold in, semi-pheno-menologically, the quark "sea" effects in the form of the Gottfried sum rule [5.24]; if one uses the notation $f^\alpha(x) = q^\alpha(x)$ $(q = u, d, s \ldots)$, then Eq. (5.68) is replaced by:

$$\frac{F_2^\alpha(x)}{x} = \frac{4}{9}[u^\alpha(x) + \bar{u}^\alpha(x)] + \frac{1}{9}[d^\alpha(x) + \bar{d}^\alpha(x)] + \frac{1}{9}[s^\alpha(x) + \bar{s}^\alpha(x)] + \ldots] \quad (5.71)$$

where the superscript α denotes the hadron species (p or n for the nucleon) and $\bar{q}(x)$ is the antiquark distribution function. Since the u, d quarks and p, n both comprise isospin doublets, we have in an obvious notation:

$$u^p = d^n \equiv u; \quad d^p = u^n \equiv d; \quad s^p = s^n \equiv s \qquad (5.71a)$$

In this notation, the conditions of zero net strangeness for a nucleon and the known charges of the proton and neutron can be expressed in the form:

$$\int_0^1 [\frac{2}{3}(u - \bar{u}) - \frac{1}{3}(d - \bar{d})] = 1; \quad \int_0^1 dx[\frac{2}{3}(d - \bar{d}) - \frac{1}{3}(u - \bar{u})] = 0$$

$$\int_0^1 dx[s(x) - \bar{s}(x)] = 0 \qquad (5.72)$$

From the first two equations of (5.72), we get:

$$\int_0^1 dx[u(x) - \bar{u}(x)] = 2; \quad \int_0^1 dx[d(x) - \bar{d}(x)] = 1 \qquad (5.73)$$

With the help of Eqs. (5.71) and (5.72), it is easy to derive the Gottfried sum rule, namely:

$$\int_0^1 \frac{dx}{x}[\nu W_2^{ep}(x) - \nu W_2^{eN}(x)] = \frac{1}{3} + \frac{2}{3}\int_0^1 dx$$

$$\{[\bar{u}(x) - \bar{d}(x)] + \text{ heavy quark "sea"}\} \qquad (5.74)$$

It is clear from the Gottfried sum rule that the surrender of isotopic neutrality of the $\bar{u}$ and $\bar{d}$ antiquarks is one way to resolve the 25% disagreement with experiment. Some support for this approach — through a recent calculation based on chiral field theory [5.25] — has shown how a sufficiently large flavor asymmetry in the light quark "sea" of the nucleon can arise when account is taken of the virtual disassociation of a quark into a quark plus pion (within the nucleon). However, the 25% discrepancy is a large effect and other explanations for the discrepancy in the Gottfried sum rule include: (1) lack of knowledge of the W_2 structure function at very low x with some possible non-perturbative surprises [5.26]; (2) failure to include the gluon — in addition to the quark — distribution inside the nucleon, whose importance is being revealed by experiments on direct, hard photon production in hadronic reactions [5.27]; (3) the factor 2 difference between the $\bar{u}$ and $\bar{d}$ masses may sufficiently alter the $(\bar{u} - \bar{d})$ distribution [see Eq. (5.74)]; and, finally (4) the crudity of the approximation made in deducing W_2^{en} from the deep inelastic electron-deuteron experiments may contribute to the discrepancy. Progress on removing the discrepancy in the Gottfried sum rule is important because its deduction follows from the basic assumptions of "Bjorken scaling" and the "quark-parton model". It seems that the Gottfried sum rule has turned out to be not only a testing ground of "Bjorken scaling" and the "quark-parton model" but also a possible probe of the asymmetry of the quark "sea", a possible measure of the gluon distribution inside the nucleon, and even a possible handle on non-perturbative QCD effects. Another set of sum rules — the Bjorken [5.28] and Ellis-Jaffe [5.29] sum rules — involving the deep inelastic scattering of polarized muons by polarized nucleons — have also led to a non-trivial discrepancy between theory and experiment, which is called the "proton spin crisis"; the present interest in the proton spin problem is so great that we discuss it as a special topic in §5.4a.

While the deep inelastic scattering of charged leptons by nucleons gave good support, on the whole, to "Bjorken scaling" and the "quark-parton model", it was highly desirable to perform similar experiments with neutrinos — where the quark current is J_μ^W (rather than J_μ^{em}) and the $W_3 (Q^2, \nu)$ must be included in addition to the W_1 and W_2 structure functions [see Eq. (5.55)] — in order to ascertain whether all deep inelastic lepton-hadron scattering process are governed by the same physics, as is expected. When we consider deep inelastic neutrino-nucleon scattering which yield charged leptons [i.e. $\nu(\bar{\nu}) + N \to \ell^- (\ell^+) + X$], Eq. (5.57) is transformed into:

$$\frac{d^2\sigma^{\nu,\bar{\nu}}}{d\Omega dE'} = \frac{G_F^2}{2\pi^2} E'^2 \left(\frac{M_W^2}{M_W^2 + Q^2}\right)^2 \left[2\sin^2\frac{\theta}{2} \, W_1^{\nu,\bar{\nu}} + \cos^2\frac{\theta}{2} \, W_2^{\nu,\bar{\nu}}\right.$$

$$\left. \mp \frac{(E + E')}{M} \sin^2\frac{\theta}{2} \, W_3^{\nu,\bar{\nu}}\right] \tag{5.75}$$

where W_i^{ν} $(i = 1, 2, 3)$ and $W_i^{\bar{\nu}}$ $(i = 1, 2, 3)$ are the three structure functions for $\nu - N$ and $\bar{\nu} - N$ scattering respectively; the W_3 structure function is included now because ν (and $\bar{\nu}$) possesses a definite helicity and the change of sign in the last term on the R.H.S. of Eq. (5.75) takes account of the change from ν to $\bar{\nu}$. With the acceptance of "Bjorken scaling", the three terms: $2MW_1^{\nu,\bar{\nu}}, \nu W_2^{\nu,\bar{\nu}}$ and $\nu W_3^{\nu,\bar{\nu}}$ are all functions of the dimensionless quantity $x = Q^2/2M\nu$ (in the deep inelastic region) and can be written:

$$2MW_1^{\nu,\bar{\nu}}(Q^2, \nu) = \bar{f}(x); \quad \nu W_2^{\nu,\bar{\nu}}(Q^2, \nu) = x\bar{f}(x);$$

$$\nu W_3^{\nu,\bar{\nu}}(Q^2, \nu) = g(x) \tag{5.76}$$

where $\bar{f}(x)$ is the analog of the Bjorken scaling function $f(x)$ for the charged lepton case [see Eq. (5.68)] and $g(x)$ is the new parity-violating Bjorken scaling function for deep inelastic neutrino-nucleon scattering.

Before deriving high energy sum rules from Eq. (5.76) for deep inelastic neutrino scattering and comparing the results with those arising from deep inelastic charged lepton scattering, it is convenient to rewrite Eq. (5.75) [using Eq. (5.76)] in terms of x and y variables [where the new variable $y = \nu/E$ is the fraction of energy transferred by the neutrino to the nucleon]; we have:

$$\frac{d^2\sigma^{\nu,\bar{\nu}}}{dx\,dy} = \frac{G_F^2 s}{\pi} x\bar{f}(x)\left[\frac{1+(1-y)^2}{2} \mp \frac{(1-y)^2}{2}\frac{g(x)}{2\bar{f}(x)}\right] \tag{5.77}$$

It is possible to show — from angular momentum conservation — that the y dependence of $\nu - q$ scattering is isotropic [since ν is lefthanded — the V-A interaction requires q to be lefthanded] and the y dependence of $\bar{\nu} - q$ scattering is proportional to $(1-y)^2$ [to suppress the backward scattering in the center of mass system since $\bar{\nu}$ is righthanded and q is lefthanded]. If follows that the y dependence of $\bar{\nu} - \bar{q}$ scattering is isotropic (like $\nu - q$ scattering) and the $\nu - \bar{q}$ scattering is proportional to $(1-y)^2$ (like $\nu - \bar{q}$ scattering). [These observations are useful when one considers the key role played by the experiment on longitudinally polarized electron-deuteron scattering [5.30] in confirming the existence of neutral weak currents in the standard gauge theory of the electroweak interaction.]

Knowledge of the y dependence of $\nu - q$ scattering and $\nu - \bar{q}$ scattering — as given above — is also very useful to obtain information about the effect of the quark "sea" (inside a nucleon) in modifying the behavior of the structure functions for deep inelastic neutrino-hadron scattering for small x. In particular, an isoscalar target (e.g. the deuteron) permits a relatively clean measurement of the ratio $\bar{q}(x)/q(x)$, where $\bar{q}(x)$ and $q(x)$ are respectively the probability distributions for $\bar{q}$ and q inside the isoscalar target at a given x. To see how this is accomplished [5.20], we write down the differential cross sections for $\nu - q$ and $\nu - \bar{q}$ scatterings; according to the instructions in the preceding paragraph, we get:

$$\frac{d^2\sigma^\nu}{dx\,dy} \sim [q(x) + (1-y)^2\bar{q}(x)]; \quad \frac{d^2\sigma^{\bar{\nu}}}{dx\,dy} \sim [\bar{q}(x) + (1-y)^2 q(x)] \tag{5.78}$$

If one compares Eq. (5.78) with Eq. (5.77), one obtains:

$$B \equiv \frac{g(x)}{\bar{f}(x)} = \frac{q(x) - \bar{q}(x)}{q(x) + \bar{q}(x)} \tag{5.79}$$

Equation (5.79) exhibits clearly the importance of the parity-violating structure function W_3 for this type of experiment and the opportunity that it offers to directly compare the parity-violating $W_3^{\nu,\bar{\nu}}$ structure functions with the parity-conserving $W_2^{\nu,\bar{\nu}}$ structure functions. The actual experiment ($Q^2 > 1$ GeV^2, $E^2 < 46$ GeV^2) gives the result [5.31]:

$$\bar{q}(x)/q(x) \longrightarrow 0 \quad \text{for } x \gtrsim 0.4; \qquad \bar{q}(x)/q(x) \longrightarrow 1 \quad \text{for } x \to 0 \qquad (5.80)$$

It is seen from (5.80) that $W_2^{\nu,\bar{\nu}}(x)$ and $W_3^{\nu,\bar{\nu}}(x)$ are very close to each other for $x > 0.4$ but that the quark "sea" begins to play an increasingly significant role as $x \to 0$. These results confirm our earlier remark in connection with the deep inelastic electron-nucleon experiments.

We have discussed some high energy sum rules that have been worked out in connection with the deep inelastic scattering of charged leptons by nucleons; some quite interesting high energy sum rules have also been obtained for the deep inelastic scattering of neutrinos by nucleons where the same experiments can be carried out with neutrinos and antineutrinos. The neutrinos are necessarily ν_μ neutrinos (because of their origin in pion decays) and interact with unpolarized protons in the reaction ν (or $\bar{\nu}$)$+p \to \mu^-(\mu^+)+X$. We comment on two sum rules — the Adler sum rule and the Gross-Llewellyn-Smith sum rules — for the deep inelastic neutrino reactions. The Adler sum rule was derived in §3.5d as an example of the ability of current algebra to yield high energy sum rules and here we merely demonstrate how readily Adler's result can be duplicated in the Bjorken scaling region. Thus, if we set the Cabibbo angle equal to zero (which is easy to rectify), and we are below the charm production threshold, the charged weak current couples to the weak isospin of the partons so that:

$$\nu(\frac{d}{u}) \to \mu^-(\frac{u}{d}) \qquad (5.81)$$

Using Eqs. (5.71)–(5.73) we get:

$$\frac{1}{x}F_2^{\nu p}(x) = 2[d(x) + \bar{u}(x)]; \quad \frac{1}{x}F_2^{\nu n}(x) = 2[u(x) + \bar{d}(x)] \qquad (5.82)$$

But, by isospin rotation, $F_2^{\nu n} = F_2^{\bar{\nu} p}$ and hence, using Eq. (5.73) (and vanishing Cabibbo angle), the Adler sum rule becomes:

$$\int_0^1 \frac{dx}{x}[F_2^{\bar{\nu} p}(x) - F_2^{\nu p}(x)] = 2 \qquad (5.83)$$

Equation (5.83) agrees with Eq. (3.107) which was derived by Adler with judicious use of current algebra and the infinite-momentum frame for the nucleon [3.31]. The important Adler sum rule for deep inelastic neutrino-hadron scattering must still be tested.

The Gross-Llewellyn-Smith sum rule [5.32] — which applies to $F_3^{\nu p}(x)$ and $F_3^{\bar{\nu}p}(x)$ — is related to the Adler sum rule if it is recognized that:

$$xF_3^{\nu p}(x) = +F_2^{\nu p}(x); \quad xF_3^{\bar{\nu}p}(x) = -F_2^{\bar{\nu}p}(x) \tag{5.84}$$

and hence, from:

$$F_3^{\nu p}(x) = 2[d(x) - \bar{u}(x)] + \ldots; \quad F_3^{\bar{\nu}p}(x) = 2[u(x) - \bar{d}(x)] + \ldots \tag{5.85}$$

it follows that:

$$\int_0^1 dx[F_3^{\nu p}(x) + F_3^{\bar{\nu}p}(x)] = 2\int_0^1 dx[u(x) - \bar{u}(x) + d(x) - \bar{d}(x)] = 6 \tag{5.86}$$

The QCD corrections to this interesting sum rule for the parity-violating structure function in deep inelastic neutrino-hadron scattering have been made and reduce the R.H.S. of Eq. (5.86) to 5.48. Agreement with experiment is good; for example, an experiment done at $Q^2 = 3$ GeV^2 yields 5.66 ± 0.40 [5.33].

In addition to the rather strong evidence for the validity of "Bjorken scaling" and the "quark-parton" hypothesis from a multiplicity of deep inelastic lepton-nucleon scattering experiments for large space-like momenta, another class of experiments — measurement of the so-called R value in $e\bar{e}$ annihilation processes in the "deep inelastic" region (corresponding to large time-like momenta) — have lent further support to "Bjorken scaling". The quantity R is defined by:

$$R = \sigma(e\bar{e} \to \text{hadrons})/\sigma(e\bar{e} \to \mu\bar{\mu}) \tag{5.87}$$

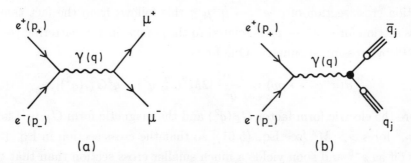

Fig. 5.6. Feynman diagrams for electron pair annihation into muon pairs and quark pairs.

The diagrams for the two types of experiment (the first type involving $\mu\bar{\mu}$ in the final state and the second type $q\bar{q}$ in the final state) are given in Fig. 5.6.

The difference between the two diagrams in Fig. 5.6 and the diagram describing electron-proton scattering (see Fig. 5.2) is that the four-momentum transfer squared, namely Q^2, is "time-like" for annihilation, whereas Q^2 is "space-like" for scattering. Since "Bjorken scaling" should apply equally well — by virtue of crossing symmetry — to time-like and space-like Q^2, the R value class of experiments provides an equally stringent test of "Bjorken scaling".

We give a brief derivation of the R value sum rule and then compare the theoretical prediction with experiment. First, we write down the total cross section for the electron pair annihilation into muon pairs, i.e. $e^+e^- \to \mu^+\mu^-$; the lowest-order result is well-known from QED and we have:

$$\sigma(e^+e^- \to \mu^+\mu^-) = \frac{4\pi\alpha_e^2}{3q^4}(1 - \frac{4m_\mu^2}{q^2})^{\frac{1}{2}}(2m_\mu^2 + q^2) \qquad (5.88)$$

where $-Q^2 = q^2 = (p_+ + p_-)^2 = s \geq 0$, with p_+ and p_- the four-momenta of the positron and electron respectively. Hence, for $s >> m_\mu^2$, one obtains:

$$\sigma(e^+e^- \to \mu^+\mu^-) \simeq \frac{4\pi\alpha_e^2}{3s} \qquad (5.89)$$

The $1/s$ dependence of the asymptotic annihilation cross section, given by Eq. (5.89), is the usual one for the annihilation of charged lepton pairs into charged lepton pairs. Obviously, $s\sigma(e^+e^- \to \mu^+\mu^-)$ is dimensionless in the region where $m_\mu^2 << s$ since then there is no longer a mass scale.

Before considering the cross section for the asymptotic (i.e. deep "time-like" inelastic) cross section for the annihilation of electron pairs into quark pairs, we note that the total cross section for the annihilation of electron-positron pairs into proton-anti-proton pairs has a much more rapid decrease with increasing s than that given by Eq. (5.89) for the asymptotic annihilation cross section of $e^+e^- \to \mu^+\mu^-$; this follows from the fact that the cross section for $e^+e^- \to p\bar{p}$ is related to that of ep elastic scattering [see Eq. (5.60)] by crossing symmetry. One finds:

$$\sigma(e^+e^- \to p\bar{p}) = \frac{4\pi\alpha_e^2}{q^2}[2M^2 G_E^2(q^2) + q^2 G_M^2(q^2)] \qquad (5.90)$$

where the electric form factor $G_E(q^2)$ and the magnetic form $G_M(q^2)$ fall off as s^{-2} for $s >> M^2$ [see Eq. (5.61)] so that the cross section in Eq. (5.90) falls off as s^{-5} and soon yields a much smaller cross section than that given by Eq. (5.89). This result is no more surprising than the behavior of the cross section for elastic $e-p$ scattering, as given by Eq. (5.60), for the same reason: the proton's "electromagnetic size" (i.e. form factor) establishes the scale.

The situation is again very different when the protons are replaced by quark-composites and the "quark-parton model" is used to derive the asymptotic cross section $\sigma(e^+e^- \rightarrow$ hadrons$)$ — regarded as an incoherent sum of electron-positron annihilations into quark-antiquark pairs and assuming that subsequently the quark pairs are completely transformed into hadrons. In this case — as was the case for deep inelastic $e-p$ scattering — we can write for the cross section in the asymptotic (deep time-like) region:

$$\sigma(e^+e^- \rightarrow \text{hadrons}) = \sum_i \sigma(e^+e^- \rightarrow q_i\bar{q}_i) \qquad (5.91)$$

where i is summed over all the "valence" quarks comprising the final hadrons. The R value, given by Eq. (5.91), refers to the asymptotic region (i.e. – $Q^2 >>$ quark masses) and becomes — in analogy to Eq. (5.70) for the deep inelastic (large space-like) lepton-hadron scattering:

$$R \equiv \frac{\sigma(e^+e^- \rightarrow \text{hadrons})}{\sigma(e^+e^- \rightarrow \mu + \mu^-)} = 3\sum_i e_i^2 + R_\tau(=1) \qquad (5.92)$$

where the R.H.S. of Eq. (5.92) now takes the color weight of 3 into account as well as the addition of 1 when the energy in the electron pair is sufficient to produce a τ lepton pair. It should be noted that the "sum rule" (5.92) is based on the "valence" quark charges in the final hadrons and takes no account of QCD corrections nor "asymptotic freedom" corrections (see below).

The simple result (5.92) for the R value has been reasonably well confirmed by experiment as shown in Fig. 5.7 [5.34] when allowance is made for the different mass thresholds for the quarks and the τ lepton. Thus, for energies below the charmed quark (c) and τ lepton thresholds, the theoretical value of R from Eq. (5.92) should be 2 (since only u, d and s are involved), while above the c and b quark (and τ lepton) thresholds, R should attain the value of $\frac{14}{3}$. As can be seen, the agreement between theory and experiment, especially above the c, b and τ thresholds, is quite good considering the assumptions that have gone into the derivation of the "time-like" sum rule. It is fair to say that the deep inelastic "space-like" and "time-like" lepton-hadron experiments give strong support to the idea that the quarks within hadrons behave as "free" point-like particles at high energies (i.e. short distances) and that some form of "asymptotic freedom" is an indispensable ingredient of the strong interaction, thereby pointing the way to non-Abelian QCD. Indeed, experimental confirmation of the logarithmic deviations from "Bjorken scaling" flowing from the asymptotic freedom property of QCD — see below — only strengthens this conclusion.

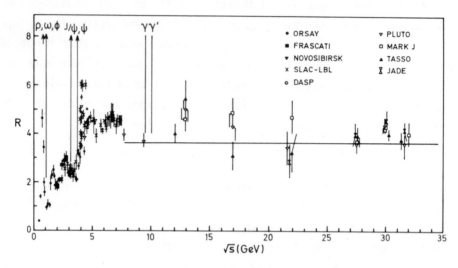

Fig. 5.7. Compilation of R-values from different e^+e^- experiments; only statistical errors are shown. The quark-parton model prediction is indicated as a solid line.

We conclude this section by citing the evidence for logarithmic departures from "Bjorken scaling" in deep inelastic lepton-hadron scattering. It was shown in §5.2b that the QCD renormalization group treatment of asymptotic freedom — with the help of the operator product expansion — leads to logarithmic corrections to the Wilson c_i coefficients [see Eq. (5.44a) for the definition of the c_i coefficients in terms of the c and b coefficients of the γ and β functions associated with the QCD renormalization group] and, hence to logarithmic corrections to the "Bjorken" structure functions. The logarithmic corrections of Eq. (5.44a) translate into logarithmic deviations from "Bjorken scaling" of the form $\ln(Q^2/Q_0^2)^k$ [$k = -c_{\mathcal{O}_i}/2b$ — see Eq. (5.44a)]. We do not work out the Wilson c_i coefficients — which, to various orders in QCD, can be accomplished by the Altarelli-Parisi method [5.19] — but state qualitatively the pattern of scaling violation. As one example, for large Q^2, $F_2(x)$ [see Eq. (5.68)] is no longer a function of x (strict "Bjorken scaling") but a function $F_2(x, Q^2)$ of Q^2 as well; it can be shown [3.26] that for $Q^2 > Q_1^2$, $F_2(x, Q_2^2) < F_2(x, Q_1^2)$ when $x \to 1$ and $F_2(x, Q_2^2) > F_2(x, Q_1^2)$ when $x \to 0$ while $\int_0^1 dx F_2(x, Q^2) = \text{const.}$ Other scaling violations in deep inelastic neutrino-proton experiments and R value measurements have definitely been seen [5.20] and are of the predicted type. On the whole, it may be said that tests of Bjorken scaling violations produce no surprises and are supportive of the asymptotic freedom property of QCD and the calculated perturbative corrections.

§5.3. Color confinement in QCD

After almost two decades of theoretical and experimental study of the property of asymptotic freedom in non-Abelian QCD, it is fair to say that asymptotic freedom is on a firm foundation and lends strong support to our belief in quarks as the basic constituents of hadronic matter (in the multi-hundred GeV region) and in the gluon-generated strong interaction among quarks as the source of the strong interaction among hadrons. However, asymptotic freedom only characterizes the short-distance (relatively weak) behavior ($\lesssim 0.1$ fermi) of the strong quark-quark interaction and is only part of the story. A full-fledged theory of the strong interaction must explain the truly strong interaction among quarks on the nucleon scale ($\gtrsim 0.1$ fermi) and be capable of describing the low energy properties of hadronic matter [meson and baryon spectrum, meson and baryon properties, etc.]. Above all, for QCD to qualify as the phenomenologically correct theory of the strong interaction, it must explain the absence of "free" quarks and "free" gluons under normal conditions. [The question of the creation of a "quark-gluon plasma" in relativistic heavy ion collisions (see §1.4a) is still a matter of speculation; it is not excluded that at such "high temperatures", deconfinement takes place.]

Detailed predictions concerning the low energy properties of hadronic matter are admittedly difficult to carry out precisely because the strong interaction among the quarks is "strong" on the nucleon scale and hence can not be treated by perturbative QCD — as in the "asymptotic freedom" region — but requires "non-perturbative" methods or models. A variety of model calculations [e.g. based on the chiral quark Lagrangian (see §5.1), the Skyrme model (see §10.5), etc.] have been performed with varying degrees of success. However, model calculations are admittedly of limited applicability and it is necessary to tackle the basic question of "quark confinement" (and "gluon confinement") to establish whether QCD can truly explain the absence of "free" quarks (and "free" gluons). In §5.3a, we comment briefly on the present theoretical status of quark confinement in QCD. Fortunately, there are quite a number of experiments on the spectroscopy of heavy quarkonia that depend for their explanation on the long-distance behavior of the $q - \bar{q}$ interaction and shed considerable positive light on the quark confinement question. The experimental evidence from heavy quarkonia spectroscopy in favor of quark confinement is summarized in §5.3b. The evidence for gluon confinement is on a less sure footing than quark confinement and is discussed in §5.4c.

§5.3a. Theoretical status of color confinement in QCD

Global $SU(3)$ color tells us that the observed hadrons (mesons and baryons) must belong to color singlet states of quark composites (a minimum of two for mesons and three for baryons) but can not, by itself, decide whether there are "free" quarks (and "free" gluons). We do expect that the gauged $SU(3)$ color theory of the strong interaction (QCD) can provide a dynamical explanation of why "free" color singlet quarks and gluons are not observed under "normal" conditions, although it is possible that some manifestation of quark deconfinement may be observed in relativistic heavy ion collisions. Assuming "normal" conditions, we review two theoretical approaches to the problem of color confinement in QCD that we choose to call: (1) 't Hooft's model of "color-electric" confinement; and (2) the lattice gauge theory (LGT) approach to color confinement. 't Hooft's model is intuitive and highly suggestive but the mathematical underpinning of his model is still a task for the future. LGT is more hopeful since, in essence, it attempts to establish — with the help of "supercomputers" — that color confinement is a rigorous consequence of the QCD Lagrangian and to draw out predictions that can be tested experimentally. We will see that LGT has had some positive successes but has not yet established that color confinement is a *sine qua non* of non-Abelian QCD. Let us turn first to 't Hooft's model of "color-electric" confinement in QCD.

(1) *'t Hooft's model of "color-electric" confinement in QCD.* 't Hooft [5.35] combines Dirac's quantization condition for magnetic monopoles (in Abelian QED) with the quantized magnetic flux tubes that are present in Type II superconductivity to develop a "color-electric flux tube" analog for non-Abelian QCD within the framework of the "dual"-transformed (Maxwell-type) Yang-Mills equations. 't Hooft points out that the total magnetic flux associated with the magnetic flux tube created in Type II superconductivity [see §10.2b], must be an integral multiple of the magnetic flux $\Phi_0 = \pi/2e$ (see §4.2a) and that it carries a finite amount of energy per unit length. The "Meissner" magnetic flux tubes which exist when the external magnetic field exceeds the "lower"' critical value of the magnetic field (H_{c_1}) in a Type II superconductor, can not break apart because the "winding number" n (in units of Φ_0) associated with each flux tube must be conserved [since the "winding number" n obeys a topological conservation law — see §10.1]. 't Hooft next imagines that a Dirac magnetic monopole (with magnetic flux $2n\Phi_0$) (see §4.6) and an anti-magnetic monopole (with magnetic

flux $-2n\Phi_0$) are immersed in a Type II superconductor. He shows that the monopole and the anti-monopole form endpoints of a "Meissner" magnetic flux tube and that the interaction between the two can be expressed in terms of the "Dirac string" (see §4.6). Since each flux tube carries a finite amount of energy per unit length, it follows that there would be a linearly rising potential pulling the monopole-anti-monopole pair together regardless of their separation. Such a behavior can be characterized as "magnetic confinement" within the Type II superconductor.

Having performed his "gedanken experiment" with Dirac's magnetic monopole inside a superconductor, 't Hooft suggests that color confinement might be understood by applying the dual transformation to the Yang-Mills equations for the "color-electromagnetic" field, namely that the "color-magnetic monopole" (which can arise in Yang-Mills theory under certain conditions) is the source of the dual "color-electromagnetic" field $\tilde{F}^{a\mu\nu}$ (just as the charge is the source of the normal electromagnetic field). Pursuing the dual transformation approach, 't Hooft [5.36] argues that the roles of magnetic and electric charges in the "color" Yang-Mills theory are reversed, that the "color-magnetic" flux tubes are converted into "color-electric" flux tubes with quantized values of "color electric" flux, ultimately producing "color-electric" confinement. This is the analogical reasoning that 't Hooft applies to non-Abelian QCD and he can take comfort in Nielsen's demonstration [5.14] that the QCD vacuum possesses the properties of a "color-dielectric". If one accepts 't Hooft's quasi-intuitive argument, it follows that electric flux lines between a positive and negative charge in QED (e.g. positronium) — as exhibited in Fig. 5.8a — are replaced by "color-electric" flux lines between $\bar{q}$

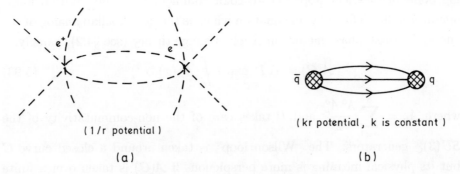

(1/r potential)   (kr potential, k is constant)

(a)   (b)

Fig. 5.8. Electric flux lines between (a): e^+ and e^- in QED (positronium) and (b): "color-electric" flux lines between $\bar{q}$ and q in QCD (heavy quarkonium) — "Coulombic" term is omitted.

and q in QCD (e.g. quarkonium) — as shown in Fig. 5.8b. From Fig. 5.8b, it is then plausible to draw the conclusion that quark and antiquark constitute the ends of a "color-Dirac string" (squeezed by the gluon "color-charges") [5.37], giving rise to the linear long-distance behavior of the $q - \bar{q}$ potential required by the heavy quarkonia experiments (see §5.3b).

't Hooft's qualitative deduction of a large-distance linear behavior for the $q - \bar{q}$ potential among quarks, thereby explaining quark confinement, is quite interesting. However, his attempt to bolster his qualitative arguments with the assertion that "color-magnetic" monopole solutions appear naturally in non-Abelian QCD and could induce Meissner-type "color-electric" flux tubes — the obverse of the magnetic flux tubes in low temperature ("electric-induced") superconductivity — is made on shaky grounds. His citation of the Georgi-Glashow model [5.38] — in which the SSB of the non-Abelian $O(3)$ group down to $U(1)_{EM}$ results in the generation of 't Hooft-Polyakov magnetic monopoles (see §10.2c) — is unconvincing because these only appear when the initial non-Abelian group is broken down spontaneously to a subgroup containing $U(1)_{EM}$ as a factor group; no such SSB mechanism is identified in QCD. 't Hooft himself recognized the incompleteness of his "color-electric" confinement model in QCD and, while he tried to define a suitable operator for "color electric" flux in QCD [5.35] — in analogy with the Wilson loop operator [5.39] serving as the QCD operator for "color-magnetic" flux — he did not succeed in finding a rigorous analytic (or topological) solution. It appears necessary to approach the color confinement problem in a more circuitous and laborious fashion — within the framework of lattice gauge theory (LGT).

(2) *Lattice gauge theory approach to color confinement in QCD.* The starting point of the LGT approach to color confinement is the "Wilson loop" operator [5.39] defined by an equation which is the non-Abelian analog of the "non-integrable phase factor" in Abelian gauge theory (see §4.2), namely:

$$\mathcal{A}(C) = \text{Tr}\, P \, \exp(i \oint_C A_\mu(x)dx^\mu) \qquad (5.93)$$

where $A_\mu = \sum_{a=1}^{8} \dfrac{\lambda^a A_\mu^a}{2}$ and P takes care of the non-commutativity of the $SU(3)_C$ generators. The "Wilson loop" is taken around a closed curve C but its physical meaning is more perspicuous if $\mathcal{A}(C)$ is taken over a finite curve $x_1 \to x_2$ [as in the original Dirac derivation of the "non-integrable phase" of a quantum mechanical wavefunction (see §4.2a)]; the operator that creates a color-charge (with representation **3**) at x_1 and a color-anti-charge (representation $\bar{3}$) at x_2 from the color-charge zero state $|\Psi_0 >$ is denoted by $M(x_1, x_2; [\mathcal{A}])$, and can be written as:

$$M(x_1; x_2[A]) = P \exp(i \int_{x_1}^{x_2} A_\mu(x)dx^\mu) \qquad (5.94)$$

The operator $M(x_1, x_2; [\mathcal{A}])$ is, in general, path-dependent and hence the energy eigenvalues corresponding to the QCD Hamiltonian must also depend on the path of integration and, in particular, on the distance R between x_1 and x_2 since:

$$HM|\Psi_0 >= E(R, \cdots)M|\Psi_0 > \qquad (5.95)$$

where $|\Psi_0 >$ is the color vacuum state and E is the energy, which is a function of R. The condition for color confinement can then be expressed simply in terms of $E(R, ...)$, thus:

$$\lim_{R \to \infty} E(R, ...) = \infty \qquad (5.96)$$

The problem now is to establish whether or not condition (5.96) is fulfilled in QCD. To decide this question at "normal" conditions, one examines the expectation value — taken over the color-vacuum state — of the evolution operator [5.40]:

$$< \Psi_0|M^\dagger e^{-HT}M|\Psi_0 >= e^{-ET} \qquad (5.97)$$

The L.H.S. of Eq. (5.98) can be expressed in terms of the functional average of the "Wilson loop" integral, as follows:

$$< \mathcal{A}(C) >= \frac{\int \cdots \int \mathcal{D}A_\mu \exp\{- \int \mathcal{L}(A)d^4x\}\mathcal{A}(C)}{\int \cdots \int \mathcal{D}A_\mu \exp\{- \int \mathcal{L}(A)d^4x\}} \qquad (5.98)$$

Equation (5.98) is a rigorous equation and, in principle, color confinement in QCD is established if $\ln < \mathcal{A}(C) >$ is asymptotically proportional to the area of the "Wilson loop" C. Actually, Wilson has proved the proportionality to the area of the "Wilson loop" in the strong coupling limit but the proof in non-Abelian QCD must still be given and, furthermore, the full proof of confinement requires the demonstration that a first-order "phase transition" is absent in the transition between the "strong coupling" and "weak coupling" (perturbation-theoretic) régimes. The second part of the proof of confinement in QCD is essential since it is known that in Abelian QED there is a non-trivial phase structure between the two régimes [1.133]. This is where the technology of LGT enters the picture because the perturbation-theoretic methods can clearly not be applied to Eq. (5.98) in the "intermediate coupling" régime [5.41] and increasingly powerful computers must be brought to bear on the numerical computations.

We do not give details of the actual LGT calculations but merely try to summarize the salient results. Having recourse to LGT offers two advantages: (1) the size of the lattice spacing a restricts the domain of momenta to a region bounded by π/a so that the kinetic energy is bounded and can be treated as a perturbation in the strong coupling limit [the virtue of putting a strong coupling theory on the lattice was already recognized by Wentzel [5.42] during the early 1940s in connection with strong coupling meson theory]; and (2) the possibility of treating the kinetic energy as a perturbation in LGT is equivalent to the method of the high-temperature expansion of the partition function in statistical mechanics. A variety of physical quantities (e.g. hadron spectrum, correlation functions, etc.) can then be studied with the path-integral formalism of Eq. (5.98) by exploiting the Monte Carlo method in computer calculations. The chief bugaboo in the application of LGT to QCD is the demonstration that the true "continuum" theory of QCD does not depend on the choice of lattice spacing a; this requirement is equivalent to the statement that there should be no first-order phase transition (in the language of statistical mechanics) in passing to the continuum.

The general strategy in LGT has been to first confirm the "area law" (and the absence of a perimeter contribution) and to then establish that there is no first-order "phase transition" in the QCD model. Thus far, calculations with an $SU(3)_C$ Yang-Mills field in a quark background field have confirmed the "area law" for the "Wilson loop" C, i.e. $W(C) = \ln <\mathcal{A}(C)> \propto$ area A of C [5.39] in the "strong coupling" limit. It is true that the proof of the "area law" for the "Wilson loop" in the "strong coupling" limit does not require having recourse to a supercomputer but, nevertheless, we give the simple derivation of the "area law" in terms of the "plaquette" language in order to communicate a sense of how the physics of QCD is translated into computer jargon. A "plaquette" is defined in terms of "Euclidean" time and space and becomes the unit area whose perimeter is the closed curve C for the "Wilson loop". The "Wilson loop" defines the "gauge links" between the vertices of the plaquette that are separated by the lattice size a and the area of the "Wilson loop" over a rectangular plaquette is $A(C) = TR$.

It can then be shown that each plaquette contributes $1/g^2$ to the "Wilson loop" correlation function $\mathcal{A}(C)$ in the "strong coupling" limit, namely [5.39]:

$$\mathcal{A}(C) \sim (\frac{1}{g^2})^{N_p} \tag{5.99}$$

where N_p is the minimum number of "plaquettes" required to cover the area of the loop $A(C)$. The area itself is given by $A(C) = a^2 N_p$; using Eq. (5.99), we get:

$$\mathcal{A}(C) \sim (g^2)^{-A(C)/a^2} = \exp\{-(\text{TR} \ln g^2)/a^2\} \qquad (5.100)$$

Equation (5.100) constitutes the "area law" for the "Wilson loop" since:

$$\lim_{T \to \infty} \mathcal{A}(C) \sim e^{-T[E(R)-2m]} \qquad (5.101)$$

From Eqs. (5.100) and (5.101), it follows that:

$$E(R) = kR; \quad k = \frac{\ln g^2}{a^2} \qquad (5.102)$$

The linear dependence of E on R, as given by Eq. (5.102), implies that the criterion for color confinement in QCD, as given by Eq. (5.97), is satisfied. This is good news for the color confinement property of QCD but, as we have already pointed out, the "area law" was anticipated in the "strong coupling" limit. The more critical test, the second condition for confinement, is the absence of a first-order phase transition (see the beginning of this section) [5.43]; unfortunately, the proof of this condition in QCD is not fully established as yet. The final confirmation of color confinement in QCD — using the LGT method — will come when the quarks do not merely constitute an external (background) source but are directly incorporated into the "Wilson loop" (i.e. the "quenched approximation" is no longer used [5.43]) and color confinement in QCD is confirmed with a realistic choice of quark masses. It must be said that a negative result on the color confinement problem in QCD will not achieve ready acceptance because of the strong experimental support for the long-distance linear potential derived from heavy quarkonium spectroscopy. This experimental evidence — to be discussed in the next section (§5.4c) — is very important, precisely because the theoretical predictions are not on a sure footing.

§5.3b. Experimental evidence for quark confinement in QCD

Indirect experimental evidence for quark confinement comes chiefly from studies of the spectroscopy of heavy quarkonia [i.e. $J/\psi = (c\bar{c})$ and $\Upsilon = (b\bar{b})$] which is made possible by the OZI rule (see §5.4b). The reason for the restriction to heavy quarkonium systems is that we wish to apply the non-relativistic potential model to the bound $q - \bar{q}$ system — with which most detailed calculations of heavy quarkonium spectroscopy have been carried out; clearly, the potential model is more trustworthy when the quarks are heavy (i.e. when their masses are large compared to Λ_{QCD}). From QCD theory, we expect the small-distance term of the non-relativistic potential $V(r)$ (see Table 1.4) to be "Coulombic" (reflecting asymptotic freedom — see §1.3d) and the large-distance term to be linear in r [reflecting the confinement property — see Eq. (5.102)]. The potential at intermediate distances is uncertain but it turns out that it is a good approximation to simply take the sum of these two contributions and to write:

$$V(r) = -\frac{4}{3}\alpha_s(Q^2)/r + kr \qquad (5.103)$$

where the factor 4/3 takes account of the color-singlet character of the $(q-\bar{q})$ wavefunction and $\alpha_s(Q^2)$ is the "running" coupling constant [and is expected to decrease somewhat when Q^2 "runs" from the J/ψ (charmonium) mass to the Υ (bottomium) mass]; k is the "tension" of the "color-electric" flux tube (see Fig. 5.8b) between q and $\bar{q}$ (fixed by experiment).

The observed energy spectra of the J/ψ and Υ systems are by now very well-determined and the ordering of the energy levels is found to be [5.44]:

$$E(1S) < E(1P) < E(2S) < E(1D)\ldots \qquad (5.104)$$

where $n = 1, 2, \ldots$ is the "radial excitation" quantum number. Figure 5.9 exhibits the different ordering of the energy levels predicted by a pure Coulombic potential and that predicted by the "QCD" form (5.103). Evidently, the confining part of the potential (kr) is needed to ensure the correct ordering of energy levels. In addition to precise energy measurements of the states, many other properties of the heavy quarkonium systems have been studied, such as radiative widths, widths for lepton pair decay, etc. [5.44] and, overall, the "QCD" potential model does surprisingly well. There is a sufficient number of such measurements that not only are the c and b quark masses and the values of α_s and k determined, but also the good agreement between theory and experiment gives powerful support to the confinement property of QCD. Table 5.2 lists a typical set of numbers for the heavy quarkonium parameters and exhibits the comparison between theory and experiment for five of the lowest energy levels of $J/\psi = (c\bar{c})$ and $\Upsilon(b\bar{b})$ [5.45]; the heavy quarkonium parameters are taken as: $m_c = 1.29$ GeV; $m_b = 4.58$ GeV; $k = 0.257$ GeV^2; $\alpha_s(m_c^2) = 0.110$ and $\alpha_s(m_b^2) = 0.105$. We see that both the ordering of the states in heavy quarkonium, as well as the numerical agreement between theory and experiment for the heavy quarkonium energy levels, support the use of the linear $q - \bar{q}$ "confinement" potential at large distances and, *a fortiori*, offer good (albeit indirect) support for color confinement in QCD.

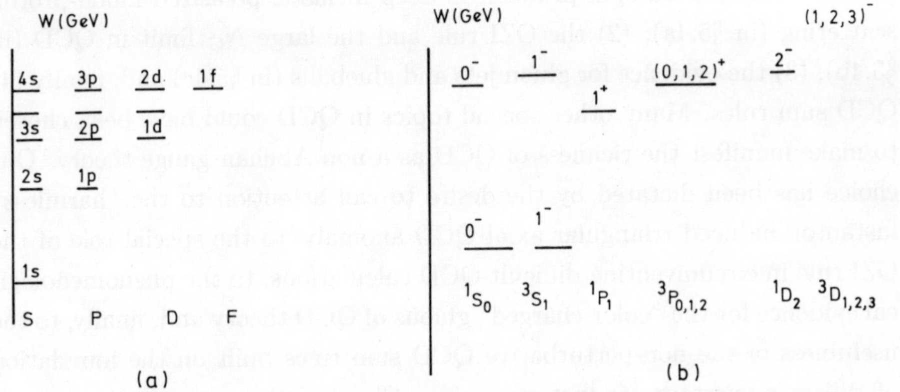

```
W(GeV)                                    W(GeV)                                         (1,2,3)⁻
                                                        1⁻
4s    3p    2d    1f                      0⁻                          (0,1,2)⁺          2⁻
                                                                 1⁺
3s    2p    1d

2s    1p

                                          0⁻         1⁻
1s

 S      P      D      F                    ¹S₀   ³S₁    ¹P₁    ³P₀,₁,₂    ¹D₂   ³D₁,₂,₃

       (a)                                                 (b)
```

Fig. 5.9. Energy level scheme using (a): a pure "Coulombic" potential $V = \alpha_s/r$ and (b): $V = -4/3\alpha_s(Q^2)/r + kr$ (k a constant).

Table 5.2: Comparison between theory and experiment for J/ψ and Υ energy levels:

first five J/ψ and Υ energy levels:					
$J/\psi = (c\bar{c})$ states		Energy in GeV	$\Upsilon = (b\bar{b})$ states		Energy in GeV
		Theory — Experiment			Theory — Experiment
1	$^3S_1$	3.097 — 3.097	1	$^3S_1$	9.460 — 9.460
2	$^3S_1$	3.686 — 3.686	2	$^3S_1$	9.908 — 10.023
3	$^3S_1$	4.123 — 4.040	3	$^3S_1$	10.230 — 10.355
4	$^3S_1$	4.492 — 4.415	4	$^3S_1$	10.501 — 10.580
1	$^1P_1$	3.488 — 3.510	1	$^1P_1$	9.779 — 9.890
(first two states fitted with m_c and k)			(first state fitted with m_b, k same as for J/ψ)		

§5.4. Special topics in QCD

In the previous sections, we focused on the basic physical content of the $SU(3)$ color gauge theory of the strong interaction (QCD) with its novel properties of asymptotic freedom and confinement. We will discuss the connection of QCD with the chiral gauge anomalies in Chapter 7, its linkages with QFD and contributions to the unification groups (both grand and partial) in Chapter 8, its paradigmatic role in the construction of preon models of quarks and leptons (in an attempt to solve the fermion generation problem) in Chapter 9 and, finally, the topological properties of non-Abelian QCD (and QFD) in Chapter 10. In the remaining sections of this chapter, we treat a variety of special topics that undertake to shed light on: (1) the nature of the proton spin problem in deep inelastic polarized muon-proton scattering (in §5.4a); (2) the OZI rule and the large N_C limit in QCD (in §5.4b); (3) the evidence for gluon jets and glueballs (in §5.4c) and, finally (4) QCD sum rules. Many other special topics in QCD could have been chosen to make manifest the richness of QCD as a non-Abelian gauge theory. Our choice has been dictated by the desire to call attention to the "harmless" instanton-induced triangular axial QCD anomaly, to the special role of the OZI rule in circumventing difficult QCD calculations, to the phenomenological evidence for the "color-charged" gluons of QCD theory and, finally, to the usefulness of the non-perturbative QCD sum rules built on the foundation of Wilson's operator product expansion. Clearly, other special topics would equally well illuminate the unusual features of the standard gauge theory of the strong interaction (QCD). The four special topics are briefly discussed in the order enumerated above.

§5.4a. Proton spin problem in deep inelastic polarized muon-proton scattering

In §5.2c, we discussed the experimental evidence from deep inelastic lepton-hadron scattering for "Bjorken scaling" and asymptotic freedom in QCD. The experiments with charged leptons (e, μ) involve unpolarized leptons [the ν and $\bar{\nu}$ experiments necessarily involve lefthanded and righthanded (longitudinally polarized) leptons] and produce good evidence for "Bjorken scaling" and the "quark-parton model". Moreover, these experiments can explain the bulk of the Gottfried high energy sum rule and give evidence for the logarithmic "Bjorken scaling" violations predicted by the asymptotic freedom property of non-Abelian QCD. We also indicated in §5.2c that some interesting experiments are underway on the deep inelastic scattering of longitudinally polarized charged leptons (in particular, muons) by longitudinally polarized protons which are testing two relevant high energy sum rules: the Bjorken sum rule and the Ellis-Jaffe sum rule. We deferred discussion of this topic to the present section and we now give a brief report of the developing situation.

Once substantial polarization of the target protons was achieved [5.30], deep inelastic polarized electron-proton experiments were performed at SLAC [5.46] in the kinematic range $x = 0.1$ to $x = 0.8$ (we use the notation of §5.2c unless otherwise indicated), to be followed by the more ambitious experiment with higher energy polarized muons (the EMC experiment at CERN [5.47]) covering the broader kinematic range $x = 0.01$ to $x = 0.7$. For this reason, we deal with the EMC experiment in what follows. The "proton spin" problem stems from the fact that the measured EMC value of the g_1 structure function (the analog of the "unpolarized" structure $F_1(x)$ — see below) integrated from $x = 0.01$ to $x = 0.7$ [and extrapolated "down" from $x = 0.01$ to $x = 0$ and "up" from $x = 0.7$ to $x = 1$] does not agree with the theoretical prediction given by the Ellis-Jaffe sum rule [5.29]. Thus, the EMC data give:

$$\int_{0.01}^{0.7} g_1^p(x)dx = 0..120 \pm 0.013 \tag{5.105}$$

If a smooth extrapolation into the unmeasured regions $x < 0.01$ and $x > 0.7$ is assumed, one gets:

$$\int_{0.7}^{1} g_1^p dx \simeq 0.001; \qquad \int_{0}^{0.01} g_1^p dx \simeq 0.002 \tag{5.105a}$$

so that the "experimental" value of $\bar{g}_1^p = \int_0^1 g_1^p(x)dx$ is:

$$\bar{g}_1^p = \int_{0}^{1} g_1^p(x)dx = 0.123 \pm 0.013 \pm 0.019 \tag{5.106}$$

at $< Q^2 > 10.7$ GeV^2.

Using the same notation for the kinematical variables as was done for writing down the hadron structure functions in connection with the deep inelastic scattering of leptons by unpolarized nucleons, we can express the difference in the cross sections for deep inelastic scattering of muons polarized anti-parallel and parallel to the spin of the target nucleon as follows [see Eq. (5.57)]:

$$\left(\frac{d^2\sigma}{dQ^2 d\nu}\right)^{\uparrow\downarrow} - \left(\frac{d^2\sigma}{dQ^2 d\nu}\right)^{\uparrow\uparrow} = \frac{4\pi\alpha^2}{E^2 Q^2}[M(E + E'\cos\theta)G_1(Q^2, \nu) - Q^2 G_2(Q^2, \nu)] \tag{5.107}$$

where $G_1(Q^2, \nu)$ and $G_2(Q^2, \nu)$ are the spin-dependent structure functions of the nucleon. In the "Bjorken scaling" limit, $G_1(Q^2, \nu)$ and $G_2(Q^2, \nu)$ are presumed to be functions of x only, thus:

$$M^2 \nu G_1(Q^2, \nu) \to g_1(x), \quad M\nu^2 G_2(Q^2, \nu) \to g_2(x) \tag{5.108}$$

We must now relate $g_1(x)$ and $g_2(x)$ to the experimentally determined quantities, then derive the Ellis-Jaffe sum rule with the help of the "quark-parton model", and finally compare the experimental results with the Ellis-Jaffe sum rule. [As we will see, the Ellis-Jaffe sum rule for the g_1^p spin function makes more assumptions in its derivation than the Bjorken sum rule which requires knowledge of both the g_q^p and g_1^n spin functions]. Since longitudinally polarized muons are scattered from longitudinally polarized nucleons, the chief experimental quantity of interest is the asymmetry parameter A which can be written:

$$A = \frac{d\sigma^{\uparrow\downarrow} - d\sigma^{\uparrow\uparrow}}{d\sigma^{\uparrow\downarrow} + d\sigma^{\uparrow\uparrow}} \qquad (5.109)$$

where, to a good approximation, we have [5.47]:

$$A = DA_1 \qquad (5.110)$$

with:

$$A_1 = \frac{\sigma_{1/2} - \sigma_{3/2}}{\sigma_{1/2} + \sigma_{3/2}}; \quad D = \frac{y(2-y)}{y^2 + 2(1-y)(1+R)} \qquad (5.110a)$$

In Eqs. (5.110) and (5.110a), D is the "depolarization factor" of the virtual photon, $\sigma_{1/2}$ ($\sigma_{3/2}$) is the "virtual" photon absorption cross section [when the projection of the total angular momentum of the photon-nucleon system along the incident lepton direction is $\frac{1}{2}$ ($\frac{3}{2}$)], R is the ratio of the longitudinal to transverse photon absorption cross sections and y, as usual, is ν/E (see §5.2). With the help of Eqs. (5.107)–(5.110a), we can express the first spin-dependent nucleon structure function $g_1(x)$ in terms of measured quantities, thus:

$$g_1(x) = A_1 g_2(x)/2x(1+R), \qquad (5.111)$$

where $g_2(x)$ is the second spin-dependent nucleon structure function. Since all the quantities on the R.H.S. of Eq. (5.111) are known or measured, we can compute the so-called nucleon (N) spin factor $\bar{g}_1^N$, which is simply $\int_0^1 g_1^N(x))dx$, that enters into the Ellis-Jaffe sum rule. The "proton spin" problem has surfaced because the measured (EMC) value of $\bar{g}_1^p$ does not agree with the theoretical value predicted by the Ellis-Jaffe sum rule (within $3\frac{1}{2}$ standard deviations).

Before making this comparison explicit, we derive the theoretical predictions for both $\bar{g}_1^p$ and $\bar{g}_1^n$ — with the help of the "quark-parton model" — in order to clarify the nature of the "proton spin" problem. Quite a few years before the Ellis-Jaffe sum rule was derived, a high energy sum rule for the difference between the proton and neutron spin-dependent structure functions was derived by Bjorken [5.28] by means of current algebra [this sum rule also follows from QCD in the "Bjorken scaling" limit] which makes few addition assumptions and relates the difference between $\bar{g}_1^p$ and $\bar{g}_1^n$ to the nucleon decay coupling constants g_V and g_A; Bjorken obtains:

$$(\bar{g}_1^p - \bar{g}_1^n) = \int_0^1 dx[g_1^p(x) - g_1^n(x)] = \frac{1}{6}|\frac{g_A}{g_V}|[1 - \frac{\alpha_s(Q^2)}{\pi}] = 0.191 \pm 0.002 \qquad (5.112)$$

where $[1 - \alpha_s(Q^2)/\pi]$ is the QCD correction [the strong coupling constant α_s is taken as 0.27 at $Q^2 \simeq 10$ GeV^2]. Unfortunately, $g_1^n(x)$ has not as yet been measured so that comparison of the experimental result given by Eq. (5.106) with the Bjorken sum rule result of Eq. (5.112) is premature. However, if one makes some plausible additional assumptions [that $SU(3)$ flavor symmetry holds and that the strange quark sea is unpolarized], one can derive separate sum rules for $\bar{g}_1^p$ and $\bar{g}_1^n$, namely the Ellis-Jaffe sum rule [5.29]:

$$\bar{g}_1^{(p,n)} = \int_0^1 g_1^{(p,n)}(x)dx = \frac{1}{12}\left|\frac{G_A}{G_V}\right|\{(1, -1) + \frac{5}{3}\frac{(3\frac{F}{D} - 1)}{(\frac{F}{D} + 1)}\} \qquad (5.113)$$

where F and D are respectively the anti-symmetric and symmetric coefficients of $SU(3)$ symmetry. Using the current values of $F/D = 0.632 \pm 0.024$ [5.48] and $|G_A/G_V| = 1.254 \pm 0.006$ [1.99], one obtains:

$$\bar{g}_1^p = \int_0^1 dx g_1^p(x) = 0.189 \pm 0.005; \quad \bar{g}_1^n = \int_0^1 dx g_1^n(x) = -0.002 \pm 0.005$$
$$(5.114)$$

Comparison of the first of Eqs. (5.114) with the measured value (5.106) yields the afore-mentioned 3.5 standard deviation discrepancy which defines the "proton spin" problem. It is seen that the fraction of the proton spin carried by the quarks is unexpectedly small within the framework of the usual "valence" quark composite picture of the nucleon. A general consensus is developing that the "proton spin" problem arising from the Ellis-Jaffe sum rule is due to the assumption that the strange quark "sea" is unpolarized. To see why this assumption is probably incorrect, we express $\bar{g}_1^p$ in "quark-parton" language, thus:

$$\bar{g}_1^p = S_\mu < p, S|\frac{1}{18}(4\bar{u}\gamma_\mu\gamma_5 u + \bar{d}\gamma_\mu\gamma_5 d + \bar{s}\gamma_\mu\gamma_5 s)|p, S >, \qquad (5.115)$$

where the matrix elements for the u, d and s quark terms are taken between proton states with definite spin S. In principle, $\bar{g}_1^p$ can be evaluated from a knowledge of three independent combinations of $(\bar{q}\gamma_\mu\gamma_5 q)$ $(q = u, d, s)$ matrix elements; these three combinations are the "invector" term (G_{IV}), the "octet" term (G_8), and the flavor-singlet combination (G_{FS}) and represent the following flavor states under $[SU(3)_F, SU(2)_I]$: $|8, 3 >, |8.1 >$ and $|1, 1 >$ respectively. We have for the "isovector" combination:

$$G_{IV} = \frac{1}{\sqrt{2}}S_\mu < p, S|\bar{d}\gamma_\mu\gamma_5 d - \bar{u}\gamma_\mu\gamma_5 u|p, S > \qquad (5.116a)$$

The "octet" combination is:

$$G_8 = \frac{S_\mu}{\sqrt{6}} < p, S|\bar{u}\gamma_\mu\gamma_5 u + \bar{d}\gamma_\mu\gamma_5 d - 2\bar{s}\gamma_\mu\gamma_5 s|p, S > \qquad (5.116b)$$

and the flavor-singlet combination is:

$$G_{FS} = \frac{S_\mu}{\sqrt{3}} < p, S|\bar{u}\gamma_\mu\gamma_5 u + \bar{d}\gamma_\mu\gamma_5 d + \bar{s}\gamma_\mu\gamma_5 s|p, S > \qquad (5.116c)$$

In the above notation, the questionable assumption underlying the sum rules in Eq. (5.114) is the statement that:

$$< p, S|\bar{s}\gamma_\mu\gamma_5 s|p, S >= 0 \qquad (5.117)$$

We now show that the first assumption that led to the Ellis-Jaffe sum rule — that $SU(3)$ flavor symmetry holds (see above) — is incompatible with the second assumption given by Eq. (5.117). The point is that when $SU(3)$ flavor symmetry holds, there is a definite relation between g_{IV} and G_8, namely:

$$G_8 = G_{IV}(\frac{3F - D}{3F + D}) \qquad (5.118)$$

As a result, G_V and G_8 completely determine $\bar{g}_1^p$ as given in Eq. (5.118). However, there is a flaw in the argument since Eq. (5.118) can not be valid within the framework of QCD because the divergence of the flavor-singlet axial vector current of Eq. (5.116c) contains the triangular axial QCD anomaly while the "octet" axial vector current of Eq. (5.116b) does not. The triangular axial QCD anomaly associated with G_{FS} is given in terms of the gluon field tensor (see §7.2b) and a non-trivial contribution from this source, in turn, leads to a non-vanishing contribution from the "strange quark sea" — in contradiction with Eq. (5.117). There is still much to sort out in the theoretical calculation and in our understanding of the QCD anomaly contribution to $\bar{g}^p(1(x))$, as measured in the deep inelastic scattering of polarized muons by polarized protons. But there is a good chance that the so-called "proton spin" problem in QCD can be resolved by the QCD counterpart [5.48] of the "harmless" triangular axial QED anomaly that explains $\pi^0 \to 2\gamma$ decay; this possibility is explored in §7.2b of the chapter on anomalies in quantum field theory.

§5.4b. Large N_C limit in QCD and the OZI rule

In §3.2d, we presented some fairly strong arguments for choosing global $SU(3)$ color to represent the new "color" degree of freedom to maintain the Pauli principle for the composite three-quark baryonic hadrons. We have also pointed out that the $SU(N)$ group has a Z_N centrum and that the hypercharges (and, *a fortiori*, the electric charges) of the constituent quarks take on the values $(0, 1, 2, ...N - 1)/N$ for the fundamental representation of $SU(N)$; the choice $N = 3$ leads to $(0, 1, 2)/3$ charges and correspondingly for

$N > 3$ (here N is the number of colors, henceforth denoted by N_C). While the gauging of $SU(3)$ color has led to the highly successful QCD theory of the strong interaction, it has been shown [5.49] that for a class of processes involving hadrons [e.g. $\pi^0 \to 2\gamma$ decay], the final answer is the same if an arbitrary N_C ($N_C \geq 3$) is assigned to each quark provided the appropriate charges — in accord with the Z_N centrum restrictions — are used. It has also been noted that for many strong interaction processes, the qualitative conclusions drawn from QCD calculations with arbitrary N_C ($N_C \geq 3$) are not altered if ($g_s^2 N_C$) is held constant. This observation [5.50] has encouraged taking the large N_C limit [i.e. choosing $SU(N_C)$ as the color gauge group and allowing $N_C \to \infty$], corresponding to the "weak coupling" limit, and examining the simplified and "rigorous" predictions of large N_C QCD for clues to the realistic case of $N_C = 3$. Calculations in the large N_C limit are especially valuable in those cases where the $N_C = 3$ calculation is basically intractable because of non-perturbative effects while the large N_C limit type of calculation is capable of giving a clean-cut proof [5.51]; we have previously mentioned two examples of large N_C QCD "proofs": (1) that the global chiral quark flavor group $SU(3)_L \times SU(3)_R$ breaks down spontaneously to $SU(3)_F$ (§5.1); and (2) that SSB can not take place for vector-like groups (see §5.1). While there are no clear-cut criteria for deciding whether a large N_C limit calculation of some QCD process leads to a useful result, we do know that the important OZI rule [3.12] is rigorously true in the large N_C limit. We discuss two well-known examples supporting the OZI rule and then recall the argument for the OZI rule in the large N_C limit. We sketch the proof of this statement in this special topics section to exhibit the type of argument that the large N_C limit provides in the absence of realistic $N_C = 3$ calculations.

The phenomenologically-inspired OZI rule, which states that "disconnected quark diagrams are suppressed relative to connected ones", has served as an excellent guiding principle in the development of strong interaction theory and exceptions to the OZI rule — which are rare — usually signify that some significant new physics is involved. The OZI rule was originally invented to explain why the dominant decay mode of the vector ϕ meson $[\phi = (s\bar{s})]$ is kaon decay, (i.e. $\phi \to K^+ K^-$), whereas the dominant decay mode of the vector ω meson $[\omega = \frac{1}{\sqrt{2}}(u\bar{u} + d\bar{d})]$ is pion decay (i.e. $\omega \to 3\pi$), even though the phase space for pion decay of the more massive ϕ meson is greater than that for the ω meson. Figure 5.10 shows how the OZI rule explains this experimental fact when the ω and ϕ mesons are represented in terms of their constituent quarks; thus Fig. 5.10a allows the $\omega \to 3\pi$ decay

process via "connected" quark diagrams whereas Fig. 5.10b shows why the
decay $\phi \to 3\pi$, involving as it does "disconnected" quark diagrams, can not
occur; on the other hand, using a quark diagram of the type shown in Fig.
5.10c, $\phi \to K^+ K^-$ can take place. It should be pointed out that there is
a small width for $\phi \to 3\pi$ decay because — in accordance with QCD —
there are gluon lines between the s and u and d quarks (see Fig. 5.10b) and
such diagrams give rise to reduced pion decay (and the deviations from an
absolute OZI rule).

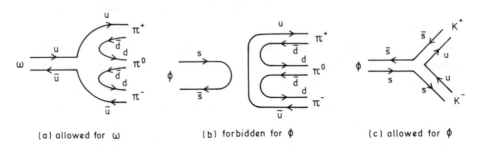

(a) allowed for ω (b) forbidden for ϕ (c) allowed for ϕ

Fig. 5.10. OZI connected and disconnected quark diagrams for ω and ϕ meson decay.

Using the same type of "OZI rule" argument, a great triumph was
achieved in explaining the preferential decay of the heavy quarkonia ψ and Υ
into c quark- and b quark- containing mesons respectively. Here, $J/\psi = (c\bar{c})$
is the analog of ϕ $(s\bar{s})$ and can decay into D^0 $(c\bar{u}) + \bar{D}^0$ $(\bar{c}u)$ provided its
mass is sufficient (which is true for the second excited state of J/ψ and all
higher ones) in accordance with Fig. 5.11a; however, J/ψ (or any of its excited states) can not decay into mesons from which a c quark is absent (see
Figs. 5.11b and 5.11c). Since the ground state of charmonium J/ψ is not
sufficiently massive to allow $D^0 + \bar{D}^0$ decay, its decay width (arising from
gluon-induced diagrams, etc.) is of the order of tens of KeV rather than
MeVs so that the observed metastability of J/ψ strongly supports the OZI
rule.

While the OZI rule is evidently not absolute, it is a good approximation
for a large number of strong interaction processes and requires some explanation. Attempts have been made to understand the origin of the OZI rule
within the framework of $N_C = 3$ QCD [e.g. it has been shown that the OZI
rule should work well for heavy quark flavors [5.52] because of unforeseen
cancellations in the matrix elements] but it is established that the OZI rule
emerges as a rigorous consequence of the application of the large N_C limit

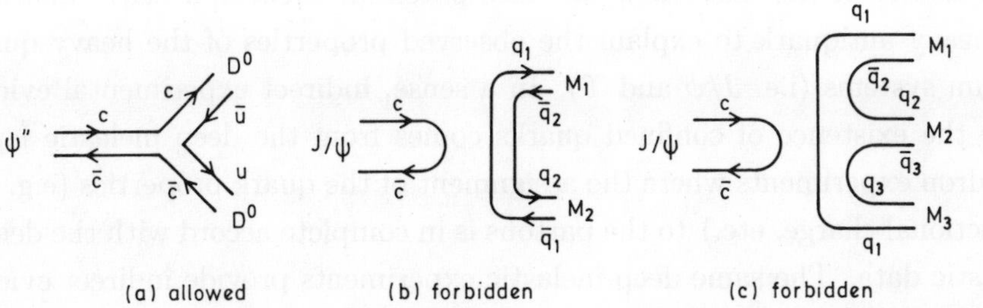

Fig. 5.11. OZI connected and disconnected quark diagrams for J/ψ and ψ'' decay.

to QCD [5.51]; for example, it has been proved that, in the large N_C limit, the transition amplitude corresponding to Fig. 5.11a is $\sim N_C^{-\frac{1}{2}}$ whereas it is $\sim N_C^{-\frac{3}{2}}$ in the case of Fig. 5.11b. The large N_C estimates for the first two schematic diagrams in Fig. 5.11 clearly tell us that the decay width for $\psi'' \to D\bar{D}$ (charmonium in the second excited state) should be substantially greater than the decay width of J/ψ (charmonium in the ground state), in agreement with experiment [$\Gamma(\psi'') = 25$ MeV against $\Gamma(J/\psi) = 0.068$ MeV [1.99]]. Similar conclusions can be drawn about the relative decay widths of the ϕ and ω mesons on the basis of the large N_C argument (i.e. the OZI rule).

In sum, we can say: (1) the OZI rule is rigorous in the $N_C \to \infty$ limit. This follows from the fact that an OZI-forbidden process involves at least two closed quark loops and hence is suppressed (completely suppressed in the large N_C limit) compared to an OZI-allowed process which receives contributions from one closed loop; and (2) since we have no way of estimating the degree of accuracy of the OZI rule for finite $N_C = 3$, we must be prepared for violations of the OZI rule; however, in view of its general success, we must try to understand the reason for OZI violation in each exceptional case. We will have occasion to call attention to the breakdown of the OZI rule in connection with "glueballs" (see §5.4c). A more indirect example of the necessity to be careful about drawing firm conclusions from large N_C QCD reasoning arises when we consider the chiral gauge anomaly-free constraints on the standard group in the large N_C limit (see §7.5).

§5.4c. Evidence for confined gluons and glueballs

A major challenge for QCD theory is to explain the failure thus far to observe "free" quarks and "free" gluons. We did discuss in §5.3b how indirect experimental evidence for quark confinement in QCD is provided by the ability of the linear long-distance potential between a heavy quark and a heavy antiquark to explain the observed properties of the heavy quarkonium systems (i.e. J/ψ and Υ). In a sense, indirect experimental evidence for the existence of confined quarks comes from the deep inelastic lepton-hadron experiments where the assignment of the quark properties (e.g. spin, fractional charge, etc.) to the partons is in complete accord with the deep inelastic data. The same deep inelastic experiments provide indirect evidence for the existence of confined gluons by virtue of the indirect observation that the quarks take up only approximately one-half the total available energy in a deep inelastic lepton-nucleon collision; the other half of the energy is transferred to the gluons inside the nucleon [5.20]. Stronger indirect evidence for the existence of confined gluons comes from the observation of three and higher multiplicity hadron jets in addition to the common two-jet phenomenon. From a study of the energy-energy and angular correlations in the multiple jet events [5.53], it is possible — with the help of perturbative QCD — to not only demonstrate the "running" of the QCD coupling constant α_s (as predicted by asymptotic freedom), but also to confirm the spin 1 and color structure of the QCD gluons. These results are highly encouraging but it is still desirable to place the confined gluon aspect of QCD on the same level of certainty as the confined quark aspect, i.e. to observe color singlet multi-gluon "glueballs" just as one observes color-singlet multi-quark hadrons. In this special topics section, we comment on the phenomenological situation with regard to "glueballs" and on the status of the lattice gauge calculations (LGT) of glueball masses and spins.

The strongest experimental evidence thus far for the existence of $J^{PC} = 2^{++}$ "glueballs" in the 2 GeV region appears to come from a careful study of the $\pi^- p \to \phi\phi n$ reaction. The argument is that one must look for resonant "glueball" states in an OZI-forbidden process (since an OZI-allowed process is quark-dominated) and that the $\pi^- p \to \phi\phi n$ reaction fulfills this condition. In an OZI-forbidden strong interaction process, Lindenbaum argues [1.126] that a favorable condition for detecting "glueballs" occurs when the strong interaction process possesses an OZI-suppressed channel with variable mass for the "disconnected" part of the diagram which is composed of the hadrons involving only new flavors of quarks. "Glueballs" with the right quantum numbers can break down the OZI suppression in the mass region where they exist and dominate the channel. Thus, OZI suppression can act as a filter for letting glueballs pass while suppressing other states. In QCD, OZI suppression can be attributed to the fact that two or more hard gluons are needed to bridge the gap in a suppressed "disconnected" or (hairpin) OZI diagram involving new types (i.e. flavors) of quarks. However, if the gluons in the intermediate state resonate to form a "glueball", the effective coupling constant (as in all resonance phenomena) becomes strong, and OZI suppression should disappear in the mass range of the "glueball". This should allow hadronic states with the "glueball" quantum numbers to form with essentially no OZI suppression.

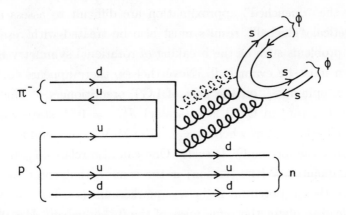

Fig. 5.12. OZI disconnected diagram for the reaction $\pi^- p \to (gb)n \to \phi\phi n$ (i.e. a doublet hairpin diagram). Two or three gluons are shown connecting the disconnected parts of the diagram depending upon the quantum numbers of the $\phi\phi$ system. Glueballs (gb) can be formed directly from the gluons.

The above is a plausible argument and apparently, the reaction $\pi^- p \to \phi\phi n$ — as shown in Fig. 5.12 — is an excellent candidate for a "glueball resonance"-dominated reaction [1.127], for the following reasons: (1) the cross section for the OZI-suppressed reaction $\pi^- p \to \phi\phi n$ of Fig. 5.12 is much larger — by a factor of 50 — than the OZI-allowed reaction $\pi^- p \to \phi K^+ K^- n$ and is much larger — by a factor of 1000 — than the OZI-allowed reaction $\pi^- p \to K^+ K^- K^+ K^- n$; (2) there is no evidence for the $f_4(2030)$ resonance which should be strongly produced through π exchange; and, finally (3) the shapes of the observed high cross section Breit-Wigner resonances with $I^G J^{PC} = 0^+ 2^{++}$ strongly suggest the three tensor "glueball" states: $gb_T(2010)$, $gb_{T'}(2300)$ and $gb_{T''}(2340)$. Lindenbaum's analysis [1.127] is extremely careful but there is still some controversy concerning his conclusions, particularly since the radiative decay $J/\psi \to \gamma\phi\phi$ has not been seen.

It would be a triumph for QCD if the tensor "glueball" states hold up under further scrutiny since the existence of "glueballs" is a direct test of the non-Abelian (self-interacting gluon) character of a Yang-Mills gauge field. Since the experimental situation is still unsettled, it is interesting to query QCD theory [with its extensive and improved lattice gauge theory (LGT) calculations carried out during the past few years] with regard to its findings in the "glueball" arena. The answer is that much effort has indeed been expended on an LGT search for "glueball" solutions of the Yang-Mills equations but primarily in the so-called "quenched" approximation (i.e. in which fermion loops are not included) [5.54]. The errors introduced by the "quenched" approximation are difficult to assess and hence the theoretical "glueball" results must also be treated with reserve [there are other problems such as the breaking of rotational symmetry by the lattice, which must be evaluated]. Nevertheless, there appears to be a convergence of opinion — on the part of LGT practitioners — that the lowest "glueball" state in QCD is a (scalar) $J^{PC} = 0^{++}$ state with a mass $M_{gb} = (10.4 \pm 0.3) \, \Lambda_{\overline{MS}}$ where $\Lambda_{\overline{MS}} \sim 150$ MeV is the modified minimum subtraction value for the QCD scale. One can also relate $M_{gb}(0^{++})$ to $\sqrt{k}$, where the quantity k (the tension) is the same k that enters into the linear part of the $q\bar{q}$ potential in heavy quarkonium [see Eq. (5.103)]; both methods lead to about the same mass of the 0^{++} glueball: $M_{gb}(0^{++}) \simeq 1.4$ GeV. The LGT calculations have followed through to the identification of a (tensor) 2^{++} "glueball" state: whose mass is $\approx 1.5 \times M_{gb}(0^{++})$, namely $M_{gb}(2^{++}) \simeq 2.1$ GeV. It is possible that Lindenbaum is seeing this 2^{++} glueball state although, thus far, there is no experimental evidence for the lighter 0^{++} glueball state predicted by LGT. In sum, there is progress in the search for a strongly-self coupled multi-gluon ("glueball") state — expected in non-Abelian QCD — but the final outcome awaits further theoretical and experimental work.

§5.4d. QCD sum rules

The non-perturbative character of all low-energy hadronic processes (of which the "glueball" problem is one example) has stimulated a number of attempts to find relations among different long-distance- (strong coupling-) dominated hadronic processes by expressing the form factors (or the structure functions or the decay amplitudes, or what have you), in terms of semi-phenomenological quantities. In a sense, dispersion theory was the first such attempt (see §1.2b), followed by current algebra (see §3.5) and chiral perturbation theory (see [5.1]). It is fair to say that these efforts were focused on the pion-nucleon interaction and were most successful at the lowest energies (soft pion limit, etc.). With the advent of QCD and the realization that asymptotic freedom could only be expected to hold in the multi-GeV region or above, the need for a semi-phenomenological method to deal with relatively low energy (~ 1 GeV), strong interaction processes within the QCD framework became apparent. One of the more significant programs in this direction — the use of QCD sum rules [5.55] — circumvents the difficulty of performing direct non-perturbative QCD calculations through a combination of Wilson's operator product expansion (OPE) and the phenomenological identification of non-perturbative quantities such as the strength of the "quark condensate", the strength of the "gluon condensate", etc.

The overall objective of the QCD sum rule method is to compute the "resonance" parameters in hadronic physics (e.g. masses, coupling strengths, etc.) in terms of QCD Lagrangian-related VEV's of "non-perturbative" higher-dimensional operators such as $< 0|\bar{q}q|0 >$ and $< 0|G_{\mu\nu}^a G_{\mu\nu}^a|0 >$, where $q, \bar{q}$ are the light quark fields and $G_{\mu\nu}^a$ is the gluon field tensor. The basic thrust is to use OPE both to determine the Wilson coefficients by means of the usual perturbative short-distance (asymptotic freedom) arguments, and to help evaluate the power corrections to the asymptotic freedom results introduced by the non-vanishing large-distance (non-perturbative) VEV's of the higher-dimensional operators in OPE. The application of Wilson's OPE to the derivation of QCD sum rules in the GeV region is quite different from its use in the deep inelastic lepton-hadron energy region where one is quite satisfied with the logarithmic (in energy) corrections to "Bjorken scaling" (see §5.2b).

The "QCD sum rule" strategy proceeds as follows: the time-ordered product of two currents, i.e. the operator product] is expanded in terms of the identity I (which contains the ordinary perturbative contributions) and the higher-dimensional terms involving the Wilson c_i coefficients $[c_i(q^2)]$ and the associated operators $\mathcal{O}_i$ thus [see Eq. (5.41)]:

$$T^j(q^2) = i \int d^4x \, e^{iqx} \, T\{j_\Gamma(x)j_\Gamma(0)\} = c_I I + \sum_i c_i(q^2)\mathcal{O}_i \qquad (5.119)$$

where:

$$j_\Gamma(x) = \bar{q}_k(x)\Gamma q_\ell(x); \quad \Gamma = 1, \gamma_5, \gamma_\mu, \text{ etc.} \qquad (5.120)$$

with k and ℓ quark flavors. The first five c_i's are the Wilson c_i coefficients of dimension 4 (two of them) and dimension 6 (three of them) [terms of higher dimension than 6 are not written down], and the $\mathcal{O}_i$'s are listed in Table 5.3; it should be noted that all the operators possess spin 0 (because of the particular choice of $T^j(q^2)$) and the Γ's are matrices taking account of the color, flavor and spinor indices of the quark fields [5.56].

Table 5.3: Leading terms in operator product expansion ($Q^2 = -q^2$).

local operator	dimension	degree of singularity	Q^2 dependence of coefficient
I	0	$\frac{1}{x^6}$	$0(1)$
$\mathcal{O}_1 = \alpha_s\, G^a_{\mu\nu} G^a_{\mu\nu}$	4	$\frac{1}{x^2}$	Q^{-4}
$\mathcal{O}_2 = m\bar{q}q$	4	$\frac{1}{x^2}$	Q^{-4}
$\mathcal{O}_3 = m(\bar{q}\sigma_{\mu\nu}\frac{\lambda_a}{2}q)\, G^a_{\mu\nu}$	6	$0(1)$	Q^{-6}
$\mathcal{O}_4 = (\bar{q}\Gamma_1 q)(\bar{q}\Gamma_2 q)$	6	$0(1)$	Q^{-6}
$\mathcal{O}_5 = f_{abc}\, G^a_{\mu\nu} G^b_{\nu\lambda} G^c_{\lambda\mu}$	6	$0(1)$	Q^{-6}

By taking the VEV's of both sides of Eq. (5.120), the unit operator I is eliminated (since it encompasses all the perturbation-theoretic terms) and one is left with the VEV's of the non-perturbative higher-dimensional terms whose evaluation is now possible because the VEV of $T^j(q^2)$ is nothing more than the two-point Green's function $\Pi^j(q^2)$ which can be related — by means of a dispersion relation — to the phenomenological parameters (masses, coupling strengths, etc.) characterizing the strong interaction process under consideration. It is straightforward to deal with most of the operators in Fig. 5.3 but the operator $< 0|\bar{q}\Gamma q\bar{q}\Gamma q|0 >$ is more complicated and has to be approximated by [5.57]:

$$< 0|\bar{q}\Gamma q\bar{q}\Gamma q|0 >= N^{-2}[(\mathrm{Tr}\ \Gamma)^2 - (\mathrm{Tr}\ \Gamma^2)](< 0|\bar{q}q|0 >)^2, \qquad (5.121)$$

where N is a normalization factor. The approximation (5.121) can be justified and one is left with $< 0|\bar{q}q|0 >$ and $< 0|G^a_{\mu\nu} G^{a\mu\nu}|0 >$ as the two basic quantities in the QCD sum rule method [the first reflecting the breakdown of chiral symmetry among the quarks and the second representing the "dilaton" responsible for the breakdown of scale invariance (see §2.3b)]. In practical terms, the QCD sum rules are finally obtained by taking the dispersion relation for $\Pi^j(q^2)$ in terms of $Im\,\Pi^j(q^2)$ and applying the Borel transform to $Im\,\Pi^j(q^2)$ [which emphasizes the low-energy resonance contributions to the cross sections associated with the currents used in OPE] in order to evaluate the phenomenological parameters.

Table 5.4: Phenomenological VEV's of some dimension 4 and 6 operators.

Operator	Value	Favored Method
$\langle 0 \mid \frac{\alpha_s}{\pi} G^{a\mu\nu} G^a_{\mu\nu} \mid 0 \rangle$	$\simeq (340 \text{ MeV})^4$	charmonium spectrum
$\langle 0 \mid m \bar{q}q \mid 0 \rangle$	$\simeq (100 \text{ MeV})^4$	current algebra
$\langle 0 \mid \bar{q}q \mid 0 \rangle$	$\simeq (250 \text{ MeV})^3$	nucleon mass
$\left[\dfrac{\langle 0 \mid \bar{q} \sigma_{\mu\nu} q G^{a\mu\nu} \mid 0 \rangle}{\langle 0 \mid \bar{q}q \mid 0 \rangle} \right]$	$\simeq 1.1 \text{ GeV}^2$	theory

We list in Table 5.4 the values deduced for the VEV's of the local operators listed in Table 5.3, which are most accessible to phenomenological determination on the basis of QCD sum rules (or current algebra [5.57]). It should be emphasized that more than one type of experiment enters into the determination of the quantities listed in Table 5.4 (the chief method is listed in Table 5.4); for example, the application of the "QCD sum rule" method to the leptonic widths of the ρ^0 and ϕ mesons — as well as to charm production in $e\bar{e}$ collisions and the charmonium spectrum — yield an accurate value of $< 0 \mid G^a_{\mu\nu} G^{a\mu\nu} \mid 0 >$. The value of $< 0 \mid \bar{q}q \mid 0 >$ given in Table 5.4 also appears well determined from current algebra in combination with "QCD sum rule" results [5.57]. However, the value of $< 0 \mid m(\bar{q}\sigma_{\mu\nu}q)G^{a\mu\nu} \mid 0 >$ given in Table 5.4 is an LGT value [5.58]; the measurements are difficult and the experimental value of this quantity [which, incidentally, is important for the calculation of the (πN) σ term — see §3.5c] varies from agreement with the LGT value in Table 5.4 down to a value five times smaller. It will be interesting to see whether LGT or the semi-phenomenological "QCD sum rule" approach prevails for this dimension six operator.

References

[5.1] J. Gasser and H. Leutwyler, *Nucl. Phys.* **B250** (1985) 465.

[5.2] S. Weinberg, *Phys. Rev.* **D11** (1975) 3583.

[5.3] N. Isgur and M.B. Wise, *Phys. Lett.* **232** (1989); *ibid.* **B237** (1990) 527.

[5.4] Y. Nambu and Jona-Lasinio, *Phys. Rev.* **122** (1961) 345; R.E. Marshak and S. Okubo, *Nuovo Cim.* **19** (1961) 1226.

[5.5] L.F. Li, *Phys. Rev.* **D9** (1974) 1723.

[5.6] S. Coleman and D.J. Gross, *Phys. Rev. Lett.* **31** (1973) 851; A. Zee, *Phys. Rev.* **D7** (1973) 1647.

[5.7] C. Vafa and E. Witten, *Nucl. Phys.* **B234** (1984) 173.

[5.8] J.I. Friedman and H.W. Kendall, *Ann. Rev. Nucl. Sci.* **22** (1972) 203.

[5.9] J.B. Kogut and L. Susskind, *Phys. Rev.* **D11** (1975) 3594; J.B. Kogut, E. Dagotto and A. Kocic, *Phys. Rev. Lett.* **60** (1987) 772.

[5.10] L. Kadanov, *Physica* **2** (1966) 263; K.G. Wilson, *Phys. Rev.* **B4** (1971) 3184.

[5.11] K.G. Wilson, *Phys. Rev.* **179** (1969) 1499.

[5.12] C.G. Callan, *Phys. Rev.* **D2** (1970) 1541; K. Symanzik, *Commun. Math. Phys.* **18** (1970) 227.

[5.13] A. Sakharov, *Memoirs*, Knopf, New York (1990), p. 84.

[5.14] N.K. Nielsen, *Am. J. Phys.* **49** (1981) 1171.

[5.15] L.D. Landau and E.M. Lifshitz, *Electrodynamics of Continuous Media*, Pergamon Press (1984).

[5.16] A. El-Khadra, G. Hockney, A. Kronfeld, and P. Mackenzie, *Phys. Rev. Lett.* **69** (1992) 729.

[5.17] W.J. Marciano, *Phys. Rev.* **D29** (1984) 580.

[5.18] D.V. Shirkov, *Renormalization Group*, ed. D. Shirkov, D. Kazakov and A. Vladimirov, World Scientific, Singapore (1988).

[5.19] G. Altarelli and G. Parisi, *Nucl. Phys.* **B126** (1977) 298.

[5.20] F.E. Close, *An Introduction to Quarks and Partons*, Academic Press, New York (1979); E. Leader and E. Perdazzi, *Gauge Theories and the 'New Physics'*, Cambridge Univ. Press, England (1982).

[5.21] C. Callan and D. Gross, *Phys. Rev. Lett.* **22** (1969) 156.

[5.22] D.A. Ross and Sachrajda, *Nucl. Phys.* **B149** (1979) 497.

[5.23] M. Virchaux, *Proc. of XXVI Intern. Conf. on High Energy Physics*, Dallas (1992).

[5.24] K. Gottfried, *Phys. Rev. Lett.* **18** (1967) 1154.

[5.25] E.J. Eichten, I. Hinchliffe and C. Quigg, FERMILAB-PUB-91/272 (1991).

[5.26] L.V. Gribov, E.M. Levin and M.G. Ryskin, *Phys. Rep.* **100** (1983) 1.

[5.27] J.F. Owens, *Rev. Mod. Phys.* **59** (1987) 465.

[5.28] J.D. Bjorken, *Phys. Rev.* **148** (1966) 1467; *ibid.*, **D1** (1970) 1376.

[5.29] J. Ellis and R. Jaffe, *Phys. Rev.* **D9** (1974) 1444.

[5.30] C.Y. Prescott *et al.*, *Phys. Lett.* **77B** (1978) 347.

[5.31] D.H. Perkins, *Contemp. Phys.* **16** (1975) 173.

[5.32] D.G. Gross and C.H. Lewellyn-Smith, *Nucl. Phys.* **B14** (1967) 337.

[5.33] F.B. Brown *et al.*, *Phys. Rev. Lett.* **65** (1990) 2491.

[5.34] D.B. MacFarlane *et al.*, *Zeit. f. Phys.* **C26** (1984) 1.

[5.35] G. 't Hooft, *Physica Scripta* **25** (1982) 133.

[5.36] G. 't Hooft, "On the phase transition towards permanent quark confinement", Utrecht preprint (1977).

[5.37] Y. Nambu, *Phys. Rev.* **D10** (1974) 4262; G. Parisi, *Phys. Rev.* **D11** (1975) 970; K.G. Wilson, *Phys. Rev.* **D1** (1974) 2445; J. Kogut and L. Susskind, *Phys. Rev.* **D9** (1974) 3501.

[5.38] H. Georgi and S.L. Glashow, *Phys. Rev. Lett.* **28** (1972) 1494.

[5.39] K.G. Wilson, *Phys. Rev.* **D10** (1974) 2445.

[5.40] B. Sakita, *Proc. XIX Intern. Conf. High Energy Phys.*, Tokyo (1978), p. 921.

[5.41] T. Applequist and H. Georgi, *Phys. Rev.* **D8** (1973) 4000; A. Zee, *Phys. Rev.* **D7** (1973) 3630.

[5.42] G. Wentzel, *Helv. Phys. Acta.* **16** (1943) 222, 551.

[5.43] M. Creutz, *Quarks, Gluons and Lattices*, Cambridge Univ. Press, England (1983); A. Ukawa, *Proc. of XXV Intern. Conf. on High Energy Phys.*, World Scientific, Singapore (1991), p. 79.

[5.44] D. Flamm and F. Choberl, *Introduction to the Quark Model of Elementary Particles*, Gordon and Breach, New York (1982).

[5.45] P. Wang, Ph.D. thesis, "Studies of Lepton and Quark Interactions", Virginia Tech (1985).

[5.46] G. Baum *et al.*, *Phys. Rev. Lett.* **51** (1983) 1135.

[5.47] M. J. Ashman *et al.*, *Nucl. Phys.* **B328** (1989) 1.

[5.48] S. Gupta, M.V.N. Murthy and J. Pasupathy, *Phys. Rev. Lett.* **39** (1989) 2547.

[5.49] V.A. Koval'chuk, *Soviet Phys. JETP Lett.* **48** (1988) 11.

[5.50] M. Bourquim *et al.*, *Zeit. f. Phys.* **C21** (1983) 27.

[5.51] E. Witten, *Annals of Phys.* **128** (1980) 363.

[5.52] P. Geiger and N. Isgur, *Phys. Rev. Lett.* **67** (1991) 1066.

[5.53] V.D. Barger and R.J.N. Philips, *Collider Physics*, Addison-Wesley, Boston (1987).

[5.54] S.R. Sharpe, *Proc. of BNL Workshop on Glueballs, Hybrids and Exotic Hadrons*, ed. S. Chung, A.I.P., New York (1989).

[5.55] M.A. Shifman, A.I. Vainshtein and V.I. Zakharov, *Phys. Rev. Lett.* **42** (1979) 297; *Nucl. Phys.* **B147** (1979) 385, 448.

[5.56] L.J. Reinders, CERN preprint TH3701 (1983).

[5.57] M.A. Shifman, A.I. Vainshtein and V.I. Zakharov, *Phys. Lett.* **77B** (1978) 80.

[5.58] H. Kremer and G. Schierholz, DESY preprint 87-024 (1987).

Chapter 6

GAUGE THEORY OF THE ELECTROWEAK
INTERACTION (QFD)

§6.1. Universal (V-A) charged current theory of weak interactions

Fermi's charged current theory of nuclear beta decay was formulated in 1934 [1.26] when only the nucleon was known to decay, there was only one family of leptons, the applicability of the gauge principle to electromagnetism was known (but Yang-Mills gauge fields were unknown), there was no parity violation (and hence chiral fermions were not invented), there were no quarks, baryon and lepton conservation were only dimly perceived, and there were no neutral weak currents. Under these circumstances, Fermi did exceedingly well when he wrote down the "local" [by "local" here is meant interacting at the same space-time point, not to be confused with "local or gauge symmetry"] weak charged current-current interaction, separated the nucleonic $(n - p)$ current from the leptonic $(e - \nu)$ current, chose the vector (V) Lorentz structure for the current [out of the five possible choices: scalar (S), vector (V), axial vector (A), tensor (T) or pseudoscalar (P)], and assumed a massless neutrino (although he considered the dramatic change of the endpoint slope of the β spectrum for a finite mass neutrino).

From the beginning, the short-range character of the weak interaction (appreciably less than the range of nuclear forces) was supported by numerous beta decay experiments [highly forbidden beta decays were a prime example [6.1]]; however, it took until the early 1980s — with the observation of the weak bosons [6.2] — to establish that the range of the weak interaction is a factor of 1000 smaller than the nuclear force range. The separation

between the hadronic (now quark) and leptonic charged currents has withstood the test of time and thus far the masslessness of ν_e is still tenable (see
Table 1.6). Fermi's assumption of a pure V charged current had to be modified when it was discovered, in the mid-1930s, that equally fast (allowed)
beta transitions occurred via "Gamow-Teller" selection rules [1.64], which
permitted an angular momentum change J of one unit in allowed β transitions — in contrast to the "Fermi" selection rules for allowed β transitions,
which mandated no change of J between the initial and final nuclear states.
It was soon realized that the "Gamow-Teller" selection rules are obtained if
the weak charged current governing β decay contains an A and/or T current, whereas "Fermi" selection rules hold if the Lorentz structure of the
weak charged current is V and/or S. Since Fermi's choice of V was not based
on an explicit gauge principle, the phenomenologically correct combination
of S *or* V and A *or* T became a highly controversial question for several
decades until parity violation entered the picture in the mid-1950's [1.72].
Within one year of the confirmation of maximal parity violation — in 1956
— the Fermi theory of beta decay gave way to the universal V-A charged
current-current theory of weak interactions [1.76].

The replacement of the parity-conserving (non-chiral) weak charged current by the maximal parity-violating [purely chiral (lefthanded)] (V-A) weak
charged current opened the door to rapid progress in weak interaction theory
in several ways:

(1) The initial derivation of the V-A current was based on the principle
of "chirality invariance". This principle takes its cue from the γ_5 invariance
of the (two-component) Weyl equation for the massless neutrino [1.75], which
exhibits manifest maximal parity violation (and CP conservation). What we
call the use of the "neutrino paradigm" in §1.2d is simply the extension of
the principle of γ_5 (or chirality) invariance to weak and electromagnetic currents involving massive fermions. Insofar as weak charged currents (which
necessarily implies two distinct flavors of fermions) are concerned, "chirality invariance" requires the weak charged current to take the chiral form
$\bar{\psi}_{iL}\gamma_\mu\psi_{jL}$ (i flavor $\neq j$ flavor) where L indicates a lefthanded chiral fermion.
On the other hand, the same principle of "chirality invariance", when it is
invoked for a neutral weak or electromagnetic current, which is necessarily a
single flavor fermion current, it only requires that the neutral fermion current
possess the Lorentz structure V+αA (α an arbitrary constant); this result is
consistent with the parity-conserving (non-chiral) V electromagnetic current
and the combination of the neutral non-chiral electromagnetic current and

neutral lefthanded weak current that enters into the standard electroweak gauge theory (see §6.2). Indeed, we will see in §6.2 that it is the common "chirality invariance" of the weak and electromagnetic currents that makes possible the mixing of the chiral $SU(2)_L$ group and the chiral $U(1)_Y$ group (Y here is the "weak" hypercharge) that spontaneously break down to the non-chiral $U(1)_{EM}$ group. The use of the term "neutrino paradigm" is, in our view, more than a rhetorical flourish. Just as the massless photon encodes the principle of "gauge invariance" in electromagnetism (which appears to be absolutely conserved), so the massless neutrino — assuming that it truly exists (see §6.4b) could encode the principle of "chirality invariance" (and CP invariance), whose violation is incurred when the fermion mass is finite; it may be reasonable to correlate the amount of chirality violation with some power of the fermion mass measured in units of the chirality breakdown scale. [One could argue that all the quarks and leptons are assumed to be massless (Weyl) fermions before the Higgs SSB mechanism operates on the electroweak group $SU(2)_L \times U(1)_Y$ to produce the (Dirac) masses of the quarks and the charged leptons and that, consequently, the measure of chirality violation is of the order of $(m_{q,\ell}/M_W)^2$. If this is so, the three charged leptons (of the three generations) and five of the six quarks (all except the t quark) are well described by the (lefthanded) V-A charged current interaction; the t quark with its mass comparable to M_W — assuming that its exists — would be the exception and could contain an admixture of (righthanded) V+A charged current.]

(2) The establishment of the universal (V-A) Lorentz structure for the weak charged current led to more serious interest in the possible existence of an "intermediate vector boson" (IVB) and in dealing with the divergent behavior of higher-order weak interactions [whether calculated with the current-current form or the current-IVB (Yukawa) form]. It was soon realized that the "unitarity limit" (when the weak interaction becomes "strong"!) is reached when the (center of mass) energy E attains the value $G_F^{-\frac{1}{2}} \sim$ 300 GeV (G_F is the Fermi coupling constant). On the other hand, the calculation of the mass difference between the long-lived and the short-lived K^0 mesons [i.e. $\Delta m(K_L^0 - K_S^0)$]–the only measured higher-(second-)order weak interaction process — leads to a "weak interaction cutoff" of the order of several GeV [1.116], thereby creating a dilemma that could only be resolved later (in 1970) by the GIM mechanism [1.86] which, in turn, was the precursor of the Kobayashi-Maskawa treatment of the difference between the mass and gauge interaction eigenstates in weak processes [6.3].

(3) The IVB hypothesis to explain the weak interaction stimulated interest in the possibility that the IVB is a gauge particle but its finite mass could not be reconciled with the thrust of the then-existing Yang-Mills gauge theory [1.51] (which seemed to require zero mass). Two early attempts that worked out the consequences of treating a pair of massive charged IVB's and the one massless photon as a triplet of vector fields interacting in different ways with "heavy" (presumably nucleons) and "light" (leptonic) fermions, are worth mentioning. In one [6.4], the two charged IVB's were postulated to break an "electric charge reflection symmetry" and the usual space reflection symmetry and the deduced consequence for the Lorentz structure of β decay was the incorrect admixture of V and T interactions; this paper was written before the V-A theory and acknowledges the surrender of the Weyl equation for the neutrino and proclaims the failure of chirality invariance for the β decay interaction. The second paper [6.5] — published shortly after the V-A theory — builds on the chirality-invariant charged current interaction and extends the $SU(2)$ Yang-Mills theory to weak interactions; the paper recognizes to some extent the implications for neutral weak currents but offers no explanation for the masses of the charged IVB's and agrees that a quantum field theory with charged massive vector fields is not renormalizable (see §4.5). A somewhat later attempt (in 1961) to treat the charged massive IVB's on a par with the photon (i.e. to somehow "unify" the weak and electromagnetic interactions) produced a fresh insight by Glashow [1.88], namely that the minimal (electroweak) group that could combine the $SU(2)_L$ weak group [6.6] and the $U(1)_{EM}$ electromagnetic group is $SU(2)_L \times U(1)_Y$, where the electric charge $Q = I_{3L} + Y/2$ (with I_{3L} the third charge of the $SU(2)_L$ weak group and Y the weak hypercharge); the 1961 $SU(2)_L \times U(1)_Y$ group — which we have called the "weak" GNN group (see §1.3c) became the gauged electroweak group of the standard model. However, at the time, the Higgs mechanism for generating massive weak bosons was still unknown [the concept of the SSB of global internal symmetry groups and the Goldstone theorem were known but the effect of gauging on the "would-be" N-G bosons (i.e. the Higgs mechanism) was several years off]. Moreover, the question of the renormalizability of such a theory was left open. It was not until the mid-1960s that the SSB Higgs mechanism [1.87] was invented to generate finite mass for (massless) gauge particles. And it was not until the early 1970's that 't Hooft [1.106] proved the renormalizability of Yang-Mills-Higgs theories with SSB generation of IVB masses.

(4) The IVB hypothesis also played an important role in the development of the concept of distinct families of hadronic (now quark) and leptonic constituents of matter. After the IVB was proposed to underpin the V-A theory, the failure to observe the $\mu \to e\gamma$ decay process acquired great urgency because the intermediation of an IVB in this decay process predicted much too large a branching ratio (compared to the decay $\mu \to e + \nu_\mu + \bar{\nu}_e$) (see §1.2d). It appeared at the time (1959) that the only way to solve the dilemma was to postulate that muon and electron neutrinos are distinct particles [1.82], a hypothesis that was shortly to be confirmed. Once the distinct second-generation neutrino was discovered, it was possible to invoke a baryon (quark)-lepton symmetry principle to argue for the existence of a second-generation baryon (or quark) that would be the counterpart of ν_μ [1.84]. The existence of ν_μ ingendered new problems as well as new physics: it drew attention to the role of the second generation s quark in strange particle physics; in particular, it heightened the puzzle with regard to the rarity of (strangeness) flavor-changing neutral currents. This puzzle was solved by means of the GIM mechanism [1.86] which could only come into play if the quark partner (i.e. the c quark) of the s quark exists. The discovery of charmonium [6.7] completed the identification of the second quark-lepton generation and thereby validated the GIM mechanism.

(5) While the V-A Lorentz structure for the charged current was not only confirmed by all hypercharge-conserving weak decays but also by the hypercharge-changing weak decays, it became clear by the early 1960s that the effective strength for the coupling of the hypercharge-violating charged current was substantially less (by a factor of 20) than that of the effective strength for the coupling of the hypercharge-conserving charged current [6.8]. The introduction of the "Cabibbo" angle [1.81] provided a phenomenological explanation of this departure from universality and opened the door to the CKM mixing matrix [6.3], wherein the weak quark current replaces the weak hadron current and inter-generational effects are taken into account. The CKM mixing matrix approach makes a clear distinction between mass and gauge interaction eigenstates in weak processes and it offers the possibility of explaining — in a natural way — the weak CP violation phenomenon when there are three generations of quarks and leptons [6.9]. Within the CKM framework, three generations of quarks require three "Cabibbo" angles $(\theta_1, \theta_2, \theta_3)$ and permit the existence of one (complex) phase shift δ whose (non-vanishing) value is responsible for the weak CP violation observed in the neutral kaon system and possibly the neutral B meson system (see §6.3f).

[Weak CP conservation can come into play for finite-mass quarks but not for finite-mass charged leptons (because of their massless neutrino partners) [6.10].] If the CKM scheme is fully validated and δ is determined, it may be possible to decide whether there is a deep connection between the structure of the CKM mixing matrix and the observed number (3) of quark generations. Thus, the existence of precisely three quark generations may not be unrelated to the fact that of finite quark mass breaks the "neutrino paradigm" (which implies "CP conservation" as well as "chirality invariance" — see point 1) and might lead one to expect weak CP violation for finite-mass quarks, and as shown by Kobayashi and Maskawa, weak CP violation can only occur as a "mass effect" for a minimum of three generations.

In view of the developments just enumerated, the time was ripe in the mid-1960s for bringing together the essential ingredients of the present-day gauge theory of the electroweak interaction except for the uncertainty of renormalizability for a non-Abelian gauge theory with SSB. The pieces were finally put in place in 1967–68 [6.11] and the standard gauge theory of the electroweak interaction that emerged is discussed in the next section (§6.2). In the same section, we preface the main discussion with a more detailed account of Glashow's early contribution to the electroweak theory.

§6.2. Gauge theory of the electroweak interaction (QFD)

We pointed out in the previous section that, while sidestepping the question of renormalizability and without knowledge of the Higgs mechanism for generating finite gauge masses, Glashow [1.88] did demonstrate that a group with the "strong" GNN structure (see §3.2b), i.e. $SU(2) \times U(1)$, is the minimal group that can accommodate both the weak and electromagnetic interactions of leptons (and baryons — quarks were not yet invented), in particular the chiral (lefthanded) character of the weak current (both charged and neutral) and the non-chiral electromagnetic current. [The "weak" GNN group put forward by Glashow bears a strong resemblance to the (global) "weak" GNN group resulting from the earlier "baryon-lepton symmetry principle" [1.84]; it is amusing that the global "weak" GNN flavor group was gauged to produce the standard electroweak gauge group, whereas the global "strong" GNN flavor group only received "global" enlargements and ultimately gave way to the gauging of color degrees of freedom as the foundation for the standard strong interaction gauge group.]

The "minimality" of an $SU(2) \times U(1)$-type group to "unify" the weak and electromagnetic interactions, can be seen as follows: we know that the two-flavor lefthanded weak charged current and its conjugate, by themselves, constitute two of the three generators of the $SU(2)_L$ group and the only question that must be settled is whether the neutral non-chiral electromagnetic current can be identified with the third generator of $SU(2)_L$ or requires a larger group. It is easy to prove that the commutator of the two lefthanded, "charge-raising" and "charge-lowering" generators — representing the two weak charged chiral currents — can not yield the electromagnetic ("neutral charge") "non-chiral" generator. Thus, if we choose for the lefthanded, "charge-raising" current (say, for the electron and its neutrino) $J_\mu^\dagger = \frac{1}{2}[\bar{\nu}_e\gamma_\mu(1 - \gamma_5)e]$, the "charge-lowering" current is J_μ; we can write for the two charges $Q^+(t)$ and $Q^-(t)$ of the $SU(2)_L$ group:

$$Q^+(t) = i \int d^3x\ J_0^\dagger(x) = \frac{1}{2} \int d^3x\ \nu_e^\dagger(1 - \gamma_5)e \quad ; \quad Q^-(t) = Q^{+\dagger} \quad (6.1)$$

If we then use the canonical equal-time anti-C.R. for fermions, we get:

$$[\psi_i^\dagger(\underline{x}, t), \psi_j(\underline{x}', t)]_+ = \delta_{ij}\ \delta^3(\underline{x} - \underline{x}') \quad (6.2)$$

and hence:

$$[Q^+(t), Q^-(t)] = 2Q^3(t); \quad Q^3(t) = \frac{1}{4} \int d^3x[\nu_e^\dagger(1-\gamma_5)\nu_e - e^\dagger(1-\gamma_5)e] \quad (6.3)$$

As expected, Q^3 is the neutral (lefthanded) chiral current and not Q^{em} [since $Q^{em} = \int d^3xe^\dagger e$] so that Q^+, Q^-, Q^{em} do not form a closed $SU(2)_L$ algebra. Hence, the chiral $SU(2)_L$ group must be enlarged to accommodate the non-chiral electromagnetic current and, *a fortiori*, the enlarged group must contain at least four generators. The minimal enlargement is the chiral, non-Abelian $SU(2)_L \times U(1)_Y$ group (where Y is the "weak" hypercharge) which is taken to break spontaneously — via the Higgs mechanism — to the unbroken, non-chiral, Abelian electromagnetic group $U(1)_{EM}$; the non-chiral electric charge Q is related to the chiral charges I_{3L} and Y of the "weak" GNN group $SU(2)_L \times U(1)_Y$ in analogy to the "strong" GNN relation (see §3.2b), thus:

$$Q = I_{3L} + \frac{Y}{2} \quad (6.4)$$

The idea of using the gauged $SU(2)_L \times U(1)_Y$ group and of generating substantial masses for the charged weak bosons by means of a Higgs mechanism was an excellent starting point for the electroweak theory but it still had to be supplemented by a suitable choice of quantum numbers [under $SU(2)_L \times U(1)_Y$] for the Higgs boson and of appropriate quantum numbers — under the same group — for the quarks and leptons of each generation.

The most economical choice was made for the Higgs boson, namely that the Higgs particle is a single complex scalar field that is a doublet under $SU(2)_L$ with "weak" hypercharge $Y = 1$; this simple choice of Higgs field, when a non-vanishing VEV is assigned to one of its neutral component fields, breaks down the chiral $SU(2)_L \times U(1)_Y$ group to the unbroken non-chiral $U(1)_{EM}$ group, which satisfies the "'weak" G_{NN} relation of Eq. (6.4). Insofar as the Weyl fermions are concerned, the simple choice of a lefthanded fermion doublet under $SU(2)_L$ for the leptons of one generation (and, subsequently, one for the quarks) plus one righthanded singlet representation (since ν_R was presumed to be absent) under $SU(2)_L$ for the leptons of one generation (and, subsequently, two righthanded singlet representations for the quarks of one generation) worked remarkably well. It should be noted that it is still not established whether the Higgs boson (or bosons) involved in the SSB of the electroweak group is achieved by an "elementary" Higgs particle or by means of a "quasi-Higgs" particle [like the "Cooper pair" responsible for breaking $U(1)_{EM}$ inside a superconductor (see §4.4a)]. The latter alternative is discussed in §6.4c but the lack of resolution of the "Higgs problem" has not vitiated the successful testing of most predictions of the standard gauge theory of the electroweak interaction carried out thus far. Not much attention was paid to the crucial papers of Weinberg and Salam in 1967–68 [6.11] because the renormalizability of what turned out to be the standard electroweak theory had not been established. When 't Hooft's proof came in 1971 [1.106], the non-Abelian electroweak gauge theory with SSB was taken seriously and henceforth was subjected to a vast array of probing tests, which it has continued to pass with flying colors.

It is our objective, in the remainder of this section, to write down the Lorentz-invariant, gauge-invariant and renormalizable Lagrangian based on the $SU(2)_L \times U(1)_Y$ group for the chiral quarks and chiral leptons of one generation [with the quantum numbers shown in Table 1.5]; the quantum numbers of the Higgs boson are taken to be (2,1) under $SU(2)_L \times U(1)_Y$. We expound in some detail the physical consequences of the rather lengthy QFD Lagrangian and then subject its predictions to a great variety of experimental tests, as detailed in succeeding sections. The agreement between theory and experiment is truly impressive (hence the standard model!) but we try to call attention, as we recount the successes, to possible loopholes and ways of "going beyond the standard model".

It is clear by now that — in contrast to unbroken, non-chiral QCD with only gauge (gluon) fields and quarks — spontaneously broken, chiral QFD possesses lepton and scalar fields in addition to quark fields; further, the quarks and leptons enter the electroweak Lagrangian with only one chirality (negative helicity). It is convenient to write the full electroweak Lagrangian in the form [2.26]:

$$\mathcal{L} = \mathcal{L}_F + \mathcal{L}_H + \mathcal{L}_G + \mathcal{L}_Y - V(\phi) \tag{6.5}$$

where $\mathcal{L}_F$ is the part of the Lagrangian involving the gauge-covariant derivatives of the fermion fields, $\mathcal{L}_H$ is the part involving the covariant derivative of the complex Higgs boson doublet ϕ, $\mathcal{L}_G$ is the "kinetic energy" contribution of the four electroweak gauge bosons, $\mathcal{L}_Y$ is the Yukawa coupling of the Higgs boson to the fermions and $V(\phi)$ is the Higgs potential. We have for $\mathcal{L}_F$:

$$\mathcal{L}_F = -i[\bar{\ell}_L \not{D}\ell_L + \bar{e}_R \not{D}e_R + \bar{Q}_L \not{D}Q_L + \bar{u}_R \not{D}u_R + \bar{d}_R \not{D}d_R]; \quad (\not{D} = \gamma_\mu D_\mu) \tag{6.6}$$

where:

$$D_\mu\ell_L \equiv (\partial_\mu - \frac{i}{2}g\tau^a \cdot W_\mu^a + \frac{i}{2}g'B_\mu)\ell_L; \quad D_\mu e_R \equiv (\partial_\mu + ig'B_\mu)e_R \tag{6.6a}$$

$$D_\mu Q_L \equiv (\partial_\mu - \frac{i}{2}g\tau_a \cdot W_\mu^a - \frac{i}{6}g'B_\mu)Q_L \tag{6.6b}$$

$$D_\mu u_R \equiv (\partial_\mu - \frac{2i}{3}g'B_\mu)u_R; \quad D_\mu d_R \equiv (\partial_\mu + \frac{i}{3}g'B_\mu)d_R; \quad (a = 1, 2, 3) \tag{6.6c}$$

In Eqs. (6.6a)–(6.6c), g and g' are the gauge coupling constants for $SU(2)_L$ and $U(1)_Y$ respectively, ℓ_L is the lefthanded lepton doublet, e_R is the righthanded charged lepton singlet, Q_L is the lefthanded quark doublet and u_R and d_R (the "up" and "down" weak isospin states respectively) are the righthanded quark singlets in each of the three generations; the W_μ^a's $(a = 1, 2, 3)$ are the three weak gauge fields associated with $SU(2)_L$ and B_μ is the one gauge field associated with the $U(1)_Y$ group. It is important to note the asymmetry in Eqs. (6.6a) and (6.6c), where $\mathcal{L}_F$ only contains the contribution from a single righthanded charged lepton whereas it contains contributions from two righthanded quarks.

The contribution of the scalar (Higgs) field to the Lagrangian, $\mathcal{H}$, is:

$$\mathcal{L}_H = (D_\mu\phi)^\dagger(D^\mu\phi) \tag{6.7}$$

where:

$$D_\mu\phi \equiv (\partial_\mu - \frac{i}{2}g\tau_a \cdot W_\mu^a - \frac{i}{2}g'B_\mu)\phi \tag{6.7a}$$

The Higgs doublet ϕ, when its VEV is non-vanishing, generates a finite mass for three of the four gauge fields, as will be seen below. The purely gauge contribution of the three non-Abelian W_μ^A gauge fields and the single Abelian B_μ gauge field to $\mathcal{L}$, $\mathcal{L}_G$, is:

$$\mathcal{L}_G = -\frac{1}{4} F_{\mu\nu}^a F^{a\mu\nu} - \frac{1}{4} B_{\mu\nu} B^{\mu\nu} \quad (a = 1, 2, 3) \tag{6.8}$$

where:

$$F_{\mu\nu}^a \equiv \partial_\mu W_\nu^a - \partial_\nu W_\mu^a + g\epsilon^{abc} W_\mu^b W_\nu^c; \quad B_{\mu\nu} \equiv \partial_\mu B_\nu - \partial_\nu B_\mu \tag{6.9}$$

One of the important assumptions that is made in constructing the standard electroweak gauge group is to assume that the same scalar (Higgs) field whose SSB is used to generate the finite masses for the three weak bosons, is utilized to generate finite (Dirac) masses for the quarks and charged leptons of the three generations. The massless neutrinos stay massless because the Higgs boson is a doublet [only the SSB due to a Higgs triplet can generate Majorana masses for the initially massless neutrinos (see §6.4b) and this possibility is not considered in the standard model]; in the absence of a neutrino contribution to $\mathcal{L}_Y$, $\mathcal{L}_Y$ becomes:

$$\mathcal{L}_Y = h_e \bar{\ell}_L \phi e_R + h_u \bar{Q}_L \tilde{\phi} u_R + h_d \bar{Q}_L \phi d_R + \text{h.c.} \tag{6.10}$$

where $\tilde{\phi} = i\tau_2 \phi^\dagger$ is (the "charge" conjugate representation of ϕ), and the h's are the "Yukawa" coupling constants to be fixed by experiment (with different values for each fermion of the three generations). When a non-zero VEV is taken for ϕ, finite Dirac masses are generated (see below).

The final step in the construction of the electroweak Lagrangian consists of adding a Higgs potential $V(\phi)$ with the typical SSB form ($\lambda > 0$, $\mu^2 < 0$):

$$V(\phi) = \mu^2 \phi^\dagger \phi + \lambda (\phi^\dagger \phi)^2 \tag{6.11}$$

As is well-known, the assumption of $\mu^2 < 0$ in Eq. (6.11) triggers the SSB of the gauge-invariant Lagrangian (6.6) (see §4.4b); we recall that the minimum of $V(\phi)$ occurs when:

$$< 0|\phi|0 >= 1/\sqrt{2} \binom{0}{v} \quad \text{where } v = \sqrt{-\mu^2/\lambda} \tag{6.12}$$

Since the Higgs doublet possesses the quantum numbers $(2,1)$ under $SU(2)_L \times U(1)_Y$, both parts of the electroweak group are broken; moreover, since:

$$\frac{1}{2}(\tau_3 + Y) < 0|\phi|0 >= 0 \tag{6.13}$$

it follows that $\frac{1}{2}(\tau_3 + Y)$ is an unbroken generator which becomes the conserved electric charge $Q = I_{3L} + Y/2$ of the unbroken group $U(1)_{EM}$ [see Eq. (6.4)].

Once the gauge-invariant Lagrangian (6.5) is spontaneously broken, we can draw a number of important conclusions that define the standard electroweak theory:

(1) The Higgs term of Eq. (6.7) generates the mass terms for the gauge bosons, namely (see §4.4b):

$$\mathcal{L}_{\text{mass}} = \frac{1}{4}g^2v^2W_\mu^+W^{-\mu} + \frac{1}{8}v^2(gW_\mu^3 - g'B_\mu)^2 \qquad (6.14)$$

where $W_\mu^\pm = \frac{1}{\sqrt{2}}(W_\mu^1 \pm iW_\mu^2)$. Equation (6.14) implies that:

$$M_W = \frac{gv}{2} \qquad (6.15)$$

and that one linear combination of the two neutral gauge fields W_μ^3 and B_μ:

$$Z_\mu \equiv \frac{1}{\sqrt{(g^2 + g'^2)}}(gW_\mu^3 - g'B_\mu) \qquad (6.16)$$

acquires the mass:

$$M_Z = \frac{1}{2}\sqrt{(g^2 + g'^2)}\,v \qquad (6.17)$$

while the orthogonal combination of W_μ^3 and B_μ, namely:

$$A_\mu = \frac{g'W_\mu^3 + gB_\mu}{\sqrt{g^2 + g'^2}} \qquad (6.18)$$

represents a massless gauge field and hence the photon.

(2) By considering the interaction of the charged gauge fields $W_\mu^\pm$ with the lepton currents, one obtains for this part of $\mathcal{L}_F$ ($\ell^\pm$ are the "up" and "down" weak isospin leptons respectively — see Table 1.5):

$$\mathcal{L}^{C.C.} = -\frac{g}{\sqrt{2}}[(\bar{\ell}_L^+\gamma^\mu\ell_L^-)W_\mu^+ + (\bar{\ell}_L^-\gamma^\mu\ell_L^+)W_\mu^-] \qquad (6.19)$$

The interaction among the charged gauge bosons and quark currents is obtained from Eq. (6.19) by means of the replacement $\ell_L^+ \to Q_L^+$ and $\ell_L^- \to Q_L^-$. The effective charged current — charged current interaction for μ decay becomes at low energies ($\ll M_W$):

$$\mathcal{L}_{eff} = (\frac{g}{\sqrt{2}})^2\{[\bar{\nu}_\mu\gamma_\lambda(1 - \gamma_5)\mu]\frac{1}{M_W^2}[\bar{e}\gamma^\lambda(1 - \gamma_5)\nu_e]\} \qquad (6.20)$$

Equation (6.20) is the original (V-A) Lagrangian in the low energy domain when one sets $G_F/\sqrt{2} = g^2/8M_W^2$; since $M_W = gv/2$ [from Eq. (6.15)], the SSB scale, v, for breaking the electroweak gauge interaction becomes $v = \sqrt{2}G_F^{-\frac{1}{2}}$, which therefore depends only on the Fermi coupling constant G_F. G_F can be determined most accurately from muon decay (after taking account of the electromagnetic radiative corrections) and turns out to be

$G_F = 1.16637(2) \times 10^{-5}$ GeV^{-2} [1.99] so that the electroweak symmetry breaking scale $\Lambda_{QFD} = v \simeq 246$ GeV. It should be emphasized that, while Λ_{QFD} is known very accurately, neither μ ($\sqrt{2}\mu$ is the Higgs boson mass) nor λ (the coupling constant in the Higgs potential) is known (see §6.3c) but only the combination $v = \sqrt{-\mu^2/\lambda}$. We note that $\Lambda_{QFD} \gtrsim 10^3 \Lambda_{QCD}$ since $\Lambda_{QCD} \sim 150$ MeV.

(3) The gauge coupling constants, g and g', of the broken $SU(2)_L$ and $U(1)_Y$ groups respectively, can be related to the gauge coupling constant e of the unbroken $U(1)_{EM}$ group, thus:

$$g = e \csc \theta_W, \quad g' = e \sec \theta_W \quad \text{or} \quad \tan \theta_W = \frac{g'}{g} \qquad (6.21)$$

where θ_W, called the "Weinberg angle", is a constant of the electroweak gauge theory, to be determined by experiment. Equation (6.21) is evident if we rewrite the fermion terms in Eq. (6.6), containing the neutral gauge fields W_μ^3 and B_μ, in terms of the $SU(2)_L \times U(1)_Y$ charges; we get for the total neutral gauge field contribution to $\mathcal{L}_F$ (dropping the Lorentz indices):

$$\mathcal{L}^{NC} \sim -[(\frac{g^2 I_{3L} - g'^2 Y/2}{\sqrt{g^2 + g'^2}})Z + \frac{gg'}{\sqrt{g^2 + g'^2}}QA] \qquad (6.22)$$

where the first term on the R.H.S. of Eq. (6.22) arises from the interaction of the neutral weak boson field Z^μ (with mass M_Z) with the neutral fermion (quark or lepton) current J_μ^{NC} and the second term from the interaction of the massless electromagnetic gauge field A^μ with J_μ^{NC}. Equation (6.21) follows from identifying Q in Eq. (6.22) with the electric charge. If the Weinberg angle θ_W is determined from experiment, the value of the $SU(2)_L$ and $U(1)_Y$ gauge coupling constants are fixed. Moreover, in the standard model of the electroweak interaction, where the SSB is induced by a Higgs doublet, the ratio of the neutral weak boson mass M_Z to the charged weak boson mass M_W is precisely $\sec \theta_W$. We will see in §6.3b that the relation $M_W = M_Z \cos \theta_W$ provides an independent method for pinning down the value of the Weinberg angle.

(4) We come now to a crucial prediction of the standard electroweak theory, namely that the neutral weak current depends only on the single parameter, $\sin^2 \theta_W$, once I_{3L} and Q [the quantum numbers under $SU(2)_L$ and $U(1)_{EM}$ respectively] are selected from Table 1.5 for the quarks and leptons of each generation (these quantum numbers are identical for all three

generations). This can be seen by writing down $\mathcal{L}^{NC}$ in terms of $\sin^2 \theta_W$:

$$\mathcal{L}^{NC} = -\frac{g}{\cos \theta_W} Z^\mu \sum_k \bar{\psi}_k \gamma_\mu (I_{3L} - Q \sin^2 \theta_W) \psi_k \qquad (6.23)$$

where k is summed over all quarks and leptons and I_{3L} and Q indicate the quantum numbers under $SU(2)_L \times U(1)_Y$ of the quark or lepton current. To put it another way, the chiral flavor assignments of all the quarks and leptons for a single generation are known from Table 1.5, and Eq. (6.23) then informs us that the same value of $\sin^2 \theta_W$ should predict the absolute cross sections for all neutral weak current processes (since g is fixed from the charged weak current processes). Since the chiral flavor assignments are identical for the three generations of quarks and leptons, the same statement holds for all three quark-lepton generations.

(5) The masses of the fermions follow from the Yukawa interaction of Eq. (6.10) when the value of v is inserted for $< 0|\phi|0 >$ provided one knows the Yukawa coupling constants h_f for the quarks and leptons of each of the three generations, thus:

$$m_f = h_f v / \sqrt{2} \qquad (6.24)$$

Since the h_f's are unknown for the quarks and charged leptons of all three generations, the Higgs-induced values of these fermion masses — as given by Eq. (6.24) — are not really predictions. Instead, the observed values of the quark and charged lepton masses — which range from $m_e = 0.5$ MeV to $m_t > 92$ GeV over the three generations (a factor of more than 2×10^5) — determine the Yukawa coupling constants. We have already remarked that the standard model assumes the absence of a righthanded neutrino so that the Dirac mass must be zero [see Eq. (6.10)]; a finite mass Majorana neutrino can be constructed in the absence of ν_R but requires a Higgs triplet (see §6.4b) [1.147].

(6) The SSB of the electroweak Lagrangian given in Eq. (6.5) makes a definite prediction for the quantity $\rho = (M_W / M_Z \cos \theta_W)^2$: ρ is predicted to be unity for the weak boson mass terms because the Higgs scalar is a weak isospin doublet [see Eqs. (6.15) and (6.18)]. Since ρ enters independently as the ratio of the neutral to charged current Lagrangian, it is possible to deduce the value of ρ from neutral weak current experiments at the same time that the parameter $\sin^2 \theta_W$ is determined. If the value of ρ derived in this way differed from unity, this would imply that representations of the Higgs field, other than doublets, contribute to the SSB of the electroweak group. In general, if arbitrary representations contribute to the SSB of the

electroweak group, ρ becomes:

$$\rho = \frac{\Sigma_i v_i^2 (I_i^2 + I_i - Y_i^2/4)}{\frac{1}{2}\Sigma_i v_i^2 Y_i^2} \tag{6.25}$$

where the summation is carried out over the representations i of the scalar fields, v_i is the VEV of the neutral member of the representation i, and Y_i and I_i are the weak hypercharge and weak isospin, respectively, of the ith representation. From Eq. (6.25), it follows that if the Higgs field is a doublet (i.e. $I = \frac{1}{2}$) and possesses $|Y| = 1$, the same value $\rho = 1$ is predicted for any number of Higgs doublets. It turns out that $\rho = 1$ within experimental error (see §6.3a).

(7) We will see in §6.3 that all the properties of the spontaneously broken gauge theory of the electroweak interaction enumerated above are in excellent agreement with experiment. However, as the total electroweak Lagrangian of Eq. (6.5) stands, there is still the problem of maintaining renormalizability of the theory because of the presence of chiral gauge anomalies due to the unbalanced numbers of lefthanded and righthanded fermion representations in the gauge interaction Lagrangian (6.5). The increasingly important role played by chiral gauge anomalies in modern-day particle physics requires a separate chapter (Chapter 7) on the subject. It suffices to state here that all three chiral gauge anomalies (in four dimensions) that are present for the quarks and leptons of each generation — the triangular axial, global $SU(2)$ and mixed gauge-gravitational anomalies — must cancel for each fermion (quark-lepton) generation in order to guarantee the renormalizability and self-consistency of the electroweak gauge theory. Actually, the requirements of chiral gauge anomaly cancellation suggest significant linkages between QCD and QFD (see §8.1) and a gratifying coherence in the present-day standard model.

(8) The theory in the minimal version defined by Eq. (6.5) (i.e. with one complex Higgs doublet) predicts the existence of a single finite mass neutral scalar (Higgs) particle, whose detection in some form ("quasi-" or elementary) is crucial to the validation of the present-day gauge theory of the electroweak interaction. Unfortunately, our ignorance of λ in the quartic term of the Higgs potential [see Eq. (6.11)] makes it impossible to predict the mass (M_H) of the Higgs boson; instead, several theoretical arguments have been advanced — to be discussed in §6.3c — to place lower and upper bounds on M_H [the resulting range of values of M_H is 10 GeV $\lesssim M_H \lesssim$ 1 TeV]. Moreover, recent experiments [6.9] have succeeded in pushing up the lower

limit on M_H to 60 GeV so that present expectation for M_H is in the range: 60 GeV$< M_H \lesssim 1$ TeV. This range of values straddles the mass of the neutral weak boson of the electroweak interaction ($\simeq 100$ GeV) and a determination of M_H is crucial for understanding the true nature of the Higgs SSB mechanism in electroweak interactions. The mysterious role of the Higgs particle in the standard electroweak theory is heightened by the fact that the measured lower bound on the t quark mass ($m_t > 92$ GeV) already exceeds M_W and allows us to contemplate the possibility that the Higgs boson is the fermion condensate $|t\bar{t}|$ (see §6.4c) which, incidentally, possesses the correct quantum numbers (2,1) under the electroweak group $SU(2)_L \times U(1)_Y$ — see Table 1.5. Attempts have been made to invoke other types of dynamical quasi-Higgs mechanisms, such as "technifermions", that might serve as the "quasi-Higgs" doublet(s) needed for the SSB of the electroweak gauge group. A brief discussion of these ideas will be given in §6.4c; here, we merely note that lack of knowledge of the Higgs boson mass and whether the Higgs is "elementary" or "quasi-" (i.e. a "fermion condensate") are unhappy deficiencies of the present-day gauge theory of the electroweak interaction. Indeed, a major reason for building the SSC is to settle this aspect of electroweak theory.

(9) Finally, the Lagrangian given in Eq. (6.5) can be used for the quarks and leptons of all three generations with the caveats that, for leptons, lepton charge conservation is distinct for each family of leptons (i.e. $L_e \neq L_\mu \neq L_\tau$) and, for quarks, the distinction is made between the gauge interaction and mass eigenstates of the quarks via the U_{CKM} quark mixing matrix [2.26]. [The U_{CKM} mixing matrix is trivial (i.e. the identity) for the leptons as long as the three generations of neutrinos are taken to be massless [6.10]]. The attractiveness of the U_{CKM} matrix approach to the quarks is that it provides a concise phenomenological formulation for the mixing angles (between the gauge and mass eigenstates of the quarks) and the single weak CP violation (phase) parameter for the observed three generations of quarks [assuming that m_t exists, for which there is a very strong theoretical expectation (see §6.4c)]. Much work during the past few years has gone into the determination of the three "Cabibbo" angles and the single weak CP violation parameter and we summarize this effort in §6.3e. Even if the outcome of this comprehensive program is favorable — it does look promising at this stage — a deeper understanding of the "fermion generation problem" may require going beyond the present U_{CKM} quark mixing matrix approach (see Chapter 9).

The nine observations enumerated above were drawn from the total electroweak Lagrangian (6.5) for one generation of quarks and leptons, supplemented by the CKM quark mixing matrix. These observations are intended to spell out the conceptual foundations for the present-day gauge theory of the electroweak interaction. We propose, in the remainder of this chapter, to first describe the highlights of electroweak phenomenology — touching upon the manifold experiments (some of them of high precision) that support the standard theory — and then to discuss some of the more probing tests that may ultimately lead to refinements in the standard model.

§6.3. Highlights of electroweak phenomenology

As stated, the standard electroweak model — defined by the total Lagrangian of Eqs. (6.5)–(6.11), the quark and lepton quantum numbers of Table 1.5, the Higgs quantum numbers $(\mathbf{2}, \mathbf{1})$ under $SU(2)_L \times U(1)_Y$, and the U_{CKM} quark mixing matrix for three generations (see §6.3e) has achieved remarkable phenomenological success. In the next several sections, we review highlights of these phenomenological successes to clarify how the key parameters of the standard electroweak model have been determined as well as to demonstrate the theory's versatility and richness of prediction. The need to "go beyond the standard model" and some of the directions that have been put forward to satisfy this need, are postponed to §6.4–§6.4c of this chapter (and to later chapters).

The highlights of electroweak phenomenology are discussed under the following six headings: §6.3a: Determination of ρ and $\sin^2 \theta_W$ from neutral weak current experiments; §6.3b: Masses and decay widths of the weak gauge bosons and the number of neutrino species; §6.3c: Properties of the Higgs boson; §6.3d: Three generations of leptons and lepton charge conservation; §6.3e: Three generations of quarks and the determination of the U_{CKM} quark-mixing matrix; and §6.3f: Weak CP violation.

§6.3a. Determination of ρ and $\sin^2\theta_W$ from neutral weak current experiments

The key to the determination of ρ and $\sin^2 \theta_W$ has been the study of weak processes involving neutral current interactions at energies substantially below 100 GeV (the mass region of the weak gauge bosons). We therefore return to Eq. (6.23) and rewrite it in the form:

$$H^{NC} = \frac{4\rho}{\sqrt{2}} G_F J_\mu^{\dagger NC} J^{\mu NC} \qquad (6.26)$$

where H^{NC} is the effective Hamiltonian, J_μ^{NC} is the neutral (electroweak) current summed over all quark and lepton flavors, G_F has replaced $\frac{g^2}{8m_W^2}$, and ρ has already been defined as $(M_W/M_Z \cos\theta_W)^2$; implicit in Eq. (6.26) is the assumption that the three weak bosons $W^\pm$ and Z possess vanishing baryon and lepton charges (i.e. $B = L = 0$ for $W^\pm$, Z). Since we limit ourselves to the class of neutrino-induced neutral current weak processes [i.e. $\nu - e$ scattering and $\nu - N$ scattering (which is a mix of $\nu - u$ and $\nu - d$ scattering)], we can write Eq. (6.26) in the form:

$$H_{NC}^\nu = 2\sqrt{2}G_F\{\bar{\nu}_L\gamma_\mu\nu_L\} \sum_f [\epsilon_L(f)(\bar{f}_L\gamma^\mu f_L) + \epsilon_R(f)(\bar{f}_R\gamma^\mu f_R)] \quad (6.27)$$

where $f = e, u, d$ and the ϵ_L's and ϵ_R's are given by:

$$\epsilon_L(\ell) = \rho(-\frac{1}{2} + \sin^2\theta_W); \quad \epsilon_R(\ell) = \rho(\sin^2\theta_W)$$

$$\epsilon_L(u) = \rho(\frac{1}{2} - \frac{2}{3}\sin^2\theta_W); \quad \epsilon_R(u) = \rho(-\frac{2}{3}\sin^2\theta_W)$$

$$(6.27a)$$

$$\epsilon_L(d) = \rho(-\frac{1}{2} + \frac{1}{3}\sin^2\theta_W); \quad \epsilon_R(d) = \rho(\frac{1}{3}\sin^2\theta_W)$$

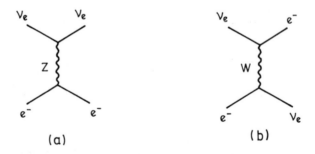

(a)   (b)

Fig. 6.1. Feynman diagrams contributing to $\nu_e - e$ scattering in the standard model; only the first Feynman diagram contributes to $\nu_\mu - e$ scattering.

The neutral current form of Eq. (6.27) [with the definitions of ϵ_L and ϵ_R given by Eq. (6.27a)] represents the contribution to the electroweak interaction between the neutrinos (both ν_e and ν_μ) and the quarks and charged leptons, generated by the neutral Z boson, as shown in Fig. 6.1a. When the neutrinos interact with charged leptons of the same family (i.e. $\nu_e - e$ scattering), the contribution of Fig. 6.1a must be augmented by the contribution of Fig. 6.1b, but with the second interaction generated by the charged W boson (and the neutral weak lepton current replaced by the charged weak lepton current). The parameters $\epsilon_L(\ell)$ and $\epsilon_R(\ell)$ of Eq. (6.27a) can be obtained directly from the neutrino (neutral-current) experiments with leptons and, if the "quark-parton" model is invoked, from the corresponding experiments on nucleons. We will see that the neutrino experiments with both charged leptons and quarks give consistent determinations of ρ and $\sin^2 \theta_W$.

The cleanest determination of $\sin^2 \theta_W$ is made by measuring the ratio $\mathcal{R}$ of the cross sections for $\nu_\mu - e$ and $\bar{\nu}_\mu - e$ scattering; this measurement is independent of ρ because one uses neutrinos and antineutrinos of the same family to make the comparison. One has, using Eq. (6.27a):

$$\mathcal{R} = \frac{\sigma(\nu_\mu e \to \nu_\mu e)}{\sigma(\bar{\nu}_\mu e \to \bar{\nu}_\mu e)} = \frac{[3\epsilon_R^2(e) + \epsilon_L^2(e)]}{[3\epsilon_L^2(e) + \epsilon_R^2(R)]} = \frac{3 - 12\sin^2\theta_W + 16\sin^4\theta_W}{1 - 4\sin^2\theta_W + 16\sin^4\theta_W}$$

$$(6.28)$$

The value for $\sin^2 \theta_W$ deduced from such experiments at CERN [6.12], including radiative corrections (with $m_t \simeq M_H \simeq 100$ GeV), is:

$$\sin^2 \theta_W = 0.232 \pm 0.012 \pm 0.008 \qquad (6.29)$$

Since the actual values of the $\nu_\mu - e$ and $\bar{\nu}_\mu - e$ scattering cross sections are also measured, one may determine the ρ parameter from the same experiments; the results are:

$$\frac{\sigma(\nu_\mu e \to \nu_\mu e)}{E_\nu} = (2.2 \pm 0.4 \pm 0.4) \times 10^{-42} \frac{\text{cm}^2}{\text{GeV}}$$

$$\frac{\sigma(\bar{\nu}_\mu e \to \bar{\nu}_\mu e)}{E_\nu} = (1.6 \pm 0.3 \pm 0.3) \times 10^{-42} \frac{\text{cm}^2}{\text{GeV}} \qquad (6.30)$$

which yields:

$$\rho = 1.10 \pm 0.08 \pm 0.13 \qquad (6.31)$$

The accuracy of the ρ determination, as given by Eq. (6.31), is not good but at least one can say that the hypothesis of a Higgs doublet SSB is consistent with the experimental data, thereby verifying the quantum numbers $(2, 1)$ of the Higgs boson under the standard electroweak group.

A more accurate determination of the ρ value is made when the "world average" is taken of a large number of deep inelastic lepton-hadron experiments using the "quark-parton" model. In this fashion, a fit of all the data ("world average" [6.13]) obtained from the deep inelastic neutrino-nucleon experiments via the process: $\nu_\mu N \to \nu_\mu X$, leads to the following values for $\sin^2 \theta_W$ and ρ:

$$\sin^2 \theta_W = 0.232 \pm 0.014 \pm 0.008; \quad \rho = 0.999 \pm 0.013 \pm 0.008 \qquad (6.32)$$

It is seen that this value of $\sin^2 \theta_W$ (and its errors) agrees well with the value obtained from the neutrino-electron scattering experiments whereas the value of ρ, if taken at face value, is very close to unity. We can say that two pillars of the standard electroweak gauge theory — the constancy of $\sin^2 \theta_W$ and the value of unity for ρ — are well confirmed by neutral current-induced neutrino scattering experiments.

It should be noted that as early as 1978, the SLAC–Yale experiment [6.14] — measuring the parity-violating asymmetry in the deep inelastic scattering of polarized electrons on deuterium — was not only a convincing blow against the mistaken atomic parity-violating experiments (in heavy nuclei) at that time (claiming the absence of such an effect [6.15]), but it also led to a value of $\sin^2 \theta_W$ (0.231) in agreement with the most recent value. This was an early triumph for the standard electroweak theory, which requires the same $\sin^2 \theta_W$ to enter the expression for the neutral electron current in all electroweak processes.

§6.3b. Masses and decay widths of the weak gauge bosons and the number of neutrino species

In the previous section, it was established that two key predictions of the standard electroweak theory are fully confirmed by experiment, namely: (1) the constancy of the Weinberg angle θ_W for quark and lepton weak processes; and (2) the value unity for ρ, which follows from the weak isospin doublet character of the Higgs boson built into the standard electroweak theory. Even more striking confirmation of the standard electroweak theory comes from recent measurements of the masses and decay widths of the weak (gauge) bosons, Z and W, especially the Z boson. Indeed, probes of the electroweak sector of the standard model — for Z (not W as yet) — are entering a new era with the advent of precision measurements in the neighborhood of the Z resonance in $e^+ e^-$ annihilation, carried out at LEP (large electron-positron collider at CERN) and, to a lesser extent, at SLC (Stanford Linear Collider). These experiments are supplemented by W boson measurements in $p\bar{p}$ colliders (at Fermilab and at CERN). The Z weak boson experiments give a precision determination of M_Z and the partial decay widths of Z; the Z partial width measurements have led to a good determination of the number of neutrino species (and, *a fortiori*, the number of quark-lepton generations) — see below. The W weak boson mass measurements have led to a new and independent determination of $\sin^2 \theta_W$ (see below) but precision measurements of the W weak boson partial decay widths have not as yet proved feasible.

We know (from §6.2) that, to lowest-order in α_e, the W and Z masses are determined in terms of α_e, G_F and θ_W as follows:

$$\text{(a)} \quad M_W^2 = \frac{\pi\alpha_e}{\sqrt{2}G_F}\frac{1}{\sin^2\theta_W} \qquad \text{(b)} \quad M_Z^2 = \frac{M_W^2}{\cos^2\theta_W} \qquad (6.33)$$

It is clear from Eq. (6.33) that independent measurements of the charged and neutral weak boson masses can give yet another determination of $\sin^2\theta_W$ — through the ratio M_W/M_Z — although this determination is not as reliable because of sizable radiative corrections (see below). Because of the large W and Z masses, the higher-order electromagnetic radiative corrections become non-trivial and it is necessary to pay attention to these corrections in order to deduce the masses. It is convenient to define θ_W in terms of physical quantities in order to remove the ambiguity due to the renormalization procedure; in addition, the common prescription [6.16] is to maintain the relation: $M_W = M_Z\cos\theta_W$, even after loop effects are included. M_W then receives radiative corrections that can be written in the form [6.17]:

$$M_W^2 = \frac{\pi\alpha_e}{\sqrt{2}G_F}\frac{1}{\sin^2\theta_W}\frac{1}{1-\Delta r} \qquad (6.34)$$

where Δr is a quantity of order α_e and contains two large terms, thus:

$$\Delta r \simeq \frac{\alpha_e}{\pi}[\frac{1}{3}\sum_f N_C^f e_f^2 \ln(M_Z^2/m_f^2) - \frac{3}{16\sin^4\theta_W}\frac{m_t^2}{M_Z^2}] \qquad (6.35)$$

In Eq. (6.35), N_C^f is the "color weight" [1 for lepton flavor and 3 for quark flavor — see Table 1.5] and the first term is a sum over fermions lighter than Z (and is logarithmic because this term takes into account the "running" of α_e). The second term in Eq. (6.35) results from the effect of "radiative corrections" on $(\rho-1)$ and becomes important if the top quark mass is large (since the Yukawa coupling is proportional to the fermion mass); this term is negative and can almost cancel the first term for $m_t \sim 250$ GeV. On the other hand, the "radiative corrections" due to the Higgs boson have only a logarithmic dependence on M_H [e.g. changing M_H from 10 to 10^3 GeV only changes Δr by about 1%] — in contrast to the power behavior of m_t. Evidently, the uncertainty in our knowledge of m_t introduces the major error in the deduction of $\sin^2\theta_W$ from Eq. (6.33b).

We will list the rather precise values of M_Z and M_W that have been measured recently but, before doing so, we write out the partial decay widths of the W and Z bosons. While there have been precision measurements of the Z boson partial decay widths, there are as yet no meaningful measurements of the partial decay widths for the W boson. Nevertheless, we also record the theoretical predictions for the W boson decays before turning to the theoretical predictions and measurements of the Z boson decays, because the comparison is instructive. One expects the decay modes of W and Z to be of two types, leptonic and hadronic; clearly:

$$W^- \longrightarrow \begin{cases} e^-\nu_e, \mu^-\bar{\nu}_\mu, \tau^-\bar{\nu}_\tau \\ \bar{u}d, \bar{c}s \end{cases} \quad [(t\bar{b}) \text{ is excluded as a final state}] \quad (6.36a)$$

$$Z^- \longrightarrow \begin{cases} \nu_e\bar{\nu}_e, \nu_\mu\bar{\nu}_\mu, \nu_\tau\bar{\nu}_\tau; \; e\bar{e}, \mu\bar{\mu}, \tau\bar{\tau} \\ u\bar{u}, d\bar{d}, c\bar{c}, s\bar{s}, b\bar{b} \end{cases} \quad [t\bar{t} \text{ is excluded as a final state}]$$

$$(6.36b)$$

The partial decay width of W to a single lepton pair is given by (neglecting the charged lepton mass):

$$\Gamma(W^- \to \ell^-\bar{\nu}_\ell) = \frac{G_F}{6\sqrt{2}\pi} M_W^3 = 229 \pm 3 \text{ MeV} \quad (6.37)$$

for $M_W = 80.6$ GeV. The total width of W, including quark pair final states and taking account of 3 colors per quark, is then (to order α_s):

$$\Gamma_W = \frac{3G_F}{2\sqrt{2}\pi} M_W^3 \left(1 + \frac{2}{3}\Delta\right) \quad (6.38)$$

where the fermion masses are neglected and Δ is the QCD correction factor for decay into quark pairs, namely:

$$\Delta = \frac{\alpha_s}{\pi} + O(\alpha_s^2) \quad (6.38a)$$

For $M_W = 80.6$ GeV and $\alpha_s = 0.12$ (see §5.2a), the total width $\Gamma_W \sim$ 2.11 GeV.

Turning to Z boson decay, the lowest-order standard model predictions for the partial leptonic decay widths of Z are:

$$\Gamma(Z \to \nu\bar{\nu}) = \frac{G_F M_Z^3}{12\sqrt{2}\pi}; \quad \Gamma(Z \to \ell\bar{\ell}) = \frac{G_F M_Z^3}{12\sqrt{2}\pi}(1 - 4s_W^2 + 8s_W^4) \quad (6.39a)$$

where $s_W = \sin\theta_W$. The quark decay widths of Z are:

$$\Gamma(Z \to u\bar{u}) = \frac{G_F M_Z^3}{12\sqrt{2}\pi}\left(3 - 8s_W^2 + \frac{32}{3}s_W^4\right)(1 + \Delta);$$

$$\Gamma(Z \to d\bar{d}) = \frac{G_F M_Z^3}{12\sqrt{2}\pi}\left(3 - 4s_W^2 + \frac{8}{3}s_W^4\right)(1 + \Delta) \quad (6.39b)$$

The use of G_F instead of $\alpha_e/\sin^2\theta_W \cos^2\theta_W$ reduces the size of the $O(\alpha_e)$ corrections, since the effect of "running" α_e is incorporated. The total width of Z for three generations of leptons and five flavors of quarks (no $t\bar{t}$) is predicted to be:

$$\Gamma_Z = \frac{G_F M_Z^3}{\sqrt{2}\pi}[(\frac{1}{2} - s_W^2 + 2s_W^4) + (\frac{5}{4} - \frac{7}{3}s_W^2 + \frac{22}{9}s_W^4)(1 + \Delta)] \quad (6.40)$$

We are now in a position to compare the theoretical predictions for M_Z and Γ_Z with the recent precision experiments. These experiments have been carried out with colliding electron-positron beams (at LEP and SLC), chiefly at the Z boson resonance, and the scanning of this resonance by means of the produced fermion pairs (both leptons and quarks) yields values of M_Z and Γ_Z. To show how this is done, we write down the differential cross section (to lowest order) for the process $e\bar{e} \to f\bar{f}$ [f is the fermion (lepton or quark)], prescribed by the standard electroweak theory, without regard to the final fermion polarizations; if θ is the fermion polar angle, the differential cross section becomes:

$$\frac{d\sigma_{f\bar{f}}}{d\cos\theta} = \frac{\pi\alpha_e^2 Q_f^2 N_C^f}{2s}(1 + \cos^2\theta) - \frac{\alpha_e Q_f N_C^f G_F M_Z^2(s - M_Z^2)}{8\sqrt{2}[(s - M_Z^2)^2 + M_Z^2\Gamma_Z^2]}$$

$$\times \{[\epsilon_R(e) + \epsilon_L(e)][\epsilon_R(f) + \epsilon_L(f)](1 + \cos^2\theta)$$

$$\times [+2[\epsilon_R(e) - \epsilon_L(e)][\epsilon_R(f) - \epsilon_L(f)]\cos\theta]$$

$$+ \frac{N_C^f G_F^2 M_Z^4 s}{64\pi[(s - M_Z^4)^2 + M_Z^2\Gamma_Z^2]}$$

$$\times [\epsilon_R^2(e) + \epsilon_L^2(e)][\epsilon_R^2(f) + \epsilon_L^2(f)](1 + \cos^2\theta)$$

$$+ 2[\epsilon_R^2(e) - \epsilon_L^2(e)][\epsilon_R^2(f) - \epsilon_L^2(f)]\cos\theta\} \quad (6.41)$$

where Q_f is the fermion electric charge, $s = E_{c.m.}^2$, and the ϵ_L's and ϵ_R's are given by Eq. (6.27a). In deriving Eq. (6.41), both diagrams in Fig. 6.2 are taken into account, and the Breit-Wigner formula has been used for the Z propagator (appropriate for an unstable particle with mass M_Z and total width Γ_Z). The first term of Eq. (6.41) represents the electromagnetic contribution [due to diagram (a) of Fig. 6.2], the third term represents the weak interaction contribution [due to diagram (b) of Fig. 6.2], and the second term receives its contribution from the interference of the electromagnetic and weak interactions. For $s \ll M_Z^2$, the first term clearly dominates but, at the Z boson pole (i.e. when $s = M_Z^2$), the second (interference) term vanishes and the third (weak interaction) term is much enhanced over the electromagnetic term and assumes the form:

$$\frac{d\sigma_{f\bar{f}}}{d\cos\theta} = \frac{N_C^f G_F^2 M_Z^4}{64\pi\Gamma_Z^2}[(a_e^2 + v_e^2)(a_f^2 + v_f^2)(1 + \cos^2\theta) + 8a_e v_e a_f v_f \cos\theta] \quad (6.42)$$

where $a = \epsilon_L - \epsilon_R$ and $v = \epsilon_L + \epsilon_R$, are, respectively, the "axial vector" and "vector" coupling constants. For recent experiments at the Z boson resonance, the form (6.42) is useful and we give in Table 6.1 the comparison between experiment [6.18] and theory for the a and v parameters of Eq. (6.42) for the leptons and quarks of the first generation; two sets of theoretical numbers are given corresponding to $m_t = 100$ and 200 GeV. The agreement between theory and experiment is fairly good but, clearly, more accuracy is required on the experimental side and knowledge of m_t would be helpful.

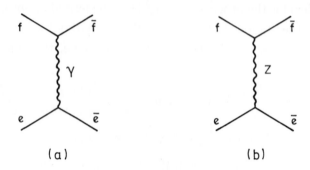

(a)　　　　　　　　　(b)

Fig. 6.2. Lowest order contributions to the reaction $e\bar{e} \rightarrow f\bar{f}$ in the standard model.

Table 6.1: a, v parameters with $m_t = 100(200)$ GeV.

Quantity	Experimental value	Standard model prediction
a_e	-0.513 ± 0.025	$-0.504(-0.509)$
a_u	0.505 ± 0.018	$0.498(0.503)$
a_d	$-0.459\ {}^{+0.078}_{-0.049}$	$-0.505(-0.509)$
v_e	-0.045 ± 0.022	$-0.036(-0.042)$
v_d	$-0.413\ {}^{+0.078}_{-0.049}$	$-0.3439(-0.353)$

We next summarize in Table 6.2 recent LEP results [6.18] for M_Z and its partial decay widths $\Gamma_Z^q, \Gamma_Z^\ell$ and Γ_{inv}, from which N_ν (the number of neutrino species) is derived; the relevant standard model predictions are also given in Table 6.2. In Table 6.2, Γ_{inv} is the "invisible" contribution to Γ_Z [i.e. the contribution of lefthanded neutrinos whose masses are sufficiently small ($\lesssim M_Z/2$ GeV) so that they can contribute to Γ_{inv}], which is equal to

$\Gamma_Z - \Gamma_Z^q - 3\Gamma_Z^\ell$; the central theoretical values are given for $m_t = 134$ *GeV* and $M_H = 100$ GeV [1.132] and the uncertainties include variations of the parameters within the bound 92 GeV$< m_t < 200$ GeV, 60 GeV$< M_H < 1$ TeV. The value of Γ_Z^ℓ tells us that the Z boson decays predominantly to quarks (i.e. hadrons), in the ratio of about 3 : 1; this is further confirmation — if that were necessary — for the three color degrees of freedom enjoyed by the quarks as compared to the color singlet leptons. Finally, the measurement of N_ν is most interesting because it is the first particle physics measurement of the number of generations [6.19] — assuming that the t quark is found — and tells us that there exist only three quark-lepton generations (provided the neutrino mass does not exceed half of the Z mass).

Table 6.2: LEP results and standard model predictions for M_Z and Γ_Z.

	M_Z (GeV)	Γ_Z (GeV)	Γ_Z^q (GeV)	Γ_Z^ℓ (GeV)	Γ_{inv} (MeV)	N_ν
LEP	91.177 ± 0.1021	2.498 ± 0.013	1.764 ± 0.014	83.70 ± 0.65	482 ± 13	2.92 ± 0.08
standard model		$2.492\,^{+0.031}_{-0.025}$	$1.741\,^{+0.025}_{-0.021}$	$83.69\,^{+0.84}_{-0.61}$	$500\,^{+4}_{-3}$	

The value $N_\nu \simeq 3$ agrees surprisingly well with the number of neutrino species deduced from the "big bang" hypothesis of the universe, i.e. $N_\nu = 2.7 \pm .2$ [6.20]. Indeed, the good agreement between the accelerator-deduced N_ν and the astrophysical deduction of the same quantity is a signal triumph for astrophysics. It should be noted that it is also possible to exploit the knowledge of the "invisible" width $\Gamma_{inv}(Z)$ to place limits on the presence of unknown decay products of Z [6.18]; e.g. it is established that there are no new "standard" quarks of the bottom or top type below 46 GeV/c^2, there is no sequential heavy lepton below 44.3 GeV/c^2, there are no squarks nor sleptons below 43 GeV/c^2, etc. This is quite a menagerie of unseen particles but, while one is constantly on the alert for evidence of the t quark and Higgs boson — which are expected on the basis of the standard model — it is useful to be on the lookout for the more hypothetical particles which might be present under a variety of speculative extensions of the standard model (e.g. supersymmetric partners — see §2.3a).

We conclude this section with mention of the status of the independent determination of $\sin^2 \theta_W$ from a comparison of the measured values of M_Z and M_W. While the present electron-positron colliders — chiefly LEP — are ideal for doing experiments at the Z boson resonance, the best — albeit not as accurate — measurements of M_W have come from the proton-anti-proton colliders at Fermilab and at CERN. The average value obtained from these experiments is: $M_W = 80.25 \pm 0.35$ GeV. This value of M_W and the value of M_Z from the LEP experiments, enable one to deduce an independent value of $\sin^2 \theta_W$; using the relation $\sin^2 \theta_W = 1 - M_W^2/M_Z^2$, one obtains $\sin^2 \theta_W = 0.225 \pm 0.007$. As noted earlier, this value of $\sin^2 \theta_W$ strongly depends on m_t [through Δr — see Eq. (6.35)] and m_t is still unknown. [Experiments to find the t quark and to measure its mass have thus far been unsuccessful and only lower limits on m_t have been established — see §6.3d.] The preliminary value of $\sin^2 \theta_W$ deduced from the ratio M_W/M_Z is satisfactory but a very accurate value must await knowledge of m_t. Actually, one can use the preliminary value of $\sin^2 \theta_W$, taken together with the precision determination of M_Z, to derive an upper bound on m_t, which turns out to be 200 GeV. [In comparison, the low-energy charged and neutral current cross sections give an upper bound on m_t of about 250 GeV (the so-called $(\rho - 1)$ constraint).]

§6.3c. Properties of the Higgs boson

The precision measurements of key parameters of the Z (and to a lesser extent) W weak gauge bosons are very gratifying. They confirm the capability of the standard electroweak gauge theory to make definite predictions concerning the masses and decay widths of the Z and W weak bosons (as well as other properties) — provided it is assumed that one or more doublet Higgs bosons (with weak hypercharge $|Y| = 1$) are responsible for the SSB of the electroweak group. Unfortunately, the standard electroweak theory is incapable of predicting the mass of the Higgs boson (and, *a fortiori*, its decay width), or even to say whether the Higgs boson is an "elementary" particle (say, with the same status as the Z boson) or a quasi-Higgs particle (e.g. a $t\bar{t}$ quark condensate). The reason for the inability of the electroweak theory to predict the Higgs mass, M_H, is easy to see at "tree level" where, using Eqs. (6.11) and (6.12), we get:

$$M_H^2 = 2\lambda v^2 \tag{6.43}$$

While the Higgs SSB scale ($v = \Lambda_{QFD}$) is known very accurately, the Higgs boson mass M_H is obviously not fixed since it depends on the unknown coupling constant λ of the quartic self-interaction in the "Higgs" potential of Eq. (6.11).

In view of our complete ignorance of the value of λ until recently, theoretical attempts were made — using plausible properties of gauged quantum field theories — to establish lower and upper bounds on λ and, hence, M_H. Recently, an experimental lower limit of 60 GeV [6.9] has been set on M_H and has replaced the theoretical lower bound as the number to be reckoned with. However, some methodological interest attaches to the theoretical argument for the lower bound on M_H and, hence, we mention it at the same time that we consider the theoretical argument for the upper bound. Both the lower and upper bounds on the Higgs mass are derived by appealing to the constraints of "vacuum stability" and "triviality" respectively, and, indeed, both bounds can be discussed by making use of renormalization group analysis at the one-loop level. For this purpose, we look at the evolution of the three coupling constants g_s, g and g' [respectively associated with the three parts of the standard group $SU(3)_C \times SU(2)_L \times U(1)_Y$], the t quark coupling constant $h = m_t\sqrt{2}/v$ to the Higgs boson [since the other fermions are too light to affect the result] and λ. The renormalization group equations for the gauge couplings g_s, g and g' are well-known, namely [6.21]:

$$\frac{d}{dt}g_s^2 = -\frac{(33 - 4N_g)}{48\pi^2}g_s^4; \qquad \frac{d}{dt}g^2 = -\frac{(22 - 4N_g - \frac{1}{2}N_H)}{48\pi^2}g^4$$

$$\frac{d}{dt}g'^2 = \frac{(\frac{20}{3}N_g + \frac{1}{2}N_H)}{48\pi^2}g'^4 \tag{6.44}$$

where N_g is the number of generations ($N_g = 3$ for standard model), N_H is the number of complex scalar doublets ($N_H = 1$ for standard model) and $t = \ln(Q^2/Q_0^2)$ [with Q^2 the four-momentum transfer squared and Q_0^2 the reference scale for renormalization]. It is to be noted that Eqs. (6.44) are independent of both h and λ, which have their own "running" equations (with the "running" equation for h not depending on λ); they are:

$$\frac{d}{dt}h = \frac{1}{16\pi^2}[\frac{9}{4}h^3 - 4g_s^2h - \frac{9}{8}g^2h - \frac{17}{24}g'^2h] \tag{6.45}$$

$$\frac{d}{dt}\lambda = \frac{1}{16\pi^2}\{12\lambda^2 + 6\lambda h^2 - 3h^4 - \frac{3}{2}\lambda(3g^2 + g'^2) + \frac{3}{16}[2g^4 + (g^2 + g'^2)^2]\} \tag{6.45a}$$

One can now derive lower and upper bounds on the Higgs mass. The lower bound [6.22] on M_H is derived by considering the behavior of the "running" coupling constant λ in the limit where M_H is small (corresponding to small λ); neglecting terms containing λ in Eq. (6.45a), one obtains:

$$\frac{d}{dt}\lambda = \beta_\lambda, \tag{6.46}$$

where:

$$\beta_\lambda = \frac{1}{16\pi^2}[-3h^4 + \frac{3}{16}(2g^4 + (g^2 + g'^2)^2)] \tag{6.47}$$

Approximating β_λ by a constant and integrating up to scales Q of the order of the Higgs field ϕ itself, produces the logarithmic expression:

$$\lambda(\phi) = \lambda(Q_0) + \beta_\lambda \ln\frac{\phi^2}{Q_0^2} \tag{6.48}$$

where $\phi^2 \equiv \phi^\dagger\phi$. The $\lambda(\phi)$ so defined can be used to specify the one-loop effective potential as follows [see Eq. (6.11)]:

$$V \simeq \mu^2\phi^2 + \lambda(\phi)\phi^4 = \mu^2\phi^2 + \lambda(Q_0)\phi^4 + \beta_\lambda\phi^4 \ln(\frac{\phi^2}{Q_0^2}) \tag{6.49}$$

The VEV, determined by the requirement:

$$\frac{\partial V}{\partial\phi}\Big|_{\phi=v/\sqrt{2}} = 0 , \tag{6.50}$$

yields for M_H:

$$M_H^2 = \frac{1}{2}\frac{\partial^2 V}{\partial\phi^2}\Big|_{\phi=v/\sqrt{2}} = \frac{-8}{v^2}V(\frac{v}{\sqrt{2}}) + \beta_\lambda v^2 \tag{6.51}$$

Since $V(0) = 0$ when there is no symmetry breaking, the requirement that the symmetry-breaking vacuum is preferred, i.e. $V(v/\sqrt{2}) < V(0)$ (the condition of "vacuum stability"), implies:

$$M_H^2 > \beta_\lambda v^2 = \frac{3}{16\pi^2 v^2}[2M_W^4 + M_Z^4 - 4m_t^4] \tag{6.52}$$

Equation (6.52) is the final result for the lower limit on the Higgs mass using the "vacuum stability" argument; when this limit was first derived, it was expected that $m_t << M_W$ and that one could use Eq. (6.52) to derive the lower bound $M_H \geq 7$ GeV. However, as we have already remarked, recent measurements have placed a lower bound of 92 GeV on m_t and evidently, this high value invalidates the use of Eq. (6.52) (whose R.H.S. becomes negative!). To correct the deficiency of the one-loop calculation that leads to Eq. (6.52), the "vacuum stability" lower bound at the two-loop level has been studied, and the overall conclusion is not too helpful, namely the lower bound on M_H can be substantially larger (increasing with the value of m_t for $m_t > 92$ GeV), but still below the observed lower limit of 60 GeV. It seems sensible to accept 60 GeV as the lower limit of the Higgs mass and to keep pushing the experimental search for the Higgs that will, in the foreseeable future, establish its existence or at least place a reliable lower limit on its mass.

The derivation of the upper limit on the Higgs mass also has its limitations but it raises important questions of principle concerning the "triviality" of the Higgs sector in the standard electroweak theory that must be addressed. Thus, consider the ungauged scalar (Higgs) theory [which is equivalent to considering the scalar ϕ^4 theory that served as a model for deriving the coupling constant (λ) renormalization constant in §4.5]; from our previous discussion (in §5.2a), we know that the ϕ^4 (or Higgs potential) theory possesses a stable "infrared" fixed point at the "origin" — like QED — and the "running" coupling constant $\lambda(Q)$ acquires the form:

$$\lambda(Q) = \frac{\lambda(v)}{1 - \lambda(v)\frac{3}{4\pi^2}\ln\frac{Q^2}{v^2}} \tag{6.53}$$

where the renormalization reference point can be taken as v, the Higgs SSB scale. Except for a numerical factor in the coefficient of $\ln Q^2/v^2$ in the denominator on the R.H.S. of Eq. (6.53), the "running" coupling constant for the Higgs sector is subject to the same "disease" as the "running" electric charge of QED which provoked Landau's conjecture that a consistent QED quantum field theory requires a zero renormalized charge, i.e. that a consistent QED suffered from "triviality" (the disease of the "Moscow zero" — see §4.5). The term "triviality" — applied to the Higgs sector [6.21] — is simply the counterpart of QED "triviality", namely the hypothesis that the coupling $\lambda(v)$ at low energy is driven to zero (implying a "trivial" non-interacting theory) if we let $Q \to \infty$ while keeping $\lambda(Q) > 0$ (as required for stability of the vacuum). However, the "trivality" problem of the Higgs sector in electroweak theory may be more serious than the corresponding problem in QED because the coupling constant λ can be quite large (if M_H is large) — much larger than α_e — and if one inserts the upper limit on M_H (1 TeV — see below), one finds a cutoff Λ — to avoid "triviality" — of the same order [6.23]; this result is in sharp contrast to the QED situation where the corresponding cutoff — to avoid Landau "triviality" — must be of the order of 10^{276} GeV. There is one obvious caveat with regard to the "triviality" problem — whether in the Higgs sector or QED — and that is that the one-loop approximation has been used in deriving the expression for the "running" coupling constant and it is conceivable that higher-loop or non-perturbative corrections can produce a stable "ultraviolet" fixed point away from the "origin", thereby preventing the growth of $\lambda(Q)$ beyond that point.

With the above caveat, a crude estimate of the upper bound on M_H can be obtained from the one-loop renormalized coupling constant, i.e. taking

$\lambda(Q) \to \infty$ in Eq. (6.53); we find:

$$[\lambda(v)]_{\max} \simeq \frac{2\pi^2}{3\ln(\frac{Q}{v})} \tag{6.54}$$

and thus, the upper bound is:

$$M_H = \sqrt{2\lambda}\, v \le \frac{2\pi}{\sqrt{3}} \frac{v}{(\ln\frac{Q}{v})^{\frac{1}{2}}} \simeq 1 \text{ TeV} \tag{6.55}$$

for $\ln\frac{Q}{v} \sim 0(1)$. This crude estimate of the upper bound on M_H agrees with numerical studies of the Higgs model on a lattice and with other approaches to the upper bound on M_H. For example, a similar bound ($\simeq 1$ TeV) [6.24] is obtained by observing that "tree-level" unitarity is violated if M_H is too large. It should be noted that an upper bound of 1 TeV corresponds to a strong coupling constant λ in the Higgs potential and may not be too reliable. Indeed, it has been shown by some authors [1.135] that, for large λ, a completely new feature may enter the strong coupling régime and lead to soliton-like solutions for the Higgs boson with newly-acquired "topological" quantum numbers like baryon and/or lepton number.

Without knowing the mass of the hypothetical Higgs boson (or whether there is more than one Higgs doublet), it is difficult to make definite statements about its production and decay mechanisms, even assuming that the Higgs boson is "elementary". However, some qualitative comments are worth making: thus, if only one (complex) Higgs scalar doublet is assumed in the electroweak theory, we need only consider one residual neutral Higgs particle (denoted by H^0) after SSB, with a mass that can range from about 60 GeV to about 1 TeV. With $M_H \le 100$ GeV, the only term [see Eq. (6.10)] that can contribute to the H^0 decay width Γ_H is the one that couples H^0 to a fermion pair, i.e. $\frac{m_f}{v}\bar{f}fH$ [if $m_t > M_H/2$, this interaction favors the decay of the H^0 Higgs into a $b\bar{b}$ pair]. This term contributes the (partial) width Γ_H^f for $H^0 \to f\bar{f}$:

$$\Gamma_H^f = \frac{G_F M_H m_f^2}{4\pi\sqrt{2}} N_C^f (1 - \frac{4m_f^2}{M_H^2})^{\frac{3}{2}} \tag{6.56}$$

where N_C^f is 1 for a lepton and 3 for a quark. When M_H is sufficiently large, the direct couplings of H^0 to the weak gauge bosons W and Z in the weak interaction Lagrangian of Eq. (6.7) contribute two-body decays at the tree level, with the decay widths given by:

$$\Gamma_H^W = \frac{G_F M_H^3}{8\pi\sqrt{2}} \sqrt{1 - \frac{4M_W^2}{M_H^2}} (1 - 4\frac{M_W^2}{M_H^2} + 12\frac{M_W^4}{M_H^4})$$

$$\Gamma_H^Z = \frac{G_F M_H^3}{16\pi\sqrt{2}} \sqrt{1 - \frac{4M_Z^2}{M_H^2}(1 - 4\frac{M_Z^2}{M_H^2} + 12\frac{M_Z^4}{M_H^4})} \qquad (6.57)$$

For $(M_H >> M_Z)$, Eq. (6.57) reduces to:

$$\Gamma_H^W \simeq 328\, M_H^3 \text{ MeV}; \quad \Gamma_H^Z \simeq 164\, M_H^3 \text{ MeV} \qquad (6.58)$$

where M_H is expressed in units of 100 GeV. The radiative corrections for the decay of H^0 into the weak bosons: W^+W^-, $2Z$ turn out to be small for the highest conceivable mass of H^0 (i.e. ~ 1 TeV). We note in Eq. (6.58) that the partial decay widths of H into the W and Z bosons are relatively small compared to the H mass because of the weak coupling of H to W and Z.

Other two-body decays of H^0, such as $H^0 \rightarrow \gamma\gamma$, gg (g stands for gluon), $Z\gamma$, do not exist at "tree level" but take place through higher-order loops; the rates are generally small. For example, the one-loop contributions to the decay $H^0 \rightarrow \gamma\gamma$ come from quarks, charged leptons and W bosons through the triangle graphs shown in Fig. 6.3. The decay mode $H^0 \rightarrow \gamma\gamma$ only becomes interesting if M_H is small compared to M_W and, then, only if the fermion is too heavy to be produced directly in the decay $H^0 \rightarrow f\bar{f}$. The other two-body decay modes of H^0: $H^0 \rightarrow gg$ and $H^0 \rightarrow Z\gamma$, become significant for a sufficiently heavy Higgs boson, e.g. the decay $H^0 \rightarrow gg$ occurs through the virtual decay of H^0 to a heavy $q\bar{q}$ pair that subsequently annihilates into two gluons. As a final remark concerning the loop-induced decay modes of H^0 : $\gamma\gamma$, gg and $Z\gamma$, it can be said that the primary dependence of their branching ratios is on m_t for $M_H \geq 100$ GeV. Unfortunately, the branching ratio for gg is never above 10%, and the branching ratios for $\gamma\gamma$ and $Z\gamma$ are at most $\sim 10^{-3}$.

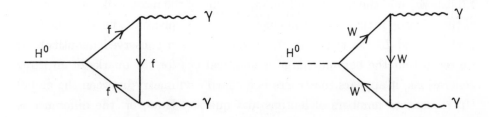

Fig. 6.3. Diagrams for the decay of a Higgs boson into two photons.

§6.3d. Three generations of leptons and lepton charge conservation

Thus far, we have considered some highlights of electroweak phenomenology from the vantage point of the structure of the electroweak interaction and the role of the gauge bosons and Higgs boson(s) in establishing the dynamics of the weak interaction. Much of the previous discussion would still hold if there was only one generation of quarks and leptons. However, it has now been established that there are precisely three generations of leptons [of the "normal" type, i.e. with neutrino masses less than $M_Z/2$ GeV (see §6.3b)] and three generations of quarks [if one accepts the existence of the t quark] and we review in this section and the next, the impact of this knowledge on our understanding of the standard electroweak gauge theory. We propose to discuss in this section (§6.3d) the lepton case and leave to the next section (§6.3e) the more complicated, albeit in certain ways better, known quark case.

The first big surprise after the "birth" of modern particle physics (ca. 1945) was the discovery of the second-generation charged lepton, the muon, in 1947; strictly speaking, the muon was discovered a decade earlier — by Anderson and Neddermeyer [1.27] — but it was initially confused with the strongly interacting "Yukawa" meson and was only given its true identification after the Bristol discovery of $\pi \to \mu$ decays [1.28]. [The important role of the muon in the development of modern particle physics — its need for its own neutrino partner, its hint of first a baryon-lepton and then a quark-lepton symmetry, etc. — was discussed in §1.2d.] It took almost thirty years before the third-generation charged lepton, the τ lepton [6.25], was discovered, and it soon replicated the muon in requiring its own distinct neutrino ($\nu_\tau \neq \nu_\mu \neq \nu_e$) and in calling for its own quark partner (the b quark). Apart from the similarities between the muon and the τ lepton, the τ lepton has the distinction of being the heaviest of the "normal" charged leptons [Table 1.6 lists m_τ and the upper limit on m_{ν_τ}]. One of the most striking properties of the τ lepton is that its lepton charge, L_τ, is "sequential", i.e. distinct from both L_μ and L_e (so that only $L_e + L_\mu + L_\tau$ is a conserved quantity). If we recall that the baryon charge is identical ($\frac{1}{3}$) for the quarks of all three generations, it appears that there is a sharp asymmetry between the global $U(1)$ quantum numbers of leptons and quarks. However, the difference is more apparent than real because the assumption of three massless neutrinos in the standard model introduces sufficient arbitrariness in the phases of the charged lepton wavefunctions that the lepton mixing matrix (U_{CKM} for leptons) reduces to the identity matrix. The greatly reduced inter-generational "baryon charge" (the counterpart of $L_e \neq L_\mu \neq L_\tau$) among the three quark doublets is reflected in the large reduction of the inter-generational values of U_{CKM} [see Eq. (6.71)] as compared to the intra-generational values of the matrix elements of U_{CKM} (see Tables 6.3 and 6.4 in the next section). We will have more to say about lepton charge conservation among the three generations of leptons but, before doing so, we sketch the "common" and "uncommon" properties of the three charged leptons (e, μ, τ) and relegate to §6.4b a similar discussion of the neutrinos of the three generations.

The "common" property of the charged leptons of the three generations is that all three possess identical quantum numbers under the standard group $SU(3)_C \times SU(2)_L \times U(1)_Y$ (see Table 1.5); indeed, it is this property which defines a generation. The fact that all three charged leptons are singlets under the $SU(3)$ color group implies the absence of a strong interaction, whereas their non-trivial quantum numbers under $SU(2)_L \times U(1)_Y$ (see Table 1.5) imply that they undergo weak and electromagnetic interactions. The three charged leptons also share the "common" property — as of now — of possessing a point-like structure down to at least 10^{-16}cm; the characteristic experiment to establish these limits is "Bhabha-type" scattering, i.e. $e\bar{e} \to f\bar{f}$ [6.26] (see §9.3). Apart from the distinctly different masses of e, μ, τ (see Table 1.5), we expect $e - \mu - \tau$ universality not only with regard to the electromagnetic interaction (e.g. the same renormalized charge for e, μ and τ) but also with respect to the weak interaction. The different mass of each charged lepton does make a difference when radiative effects are present; for example, the larger mass of μ is responsible for a substantially larger effect on its anomalous magnetic moment [due to the virtual hadron pair contribution (see §4.5a)] than on the anomalous magnetic moment of e.

Insofar as $e - \mu - \tau$ universality in the weak interaction is concerned, we know that $e - \mu$ universality is good to better than 0.3% (i.e. $g_\mu/g_e = 0.9970 \pm 0.0023$ [6.9]), where g_μ and g_e are the respective μ and e coupling strengths; the τ lepton fits into the picture of $e - \mu - \tau$ universality in the weak interaction in view of the recent measurement of the τ lifetime, namely $2.96.8 \pm 3.2 \times 10^{-13}$ sec [6.9], which agrees well with the calculated value of $2.87 \pm 0.04 \times 10^{-13}$ sec [6.27] [using the branching ratios for $\tau \to e\nu_\tau\bar{\nu}_e$ and $\tau \to \mu\nu_\tau\bar{\nu}_\mu$ and $e - \mu$ universality]. With regard to other aspects of $e - \mu - \tau$ universality in the weak interaction, the experimental results are generally supportive of universality; for example, the Michel parameter (see §1.2) in the decay $\tau \to \ell\nu_\tau\bar{\nu}_\ell$ has been measured to be 0.78 ± 0.05 for $\ell = e$ and 0.72 ± 0.08 for $\ell = \mu$, consistent with the V-A coupling for the charged currents [6.27]. There is one puzzle in τ decay which does not bear on the question of $e - \mu - \tau$ universality but has to do with the one-charged particle decay modes of

τ. The problem is that there is a few percent disagreement between the exclusive and "topological" (inclusive) experimental determinations of one-charge particle branching ratios from the τ decay. The measured sum of the exclusive one-charge particle branching ratios is $78.3 \pm 1.9\%$ [6.28] and, allowing for a theoretical estimate of the unmeasured one-charge particle modes, namely 2.2%, the "exclusive" total is $80.5 \pm 1.9\%$ — disagreeing with the value of the directly-observed "topological" one-charge particle branching fraction of $84.8 \pm 0.04\%$ [6.28]. It is not clear whether the small disagreement is indicative of the presence of a missing decay mode or a systematic bias in previous measurements of the exclusive hadronic branching ratios.

We return to what, in our view, is the deepest mystery about $e - \mu - \tau$ universality and that is the role of lepton charge conservation in lepton-related processes. On the one hand, L_e, L_μ and L_τ are interchangeable in balancing off baryon charge conservation in processes depending upon $B - L$ [e.g. cancellation of the chiral gauge anomalies (see §7.5) and the "residual" global $B - L$ charge conservation in instanton (or sphaleron)-induced B and L symmetry breakdowns in electroweak processes (see §10.3a)]. On the other hand, it seems that only the sum $(L_e + L_\mu + L_\tau)$ enjoys global lepton charge conservation and that separate global lepton charge conservation for each generation is only guaranteed through its participation in the sum of the three generational lepton charges. Curiously enough, if the three neutrinos possess mass, we would be led to an U_{CKM} lepton mixing matrix (similar to the U_{CKM} quark mixing matrix) and we would assign the same lepton charge to each lepton (just as we assign the same baryon charge of $\frac{1}{3}$ to each quark). Since there is no evidence of finite mass for any of the neutrinos, it is conceivable that the separate generational lepton charges are not conserved and that one generational lepton charge is converted into another through an SSB mechanism which generates a N-G boson (called a "majoron") in the process.

At least two "majoron" scenarios have been proposed in which the "majorons" should manifest themselves in physical processes; they both hypothesize an interaction between a Majorana mass term [which requires $\Delta L = 2$ (see §6.4b)] and a scalar field [which carries $L = 2$ (to conserve L in the Lagrangian)] but whose VEV is taken as non-vanishing to generate the "majoron". In one case, Gelmini et al [1.164], the scalar field is taken as a singlet under $SU(2)_L \times U(1)_Y$ which carries $L = 2$, and in the other case, Chicashige et al [1.164], it is taken as a triplet Higgs field which also carries $L = 2$. The first "majoron" scenario is unverifiable because of the miniscule effective coupling of the "majoron" to matter that must be built into the theory. The second "majoron" scenario is, in principle, verifiable, through the prediction of a double β decay process accompanied by a "majoron"; the trouble here is that the production of observable effects by the triplet Higgs field predicts a value of ρ that can not be easily reconciled with the small error in the $\rho = 1$ determination (see §6.3a). The *raison d'etre* for the "majoron" disappears if the observed neutrinos are massive and we are compelled to have recourse to the lepton mixing matrix description. If we turn from the leptons to the quarks of the three generations, we know that the quark masses are all finite and that the same global baryon charge conservation can be maintained for all three quark generations at the price of introducing the U_{CKM} mixing matrix; we present in the next section sufficient detail concerning the U_{CKM} quark mixing matrix to see how the generational distinction is accomplished for quarks in the standard model.

§6.3e.   Three generations of quarks and the determination of the U_{CKM} mixing matrix

The V-A charged current theory was formulated in 1957 but it was not until 1963 that Cabibbo put forth his "weak" mixing angle (i.e. the "Cabibbo angle") proposal to explain the reduced transition probabilities — by a factor of 20 on the average — for $\Delta Y = 1$ semi-leptonic weak transitions as compared to $\Delta Y = 0$ semi-leptonic weak transitions [1.81]. The introduction of the Cabibbo angle provided a systematic accounting for the difference between $\Delta Y = 1$ and $\Delta Y = 0$ hadronic (V-A) charged currents. Within a few years, the Cabibbo angle approach to explaining the suppression of $\Delta Y = 1$ with respect to $\Delta Y = 0$ weak processes was re-expressed in terms of s and d quarks; more precisely, the weak charged current of low-energy phenomenology, J_μ^{CC}, was described by the quark-model transcription of the "Cabibbo" current (θ is the Cabibbo angle), thus:

$$J_\mu^{CC} \sim \bar{u}_L \gamma_\mu d_L \quad \longrightarrow \quad J_\mu^{CC} \sim \bar{u}_L \gamma_\mu d_L^\theta \tag{6.59}$$

with:

$$d_L^\theta = d_L \cos\theta + s_L \sin\theta \tag{6.59a}$$

The Cabibbo form (6.59a) for the quark current could account for the strength hierarchy problem but it brought to the fore the problem of higher-order divergences in the V-A theory. Thus, Eq. (6.59) implies that the product of two charged quark currents generates an "effective" weak neutral quark current contribution to the weak Lagrangian $\mathcal{L}^{NC}$ of the form:

$$\mathcal{L}^{NC} \sim \sin\theta\cos\theta \bar{d}_L \gamma_\mu s_L \tag{6.60}$$

Equation (6.60) gives rise (in order G_F^2 — neutron decay is of order G_F) to $\Delta Y \neq 0$ neutral current processes such as the $\Delta Y = 1$ $K_L \to \mu\bar{\mu}$ and $K^\pm \to \pi^\pm\nu\bar{\nu}$ decays, as well as the $\Delta Y = 2$ mass difference effect between K_L and K_S [denoted by $\Delta m(K_L - K_S)$]. These are now called flavor-changing neutral current (FCNC) processes and take place in the same order (first order in G_F) — via the neutral weak quark currents — as the non-FCNC processes arising from the charged weak quark currents. In the absence of neutral currents, as in the V-A theory, FCNC processes such as $K_L \to \mu\bar{\mu}$ take place in higher orders of G_F and yield divergent answers. Consequently, in the former V-A theory, one could not make predictions concerning FCNC processes but only use examples of them to deduce upper limits on Λ_{weak} (where Λ_{weak} is the cutoff in the higher-order calculation); thus, the observed value of $\Delta m(K_L - K_S)$ could only be explained if Λ_{weak} is quite small (a few GeV) [6.29], as if to simulate the GIM mechanism (see below).

By 1967–68 [1.88], the gauge theory of the electroweak interaction had been proposed, but, at first, its prediction of weak neutral currents on a par with weak charged currents seriously aggravated the FCNC problem because the neutral current $d_L^\theta \gamma_\mu d_L^\theta$ can give rise to FCNC in first-order through the mediation of the neutral (Z) weak boson [which now exists in addition to charged (W) weak bosons]. In order to suppress a first-order neutral current-induced FCNC process, the GIM mechanism was introduced [1.86], whereby the orthogonal (Cabibbo-type) linear combination (θ is still the Cabibbo angle)

$$s_L^\theta = -d_L \sin\theta + s_L \cos\theta \tag{6.61}$$

replaces s_L in all charged and neutral weak currents. The contribution of the neutral currents $\bar{d}_L^\theta \gamma_\mu d_L^\theta$ and $\bar{s}_L^\theta \gamma_\mu s_L^\theta$ to the neutral part of the weak Lagrangian then becomes:

$$\mathcal{L}^{NC} \sim g(\bar{d}_L^\theta \gamma_\mu d_L^\theta + \bar{s}_L^\theta \gamma_\mu s_L^\theta)\, Z \sim g(\bar{d}_L \gamma_\mu d_L + \bar{s}_L \gamma_\mu s_L)\, Z \tag{6.62}$$

which no longer contains a cross term between $\bar{d}_L$ and s_L of the form (6.60) [note that the coupling is g and not $G_F \sim g^2$] and hence there is complete cancellation of FCNC processes. Moreover, in order to construct a charged weak current with s_L^θ, a fourth quark (called charm c) must be introduced to complete a "second-generation" $SU(2)$ weak isospin doublet (c, s^θ) [in analogy to the "first-generation" weak isospin doublet (u, d^θ)] so that the counterpart of $J_\mu^{CC} \sim \bar{u}_L \gamma_\mu d_L^\theta$ [see Eq. (6.59)] is:

$$J_\mu^{CC} \sim \bar{c}_L \gamma_\mu s_L^\theta \qquad (6.62a)$$

Using the two Cabibbo-type charged currents J_μ^{CC} of Eqs. (6.59) and (6.62a) — as shown in Fig. 6.4 — it is easy to see that the contribution to FCNC process vanishes if the masses of the u and c quarks are equal and only depends on $(m_c - m_u/M_W)^2$ if the u and c quark masses are unequal. Thus, the GIM mechanism eliminates the entire neutral current contribution to an FCNC process and only allows it to take place as a correction to the second-order charged-current transition. These consequences of the GIM mechanism hold for all FCNC weak processes and are simply reflective of the unitary property of the U_{CKM} mixing matrix for the special case of two generations (i.e. $N_g = 2$ — see below).

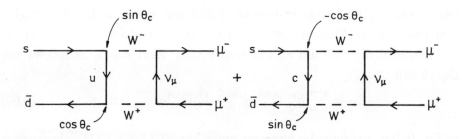

Fig. 6.4. Feynman diagrams showing operation of GIM mechanism in $K_L \to \mu\bar{\mu}$ decay.

The success of the GIM mechanism in eliminating FNNC weak processes for two quark-lepton generations inspired its generalization to the Kobayashi-Maskawa quark mixing matrix [6.3] [the KM quark mixing matrix is now called the CKM mixing matrix to incorporate Cabibbo's contribution] for an arbitrary number of quark generations. Thus, suppose N_g is the number of generations; if $N_g > 2$, the 2×2 Cabibbo unitary matrix — defined by Eqs. (6.59a) and (6.61) — must be enlarged to the $N_g \times N_g$ U_{CKM} unitary matrix. To understand how the U_{CKM} mixing matrix for the quarks arises in the general case, we recall that the Yukawa term in the Lagrangian [Eq. (6.10)] contains the weak eigenstates of the quarks before SSB takes place and the quark mass eigenstates are determined by the Yukawa coupling after SSB. Since we know from experiment that there is generational mixing among the quarks, we must allow for the possibility that the gauge and mass eigenstates of the quarks are different. We proceed to deduce the U_{CKM} matrix for the general case of N_g generations before turning to the realistic case of $N_g = 3$.

One starts with the weak representations of N_g generations of quark doublets [we replace q_i^+ and q_i^- of Table 1.5 by p_i and n_i respectively]:

$$Q_{iL}^{(0)} \equiv \begin{pmatrix} p_i^{(0)} \\ n_i^{(0)} \end{pmatrix}_L (2, \tfrac{1}{3}); \quad p_{iR}^{(0)} (1, \tfrac{4}{3}) \quad n_{iR}^{(0)} (1, -\tfrac{2}{3}) \quad (i = 1, 2, \ldots N_g) \quad (6.63)$$

The quark quantum numbers under $SU(2)_L \times U(1)_Y$ are indicated in the parentheses (see Table 1.5), and the superscript zero implies that the fields are weak — not mass — eigenstates. The weak charged current interaction Lagrangian for these fields is purely chiral (involving only the lefthanded quarks) and is given by:

$$\mathcal{L}^{CC} = \frac{-g}{\sqrt{2}} W^{+\mu} \sum_i \{ \bar{p}_{iL}^{(0)} \gamma_\mu n_{iL}^{(0)} \} + \text{h.c.} \quad (i = 1, 2, \ldots N_g) \quad (6.64)$$

Up to this point, all the quarks are massless but then the basic assumption is made that the acquisition of a VEV by the Higgs doublet generates Dirac masses; the formal statement is that one starts with a Lagrangian that contains the Yukawa coupling of the Higgs doublet to a combination of lefthanded doublet and righthanded singlet Weyl states, as follows [see Eq. (6.10)]:

$$\mathcal{L}_Y = \sum_{i,j} [h_{ij}^p \, \bar{Q}_{iL}^{(0)} \tilde{\phi} p_{jR}^{(0)} + h_{ij}^n \bar{Q}_{iL}^{(0)} \phi n_{jR}^{(0)} + \text{h.c.}] \quad (6.65)$$

In Eq. (6.65), $\mathcal{L}_Y$ must be invariant under the $SU(2)_L \times U(1)_Y$ electroweak group and it is the Y conservation that necessitates the assignment of $Y = 1$ to the Higgs doublet. Once the VEV is assigned to ϕ in Eq. (6.65), $\mathcal{L}_Y$ translates into the sum of mass terms for the quarks, [see Eqs. (6.10) and (6.24)], thus:

$$\mathcal{L}_{\text{mass}} = \sum_{q=p,n} \bar{q}_{iL} M_{ij}^{(q)} q_{jR} \quad \text{with } M_{ij}^{(q)} = \frac{v}{\sqrt{2}} h_{ij}^{(q)} \quad (6.66)$$

The mass matrices in Eq. (6.66) clearly mix the weak eigenstates of different generations (and chiralities) and, to deduce the mass eigenstates, it is necessary to diagonalize the mass matrices by means of bi-unitary transformations as follows:

$$U_L^{(q)} M^{(q)} U_R^{(q)\dagger} = M_{\text{diag}}^{(q)} \quad (6.67)$$

where M_{diag}^q is the diagonal mass matrix. The mass eigenstates can now be written as:

$$p_{L,R} = U_{L,R}^{(p)} p_{L,R}^0; \qquad n_{L,R} = U_{L,R}^{(n)} n_{L,R}^{(0)} \quad (6.68)$$

Using Eq. (6.68), the generalized charged current interaction can be rewritten in terms of the mass eigenstates as follows:

$$\mathcal{L}^{C.C.} = \frac{-g}{\sqrt{2}} W_\mu^+ (\bar{p}_L \gamma_\mu U_L^{(p)\dagger} U_L^{(n)} n_L) + \text{h.c.} \tag{6.69}$$

In Eq. (6.69), the product of the two unitary matrices $U_L^{(p)\dagger}$ and $U_L^{(n)}$ is also a unitary matrix, which is called the CKM mixing matrix U_{CKM}. The properties of U_{CKM} (as a function of N_g) are interesting: (1) U_{CKM} can be parameterized by N_g^2 parameters, of which $N_g(N_g - 1)/2$ are real "Cabibbo" angles so that the number of remaining phases is $N_g(N_g + 1)/2$; (2) since the charged current involves N_g weak isospin "up" (p_i^0) quarks and N_g weak isospin "down" (n_i^0) quarks and one overall phase is arbitrary, one may remove another $(2N_g - 1)$ phases from U_{CKM} by appropriately redefining the quark fields. This leaves $\frac{1}{2}(N_g - 1)(N_g - 2)$ as the final number of physically significant phases that may be identified with weak CP violation; (3) for two generations (i.e. $N_g = 2$), it follows that there is one "Cabibbo" angle and zero phases so that U_{CKM} can be written as the real GIM-type matrix:

$$U_{CKM} = \begin{pmatrix} \cos\theta & \sin\theta \\ -\sin\theta & \cos\theta \end{pmatrix} \tag{6.70}$$

Equation (6.70) obviously gives rise to the two gauge doublets of quarks: $\begin{pmatrix} u \\ d^\theta \end{pmatrix}$ and $\begin{pmatrix} c \\ s^\theta \end{pmatrix}$; and, finally (4) when we move on to three generations of quarks (i.e. $N_g = 3$) — the number required by the recent determination of the number of neutrino species (see §6.3b) — we find that there are three "Cabibbo" angles (denoted by $\theta_1, \theta_2, \theta_3$) and one phase (denoted by δ), which can accommodate the observed weak CP violation.

Let us focus on U_{CKM} for $N_g = 3$ and ascertain whether measurements of the four U_{CKM} parameters — $\theta_1, \theta_2, \theta_3$ and δ — provide a self-consistent picture within the framework of the U_{CKM} mixing matrix approach. For this purpose, we take the u, c and t quarks as pure mass states — which is always possible if there is no mass degeneracy — and define U_{CKM} in terms of the (d, s, b) mass eigenstates and the (d', s', b') (weak) gauge interaction eigenstates, thus:

$$U_{CKM} = \begin{pmatrix} U_{ud} & U_{us} & U_{ub} \\ U_{cd} & U_{cs} & U_{cb} \\ U_{td} & U_{ts} & U_{tb} \end{pmatrix} \tag{6.71}$$

In principle, the elements of U_{CKM} in Eq. (6.71) can be directly determined from experiment as will be discussed below. However, in order to exhibit the role of the single phase δ in establishing whether weak CP violation can be understood within the U_{CKM} framework, it is useful to write down the original expression proposed for U_{CKM} in terms of $\theta_i (i = 1, 2, 3)$ and δ [6.3], thus:

$$U_{CKM} \equiv \begin{bmatrix} c_1 & s_1c_2 & s_1s_2 \\ -s_1c_3 & c_1c_2c_3 - s_2s_3e^{-i\delta} & c_1s_2c_3 + c_2s_3e^{-i\delta} \\ -s_1s_3 & c_1c_2s_3 + s_2c_3e^{-i\delta} & c_1s_2s_3 + c_2c_3e^{-i\delta} \end{bmatrix} \qquad (6.72)$$

where $c_i = \cos\theta_i$ $(i = 1, 2, 3)$ and $s_i = \sin\theta_i$ (i=1,2,3). In terms of the three "Cabibbo" angles $\theta_1, \theta_2, \theta_3$ and the phase δ, the conditions for CP violation with three generations of quarks become: (a) $\sin\delta \neq 0$; (b) all mixing ("Cabibbo") angles, given by s_1, s_2, s_3, differ from zero; and (c) there is no mass degeneracy among the three "up" or among the three "down" quarks.

The above three conditions can be summarized in one equation, which must be fulfilled if CP violation is to occur:

$$A[(M^p)^2] \cdot B[(M^n)^2] \cdot J \neq 0, \qquad (6.73)$$

where:

$$A[(M^p)^2] = (m_t^2 - m_c^2)(m_c^2 - m_u^2)(m_t^2 - m_u^2)$$
$$B[(M^n)^2] = (m_b^2 - m_s^2)(m_s^2 - m_d^2)(m_b^2 - m_d^2)$$
$$J = c_1c_2c_2s_1^2s_2s_3\sin\delta \qquad (6.74)$$

with J the "Wronskian" of Eq. (6.72). Condition (6.74) can be restated in another way, which is independent of the parameterization of U_{CKM}; choose any basis for the quarks, and assume that the mass matrices in this basis are M^p and M^n. Then the L.H.S. of Eq. (6.73) equals (up to a possible sign) Im$\{\det[M^pM^{p\dagger}, M^nM^{n\dagger}]\}/2$. Thus, in an arbitrary basis, CP is violated in the three-generation standard model if and only if [6.30]:

$$\text{Im}\{\det[M^pM^{p\dagger}, M^nM^{n\dagger}]\} \neq 0 \qquad (6.75)$$

The quantity J, the function of mixing angles and phase [defined by Eq. (6.74)], can be written in a form which is explicitly parameterization-independent (we write U for U_{CKM}):

$$|J| = |\text{Im}(U_{ij}U_{lk}U_{ik}^*U_{lj}^*)| \qquad (6.76)$$

for any choice of i, j, k, l.

Different representations of U_{CKM} have proved to be useful as increasingly precise experiments on the determination of the U_{CKM} matrix elements have been carried out. The objective throughout has been to perform the full roster of experiments necessary to decide whether the three-generation U_{CKM} is the correct framework for encoding both the quark mixing (between the gauge interaction and mass eigenstates) and the weak CP violation among the three generations of quarks. Ideally, the direct observation of the t quark and its interactions with the other quarks defining the U_{CKM} matrix, would permit the full determination of the U_{CKM} matrix, and enable one to judge whether the information contained therein can serve as an important clue to the solution of the "fermion generation problem" (why there are precisely three quark-lepton generations) at the standard model level (see §9.1). While the failure to directly observe the t quark is frustrating, the unitarity of the U_{CKM} matrix provides three "unitarity conditions" that permit a determination of the absolute values of the matrix elements of U_{KCM}, as given by Eq. (6.71). The values of the six matrix elements deduced from carefully selected experiments, are given in Table 6.3 and the remaining three "unitarity-constrained" matrix elements are listed in Table 6.4. We note that the last "direct" matrix element listed in Table 6.3 is the least accurate and likewise for the last two "indirect" matrix elements listed in Table 6.4.

Table 6.3: Direct experimental determination of six U_{CKM} matrix elements.

CKM matrix element	Value	Experiments
$\lvert U_{ud}\rvert$	0.9744 ± 0.0010	$n \to p\,e^-\bar{\nu}_e$ $\pi^- \to \pi^0\,e^-\bar{\nu}_e$
$\lvert U_{us}\rvert$	0.2205 ± 0.0018	K_{e3} decays hyperon semileptonic decays
$\lvert U_{cd}\rvert$	0.204 ± 0.017	$\nu_\mu N \to \mu^- cX$ $(\nu_\mu d \to \mu^- c)$ $D^0 \to \pi^- e^+ \nu_e$
$\lvert U_{cs}\rvert$	1.02 ± 0.18	$\nu_\mu s \to \mu^- c$ $D^0 \to K^- e^+ \nu_e$
$\lvert U_{cb}\rvert$	0.044 ± 0.009	$\bar{B} \to D\ell\bar{\nu}_\ell$
$\lvert U_{ub}/U_{cb}\rvert$	0.11 ± 0.05	$B \to X_u\,\ell\nu_\ell$ $B \to X_c\,\ell\nu_\ell$

From Eq. (6.71) and Tables 6.3 and 6.4, we can conclude — despite the errors — that U_{CKM} has a nice hierarchical pattern. To a good approxima-

Table 6.4: Indirect unitarity determination of the U_{CKM} matrix elements involving the t quark.

U_{ti} $(i = b, s, d)$	Ranges	Unitarity constraints
$\|U_{tb}\|$	$0.9986 - 0.9994$	$\|U_{ub}\|^2 + \|U_{cb}\|^2 + \|U_{tb}\|^2 = 1$
$\|U_{ts}\|$	$0.034 - 0.055$	$U_{us}^* U_{ub} + U_{cs}^* U_{cb} + U_{ts}^* U_{tb} = 0$
$\|U_{td}\|$	$0.002 - 0.020$	$U_{ub}^* U_{ud} + U_{cb}^* U_{cd} + U_{tb}^* U_{td} = 0$

tion, this hierarchical pattern is recaptured by writing U_{CKM} as a function of the Cabibbo angle ($\lambda = \sin\theta_c = 0.22$), as follows:

$$U_{CKM} \sim \begin{pmatrix} 1 & \lambda & \lambda^3 \\ -\lambda & 1 & \lambda^2 \\ \lambda^3 & -\lambda^2 & 1 \end{pmatrix} \tag{6.77}$$

A rigorous parameterization of U_{CKM} which displays this pattern in a simple way is [6.31]:

$$U_{CKM} = \begin{pmatrix} c_{12}c_{13} & s_{12}c_{13} & s_{13}e^{-i\delta} \\ -s_{12}c_{23} - c_{12}s_{23}s_{13}e^{i\delta} & c_{12}c_{23} - s_{12}s_{23}s_{13}e^{i\delta} & s_{23}c_{13} \\ s_{12}s_{23} - c_{12}c_{23}s_{13}e^{i\delta} & -c_{12}s_{23} - s_{12}c_{23}s_{13}e^{i\delta} & c_{23}c_{13} \end{pmatrix} \tag{6.78}$$

where $c_{ij} \equiv \cos\theta_{ij}$ and $s_{ij} \equiv \sin\theta_{ij}$. The values of and limits on the elements of the U_{CKM} matrix, given in Tables 6.3 and 6.4, can be translated into values of the three parameters s_{12}, s_{23}, and s_{13} (δ is still undetermined) as shown in Table 6.5. It is seen from Table 6.5 that s_{13} is definitely small compared to s_{23} but the error is still very large and, as already mentioned, δ is still unknown. Clearly, much work must still be done to completely determine the parameters of U_{CKM} but it is encouraging that the substantial body of experimental data already available is consistent in all respects with the U_{CKM} mixing matrix approach. It is interesting to note — from Tables 6.3 and 6.4 and the schematic hierarchical pattern shown in Eq. (6.77) — that the flavor diagonal matrix elements U_{ud}, U_{cs} and U_{tb} of U_{CKM} are all close to unity, which jibes with our earlier remark that U_{CKM} in the quark sector is, in a sense, playing a similar role to that of $L_e + L_\mu + L_\tau = 1$ in the lepton sector of the standard model.

Some additional constraints on U_{CKM} — albeit not altering the above conclusions — have been derived from the measured weak CP violation in the neutral kaon system and from the large (charge conjugation) mixing observed in the neutral B meson system (to be discussed in the next section);

Table 6.5: Measured parameters in the "hierarchical" representation of U_{CKM}.

Parameter	Value
s_{12}	0.2205 ± 0.0018
s_{23}	0.044 ± 0.009
s_{13}/s_{23}	0.11 ± 0.05

for making these comparisons, two alternative statements of the hierarchical nature of U_{CKM} have proved to be useful. The first is the rewriting of the phenomenological hierarchy of U_{CKM}, expressed by Eq. (6.78), in terms of the Wolfenstein parameters [6.32], thus:

$$s_{12} = \lambda; \quad s_{23} = A\lambda^2; \quad s_{13} = A\sqrt{\rho^2 + \eta^2}\lambda^3; \quad \delta = \tan^{-1}(\frac{\eta}{\rho}) \qquad (6.79)$$

Equation (6.79) contains the four parameters λ, A, ρ and η and leads (in order λ^3) to another parameterization of U_{CKM} [6.32]:

$$U_{CKM} \simeq \begin{pmatrix} 1 - \frac{\lambda^2}{2} & \lambda & A\lambda^3(\rho - i\eta) \\ -\lambda & 1 - \frac{\lambda^2}{2} & A\lambda^2 \\ A\lambda^3(1 - \rho - i\eta) & -A\lambda^2 & 1 \end{pmatrix} \qquad (6.80)$$

The usefulness of (6.80) is seen in §6.3f. The second observation is that the unitarity of U_{CKM} — when account is taken of its hierarchical pattern — implies a simple relation among the matrix elements:

$$U_{ub}^* + U_{td} = \lambda U_{cb}^* \qquad (6.81)$$

Equation (6.81) is shown schematically in Fig. 6.5, where the three sides of the triangle correspond to the three complex quantities U_{ub}^*, U_{td} and λU_{cb}^*; this triangle has been dubbed the "unitarity triangle" and its usefulness will become apparent in §6.3f.

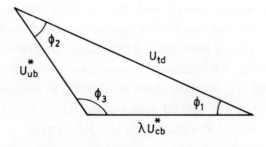

Fig. 6.5. The unitarity triangle of U_{CKM}.

A final remark about the U_{CKM} mixing matrix is worth making: we recall that J [as defined in Eqs. (6.74) and (6.76)] is essentially the measure of weak CP violation and it is evident that, within present experimental accuracy, the phase angle δ need not be small in view of the rapid decrease of the three "Cabibbo" angles for the successive generations. This could be a satisfactory state of affairs since it would correlate the "weakness" of CP violation in hadrons with the rapid decrease in the values of the "Cabibbo" angles which, in turn, would correlate inversely with the rapid increase in the quark masses in successive generations. Moreover, it opens up the possibility that the phase angle δ is of order unity and may have topological significance so that the phenomenological value $N_g = 3$ is a rather deep — but as-yet-unexplained — property of the standard model with the U_{CKM} quark mixing matrix.

§6.3f. Weak CP violation

We commented in §6.1 that the concept of a "neutrino paradigm" (suggested by massless neutrinos in all three generations) encompasses two important invariances — chirality (γ_5) invariance and CP invariance — and can serve as a useful guiding principle in dealing with the properties and interactions of finite-mass fermions. Because of the flavor structure of the standard model before SSB takes place, the principle of chirality invariance predicts a (V-A) Lorentz structure for the two-flavor charged weak current and an arbitrary admixture of V and A Lorentz structure for the one-flavor neutral weak and/or electromagnetic current; after SSB, because of the large SSB scale, it is expected that the corrections to the "neutrino paradigm" predictions are small [2.12]. By the same token, when the SSB of the electroweak group gives rise to the finite quark masses, a small departure from CP invariance (i.e. weak CP violation) is expected to occur and it is possible that weak CP violation [6.33] will find its explanation within the framework of the U_{CKM} quark mixing matrix and the observed number of three quark generations. For this and many other reasons, renewed experimental efforts are being made to obtain more information concerning the salient properties of weak CP violation. We comment on several questions that are under close scrutiny at present: (1) whether there is "direct", in addition to "indirect", weak CP violation in the neutral kaon system (these terms are defined below). "Indirect" weak CP violation in the K^0 system — as measured by the parameter ϵ [see Eq. (6.86)] — is well-established [6.33] but the status of "direct" weak CP violation — as measured by the parameter ϵ' [see Eq.

(6.86)] — is still unresolved; (2) whether there is weak CP violation at all — "indirect" or "direct" — in rare K_L decays such as $K_L \to \pi^0 e\bar{e}$; (3) whether there is weak CP violation in other neutral inter-generational meson systems such as the neutral B meson system; and, finally, (4) whether there is any hope to test the standard model — through weak CP violation effects — by continuing the prodigious efforts to measure the electric dipole moments of neutron [and charged leptons (e and μ)]. We treat questions (1), (3) and (4) in this section under the headings: (a) neutral kaon system; (b) neutral B meson system; and (c) electric dipole moments of neutron and charged leptons while (2) is discussed in the next section (§6.4a).

(a) *Neutral kaon system*

The discovery of weak CP violation in the neutral kaon system in 1963 [6.33] was completely unexpected, coming so soon after the application of chirality invariance — with its built-in CP invariance — to provide the rationale for the highly successful V-A theory [1.80]. [When the "chirality invariance" derivation of the effective low-energy V-A theory was proposed in 1957, there was no electroweak gauge theory and no Higgs mechanism and the finite-mass effects on the V-A Lorentz structure were simply ignored.] The V-A theory predicts maximal parity violation and maximal charge conjugation violation but — consistent with the "neutrino paradigm" — CP invariance. Although the observation of a small (0.1% in amplitude) departure from CP invariance in a single physical system (i.e. the neutral kaon system) does not affect the successful application of the V-A theory to all other weak processes (not involving the neutral kaon), the presence of weak CP violation in the neutral kaon system still has to be understood. With advances in experimental technique, there has been a recent surge of activity to make improved measurements of the weak CP violation parameters, ϵ and ϵ', in the neutral kaon system and we report on the present status of these efforts.

We start by reminding ourselves that linear superpositions of K^0 and $\bar{K}^0$ give rise to two mass eigenstates with different lifetimes [6.34]. If CP invariance holds and if we write $K_S = \frac{1}{\sqrt{2}}(K^0 + \bar{K}^0)$ and $K_L = \frac{1}{\sqrt{2}}(K^0 - \bar{K}^0)$, K_S possesses the eigenvalue $CP = +1$, can decay into 2π and is short-lived, whereas K_L has eigenvalue $CP = -1$, can not decay into 2π, and is long-lived. If we denote the masses and decay widths of K_S and K_L by M_S, Γ_S and M_L, Γ_L respectively, it turns out that K_L is slightly heavier than K_S, i.e. $\Delta m(K_L - K_S) = 3.5 \times 10^{-6}$ eV, which is $0.44\Gamma_S$. With CP invariance, it was expected that the decay process $K_L \to 2\pi$ would never take place, in contrast to the dominant 2π decay mode of K_S. The surprise of weak CP violation in the neutral kaon system came when the decay mode $K_L \to 2\pi$ was seen with a finite transition amplitude ($\sim 2 \times 10^{-3}$) compared to $K_S \to 2\pi$. With CP violation, the description of the neutral kaon system becomes more complicated, requiring the introduction of the parameters ϵ and ϵ'. After recalling the definition of the ϵ and ϵ' parameters, we continue with our status report.

It is well-known that the decays $K^0 \to 2\pi$ can be described by two parameters [6.34]:

$$\eta_{+-} = \frac{< \pi^+\pi^- |H_w|K_L >}{< \pi^+\pi^- |H_w|K_S >}; \quad \eta_{00} = \frac{< \pi^0\pi^0 |H_w|K_L >}{< \pi^0\pi^0 |H_w|K_S >} \tag{6.82}$$

where H_w is the weak interaction Hamiltonian. With isospin decomposition of the final 2π state into its $I = 0$ and $I = 2$ parts, one has:

$$\eta_{+-} = (\epsilon_0 + \epsilon_2)/(1 + \frac{\omega}{\sqrt{2}}); \quad \eta_{00} = (\epsilon_0 - 2\epsilon_2)/(1 - \sqrt{2}\omega) \tag{6.83}$$

where:

$$\epsilon_2 = \frac{< I = 2|H_w|K_L >}{< I = 0|H_w|K_S >}; \quad \epsilon_0 = \frac{< I = 0|H_w|K_L >}{< I = 0|H_w|K_S >}$$

$$\omega = \frac{< I = 2|H_w|K_S >}{< I = 0|H_w|K_S >} \tag{6.84}$$

The quantity ω can be neglected in Eq. (6.83) because of the observed $\Delta I = \frac{1}{2}$ rule for CP-conserving decay [6.34]. It is then possible to parameterize the $K_L^0 \to 2\pi$ amplitudes as:

$$< n|H_w|K^0 >= A_n e^{i\delta_n} \tag{6.85}$$

where δ_n is the $\pi\pi$ phase shift in the $I = n$ channel coming from the final state interactions. A commonly adopted phase convention is to choose A_0 to be real, i.e. Im $A_0 = 0$, which leads to:

$$\epsilon_0 = \epsilon; \quad \epsilon_2 = \frac{i}{\sqrt{2}}e^{i(\delta_0 - \delta_2)}\frac{\text{Im}A_2}{A_0} \equiv \epsilon' \tag{6.86}$$

and consequently:

$$\eta_{+-} = \epsilon + \epsilon'; \quad \eta_{00} = \epsilon - 2\epsilon' \tag{6.87}$$

Since ϵ'/ϵ is small compared to one, it follows that $\mathcal{R}e(\epsilon'/\epsilon) \simeq \frac{1}{6}(1 - R)$ where $R = |\eta_{00}/\eta_\pm|^2$.

In sum, the parameter ϵ measures the CP-violating "mass-mixing" between the orthogonal combinations of K^0 and $\bar{K}^0$ and its non-vanishing value $|\epsilon| = 2.275 \pm 0.021 \times 10^{-3}$ [$\eta_\pm = 2.275 \pm 0.021 \times 10^{-3} e^{i(44.6 \pm 1.2^0)}$] measures the so-called "indirect" CP violation. The second parameter, ϵ', measures the relative contribution of the final state $I = 2$ and $I = 0$ amplitudes in $K_\ell \rightarrow 2\pi$ decay so that $|\epsilon'|$ becomes a measure of so-called "direct" CP violation in the neutral kaon system. Unfortunately, while "indirect" CP violation in the neutral kaon system is definitely established, the existence of "direct" CP is still in controversy since there are two somewhat contradictory experiments [6.27]: one experiment at CERN gives $2.3 \pm 0.7 \times 10^{-3}$ for $\mathcal{R}e(\epsilon'/\epsilon)$ while the other experiment at Fermilab yields $0.60 \pm 0.69 \times 10^{-3}$ for $\mathcal{R}e(\epsilon'/\epsilon)$, which is consistent with $\epsilon' = 0$, i.e. no "direct" CP violation. Thus, these two experiments leave the existence of ϵ' in limbo. This is unfortunate because the theoretical calculations of ϵ and ϵ', within the framework of the standard model, depend on two different types of diagrams: the "box" diagrams (shown in Fig. 6.6) dominate the calculation for ϵ, whereas the "penguin" diagrams (shown in Fig. 6.7) are the chief contributors to the calculation of ϵ'. We do not write out the complicated expression for ϵ [6.35] except to note that a key parameter [which measures the "short distance" contribution to $\Delta m(K_L - K_S)$] is poorly known. It is helpful that the "long distance" effects — arising from virtual transitions such as $K_L \rightarrow \pi^0, \eta$: $K_S \rightarrow 2\pi$, which are important for the calculation of $\Delta m(K_L - K_S)$ (where the real part of the box diagram shown in Fig. 6.6 is needed) — are not important for the calculation of ϵ (which receives contributions from the imaginary part of the box diagrams). Not unexpectedly, the theoretical value of ϵ is affected by the magnitude of m_t; indeed, for $m_t > 60$ GeV, a good approximation to ϵ — that exhibits clearly the dependence on m_t — can be written in terms of the mixing parameters A, ρ and η of the Wolfenstein parameterization [see Eq. (6.80)]:

$$|\epsilon| \sim 2.2 \times 10^{-3} A^2 \eta B_K \{1 + 0.5 A^2 (1 - \rho)(\frac{m_t}{50 \text{ GeV}})^2\} \qquad (6.88)$$

where the B_K parameter $(= 0.7 \pm 0.2)$ is the ratio between the "short-distance" contribution to $\Delta m_K [= \Delta m(K_L - K_S)]$ and its value in the "vacuum insertion" approximation [6.36]. Since the value of $|\epsilon|$ is known,

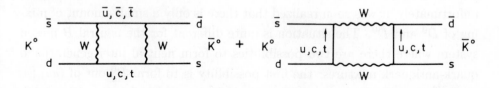

Fig. 6.6. "Box" diagrams contributing to the ϵ parameter in $K_L \rightarrow 2\pi$ decay.

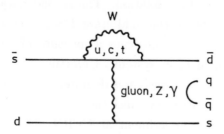

Fig. 6.7. "Penguin" diagram contributing to the ϵ' parameter in $K_L \to 2\pi$ decay.

Eq. (6.88) becomes a valuable constraint equation — via m_t — on the combination $U_{td}^* U_{ts}$ of the U_{CKM} matrix elements.

Insofar as the theoretical calculation of ϵ' is concerned, it matches the ϵ' experiments in difficulty and inconclusiveness. We have already pointed out that the major contribution to ϵ' — in contrast to ϵ — is generated by the so-called "penguin" diagrams shown in Fig. 6.7 for $m_t < M_W$ [6.36]. However, if $m_t > M_W$, it is necessary to include not only the contribution from the "gluon penguin" diagram in Fig. 6.7 but also contributions from "electroweak penguins" (Z and γ). The "long-distance" effects also play a role in the ϵ' calculation. It is fair to say that the experimental measurements and theoretical calculations of ϵ' are inconclusive with regard to the existence of "direct" weak CP violation in the neutral kaon system. For this reason, interest has turned in recent years to studies of the neutral B system because larger effects are expected.

(b) *Neutral B meson system*

It is easy to see from the structure of U_{CKM} [see Eq. (6.71)] that weak CP violation can only occur in a neutral meson system wherein the neutral meson (and its antiparticle) are constructed out of quarks belonging to different generations. Hence, it is not surprising that weak CP violation was first discovered in the neutral kaon system. When the c quark was discovered, one expected to see and one soon did see, the mixing of the neutral inter-generational quark-anti-quark combinations of $D^0 = (\bar{u}c)$ and $\bar{D}^0 = (u\bar{c})$; unfortunately, it was soon realized that there is only a small amount of mixing of D^0 and $\bar{D}^0$. The situation is quite different for the neutral B meson system where there are two possibilities to form neutral inter-generational quark-antiquark mixtures: the first possibility is to form B_d^0 out of $(\bar{d}b)$ [so that $\bar{B}_d^0 = (d\bar{b})$] and the second is to form B_s^0 out of $(\bar{s}b)$ [so that $\bar{B}_s^0 = (s\bar{b})$]. Fortunately, two experimental groups (ARGUS and CLEO) [6.37] have reported a high degree of mixing between the B_d^0 and $\bar{B}_d^0$ mesons (of the order of 20%). This finding has triggered intense interest in searching for CP violation in the B_d^0 system, including discussion of the desirability and feasibility of a B meson "factory". We comment briefly on the $(B_d^0 - \bar{B}_d^0)$ mixing situation.

The $(B_d^0 - \bar{B}_d^0)$ mixing can be described in the usual way by a mass matrix with non-zero off-diagonal elements written in the form:

$$\begin{bmatrix} M - \frac{1}{2}i\Gamma & M_{12} - \frac{1}{2}i\Gamma_{12} \\ M_{12}^* - \frac{1}{2}i\Gamma_{12}^* & M - \frac{1}{2}i\Gamma \end{bmatrix} \tag{6.89}$$

By diagonalizing this matrix, one obtains the mass eigenstates, M_{B_S} and M_{B_L}, with definite widths Γ_S and Γ_L; if M and Γ are the average mass and width and one defines: $Q \equiv \sqrt{(M_{12} - \frac{1}{2}i\Gamma_{12})(M_{12}^* - \frac{1}{2}i\Gamma_{12}^*)}$, then $\Delta m_B = M_{B_L} - M_{B_S} = 2\,\mathrm{Re}Q$ and $\Delta\Gamma = -4\,\mathrm{Im}Q$. It turns out — in contrast to the $(K^0 - \bar{K}^0)$ system — that Δm_{B_S} is much larger than $\Delta\Gamma$ so that the lifetime difference between $\tau_{B_L} - \tau_{B_S}$ is small. It can be shown [6.35] that the $(B_q^0 - \bar{B}_q^0)$ ($q = d$ or s) mixing parameter $x_q \equiv \Delta m/\Gamma$ is given by:

$$x_q = \frac{G_F^2}{6\pi^2} M_B f_B^2 B_B \tau_B M_W^2 |U_{ts}U_{tq}|^2 \alpha(\eta) \tag{6.90}$$

where $\alpha(\eta)$ is a numerical function of $\eta = (m_t/M_W)^2$ and the B_B parameter is now the ratio between the "short distance" contribution to Δm_B and its value in the "vacuum insertion" approximation. Using Eq. (6.90) and the combined result for x_d, obtained by ARGUS and CLEO, i.e. $x_d = 0.66\pm0.11$ [6.37], one obtains another constraint on $U_{ts}U_{td}^*$ and m_t, from which one derives the ratio $x_s/x_d \simeq |U_{ts}|^2/|U_{td}|^2$, which yields $x_s \geq 3$. The last result is most interesting because it implies an even larger mixing for the B_s^0 system than for the B_d^0 system. The great challenge now is to detect CP violation in the B^0 system (whether B_d^0 or B_s^0) and to establish whether the presumed weak CP violation is fully explained within the framework of the U_{CKM} mixing matrix approach for three generations.

We conclude this section on weak CP violation by spelling out the connection between the weak CP-violating parameters associated with selected decay modes of the B^0 meson and the angles of the "unitarity triangle" of Fig. 6.5. To derive this relationship, we note that a CP-violating process requires interference between the particle and antiparticle amplitudes (with different phases) and this can come about in the neutral B meson system through $B^0 - \bar{B}^0$ mixing. Because of this mixing, the decay rate of a time-evolved initially pure $B^0(\bar{B}^0)$ state into a CP-eigenstate, f, is given by the well-known formula [6.34]:

$$\Gamma(B^0_{phys}(t) \to f) \propto e^{-\Gamma t}\,[1 - \mathrm{Im}\,\bar\lambda\,\sin(\Delta m_B)]$$

$$\Gamma(\bar B^0_{phys}(t) \to \bar f) \propto e^{-\Gamma t}\,[1 + \mathrm{Im}\,\bar\lambda\,\sin(\Delta m_B) \tag{6.91}$$

where Γ is the total width and $\mathrm{Im}\,\bar\lambda$ is defined by means of the CP-violating asymmetry quantity $\alpha_f(t)$ (through the interference term $\mathrm{Im}\,\bar\lambda$), thus:

$$\alpha_f(t) = \frac{\Gamma(B^0_{phys}(t) \to f) - \Gamma(\bar B^0_{phys}(t) \to \bar f)}{\Gamma(B^0_{phys}(t) \to f) + \Gamma(\bar B^0_{phys}(t) \to \bar f)} \tag{6.92}$$

The CP-asymmetry $\alpha_f(t)$ (or, equivalently, $\mathrm{Im}\,\bar\lambda$) measures different combinations of the phases entering into U_{CKM}. It can now be shown that $\mathrm{Im}\,\bar\lambda$ for a selected B^0 meson decay can be simply related to an angle of the "unitarity triangle" of Fig. 6.5; in particular, $\mathrm{Im}\,\bar\lambda = \sin(2\phi_i)$ where $\phi_i(i = 1, 2, 3)$ denotes one of the three angles of the "unitarity triangle" of Fig. 6.5. A possible choice of three decay modes of B^0 mesons (B^0_d and B^0_s) is given in Table 6.6, which also lists the three "unitarity" angles corresponding to the three decay modes shown (see Fig. 6.5).

Table 6.6: Weak CP violation for neutral B mesons and "unitarity triangle" for U_{CKM}.

Decay modes	$\mathrm{Im}\,\bar\lambda$
$B^0_d \to \psi K_S$	$\sin(2\phi_1)$
$B^0_d \to \pi^+\pi^-$	$\sin(2\phi_2)$
$B^0_s \to \rho K_S$	$\sin(2\phi_3)$

Considerable effort is going into verifying the correctness of the "unitarity" triangle as a clearcut test of the basic assumptions underlying the use of U_{CKM} for three quark generations (i.e. unitarity and the hierarchical pattern). Hence, it seems worthwhile to sketch the derivation of one of the results listed in Table 6.6. We focus on the decay $B^0_s \to \rho K_S$, and enumerate below — with the help of Fig. 6.8 — the matrix elements that contribute to the final result $\mathrm{Im}\,\bar\lambda(\rho K_S) = \sin 2\phi_3$ (see Table 6.6). From Fig. 6.8, we can say that:

(1) The direct decay $\bar b \to u\bar u d$ is dominated by the W-exchanged "tree-level" diagram which gives:

$$\bar\lambda \propto \left(\frac{X}{X^*}\right) \qquad \text{where } X(\bar b \to u\bar u d) = U_{ub}U^*_{ud} \tag{6.93}$$

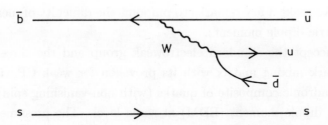

Fig. 6.8. Diagram contributing to $B_s \to \rho K_s$ decay.

(2) The mixing in the B_S^0 system is dominated by a "box" diagram with virtual t quarks (see Fig. 6.6); this yields:

$$\bar{\lambda} \propto \left(\frac{Y}{Y^*}\right) \quad \text{where} \quad Y = U_{tb}^* U_{ts} \tag{6.94}$$

(3) The mixing in the K^0 system is dominated by "box" diagrams with c quarks since B_s produces a $\bar{K}^0$ meson, $\bar{B}_s^0$ produces a K^0 meson and interference is possible only with $K^0 - \bar{K}^0$ mixing (see Fig. 6.6); this gives:

$$\bar{\lambda} \propto \left(\frac{Z}{Z^*}\right) \quad \text{where} \quad Z = U_{cd} U_{cs}^* \tag{6.95}$$

Combining all three expressions, the final result is:

$$\text{Im } \bar{\lambda}(B_s \to \rho K_s) = \text{Im } \left(\frac{U_{ub}U_{ud}^*}{U_{ub}^* U_{ud}}\right)\left(\frac{U_{tb}^* U_{ts}}{U_{tb}U_{ts}^*}\right)\left(\frac{U_{cd}^* U_{cs}}{U_{cs}^* U_{cs}}\right) \simeq \text{Im } \frac{A\lambda^3(\rho - i\eta)}{A\lambda^3(\rho + i\eta)}$$

$$= \frac{-2\eta\rho}{\rho^2 + \eta^2} \simeq \sin 2\phi_3 \tag{6.96}$$

This is the result listed in Table 6.6; similar derivations can be given for the other two decay modes of the B^0 meson in terms of the angles of the U_{CKM} "unitarity triangle". Obviously, it will take time to carry out the necessary measurements on the B^0 mesons and, hopefully, some sort of B "meson factory" will come to the rescue.

(c) *Electric dipole moments of neutron and charged leptons*

A third test of the U_{CKM} quark mixing matrix approach to the explanation of weak CP violation proceeds in quite a different direction from the previous two tests and consists of measuring the electric dipole moment of the neutron and, to a less promising extent, the electric dipole moment of a charged lepton (e or μ). Like the previous two tests of weak CP violation — the measurement of ϵ'/ϵ for the neutral K system and the determination of the ϵ's and ϵ''s for the two neutral B systems — the third test is also very difficult. We make a few remarks to pinpoint the difficulty of measuring any of the electric dipole moments.

If we accept the standard electroweak group and the three-generation U_{CKM} quark mixing matrix with its provision for weak CP violation, we expect a hadronic composite of quarks (with non-vanishing spin) to possess an electric dipole moment (EDM) at some level. The presumed existence of the θ vacuum in QCD creates the possibility of a "strong CP violation"-induced electric dipole moment but the theory of "strong CP violation" (see §10.3c) does not predict the value of the θ parameter and, in actual fact, the observed upper limit on EDM of n (denoted by d_n) is used to determine the upper limit on the "strong CP" θ parameter. It is interesting to predict d_n on the basis of the standard model and U_{CKM} — see below. With regard to possible EDM's for the charged leptons, we reiterate an earlier point that, if the neutrinos of the three generations are massless (which is a basic assumption of the standard model), there is no three-generation U_{CKM} lepton mixing matrix and hence no CP violation and none of the charged leptons can acquire an EDM in this way; only if two of the three neutrinos are massive is it possible to construct a U_{CKM} lepton mixing matrix and realize an EDM for a lepton. Consequently, in discussing the possibility of a leptonic EDM, one implicitly assumes a departure from the standard model, whether in accepting at least two massive neutrinos, or allowing for more than one Higgs doublet with complex Yukawa coupling to the leptons, or some other modification of the basic standard model assumptions. Since we have very little guidance as to the choice of departure from the standard model that could yield non-vanishing EDM's for leptons, we limit our brief comments here to hadronic EDM's that are in accord with the standard model plus U_{CKM}.

Within the above limitation, the nucleon is of primary interest and, in particular, the neutron — where a direct measurement of d_n has been pursued for decades [6.38] and increasingly so in the last few years [6.39]. An upper limit on d_p has been obtained from a measurement on atomic hydrogen but this upper limit is four orders of magnitude higher than the upper limit on d_n (see Table 6.7 below). In reporting on EDM (n), it is useful to recall the P- and CP-violating interaction that describes the EDM of a $J = \frac{1}{2}$ particle; if $\psi(x)$ is the Dirac field and $F_{\mu\nu}$ is the electromagnetic field tensor, then the "effective" interaction must be of the form:

$$if(q^2)\bar{\psi}(p_2)\sigma_{\mu\nu}\gamma_5\psi(p_1)F^{\mu\nu} \tag{6.97}$$

where $f(q^2)$ is the form factor and $q^2 = (p_2 - p_1)^2$. The quantity d_n corresponds to the static limit of the form factor $f(q^2)$, i.e.:

$$d_n = f(0) \tag{6.98}$$

The requirement of renormalizability rules out the possibility of a fundamental (lowest-order) coupling of the type shown in Eq. (6.93). However, such interactions can arise as a result of quantum loop corrections and it then follows that the induced EDM is finite and calculable. The painstaking search for an EDM of the neutron has recently succeeded in reducing the upper limit on d_n to $< 10^{-25} e$ cm [6.39] (see Table 6.7), an upper limit that is easily consistent with the standard model prediction of the order of $10^{-33} e$ cm [6.40].

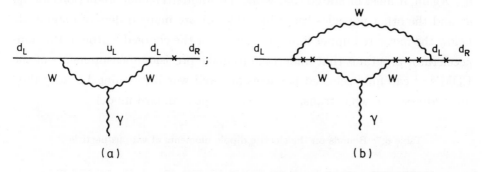

Fig. 6.9. One- and two-loop contributions to the electric dipole moment of the neutron (a): one loop; (b) two-loop.

It is unfortunate and somewhat accidental that the "standard model" prediction for d_n is so miniscule. We show in Fig. 6.9 the "one-loop" (diagram (a)) and "two-loop" (diagram (b)) standard model contributions to d_n. Figure 6.9a exhibits the cancellation of the U_{CKM} phases at the two W vertices so that the "one-loop" contribution vanishes (there is an external mass insertion in diagram (a) of Fig. 6.9 — denoted by a cross — because the W vertex conserves chirality); one can say that the "one-loop" contribution in the standard model vanishes due to the unitarity of U_{CKM}, i.e. $U_{CKM} U_{CKM}^* = 1$). At the "two-loop" level, there is a greater number of quark mass insertions — as is manifest in diagram (b) of Fig. 6.9 — but the sum over all contributions surprisingly still cancels [6.40]. However, it has been argued [6.40] that the gluon radiative corrections must be taken into account and that, at the "two-loop" level, they yield a non-vanishing result, estimated to be:

$$d_n \sim e m_q \frac{G_F \alpha_e \alpha_s}{\pi^4} \frac{m_t^2 m_s^2}{M_W^4} J \sim 10^{-33} e \text{ cm} \qquad (6.99)$$

where J is defined by Eq. (6.76) and it has been assumed that $m_t \simeq 100$ GeV. This is discouraging for observing d_n in the foreseeable future or one can state this outcome more optimistically: the detection of d_n in the "foreseeable future" would be powerful evidence for the incompleteness of the standard electroweak gauge group $SU(2)_L \times U(1)_Y$ with a single Higgs SSB doublet and three massless neutrinos.

We conclude this part of §6.3f with a table listing the bounds on the electric dipole moments of a half dozen or so particles. We note in this table the much larger upper limit on d_p (than on d_n) and the much smaller limit on d_e. Again, it must be stated that while the standard model predictions for d_p, d_e and the other particles listed in Table 6.7 are many orders of magnitude below the measured upper limits (e.g. zero for the charged leptons), the non-accelerator programs that undertake to make precision measurements of the EDM's of carefully selected particles are well worth pursuing because they are addressed to what transpires "beyond the standard model".

Table 6.7: Bounds on the electric dipole moments of various particles.

Particle	EDM (e cm)	Remarks
neutron	$d_n = -(0.3 \pm 0.5) \times 10^{-25}$	direct measurement
electron	$d_e = (-2.7 \pm 8.3) \times 10^{-27}$	atomic experiment (^{205}Tl)
neutrinos	$\|d_\nu\| \lesssim 10^{-20}$	$\nu e \to \nu e$
	$\lesssim 2 \times 10^{-22}$	astrophysics
	$\lesssim 2 \times 10^{-27}$	primordial nucleosynthesis
proton	$\|d_p\| < 2 \times 10^{-21}$	atomic experiment
muon	$\|d_\mu\| < 7.3 \times 10^{-19}$	$(g_\mu - 2)/2$
tau	$\|d_\tau\| < 10^{-16}$	$e^+ e^- \to \tau^+ \tau^-$
W boson	$\|d_W\| < 10^{-19}$	induced d_n

§6.4. Special topics in QFD

In the preceding sections (§6.3a–§6.3f), we have discussed highlights of electroweak phenomenology and have found that increasingly precise experiments continue to confirm the standard gauge theory of the electroweak interaction in all cases where definitive predictions have been made. Despite these successes for standard QFD, there is an increasing consensus of views that departures from the standard model will ensue in the multi-TeV region and that increasing attention should be given to the clues to the new physics that beckon beyond the standard model. We propose, in concluding this chapter on QFD, to consider three "special topics" (the selection is somewhat arbitrary) that are being studied intensively at the present time — either experimentally or theoretically or both — and that purport to tell us in what directions departures from the standard model are taking place, if indeed there is new physics "beyond the standard model". In §6.4a, we consider "rare weak decays" which promises to shed further light on the U_{CKM} quark mixing matrix and weak CP violation. In §6.4b, we tackle the problem of whether the neutrinos of all three generations are massless and, if not, whether the masses are of the Majorana or Dirac type or, possibly, some admixture of the two; in that section, we also discuss the inconclusive evidence thus far for "free" neutrino oscillations and the popular hypothesis (the "MSW mechanism" [6.41]) that "electron-induced" neutrino oscillations can solve the "solar neutrino problem". Finally, as the last of the special topics in this chapter, we discuss in §6.4c the important question of whether the Higgs boson is "elementary" or a quasi-Higgs particle; if the latter, we consider — along the lines of the Nambu-Jona-Lasinio (NJL) mechanism (see §6.4c) — whether the quasi-Higgs scalar excitation originates from the same dynamical chiral symmetry-breaking mechanism that gives rise to t quarks of large mass and to the $t\bar{t}$ condensates that are identified with the quasi-Higgs bosons.

§6.4a. Rare kaon decays

The study of rare kaon decays has played a pivotal role in the formulation of the standard model of electroweak interactions; for example, the failure to observe $K_L \rightarrow \mu\bar{\mu}$ and other rare decay modes of charged and neutral kaons led to the GIM mechanism (see §6.3e) and, ultimately, to the U_{CKM} quark mixing matrix. In recent years, the search for rare kaon decays has attracted renewed attention due to the possibility of performing significantly improved experiments; in addition, the prospect of an unexpectedly large t quark mass has created the possibility that more sensitive studies of rare meson decays (chiefly rare kaon decays in the near future) will provide deeper insight into the role of the most massive quark in validating the standard model and possibly indicating departures from it. Rare neutral kaon decays such as: $K_L \to \mu\bar{\mu}$, $K_L \to \pi^0 e^+ e^-$ and $K_L \to \pi^0 \nu\bar{\nu}$, and rare charged kaon decays such as: $K^+ \to \pi^+ \nu\bar{\nu}$ and $K^+ \to \pi^+ l\bar{l}$ are all examples of flavor-changing neutral current (FCNC) weak processes and are forbidden to lowest order by the unitarity of U_{CKM} (see §6.3e); however, they can occur radiatively through the typical "one-loop" diagrams shown in Fig. 6.10. These diagrams are sensitive to the virtual (massive) t quark so that the standard model predictions for these K decays depend not only on the values of the U_{CKM} parameters but also on m_t. There are in addition several "exotic" decays such as $K_L \to \mu\bar{e}$ and $K^+ \to \pi^+ \mu\bar{e}$ that are forbidden in the standard model (e.g. by "intra-generational" lepton charge conservation) and observation of such decays would be clear evidence for "new physics"; discussion of "exotic" decays is deferred until §8.4b when we discuss unification models. In this section, we confine ourselves to brief remarks concerning three crucial rare K decays that can test the standard U_{CKM} model, namely: (1) $K^+ \to \pi^+ \nu\bar{\nu}$; (2) $K_L \to \pi^0 e\bar{e}$; and (3) $K_L \to \pi^0 \nu\bar{\nu}$. We consider each of these in turn.

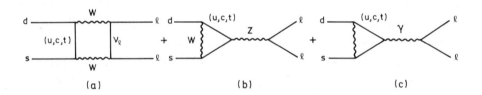

Fig. 6.10. Neutral kaon decays due to (radiative) one-loop diagrams.

(1) $\underline{K^+ \to \pi^+ \nu\bar{\nu}}$: This decay is a prime target of current rare kaon decay experiments and a favorite of theorists. It is rather insensitive to "long-distance" effects and can, therefore, be reliably calculated. In addition, it is sensitive to m_t and the U_{CKM} matrix element U_{td}. The branching ratio expected in the standard model is $(1.2 - 4.0) \times 10^{-10}$ [6.42] for m_t ranging from 92 to 250 GeV. The current experimental upper limit of the decay branching ratio is 5×10^{-9} which is not terribly far from the standard model prediction.

(2) $\underline{K_L \to \pi^0 e^+ e^-}$: This decay has recently also attracted much attention theoretically and experimentally since it may directly test the mechanism of "direct" weak CP violation in the standard U_{CKM} model. It receives three types of contributions: (a) a CP-conserving one through a two-photon intermediate state, i.e. $K_L \to \pi^0 \gamma\gamma \to \pi^0 e^+ e^-$; (b) an "indirect" weak

CP-violating one, induced by the mixing of the K^0 and $\bar{K}^0$ states characterized by the ϵ parameter, i.e. $K \to \pi^0 \gamma^* \to \pi^0 e^+ e^-$, where γ^* is a virtual photon in an effective $J = 1$ CP-even state; and (c) through the "direct" weak CP-violating decay of $K_L \to \pi^0 \gamma^* \to \pi^0 e^+ e^-$. The branching ratio of the CP-conserving part of $K_L \to \pi^0 e^+ e^-$, which depends on the decay $K_L \to \pi^0 \gamma\gamma$, is estimated to be of order 10^{-13}. The branching ratio due to the "indirect" weak CP-violating contribution is estimated to be 2×10^{-12}, whereas the "direct" weak CP contribution — coming from the short-distance electroweak loop effects and therefore depending sensitively on m_t — yields $10^{-12} - 10^{-11}$ [6.43]. The present experimental upper limit on the branching ratio of $K_L \to \pi^0 e\bar{e}$ decay is 5.5×10^{-9}, with the prospect of reaching the standard model level of about 10^{-11} in a few years. There is a glimmer of hope that the measurement of this decay can shed light on the magnitude of the "direct" weak CP-violating amplitude.

(3) $\underline{K_L \to \pi^0 \nu\bar{\nu}}$: This decay is similar to the previous one but it only receives contributions from the weak CP-violating amplitudes since there is no two-photon intermediate CP-conserving state. As a result, the decay is most interesting because the branching ratio of the "direct" weak CP-violating contribution is much larger than that of the "indirect" one because of the large t quark mass. Thus, this decay provides a very clean test for weak CP violation in the standard model but, considering that the present measured upper limit on the branching ratio of this decay is 7.6×10^{-3} [6.44]; clearly a theoretically useful measurement of this rare neutral kaon decay model is a long way off.

In sum, the intense program under way at BNL and other laboratories on rare kaon decays is pushing down the upper limits on the branching ratios on many of these FCNC weak decays. A "kaon factory" could obviously do better in this arduous program. Meanwhile, one can only say that the

Table 6.8: Upper limits on the branching ratios of rare kaon decays.

Decay	Upper limit (90% C.L.)
$K^+ \to \pi^+ \nu\bar{\nu}$	5×10^{-9}
$K_L \to \pi^0 e^+ e^-$	5.5×10^{-9}
$K_L \to \pi^0 \nu\bar{\nu}$	7.6×10^{-3}
$K_L \to \mu^\pm e^\mp$	6.1×10^{-11}
$K^+ \to \pi^+ \mu^+ e^-$	2.1×10^{-10}

standard model is remarkably resilient and that no evidence has been found for deviations from the standard model. Perhaps, the most convincing way to register this last point is to present in Table 6.8 the present upper limits on the branching ratios of several of the rare kaon decays that have been studied. Table 6.8 [6.44] will also be useful later in connection with our discussion of a number of unification attempts to "go beyond the standard model" (see §8.4b).

§6.4b. Neutrino masses, neutrino oscillations and the solar neutrino problem

Up to this point, in our extended discussion of the standard gauge theory of electroweak interactions, we have assumed massless neutrinos for all three quark-lepton generations. The two-component Weyl equation satisfied by a massless neutrino is manifestly chirality-invariant and CP invariant and both of these properties of Weyl fermions define what we have called the "neutrino paradigm": the chirality invariance property of the Weyl fermion played a key role in the construction of the universal V-A theory of weak interactions [1.80] and the small (weak) departure from CP invariance among quarks may be a relic of the "Weyl" quarks in the electroweak theory after the Higgs SSB mechanism comes into play. We also noted that it is the masslessness of the neutrinos of the three generations that wipes out the U_{CKM}-type of mixing matrix for leptons and that a "U_{CKM}" lepton mixing matrix would reappear if at least two of the neutrino masses are finite. There are many other reasons — in particle physics, astrophysics and cosmology — for wishing to have finite (albeit small) masses for the lefthanded neutrinos that we know. On the other hand, there are arguments based on the chiral gauge anomaly-free conditions in four dimensions for preferring the standard model with its massless neutrinos.

Unfortunately, for those who strongly prefer finite-mass neutrinos [to resolve some still-unsolved problem (such as the "solar neutrino problem" or dark matter)], there is no convincing evidence, as of now, that any one of the three known neutrinos possesses a finite mass despite many years of hard and skillful experimental work. We do know upper limits on the masses of ν_e, ν_μ and ν_τ and these upper limits are listed in Table 1.6 and the applicable experiments are noted in Eq. (6.100) below:

$$m(\nu_e) < 7.3\text{eV} \quad (^3\text{H decay}); \quad m(\nu_\mu) < 270 \text{ KeV} \quad (\pi \to \mu\nu \text{ decay});$$

$$m(\nu_\tau) < 35 \text{ MeV} \quad (\tau \to 5\pi + \nu_\tau \text{ decay}) \tag{6.100}$$

These upper limits on the neutrino masses are quite appreciable but they do serve a useful purpose in theoretical model building. Obviously, the actual values of neutrino masses constitute the real goal and it is quite conceivable that the vast effort being expended now to pin down the values of the neutrino masses will succeed. If a finite mass is established for any one of the three neutrinos, a major modification will be required in the standard model and it is therefore incumbent upon us to review the status of the neutrino mass search, whether implemented directly through kinematical measurements, or indirectly by means of a finite-mass-related property of neutrinos such as neutrino oscillations. We carry out this review under three headings: (1) Dirac versus Majorana finite-mass neutrinos and direct searches for their presence; (2) theory and experimental status of free neutrino oscillations; and (3) the solar neutrino problem and induced neutrino oscillations.

(1) *Dirac versus Majorana finite-mass neutrinos and direct searches*
    for their presence

In the standard $SU(2)_L \times U(1)_Y$ electroweak theory, the lefthanded (L) neutrino of each $SU(2)_L$ doublet is massless provided that there is no $SU(2)_L$ righthanded (R) singlet (ν_R) partner present to generate a Dirac mass with the help of a Higgs boson doublet; a Majorana mass for the neutrino requires a Higgs triplet (see §7.7 and §8.4) although the experimental finding that the ρ parameter (see §6.3a) is very close to unity leaves little room for this possibility at the standard model level. However, an enlarged electroweak group [e.g. the left-right-symmetric group $SU(2)_L \times SU(2)_R \times U(1)_{B-L}$ — see §8.4] can easily accommodate a Higgs triplet that generates a pair of Majorana neutrinos — one light L neutrino and one heavy R neutrino — through the so-called "see-saw" mechanism [6.45]; hence, we consider both the Dirac and Majorana possibilities for the neutrino mass. Insofar as a massive Dirac neutrino is concerned, if ν_R is present, the Dirac mass term for the neutrino is simply: $m_D \bar{\nu}_L \nu_R +$ h.c.; this procedure works for all the charged particles in the standard model (i.e. all the quarks and charged leptons) through the use of the Higgs SSB mechanism to generate the effective Yukawa coupling constants and, *a fortiori*, the Dirac fermion masses. Since there is no evidence for the existence of ν_R and a Dirac mass for the neutrino possesses no built-in explanation for its smallness compared to the charged particle masses, the alternative of postulating the existence of a Higgs triplet has, as we have already remarked, also been considered; the Higgs triplet — through its Yukawa coupling to leptons — can provide a Higgs SSB mechanism for generating a Majorana mass term of the form: $m_M \nu_L^T C \nu_L + $h.c. [where ν_L^T is the transpose of ν_L and C is the charge conjugation operator] and a natural explanation for the smallness of the Majorana masses of the three lefthanded neutrinos is provided in some extensions of the standard model (e.g. the left-right-symmetric model — see §8.4b). In either case, whether the finite mass of the neutrino has a Dirac or a Majorana origin, we would have to deal with "new physics" beyond the standard model, but in different ways.

In order to provide a common language for the Dirac and Majorana neutrino mass possibilities, we recall the most general expression for the neutrino mass (we only consider one generation for simplicity) that can arise after the SSB of the electroweak group $SU(2)_L \times U(1)_Y$ [6.46]:

$$\mathcal{L}_\nu = -m_D[\bar{\nu}_L \nu_R + \bar{\nu}_R \nu_L] - \frac{1}{2} m_M^L [\nu_L^T C \nu_L + \bar{\nu}_L C \bar{\nu}_L^T] - \frac{1}{2} m_M^R [\nu_R^T C \nu_R + \bar{\nu}_R C \bar{\nu}_R^T]$$

(6.101)

If we make use of the identities:

$$\bar{\nu}_L \nu_R = \frac{1}{2}[\bar{\nu}_L \nu_R + \bar{\nu}_L^c \nu_R^c]; \quad \nu_R^T C \nu_R = \bar{\nu}_L^c \nu_R; \quad \bar{\nu}_L C \bar{\nu}_L^T = \bar{\nu}_L \nu_R^c \quad (6.102)$$

Equation (6.101) can be rewritten in the familiar form:

$$\mathcal{L}_\nu = \frac{1}{2}(\bar{\nu}_L, \bar{\nu}_L^c) m_\nu \begin{pmatrix} \nu_R^c \\ \nu_R \end{pmatrix} + \text{h.c.}$$

(6.103)

where the mass matrix m_ν is given by:

$$m_\nu = \begin{pmatrix} m_M^L & m_D \\ m_D & m_M^R \end{pmatrix}$$

(6.103a)

Both Eq. (6.101) and Eq. (6.103a) tell us that we can think of a neutrino as having both a Dirac as well as a Majorana contribution to its mass. It is interesting to point out that an experiment can be "neutral" as to the Dirac or Majorana character of the neutrino mass that is measured or it can be designed to measure either the Dirac or Majorana component (assuming that both exist) of the total mass. Thus, the experiment seeking to determine m_{ν_e} from tritium decay: $^3\text{H} \rightarrow {}^3\text{He} + e^- + \bar{\nu}_e$ can not distinguish between the Dirac and Majorana contributions; thus, the upper limit on the mass of ν_e given in Eq. (6.100) comes from a "world average" of tritium measurements [6.47] and we do not know whether the upper limit refers to a Dirac or a Majorana mass. On the other hand, the distinction between the Dirac and Majorana contributions to the neutrino mass can be achieved through a suitable design of the experiment; for example, one can decide between a Dirac and Majorana neutrino by looking for the two phenomenologically different double beta decay processes: (a) double beta decay accompanied by two Dirac neutrinos and satisfying lepton charge conservation, and (b) neutrinoless double beta decay, which is only possible for Majorana neutrinos involving a lepton charge change of two units. Figure 6.11 shows the diagrams for the two cases. The two- (Dirac) neutrino process can proceed with massless (or finite mass) neutrinos and has finally been detected in the double beta decay ^{82}Se $\rightarrow$ Kr^{82} + $2e^-$ + $2\bar{\nu}_e$, with a lifetime of 10^{20} year [6.48]; however, despite much effort devoted to searching for neutrinoless double beta decay (Majorana neutrinos must have finite mass in order to be distinguishable from Dirac neutrinos [6.34]) from the most feasible candidate ^{76}Ge (proceeding through ^{76}Ge $\rightarrow$ ^{76}Se + $2e^-$), the process has not been observed and the lower limit on the lifetime is 8×10^{23} year [6.48]. This last result places an upper limit on the Majorana contribution to m_{ν_e} of several eV. This value is consistent with the 7.3 eV upper limit quoted in Eq. (6.100) [1.99].

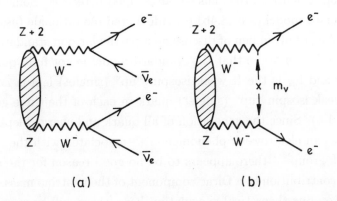

Fig. 6.11. Possible double beta decay processes: a) $2\beta(2\nu)$; b) $2\beta(0\nu)$.

To study the mass matrix of Eq. (6.103a), the most popular model is the "see-saw" mechanism [6.45], which can be realized in a natural way in some extensions of the standard model [e.g. the left-right symmetric model — see §8.4b]. In the "see-saw" model, one assumes a hierarchical structure for m_ν as follows:

$$m_M^R \equiv M >> m_D << m_M^L \simeq 0 \qquad (6.104)$$

where M is a large mass scale ($>> \Lambda_{QFD}$) defined by the theory "beyond the standard model". The mass matrix in Eq. (6.103a) then becomes:

$$m \simeq \begin{pmatrix} 0 & m_D \\ m_D & M \end{pmatrix} \qquad (6.105)$$

and its diagonalization immediately yields two Majorana mass eigenstates, ν_1 and ν_2, with:

$$\nu_L \simeq \nu_{1L}^c + \frac{m_D}{M}\nu_{2L}^c; \quad \nu_R \simeq \nu_{2R} - \frac{m_D}{M}\nu_{1R} \qquad (6.106)$$

The physical masses corresponding to the two eigenstates are:

$$m_1 \simeq \frac{m_D^2}{M} \ll m_D; \quad m_2 \simeq M \qquad (6.107)$$

Consequently, if m_D is of the same order as the mass of a quark or a charged lepton, we see that the L neutrino mass is reduced by the factor (m_D/M) and the R neutrino mass becomes large ($\sim M$). Clearly, this method can easily be generalized to N_g families.

One might inquire into the rationale for the assumption contained in Eq. (6.104), which leads to a phenomenologically acceptable mass for the L neutrino (if M is chosen large enough) and a phenomenologically desirable large mass for the R neutrino that has so far escaped detection. Some extensions of the standard model (see §8.4b) provide a credible rationale (assuming the existence of ν_R) but, in lieu of discussing a particular non-standard model at this point, one can argue that the standard model is not fully quark-lepton symmetric and ν_R is the R weak isospin "up" (singlet) lepton counterpart of the R weak isospin "up" (singlet) quark in each of the three generations (see Table 1.5). Since the generation of all quark and charged lepton masses is treated as an electroweak phenomenon — associated with the SSB of the electroweak group — there appears to be no good reason for the absence of ν_R and its contribution to a Dirac component of the neutrino mass. However, we must stop our theoretical speculation here [there will be enough of this in Chapters 8 and 9] and turn to experiment in order to ascertain whether there is any clearcut evidence for the finiteness of the neutrino mass (Dirac or Majorana) or any other "non-standard property" of the neutrino (e.g. an appreciable magnetic moment).

One powerful experimental method for establishing the finiteness of neutrino mass — other than direct measurement — is the detection of neutrino oscillations. We comment briefly on the status of neutrino oscillations which thus far have yielded a negative result in laboratory experiments — under heading (2) — and then briefly review the current situation — under heading (3) — with regard to the extra-terrestrial evidence for matter-induced neutrino oscillations as an explanation of the "solar neutrino problem".

(2) *Theory and experimental status of neutrino oscillations*

The assumption of finite neutrino masses — which goes beyond the standard model — opens up the door to an interesting phenomenon suggested more than thirty years ago by Pontecorvo [6.49]. With finite neutrino masses, the neutrino mass eigenstates need not be identical with the weak eigenstates — similar to the quark situation — and may give rise to neutrino oscillations among the weak states (unlike the electrically charged quarks that can not undergo "quark oscillation"). The basic idea of ν oscillations can be explained by using the two generations of neutrinos; the extension to three generations is straightforward. We write the two weak states ν_e and ν_μ as mixtures of the mass eigenstates ν_1 and ν_2:

$$|\nu_e> = \cos\theta\,|\nu_1> + \sin\theta|\nu_2>; \qquad |\nu_\mu> = -\sin\theta\,|\nu_1> + \cos\theta\,\nu_2>$$
$$(6.108)$$

If at $t = 0$, ν_e is produced, then at time t one has:

$$|\nu(t)> = \cos\theta\,|\nu_1(t)> + \sin\theta\,|\nu_2(t)>$$
$$= \cos\theta\,e^{-iE_1 t}\,|\nu_1(0)> + \sin\theta\,e^{-iE_2 t}\,|\nu_2(0)>$$
$$(6.109)$$

where:

$$E_2 - E_1 \simeq (m_2^2 - m_1^2)/2p \qquad (6.109a)$$

assuming $\Delta m^2 \equiv (m_2^2 - m_1^2) << p^2$ (p is the momentum). Clearly, the state $\nu(t)$ of Eq. (6.109) is no longer purely ν_e and has a non-zero projection on the state ν_μ given by:

$$<\nu_\mu|\nu(t)> = -\sin\theta\cos\theta e^{-iE_1 t} + \cos\sin\theta e^{-iE_2 t} \qquad (6.110)$$

The probability that the ν_e state oscillates in time t to the ν_μ state is given by:

$$P_{\mu e} = |<\nu_\mu|\nu(t)>|^2 = \frac{1}{2}\sin^2 2\theta[1 - \cos(E_1 - E_2)t] = \sin^2 2\theta \sin^2(\frac{\Delta m^2 \ell}{4p})$$
$$(6.111)$$

where ℓ is the path traversed. Equation (6.111) tells us that, given fixed neutrino momentum from a source, longer paths ℓ are required to detect smaller values of Δm^2. For the purposes of orientation, it should be noted that, for a neutrino momentum p of 1 MeV and a mass squared difference Δm^2 of 1 eV^2, the "oscillation length" ℓ_ν is 2.5 meters.

A large number of experiments over many years have been performed searching for some evidence of neutrino oscillations [6.50], which could at least settle the question of the finite mass of the neutrino if not its Dirac or Majorana character. Occasionally, a claim has been made that ν oscillations have been found but, in all cases thus far, the claim was repudiated by other more accurate measurements. In Table 6.9, we list several typical ν oscillation experiments that were carried out, and the limits that were found for $\sin^2 2\theta$ and Δm^2 [6.50]. It is seen from Table 6.9 that the assortment of experiments performed in the laboratory find no evidence for neutrino oscillations, with upper limits on Δm of the order of a few tenths eV and on the mixing angle θ of a few tenths of a radian. In sum, the results of Table 6.9 (and other laboratory neutrino experiments that are not listed) are consistent with massless neutrinos and one might say that the resiliency of the standard model is again reaffirmed by experiments on ν oscillations carried out in the laboratory. There is one caveat to this conclusion, however, because all laboratory ν oscillation experiments are "free" — in the sense that the oscillations have not been "induced" by some extrinsic physical mechanism (i.e. extrinsic to the neutrino system). Curiously enough, the "solar neutrino problem", which has been around for more than two decades and consists of the discrepancy between theory and observation in the detection of the neutrinos emitted during the course of the thermonuclear processes taking place in the solar interior, is currently a prime candidate for being explained in terms of "induced" ν oscillations. Indeed, the "induced" ν oscillation mechanism for solving the "solar neutrino problem" is receiving so much attention at present that we discuss it as a good — but not conclusive — example of the striking physical consequences that can result from ν oscillations.

Table 6.9: Upper limits on Δm^2 and $\sin^2 2\theta$ from laboratory ν oscillation experiments.

Limits on Δm^2 and $\sin^2 2\theta$		Experiment
$\Delta m^2 < 15$ eV^2	$(\sin^2 2\theta = 1)$	$\nu_\mu N \to \mu^- X$
$\sin^2 2\theta < 0.02$	$(\Delta m^2 \simeq 10$ eV$^2)$	
$\Delta m^2 < 0.20$ eV^2	$(\sin^2 2\theta = 1)$	$\nu_\mu \to \nu_e$
$\sin^2 2\theta < 0.04$	$(\Delta m^2 \simeq 2$ eV$^2)$	
$\Delta m^2 < 0.49$ eV^2	$(\sin^2 2\theta = 1)$	$\pi \to \mu\nu$
$\sin^2 \theta < 0.028$	$(\Delta m^2 = 2$ eV$^2)$	
$\Delta m^2 < 0.14$ eV^2	$(\sin^2 2\theta = 1)$	$\nu_\mu \to \mu^- X$
$\sin^2 2\theta < 0.02$	$(\Delta m^2 \simeq 2.2$ eV$^2)$	$\nu_e N \to e^- X$

(3) *Solar neutrino problem*

Under the heading (2), we focused on ν oscillation experiments with laboratory-produced neutrinos. It is evident from Eq. (6.111) that Δm^2 decreases as ℓ increases and solar-produced neutrinos detected on earth can be sensitive to oscillations associated with much smaller values of Δm^2 than are possible in laboratory experiments; i.e. Δm^2 can be as small as 10^{-11}eV^2. That is why the "solar neutrino problem" has become so interesting to particle physics, although the astrophysical implications are equally important.

Let us place the "solar neutrino problem" in historical context. By the late 1930s, Bethe had established that the two major thermonuclear sources of energy in "normal" main sequence stars are the proton-proton set of reactions [6.51] and the carbon cycle [6.52]. Some model calculations (particularly the "point-convective" model) were already indicating that the carbon cycle is not the primary source of energy in the sun (as it is for more luminous main sequence stars like Sirius A) but that the proton-proton set of reactions is a serious competitor [6.53]. It was then established that the helium composition of main sequence stars is appreciable [later confirmed by the "big-bang" hypothesis of the universe [6.54], thereby lowering central temperatures and moving forward the proton-proton set of reactions as the dominant thermonuclear source of energy in the sun [6.55]]. The confident belief that astrophysical theory could predict the neutrino flux (and spectrum) from the proton-proton set of reactions in the sun encouraged experimental attempts to measure the solar neutrino flux. The earliest measurements of the neutrino flux from the sun by Davis and collaborators [6.56] were performed with a large chlorine detector from which neutrino-induced A^{37} — via the inverse β reaction $\nu_e + Cl^{37} \rightarrow e^- + A^{37}$ — was extracted; the average neutrino capture rate was found to be 2.07 ± 0.25 SNU (one SNU represents 10^{-36} interactions per second per atom), in contrast to the (most recent) prediction of 7.9 ± 2.6 SNU [6.57]. The discrepancy between theoretical prediction and observation of the solar neutrino flux — which has gained urgency in recent years — is known as the "solar neutrino problem".

There are three ways — not mutually exclusive (all of which have been tried and are continuing) — of resolving the "solar neutrino problem": (a) to perform improved and more discriminating experiments on the detection of solar neutrinos; (b) to improve or even alter the assumptions underlying the calculation of the temperature-density distribution in the solar interior; and (c) to assume that the ν_e's are converted into "sterile" ν_μ's or ν_τ's, by means of some neutrino oscillation mechanism — during their passage through the sun — with the "sterile" ν_μ's and/or ν_τ's then escaping detection on earth. We consider each in turn:

(a) With regard to the first approach, the vigorous experimental program under way to detect solar neutrinos is trying to take advantage of the fact that the solar neutrino flux possesses a dispersed energy spectrum due to the variety of contributions that it receives from the different thermonuclear reactions participating in the solar proton-proton chain exhibited in Table 6.10 [6.57]. The neutrino energy is continuous (because of the continuous β spectrum) for reactions $1, 4$ and 10 (with very different cutoff energies), and discrete for reactions 2 and 7 (because of "K capture"). The predicted neutrino fluxes corresponding to the different ν_e-producing reactions are given in Table 6.11. With the help of Tables 6.10 and 6.11, it is possible to show that the rare but high-energy B^8 neutrinos (reaction 10) is the dominant neutrino source in the Cl^{37} experiment, with the Be^7 neutrinos (reaction 7) contributing about 15% and the other more abundant neutrino sources (from reactions 1 and 2) less important. On the other hand, the much more recent gallium experiments — where the detecting reaction is $^{71}Ga + \nu_e \rightarrow {}^{71}Ge + e^-$ — have a detection threshold of 0.233 MeV, which should definitely favor reaction 1 as its source of solar neutrinos. The latest result of the gallium experiments [6.58] is $63 \pm 16\%$ of the predicted value, which is substantially below the theoretical expectation of 135 SNU although the large experimental error must be kept in mind and the "solar neutrino problem" can still disappear. A second recent experiment — the Kamiokande II experiment [6.59] — which detects neutrino events via the Cerenkov radiation emitted by electrons from $\nu_e - e$ scattering, can distinguish between the high energy B^8 neutrinos and the the lower energy pp neutrinos (see Table 6.11) and gives lower values of the B^8-induced neutrino flux than is predicted by theory.

In sum, neither the B^8-dependent (i.e. the Cl^{37} and Kamiokande II experiments) nor the pp-dependent (i.e. gallium experiments) solar neutrino experiments are in good accord with our theoretical understanding of the central temperature of the sun T_c (which is the decisive parameter determining the solar neutrino flux). The gallium experiments will be decisive — because they are less dependent on our knowledge of T_c than the other experiments — in determining the seriousness — if any — of the "solar neutrino problem". While awaiting the final outcome of the gallium experiments, it seems worthwhile to comment briefly on the accuracy of the theoretical prediction of T_c and then to sketch the "induced" neutrino oscillation mech-

anism that has been suggested to solve the "solar neutrino problem", if it
persists.

Table 6.10: Nuclear reactions in central region of the sun.

Reaction	No.	Neutrino energy (MeV)
$p + p \rightarrow H^2 + e^+ + \nu_e$	1	≤ 0.420
$p + e^- + p \rightarrow H^2 + \nu_e$	2	1.442
$H^2 + p \rightarrow He^3 + \gamma$	3	$--$
$H^3 + p \rightarrow He^4 + e^+ + \nu_e$	4	≤ 18.77
$He^3 + He^3 \leftrightarrow \alpha + 2p$	5	$--$
$He^3 + He^4 \rightarrow Be^7 + \gamma$	6	$--$
$Be^7 + e^- \rightarrow Li^7 + \nu_e$	7	0.861 (90%)
$Li^7 + p \rightarrow 2\alpha$	8	$--$
$Be^7 + p \rightarrow B^8 + \gamma$	9	$--$
$B^8 \rightarrow Be^{8*} + e^+ + \nu_e$	10	< 15
$Be^{8*} \rightarrow 2\alpha$	11	$--$

Table 6.11: Neutrino fluxes from the sun.

Source	Reaction	Flux (10^{10} cm^{-2}s^{-1})
pp	1	$6.0(1 \pm 0.02)$
pep	2	$0.014(1 \pm 0.05)$
$H^3 p$	4	8×10^{-7}
^7Be	7	$0.47(1 \pm 0.15)$
^8B	10	$5.8 \times 10^{-4}(1 \pm 0.37)$

    (b) With regard to the astrophysical calculations of the solar neutrino
flux, it must be emphasized that the theoretical prediction is very sensitive to T_c, especially for the B^8 neutrinos. The astrophysicists claim [6.57]
that present knowledge of the radiative opacity and other basic assumptions
(e.g. chemical homogeneity, etc.) are adequate to obtain accurate solutions
to the standard equations of stellar equilibrium for the sun. They state
that the small theoretical error in the predicted neutrino flux derives from
incomplete knowledge of the relevant low-energy nuclear cross sections and
the rough estimate of the primordial abundance of heavy elements. These

uncertainties are supposed to be incorporated in the 37% error given for the B^8 neutrinos in Table 6.11. Some theoretical attempts have been made to modify one or more assumptions underlying the "standard astrophysical model" calculation of the solar neutrino flux; apart from some changes in the underlying assumptions (e.g. greater convective instability or more rapid element diffusion), there have been many exotic proposals [e.g. the existence of a black hole in the central core of the sun [6.60]) or the existence of weakly interacting massive particles (WIMPS) [6.54] that — through their enhanced transport of energy outward — could reduce T_c. We do not discuss these many speculations but, instead, turn to the proposed "induced"-neutrino oscillation explanation of the problematic "solar neutrino problem".

(c) *"Induced"-neutrino oscillations in the sun*: If one accepts the astrophysical calculation of the solar neutrino flux and the lower than expected ν_e flux detected on earth, then, somehow, a substantial fraction of the ν_e's must disappear en route from the center of the sun to the earth. At least three possibilities have been suggested: (1) ν_e oscillates to ν_μ or ν_τ in the "vacuum" between sun and earth. Assuming $m(\nu_e) << m(\nu_\mu)$, a value of $m(\nu_\mu) \gtrsim 10^{-5}$eV could account for the observed suppression of ν_e's reaching the earth provided there is an improbably large mixing angle (see Table 6.9); (2) ν_e could have its spin flipped to a righthanded non-interacting ν_e in traversing the magnetic fields of several kilogauss known to exist in the solar surface. However, the (Dirac) magnetic moment of ν_e would have to possess the improbably large value of 10^{-10} to 10^{-11} Bohr magnetons [6.60]; and, finally (3) a substantial fraction of the ν_e's could become "sterile" (e.g. be converted into ν_μ's or ν_τ's) during their passage through the solar interior. An attractive model along these lines has recently been developed, called the MSW mechanism [6.41]. The MSW mechanism attributes the reduction in the detectable solar ν_e flux reaching the earth as essentially due to the difference between the $\nu_e - e$ elastic scattering cross section and the $\nu_\mu - e$ and/or $\nu_\tau - e$ scattering cross sections; the point is that both charged and neutral weak currents are involved in the $\nu_e - e$ scattering whereas $\nu_\mu - e$ and $\nu_\tau - e$ scatterings occur only as a result of neutral weak currents (see Fig. 6.1). The additional contribution to $\nu_e - e$ scattering inside the solar interior, adds an additional "phase" to the ν_e component of the neutrino wavefunction that is not available to the ν_μ and ν_τ components (assuming that either or both are mixed with ν_e).

Since the possibility (3) mentioned above is more promising than the other two, we sketch a few details of the MSW calculation in order to appreciate its role as a possible solution to the "solar neutrino problem". We write down the Hamiltonian in matter for a mix of two neutrinos as follows [6.41]:

$$H_{\text{matter}} = H_{\text{vac}} + \sqrt{2} G_F n_e \begin{pmatrix} 1 & 0 \\ 0 & 0 \end{pmatrix} \tag{6.112}$$

where:

$$H_{\text{vac}} = \frac{1}{2p} \begin{pmatrix} m_1^2 \cos^2 \theta + m_2^2 \sin^2 \theta & (m_2^2 - m_1^2) \sin \theta \cos \theta \\ (m_2^2 - m_1^2) \sin \theta \cos \theta & m_1^2 \sin^2 \theta + m_2^2 \cos^2 \theta \end{pmatrix} \tag{6.112a}$$

with m_1 and m_2 the masses of the ν_1 and ν_2 neutrinos, θ the mixing angle between ν_1 and ν_2, and n_e the electron density in the matter of the solar interior. From Eq. (6.112), one obtains a different expression [compared to vacuum — see Eq. (6.111)] for the probability $\bar{P}$ (barred quantities refer to matter, unbarred to vacuum), namely:

$$\bar{P} = \sin^2 2\bar{\theta} \, \sin^2 \frac{\Delta \bar{m}^2 \ell}{4p} \tag{6.113}$$

where:

$$\sin^2 2\bar{\theta} \simeq \sin^2 2\theta \left[\frac{\Delta m^2}{\Delta \bar{m}^2}\right]; \quad (\Delta \bar{m}^2) = [(D - \Delta m^2 \cos 2\theta)^2 + (\Delta m^2 \sin 2\theta)^2]^{\frac{1}{2}} \tag{6.113a}$$

with:

$$D = 2\sqrt{2} G_F n_e p \tag{6.113b}$$

Equation (6.113) has an interesting property for $D \neq 0$ (i.e. $n_e \neq 0$): the two mass eigenstates in matter are:

$$m_{\mp}^2 = \frac{1}{2}[m_1^2 + m_2^2 + D \mp (\Delta \bar{m}^2)] \tag{6.114}$$

(otherwise, for $D = 0$, $m_- = m_1$, $m_+ = m_2$). It follows that, in matter, the two-neutrino system exhibits a resonance phenomenon when the matter mixing angle factor $\sin^2 2\bar{\theta}$ goes to unity. This occurs at a critical density, n_e^{crit}, which depends on the momentum of the neutrinos, defined by the equation:

$$D = 2\sqrt{2} G_F n_e^{\text{crit}} p = \Delta m^2 \cos 2\theta \tag{6.114a}$$

Equation (6.114a) tells us that n_e^{crit} depends on the vacuum mixing angle θ and that, at n_e^{crit}, a level-crossing phenomenon takes place, causing the neutrino mass eigenstates to switch identities, i.e. a ν_e state is completely converted into a "sterile" ν_μ or ν_τ state, depending on the value of Δm^2. This level-crossing phenomenon is exhibited in Fig. 6.12 and is responsible for the MSW mechanism.

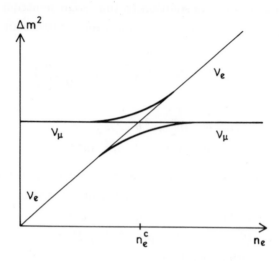

Fig. 6.12. Neutrino level crossing at the critical electron density inside matter.

Applied to the "solar neutrino problem" — assuming it exists — the MSW effect offers the possibility of eliminating the bulk of the ν_e's produced in the central core of the sun as they pass through the "critical" density region, provided the passage is "adiabatic", i.e. the change in the matter mixing angle $\bar{\theta}$ is slow compared to the frequency of neutrino oscillations (in vacuum). The adiabatic constraint requires that:

$$\frac{\Delta m^2 \sin^2 2\theta}{\cos 2\theta} > 2 \times 10^{-8} \text{eV}^2 \tag{6.115}$$

On the other hand, we can get a good upper limit on Δm^2 [from Eq. (6.114)] by choosing $p = p_{\max} \simeq 14$ MeV for the ^{8}B reaction and setting the critical density n_e^{crit} equal to that prevailing at the center of the sun (θ is expected to be small), namely $\Delta m^2_{\max} = 10^{-4}$ eV^2. Combining these constraints, we can say that the operation of the MSW mechanism in the sun requires $\Delta m \simeq \Delta^{-3}$ eV (which an uncertainty of a factor 10). This range of values for Δm is compatible with a Majorana neutrino mass, ν_μ, generated by the "see-saw" mechanism and would support the $SO(10)$ type of GUT model (see §8.4b). However, we must recognize that there is no independent evidence for finite-mass neutrinos (let alone Majorana ones) and it is premature to argue, in our view, that the "solar neutrino problem" has been solved and that the MSW mechanism constitutes evidence for the existence of finite-mass Majorana neutrinos and the correctness of $SO(10)$ GUT.

§6.4c. Dynamical symmetry-breaking origin of t quark and Higgs masses

As of this writing, one of the big surprises in QFD is the unusually large mass indicated for the heaviest member of the third generation of fermions, namely the t quark, with $m_t > 92$ GeV. The existence of the t quark is not seriously in doubt — in view of the need for freedom from the three chiral gauge anomalies (triangular, global $SU(2)$ and mixed gauge-gravitational — see Chapter 7), as well as the strong phenomenological support given by the B meson decays for the completion of a weak third generation $SU(2)_L$ quark doublet [6.61]. However, the large t quark mass raises questions about treating the origin of the t quark mass on a par with that of all the other known quarks and leptons of the three generations, as is done in the standard model through the Yukawa coupling (with arbitrary coefficients) of the same "elementary" Higgs boson to the massless (Weyl) fermions of the three generations. The mystery of the large t quark mass has combined with the ever-present alternative hypothesis that the Higgs boson is a fermion condensate, to revive interest in the possibility that there is a common dynamical symmetry-breaking origin of both the t quark and Higgs masses. The concept of a dynamical symmetry-breaking mechanism for fermion mass generation has a long history, going back more than three decades (see §1.2e). In particular, Nambu and Jona-Lasinio (NJL) [1.92] developed a model for fermion mass generation that took its cue from the BCS theory of low temperature superconductivity [1.90] and is capable of making predictions about mass excitations for bosons as well as fermions. Since, effectively, the NJL method bypasses explicit use of the Higgs SSB mechanism, its reformulation, within the QCD framework, may yield a theory that possesses the virtues of the old "technicolor" theory [1.154] (with the technifermions replaced by the t quark) and none of its deficiencies [e.g. the presence of unwanted FCNC processes]. We propose, in this section, to give a brief review of the current status of NJL theory after recalling its original formulation.

In the original NJL papers [1.92], a Lagrangian is written down for massless fermion fields that contains a chiral-invariant four-fermion interaction. The four-fermion interaction is used to generate masses for the chiral fermions by means of a dynamical symmetry-breaking mechanism patterned after the "energy gap" generated by "Cooper pairs" of electrons in low temperature superconductivity (see §1.2e and §4.4a). The key to the BCS microscopic theory of superconductivity is the existence of the "quasi-Higgs" "Cooper pair" — produced by the phonon-mediated electron-electron interaction — which is responsible for the "energy gap" Δ that is a measure of the finite energy difference between the lower superconductive ground state and the normal state. The size of the "energy gap" Δ must be calculated in a non-perturbative fashion and is actually carried out in the BCS theory by a self-consistent Hartree-Fock calculation based on the phonon-mediated electron-electron interaction. The self-consistent Hartree-Fock calculation not only predicts Δ but, in addition, gives rise to "collective" excitations that restore the electromagnetic gauge invariance of the theory (broken by the quasi-Higgs "Cooper pairs") as well as to scalar-type excitations with energy 2Δ. It is intriguing to see how the successful BCS ideas of superconductivity were translated into the NJL model in particle physics three decades ago [1.92], and to comment on the possible relevance of a "modernized" NJL theory to present-day QFD.

To simplify the mathematics, we recall the first NJL paper [1.92] which starts with the one-flavor global chiral quark group $U(1)_L \times U(1)_R$ and assumes a simple chiral-invariant form for the four-fermion interaction term, thus:

$$\mathcal{L} = -\bar{\psi}\gamma_\mu\partial_\mu\psi + g_0[(\bar{\psi}\psi)^2 - (\bar{\psi}\gamma_5\psi)^2] \qquad (6.116)$$

where the unrenormalized mass is taken as zero and g_0 is the unrenormalized coupling constant. The Lagrangian given by Eq. (6.116) is invariant under the global chiral quark group $U(1)_L \times U(1)_R$ and the goal is to derive an expression for the mass generated by the four-fermion interaction [in Eq. (6.116)] by means of a Hartree-Fock procedure similar to the calculation of the energy gap Δ in superconductivity — when carried through in quantum field-theoretic-language. If one lets $\Sigma(p, m, g, \Lambda)$ be the unrenormalized 1PI self-energy of the quark (see §4.5) — derived from Eq. (6.116) and expressed in terms of the physical mass m, the physical coupling constant g and a cutoff Λ — the physical Dirac particle must satisfy the equation:

$$\gamma \cdot p - \Sigma(p, m, g, \Lambda) = 0 \qquad (6.117)$$

for $\gamma \cdot p - m = 0$. Hence:

$$m = \Sigma(p, m, g, \Lambda)\,|_{\gamma \cdot p - m = 0} \qquad (6.118)$$

Further, g must be related to the bare coupling g_0 by an equation of the type (see §4.5):

$$g/g_0 = \Gamma(m, g, \Lambda) \qquad (6.119)$$

Application of straightforward perturbation theory to Eqs. (6.118) and (6.119) does not lead to a finite physical mass (i.e. $m \neq 0$); however, one can generate a finite mass for the Dirac particle if one invokes some form of non-perturbative approach which, in the NJL model, is precisely the generalized Hartree-Fock procedure (as in BCS theory).

To apply the Hartree-Fock procedure to the chirally-invariant Lagrangian [6.116], one writes it in the form: $\mathcal{L} = \mathcal{L}_0 + \mathcal{L}_I$, where $\mathcal{L}_0$ is the first term on the R.H.S. of Eq. (6.116) (the "free Lagrangian") and $\mathcal{L}_I$ is the second term on the R.H.S. of Eq. (6.116) (the "interaction Lagrangian"). If the "self-energy" is denoted by $\mathcal{L}_m$ (i.e. $\mathcal{L}_m = -m\bar{\psi}\psi$), then one can write:

$$\mathcal{L} = (\mathcal{L}_0 + \mathcal{L}_m) + (\mathcal{L}_I - \mathcal{L}_m) = \mathcal{L}_0' + \mathcal{L}_I' \qquad (6.120)$$

The trick now is to diagonalize $\mathcal{L}_0'$ (not $\mathcal{L}_0$), to treat $\mathcal{L}_I'$ as a perturbation (not $\mathcal{L}_I$), and to determine $\mathcal{L}_m$ from the requirement that $\mathcal{L}_I'$ does not yield additional "self-energy" effects. This procedure leads to Eq. (6.118). More explicitly, if one introduces the propagator $S_F^m(x)$ for the Dirac particle with mass m, and uses the lowest-order "bubble" diagram shown in Fig. 6.13 (see Fig. 4.2), the expression for Σ can be written in the form (see §4.5 and [1.92]):

$$\Sigma = 2g_0[\mathrm{Tr}\ S_F^{(m)}(0) - \gamma_5\mathrm{Tr}\ S_F^{(m)}(0)\gamma_5 - \frac{1}{2}\gamma_\mu\mathrm{Tr}\ S_F^{(m)}(0)$$
$$+ \frac{1}{2}\gamma_\mu\gamma_5\mathrm{Tr}\ \gamma_\mu\gamma_5 S_F^{(m)}(0)] \qquad (6.121)$$

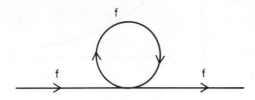

Fig. 6.13. Lowest-order "bubble" diagram used in NJL's Hartree-Fock calculation of fermion mass.

Since Eq. (6.121) gives rise to a quadratic divergence, NJL uses a relativistic cutoff and obtains in momentum space:

$$\Sigma = -\frac{8g_0 i}{(2\pi)^4} \int \frac{m}{p^2 + m^2 - i\epsilon} d^4p\ F(p,\Lambda) \qquad (6.122)$$

where $F(p,\Lambda)$ is the cutoff factor and the cutoff scale Λ can be thought of as the (large) mass of an intermediate boson mediating the four-fermion interaction. From Eq. (6.118), it follows that:

$$m = -\frac{g_0 mi}{2\pi^4} \int \frac{d^4 p}{p^2 + m^2 - i\epsilon} F(p, \Lambda) \tag{6.123}$$

If one factors out the trivial (perturbative) solution (i.e. $m = 0$), one gets for the non-trivial ("non-perturbative") solution (taking the invariant cutoff at $p^2 = \Lambda^2$):

$$\frac{2\pi^2}{g_0 \Lambda^2} = 1 - \frac{m^2}{\Lambda^2} \ln(\frac{\Lambda^2}{m^2} + 1) \tag{6.124}$$

The L.H.S. of Eq. (6.124) is plotted as a function of (m^2/Λ^2) in Fig. 6.14 and it is seen that a physically meaningful solution only exists if $g_0 > 0$ and that m can not be expanded in powers of g_0 [i.e. the solution is non-analytic as befits a solution with an essential singularity]. It can further be shown that the non-trivial (non-perturbative) finite m solution m_F has an associated ground state (vacuum) energy that is lower than the vacuum energy associated with the trivial (perturbative) $m = 0$ solution [this is the analog of the "Cooper pair" state in the BCS theory where the "Cooper pair" state is lower than the normal state by the amount Δ].

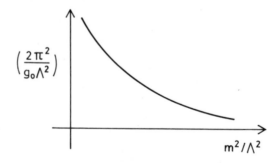

Fig. 6.14. Plot of the non-perturbative NJL solution as a function of coupling constant and cutoff.

The existence of the stable finite mass vacuum is already a very interesting qualitative result of the NJL model, demonstrating the possibility of dynamical symmetry breaking of a global quark flavor group (for massless quarks and a chirality-invariant four-quark interaction) into a global non-chiral quark flavor group for finite mass quarks. What adds further interest to the NJL model is that it predicts "collective" $q\bar{q}$ pair excitations that are interpretable as pseudoscalar and scalar mesons whose masses are fixed relative to the fermion mass m (at least in lowest approximation). More precisely, it is proved within the framework of the NJL model, that the pseudoscalar $q\bar{q}$ bound-state possesses a pole at $q^2 = 0$ (i.e. $m_{PS} = 0$) and that the lowest (massive) scalar $q\bar{q}$ bound state has its pole at $q^2 = 4m^2$ (i.e. $m_S = 2m_F$) — in complete analogy to the BCS theory.

The predictions of the simplest NJL model [corresponding to the global chiral quark symmetry group $U(1)_L \times U(1)_R$] — that a finite mass (m_F) fermion exists and that the pseudoscalar $q\bar{q}$ pair excitation m_{PS} vanishes while the mass of the scalar boson is $2m_F$ — persist when $U(1)_L \times U(1)_R$ is enlarged to $SU(2)_L \times SU(2)_R$ by introducing the isospin operators into Eq. (6.116) so that the $SU(2)_L \times SU(2)_R$-chiral-invariant Lagrangian becomes:

$$\mathcal{L} = -\bar{\psi}\gamma_\mu\psi + g_0[(\bar{\psi}\psi)^2 - \sum_{i=1}^{3}(\bar{\psi}\gamma_5\tau_i\psi)^2] \tag{6.125}$$

The replacement of Eq. (6.116) by Eq. (6.125) does not alter the qualitative conclusions of the $U(1)_L \times U(1)_R$ model except that now there are three massless $q\bar{q}$ excitations (i.e. "pions") instead of one, etc. [This reasoning led Nambu [1.93], prior to the advent of the Goldstone theorem (see §1.2e), to suggest a dynamical symmetry-breaking origin of the collective massless pseudoscalar excitations that he identified with the observed pions.]

At this point, one might inquire as to the relevance of the NJL method to the standard gauge theory of the electroweak interaction and, in particular, its bearing on the existence of a large t quark mass and a commensurately large quasi-Higgs mass. Our first comment is to point out that the Higgs boson serves two purposes: (1) the Higgs doublet generates the finite masses for the three weak bosons $(W^\pm, Z)$ in the course of spontaneously breaking $SU(2)_L \times U(1)_Y$ to $U(1)_{EM}$; and (2) the Higgs doublet generates finite masses for the quarks and charged leptons (of the three generations) through the Yukawa couplings. The NJL method changes the emphasis to the dynamical primacy of the Higgs boson and aims to produce a composite (quasi-) Higgs boson state [treated as a $t\bar{t}$ condensate] by means of the same dynamical symmetry-breaking mechanism that is responsible for generating the mass of the t quark. It is then the turn of the quasi-Higgs boson to generate the finite masses for the three electroweak bosons $(W^\pm, Z)$. This goes further than the BCS theory wherein the scalar "Cooper pair"-induced quasi-particle excitation — whose energy is predicted to be twice the energy gap Δ — is merely an excited state of the superconductive system.

In a series of recent papers [6.62], Nambu has tried to clarify his BCS-inspired idea that chiral symmetry breaking can produce the t quark condensate in the first place and that the scalar t quark pair excitation can then be identified with the quasi-Higgs boson. This courageous leap from the BCS theory to electroweak theory is justified by Nambu by borrowing from the old "bootstrap hypothesis" (see §1.2b) and arguing that just as the ρ meson is a $\pi\pi$ resonance in the s channel due to the exchange of ρ in the t channel, so the quasi-Higgs boson can be regarded as a $t\bar{t}$ condensate (resonance) in the s channel due to the exchange of the Higgs boson in the t channel. The connection with "bootstrapping" is made plausible by recalling that the self-energy of the fermion — which enters into the Hartree-Fock self-consistent solution — can be obtained from the scattering process by contraction in various channels: t channel contraction gives rise to the "tadpole" contribution while s channel contraction gives rise to the "exchange" diagram. Nambu proceeds to evaluate the "tadpole" and "exchange" contributions to the "energy gap" (t quark mass) by treating the VEV of the Higgs field, v, as a dynamical quantity, the so-called "tadpole potential". The "tadpole potential" is supposed to act on the fermion and to give it mass due to the Higgs exchange with the zero-point fermion, Higgs and gauge fields. Since the "tadpoles" themselves have a "strength" proportional to v, one obtains an "energy gap" equation for v with quadratically divergent coefficients which must be cut off at some energy scale $\Lambda >> \Lambda_{QFD}$ (since the electroweak interaction must be regarded as an effective theory at the scale Λ_{QFD}). Furthermore, since the "bootstrap" hypothesis, so the argument goes, implies that a (low-energy) effective theory must not depend on quadratically divergent coefficients, the divergent contribution to the "energy gap" equation must vanish. Accepting the hypothesized requirement that the quadratic divergences among the "tadpole" contributions from the fermion, Higgs and electroweak gauge fields cancel each other [6.63], Nambu [6.64] obtains two equations relating m_t to M_H with a logarithmic dependence on the cutoff energy Λ, namely:

$$4m_t^2 = M_H^2 + (2M_W^2 + M_Z^2);$$

$$M_H^2 v^2 = (1/16\pi^2)([(12m_t^2 - 3M_H^2 - 6M_W^2 - 3M_Z^2) - 12m_t^4 \ln(\Lambda/m_t)^2$$
$$+ (3/2)M_H^4 \ln(\Lambda/M_H)^2 + 6M_W^4 \ln(\Lambda/M_W)^2 + 3M_Z^4 \ln(\Lambda/M_Z)^2)].$$
$$(6.126)$$

The set of inequalities — on m_t and M_H — obtained from Eq. (6.126) — which turn out to be insensitive to Λ — are:

$$m_t > 120 \text{ GeV}; \quad M_H > 200 \text{GeV} \qquad (6.127)$$

It is seen from Eq. (6.127) that (M_H/m_t) is expected to be smaller than the "canonical" NJL value of 2 that is drawn from the BCS theory.

Clearly, much work remains to be done in applying the "modernized" NJL approach to the t quark-Higgs boson problem in the standard theory of the electroweak interaction. Indeed, a number of attempts have been made to place the compositeness condition for the Higgs boson (to serve as a $t\bar{t}$ condensate) on a more rigorous mathematical foundation. In particular, Bardeen *et al.* [6.65] have shown that the magical ratio 2 for (M_H/m_t) in the original NJL calculation is due to their use of the "bubble approximation" [see Fig. 6.13], which is equivalent to taking the large N_C limit (i.e. the vanishingly small coupling constant limit — see §5.4b) of the standard electroweak theory when it is assumed that the top quark condensate is produced by an attractive four-fermion interaction. Bardeen *et al.* go further [6.65] in their analysis of the gauged electroweak Lagrangian when the Higgs sector is eliminated in favor of local, attractive four-fermion interactions — *a la* the NJL model — which ultimately produce the composite Higgs boson and enable predictions to be made of the m_t and M_H masses. We do not enter into the details of the Bardeen *et al.* modification of the NJL approach except to note that the authors relate the compositeness condition for the quasi-Higgs boson to the renormalization constant derived from the renormalization group equations for the complete standard model (without the Higgs sector at high energy [6.65]); their basic result is that m_t is of the order of 200 GeV and $M_H > m_t$ but by a factor only slightly larger than unity. A successful "modernized" NJL solution to the t quark-Higgs boson problem in the standard model would hopefully obviate the need for "technicolor" theory and its attendant difficulties [1.156]. However, the explanation of the much smaller masses of the other quarks and of the charged leptons of the three generations, as well as the "fermion generation problem" itself, would still remain.

References

[6.1] E.J. Konopinski and G.E. Uhlenbeck, *Phys. Rev.* **60** (1941) 308; R.E. Marshak, *Phys. Rev.* **61** (1942) 431.

[6.2] G. Arnison *et al.*, *Phys. Lett.* **122B** (1983) 103.

[6.3] M. Kobayashi and K. Maskawa, *Prog. Theor. Phys.* **49** (1975) 652.

[6.4] J. Schwinger, *Annals of Phys.* **2** (1957) 407.

[6.5] S. Bludman, *Nuovo Cim.* **9** (1958) 433.

[6.6] T.D. Lee and C.S. Wu, *Ann. Rev. Nucl. Sci.* , Palo Alto (1966) 381.

[6.7] J.J. Aubert *et al.*, *Phys. Rev. Lett.* **33** (1974) 1404; J.E. Augustin *et al.*, *Phys. Rev. Lett.* **33** (1974) 1406.

[6.8] W. Braunschweig *et al.*, *Zeit. f. Phys.* **C37** (1988) 171.

[6.9] L. Rolandi, *Proc. of XXVI Intern. Conf. on High Energy Phys.*, Dallas (1992).

[6.10] R.E. Marshak and R.N. Mohapatra, Festschrift in honor of M. Goldhaber's 70th birthday, *Trans. N.Y. Acad. Sci.* **40** (1980) 124.

[6.11] S. Weinberg, *Phys. Rev. Lett.* **19** (1967) 1264; A. Salam, *Elementary Particle Theory*, ed. N. Swartholm, Almquist and Wissell, Stockholm (1968).

[6.12] Charm-II Collaboration, D. Geiregat *et al.*, *Phys. Lett.* **232B** (1989) 539.

[6.13] U. Amaldi *et al.*, *Phys. Rev.* **D36** (1987) 1385.

[6.14] M.J. Alguard *et al.*, *Phys. Rev. Lett.* **41** (1978) 70.

[6.15] J.H. Hollister, G.R. Apperson, L.L. Lewis, T.P. Emmons, T.G. Vold, and E.N. Fortson, *Phys. Rev. Lett.* **46** (1981) 643; L.M. Barkov and M.S. Zolotorev, *Phys. Lett.* **B85** (1979) 308.

[6.16] A. Sirlin, *Phys. Rev.* **D22** (1980) 971; *ibid.* **D29** (1984) 89.

[6.17] W. Marciano and A. Sirlin, *Phys. Rev.* **D12** (1980) 2695; *ibid.* **D29** (1984) 945; M. Consoli, S. Lo Presti, and L. Maiani, *Nucl. Phys.* **B223** (1983) 474; Z. Hioki, *Prog. Theor. Phys.* **68** (1982) 2134; *Nucl. Phys.* **B229** (1983) 284.

[6.18] J.R. Carter, *Proc. Joint Intern. Lepton–Photon Symp. and Europhys. Conf. on High Energy Phys.*, Geneva (1992).

[6.19] G.J. Feldman, *Proc. Intern. Symp. on Lepton and Photon Interactions at High Energies*, ed. M. Riordan, World Scientific, Singapore (1990), p. 225.

[6.20] D.N. Schramm, *Proc. of XXV Intern. Conf. on High Energy Physics*, ed. K.K. Phua and Y. Yamaguchi, World Scientific, Singapore (1991).

[6.21] Cf. J.F. Gunion *et al.*, *The Higgs Hunter's Guide*, Addison-Wesley (1990); *Standard Model Higgs Boson*, ed. M. Einhorn, North-Holland (1991).

[6.22] K.G. Wilson, *Phys. Rev.* **B4** (1971) 3184; K.G. Wilson and J. Kogut, *Phys. Rev.* **12** (1974) 75.

[6.23] M.A.B. Beg, *Proc. of Workshop on Higgs Particles*, Erice, Italy (1989).

[6.24] B.W. Lee, C. Quigg and G.B. Thacker, *Phys. Rev. Lett.* **38** (1977) 883; *Phys. Rev.* **D16** (1977) 1519.

[6.25] M.L. Perl *et al.*, *Phys. Rev. Lett.* **35** (1975) 1485.

[6.26] W. Bartel *et al.*, *Zeit. f. Phys.* **C30** (1986) 371.

[6.27] P. Drell, *Proc. XXVI Intern. Conf. on High Energy Phys.*, Dallas (1992).

[6.28] K. Riles, *Proc. XXVI Intern. Conf. on High Energy Phys.*, Dallas (1992).

[6.29] R.N. Mohapatra and R.E. Marshak, *Phys. Rev. Lett.* **20** (1968) 1081.

[6.30] C. Jarlskog, *CP Violation*, ed. C. Jarlskog, World Scientific, Singapore (1989).

[6.31] H. Harari and M. Leurer, *Phys. Lett.* **181B** (1986) 123.

[6.32] L. Wolfenstein, *Phys. Rev. Lett.* **51** (1983) 1945.

[6.33] J.H. Christenson, J.W. Cronin, V.L. Fitch, and R. Turlay, *Phys. Rev. Lett.* **13** (1964) 138.

[6.34] R.E. Marshak Riazuddin and C.P. Ryan, *Theory of Weak Interactions in Particle Physics*, Wiley-Interscience, New York (1969).

[6.35] T. Inami and C.S. Lim, *Prog. Theor. Phys.* **65** (1981) 297.

[6.36] F.J. Gilman and M.B. Wise, *Phys. Lett.* **83B** (1979) 83.

[6.37] ARGUS Collaboration, H. Albrecht *et al.*, *Phys. Lett.* **B192** (1987) 245; CLEO Collaboration, R. Fulton *et al.*, *Phys. Rev. Lett.* **64** (1990) 16.

[6.38] E.M. Purcell and N.F. Ramsey, *Phys. Rev.* **78** (1950) 807.

[6.39] K.F. Smith *et al.*, *Phys. Lett.* **B234** (1990) 191.

[6.40] E.P. Shabalin, *Sov. Phys. Usp.* **26** (1983) 2971.

[6.41] L. Wolfenstein, *Phys. Rev.* **D17** (1978) 2369; *ibid.* **D20** (1979) 2634; S.P. Mikheyev and A. Yu. Smirnov, *Nuovo Cim.* **9C** (1986) 17; H.A. Bethe, *Phys. Rev. Lett.* **56** (1986) 1305.

[6.42] G. Belanger and C.Q. Geng, *Phys. Rev.* **D43** (1991) 140.

[6.43] G. Ecker, A. Peck and E. deRafael, *Phys. Lett.* **237** (1990) 481.

[6.44] W. Molzon, *Proc. of Joint Intern. Lepton–Photon Symp. and Europhys. Conf. on High Energy Phys.*, Geneva (1992).

[6.45] M. Gell-Mann, P. Ramond and R. Slansky, *Supergravity*, ed. D.Z. Freedman *et al.*, North-Holland (1979).

[6.46] F.J. Gilman and Y. Nir, *Ann. Rev. Nucl. Part. Sci.* **40**, 213 (1990).

[6.47] S.R. Elliott *et al.*, *Phys. Rev. Lett.* **59** (1987) 2020.

[6.48] F.T. Avignone and R.L. Brodzinski, Univ. of South Carolina Review on Double-Beta Decay (1988).

[6.49] B. Pontecorvo, *Soviet Phys. JETP* **6** (1958) 429.

[6.50] R.D. Peccei, *Proc. of the Flavor Symposium*, Peking University, Beijing, (1988).

[6.51] H.A. Bethe and C.L. Critchfield, *Phys. Rev.* **54** (1938) 248.

[6.52] H.A. Bethe, *Phys. Rev.* **55** (1939) 434.

[6.53] R.E. Marshak and H.A. Bethe, *Bull. Amer. Phys. Soc.* **14**, No. 3 (1939).

[6.54] E. Kolb and M. Turner, *The Early Universe: Reprints*, Addison-Wesley, Boston (1988).

[6.55] M. Schwartzschild, *Structure and Evolution of Stars*, Princeton Univ. Press (1958).

[6.56] J.K. Rowley, B.T. Cleveland and R. Davis, Jr., *Solar Neutrinos and Neutrino Astronomy*, AIP Conference Proc. No. 126, ed. M.L. Cherry, W.A. Fowler and K. Lande, A.I.P., New York (1985).

[6.57] J.N. Bahcall *et al.*, *Rev. Mod. Phys.* **54** (1982) 767; J.N. Bahcall and R.K. Ulrich, *Rev. Mod. Phys.* **60** (1988) 297.

[6.58] P. Anselmann *et al.* (Gallex collaboration), *Phys. Lett.* **B285** (1992) 376, 390.

[6.59] V.N. Gavrin, *Proc. of 14th Intern. Conf. on Neutrino Phys. and Astrophys.*, ed. K. Winter, North-Holland (in press).

[6.60] M.B. Voloshin, M.I. Vysotsky, and L.B. Okun, *Sov. JETP* **64** (1986) 446.

[6.61] G. Kane and M. Peskin, *Nucl. Phys.* **B195** (1982) 29; P. Avery *et al.*, *Phys. Rev. Lett.* **53** (1984) 1309.

[6.62] Y. Nambu, *Proc. XI Warsaw Symposium on Elementary Particle Phys.*, ed. Z.A. Adjuk *et al.*, World Scientific, Singapore (1987).

[6.63] M. Veltman, *Acta. Phys. Polon.* **B12** (1981) 437.

[6.64] Y. Nambu, Univ. of Chicago preprint EFI90-46 (1990).

[6.65] W.A. Bardeen, C.T. Hill and M. Lindner, *Phys. Rev.* **D41** (1990) 1647; W.A. Bardeen, Fermilab-Conf.-90/269-T (1990).

Chapter 7

CHIRAL GAUGE ANOMALIES
IN THE STANDARD MODEL

§7.1. Chiral anomalies in four-dimensional quantum field theory

In the previous chapters, we have shown how the seminal concepts of gauge invariance (both for Abelian and non-Abelian global internal symmetry groups), chirality invariance (both for charged and neutral fermion currents) and spontaneous symmetry breaking (SSB) — due to "quasi-" or "elementary" Higgs scalars — have provided the foundation for the highly successful standard model of the strong and electroweak interactions for three generations of quark and lepton doublets. The standard model succeeds because it is a gauged quantum field theory that maintains its renormalizability and self-consistency despite properties that, at first sight, militate against a well-behaved theory: (1) in the standard gauge theory of the strong interaction (QCD), the asymptotic freedom property of non-Abelian $SU(3)$ color eliminates the "Landau singularity" problem [present in QED at high energies — see §5.2a] and gives a credible status to perturbative QCD at short distances; (2) in the standard (spontaneously broken) gauge theory of the electroweak interaction (QFD), the ultraviolet divergences that would normally accompany a quantum field theory with a massive, charged intermediate vector boson (IVB) are moderated by the SSB origin of the IVB mass; and (3) of equal importance, in a gauged quantum field theory with chiral fermion interactions, the insistence on simultaneous gauge invariance, chirality invariance and general covariance could lead to chiral gauge anomalies that destroy the renormalization and/or self-consistency of the theory. Fortunately, it turns out that the standard electroweak group

413

$SU(2)_L \times U(1)_Y$, with the quark and lepton quantum numbers observed for each family (see Table 1.5), is free of all three known chiral gauge anomalies in four dimensions. This good fortune must be understood because, in many ways, the absence of "harmful" chiral gauge anomalies for the standard group is the most striking feature of the standard model's success.

The appearance of chiral gauge anomalies in relativistic quantum field theory was unexpected but should not have been. Essentially, anomalies arise in quantum field theory when the symmetries of classical field theory are broken by the quantum fluctuations inherent in a quantum field theory [7.1]. In a sense, the entire renormalization program in quantum gauge theories represents the outcome of calculating the "quantum mechanical" anomalies present in a classical gauge theory with scale invariance; thus, whether it is Abelian QED with massless fermions or non-Abelian QCD with massless fermions, the classical Lagrangian in both cases possesses the property of scale invariance (and conformal invariance — see §2.3b) because the gauge fields are massless and the coupling constants are dimensionless. However, "quantum mechanical" renormalization introduces a finite renormalization scale for both unbroken QED and unbroken QCD and breaks scale invariance (and, *a fortiori*, conformal invariance) in the process. It is true that some form of "residual" scale invariance remains in both QED and QCD by virtue of the renormalization group properties of the quantum field theory in question but the departure from scale invariance for Abelian QED (with its "infrared slavery") is very different as compared to non-Abelian QCD (with its "ulraviolet" asymptotic freedom). A further example of the "quantum mechanical" breaking of a symmetry in a classical gauge theory with massless fermions is associated with the triangular chiral gauge anomaly. The "quantum fluctuations" resulting from triangular fermion loop corrections in the renormalization process — break down classical chirality invariance and lead to the so-called chiral gauge anomalies. These anomalies in the axial vector current part of the chiral fermion interaction with the gauge field eliminate the axial vector Ward-Takahashi identities and destroy the renormalizability of the quantum field theory. These "harmful" chiral gauge anomalies must therefore cancel out in order to maintain the renormalizability of the theory (as we will see, this applies to the chiral gauge electroweak part of the standard model but not to the vector-like QCD part).

There are actually three chiral gauge anomalies in four dimensions (all of which will be discussed in this chapter) and, to be precise, the afore-mentioned chiral gauge anomaly is also called the triangular axial anomaly, which also appears in a "harmless" form. Thus, if the axial vector current (due to the massless fermions), whether part of a chiral fermion current or by itself, is not coupled to the gauge field, the conflict between the demands of chirality invariance and gauge invariance in a quantum field theory gives rise to the "harmless" triangular axial anomaly, whose physical consequences can be quite significant; two examples of "harmless", albeit useful, triangular axial anomalies are: (1) the global QED anomaly (also called the ABJ anomaly [1.142] — see below), which results from expressing the pion field in terms of the divergence of the axial vector quark current (i.e. using PCAC) at one of the three vertices of the fermion (quark) triangle and allowing QED vector gauge currents to interact with photons at the other two vertices; and (2) the "harmless" triangular axial anomaly in QCD (i.e. the global QCD anomaly) where the axial vector quark current enters at one vertex of the fermion triangle and the two vector quark currents interact with gluons at the other two vertices. The "harmless" global QED anomaly is best known for its significant role in the calculation of the $\pi^0 \to 2\gamma$ decay width in the soft pion limit and is discussed in some detail in §7.2a. The counterpart of the "harmless" global QED anomaly in QCD, namely the "harmless" global QCD anomaly, has found useful application in strong interactions, of which several examples are discussed in §7.2b.

The situation is very different when an axial vector (or chiral) fermion current is coupled to a gauge field(s), as in QFD; in that case the resulting triangular axial anomalies — henceforth called chiral gauge anomalies because they are no longer "global" — must be absent in order to carry through the renormalization of the gauge theory with the help of the axial vector, in addition to the vector, Ward–Takahashi identities. It turns out that the required triangular chiral gauge anomaly cancellation in QFD is made possible by the existence of both quarks and leptons in appropriate representations of the electroweak group. The general conditions for the cancellation of triangular chiral gauge anomalies for a variety of chiral gauge groups and the specific application to QFD, will be worked out in §7.2c. [It is worth noting that it is the "triangular" fermion loop which gives rise to the triangular chiral gauge anomaly in four dimensions; for higher-dimensional chiral gauge theories (i.e. $D > 4$ space-time dimensions), the primary chiral gauge anomaly is due to a fermionic loop with $(\frac{D}{2}+1)$ vertices (e.g. "square" fermion loop for $D = 6$, "hexagonal" fermion loop for $D = 10$, etc.) [7.2].]

As we will see, the triangular chiral gauge anomaly plays a crucial role in the standard model (and in all attempts to enlarge the standard group — see

§7.7), but it happens that in addition two other chiral gauge anomalies have been identified in four dimensions that have important consequences for the standard group [albeit increasingly less so as the standard group is enlarged — see §7.7]. The additional two chiral gauge anomalies are: the (1) "non-perturbative" global $SU(2)$ chiral gauge anomaly which must be absent to maintain the self-consistency of the chiral gauge theory; (2) the "perturbative" mixed gauge-gravitational anomaly, which — like the "perturbative" triangular chiral gauge anomaly — must be cancelled to guarantee renormalizability of the chiral gauge theory. In §7.3, we explain the origin of the "non-perturbative" global $SU(2)$ anomaly, which arises if an axial vector (or chiral) current is coupled to a non-Abelian gauge field (wherein topologically non-trivial gauge transformations are necessarily present [1.144]). The global $SU(2)$ anomaly must be absent in order to maintain the mathematical self-consistency of the theory; this anomaly is discussed in §7.3. Furthermore, there is a connection between the absence of the non-perturbative global $SU(2)$ anomaly and the cancellation of the perturbative triangular axial anomaly for some of the physically interesting Lie groups and these matters are also considered in some detail in §7.3. Finally, the global $SU(2)$ anomaly introduces us to the difficulty of defining — in a gauge-invariant way — the fermion path integral in a background non-Abelian gauge field and the value of the Atiyah-Singer "index theorem" in shedding light on the topologically non-trivial properties of non-Abelian chiral gauge theory (see §10.4a).

The last of the three chiral gauge anomalies in four dimensions, the mixed gauge-gravitational anomaly, is rather special for $D = 4$ space-time dimensions. Thus, while there are no purely gravitational anomalies in four dimensions [such anomalies only occur in $D = 4k + 2$ $(k \geq 0)$ space-time dimensions [7.3]], the mixed gauge-gravitational anomaly does show up in four dimensions and imposes an important constraint on non-semi-simple gauge groups like the standard group. The perturbative mixed gauge-gravitational anomaly (which we usually called the "mixed anomaly") is structurally quite similar to the perturbative triangular chiral gauge anomaly: in the case of the "mixed anomaly", an axial vector (or chiral) current is coupled to a gauge field at one of the vertices and two symmetric tensor currents are coupled to gravitons at the other two vertices of the triangular fermion loop [1.145]. While the triangular chiral gauge anomaly must be cancelled in order to guarantee simultaneous chirality conservation and gauge invariance of the quantum field theory, the "mixed anomaly" must be absent in order to ensure simultaneous gauge invariance and general covariance of the quantum field theory [7.3]. The nature of the "mixed anomaly" and its consequences for QFD are discussed in §7.4.

In sum, the three chiral gauge anomalies in four dimensions [the triangular, the global $SU(2)$ and the "mixed"] must all be absent in the standard model and, quite remarkably, all three anomaly-free constraints are satisfied by the standard group with the quark and lepton Weyl representations given in Table 1.5. Once it is shown that the standard model passes all three chiral gauge anomaly-free tests, it is natural to turn the question around and to inquire into the extent to which the Weyl fermion representations of the standard group are determined by the three chiral gauge anomaly-free constraints. This question is considered in §7.5 where it is proved that the cancellation of all three chiral gauge anomalies is sufficient — with the assumption of minimality — to fix the phenomenologically correct representations and (quantized) hypercharges of the quarks and leptons of the three generations. In this proof, the "mixed" anomaly-free constraint plays a key role because, without it, quantization of hypercharge does not occur. This last result suggests the relevance of the "mixed" gauge-gravitational anomaly at much lower scales than the Planck scale (where quantum gravity must play a role), namely at the electroweak scale Λ_{QFD} [i.e. the long wave length limit for the "mixed" anomaly — see §7.4].

We go further in §7.6 and point out that the three chiral gauge anomaly-free constraints plus the Higgs SSB breaking of the electroweak group in the lepton sector are sufficient to ensure electric charge quantization (ECQ) of all quarks and leptons in each generation (a fact of nature previously thought to require grand or partial unification at a scale much higher than the electroweak scale [3.26]). Finally, in §7.7, we also consider the role of chiral gauge anomalies (and charge quantization — both hyper- and electric) in enlarged chiral gauge groups up through GUT groups. The interesting result is established that the standard gauge group with its "standard" quantum numbers for the chiral fermions in each generation maximizes the use of the three chiral gauge anomaly-free constraints and the charge quantization conditions.

§7.2. Perturbative triangular axial and chiral gauge anomalies in quantum field theory

The existence of a theoretical problem in the calculation of $\pi^0 \rightarrow 2\gamma$ decay was first noticed in 1969 by Adler, Bell and Jackiw (ABJ) [1.142], when an attempt to compute the π^0 decay width by means of current algebra plus soft pion techniques (see §3.5) gave a null result — in disagreement with experiment. The solution to the $\pi^0 \to 2\gamma$ problem was made possible by the discovery of the global $[U(1)_A]$ triangular axial anomaly in the calculation of the $\pi^0 \to 2\gamma$ decay width. The ABJ anomaly is important for several reasons: (1) it turns out that the entire contribution to the width for $\pi^0 \to 2\gamma$ decay is due to the anomaly term and the excellent agreement with experiment provides the most direct confirmation of the existence of the $U_A(1)$ triangular axial anomaly for an Abelian gauge group. The axial vector current in the ABJ calculation of $\pi^0 \to 2\gamma$ decay is associated with a global chiral symmetry of the theory and, hence, is "harmless", i.e. the non-vanishing global QED anomaly in $\pi^0 \to 2\gamma$ decay is measured, not cancelled. A similar situation obtains in QCD wherein the "flavor-singlet" [in $SU(2)$ or $SU(3)$ flavor space] axial vector current gives rise to the "harmless" global QCD anomaly; (2) the "harmless" global QED and QCD anomalies are both $U(1)_A$ (axial) anomalies but the structure of the $U(1)_A$ QED anomaly is determined by the topologically trivial Abelian gauge fields of QED whereas the counterpart $U(1)_A$ QCD anomaly is determined by the topologically non-trivial non-Abelian gauge fields of QCD. The physical consequences of this difference are discussed in §7.2b; (3) the triangular axial anomalies are not "harmless" when the axial vector (or chiral) fermionic currents are all coupled to gauge fields (Abelian and/or non-Abelian) and an odd number of such currents is axial vector (or chiral). The primary features of the global triangular axial anomaly — its dependence on the product of the (antisymmetric tensor) field and its dual field, and its independence of the fermion mass in the triangular fermion loop [see Eq. (7.15)] — are retained in the triangular chiral gauge anomaly; this structure is responsible for the breakdown of the renormalizability of the theory when only axial vector (or chiral) gauge currents are present at all three vertices of the triangular fermionic loop and necessitate the cancellation of these anomalies. The fermion mass independence of the chiral gauge anomalies permits the decision (with regard to cancellation) to be made solely in terms of the representation matrices for the chiral fermions at the three current vertices and without taking account of the dynamics of the quark and lepton interactions and, finally; (4) since the gauged electroweak group requires the coupling of chiral fermion currents to gauge fields, one expects triangular chiral gauge anomalies in the electroweak theory and the need for their cancellation. It happens that the observed L and R quark-lepton representations of the electroweak group (satisfying Table 1.5), plus the mass independence of the triangular anomaly, conspire to produce the necessary cancellation and thereby remove the threat to the renormalizability of electroweak gauge theory.

We elaborate points (1)–(4) below. We give a somewhat detailed account, in §7.2a, of the prime example of the highly successful application — to the $\pi^0 \to 2\gamma$ decay width — of the "harmless" triangular QED. We then sketch in §7.2b three examples of the application of the "harmless" triangular QCD anomaly: (a) its role in restoring isospin invariance for the two nucleonic multi-quark states despite the substantial breakdown of isospin invariance for the u and d component quarks; (b) its contribution to the mass of the η' meson [one part of the so-called $U(1)$ problem in QCD — see §10.3b]; and (3) its potentially valuable role in helping to resolve the "proton spin" puzzle surrounding the deep inelastic scattering of polarized muons by polarized protons. In §7.2c, we deal with the conditions for cancellation of triangular chiral gauge anomalies in order to preserve the renormalizability of the quantum field theory under consideration, in particular QFD.

§7.2a. Harmless global triangular axial QED anomaly

We first treat the ABJ calculation of the $\pi^0 \to 2\gamma$ decay width in terms of the three currents at the vertices of the triangular fermionic loop: the first is the "global" axial vector quark current (whose divergence represents the pion field via PCAC — see §3.5) and the other two are vector quark currents coupled to the gauge field (photon) of QED. ABJ discovered that straightforward application of the customary regularization methods of quantum field theory to the $\pi^0 \to 2\gamma$ transition amplitude gives rise to a conflict between the requirements of gauge invariance and chirality conservation as certified, for example, by the vanishing of the divergence of the axial vector current (i.e. the "axial density") in the limit of zero fermion mass. Since gauge invariance must be maintained, this discordance necessitates the addition of an anomalous term to the divergence of the axial vector current. Let us see how this comes about.

We recall that when $m = 0$ for the fermions in a Dirac field theory, the Lagrangian $\mathcal{L}$ is invariant under a chirality transformation, i.e. $\mathcal{L}$ is invariant under the (global) chirality transformation (α is constant — see §3.3):

$$\psi(x) \to e^{-i\alpha\gamma_5}\psi(x) \tag{7.1}$$

In accordance with Noether's theorem, the axial vector current is conserved:

$$\partial_\mu J_5^\mu = 0 \tag{7.2}$$

where:

$$J_5^\mu = \bar\psi\gamma^\mu\gamma_5\psi \tag{7.2a}$$

For $m \neq 0$, the "naive" approach yields the operator identity:

$$\partial_\mu J_5^\mu = 2mj_5 \tag{7.3}$$

where m is the fermion mass, and j_5 is the "axial density":

$$j_5 = i\bar\psi\gamma_5\psi \tag{7.4}$$

Since J_5^μ is the product of the field $\bar\psi(x)$ and $\psi(x)$ at the same space-time point, regularization is required and, after this is carried out in a gauge-invariant manner, Eq. (7.3) must be replaced by:

$$\partial_\mu J_5^\mu = 2mj_5 + \frac{\alpha_e}{2\pi}F_{\mu\nu}\tilde{F}^{\mu\nu} \tag{7.5}$$

where $F_{\mu\nu} = \partial_\mu A_\nu - \partial_\nu A_\mu$ (A_μ is the electromagnetic gauge field) is the (antisymmetric tensor) electromagnetic field, $\tilde{F}^{\mu\nu} = \frac{1}{2}\epsilon^{\mu\nu\rho\sigma}F_{\rho\sigma}$ is the dual tensor field, and α_e is the fine-structure constant. We note that the anomalous term in Eq. (7.5) is the product of the electromagnetic field tensor and its dual; if it is recalled that the electric current is the source of $F_{\mu\nu}$ and that a "magnetic current" (associated with a hypothetical Dirac magnetic monopole — see §4.6) is the source of $\tilde{F}_{\mu\nu}$, one can begin to anticipate some of the topological non-Abelian strictures of the anomalous term that will be explored in §10.3. In any case, the result is that the axial vector current is not conserved in the massless limit — the "naive" classical result — but requires the introduction of the triangular axial anomaly; this triangular axial anomaly is called the perturbative triangular anomaly because it originates from the quantum effects associated with local transformations — described by Feynman diagrams containing a triangular fermion loop (in lowest order) to one of whose vertices an axial vector current is attached.

To explain the origin of the anomalous term in Eq. (7.5), one can start with the three-point Green's function $T_{\mu\nu\lambda}(k_1, k_2, q)$ — needed to calculate the transition amplitude for $\pi^0 \rightarrow 2\gamma$ decay — as follows (see Fig. 7.1):

$$T_{\mu\nu\lambda}(k_1, k_2, q) = i\int d^4x_1 d^4x_2 < 0|T(V_\mu(x_1)V_\nu(x_2)A_\lambda(0))|0 > e^{ik_1\cdot x_1 + ik_2\cdot x_2} \tag{7.6}$$

where $V_\mu(x_1)$ and $V_\nu(x_2)$ are the two vector quark currents coupled to the gauge (photon) field, and $A_\lambda(0)$ is the axial vector quark current whose divergence represents π^0 [see §3.5a]. To lowest-order, the two Feynman diagrams

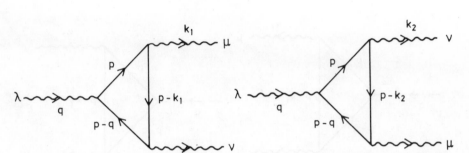

Fig. 7.1. Lowest-order contributions to the Green's function $T_{\mu\nu\lambda}$ in $\pi^0 \to 2\gamma$ decay.

in Fig. 7.1 contribute to $T_{\mu\nu\lambda}$ (and obey "crossing symmetry") and one gets:

$$T_{\mu\nu\lambda} = i \int \frac{d^4p}{(2\pi)^4} (-1) \{ \text{Tr}[\frac{i}{\not{p} - m} \gamma_\lambda \gamma_5 \frac{i}{(\not{p} - \not{q}) - m} \gamma_\nu \frac{i}{(\not{p} - \not{k}_1) - m} \gamma_\mu]$$

$$+ \begin{pmatrix} k_1 \leftrightarrow k_2 \\ \mu \leftrightarrow \nu \end{pmatrix} \} \tag{7.7}$$

One also needs the three-point Green's function where the "axial density" $P(0)$ [defined by Eq. (7.4)] replaces the axial vector current $A_\lambda(0)$, namely:

$$T_{\mu\nu}(k_1, k_2, q) = i \int d^4x_1 d^4x_2 < 0|T(V_\mu(x_1)V_\nu(x_2)P(0))|0 > e^{ik_1 \cdot x_1 + ik_2 \cdot x_2} \tag{7.8}$$

The lowest-order (Feynman) contribution to the three-point Green's function $T_{\mu\nu}(k_2, k_2, q)$ becomes (see Fig. 7.2):

$$T_{\mu\nu} = i \int \frac{d^4p}{(2\pi)^4} (-1) \{ \text{Tr}[\frac{i}{\not{p} - m} \gamma_5 \frac{i}{(\not{p} - \not{q}) - m} \gamma_\nu \frac{i}{(\not{p} - \not{k}_1) - m} \gamma_\mu]$$

$$+ \begin{pmatrix} k_1 \leftrightarrow k_2 \\ \mu \leftrightarrow \nu \end{pmatrix} \} \tag{7.9}$$

The imposition of gauge invariance yields the following conditions on $T_{\mu\nu\lambda}$:

$$k_1^\mu T_{\mu\nu\lambda} = k_2^\nu T_{\mu\nu\lambda} = 0 \tag{7.10}$$

The axial vector "Ward-Takahashi-identity" relation between the two Green's functions $T_{\mu\nu\lambda}$ and $T_{\mu\nu}$ is:

$$q^\lambda T_{\mu\nu\lambda} = 2m T_{\mu\nu} \tag{7.11}$$

and when $m = 0$, $q^\lambda T_{\mu\nu\lambda} = 0$, which is the chirality invariance condition that is expected to hold. However, if one carries out the formal manipulations to check Eqs. (7.10) and (7.11), one is confronted with a linear divergence in

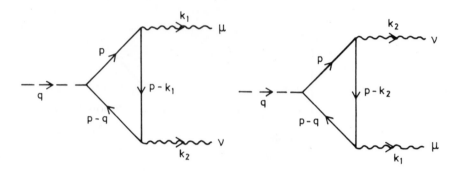

Fig. 7.2. Lowest-order contributions to the Green's function $T_{\mu\nu}$ in $\pi^0 \to 2\gamma$ decay.

the loop momentum p [see Figs. 7.1–7.2 and §4.5a], and the answer depends on the choice of "routing" of p [3.26]. Thus, if one reroutes from p to $p + a$, where a is some linear combination of k_1 and k_2, i.e. $a = \alpha k_1 + (\alpha - \beta)k_2$ (α, β are constants), one obtains for Eqs. (7.10) and (7.11):

$$k_1^\mu T_{\mu\nu\lambda} = -k_2^\mu T_{\mu\nu\lambda} = (\frac{1+\beta}{8\pi^2})\epsilon_{\nu\lambda\sigma\rho}k_1^\sigma k_2^\rho \qquad (7.12)$$

and:

$$(k_1 + k_2)^\lambda T_{\mu\nu\lambda} = 2mT_{\mu\nu} - \frac{(1-\beta)}{4\pi^2}\epsilon_{\mu\nu\sigma\rho}k_1^\sigma k_2^\rho \qquad (7.13)$$

Equation (7.12) tells us that gauge invariance is violated unless $\beta = -1$ whereas Eq. (7.13) requires $\beta = 1$ to maintain chirality conservation. Since gauge invariance is essential to derive the (vector) Ward-Takahashi identities required to prove the renormalizability of QED — we choose $\beta = -1$ in Eq. (7.12). But then Eq. (7.13) becomes:

$$q^\lambda T_{\mu\nu\lambda} = 2mT_{\mu\nu} - \frac{1}{2\pi^2}\epsilon_{\mu\nu\sigma\rho}k_1^\sigma k_2^\rho \qquad (7.14)$$

Manifestly, Eq. (7.14) no longer vanishes in the chiral limit $m \to 0$, and we conclude that chirality conservation is violated by an anomaly associated with the triangular fermionic loop graph when gauge invariance is enforced in the regularization process. Indeed, from Eq. (7.14), it follows that, in the chiral ($m = 0$) limit, $T_{\mu\nu\lambda}$ behaves as:

$$T_{\mu\nu\lambda} \xrightarrow[m \to 0]{} \frac{i}{2\pi^2}\epsilon_{\mu\nu\sigma\rho}k_1^\sigma k_2^\rho \frac{q_\lambda}{q^2 + i\epsilon} \qquad (7.15)$$

Thus, the entire contribution to $T_{\mu\nu\lambda}$ in the massless limit comes from the axial anomaly that manifests itself as a pole at $q^2 = 0$ (the "axial anomaly pole").

The great importance of the triangular axial anomaly and the "axial anomaly pole" is that the higher-order radiative corrections [due to (vector) photon loops of increasing complexity] do not alter the structure of the lowest-order calculations of the axial anomaly term given in Eq. (7.5) and the "axial anomaly pole" given in Eq. (7.15). [Parenthetically, the structures of Eqs. (7.5) and (7.15) are not altered when the corresponding renormalized quantities are inserted. For example, the Z_3^2 "charge renormalization" function (in QED) associated with α_e [in Eq. (7.5)] is just cancelled by the Z_3^{-2} term connected with $F_{\mu\nu}\bar{F}_{\mu\nu}$ [the Z_2 and Z_1 renormalization functions do not enter the picture by virtue of gauge invariance (see §4.5a)]; moreover, the "axial anomaly pole" structure in Eq. (7.15) is maintained because the pole structure of a Green's function is not altered by renormalization.] These remarks justify the common statement that the triangular axial anomaly is a "short wavelength" (i.e. a high energy) effect [since all the ultraviolet divergences of the radiative corrections are taken care of by renormalization]. On the other hand, the presence of the "axial anomaly pole" clearly indicates that the triangular axial anomaly is also a "long wavelength" (i.e. low energy) phenomenon. The compatibility of the "long" and "short wavelength" features of the triangular axial anomaly is, of course, possible because of the mass-independence of this term (i.e. there is no scale).

The simultaneous existence of "long" and "short wavelength" limits for the triangular axial anomaly explains its importance in quantum field theories: the short wavelength limit leads — for chiral gauge theories — to the need for cancellation to preserve the renormalizability of the theory and its "long wavelength" limit justifies the 't Hooft "anomaly matching condition" in composite fermion models [1.159] (see §9.5). The wide-ranging constraining role played by the triangular axial anomaly has stimulated a number of attempts to probe more deeply into its origin; the most illuminating observation appears to be [7.1] that the "negative energy sea" associated with the Dirac equation can not be separately defined, in a gauge-invariant way, for L and R fermions despite the fact that the "classical" Dirac equation itself does separate into L and R chiral parts; this confirms the earlier statement that the triangular axial anomaly is a purely quantum mechanical effect.

We next comment on the positive role played by the triangular axial anomaly of Eq. (7.5) in explaining the $\pi^0 \to 2\gamma$ decay width — before turning to a second interesting application — in QCD. We recall that the triangle graphs in Fig. 7.1 define the three-point Green's function of Eq. (7.6), where q denotes the four-momentum of π^0, and k_1 and k_2 are the four-momenta of the two photons. It can now be shown that, in the soft-pion limit $(q^\mu \to 0)$, the first term on the R.H.S. of Eq. (7.14) vanishes [since $T_{\mu\nu\lambda}$ does not have any intermediate states (degenerate with the vacuum) that couple to the vacuum through the axial vector current] and the only contribution to $\pi^0 \to 2\gamma$ decay comes from the anomalous term in Eq. (7.14). A straightforward calculation of the decay amplitude Γ_{π^0} for $\pi^0 \to 2\gamma$, making use of the "axial anomaly" pole, then yields:

$$\Gamma_{\pi^0} = \frac{N_C e^2 D}{2\pi^2 f_\pi} \tag{7.16}$$

where N_C is the number of colors, D is the "axial anomaly coefficient" [depending on the "charges" at the three vertices of the triangle graph (see Fig. 7.1)], and f_π is the amplitude for the weak decay process $\pi \to \mu\nu$ (see §3.5a). To calculate D, one assumes the simple quark model (without color), where the vector (electromagnetic) and axial vector (π^0-related) currents are respectively:

$$V_\mu(x) = \bar{q}(x)\gamma_\mu Q q(x) \qquad (Q \text{ is the electric charge}) \tag{7.17a}$$

$$A_\mu^3(x) = \bar{q}(x)\gamma_\mu\gamma_5 \frac{\lambda_3}{2} q(x) \qquad [\lambda_3 \text{ is the third generator of } SU(3)_F] \tag{7.17b}$$

with:

$$Q = \frac{1}{3}\begin{pmatrix} 2 & & \\ & -1 & \\ & & -1 \end{pmatrix} \quad \text{and} \quad \lambda_3 = \begin{pmatrix} 1 & & \\ & -1 & \\ & & 0 \end{pmatrix} \tag{7.17c}$$

Using Eqs. (7.17), D takes on the value;

$$D = \frac{1}{2}\text{Tr}(\{Q,Q\}\frac{\lambda_3}{2}) = \frac{1}{6} \tag{7.18}$$

and, if one inserts $N_C = 3$ into Eq. (7.16), the theoretical value of the π^0 aplitude $\Gamma_{\pi^0}^{\text{th}}$ is 0.0369 m_π^{-1} — in excellent agreement with the experimental value $\Gamma_{\pi^0}^{\text{exp}} = 0.0375\ m_\pi^{-1}$. It is to be noted: that (1) without color, there would be a factor 3 discrepancy between the theoretical and experimental values for the $\pi^0 \to 2\gamma$ decay amplitude [strictly speaking, the number of colors N_C is not fixed by the $\pi^0 \to 2\gamma$ measurement because the charge of each quark must be redefined for N_C colors and agreement with the $\pi^0 \to 2\gamma$ measurement will result for arbitrary N_C [5.49]; this is no longer true for other QCD-related measurements such as the R value in electron-positron annihilation (see §5.2c) where $N_C = 3$ is the preferred value]; and (2) the strange (s) quark does not contribute to the result as is evident from Eqs. (7.17c) and (7.18); this is not surprising since, in the "valence" approximation, the pion is composed only of u and d quarks. In any case, the good agreement found for the $\pi^0 \rightarrow 2\gamma$ decay amplitude clearly supports the reality of the triangular axial anomaly introduced by fermion loops in QED. It should not be surprising that there is a counterpart "harmless" global triangular axial anomaly in QCD — which we treat in the next section.

§7.2b. Harmless global triangular axial QCD anomaly

We have seen that the global QED anomaly that arises in connection with $\pi^0 \rightarrow 2\gamma$ decay is not only "harmless" but completely accounts for the observed decay width to an impressive accuracy. The QCD anomaly equation is a straightforward translation of the QED anomaly equation, as follows:

$$\partial^\mu (\bar{q} \gamma_\mu \gamma_5 q) = 2i m_q (\bar{q} \gamma_5 q) + \frac{g_s^2}{8\pi^2} \mathrm{Tr}\ G_{\mu\nu} \tilde{G}^{\mu\nu} \qquad (7.19)$$

where $G_{\mu\nu}^a$ is the gluon field tensor ($\tilde{G}^{a\mu\nu}$ is its dual) and it is understood that the trace is taken over the light quark flavors (2 or 3) so that we have a flavor singlet $U(1)_A$ QCD anomaly in $SU(2)$ or $SU(3)$ flavor space. The global QCD anomaly in Eq. (7.19) is equally "harmless" and some of its physical consequences are of comparable importance to those of the global QED anomaly. It is true that — in contrast to the global QED anomaly — the nature and importance of the global QCD anomaly could only be appreciated after the formulation of QCD itself [1.89] and the discovery of the instanton [1.136]; indeed, it was soon recognized that the instanton-related, topologically non-trivial QCD anomaly possesses non-zero matrix elements at zero momentum [7.4] [thereby implying the non-conservation of the flavor-singlet axial vector current without the generation of an N-G boson [1.161]] in contrast to the topologically trivial global QED anomaly. Further studies of the global QCD anomaly demonstrated its flavor-invariance-saving role in QCD, its predominant role in explaining the two parts of the "$U(1)$ problem in QCD" (see §10.3b) — the occurrence of $\eta \rightarrow 3\pi$ decay and the large magnitude of the η' meson mass — and its general usefulness in other QCD problems such as the recent "proton spin" problem arising in connection with the deep inelastic scattering of polarized muons by polarized protons.

In this section, we propose to comment briefly on these three applications of the global QCD anomaly. With regard to the flavor-invariance-saving role of the global QCD anomaly, we limit ourselves to two quark flavors (u and d) and show how the flavor-singlet QCD anomaly restores the "isospin-invariant" behavior for the nucleonic isospin doublet, which would otherwise be grossly violated because of the substantial mass difference between the u and d quarks. In sketching this proof, we assume that the "valence" quark composition of the two nucleonic isospin states p and n is (uud) and (udd) respectively and that the ratio of the current masses (see §3.5c) is $m_d/m_u \simeq 1.8$ [i.e. $(m_d - m_u)/(m_d + m_u) \sim 30\%$] so that there is substantial violation of isospin invariance for the u, d doublet. Let us see how the triangular axial QCD anomaly (for two quark flavors) reduces the expected isospin violation to one more consonant with the mass difference between proton and neutron. We write down the axial vector matrix element in β decay:

$$< p|\bar{u}\gamma_\mu\gamma_5 d|n >=< p|\bar{u}\gamma_\mu\gamma_5 u - \bar{d}\gamma_\mu\gamma_5 d|p >=< p|A_\mu^{I=1}|p > \qquad (7.20)$$

Next, we compute the "naive" divergence of the axial vector quark current, which is:

$$\partial_\mu[\bar{q}\gamma_\mu\gamma_5 q] = 2im_q\bar{q}\gamma_5 q \qquad (7.21)$$

It follows that the divergence of $A_\mu^{I=1}$ is given by:

$$\partial_\mu A_\mu^{I=1} = i(m_u + m_d)[\bar{u}\gamma_5 u - \bar{d}\gamma_5 d] + i(m_u - m_d)[\bar{u}\gamma_5 u + \bar{d}\gamma_5 d] \qquad (7.22)$$

The isovector, axial vector current $A_\mu^{I=1}$ does not give rise to a QCD anomaly whereas the isosinglet, axial vector current $A_\mu^{I=0} = (\bar{u}\gamma_\mu\gamma_5 u)+(\bar{d}\gamma_\mu\gamma_5 d)$ does; if we ignore the anomalous gluonic contribution to the divergence of $A_\mu^{I=0}$, we get:

$$\partial_\mu A_\mu^{I=0} = 2im_u(\bar{u}\gamma_5 u) + 2im_d(\bar{d}\gamma_5 d) = i(m_u + m_d)[\bar{u}\gamma_5 u + \bar{d}\gamma_5 d]$$
$$+ i(m_u - m_d)[\bar{u}\gamma_5 u - \bar{d}\gamma_5 d] \qquad (7.23)$$

Using charge symmetry, we obtain:

$$< p|\bar{u}\gamma_5 u + \bar{d}\gamma_5 d|p >=< n|\bar{u}\gamma_5 u + \bar{d}\gamma_5 d|n > \qquad (7.24)$$

and consequently:

$$< N|\partial_\mu A_\mu^{I=0}|N >_{I=1} =< p|\partial_\mu A_\mu^{I=0}|p > - < n|\partial_\mu A_\mu^{I=0}|n >$$
$$= \frac{(m_u - m_d)}{(m_u + m_d)} < N|\partial_\mu A_\mu^{I=1}|N >_{I=1} \qquad (7.25)$$

where the subscript $I = 1$ on the nucleon matrix elements denotes the difference between proton and neutron matrix elements. It is evident from Eq. (7.25) that an appreciable violation (of the order of 30%) is expected in the divergence of the axial vector current if the triangular axial QCD anomaly is not taken into account. However, it can be shown [7.5] that the gluon anomaly term in Eq. (7.19) just cancels out the large isospin-violating effect of order $[(m_u - m_d)/(m_u + m_d)]$ and reduces the isospin-violating term in the two-flavor-singlet axial vector current to a much smaller term, of order $(m_u - m_d)/\Lambda_{QCD}$. This is an example of the important role played by the "harmless" global QCD anomaly for two light quark flavors.

The "harmless" three-flavor QCD anomaly is a simple generalization of the two-flavor QCD anomaly where the three-flavor-singlet axial vector current again satisfies Eq. (7.19) but now:

$$(\bar{q}\gamma_\mu\gamma_5 q) = \frac{1}{\sqrt{3}}[\bar{u}\gamma_\mu\gamma_5 u + \bar{d}\gamma_\mu\gamma_5 d + \bar{s}\gamma_\mu\gamma_5 s] \tag{7.26}$$

With the help of Eq. (7.26), we next discuss the attempt to utilize the three-flavor QCD anomaly to deal with one part of the "$U(1)$ problem" in QCD (see §10.3b), namely the large mass of the η' meson, and conclude this section with some remarks about the effort to resolve the recently uncovered "proton spin" problem by means of the three-flavor QCD anomaly. Insofar as the η' mass is concerned, it can be shown that without the QCD anomaly, the mass of the η_{ns} ($I = 0$ "non-strange" pseudoscalar) meson can only be slightly larger than the mass of the pion, i.e. $m_{\eta_{ns}} \leq \sqrt{3}m_\pi$. [We employ the notation $\eta_{ns} = \sqrt{\frac{2}{3}}\eta_1 + \frac{1}{\sqrt{3}}\eta_8$, $\eta_s = \frac{1}{\sqrt{3}}\eta_1 - \sqrt{\frac{2}{3}}\eta_8$, where η_1 is the $SU(3)$ flavor singlet pseudoscalar meson, η_8 the eighth component of the $SU(3)$ flavor octet pseudoscalar meson, and η_s is the $I = 0$ "strange" pseudoscalar meson.] In terms of our notation, we can write (see §3.5c):

$$\begin{aligned} F_{\eta_{ns}} m_{\eta_{ns}}^2 &= <0|\partial_\mu J_\mu^5|\eta_{ns}> \\ &= \frac{2\bar{m}}{F_{\eta_{ns}}}\sqrt{\frac{2}{3}} <s_0> +2 <0|\frac{\alpha_s}{4\pi}\,\mathrm{Tr}(G_{\mu\nu}\tilde{G}_{\mu\nu})|\eta_{ns}> \end{aligned} \tag{7.27}$$

where $s_0 = \bar{q}\frac{\lambda_0}{2}q$; correspondingly:

$$F_\pi m_\pi^2 = \frac{2\bar{m}}{F_\pi}\sqrt{\frac{2}{3}} <s_0> \tag{7.28}$$

with $2\bar{m} = (\frac{m_s - \bar{m}}{m_K^2 - m_\pi^2})m_\pi^2$. Hence:

$$m_{\eta_{ns}}^2 = \frac{F_\pi^2}{F_{\eta_{ns}}^2}m_\pi^2 + \frac{2}{F_{\eta_{ns}}} <0|\frac{\alpha_s}{4\pi}\,\mathrm{Tr}(G_{\mu\nu}\tilde{G}_{\mu\nu})|\eta_{ns}> \tag{7.29}$$

If the anomaly term in Eq. (7.29) is neglected, we obtain the equality: $F_{\eta_{ns}}^2 m_{\eta_{ns}}^2 = F_\pi^2 m_\pi^2$, which, on using the inequality $F_{\eta_{ns}} \geq F_\pi/\sqrt{3}$, yields the inequality $m_{\eta_{ns}} \leq \sqrt{3}m_\pi$ [5.2]. The last inequality became a controversial part of the $U(1)$ problem in QCD [7.6] and the resolution only came when it was realized that the anomaly term had been neglected. Taking into account the QCD anomaly, Eq. (7.29), in the soft pion limit, becomes [7.7]:

$$m^2_{\eta_{ns}} = \frac{2}{F_{\eta_{ns}}} < 0|\frac{\alpha_s}{4\pi}\text{Tr}(G_{\mu\nu}\tilde{G}_{\mu\nu})|\eta_{ns} > \tag{7.30}$$

But we also have (with $J^5_{s\mu} = \bar{s}i\gamma_\mu\gamma_5 s$):

$$\sqrt{2}F_{\eta_s}m^2_{\eta_s} \equiv < 0|\partial_\mu J^5_{s\mu}|\eta_s >= 2m_s < 0|\bar{s}i\gamma_5 s|\eta_s >$$
$$+ 2 < 0|\frac{\alpha_s}{4\pi}\text{Tr}(G_{\mu\nu}\tilde{G}_{\mu\nu}|\eta_s > \tag{7.31}$$

Comparison of the first term on the R.H.S. of Eq. (7.31) with the corresponding relation for the K meson [in the soft pion limit] gives (see §3.5c):

$$\sqrt{2}F_K m^2_K \equiv < 0|\partial_\mu(\bar{s}i\gamma_\mu\gamma_5 d)|K >= m_s < 0|\bar{s}i\gamma_5 d|K > \tag{7.32}$$

Further, in the $SU(3)$ limit:

$$< 0|\bar{s}i\gamma_5 d|K > \simeq < 0|\bar{s}i\gamma_5 s|\eta_s > \tag{7.33}$$

and hence:

$$2m_s < 0|\bar{s}i\gamma_5 s|\eta_s > \simeq 2\sqrt{2}F_K m^2_K \tag{7.34}$$

so that:

$$m^2_{\eta_s} = 2(F_K/F_{\eta_s})m^2_K + \frac{2}{\sqrt{2}F_{\eta_s}} < 0|\frac{\alpha_s}{4\pi}\text{Tr}(G_{\mu\nu}\tilde{G}_{\mu\nu}|\eta_s > \tag{7.35}$$

Noting that $\eta_8 = (\eta_{ns} - \sqrt{2}\eta_s)/\sqrt{3}$, we can rewrite Eq. (7.30) as:

$$m^2_{\eta_{ns}} = \frac{F_{\eta_8}}{F_{\eta_{ns}}}m^2_{\eta_8} + \frac{2\sqrt{2}}{F_{\eta_{ns}}} < 0|\frac{\alpha_s}{4\pi}\,\text{Tr}(G_{\mu\nu}\tilde{G}_{\mu\nu}|\eta_s > \tag{7.36}$$

If we next use the relation:

$$m^2_{\eta'} + m^2_\eta = m^2_{\eta_s} + m^2_{\eta_{ns}} \tag{7.37}$$

together with Eqs. (7.35) and (7.36), we get:

$$m^2_{\eta'} \simeq 2m^2_K - m^2_\eta + A^2 \tag{7.38}$$

where, for simplicity, we have put $F_{\eta_8} \simeq F_K \simeq F_{\eta_{ns}} \simeq F_\pi$ so that A^2 is the anomaly contribution. Equation (7.38) is the final result and tells us that since $(2m^2_K - m^2_\eta) = 0.19$ GeV^2 while $m^2_{\eta'} = 0.92$ GeV^2, most of the mass of the η' meson arises from the global QCD anomaly. More precisely, it can be shown by further current algebra manipulations that $A^2 \geq 0.37$ GeV^2 so that $m_{\eta'}$ is at least 720 MeV, about 25% less than the experimental value of 958 MeV [7.7]. In sum, just as the global QED anomaly corrected for the deficiencies of the straightforward current algebra calculation of $\pi^0 \to 2\gamma$

decay when the divergence of an axial vector quark current is involved, so the global QCD anomaly performs a similar significant role in correcting for deficiencies in current algebra calculations of strong interaction processes in which the coupling of the quark currents to gluon fields is involved.

Finally, we comment on a possibly useful role for the global QCD anomaly in connection with the "proton spin" problem. As we indicated in §5.4a, there is a 3.5 standard deviation disagreement between the Ellis-Jaffe sum rule and the EMC experimental result on the deep inelastic scattering of polarized muons by polarized protons. This discrepancy has caused much concern and drawn a great deal of theoretical interest to explain the disagreement. The most popular proposal to remove the disagreement is to invoke the three-flavor global QCD anomaly to provide gluonic, in addition to quark, contributions to the proton spin function. We recall that the experimental result for the proton spin function is already a harbinger of trouble since the measured value of 0.126 ± 0.018 [see Eq. (5.106)] is close to zero, contrary to the "naive" predictions of the "valence" quark model. Moreover, the Ellis-Jaffe high energy sum rule — derived on the basis of light-cone algebra, $SU(3)$ flavor symmetry and an unpolarized "strange quark sea" — yields 0.189 ± 0.005 for the proton spin function under the conditions of the EMC experiment. It is this discrepancy which constitutes the "proton spin" problem and has led to a reexamination of the Ellis-Jaffe assumptions, in particular their neglect of the three-flavor global QCD anomaly.

Apart from the failure to consider the global QCD anomaly in deriving the Ellis-Jaffe sum rule, it is generally agreed that the assumption of an unpolarized "strange quark sea" is the weakest link in the derivation; indeed, acceptance of the experimental measurement of the proton spin function and the Ellis-Jaffe sum rule without consideration of the global QCD anomaly, implies a non-trivial contribution of the "strange quark sea" to the proton spin function. One finds (see §5.4a) that the s quark "sea" contributes an appreciable amount (almost 40% of the d quark) to the proton spin function. Prima facie, this result is disconcerting because it violates the OZI rule [which requires the s quark contribution to involve "disconnected diagrams" in contrast to the u and d quarks which contribute through "connected diagrams"]; however, one can argue that an appreciable polarization of the s quark "sea" vitiates the Ellis-Jaffe sum rule unless an alternative explanation — such as the global QCD anomaly — is found to replace the s quark "sea" contribution to the proton spin function. Unfortunately, evaluation of the QCD gluon anomaly contribution to the proton spin function becomes the sticking point:

a variety of calculations has been performed [7.8] — from treating the QCD anomaly as a correction to the Goldberger-Treiman relation (see §3.5a) to trying to evaluate it within the framework of the Ellis-Jaffe sum rule — but, in most calculations, the QCD anomaly accentuates the need for the "strange quark sea". More theoretical work has to be done to evaluate the gluon (anomaly) contribution to the proton spin function and it is necessary to perform deep inelastic scattering experiments on polarized muons by polarized deuterons in order to extract the neutron spin function and thereby test the Bjorken sum rule [for the difference between the proton and neutron spin functions — as given by Eq. (5.112)] in addition to the Ellis-Jaffe sum rule.

§7.2c. Harmful triangular chiral gauge anomalies and renormalizability of the standard model

We have seen that the "harmless" (global) triangular axial anomalies can be very useful in the calculation of certain types of QED and QCD processes involving global axial vector quark currents. However, the triangular axial anomaly can pose a threat to the renormalizability of gauge theory if the axial vector current is coupled directly to the gauge field(s) of the theory. Since only vector (or vector-like) currents are coupled to the photon (gauge) field in QED and to the gluon gauge fields in QCD, the problem of renormalizability does not arise for either QED or QCD. The situation is quite different when one or three (an odd number) of axial vector (or chiral fermion currents) is coupled to gauge fields at the vertices of the triangular fermion loop — and even more complicated when some of the gauge fields are non-Abelian in character. In general, for non-Abelian gauge groups in $D = 4$ "space-time" dimensions, n-point Green's functions with an odd number of axial vector couplings — up to five — contribute anomalous terms to the axial density. However, the triangular chiral gauge anomaly is basic and its absence implies the absence of all other anomalous diagrams [7.2]. In the standard model, the electroweak group $SU(2)_L \times U(1)_Y$ couples (lefthanded) chiral fermion currents to both Abelian [through $U(1)_Y$] and non-Abelian [through $SU(2)_L$] gauge fields and it is necessary to confront the question of "harmful" chiral gauge anomalies in QFD.

We have already written down in Eq. (7.19) the non-Abelian generalization of Eq. (7.5) but it is worthwhile to exhibit the few explicit steps that lead to Eq. (7.19). We start with the generalization of Eq. (7.14), which is:

$$q^\lambda T^{abc}_{\mu\nu\lambda} = 2m T^{abc}_{\mu\nu} - \frac{1}{2\pi^2}\epsilon_{\mu\nu\sigma\rho}k_1^\sigma k_2^\rho D^{abc} \tag{7.39}$$

where $T^{abc}_{\mu\nu\lambda}$ and $T^{abc}_{\mu\nu}$ are the counterparts of Eq. (7.6) and (7.8) respectively:

$$T^{abc}_{\mu\nu\lambda} = i \int d^4x_1 d^4x_2 < 0 = |T(V^a_\mu(x_1)V^b_\nu(x_2)A^c_\lambda(0))|0 > \cdot e^{ik_1 \cdot x_1 + ik_2 \cdot x_2}$$

(7.40)

$$T^{abc}_{\mu\nu} = i \int d^4x_1 d^4x_2 < 0|T(V^a_\mu(x_1)V^b_\nu(x_2)P^c(0))|0 > e^{ik_1 \cdot x_1 + ik_2 \cdot x_2}$$ (7.41)

with:

$$V^a_\mu = \bar\psi T^a \gamma_\mu \psi, \quad A^a_\mu = \bar\psi T^a \gamma_\mu \gamma_5 \psi, \quad P^a = \bar\psi T^a \gamma_5 \psi$$ (7.42)

In Eq. (7.42), T^a is the representation matrix corresponding to the $SU(N)$ flavor group $[a = 1, 2, \ldots (N^2 - 1)]$ and satisfying the usual C.R. (f^{abc} are the structure constants)

$$[T^a, T^b] = if^{abc}T^c$$ (7.43)

The quantity D^{abc} depends only on the T^a's, as follows:

$$D^{abc} = \frac{1}{2}\text{Tr}(\{T^a, T^b\}, T^c)$$ (7.44)

The counterpart of Eq. (7.5) for the divergence of the $U(1)$ axial vector current, in the presence of non-Abelian gauge fields, is:

$$\partial_\mu J^\mu_5 = 2mj_5 + \frac{\alpha}{2\pi}\text{Tr}\ (F_{\mu\nu}\tilde{F}^{\mu\nu})$$ (7.45)

where:

$$\text{Tr}(F_{\mu\nu}\tilde{F}^{\mu\nu}) = F^a_{\mu\nu}\tilde{F}^{a\mu\nu}; \quad F^a_{\mu\nu} = \partial_\mu A^a_\nu - \partial_\nu A^a_\mu + gf_{abc}A^b_\mu A^c_\nu$$ (7.46)

with $g(\alpha = g^2/4\pi)$ the non-Abelian gauge constant. In QCD, the non-Abelian field tensor $F_{\mu\nu}$ is denoted by $G_{\mu\nu}$ and g by $g_s(\alpha_s = \frac{g^2_s}{4\pi})$, and it is then obvious that Eq. (7.36) is identical with Eq. (7.19).

We have stated repeatedly that the presence of a $U(1)_A$ triangular chiral gauge anomaly (for an Abelian or non-Abelian quantum field theory) destroys the renormalizability of the gauge theory unless the chiral gauge anomaly is cancelled. Before deriving the general conditions for the cancellation of $U(1)_A$ anomalies, it is instructive to give one explicit example of how the triangular fermion loop responsible for the anomaly exhibits its pathological character. We choose a simple example from QFD exhibiting the unphysical behavior of a scattering amplitude at high energy. Thus, consider the scattering of a photon by an electron via the weak neutral Z boson in QFD, where the Feynman diagram is shown in Fig. 7.3. It can be proved that unitarity is violated by the scattering amplitude (in fourth order), arising from the chiral fermion coupling of the electron loop at the Z gauge boson vertex; the unitary bound is also violated if the photons are replaced by Z

bosons so that there are three axial vector currents (an odd number!). To avoid this catastrophe, other fermion (ABC) loops (see Fig. 7.3) must be included (i.e. the neutrino and the quarks of the same generation — see below) to cancel the electron loop and guarantee unitarity. Many other physical examples can be given of unacceptable predictions of non-Abelian QFD if the chiral fermion loops contributing to the $U(1)_A$ are not cancelled. The only way to guarantee cancellation of the $U(1)_A$ anomalies, so as to maintain the renormalizability of QFD, is to have more than one type of chiral fermion loop so that cancellation can occur.

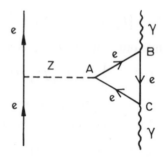

Fig. 7.3. A scattering amplitude for an electron and a photon in fourth order, taking account of the chiral interaction with Z.

In the general case, the crucial point is that since $U(1)_A$ anomalies are independent of the fermionic mass, it is possible to write down general conditions for the cancellation of these anomalies exclusively in terms of the properties of the transformation matrices at the three vertices of the fermion triangle loop (see Fig. 7.3). In QFD, there are, in fact, other charged leptons and quarks that have to be included, and it turns out that the triangular anomaly-free condition is satisfied for the afore-mentioned $e - \gamma$ scattering process. Indeed, all strong and electroweak processes governed by the standard model satisfy the $U(1)_A$ (triangular axial) anomaly-free conditions: the strong interaction processes because the fermionic quarks are in a vector-like representation (see below) and the electroweak processes (whether they involve leptons or quarks) because — as in the case of the $e-\gamma$ scattering process — the chiral (Weyl) leptons and quarks of each generation possess the correct quantum numbers — under the electroweak chiral $SU(2)_L \times U(1)_Y$ group (see Table 1.5) — so that triangular anomaly cancellation takes place. [It would appear at first sight in Table 1.5, that there is an asymmetry in the $SU(2)_L \times U(1)_Y$ quantum numbers as between quarks and leptons (since there is no righthanded neutrino ν_R) so that the triangular anomaly cancellation might be thought to be accidental; however, ν_R, apart from the fact that it is a color singlet, is also "inert" with respect to the $SU(2)_L$ and $U(1)_Y$ quantum numbers (which are 1 and 0 respectively — see Table 1.5) and, consequently, does not contribute to any of the triangular anomaly-free equations (see §7.5).]

Let us now work out the conditions for the cancellation of $U(1)_A$ triangular chiral gauge anomalies for the general (compact) Lie gauge group G and show that the triangular anomalies cancel for the electroweak group. For this purpose, it is convenient to rewrite the Lagrangian in the general case for the interaction of the L and R chiral currents with the gauge bosons A_μ^a of the group G, thus:

$$\mathcal{L}^{\text{int}} = gA_\mu^a(\bar{\psi}_L\gamma^\mu T_L^a\psi_L + \bar{\psi}_R\gamma^\mu T_R^a\psi_R) \tag{7.47}$$

where the lefthanded T_L^a and righthanded T_R^a need not transform in the same way under G, i.e. $T_L^a \neq T_R^a$ (as is the case for the electroweak group). We can now rewrite the triangular anomaly Eq. (7.45) (when $m = 0$) for the lefthanded and righthanded chiral currents, J_L^μ and J_R^μ respectively, as follows:

$$\partial_\mu J_{L,R}^\mu = \partial_\mu(\bar{\psi}_{L,R}\gamma^\mu\psi_{L,R}) = \frac{i}{8\pi^2}\text{Tr}\,(F_{L,R})_{\mu\nu}\tilde{F}_{L,R}^{\mu\nu} \tag{7.48}$$

where $F_{L\mu\nu} = F_{\mu\nu}^a T_L^a$, with a corresponding equation for the righthanded chiral current J_R^μ. Earlier, we expressed the triangular anomaly in terms of the D^{abc} coefficients [see Eq. (7.44)], which depend only on the transformation matrices for the gauge groups involved. Since there is no mixing of T_L^a and T_R^a in Eq. (7.44), and since $D_L^{abc} = -D_R^{abc}$, we can write the general triangular anomaly-free condition in the form:

$$D_L^{abc} - D_R^{abc} = 0 \tag{7.49}$$

where:

$$D_L^{abc} = \frac{1}{2}\text{Tr}(\{T_L^a, T_L^b\}, T_L^c); \quad D_R^{abc} = \frac{1}{2}\text{Tr}(\{T_R^a, T_R^b\}, T_R^c) \tag{7.49a}$$

But $D_{L,R}^{abc}$ is totally symmetric in a, b, c so that:

$$\text{Tr}(\{T^a, T^b\}T^c) \sim \text{Tr}[T^aT^bT^c]_{\text{sym}} \tag{7.50}$$

Thus, the theory is free of triangular anomalies if, for all values of a, b, c:

$$\text{Tr}(T_L^aT_L^bT_L^c)_{sym} = \text{Tr}(T_R^aT_R^bT_R^c)_{sym} \tag{7.51}$$

Condition (7.51) can be satisfied in a variety of ways for the (compact) Lie gauge groups that are of interest in particle physics. We enumerate the different ways in which triangular chiral gauge anomaly cancellation can take place and identify the particular Lie group or class of Lie groups for which that mode of cancellation is applicable [this brief overview will be useful in connection with unification groups (see Chapter 8)]. We explain, in the process, why the gauge theories of the strong and electroweak interactions pass the test of triangular anomaly cancellation. The ways of achieving triangular anomaly cancellation for simple Lie gauge groups are as follows (the extension to non-simple groups is straightforward):

(1) The triangular anomalies caused by lefthanded fermions are cancelled by the triangular anomalies caused by the righthanded fermions. Evidently, this is the case if $T_L^a = T_R^a$ for all a, i.e. if the theory is a "vector-like" gauge theory. An example of a "vector-like" theory is QCD where both the lefthanded and righthanded quarks transform as triplet representations under the color group $SU(3)_C$. The Abelian gauge theory QED is also an example of a "vector-like" theory and is triangular anomaly-free.

(2) $D_L^{abc} = D_R^{abc} = 0$ even if $T_L^a \neq T_R^a$ for all a. In this case, the fermion representation is said to be "safe", i.e. the triangular anomalies cancel separately for the lefthanded and righthanded fermions despite the non-equality of $T_L^a = T_R^a$ (for all a). This is the case, for example, if the fermion representations are real (or pseudo-real). For a real representation, i.e. a representation which is equivalent to its complex conjugate, one has:

$$T^a = -U^{-1}T^{a*}U \tag{7.52}$$

where U is an arbitrary unitary matrix. Consequently, we have:

$$D^{abc} = \frac{\text{Tr}}{2}(\{T^a, T^b\}T^c) = -\frac{\text{Tr}}{2}(\{T^{a*}, T^{b*}\}T^{*c})$$

$$= -\frac{\text{Tr}}{2}(\{T^a, T^b\}T^c) = -D^{abc} \tag{7.53}$$

i.e. $D_L^{abc} = D_R^{abc} = 0$. Thus, real (or pseudo-real) representations do not produce triangular anomalies and are "safe"; since the $SU(2)$ group only possesses real or pseudo-real representations, it follows that all $SU(2)$ representations are individually triangular anomaly-free (i.e. "safe").

The converse of the above statement is not true, i.e. "safe" representations are not necessarily real (or pseudo-real). Thus, for $N \geq 3$, the $SU(N)$ class of groups has some representations that are real [e.g. **8** and **27** of $SU(3)$], some that are complex but are triangular anomaly-free [e.g. **10** of $SU(3)$, because of a special cancellation of phases], and some that are complex and separately give rise to triangular anomalies [e.g. the fundamental representation of $SU(N)$ for $N \geq 3$] but, which, when combined with other representations, can achieve triangular anomaly cancellation [e.g. the combination of the antisymmetric **10** plus the conjugate fundamental representation $\bar{\bf 5}$ in $SU(5)$ yields triangular anomaly cancellation]. The symplectic groups $Sp(n)$ $(n \geq 1)$ $[Sp(1) = SU(2)]$ all have real (or pseudo-real) representations and are therefore all "safe". The $SO(N)$ series of groups is more complicated: $SO(2k+1)$ $(k \geq 1)$ and $SO(4k)$ $(k > 2)$ have only real (or pseudo-real) representations and hence are all "safe"; on the other hand, $SO(4k+2)$ $(k \geq 2)$ possess complex ("spinorial") representations — of dimension 2^{2k} — as well as real representations. The real representations of $SO(4k+2)$ are necessarily "safe", but it also turns out that the complex "spinorial" representations are all triangular anomaly-free, i.e. "safe". Of the five exceptional compact Lie groups, four: G_2, F_4, E_7 and E_8, have only real (or pseudo-real) representations and are therefore all "safe"; one, E_6, possesses complex representations, all of which are triangular anomaly-free and therefore "safe". It is intriguing that the only simple Lie groups that possess "unsafe" representations are the unitary groups $SU(N)$ with $N \geq 3$; this property is exploited in constructing larger groups than the standard group.

We must now convince ourselves that the standard gauge group [i.e. $SU(3)_C \times SU(2)_L \times U(1)_Y$], governing the strong and electroweak interactions of quarks and leptons within each generation, is free of triangular anomalies and hence fully renormalizable. We have already noted that the simple gauge $SU(3)_C$ subgroup of the standard group is "safe" because it is vector-like. The non-semi-simple electroweak gauge group $SU(2)_L \times U(1)_Y$ is chiral and hence possesses triangular chiral gauge anomalies which must be cancelled to ensure renormalizability of the theory. We show that the observed quark and lepton representations, when due account is taken of the three colors for the quarks, combine to yield triangular anomaly cancellation for the electroweak group for each generation. To see this, we remind ourselves that the lefthanded and righthanded representations of the Weyl fermions of one generation, under $SU(3)_C \times SU(2)_L \times U(1)_Y$ [$SU(3)_C$ has been included to differentiate between quarks and leptons], are those given in Table 1.5; again note that there is no ν_R in the standard model. Since the $SU(2)$ representations in Table 1.5 are real (or pseudo-real), the triangular anomaly associated with $SU(2)_L$ currents (at the three vertices) must vanish; this can also be seen from [see Eq. (7.44)]:

$$\text{Tr}(\{\tau^a, \tau^b\}\tau^c) = 2\delta^{ab}\,\text{Tr}(\tau^c) = 0 \qquad (7.54)$$

When two of the fermion currents are $SU(2)_L$ and one $U(1)_Y$, we get (since every member of an $SU(2)_L$ multiplet has the same hypercharge):

$$\text{Tr}(\{\tau^a, \tau^b\}Y) = 2\delta^{ab}\,\text{Tr}\,Y|_{\text{fermion doublets}} = 0 \qquad (7.55)$$

When all three currents are associated with $U(1)_Y$, we get (from Table 1.5):

$$\sum_{\text{leptons}} (Y_L^3 - Y_R^3) + 3 \sum_{\text{quarks}} (Y_L^3 - Y_R^3) = 0 \qquad (7.56)$$

Thus the QFD is triangular anomaly-free and consequently, renormalizable. It is important to note that freedom from triangular chiral gauge anomalies for the electroweak group is due to the presence of both quarks and leptons in each generation, and because the three color degrees of freedom of each quark balance off the fractional hypercharges of the quarks as compared to the leptons. It can be said that the strongest theoretical argument for the existence of the t quark (there are experimental arguments from b quark decays — see [6.61]) is the need for the third generation of quarks and leptons to satisfy the triangular anomaly-free constraints.

Before leaving the subject of the $U(1)_A$ triangular chiral gauge anomaly, we rewrite the $U(1)_A$ anomalies for an Abelian gauge group [see Eq. (7.5)] and a non-Abelian gauge group [see Eq. (7.45)], and comment on the physical meaning of the alternate expressions for later use in Chapter 10. Consider first Eq. (7.5) for an Abelian gauge group, which can be rewritten in terms of the divergence of a four-current, to wit:

$$2F_{\mu\nu}\tilde{F}^{\mu\nu} = \partial^\mu K_\mu \qquad (7.57)$$

where:

$$K_\mu = 4\epsilon_{\mu\nu\alpha\beta}\left(A_\nu \partial_\alpha A_\beta + \frac{2}{3} A_\nu A_\alpha A_\beta\right) \qquad (7.57a)$$

One can then define a "new" axial vector current [see Eq. (7.5) — with α_e replaced by a general coupling constant α]:

$$\tilde{J}_5^\mu \equiv J_5^\mu - \frac{\alpha}{2\pi} K^\mu \qquad (7.58)$$

that satisfies the condition:

$$\partial_\mu \tilde{J}_5^\mu = 2m j_5 \qquad (7.59)$$

so that $\tilde{J}_5^\mu$ is conserved in the limit $m \to 0$. However, the "new" current, $\tilde{J}_5^\mu$, is not gauge invariant, and hence can not be coupled to physical fields, nor lead to the axial vector Ward-Takahashi identities that are required for a renormalizable theory. However, the "new" axial charge $\tilde{Q}_5 = i \int \tilde{J}_5^0 d^3 x$ which, by virtue of Eq. (7.59), is a constant of motion (in the limit $m \to 0$), is gauge invariant; that is:

$$\tilde{Q}_5 = i \int d^3 x \tilde{J}_5^0 = \int d^3 x \{ \psi^\dagger \gamma_5 \psi - \frac{\alpha_e}{\pi} \underline{A} \cdot \nabla \times \underline{A} \} \qquad (7.60)$$

is gauge invariant and can be given a physical meaning — as the generator of the chiral transformation — although this is not sufficient to validate the axial vector Ward-Takahashi identities needed for a renormalizable theory.

The $U(1)_A$ anomaly term in Eq. (7.45) for non-Abelian gauge groups can similarly be rewritten in terms of a divergence of an axial vector current K_μ so that the "new" axial vector current $\widetilde{\widetilde{J}}_5^\mu$ is $J_5^\mu - \frac{\alpha}{2\pi} \tilde{K}^\mu$, where $\tilde{K}^\mu$ (the so-called Chern-Simons current — see §10.3) possesses the same form as that given for K^μ in Eq. (7.57a) except that now $A_\nu = A_\nu^a T^a$ ($T^a = T_L^a$ or T_R^a). While $\tilde{K}^\mu$ appears to be a straightforward generalization of K^μ, the physical consequences are quite different for the Abelian and non-Abelian cases because the "non-Abelian" $\tilde{K}^\mu$ derives its "hard" character (i.e. contributes on the surface at ∞) from the topologically non-trivial gauge transformations (or, equivalently, the instanton solutions — see §10.3) present in non-Abelian gauge groups. One immediate consequence is that, whereas the "new" axial charge $\tilde{Q}_5$ for the Abelian case [see Eq. (7.60)] is gauge invariant, this is not true for the "new" axial charge for the non-Abelian case, namely:

$$\widetilde{\widetilde{Q}}_5 = i \int d^3 x \widetilde{\widetilde{J}}_5^0 \qquad (7.61)$$

The "non-Abelian" topologically non-trivial charge $\widetilde{\widetilde{Q}}_5$ of Eq. (7.61) is no longer gauge invariant although it can still serve as a "chirality" generator — because it is a "constant of motion" within the framework of the θ vacuum (see §10.3). This is in contrast to the $U(1)_A$ anomaly for the Abelian case which carries no topological significance. It should be emphasized, however, that the $U(1)_A$ triangular chiral gauge anomaly — whether it is due to Abelian or non-Abelian gauge groups — makes it impossible to construct axial vector Ward-Takahashi identities and, consequently, interferes with the renormalizability of the quantum field theory in question. The only way to ensure the renormalizability of a quantum field theory is to have triangular chiral gauge anomaly cancellation for both Abelian and non-Abelian groups.

§7.3. Non-perturbative global $SU(2)$ chiral gauge anomaly and self-consistency of the standard model

We have seen that a necessary condition for the renormalizability of a chiral gauge theory like QFD is the cancellation of all triangular axial anomalies. As we have remarked, the triangular anomaly is a perturbative anomaly because only the first term is retained in the perturbative expansion of the gauge transformation, i.e. $U(x) = \exp\ i\epsilon^a(x)T^a \simeq 1 + i\epsilon^a(x)T^a + \dots$ [as before, the T^a's are the generators (transformation matrices) of the group in question and the $\epsilon^a(x)$'s are the local phase functions associated with the T^a's]; consequently, it is possible to derive the triangular anomaly by means of perturbation theory (i.e. Feynman diagrams). We have also pointed out that the perturbative triangular chiral gauge anomaly possesses the very special property that the structure of the lowest-order perturbative term already contains the essential features (e.g. independence of fermion mass in fermion loops) of the quantum corrections (to all orders in perturbation theory) to the classical prediction for axial vector current conservation in the massless fermion limit; it is for this reason that the perturbative triangular anomaly is both a "short wavelength" as well as a "long wavelength" phenomenon. The question that is now being posed is whether a chiral gauge theory — Abelian or non-Abelian — in (3+1)-space-time dimensions develops any new mathematical ambiguities or inconsistencies when the gauge transformations are not infinitesimally close to the identity.

The answer to the last question is in the affirmative for a non-Abelian chiral gauge theory in four dimensions, which develops an inconsistency when the full-fledged ("large"), global gauge transformation $U(x) = \exp[i\epsilon^a(x)T^a]$ is used to explore gauge parameter space [and not merely the infinitesimal form $i\epsilon^a(x)T^a$]. The anomaly that arises in four-dimensional chiral gauge theory for a "non-infinitesimal" gauge transformation is called the global $SU(2)$ anomaly [1.144] since $SU(2)$ is the smallest non-Abelian group for which this anomaly develops. By its very nature, the global $SU(2)$ anomaly can not be calculated by means of (perturbative) Feynman diagrams but must be identified by means of non-perturbative (topological) methods. Consequently, this new anomaly is also termed the non-perturbative global $SU(2)$ chiral gauge anomaly. We henceforth refer to it as the global $SU(2)$ anomaly.

The global $SU(2)$ anomaly is interesting in several respects: (1) while the global $SU(2)$ anomaly occurs for a chiral gauge theory for the same basic reason as does the triangular anomaly (the effect depends crucially upon the properties of the γ_5 operator), it differs from the triangular anomaly in requiring the full exploration of gauge field space — thereby allowing topologically non-trivial gauge transformations to play a decisive role (this is not possible with infinitesimal gauge transformations); (2) the conventional gauge theory of the electromagnetic interaction is based on the $U(1)$ group and is topologically uninteresting — except for the ad hoc introduction of Dirac magnetic monopoles (see §4.6). The situation changes drastically for the standard model, which is based on non-Abelian groups, and necessarily involves topologically non-trivial gauge transformations. While the global $SU(2)$ anomaly is "harmless" for "vector-like" non-Abelian QCD (see point 4 below), it must be taken very seriously for chiral non-Abelian QFD; (3) the presence of the global $SU(2)$ anomaly is disastrous for a chiral gauge theory and must be cancelled to achieve a viable theory, i.e. a well-defined theory without mathematical inconsistencies. In particular, Witten [1.144] has shown that the presence or absence of the global $SU(2)$ anomaly in an $SU(2)$ chiral gauge theory is correlated with the oddness or evenness in the number of lefthanded fermion doublets in the theory. Thus, QFD emerges unscathed from the global $SU(2)$ anomaly constraint because the total number of left-handed fermion doublet representations in the electroweak theory is even [despite the fact that the numbers of chiral quark and chiral lepton doublets are separately odd]. This apparent accident must be understood; (4) the condition for the absence of the global $SU(2)$ chiral gauge anomaly can be generalized to any gauge group of which $SU(2)$ is a subgroup. This opens the door to the imposition of freedom from the global $SU(2)$ anomaly as a second constraint (the first being freedom from the triangular anomaly) in model building and is especially useful in fixing the representations of the standard group (as we will see in §7.5); and, finally (5) it is reasonable to inquire whether there is any relationship between the two "freedoms" — freedom from the triangular anomaly and freedom from the global $SU(2)$ anomaly — in the construction of chirally gauged quantum field theories. It turns out that, for certain well-defined group properties, the absence of the perturbative triangular anomaly guarantees the absence of the non-perturbative global $SU(2)$ anomaly (see §10.4b).

Let us clarify some of the points enumerated above. We have remarked that the global $SU(2)$ anomaly is a consequence of the careful treatment of the topologically non-trivial gauge transformations in the $SU(2)$ group and this statement must be explained. The explanation can be given at various levels of mathematical sophistication and we start by appealing to

Witten's "intuitive" approach [1.144] in this chapter and return later to the same question in a more general topological context — within the framework of the Atiyah-Singer "index theorem" [7.9] — in §10.4a. What I call Witten's "intuitive" approach consists of using the path integral method to study the mathematical properties of the chiral fermion determinant for a single doublet of Weyl chiral fermions in the presence of background (classical) $SU(2)$ gauge fields. More precisely, Witten starts with the (Euclidean) path integral Z for a single doublet of Weyl fermions ψ in the presence of the background $SU(2)$ gauge field: $A_\mu = \sum_{a=1}^{3} \tau^a A_\mu^a$, as follows:

$$Z = \int d\psi d\bar{\psi} \int dA_\mu \exp -\{ \int d^4x [(\frac{1}{2g^2})\mathrm{Tr}\, F_{\mu\nu}^2 + \bar{\psi} i \displaystyle{\not}D \psi] \} \qquad (7.62)$$

where:

$$F_{\mu\nu} = \sum_{a=1}^{3} \tau^a F_{\mu\nu}^a \qquad (7.62a)$$

with $F_{\mu\nu}^a = \partial_\mu A_\nu^a - \partial_\nu A_\mu^a + ig\epsilon_{abc} A_\mu^b A_\nu^c$, $\displaystyle{\not}D = \gamma_\mu D_\mu$ and D_μ the covariant derivative $(\partial_\mu + igA_\mu)$.

The first step in Witten's proof is to perform the fermion integration over a single Dirac doublet in Eq. (7.44); one gets:

$$\int (d\psi d\bar{\psi})_{\mathrm{Dirac}} \exp(\bar{\psi} i \displaystyle{\not}D \psi) = (\det i\displaystyle{\not}D) \qquad (7.63)$$

One then uses the "intuitive" argument that, since each Dirac fermion is equivalent to two Weyl fermions, the fermion integration over a single Weyl doublet yields:

$$\int (d\psi d\bar{\psi})_{\mathrm{Weyl}} \exp(\bar{\psi} i \displaystyle{\not}D \psi) = (\det i\displaystyle{\not}D)^{\frac{1}{2}} \qquad (7.64)$$

i.e. the fermion integration for the single Weyl doublet yields the square root of the integral over a single Dirac doublet. In going from Eq. (7.63) to (7.64), a crucial complication is introduced: whereas the R.H.S. of Eq. (7.63) has an unambiguous value (for a single Dirac fermion doublet) and leads to complete mathematical consistency (with the help of Pauli-Villars regulators) in the calculation of the expectation values of physical operators, the situation is quite different for Eq. (7.64); for Eq. (7.64), the ambiguity in the square root creates ambiguity in the evaluation of expectation values of physical operators when the gauge group contains topologically non-trivial gauge transformations (e.g. instantons with non-vanishing "winding number" — see §10.3). Thus, if one considers the $SU(2)$ gauge group, it can be shown (see §10.4a) that the application of the topologically non-trivial $SU(2)$ gauge transformation [denoted by $U(x)$] on A_μ, i.e. $A_\mu^U \to U^{-1} A_\mu U - iU^{-1} \partial_\mu U$ [see Eq. (4.19)], changes the sign of the fermion integration over a single Weyl doublet, i.e.:

$$[\det i\not{D}(A_\mu^U)]^{\frac{1}{2}} = -[\det i\not{D}(A_\mu)]^{\frac{1}{2}} \qquad (7.65)$$

The proof of Eq. (7.65) can be given by means of the Atiyah-Singer "index theorem" (see §10.4a); here, we accept Eq. (7.65), continue with Witten's "intuitive" derivation of the global $SU(2)$ chiral gauge anomaly and draw out its implications for the standard electroweak group and its possible enlargements. After the fermion integration over a single Weyl doublet, Eq. (7.62) becomes:

$$Z = \int dA_\mu (\det i\not{D})^{\frac{1}{2}} \exp(-\frac{1}{2g^2} \int d^4x \, \mathrm{Tr} \, F_{\mu\nu}^2) \qquad (7.66)$$

with the built-in ambiguity of sign carried by the factor $(\det i\not{D})^{\frac{1}{2}}$. This result implies that the contribution of any gauge field A_μ is cancelled by the equal and opposite contribution of A_μ^U — by virtue of Eq. (7.65) — so that the expectation values of all gauge-invariant physical operators are indeterminate and the theory can not claim mathematical consistency. It is evident that the last statement about mathematical inconsistency applies to any chiral gauge quantum theory with an odd number of Weyl fermion doublets, but not with an even number (since $2n$ Weyl fermion doublets are equivalent to n Dirac doublets). The lack of definition of the $SU(2)$ gauge theory in the presence of an odd number of Weyl fermion doublets constitutes the global $SU(2)$ chiral gauge anomaly, is unacceptable and must be eliminated. As we have implied, the remedy is fairly obvious: the global $SU(2)$ anomaly is absent when the theory contains an even number of Weyl fermion doublets since the fermion integration over every pair of Weyl doublets no longer contains the troublesome square root in Eq. (7.46), and the path integral Z is invariant under a topologically non-trivial $SU(2)$ gauge transformation. In sum, the global $SU(2)$ anomaly is absent for any number of non-chiral (Dirac) fermions and only arises for an odd number of chiral (Weyl) fermions.

The earlier statements made under points (1)–(3) are now clear [points (4)–(5) will be clarified in §10.4a and §10.4b] and we are in a position to spell out the status of the global $SU(2)$ anomaly for the standard model. For the $SU(3)_C$ part of the standard group, no global $SU(2)$ anomaly is present — even though an $SU(2)$ group is a subgroup of $SU(3)_C$ — because the coupling of the quark currents to the gluonic gauge fields is non-chiral (vector-like) so that, effectively, only Dirac fermions are involved. The situation is different for the electroweak subgroup of the standard group, $SU(2)_L \times U(1)_Y$, since it manifestly contains $SU(2)$ chiral (lefthanded) flavor doublets so that we must check whether the total number of weak isospin doublets — for the quarks and leptons of a single generation — is even or odd. Curiously enough, we have the fortunate circumstance that the total number is 4, of which 3 are lefthanded chiral quark doublets (3 colors per quark) and 1 is a lefthanded lepton doublet (one "color" per lepton). It is intriguing that for each generation of quarks and leptons, the numbers of chiral (weak isospin) doublets are separately odd but the total number is even so that the global SU(2) anomaly is absent and the mathematical consistency of QFD is maintained.

Just as in the case of the triangular chiral gauge anomaly — where both quarks and leptons are needed to achieve cancellation — the absence of the global SU(2) anomaly for the chiral gauge group $SU(2)_L \times U(1)_Y$ [the group that is nature's choice for the electroweak interaction — at least up to several hundred GeV] requires the simultaneous presence of quark and lepton chiral doublets. It appears that for both types of chiral gauge anomalies — the triangular and the global $SU(2)$ — a linkage between the quark and lepton worlds is mandated by the combined requirements of gauge invariance and chirality invariance, suggesting a deeper connection between the two types of chiral gauge anomalies than might have been expected. Indeed, a rigorous relation between the perturbative triangular anomaly and the non-perturbative global $SU(2)$ anomaly has been established for simple gauge groups G that satisfy the condition $\Pi_4(G) = 0$ (Π_4 is the fourth homotopy group — see §10.4c) and $G \supset SU(2)$; it turns out that if the representations of G yield freedom from the triangular anomaly, freedom from the global $SU(2)$ anomaly is guaranteed [a proof of this last result is given in §10.4b]. An important condition of the proof is that the gauge group must be simple and only then can it be shown that triangular anomaly cancellation for a choice of Weyl fermion representations implies the absence of the global $SU(2)$ anomaly. Since the standard group $SU(3)_C \times SU(2)_L \times U(1)_Y$ is not simple, freedom from the triangular anomaly and freedom from the global $SU(2)$ anomaly impose independent constraints on the possible quark and lepton representations of the standard group and, as will be seen, contribute importantly to the unique determination of the observed quark and lepton quantum numbers for the standard gauge group. Whether the path "beyond

the standard model" will lead to grand unification (a simple GUT group) —
to which the theorem, that triangular anomaly cancellation implies absence
of the global $SU(2)$ anomaly, applies — is problematic at this stage (this
problem will be more fully discussed in Chapter 8) and so it seems sensible
to consider (in §7.5) the separate constraints imposed by the triangular and
global $SU(2)$ anomaly-free conditions on the standard model. Before doing
so, we examine the status (in the next section) of the third known chiral
gauge anomaly in four dimensions, namely the mixed gauge-gravitational
anomaly.

§7.4. Perturbative mixed gauge-gravitational anomaly and its cancellation in the standard model

The perturbative mixed (chiral) gauge-gravitational anomaly has been
around for a couple of decades [1.145] but it was only realized in the 1980s
[7.3] that the existence of a "long wavelength limit" (as well as a "short wavelength" limit) for this perturbative anomaly justified taking it seriously at the
electroweak level. The perturbative mixed gauge-gravitational anomaly was
derived [1.145] in close analogy to the perturbative triangular chiral gauge
anomaly by replacing the two vector (electromagnetic) currents interacting
with the two photons [at two vertices of the triangular fermionic loop —
see Fig. 7.1] by two symmetric tensor (gravitational) currents interacting
with two graviton fields [at two of the vertices of the triangular fermionic
loop — see Fig. 7.4]. One finds, not surprisingly, that the divergence of an
axial vector (or chiral) current coupled to a gauge field at the third vertex
(see Fig. 7.4) acquires an anomaly, which is called the mixed (chiral) gauge-
gravitational anomaly; strictly speaking, in the case of the mixed anomaly,
the so-called "sea-gull" term [diagram (c) of Fig. 7.4] also contributes to
the mixed anomaly. It should be pointed out that if the axial vector (or
chiral) current is not coupled to a gauge field, the mixed chiral gauge-
gravitational anomaly becomes a mixed "global" axial-gravitational anomaly
in the same sense that the original ABJ anomaly is a "global" triangular axial
anomaly [in contradistinction to the triangular chiral gauge anomaly present
in QFD]. Parenthetically, the mixed "global" axial-gravitational anomaly
arises from the need to reconcile chirality conservation and general covariance and is hence "harmless" like the original ABJ anomaly; however, it
would be extremely difficult to test the predictions of the mixed "global"
axial-gravitational anomaly, unlike the ABJ anomaly, which is severely and
successfully tested through the $\pi^0 \rightarrow 2\gamma$ decay process. Since we are inter-

ested in drawing conclusions from the additional requirement of cancellation of the mixed (chiral) gauge-gravitational anomaly present in QFD, we assume in what follows that the axial vector current is coupled to a gauge field.

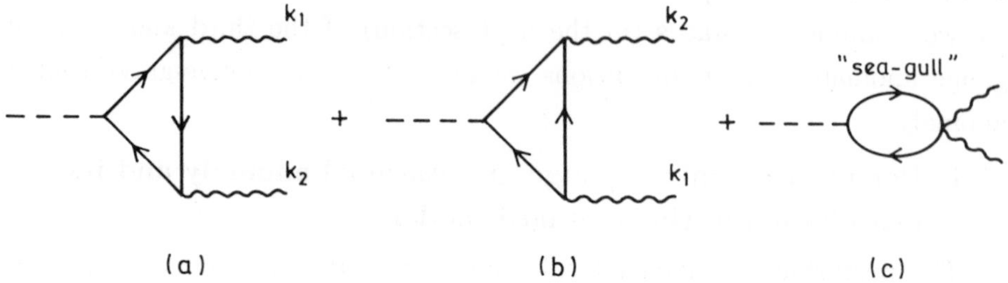

Fig. 7.4. One-fermion loop contributions to the mixed gauge-gravitational anomaly in four space-time dimensions.

The existence of the mixed gauge-gravitational anomaly in four dimensions is of particular interest because there is no pure (one-loop) gravitational anomaly in four dimensions. Pure gravitational anomalies do occur for Weyl fermions (of spin 1/2 or spin 3/2) in $4k + 2$ ($k \geq 0$) dimensions [7.3] [e.g. for $k = 2$ ($D = 10$), the "hexagon diagram" with six external gravitons is anomalous and is of interest in superstring theory (see §1.4c)]. However, in four dimensions, the only anomaly involving external gravitons is the mixed anomaly with one axial vector (or chiral) fermion current and two gravitational currents. If the axial vector (or chiral) fermion current is J_5^μ, the analog of the "triangular" anomalous equation (7.5) (in the massless fermion limit) becomes:

$$\partial_\mu J_5^\mu = -\frac{n_f}{192\pi^2} R_{\mu\nu\sigma\tau} \tilde{R}^{\mu\nu\sigma\tau} \tag{7.67}$$

where n_f is the number of fermion flavors, $R_{\mu\nu\sigma\tau}$ is the Riemann-Christoffel curvature tensor of Einsteinian gravity [7.10] and $\tilde{R}^{\mu\nu\sigma\tau} = \frac{1}{2}\epsilon^{\mu\nu\alpha\beta} R_{\alpha\beta}^{\sigma\tau}$ is the dual Riemann-Christoffel tensor. Evidently, Eq. (7.67) possesses a similar structure to that of the triangular anomaly defined by Eq. (7.45), since the Reimann and Maxwell tensors play comparable roles in representing the photon and graviton fields respectively.

The analogy between the $U(1)$ mixed and $U(1)_A$ triangular anomalies goes even deeper [assuming that the $U(1)_A$ triangular anomaly is due to non-Abelian gauge fields]. In both cases: (1) the anomaly terms have dimension L^{-4} (L is length) so that the space-time integrals are dimensionless (i.e. independent of the mass) and are connected to topological invariants [non-Abelian "internal symmetry instantons" for the triangular anomaly and "gravitational instantons" for the mixed anomaly — second ref. [1.145]; (2) both anomalies are singlets of all internal symmetries (color, flavor); (3) both anomaly terms can be written as exact divergences with similar properties; (4) both anomaly terms are pseudoscalar and even under charge conjugation; and, finally (5) both the mixed anomaly and the triangular anomaly possess "short wavelength" as well as "long wavelength" limits. However, the "short wavelength" limit of each anomaly has a different physical meaning, which explains the somewhat different reasons for the required cancellation of the two anomalies in order to maintain viable quantum field theories. As we have seen, the basic reason for requiring the cancellation of the triangular anomaly is to simultaneously maintain chirality conservation and gauge invariance for a renormalizable quantum field theory with massless fermions (hence the need for a "short wavelength" limit). In the case of the mixed anomaly, it is mandated that chirality conservation, gauge invariance and general covariance must be simultaneously maintained. It is true that in the Planckian "short wavelength limit" — where gravitational fluctuations are as significant as gauge fluctuations — the unknown effects of quantum gravity have to be considered; however, if we only consider the "quantum fluctuations" at the electroweak level, we are only paying attention to the "long wavelength limit" of a mixed gauge-gravitational anomaly where quantum gravity is unimportant. Thus, our ignorance about quantum gravity should not alter the conclusion drawn about the need for cancellation of the mixed anomaly at the relatively "long wavelength" electroweak scale.

With the last caveat, we state the result of applying the mixed (chiral) gauge-gravitational anomaly to the standard electroweak group $SU(2)_L \times U(1)_Y$. It is evident that since the chiral gauge electroweak group has only one vertex of the fermionic loop diagram at its disposition, so to speak, the $SU(2)_L$ part of the electroweak group does not contribute and hence (see Fig. 7.4), the mixed anomaly can only arise from the $U(1)_Y$ part. Since the mixed anomaly arises from a triangular fermion loop with one hypercharge current and two gravitational currents, it must be proportional to Tr Y. Consequently, the condition for the renormalizability of the $SU(2)_L \times U(1)_Y$ group coupled to gravity is:

$$\text{Tr } Y = 0 \qquad (7.68)$$

This mixed anomaly-free condition [to be distinguished from the triangular anomaly-free condition $\text{Tr } Y^3 = 0$] is non-trivial and places a valuable constraint — in conjunction with the triangular and global $SU(2)$ anomaly-free conditions — on a non-semi-simple chiral gauge theory such as QFD (see below). It happens that the condition $\text{Tr } Y = 0$, summed over the hypercharges of one generation of quarks and leptons (see Table 1.5) and, taking account of the three color degrees of freedom for each quark, is satisfied for the standard electroweak group; we note that taking cognizance of the different "color weights" of the chiral quarks and leptons in the mixed anomaly-free condition $\text{Tr } Y = 0$, amounts to taking cognizance of $SU(3)_C$ for the mixed anomaly-free condition. The overall conclusion is that the standard group with the observed quantum numbers of the quarks and leptons for all three generations (assuming that the not-as-yet-observed t quark is no exception) is free of the three known chiral gauge anomalies in four dimensions (triangular, global $SU(2)$ and mixed).

This result is gratifying and reassures us with regard to the renormalizability, chirality conservation (for massless quarks), self-consistency and general covariance of the standard model at scales appreciably below the Planck scale. However, it is not clear why the standard model is so compliant with these powerful anomaly-free constraints and we must try to understand why all three four-dimensional anomalies are cancelled in the standard group $SU(3) \times SU(2) \times U(1)$. For this purpose, we turn the question around and examine, in the next section, what quark and lepton representations are possible when all three anomaly-free conditions are imposed on the standard group [we distinguish between the standard group $SU(3) \times SU(2) \times U(1)$ and the standard model $SU(3)_C \times SU(2)_L \times U(1)_Y$ where the representations and quantum numbers are those given in Table 1.5].

§7.5. Anomaly-free constraints on Weyl representations of the standard group

In this section, we inquire into the extent to which the observed Weyl fermion representations and their hypercharges are determined by the combination of the triangular, global $SU(2)$ and mixed anomaly-free conditions. We will show that starting with the non-semi-simple standard group $SU(3) \times SU(2) \times U(1)$, the imposition of the three anomaly-free conditions in four dimensions on this group leads uniquely — except for the hypercharge scale — to the correct phenomenological result for the minimal number of (lefthanded) Weyl representations of $SU(3)_C \times SU(2)_L$ and their $U(1)_Y$ charges. This is a welcome result, emphasizing the power of anomalies in quantum field theory, and suggests that we have at least a partial handle on the non-perturbative aspects of the standard model [through the global $SU(2)$ anomaly and the "short wavelength" limit of the triangular anomaly] and some clue as to the effect of Einsteinian gravity [through the "long wavelength" limit of the mixed gauge-gravitational anomaly]. We describe in this section how the three anomaly-free conditions constrain the Weyl representations and quantum numbers under the standard group $SU(3) \times SU(2) \times U(1)$ so as to yield a unique minimal set of quantum numbers that are identical with those of the standard model fermions. We then comment (in §7.6) on further ramifications of the anomalies approach with respect to electric charge quantization when the three chiral gauge anomaly-free conditions are combined with the Higgs mechanism for spontaneously breaking the group $SU(3)_C \times SU(2)_L \times U(1)_Y$ down to the unbroken non-chiral $SU(3)_C \times U(1)_{EM}$ group. In the concluding section (§7.7), we consider briefly the anomaly-free constraints on chiral gauge groups that go "beyond the standard model" and the increasing degree of redundancy that is encountered in the anomaly-free constraints as the standard chiral gauge group is enlarged.

The demonstration of the predictive power of the three chiral gauge anomaly-free conditions for the Weyl fermion representations (and quantum numbers) of the standard model depends on taking seriously all three anomaly-free conditions, working only with Weyl (chiral) fermion states, accepting the condition that only Dirac masses of the quarks and leptons can be generated via the SSB of the standard group, and adopting the principle of minimality [so that only the most economical solution, i.e. the smallest number of Weyl fermions, is accepted although any number of identical "copies" (generations) are allowed]. Keeping these points in mind, we start by allowing an arbitrary number of (lefthanded) Weyl fermion representations under the group $SU(3) \times SU(2) \times U(1)$, as in Table 7.1. In Table 7.1, the integers j, k, l, m, n and p, and the $U(1)$ charges — Q_i, Q'_i, $\bar{Q}_i$, q_i and $\bar{q}_i$ — are all arbitrary. Adding arbitrary non-fundamental and singlet representations to Table 7.1 do not change the conclusions. It should be emphasized that we are focussing on the standard group before the SSB Higgs mechanism is operative; it is for this reason that we speak of Weyl fermion representations and their chiral hypercharges. The uniqueness statement about the representations under the standard group, as determined by the three anomaly-free conditions, applies to the Weyl fermion representations and their chiral hypercharges. It is only after the precise set of minimal

Weyl fermion representations and their chiral hypercharges are determined that we can assign them to the initial $SU(3) \times SU(2) \times U(1)$ group and recognize that the imposition of the three anomaly-free constraints has fixed the $SU(3) \times SU(2) \times U(1)$ group as the standard group of strong and electroweak interactions, i.e. $SU(3)_C \times SU(2)_L \times U(1)_Y$. And it is only when the electroweak part of the standard group, $SU(2)_L \times U(1)_Y$, is spontaneously broken to $U(1)_{EM}$, that we can speak of the electric charges of the quarks and leptons and the quantization of electric charge. This will become clear as we proceed.

Table 7.1: Possible Weyl fermion representations under the $SU(3) \times SU(2) \times U(1)$ group.

$SU(3)$	$SU(2)$	$U(1)$
3	2	$Q_i\ (i = 1,2,3,\dots,j)$
3	1	$Q'_i\ (i = 1,2,3,\dots,k)$
$\bar{3}$	1	$\bar{Q}_i\ (i = 1,2,3,\dots,l)$
$\bar{3}$	2	$\bar{Q}'_i\ (i = 1,2,3,\dots,m)$
1	2	$q_i\ (i = 1,2,3,\dots,n)$
1	1	$\bar{q}_i\ (i = 1,2,3,\dots,p)$

Using the representations listed in Table 7.1, we find that freedom from the triangular anomalies leads to the following equations:

$$[SU(3)]^3 : \sum_{i=1}^{j} 2 + \sum_{i=1}^{k} 1 - \sum_{i=1}^{\ell} 1 - \sum_{i=1}^{m} 2 = 0, \tag{7.69a}$$

$$[SU(3)]^2 U(1) : 2\sum_{i=1}^{j} Q_i + \sum_{i=1}^{k} Q'_i + \sum_{i=1}^{\ell} \bar{Q}_i + 2\sum_{i=1}^{m} \bar{Q}'_i = 0 \tag{7.69b}$$

$$[SU(2)]^2 U(1) : 3\sum_{i=1}^{j} Q_i + 3\sum_{i=1}^{m} \bar{Q}'_i + \sum_{i=1}^{n} q_i = 0, \tag{7.69c}$$

$$U(1)^3 : 6\sum_{i=1}^{j} Q_i^3 + 3\sum_{i=1}^{k} Q_i'^3 + 3\sum_{i=1}^{\ell} \bar{Q}_i^3 + 6\sum_{i=1}^{m} \bar{Q}_i'^3 + 2\sum_{i=1}^{n} q_i^3 + \sum_{i=1}^{p} \bar{q}_i^3 = 0. \tag{7.69d}$$

The global $SU(2)$ anomaly-free condition leads to:

$$3j + 3m + n = N \tag{7.70}$$

where N is an even integer. Finally, the third (mixed) anomaly-free condition $\mathrm{Tr}\, Y = 0$ yields — with the help of Eq. (7.51b):

$$2q_q + \bar{q}_1 = 0 \tag{7.71}$$

We now seek the minimal solution that satisfies Eqs. (7.69)–(7.71) plus the extremely plausible condition — since we are using only Weyl fermion representations — that no Dirac fermion mass is generated without the SSB of the $G_{321} = SU(3) \times SU(2) \times U(1)$ symmetries. Equation (7.70) is an especially restrictive condition and, even when we set $j = m = 0$, the lowest value of n is 2 and leads to 10 Weyl fermions in the representations: $(\mathbf{1}, \mathbf{2}, q)$, $(\mathbf{1}, \mathbf{2}, -q)$, $(\mathbf{3}, \mathbf{1}, Q)$, and $(\bar{\mathbf{3}}, \mathbf{1}, -Q)$ under G_{321}; however, the first two and last two representations can combine to generate Dirac fermion masses without the SSB of G_{321}. Hence, we must choose non-vanishing values for j and n and the minimum choice is $j = n = 1$; sticking with the "minimality" condition, we select $m = 0$ and are then compelled to take $l = 2$, $k = p = 0$. This combination of chiral hypercharges yields 14 Weyl fermions in the representations: $(\mathbf{3}, \mathbf{2}, \mathbf{0})$, $(\bar{\mathbf{3}}, \mathbf{1}, \bar{Q})$, $(\bar{\mathbf{3}}, \mathbf{1}, -\bar{Q})$ and $(\mathbf{1}, \mathbf{2}, \mathbf{0})$. The trouble with this set of 14 Weyl fermions is that the $(\mathbf{1}, \mathbf{2}, \mathbf{0})$ doublet can not acquire a Dirac mass even with the SSB of G_{321}. Hence, still accepting "minimality", we are compelled to move on to 15 Weyl fermions by taking $p = 1$, i.e. adding the state $(\mathbf{1}, \mathbf{1}, \bar{q})$; we can then generate a Dirac fermion mass out of the Weyl representations $(\mathbf{1}, \mathbf{2}, q)$ and $(\mathbf{1}, \mathbf{1}, \bar{q})$ with an appropriate Higgs.

For clarity, we rewrite in an obvious notation the surviving anomaly-free constraint equations for these 15 Weyl fermions [corresponding to $j = 1, k = 0, l = 2, m = 0, n = 1, p = 1$]:

$$2Q + \bar{Q}_1 + \bar{Q}_2 = 0; \quad 3Q + q = 0 \tag{7.72}$$

$$6Q^3 + 3\bar{Q}_1^3 + 3\bar{Q}_2^3 + 2q^3 + \bar{q}^3 = 0 \tag{7.73}$$

$$2q + \bar{q} = 0 \tag{7.74}$$

Strictly speaking, Eqs. (7.72)–(7.74) possess two solutions: the "standard model" solution [1.146] and the so-called "bizarre" solution [7.11]. The "standard model" solution is given by:

$$Q = -\frac{1}{3}q, \quad \bar{Q}_1 = \frac{4}{3}q, \quad \bar{Q}_2 = -\frac{2}{3}q, \quad \bar{q} = -2q \tag{7.75}$$

The "bizarre" solution is very different; one gets:

$$Q = q = \bar{q} = 0, \quad \bar{Q}_1 = -\bar{Q}_2 \tag{7.76}$$

Equation (7.75) is a very interesting result: it tells us that, except for the scale which can be adjusted by altering the $U(1)$ gauge coupling constant, the charges of all 15 Weyl fermion states are fixed and correspond precisely to the observed hypercharges of one generation of quarks and leptons. Furthermore, the "bizarre" solution must be excluded: apart from the trivialization of the mixed anomaly-free condition [see Eq. (7.74)], the "inert" state $(1, 1, 0)$ is a non-chiral representation which violates a key assumption [that only Weyl (chiral) representations of G_{321} are allowed] in this derivation. In passing, it is interesting to note that the solution of Eqs. (7.72)–(7.74) in the large N_C limit (i.e. when $N_C \to \infty$ instead of $N_C = 3$ — see §5.4b) becomes a "half-bizarre" solution in the sense that $Q = 0$ and $\bar{Q}_1 = -\bar{Q}_2$ [see Eq. (7.76)] but, as might be expected, there is no condition on q and $\bar{q}$ (the leptonic sector!). This is a curious result and suggests that one must be careful about large N_C results (see §5.4b) when the leptonic and quark sectors are both involved in the process.

Apart from some fine points, we can conclude that "minimality" and the application of the three chiral gauge anomaly-free conditions in four dimensions to the Weyl representations of the standard group yield — except for the hypercharge scale — a unique set of Weyl fermion representations and quantized (chiral) hypercharges that coincide with those of the observed quarks and leptons of one generation; obviously, the number of "copies" of the 15 Weyl states (i.e. the number of generations) is not determined through the application of the three anomaly-free constraints in four dimensions. If we go further and choose the normalization of the lefthanded lepton doublet hypercharge $q = -1$ — in anticipation that this value will correspond to zero electric charge for the neutrino after the SSB of the electroweak group — we obtain the familiar quantum numbers of the 15 Weyl states for one family of quarks and leptons (listed in Table 1.5). We are now in a position to treat the Higgs spontaneous breaking of the electroweak group and to show how — with the help of the three chiral gauge anomaly-free constraints — electric charge quantization can be achieved within the framework of the standard model.

§7.6. Higgs breaking of the electroweak group and electric charge quantization (ECQ)

We have seen in the previous section that imposition on the standard group of the three chiral gauge anomaly-free constraints in four dimensions [triangular, global $SU(2)$ and mixed] leads — under the condition of "minimality" — to the correct "one-generation" quantum numbers of the quarks and leptons plus chiral hypercharge quantization; the mixed chiral gauge-gravitational anomaly-free constraint is indispensable to obtain the welcome feature of chiral hypercharge "quantization". In this section, we show how — building on the three anomaly-free conditions — we can secure electric charge quantization (ECQ) at the electroweak scale with a simple assumption on the Higgs breaking of the electroweak group in the lepton sector [doublet under $SU(2)_L$ but arbitrary hypercharge Y_ϕ]. This can be seen as follows: in the standard model, the fermion masses (both quark and lepton) are generated by means of the Yukawa Lagrangian [see Eq. (6.10)]:

$$\mathcal{L}_Y = h_u \bar{q}_L \tilde{\phi} u_R + h_d \bar{q}_L \phi d_R + h_e \bar{l}_L \phi e_R + \text{h.c.} \qquad (7.77)$$

where the h's are the Yukawa coupling constants, $\tilde{\phi} = i\tau_2 \phi^*$, and ϕ is the Higgs doublet possessing the quantum numbers, $(\mathbf{1}, \mathbf{2}, Y_\phi)$ under the standard group $SU(3)_C \times SU(2)_L \times U(1)_Y$.

In utilizing Eq. (7.77) to generate the (Dirac) fermion masses in the standard model, the hypercharge quantum number of the Higgs doublet responsible for the SSB of the electroweak group is chosen to achieve phenomenological agreement. However, this assignment is not necessary for the purposes of demonstrating electric charge quantization. Thus, consider that the electric charge Q associated with the unbroken $U(1)_{EM}$ group assumes the more general form [7.12]:

$$Q = I_{3L} - YI_{3\phi}/Y_\phi \qquad (7.78)$$

where $I_{3\phi} = -\frac{1}{2}$ for the component of the Higgs (weak isospin) doublet that acquires a VEV The invariance of Eq. (7.77) under the global hypercharge gauge transformation then yields the following hypercharge relations [we now use the hypercharge notation appropriate to the standard group, wherein $Q \equiv Y_q$, $\bar{Q}_1 \equiv -Y_u$, $\bar{Q}_2 \equiv -Y_d$, $q = Y_l$, $\bar{q} = -Y_e$ (see Eq. (7.75))]:

$$Y_u = Y_q + Y_\phi; \quad Y_d = Y_q - Y_\phi; \quad Y_e = Y_l - Y_\phi \qquad (7.79)$$

The sum of the first two relations in Eq. (7.79) duplicates the triangular anomaly-free condition of Eq. (7.72) and the third relation of Eq. (7.79), when combined with the mixed anomaly-free condition of Eq. (7.74), imposes the constraint $Y_\phi = -Y_\ell$ [Y_ℓ is the scale in terms of which all the Weyl fermion hypercharges, corresponding to the standard group, were expressed in Eq. (7.75)]. When the relation $Y_\phi = -Y_\ell$ is substituted into Eq. (7.78) for Q and the values of the hypercharges Y (expressed in units of Y_ℓ) are taken from Eq. (7.75), we obtain the correct quantized electric charges for the quarks and leptons of one generation, i.e. $Q_u = \frac{2}{3}$, $Q_d = -\frac{1}{3}$, $Q_\nu = 0$ and $Q_e = -1$. Clearly, in the standard model, the chiral gauge anomaly-free conditions (including the mixed anomaly-free condition) play an essential role at the electroweak level when the Higgs SSB mechanism is employed to obtain the unbroken $U(1)_{EM}$ group and to generate the (Dirac) fermion masses (of a single generation).

It is interesting to note that the relations between the quark hypercharges and the Higgs hypercharge given by Eq. (7.79) are not needed to prove ECQ at the electroweak level; the point is that the use of all three anomaly-free conditions is sufficiently powerful in the case of the standard group (with 15 Weyl states) to determine all quantized quark and lepton hypercharges (in units of Y_ℓ) and to secure ECQ solely by taking account of the Yukawa coupling of the Higgs particle with the leptons [see Eq. (7.79)]. This feature is replicated for the enlarged left-right-symmetric (LRS) group (see below) and one may be tempted to speculate that this opens the door to separate Yukawa couplings of distinct Higgs doublets to quarks and leptons [7.13]. However, such speculation is premature and it is best to reiterate the essential result that the application of all three anomaly-free conditions to the Weyl fermion representations of the standard group is not only capable — with the assumption of "minimality" — to predict a unique and phenomenologically correct set of quantum numbers for one family of quarks and leptons but also, in combination with the Higgs SSB mechanism, to predict ECQ at the electroweak scale. This result for the standard group is to be contrasted with the increasing redundancy of the global $SU(2)$ and mixed anomaly-free conditions and the increasing loss of automatic ECQ, that characterize groups larger than the standard group — to be discussed in the next section.

§7.7. Four-dimensional chiral gauge anomaly constraints beyond the standard model

It is interesting to inquire into the transformed status of the three anomaly-free conditions, and the transformed role of the Higgs SSB mechanism and ECQ, when the standard group is enlarged, in successive stages, to a simple GUT group. We will notice the increasing redundancy of the anomaly-free conditions [first of the mixed anomaly-free condition and then of the global $SU(2)$ anomaly-free condition] as the electroweak group is enlarged in progressive steps by first adjoining a $U(1)$ [or $SU(2)$ group] to the standard group, then jumping to the semi-simple Pati-Salam group [1.148] and, finally, reaching a GUT group (as a prelude to Chapter 8).

In describing the changing status of the anomaly-free conditions as the standard group is enlarged, we first point out that the (Abelian) $U(1)_Y$ hypercharge part of the standard group is crucial for the effectiveness of the constraints imposed by the mixed and triangular anomaly-free conditions. Thus, the fact that the standard group contains the $U(1)$ group (so that it is not even semi-simple) is responsible for the power of the three anomaly-free conditions and it is instructive to check what happens when the standard group is enlarged to contain more than one $U(1)$ group, and then to identify the reduced role of the three anomaly-free conditions when the non-semi-simple standard group is enlarged to a semi-simple or simple group. We start by summarizing the consequences for the chiral hypercharge quantum number determinations and ECQ when the standard group is enlarged to the gauge group $G_{3211}[\equiv SU(3)_C \times SU(2)_L \times U(1)_Y \times U(1)_{Y'}]$, and a righthanded (Weyl) neutrino (ν_R) is added to the 15 Weyl fermion states of the standard model. We first show that it is not possible to validate ECQ when the standard group $G_{321}[= SU(3)_C \times SU(2)_L \times U(1)_Y]$ is retained and the number of Weyl fermion states of G_{321} is increased from 15 to 16 (i.e. ν_R is added). The 16th Weyl state (ν_R) possesses the quantum numbers $(1, 1, Y_\nu)$ under $SU(3)_C \times SU(2)_L \times U(1)_Y$, allowing Y_ν to be arbitrary. The triangular and mixed anomaly-free conditions of Eqs. (7.69) and (7.71) become:

$$2Y_q - Y_u - Y_d = 0; \quad 3Y_q + Y_l = 0 \tag{7.80a}$$

$$6Y_q^3 - 3Y_u^3 - 3Y_d^3 + 2Y_l^3 - Y_\nu^3 Y_e^3 = 0; \quad 6Y_q + 2Y_l - 3Y_u - 3Y_d - -Y_\nu - Y_e = 0 \tag{7.80b}$$

It is immediately noted that the additional hypercharge Y_ν introduces an ambiguity into Eqs. (7.80), which is not resolved even after allowing the Higgs doublet to generate a Dirac mass for ν_R. It is also easy to see that the hypercharge Y_ϕ (associated with the Higgs doublet) yields a new relation $Y_\nu = Y_l + Y_\phi$ [see Eq. (7.79) for notation] but, unfortunately, this relation already follows from the relation $2Y_l - Y_e - Y_\nu = 0$ [which results from a combination of Eqs. (7.80a) and (7.80b) and the relation $Y_e = Y_l - Y_\phi$ of Eq. (7.79)]. Since one more parameter (Y_ν) is added to the five previous parameters (Y_q, Y_u, Y_d, Y_l, Y_e) and no new constraints are provided by the gauge invariance of the Yukawa coupling of the Higgs doublet to the Weyl fermions, it is not possible to deduce the electric charges of the quarks and leptons of a single generation; hence, ECQ is lost at the electroweak scale [7.14]. One can try introducing a triplet Higgs boson (in addition to the doublet) to generate a Majorana mass for ν_R but the structure of a Majorana mass term, $\nu_R^T C^{-1} \nu_R$, does not lead to ECQ. To achieve ECQ — even with a Majorana mass for ν_R — requires enlarging the standard gauge group G_{321} by adding one more $U(1)$ gauge group.

The presence of two $U(1)$ gauge groups [which we denote by $U(1)$ and $U(1)'$ with associated hypercharges Y and Y' respectively] gives rise to two more triangular anomaly-free conditions. Consequently, the number of additional relations among the Y's and Y''s [provided one makes use of both triangular anomaly-free conditions involving $U(1)_Y$ and $U(1)'_{Y'}$] is sufficient to fix the two $U(1)$ groups as: $U(1)_{B-L} \times U(1)_R$ [where $U(1)_R$ is defined by assigning the same $U(1)_R$ charges to the R (weak isospin) "up" and "down" quark and lepton and the same $U(1)_R$ charges (but opposite to the "up" charge) to the righthanded "down" quark and "down" lepton respectively]. One way to recapture ECQ is to insist on a Majorana mass for ν_R. Implicit in this derivation is the use of the "weak" GNN relation (see §1.2d): $Q = I_{3L} + Y_R/2 + Y_{B-L}/2$, corresponding to the group $G_{3211}[\equiv SU(3)_C \times SU(2)_L \times U(1)_R \times U(1)_{B-L}]$, as well as the use of the mixed anomaly-free condition and the relations among the hypercharges imposed by the Higgs SSB mechanism giving rise to a Majorana mass for ν_R.

We have learned from the above example of the enlarged G_{3211} group that full use of the three anomaly-free conditions, plus insistence on a Majorana mass for ν_R, make it possible to achieve uniqueness in the quantum numbers of the 16 Weyl fermions under the enlarged G_{3211} group, as well as ECQ. The enlarged non-semi-simple group G_{3211} is of potential interest as a subgroup of $SO(10)$ GUT, as are the non-semi-simple group $G_{3221}[\equiv SU(3)_C \times SU(2)_L \times SU(2)_R \times U(1)_{B-L}]$ and the semi-simple group G_{422} [$\equiv SU(4)_C \times SU(2)_L \times SU(2)_R$] (see §8.4b). We briefly comment on the constraints imposed by the three anomaly-free conditions plus the SSB Higgs mechanism (with EQC) as we move up through the G_{3221} and G_{422} subgroups of $SO(10)$ GUT; we then describe the dramatic change in the potency of the anomaly-free conditions at the grand unification level — whether $SO(10)$ or one of the other GUT groups.

With regard to the left-right-symmetric (LRS) group G_{3221}, it is easily shown that — with "minimality" — the Weyl fermion representations are those listed in Table 7.2. It is obvious from Table 7.2 that there are even numbers of lefthanded and righthanded (weak) doublets so that the global $SU(2)$ anomaly-free condition is automatically satisfied. The relevant triangular anomaly-free conditions are:

$$\text{Tr}\,[SU(3)_C]^2 U(1)_{B-L} = 0 \Rightarrow \qquad Y_q + Y_{\bar{q}} = 0$$
$$\text{Tr}\,[SU(2)_L]^2 U(1)_{B-L} = 0 \Rightarrow \qquad 3Y_q + Y_\ell = 0 \qquad (7.81)$$
$$\text{Tr}\,[SU(2)_R]^2 U(1)_{B-L} = 0 \Rightarrow \qquad 3Y_{\bar{q}} + Y_{\bar{\ell}} = 0$$

Table 7.2: Weyl fermion representations under the left-right-symmetric (LRS) group.

Particles	$SU(3)_C$	$\times$	$SU(2)_L$	$\times$	$SU(2)_R$	$\times$	$U(1)_{B-L}$
$q_L \begin{pmatrix} u \\ d \end{pmatrix}_L$	3		2		1		Y_q
$\bar{q}_L = \begin{pmatrix} \bar{u} \\ \bar{d} \end{pmatrix}_L$	$\bar{3}$		1		2		$Y_{\bar{q}}$
$\ell_L = \begin{pmatrix} \nu \\ e \end{pmatrix}_L$	1		2		1		Y_ℓ
$\bar{\ell}_L = \begin{pmatrix} \bar{\nu} \\ \bar{e} \end{pmatrix}_L$	1		1		2		$Y_{\bar{\ell}}$

The solution of Eqs. (7.81) is:

$$Y_{\bar{q}} = -Y_q, \quad Y_{\bar{\ell}} = -Y_\ell, \quad Y_q = -Y_\ell/3 \qquad (7.82)$$

The mixed anomaly-free condition is automatically satisfied because of the manifest left-right symmetry in the electroweak sector. It is clear from Eq. (7.82) that the new feature of left-right symmetry (LRS) in the representations of the group produces the redundancy of the mixed anomaly-free condition — in contrast to the situation for the standard group. Nevertheless, the constraints imposed by the triangular [and global $SU(2)$] anomaly-free condition are sufficient to fix the hypercharges (up to a common scale) to the same extent as was the case for the standard group with a non-redundant mixed anomaly-free condition. The question of ECQ must still be answered and, it turns out, that a Majorana mass for ν_R again secures ECQ for the LRS model whereas a Dirac mass for ν_R does not. Thus, as in the case of the G_{3211} group, if one tries to generate Dirac masses for the quarks and leptons in the LRS model, the simplest way is to choose a Higgs boson with the quantum numbers $(\mathbf{1}, \mathbf{2}, \mathbf{2}, Y_H)$ under G_{3221}. But then $Y_H = 0$ and no additional relation between Y_q and Y_l emerges from the gauge invariance of the Yukawa coupling of the Higgs particle to the quarks and leptons — in addition to the relation in Eq. (7.82) — and hence there is no ECQ.

The situation changes for the LRS group when one insists on a Majorana mass for ν_R (or ν_L, for that matter), which requires the use of two

Higgs bosons (weak triplets) with the quantum numbers $(\mathbf{1}, \mathbf{3}, \mathbf{1}, Y_\Delta)$ and $(\mathbf{1}, \mathbf{1}, \mathbf{3}, Y_\Delta)$ under G_{3221}; the gauge invariance of the resulting Yukawa Lagrangian then implies:

$$2Y_l + Y_\Delta = 0 \tag{7.83}$$

and the electric charge formula:

$$Q = I_{3L} + I_{3R} + \frac{Y}{Y_\Delta} \tag{7.84}$$

Equations (7.83) and (7.84) tell us that $Y_q = \frac{1}{3}$ and $Y_\ell = -1$ [so that $U(1)_Y$ in Table 7.2 is really $U(1)_{B-L}$]. Thus, for the LRS group, the imposition of the conditions of freedom from triangular and global $SU(2)$ anomalies plus a Higgs-generated Majorana mass for the neutrino are sufficient to fix the quantum numbers of the minimal number of Weyl fermion representations and to achieve ECQ [7.12]. The new feature — compared to the G_{3211} case — is that the mixed anomaly-free condition is guaranteed by the LRS property; that is, the mixed anomaly-free condition for the LRS group: $3Y_q + 3Y_{\bar{q}} + Y_\ell + Y_{\bar{\ell}} = 0$ is an immediate consequence of the triangular anomaly-free conditions, which, in turn, reflect the LRS of the G_{3221} group.

The next interesting step — after G_{3221} — in the "enlargement" chain of groups leading to $SO(10)$ GUT (see §8.4a) — is the Pati-Salam group $G_{422}[\equiv SU(4)_C \times SU(2)_L \times SU(2)_R]$ with its sixteen (minimal) Weyl states $[(\mathbf{4}, \mathbf{2}, \mathbf{1})$ plus $(\bar{\mathbf{4}}, \mathbf{1}, \mathbf{2})]$ under G_{422}. It is obvious from the manifest LRS of G_{422} that the mixed anomaly-free condition is redundant. Thus, as one proceeds beyond G_{3211}, the manifest LRS of the Weyl states leads to a redundant mixed anomaly-free condition, as was the case for the G_{3221} group. Hence the LRS groups larger than G_{3211} [namely the intermediate steps in the chain to $SO(10)$ GUT: G_{3221} and G_{422} — see §8.4a] are not explicitly constrained by the mixed anomaly-free condition; instead, freedom from the mixed anomaly is built into these larger groups.

A further redundancy develops — the redundancy of the global $SU(2)$ anomaly-free condition — when one enlarges the semi-simple G_{422} group even further — to one of the (simple) GUT groups. [It is not necessary that the GUT group be $SO(10)$ — the following arguments about the redundancy of both the mixed and global $SU(2)$ anomaly-free conditions hold for any GUT group.] In the case of any one of the GUT groups, it can be shown that triangular anomaly cancellation is sufficient to guarantee the absence of the global $SU(2)$ anomaly while the "simple" character of the GUT group suffices to ensure freedom from the mixed anomaly. It will be explained in

§8.3 why the simple groups $SU(N)$ $(N \geq 5)$, $SO(4k+2)$ $(k \geq 2)$ and E_6 are the only candidate GUT groups. Further, it will be proved in §10.4b that the absence of the global $SU(2)$ anomaly is guaranteed by the triangular anomaly-free condition as long as the $SU(2)$ group is embedded in a simple group G with the property $\Pi_4(G) = 0$ (Π_4 is the fourth homotopy group). Since all candidate GUT groups satisfy the conditions of this theorem, the global $SU(2)$ anomaly-free condition is redundant in determining the Weyl fermion representations of a GUT group. The mixed anomaly-free condition is also redundant for a GUT group, albeit for another reason. The point is that freedom from the mixed anomaly requires the condition Tr $Y = 0$ to be satisfied for each fermion generation, and this follows automatically for a GUT group because the hypercharge Y is a generator of the group and must be traceless. These are the reasons why it has been unnecessary — in studying the three most interesting GUT groups [$SU(5)$, $SO(10)$, and E_6 — see Chapter 8] to invoke the global $SU(2)$ and mixed gauge-gravitational anomaly-free conditions — in addition to the triangular anomaly-free condition — because the triangular anomaly-free condition has always been satisfied in the choice of representation(s) of the GUT groups that have been studied (see Chapter 8), e.g. the triangular anomaly-free pair of representations **5̄ + 10** of SU(5), the triangular anomaly-free spinorial representation **16** of $SO(10)$, and the triangular anomaly-free fundamental representation **27** of E_6.

A final remark should be made concerning the status of electric charge quantization in the case of a simple GUT group, namely that ECQ is automatic for a GUT group despite the loss of the independent constraint imposed by the mixed anomaly-free condition. This follows from the fact that the generators of a (non-Abelian) GUT group are non-commutative and consequently possess discrete eigenvalues whereas a free-standing $U(1)$ group possesses a continuous set of eigenvalues. It is therefore only necessary to express the electric charge operator Q — for a GUT group — as a linear combination of appropriate traceless generators (see §8.3) to secure ECQ; a specific Higgs SSB mechanism need not be spelled out although it is always understood that electric charge refers to the $U(1)$ charge of the unbroken $SU(3)_C \times U(1)_{EM}$ group. Indeed, this has been the conventional wisdom concerning ECQ, and one of the key arguments for grand unification. The fact is that ECQ can be established at the electroweak scale — within the framework of the non-semi-simple standard group — by fully exploiting the three anomaly-free conditions in four dimensions and the Higgs mechanism.

Thus far, our discussion in this section has focused on the constraints imposed by the anomaly-free conditions on the quantum numbers of one generation of quarks and leptons. No constraint has emerged on the number of "copies" of a quark-lepton family with the same set of quantum numbers (i.e. generations) and neither has there been any explanation why the observed number of generations at the electroweak scale is precisely three. Nor is more light shed — from the anomalies standpoint — on the "fermion generation problem" when we adjoin a "family" gauge group like $SU(2)_f$ or $SU(3)_f$ to the standard group G_{321} or one of its enlargements. We have already seen that, in essence, adjoining $SU(2)_f$ to the standard group (to give G_{3221}) and applying the chiral gauge anomaly-free conditions in a straight-forward fashion, simply gives rise to the one-family LRS theory. Nor is greater success achieved by adjoining the family gauge group $SU(3)_f$ to the standard group G_{321}; in this case, we get the enlarged group $G_{321} \times SU(3)_f$. When one applies the anomaly-free conditions to $G_{321} \times SU(3)_f$ and invokes "minimality" — as we did with the standard group — it is easily shown that the Weyl representations are those listed in Table 7.3 [7.14]. It is seen that the chiral fermion representations in Table 7.3 define the "chiral color" group $G_{3321}[\equiv SU(3)_{CL} \times SU(3)_{CR} \times SU(2)_L \times U(1)_Y]$, so that adjoining the $SU(3)$ family group onto the standard group, converts the resulting group into the one-family "chiral color" model (with its additional exotic fermion representations) [7.15]. Unfortunately, we have no explanation of the fermion

Table 7.3: The minimal Weyl fermion quantum numbers in the "chiral-color" model.

Particles	$SU(3)_{CL}$ ×	$SU(3)_{CR}$ ×	$SU(2)_L$ ×	$U(1)_Y$
$\binom{u}{d}_L$	3	1	2	$\frac{1}{3}$
$\bar{u}_L$	1	$\bar{3}$	1	$-\frac{4}{3}$
$\bar{d}_L$	1	$\bar{3}$	1	$\frac{2}{3}$
$\binom{\nu}{e}_L$	1	1	2	-1
$\bar{e}_L$	1	1	1	2
Q_{1L}	$\bar{3}$	1	1	Q
$\bar{Q}_{1L}$	1	3	1	$-Q$
Q_{2L}	$\bar{3}$	1	1	$-Q - \frac{2}{3}$
$\bar{Q}_{2L}$	1	3	1	$Q + \frac{2}{3}$

generation problem on the basis of the enlarged "chiral color" group. It is possible to start with the standard group and adjoin an unbroken family gauge group, say $SU(2)_f$ [7.16], assume that the $SU(2)_L$ gauge group is unbroken at the $SU(2)_f$ "confining" scale [7.17], apply the three chiral gauge anomaly-free constraints on the Weyl representations and, finally, invoke the condition of asymptotic freedom for the unbroken gauge groups. One then finds that, with "minimality", the anomaly-free constraints limit the number of Weyl representations of the enlarged group to 23 (an increase of 8 over the standard group) and an upper limit (3) on the number of possible generations [7.18]. This is an amusing result but the price paid is eight "exotic" fermions (sarks!) and several ad hoc assumptions.

From a conceptual point of view, it is not clear whether the fermion generation problem can be solved at the electroweak scale but, in view of the success of the three known chiral gauge anomaly-free conditions in fixing the representations and hypercharges of the 15 Weyl fermion states of the standard group, one might still entertain the hope that phenomena at a much larger mass scale (i.e. the Planck scale) need not be invoked and that some imaginative — albeit more modest — enlargement of the standard group can determine the observed number of fermion generations. Or, it is conceivable that if "weak CP violation" is fully explained within the framework of the U_{CKM} quark mixing matrix for three quark-lepton generations (see §6.3e), a clue will be found by probing nature's choice of the minimal number of generations that can sustain "weak CP violation". Or, it is possible that an explanation of the quite unexpected size of the t quark mass will explain why the generation concept itself breaks down beyond three generations. Of course, more speculative approaches to the fermion generation problem have been invented: such as the composite preon models of quarks and leptons and the very ambitious superstring theory (TOE) mentioned in §1.4c. In view of the lack of progress thus far on the fermion generation problem, we postpone further discussion of this vital problem until Chapter 9, where we do give a status report on the higher symmetry and preon model approaches to the "fermion generation problem".

References

[7.1] R.J. Jackiw, in *Proc. of the Symposium on Anomalies, Geometry Topology*, ed. W.A. Bardeen and A.R. White, World Scientific, Singapore (1985).

[7.2] W.A. Bardeen, *Phys. Rev.* **184** (1969) 1848.

[7.3] L. Alvarez-Gaumé and E. Witten, *Nucl. Phys.* **B234** (1983) 269.

[7.4] E. Witten, *Nucl. Phys.* **B176** (1976) 269.

[7.5] D.J. Gross, S.B. Treiman and F. Wilczek, *Phys. Rev.* **D19** (1979) 2188.

[7.6] T.P. Cheng and L.F. Li, *Phys. Rev. Lett.* **62** (1989) 1441.

[7.7] Riazuddin and R.E. Marshak, Virginia Tech preprint 91/1.

[7.8] Riazuddin and Fayazuddin, *Phys. Rev.* **D41** (1989) 3517; R.D. Carlitz and A.V. Manohar, preprint UCSD/PTH 90-32, Univ. of Calif. (San Diego) 1990; R.L. Jaffe and A. Manohar, *Nucl. Phys.* **B337** (1990) 509.

[7.9] M. Atiyah and I. Singer, *Ann. Math.* **87** (1968) 484.

[7.10] C.W. Misner, K.S. Thorne and J.A. Wheeler, *Gravitation*, W.H. Freeman, New York (1973).

[7.11] J.A. Minahan, P. Ramond and R.C. Warner, *Phys. Rev.* **D41** (1990) 716; C.Q. Geng and R.E. Marshak, *Phys. Rev.* **D41** (1990) 717.

[7.12] K.S. Babu and R.N. Mohapatra, *Phys. Rev. Lett.* **63** (1989) 938.

[7.13] H.M. Georgi, E.E. Jenkins and E.H. Simmons, *Phys. Rev. Lett.* **62** (1989) 2789.

[7.14] C.Q. Geng, *Phys. Rev.* **D39** (1989) 2402.

[7.15] P.H. Frampton and S.L. Glashow, *Phys. Lett.* **B190** (1987) 192; *ibid.*, *Phys. Rev. Lett.* **58** (1987) 2168.

[7.16] P.H. Frampton and Y.J. Ng, *Phys. Rev.* **D42** (1990) 3242.

[7.17] L.E. Abbott and E. Farhi, *Phys. Lett.* **101B** (1981) 69.

[7.18] P.H. Frampton, preprint IFP-390-UNC, Univ. of North Carolina Chapel Hill (1991).

Chapter 8

UNIFICATION OF STRONG AND
ELECTROWEAK INTERACTIONS

§8.1. Linkages between QCD and QFD

The standard model of quark and lepton interactions — as discussed in Chapters 5–7 — is defined by the product of two gauge groups: the unbroken non-Abelian vector-like $SU(3)_C$ group of the strong interaction among quarks (QCD) and the spontaneously broken non-semisimple gauge group $SU(2)_L \times U(1)_Y$ of the electroweak interaction among chiral quarks and leptons (QFD). The standard model gives a highly successful account — at least up to 100 GeV — of the properties of three of the four fundamental interactions in nature [strong, weak and electromagnetic — the gravitational interaction enters only peripherally in the Λ_{QFD} range through the mixed gauge-gravitational anomaly (see §7.4)] among the three generations of quark and lepton flavor doublets. At the standard group level, the sharp distinction between quarks and leptons is expressed through the very different representations of $SU(3)_C$ to which quarks and leptons belong. The **3** and **3̄** $SU(3)_C$ representations (see Table 1.5), describing the quarks and antiquarks of each generation, signal the strong, vector-like, and confining color interaction among quarks whereas the color singlet representations for the leptons imply the absence of the strong interaction among leptons. The unbroken non-Abelian character of $SU(3)_C$ gives rise to the crucial features of asymptotic freedom and confinement in QCD that are invoked to explain the phenomenology of the quark world and, *a fortiori*, the hadron world. The big stumbling block to the acceptance of a gauge theory of the strong interaction — its small finite range — was overcome by recognizing that the combination

461

of finite renormalization scale, asymptotic freedom and confinement leads to a finite QCD scale [$\Lambda_{QCD} \simeq 150$ MeV (see §5.2a) of the order of the pion mass] despite the fact that the gauge coupling constant g_s is dimensionless, the gluons are massless and the quarks can be taken as massless without changing the basic result.

While QCD ignores the lepton world, QFD treats the quark and lepton worlds on a more or less equal "chiral flavor" footing [ν_R is the exception — see below]. The quark and lepton chiral currents couple in a symmetrical fashion with the electroweak gauge fields and the same Higgs SSB mechanism gives rise to the large masses of the W and Z bosons that mediate the extremely short-range electroweak interaction among quarks and leptons. The Higgs mechanism responsible for the W and Z boson masses is also presumed to generate the finite Dirac masses of the six quarks and three charged leptons by means of Yukawa couplings of a single Higgs doublet to these nine fermions. This is where the exception to "full" chiral quark — chiral lepton — symmetry enters the picture; one simply postulates the absence of a righthanded neutrino in each of the three fermion generations, which ensures that the three (lefthanded) neutrinos possess vanishing Dirac mass. The absence of ν_R formally destroys "full" chiral quark-chiral lepton symmetry but, for most purposes [e.g. satisfying the three chiral gauge anomaly-free constraints in four dimensions (see §7.5)], this ν_R is hardly missed because its quantum numbers [$(1,1,0)$ under $SU(3)_C \times SU(2)_L \times U(1)_Y$ (see Table 1.5)] correspond to complete "inertness" in its strong and electroweak interactions. On the other hand, a finite Majorana mass (see §6.4b) for the observed neutrinos would strongly modify the concept of "full" quark-lepton symmetry and lead to substantially "new physics". This significant difference between Dirac and Majorana neutrinos of finite mass is reflected in the partial and grand unification groups, as we will see.

While QFD manifestly establishes connections between the quark and lepton worlds (even if ν_R is absent), QCD apparently applies only to quarks and it would appear that attempts to "unify" QCD and QFD are likely to fail. However, there are some identifiable "linkages" between QCD and QFD that are supportive of the unification idea, to which we draw attention before spelling out the usual arguments for "partial" and/or "grand" unification models. The first prescient observation of a possible "linkage" between the strong and weak interactions (QCD and QFD were still unborn) — we call this Linkage I — occurred during the early 1960s [3.14] when the algebra of vector and axial vector "strong" currents was proposed and these vector and axial vector currents were identified with the physical weak currents in the V-A theory. The vector and axial vector "strong" currents soon were translated into the chiral quark currents associated with global $SU(3)_L \times SU(3)_R$ and a rationale was provided for the many successful calculations of weak (and electromagnetic) processes based on current algebra, PCAC and soft pion techniques. In modern parlance, Linkage I stems from the fact that the (u, d, s) quarks are light compared to Λ_{QCD} so that the massless approximation for these three quarks in the QCD Lagrangian (see §5.1) can be exploited to generate the global chiral quark flavor symmetry $SU(3)_L \times SU(3)_R$ of QCD. Clearly, <u>Linkage I</u> was the first crucial step that began to bridge the gap in a well-defined way between the strong and weak interactions.

A second linkage between QCD and QFD follows from the quasi-Higgs nature of the $q\bar{q}$ condensates produced by the confining color interaction in QCD. These $q\bar{q}$ condensates do not lead to the SSB of the color gauge group $SU(3)_C$ [because the singlet color state of the $q\bar{q}$ condensate is the "most attractive channel" — see §5.1] but it does give rise to the SSB of $SU(3)_L \times SU(3)_R$ down to the global vector quark flavor group $SU(3)_F$, accompanied by an N-G octet of pseudoscalar bosons ("pions"). It is tempting to identify this octet of "pions" with the physical octet of pseudoscalar bosons $(\pi^{\pm}, \pi^0; K^{\pm}, K^0, \bar{K}^0; \eta)$ except that the masses of the physical pseudoscalar octet are not identically zero (no more than the u, d, s quarks are massless). At this point, QFD comes to the rescue and attributes the small but finite masses of the three light quarks to the SSB of the electroweak group $SU(2)_L \times U(1)_Y$ so that the three light quark masses depend on the SSB scale Λ_{QFD} of the electroweak group (see §6.2). These finite quark masses then convert the octet of massless (N-G) "pions" into finite mass (pseudo-N-G) "pions", whose masses depend, apart from the Yukawa coupling constants h_q, on the "geometric mean" of Λ_{QCD} and Λ_{QFD}, thus (see §3.5c):

$$m(\text{"pions"}) \simeq h_q \sqrt{\Lambda_{QFD} \, \Lambda_{QCD}} \tag{8.1}$$

Equation (8.1) constitutes Linkage II between QCD and QFD and is quite remarkable since $\Lambda_{QFD} \gtrsim 10^3 \, \Lambda_{QCD}$ and the mass of the lightest hadron (i.e. the strongly interacting "pion") is related not only to Λ_{QCD} but also to the Higgs SSB scale Λ_{QFD}.

Linkages I and II may be thought of as QCD-induced linkages between QCD and QFD. However, the next three linkages (Linkages III–V) may be termed QFD-induced because they all flow from the three known chiral

gauge anomaly-free constraints in four dimensions: the triangular anomaly
(Linkage III), the global $SU(2)$ anomaly (Linkage IV) and the mixed gauge-
gravitational anomaly (Linkage V) (see §7.2–§7.4). Linkage III tells us that
the cancellation of the perturbative triangular chiral gauge anomalies of the
electroweak group is made possible by the presence of both quarks and leptons — with very different representations under the non-chiral $SU(3)$ color
group but very similar representations under the chiral $SU(2)_L \times U(1)_Y$ electroweak group. The quantum numbers of the quarks and leptons (in each
generation) under the $SU(2)_L \times U(1)_Y$ electroweak group are not identical
(i.e. there is not "full" quark-lepton symmetry — see §8.4); however, the
differences are such — due to the "inert" character of the ν_R quantum numbers — that the cancellation of the triangular chiral gauge anomalies in the
standard model amounts to "full" quark-lepton symmetry. A second contributing factor to Linkage III is that the different color weights for quarks
and leptons in each generation (3 for quark, 1 for lepton — see Table 1.5)
are precisely the factors needed to balance off the fractional versus integral
(weak) hypercharges of quarks and leptons (respectively) in order to achieve
cancellation of the triangular chiral gauge anomalies (see §7.2c).

Equally curious, and hopefully as significant, is the way Linkage IV is accomplished: through the apparently fortuitous even sum (4) of the separate
odd numbers (3 and 1 respectively) of $SU(2)$ lefthanded doublet representations of quarks and leptons. A total even number of lefthanded $SU(2)$
doublet Weyl fermion representations is required to achieve cancellation of
the global $SU(2)$ chiral gauge anomaly of the standard electroweak group. It
is gratifying that the odd fundamental (3) quark representation of $SU(3)_C$
and the odd singlet lepton representation under $SU(3)_C$, conspire to cancel the global SU(2) anomaly; this saves the self-consistency of electroweak
theory. Linkage IV is clearly of a different character from Linkage III — as
might be expected — since it relates to the non-perturbative, topologically
non-trivial global $SU(2)$ anomaly while Linkage III relates to the perturbative, topologically trivial triangular anomaly. As in the case of Linkage III,
Linkage IV appears be an argument for a deeper connection between the
quark and lepton worlds.

Finally, Linkage V — which imposes an important constraint on the standard electroweak group and follows from the requirement of freedom from
the perturbative mixed gauge-gravitational anomaly — seems to be telling
us, in some subtle way, that both quarks and leptons are needed to reconcile (Einsteinian) general covariance with chirality conservation and gauge

invariance, even at the electroweak level. We have seen in §7.4 that Linkage V derives from the condition Tr $Y = 0$ for each generation and that the mixed gauge-gravitational anomaly-free condition, in combination with the other two chiral gauge anomaly-free conditions in four dimensions, mandates hypercharge quantization (see §7.5) and, after the Higgs breaking of $SU(2)_L \times U(1)_Y$, electric charge quantization as well (see §7.6). Linkage V seems to be going further than Linkages III and IV in not only indicating a deeper connection between the quark and lepton worlds, but also in suggesting the compatibility of a future TOE (including gravity) with some form of quark-lepton unification.

It is evident that all five linkages enumerated above between the quark and lepton sectors and between the strong and electroweak interactions result from the entry of chiral fermions into the standard model and express the need to "go beyond the standard model". What is far from clear is whether going "beyond the standard model" is best implemented by following the route of "partial" and/or "grand" unification, or by ascribing a common origin (e.g. by means of preons) to quarks and leptons subject to a new confining gauge interaction, or by starting with non-point-like entities (e.g. "superstrings") in a higher number of space-time dimensions that compactify in a well-defined way to our four-dimensional world, or, finally, by probing more deeply into the basic foundations of the standard gauge theory itself (perhaps along topological lines) to discover some hidden symmetries that can provide answers to the unsolved problems of the standard model. Because of the paucity of experimental data in the multi-hundred GeV region, we would be justified to postpone further discussion of the "new" physics until the higher energy accelerators (just completed or under construction) begin to produce definitive departures from the standard model. However, theoretical speculations concerning the "new" physics have been so productive of new concepts in particle physics — some of which may even turn out to be correct and some of which may be useful in other branches of physics — that we propose to discuss some of these speculative expeditions "beyond the standard model".

The remaining sections of this chapter are devoted to the traditional (as of the 1980s!) unified theories ["partial unification" and "grand unification" — except for a section (§8.4c) on "non-traditional" $SO(10)$ GUT (i.e. $SO(10)$ GUT with D parity)]. In the following chapter (Chapter 9), we review the status of the composite preon approach to unification, with particular emphasis on the "fermion generation problem". We choose not to expand on the higher-dimensional (e.g. Kaluza-Klein and superstring) approaches to TOE beyond the comments offered within a historical context in §1.4c. The TOE field is changing too rapidly and is best left to the active practitioners to share their hopes and expectations with the rest of us. However, we do undertake a final expository chapter (Chapter 10) on the topological conservation laws associated with solitons, instantons and certain types of spontaneously broken gauge theories; in that chapter, we do draw illustrations of the power of topological conservation laws from as-yet-unobserved phenomena in particle physics (e.g. 't Hooft-Polyakov magnetic monopoles and sphalerons) and from observed topological effects in lower-dimensional condensed matter physics.

§8.2. Arguments for unification of QCD and QFD

In the previous section of this chapter, we searched for clues — from the standard gauge theory — to a closer interrelationship between the $SU(3)_C$ (QCD) and $SU(2)_L \times U(1)_Y$ (QFD) parts of the standard group and to a more intimate interdependence of the quark and lepton representations and of the strong and electroweak interactions. We found such clues, which we termed "linkages" between QCD and QFD and, at the end of the last section, we mentioned several possible directions in which these linkages could be built upon to formulate a more unified theory of the interactions among quarks and leptons. In this and the following four sections, we undertake to describe the traditional "unification" (partial and grand) program which, in many ways, is the most conservative of the possible extensions of the standard model since it makes no claim of incorporating gravity nor of accounting for the correct number of quark-lepton families that are observed.

The traditional unification program, broadly speaking, attempts to "go beyond the standard model" by enlarging the standard group and, through such enlargement, to achieve a maximum reduction in the number of arbitrary parameters associated with the standard group [8.1]. It is hoped that the enlarged group can predict the observed values of some of the arbitrary parameters in the standard model and also make new predictions that can be tested by experiment. We call the enlarged "unification" group a "partial unification theory" (PUT) if the enlarged group is not a simple group (i.e. it is semisimple or even non-semisimple) and a "grand unification theory" (GUT) if the enlarged group is simple. One expects the GUT group to be more satisfactory since: (1) it should lead to the maximum reduction in the number of arbitrary parameters associated with the standard group; (2) the large unification mass (scale) required by a simple (GUT) group brings the theory within range of the Planck energy ($\sim 10^{19}$ GeV) and makes it more likely that some kind of "super-grand unified" theory (a TOE) — unifying the successful GUT theory with Einsteinian gravity — can be constructed; and (3) a simple grand unification group (i.e. a GUT group) may possess "hidden" symmetries (and/or "inner automorphisms") which do not apply to semi-simple or non-semi-simple partial subgroups of a GUT group [as in the case of $SO(10)$ — see §8.4c]. In view of these arguments, we focus in this section on the grand unification approach and give additional reasons for pursuing the GUT program, concluding with a summary list of properties that a viable GUT must possess. We then turn — in §8.3 — to a discussion of the most appealing (lowest rank) $SU(5)$ GUT group. The disappointing performance of $SU(5)$ GUT [e.g. the failure to observe the predicted proton decay and the discrepancy between the observed and predicted values of $\sin^2 \theta_W$] has been regarded by some as destructive of the entire unification program. However, we believe that the attractive features of the unification program justify a fairly detailed examination of the $SO(10)$ GUT option and, to a lesser extent, the E_6 grand unification option. Consequently, in §8.4, we consider the next higher rank $SO(10)$ GUT model [the rank of $SO(10)$ is 5, in contrast to rank 4 for $SU(5)$] and, along the way to $SO(10)$ [which contains a local $U(1)_{B-L}$ symmetry], we discuss other "$B-L$" groups such as the Pati-Salam group $SU(4)_C \times SU(2)_L \times SU(2)_R$ and the LRS group $SU(2)_L \times SU(2)_R \times U(1)_{B-L}$ (see §8.4a and §8.4b) and search for phenomenological tests of the intermediate mass scales associated with the "partial unification groups". The following section (§8.4c) is devoted to a brief review of the consequences of taking explicit account of an inner automorphism — so-called D parity — in $SO(10)$ GUT. Finally, we treat briefly (in §8.5) the "superstring-inspired" E_6 GUT model and point out its limitations within the context of the unification program. It should be reemphasized that all three GUT models [$SU(5)$, $SO(10)$ and E_6] are one-fermion-generation models and the three-fold replication of fermion generations must still be explained. The "fermion generation problem" will be discussed — within the framework of enlarged GUT groups — in the first section of Chapter 9 where it will be shown that further enlargement of $SU(5)$, $SO(10)$ or E_6 GUT does not provide a solution of the "fermion generation problem". It is this failure that gave impetus to the composite preon attempt to resolve the fermion generation problem, as is explained in the remainder of Chapter 9.

We now turn to the arguments for pursuing the GUT approach to unify quarks, leptons and the gauge groups defining their interactions. Let us start with "negative" arguments in favor of seeking more unification than has been achieved by the standard model. The fact is that, despite its great successes, the standard model is not particularly elegant with its approximately twenty free parameters: three gauge coupling constants $[g_s, g$ and g' or, equivalently, g_s, e and $\sin\theta_W$ — see §6.2 for notation], two Higgs parameters v and λ (v is determined by G_F but λ is unknown), nine Yukawa coupling constants h_f [determined by the empirical masses of the six quarks and three charged leptons of the three generations — the neutrinos are taken as massless in the standard model], four CKM parameters [three "Cabibbo" angles $(\theta_1, \theta_2, \theta_3)$ and one CP-violating phase angle δ] and, finally, one "strong CP"-violating θ-parameter, for a total of nineteen free parameters. To this "negative" argument for the pursuit of the GUT program, there was added — until recently — the further "negative" argument that an explanation of electric charge quantization could not be achieved within the framework of the standard model. However, we have seen in §7.6 that the imposition of all three chiral gauge anomaly-free conditions plus minimal use of the Higgs SSB mechanism do enable us to explain both hypercharge and electric charge quantization at the electroweak level. Consequently, there is one less "negative" argument — 19 instead of 20! — to "go beyond the standard model" since there is no need for a GUT group to explain ECQ.

More to the point, there are several "positive" arguments for seeking a unified gauge theory of the strong and electroweak interactions without gravity. There is the usual argument that the gravitational coupling constant is still much smaller than the strong and electroweak coupling constants at acceptable GUT scales. But, perhaps, the most striking "positive" argument for first pursuing unification of the strong and electroweak interactions without gravity is the realization that both the strong and electroweak interactions are mediated by vector (gauge) fields [giving rise to both positive and negative "charges" and, *a fortiori*, both repulsive and attractive forces] whereas the gravitational interaction is mediated by a (symmetric second-rank tensor) graviton field (giving rise only to an attractive interaction). Furthermore, the gauge symmetries in QCD and QFD relate to internal symmetries whereas the symmetries in Einsteinian gravity relate directly to space-time. This clear-cut differentiation between the strong and electroweak interactions, on the one hand, and the gravitational interaction, on the other, must somehow be bridged in TOE but, until a successful TOE emerges, it is worth reviewing in concrete terms the more modest GUT program which omits gravity.

Granting then that the strong and electroweak parts of the standard group can be unified into a GUT group without gravity — at scales substantially below the Planck scale — it is possible to exploit the observation that two of the three gauge groups of the standard group, i.e. $SU(3)_C$ and $SU(2)_L$, are non-Abelian and hence are asymptotically free whereas the third gauge group, $U(1)_Y$ [closely related to unbroken $U(1)_{EM}$], is Abelian and not asymptotically free. This tells us that the renormalized coupling constants for $SU(3)_C$ and $SU(2)_L$ decrease with increasing energy whereas the renormalized coupling constant for $U(1)_Y$ increases with increasing energy, thereby creating the possibility of a crossover ("unification") at sufficiently high energy. This "crossover" looks promising because, at low energies, the $SU(2)_L$ gauge coupling constant is smaller than that of $SU(3)_C$, and its rate of decrease (with increasing energy) is also smaller [because the dimension of $SU(2)$ is smaller than that of $SU(3)$ — see §5.2a]. More precisely, it is known from §5.2a that the "running" coupling constants for $U(1)_Y$, $SU(2)_L$ and $SU(3)_C$ groups are [see Eq. (5.24)]:

$$\alpha_i^{-1}(Q) = \alpha_i^{-1}(\mu) + 4\pi b_i \ln \frac{Q^2}{\mu^2} \qquad (8.2)$$

where $\alpha_i = g_i^2/4\pi$ $(i = 1, 2, 3)$ runs over the $U(1)_Y$, $SU(2)_L$ and $SU(3)_C$ groups respectively. The b_i's of Eq. (8.2) are the coefficients of the respective β functions — taken in one-loop approximation — of $U(1)_Y$, $SU(2)_L$ and $SU(3)_C$ (see §5.2a):

$$b_1 = -2n_f/24\pi^2; \quad b_2 = (22 - 2n_f)/12\pi; \quad b_3 = (33 - 2n_f)/12\pi \qquad (8.3)$$

where n_f is the number of fermion flavors. We have plotted in Fig. 8.1 the values of the three properly normalized "running" coupling constants as a function of $\ln(Q/\text{GeV})$, where $\alpha_1 = 5/3 \frac{g'^2}{4\pi}$, $\alpha_2 = \frac{g^2}{4\pi}$ and $\alpha_3 = \frac{g'^2}{4\pi}$ retain their usual definitions; $SU(5)$ GUT numbers have been used in drawing the lines of Fig. 8.1 and the failure of the three lines to meet at a point must be taken seriously. At this stage, we do not commit ourselves with regard to the "mismatch" of $SU(5)$ GUT (see §8.3) [8.2] so that we can make some qualitative points.

Figure 8.1 shows clearly that the rate of approach to asymptotic freedom, i.e. the rate of decrease with Q of the non-Abelian gauge coupling

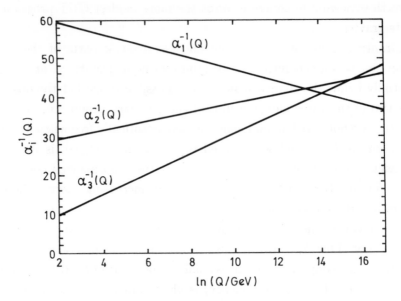

Fig. 8.1. Schematic view of the "running" gauge coupling constants in QCD and QFD and their possible unification.

constant α_3 — compared to α_2 — is greater, the larger the dimension of the non-Abelian group [$SU(3)_C$ compared to $SU(2)_L$]. When this is combined with the fact that the coupling constant α_1 increases with Q for the Abelian group, $U(1)_Y$, the experimental ordering of the coupling constants at low energies ($\alpha_1 < \alpha_2 < \alpha_3$) is quite compatible with the objective of a GUT group, which is to achieve unification (with a single coupling constant α) at some high energy, M_U. Further, we can say that because the "running" of the coupling constants with Q is logarithmic and thus very slow (which is a reflection of the renormalizability of the theory), the scale of unification will indeed be very high, $M_U \gtrsim 10^{14}$ GeV. This feature of the standard model gives "positive support" to the grand unification idea, namely that a larger simple ("grand unified") group, with one gauge coupling constant, could exist that breaks down spontaneously (via a Higgs-type mechanism) directly or, through a series of stages, to the standard group $SU(3)_C \times SU(2)_L \times U(1)_Y$ ($\equiv G_{321}$) which, in turn, breaks down to the (unbroken) group $SU(3)_C \times U(1)_{EM}$ ($\equiv G_{31}$). Thus, a GUT group certainly reduces the number of gauge coupling constants, and, moreover, requires its (irreducible) representations to contain both quark and lepton fields (in contrast to the standard group), thereby opening up the possibility of transitions between the quark and lepton worlds (e.g. proton decay).

We next mention two key properties that a viable grand unification theory ("candidate" GUT group) of the strong and electroweak interactions must possess; these are: (1) if the GUT group is denoted by G, G must contain G_{321} as a subgroup and must therefore be a chiral group [since the $G_{21} \equiv SU(2)_L \times U(1)_Y$ part of G_{321} is a chiral group]. This means that G must contain a complex representation or G_{321} can never be a subgroup of G. As pointed out in §4.3, this requirement restricts the choice of G to $SU(N)$ $(N \geq 3)$, $SO(4k+2)$ $(k \geq 2)$ and E_6; and (2) the representations of G must be chosen in such a way that there are no triangular chiral gauge anomalies; we have noted in §7.7 (see also §10.4b) that the absence of the triangular anomaly automatically guarantees the absence of the global $SU(2)$ anomaly in a GUT group and that, furthermore, the mixed gauge-gravitational anomaly is always absent in a GUT group because the hypercharge is one of its generators. Thus, the triangular anomaly-free constraints must be satisfied by the representation(s) of the GUT group to ensure a renormalizable and self-consistent theory. Since the fundamental representation of an $SU(N)(N \geq 3)$ group is not triangular anomaly-free, at least two irreducible representations must be used for an $SU(N)(N \geq 3)$ GUT model in order to achieve triangular anomaly cancellation. On the other hand, the spinoral representations of $SO(4k+2)$ $(k \geq 2)$, and the fundamental representation of E_6 are themselves triangular anomaly-free and one can construct $SO(4k+2)(k \geq 2)$ and E_6 GUT theories with single complex representations.

Apart from the two essential requirements just mentioned for a "candidate" GUT group, several other considerations must be taken into account: (1) as is well-known (see §7.5), the standard group — with only massless (lefthanded) neutrinos — possesses 15 Weyl fermion states per generation, for a total of 45 for the three generations. If one adds a right-handed neutrino — as is suggested by "full" quark-lepton symmetry (see §8.4 below) — there are sixteen Weyl fermion states per generation (or 48 in three generations) that must be fitted into the representations of the GUT group G. If one does not try to unify all three generations within the same GUT group, one can obviously work with lower-dimensional GUT groups and lower-dimensional representations. Actually, most GUT theories are "one-generation" models (see §9.1); (2) since G must break down spontaneously to G_{321}, a Higgs mechanism or some dynamical breaking scheme must be invoked in searching for an acceptable GUT model. Thus far, no one has fully succeeded in giving a dynamical explanation of the SSB of the electroweak group [e.g. in terms of fermion condensates (see §6.4c)] so that the standard model is compelled to settle for an "elementary" Higgs mechanism. [In many GUT models, the "elementary" Higgs representations are chosen so as to simulate the quantum numbers of fermion condensates formed out of the basic representations of the model.] The same approach has been followed in the grand unification program and this has introduced considerable arbitrariness in the choice of the Higgs representations of G since, in general, more than one SSB is involved (other than the SSB of G_{321} to G_{31}). In deciding the SSB pattern induced by the choice of "elementary" Higgs representations, it is plausible to use the so-called "Michel conjecture" [8.3], which states that a renormalizable Higgs potential can only develop vacuum expectation values (VEV's) if it breaks down the initial group to its maximal "little" subgroup and, finally; (3) the minimum rank, i.e. the number of commuting observables [or, equivalently, the number of generators that can be simultaneously diagonalized] of the GUT group must be at least 4, since it must at least equal the rank of the standard group G_{321}. We have previously remarked that the "candidate" GUT groups are comprised in the class of $SU(N)$ ($N \geq 3$) special unitary groups, the class of $SO(4k+2)$ ($k \geq 2$) orthogonal groups and the E_6 exceptional group. Since the rank of an $SU(N)$ group is $(N-1)$, $SU(5)$ is the minimum rank GUT group of this class. Correspondingly, $SO(10)$ enjoys the minimum rank of the orthogonal class of groups — namely 5 — and E_6 possesses rank 6 (as indicated). Indeed, $SU(5)$, $SO(10)$ and E_6 are the three GUT groups that have been seriously studied for their phenomenological implications and we now discuss each of them in turn.

§8.3. $SU(5)$ GUT and proton decay

We describe what is called "minimal" $SU(5)$ GUT in some detail, even though it is in trouble with experiment, because it is the lowest rank GUT model (with the smallest number of fermions and Higgs bosons) and provides a good reference point for non-minimal $SU(5)$ GUT and the higher rank GUT theories. As a matter of notation, we express the Weyl fermion content of each GUT model purely in terms of lefthanded representations; this has not always been done in our discussion of the standard model but will be followed without exception for the GUT models. Thus, instead of writing f_R (where f_R is a righthanded Weyl fermion in a certain representation), we write f_L^c (where f_L^c is the lefthanded charge conjugate of the Weyl fermion). This means that the fifteen Weyl fermion states of the first generation have the following quantum numbers under the standard group $SU(3)_C \times SU(2)_L \times U(1)_Y$:

Table 8.1: Left-handed fermion representations under $G_{321} = (SU(3)_C \times SU(2)_L) \times U(1)_Y$ for each generation.

$(u, d)_L$	(u_L^c)	(d_L^c)	$(\nu, e)_L$	(e_L^c)
$(3, 2, \frac{1}{3})$	$(\bar{3}, 1, -\frac{4}{3})$	$(\bar{3}, 1, \frac{2}{3})$	$(1, 2, -1)$	$(1, 1, 2)$

We have pointed out that two irreducible representations of $SU(5)$ are needed to satisfy the triangular anomaly-free condition and it is convenient to use the fundamental $\bar{5}$ representation and the two-index anti-symmetric representation 10 of $SU(5)$. The choice of $\bar{5} + 10$ yields the decompositions under G_{321} shown in Eq. (8.4):

$$\bar{5} = (\bar{3}, 1, \frac{2}{3}) + (1, 2, -1); \qquad 10 = (3, 2, \frac{1}{3}) + (\bar{3}, 1, -\frac{4}{3}) + (1, 1, 2) \quad (8.4)$$

in complete agreement with the quantum numbers of Table 8.1 needed to define one generation of quarks and leptons in the standard model. It is useful at this point to exhibit the rank 4 character of $SU(5)$ GUT and the physical meaning of the relevant generators. We do not write out all 24 generators of $SU(5)$ but only record the four diagonalized generators that establish rank 4 for $SU(5)$. If we recall the notation λ_i $(i = 1, 2, \ldots 8)$ for the eight generators of the $SU(3)$ group (see §3.2c), and denote the 24 generators of $SU(5)$ by Λ_i $(i = 1, \ldots 24)$, the four diagonal generators are:

$$\Lambda_3 = \begin{pmatrix} & & 0 & 0 \\ \lambda_3 & & 0 & 0 \\ & & 0 & 0 \\ 0 & 0 & 0 & 0 & 0 \\ 0 & 0 & 0 & 0 & 0 \end{pmatrix}, \quad \Lambda_8 = \begin{pmatrix} & & 0 & 0 \\ \lambda_8 & & 0 & 0 \\ & & 0 & 0 \\ 0 & 0 & 0 & 0 & 0 \\ 0 & 0 & 0 & 0 & 0 \end{pmatrix} \quad (8.5)$$

$$\Lambda_{23} = \begin{pmatrix} & & 0 & 0 \\ 0 & & 0 & 0 \\ & & 0 & 0 \\ 0 & 0 & 0 & \\ 0 & 0 & 0 & \tau_3 \end{pmatrix}, \quad \Lambda_{24} = \frac{2}{\sqrt{15}} \begin{pmatrix} 1 & 0 & 0 & 0 & 0 \\ 0 & 1 & 0 & 0 & 0 \\ 0 & 0 & 1 & 0 & 0 \\ 0 & 0 & 0 & -\frac{3}{2} & 0 \\ 0 & 0 & 0 & 0 & \frac{3}{2} \end{pmatrix}$$

From Eq. (8.5), it is seen that the first two generators refer to $SU(3)_C$ [i.e. break Λ_3 is the third component of the "color spin" and Λ_8 is the "color hypercharge" when properly normalized] and the last two refer to $SU(2)_L$ and $U(1)_Y$ respectively [i.e. Λ_{23} represents the third component of the weak isospin I_{3L} and Λ_{24} the weak hypercharge Y when properly normalized].

Since the $SU(5)$ representations mix the quark and lepton states, it is interesting at this point to raise the question as to whether electric charge quantization (ECQ) is automatically predicted by the non-Abelian $SU(5)$ group; the answer is in the affirmative if it is recalled that the electric charge Q ($Q = I_{3L} + Y/2$) can be written in terms of Λ_{23} and Λ_{24}, as $Q = \frac{\Lambda_{23}}{2} - \sqrt{\frac{5}{2}}\Lambda_{24}$; for the representation $\bar{5}$ of SU(5), this gives:

$$Q(\bar{5}) = \begin{bmatrix} \frac{1}{3} & & & & \\ & \frac{1}{3} & & & \\ & & \frac{1}{3} & & \\ & & & -1 & \\ & & & & 0 \end{bmatrix} \tag{8.6}$$

It should be noted that expressing Q in terms of Λ_{23} and Λ_{24} and, *a fortiori*, obtaining the expression (8.6) for $Q(\bar{5})$, is possible because it has been implicitly assumed that $SU(5)$ breaks down to G_{321} which, in turn, breaks down to G_{31}. With this understanding, it is easy to derive ECQ from Eq. (8.6); one merely argues that, as a generator of $SU(5)$, Tr $Q(\bar{5}) = 0$ and hence:

$$3Q_{d^c} + Q_{e^-} = 0 \tag{8.7}$$

where the factor 3 on the L.H.S. of Eq. (8.7) is the color weight of $SU(3)_C$. Equation (8.7) reestablishes the quantization relationship between the electric charges of quarks and leptons (i.e. ECQ) that was derived in the standard model with the help of the chiral gauge anomaly-free conditions (see §7.6). In essence, when the chiral gauge anomaly-free conditions are used in conjunction with the non-semi-simple electroweak group, it is possible to derive ECQ without appealing to the "quantization property" of a non-Abelian group [3.26].

Before considering the Higgs mechanism that is responsible for the SSB of $SU(5)$ down to G_{321}, it is instructive to demonstrate that the combination of $\bar{5} + 10$ as the two $SU(5)$ representations of the (lefthanded) Weyl fermion states of one generation respects the requirement of freedom from triangular anomalies; we know that each generation is triangular anomaly-free with respect to G_{321} but the freedom from triangular chiral gauge anomalies must also hold for the fermionic couplings to all the remaining $SU(5)$ gauge bosons (of which there are 12 — see below). The proof of this statement follows from Eq. (7.51) if one exploits the fact that $D(R)$ [the anomaly coefficient due to a given fermion representation R] is independent of the generators and is normalized to unity for the fundamental representation. Hence one can choose a simple generator of $SU(5)$ such as electric charge Q to prove the validity of the triangular anomaly-free constraint; thus, let $T^a = T^b = T^c =$

Q [in Eq. (7.51)] and one finds immediately:

$$\frac{D(\bar{5})}{D(10)} = \frac{\text{Tr } Q^3(\bar{5})}{\text{Tr } Q^3(10)} = \frac{3(\frac{1}{3})^3 + (-1)^3 + 0^3}{3(-\frac{2}{3})^3 + 3(\frac{2}{3})^3 + 3(-\frac{1}{3})^3 + 1^3} = -1 \qquad (8.8)$$

so that:

$$D(\bar{5}) + D(10) = 0 \qquad (8.9)$$

It is clear from Eqs. (8.8) and (8.9) that, while the triangular anomaly coefficient $D(R)$ is separately non-vanishing for the $\bar{5}$ and 10 irreducible representations of $SU(5)$, the combination $\bar{5}$ and 10 ensures triangular anomaly cancellation. One expects from the "even-odd" rule (see §10.4b) that, while the number of Weyl fermion (weak isospin) doublets is separately odd for $\bar{5}$ and 10 $SU(5)$, the total number of Weyl fermion doublets is even for the combination $\bar{5}$ and 10 (as required by the "even-odd" rule); this is confirmed when one examines Eq. (8.4) where the $\bar{5}$ representation contains one weak fermion doublet and the 10 representation three Weyl fermion doublets, for a total of four — precisely the number of Weyl fermion doublets found for the standard model (see Table 8.1). On the other hand, the mixed gauge-gravitational anomaly-free constraint should be obeyed separately for each irreducible representation of $SU(5)$ since Tr Y vanishes separately for the $\bar{5}$ and 10 representations, which is also seen to be true in Eq. (8.4). These simple observations should help clarify the "redundancy" discussion in §7.7 *vis-à-vis* the three chiral gauge anomalies in four dimensions.

We can now turn to the SSB of $SU(5)$ GUT down to G_{321}; we assume "elementary" Higgs fields since we are only interested in the group-theoretic properties of SSB. From group theory [8.4], we know that the rank 4 $SU(5)$ group can only break down to two maximal (rank 4) subgroups containing $SU(n)$: $SU(4) \times U(1)$ and $SU(3) \times SU(2) \times U(1)$. The first such maximal subgroup is of no interest because it can not break down further to G_{321} and so the only viable path is the direct SSB of $SU(5)$ to the standard group G_{321}. The lowest-dimension Higgs representation that can serve this purpose (and satisfy the "Michel conjecture") is the (adjoint) 24 representation Φ of $SU(5)$, whose decomposition in terms of the irreducible representations of G_{321}, is:

$$\mathbf{24} = (\mathbf{8},\mathbf{1},0) + (\mathbf{1},\mathbf{3},0) + (\mathbf{1},\mathbf{1},0) + (\mathbf{3},\mathbf{2},-\frac{5}{3}) + (\bar{\mathbf{3}},\mathbf{2},\frac{5}{3}) \qquad (8.10)$$

The fact that one term in the decomposition of 24 is $(\mathbf{1},\mathbf{1},0)$ implies that G_{321} is a maximal *little* group of $SU(5)$, thereby satisfying the "Michel conjecture". The choice of VEV of the Higgs scalar Φ that gives $SU(5) \rightarrow$

G_{321} is:

$$< \Phi > = V \begin{pmatrix} 1 & & & & \\ & 1 & & & \\ & & 1 & & \\ & & & -\frac{3}{2} & \\ & & & & -\frac{3}{2} \end{pmatrix} \tag{8.11}$$

where V is the SSB scale from $SU(5)$ to G_{321}.

Thus far, we have considered the SSB of $SU(5)$ to G_{321}; but we must not forget that there is a final step in the SSB path of $SU(5)$, namely: $G_{321} \rightarrow G_{31}$ which is accomplished by the fundamental **5** representation of $SU(5)$. From Eq. (8.4), it is evident that it is the $(\mathbf{1}, \mathbf{2}, 1)$ decomposite which specifies the quantum numbers of the Higgs **5** representation that breaks the $SU(2)_L \times U(1)_Y$ group down to $U(1)_{EM}$ [but leaves unbroken the $SU(3)_C$ group]. If we denote the **5** representation of $SU(5)$ by ϕ, we have:

$$< \phi > = \begin{pmatrix} 0 \\ 0 \\ 0 \\ 0 \\ \frac{v}{\sqrt{2}} \end{pmatrix} \tag{8.12}$$

where v is the SSB scale from G_{321} to G_{31}. The complete SSB pattern for the $SU(5)$ GUT model is consequently (in an obvious notation) [8.5]:

$$SU(5) \xrightarrow[M_U]{24} G_{321} \xrightarrow[M_W]{5} G_{31} \tag{8.13}$$

where we have indicated the Higgs representations above the arrows, and the gauge boson mass scales below the arrows.

With the fermion and Higgs representations chosen, we can draw out the physical consequences of the $SU(5)$ GUT model. First, we note that from Eqs. (8.2) and (8.3), it is straightforward to derive expressions for M_U and $\sin^2 \theta_W$, namely [8.5]:

$$\ln(M_U/M_W) = (\pi/11)[\frac{1}{\alpha_e(M_W)} - \frac{8}{3\alpha_s(M_W)}] \tag{8.14}$$

$$\sin^2 \theta_W = \frac{3}{8} - \frac{55}{24\pi}\alpha_e(M_W) \ln(M_U/M_W) \tag{8.15}$$

These equations yield [8.6]:

$$M_U = 2.0^{+2.1}_{-1.0} \times 10^{14}\text{GeV}; \quad \sin^2 \theta_W = 0.214^{+0.003}_{-0.004} \tag{8.16}$$

where $\alpha_e(M_W) = \frac{1}{125}$, $\alpha_s(M_W) \simeq 0.12$ and $\Lambda_{\overline{MS}} = 150$ MeV have been used (see §5.2a). As expected from the logarithmic energy dependence of the "running" coupling constants, the value of M_U is many orders of magnitude greater than M_W. More importantly, the value of $\sin^2 \theta_W$ predicted by $SU(5)$ GUT can be compared with the latest experimental value which is [6.12]:

$$\sin^2 \theta_W (M_W)_{\text{exp}} = 0.232 \pm 0.012 \pm 0.008 \qquad (8.17)$$

The value given in (8.17) is for $m_t = 45$ GeV and $M_H = 100$ GeV, with the discrepancy between theory and experiment increasing for larger m_t (which is already > 92 GeV). The radiative corrections have been taken into account for both the theoretical and experimental values. The disagreement between theory and experiment with regard to $\sin^2 \theta_W$ is reflective of the fact that the three lines do not meet at a single point in Fig. 8.1, as they should for $SU(5)$ GUT. Despite the recent LEP measurements of the relevant parameters in the "running" equations to draw Fig. 8.1 [8.2], questions may still be raised about the claimed accuracies in the value of $\sin^2 \theta_W$ and it behooves us to turn to another relatively clear-cut theoretical prediction of a measurable quantity, namely proton decay.

To make the comparison between theory and experiment for the proton decay lifetime, we must look at the gauge bosons contained in the adjoint representation of the $SU(5)$ group and identify their quantum numbers under the standard group G_{321}. From the SSB path of $SU(5)$ GUT shown in Eq. (8.13), we note that twelve of the gauge bosons must acquire mass at the grand unification scale, M_U, and three of the remaining twelve gauge bosons must acquire mass at the electroweak scale Λ_{QFD}. This can be seen more explicitly from the decomposition of the **24** representation of $SU(5)$ under G_{321} in Eq. (8.10) where, clearly, the first three terms represent the twelve gauge bosons that we already know: the first term $(\mathbf{8}, \mathbf{1}, 0)$ represents the eight massless gluons, the second term $(\mathbf{1}, \mathbf{3}, 0)$ corresponds to the three weak gauge bosons [that acquire masses after the SSB of G_{321} down to G_{31} via $< \phi > \neq 0$ [see Eq. (8.12)]] and the third term $(\mathbf{1}, \mathbf{1}, 0)$ is the massless photon. The fourth and fifth terms of Eq. (8.10) represent the twelve new gauge bosons that are called "lepto-quarks" because they are color triplets as well as flavor doublets; they are usually denoted by X_μ^α and Y_μ^α ($\alpha = 1, 2, 3$ color). It is these gauge bosons that acquire superheavy mass $M_U = g(M_U)V$ [where V is the VEV of the Φ Higgs — see Eq. (8.11) — and $g(M_U)$ is the "unified" gauge coupling constant evaluated at M_U (see Fig. 8.1)]. The twelve superheavy "lepto-quarks" give rise to the new physics "beyond the standard model", among which proton decay has been the object of an intensive but thus far fruitless experimental search. Proton decay can take place because $SU(5)$ GUT eliminates the distinction between quarks and leptons by placing quarks and leptons in the same irreducible — both $\bar{5}$ and 10 — $SU(5)$ representations; hence conversion of quark into lepton states becomes possible. Observation of proton decay became the overriding test for $SU(5)$ GUT because it would violate the two global conservation laws of baryon charge and lepton charge, which had held up extremely well ever since they were proposed (see §3.1). Despite the possibility of B and L violation within the framework of "minimal" $SU(5)$ GUT, it was realized at the outset that global $B - L$ conservation would have to be maintained and that the dominant decay process obeying this constraint is $p \to e^+\pi^0$ (or $p \to \bar{\nu}_e\pi^+$) from simple phase space considerations [8.7].

To compute the lifetime for proton decay, one writes down that part of the Lagrangian which involves the exchange of the "lepto-quark" gauge bosons $X(Q = 4/3)$ and $Y(Q = -1/3)$ between the quarks and leptons of the first generation and which is invariant under the standard $SU(3)_C \times SU(2)_L \times U(1)_Y$ group, so [8.8]:

$$\mathcal{L}_{X,Y} = \frac{ig_5}{\sqrt{2}} X^\alpha_\mu (\epsilon_{\alpha\beta\gamma}\bar{u}^c_{\gamma L}\gamma^\mu u_{\beta L} + \bar{d}_\alpha\gamma^\mu e^c_L) + \bar{d}_{\alpha R}\gamma^\mu e^c_R) + \text{h.c.}$$

$$+ \frac{ig_5}{\sqrt{2}} Y^\alpha_\mu (\epsilon_{\alpha\beta\gamma}\bar{u}^c_{\beta L}\gamma^\mu d_{\gamma L} + \bar{u}_{\alpha L}\gamma^\mu e^c_L + \bar{d}_{\alpha R}\gamma^\mu \nu^c_R) + \text{h.c.} \tag{8.18}$$

We see from Eq. (8.18) that:

$$X \to u^c u^c; \quad X \to ed; \quad Y \to d^c u^c; \quad Y \to eu; \quad Y \to \nu d \tag{8.18a}$$

We can use Eq. (8.18a) to obtain baryon-violating decays of the proton, i.e. $p \to e^+ + \pi^0$ or $\bar{\nu} + \pi^+$, as seen in Fig. 8.2. The lepto-quark X gives rise to the baryon-violating decay $p \to e^+ + \pi^0$ whereas the lepto-quark Y can induce both $p \to e^+ + \pi^0$ and $p \to \bar{\nu} + \pi^+$ decay. Equation (8.18) and Fig. 8.2 make explicit the fact that the "lepto-quarks" X $(Q = -\frac{4}{3})$ and Y $(Q = -\frac{1}{3})$ possess the same quantum number $B - L = -\frac{2}{3}$ [8.9] so that the initial and final states of the decaying proton always have the quantum numbers $B - L = +1$; in other words, proton decay is governed by an exact global symmetry, namely $(B - L) = 1$. It has been demonstrated [8.7] that global $B - L$ conservation holds exactly for $SU(5)$ GUT while, for $SO(10)$ GUT and E_6 GUT, it only holds to lowest order in (M_W/M_U); on the other hand, it is possible to gauge $B - L$ symmetry in rank 5 $SO(10)$ GUT and rank 6 E_6 GUT, in contrast to rank 4 $SU(5)$ GUT (see §8.4a).

Returning to the calculation of the decay process $p \to e^+\pi^0$ $(p \to \bar{\nu}_e +\pi^+$ is much more difficult to detect), we can use the Lagrangian (8.17) to write

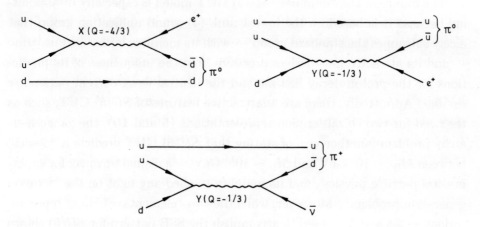

Fig. 8.2. Some mechanisms for proton decay in the $SU(5)$ GUT model of unification.

down the lifetime which is basically:

$$\tau_p \sim M_U^4/\alpha_5^2(M_U)m_p^5 \sim 10^{29} \text{ yr} \tag{8.19}$$

where α_5 is the $SU(5)$ gauge constant. Since there a unique SSB path from the $SU(5)$ GUT group to the standard G_{321} group, intermediate mass scales between M_W and M_U must be absent and, in principle, the theoretical prediction for τ_P is unequivocal and testable. There are, of course, radiative corrections due to the exchanges of the $SU(3)_C \times SU(2)_L \times U(1)_Y$ gauge bosons but these can be evaluated to reasonable accuracy by means of renormalization group techniques. Using the values of M_U and $\sin^2 \theta_W$ given by Eqs. (8.14) and (8.15), one obtains the theoretical prediction for the partial lifetime τ (for $p \to e^+\pi^0$):

$$\tau_p^{\text{theor}} \simeq 3.7 \times 10^{29\pm0.7}[M_U/(2 \times 10^{14}\text{GeV})]^4 \text{ yr} \tag{8.19a}$$

The uncertainty in the theoretical prediction given in Eq. (8.19a) is rather conservative and the poor agreement with the latest experimental lower limit, $\tau_P^{\text{exp}} \geq 6 \times 10^{32}$ years [8.6] rules out "minimal" $SU(5)$ GUT. The failure of minimal $SU(5)$ GUT in its key prediction of proton decay is supported by the discrepancy between the $SU(5)$ GUT prediction of $\sin^2 \theta_W$ [or, equivalently, the failure of the "running" equations for α_1, α_2 and α_3 to intersect at a single point — see Fig. 8.1] (see above). Vigorous attempts have been made to save $SU(5)$ GUT by going to a "non-minimal" version [e.g. by introducing a large number of Higgs [8.6] or supersymmetry (see below)], but all of these "life-saving" measures are quite ad hoc.

The failure of the "minimal" $SU(5)$ GUT model is especially disappointing because it is based on the lowest rank (4) grand unification group that nicely subsumes the standard group — with its massless lefthanded neutrino — and its attractiveness derives precisely from the uniqueness of its predictions for the proton decay lifetime and the neutral weak current parameter $\sin^2 \theta_W$. Admittedly, there are unattractive features of $SU(5)$ GUT, such as the need for two chiral fermion representations ($\bar{5}$ and 10), the gauge hierarchy problem [another way of stating that $SU(5)$ GUT predicts a "desert" between $M_W \sim 10^2$ GeV and $M_U \sim 10^{14}$ GeV — a dismal prospect for experimental particle physics], and its inability to shed any light on the "fermion generation problem". Moreover, while the two "elementary" Higgs representations — 24 and 5 — used to accomplish the SSB pattern for $SU(5)$ shown in Eq. (8.12), can be thought of as two-fermion condensates (quasi-Higgs) constructed out of the $\bar{5}$ and 10 fermion representations, there is no clue as to whether the dynamical symmetry breaking approach to the $SU(5)$ Higgs — along the lines being attempted to generate a quasi-Higgs doublet out of the $t\bar{t}$ condensate in the standard model (see §6.4c) — is feasible. Nevertheless, the failure of $SU(5)$ GUT does not necessarily invalidate the grand unification hypothesis for other simple groups and, indeed, it calls for more careful study of the higher rank GUT options. In examining such possibilities, it is incumbent upon us to find additional "positive" arguments (physical and/or mathematical) to justify serious consideration of these higher rank GUT groups. We propose, in the remaining sections, to review "positive" arguments on behalf of $SO(10)$ GUT and E_6 GUT and to comment on some distinctive features of these higher rank GUT groups *vis-à-vis* $SU(5)$ GUT. In the process, we will have occasion to deal with some more "new physics".

§8.4. B-L local symmetry, partial unification and $SO(10)$ GUT

We know that color-flavor grand unification can only be achieved with one of the groups $SU(N)$ ($N \geq 3$), or $SO(4k+2)$ ($k \geq 2$), or E_6 (see §8.2). We also know that the lowest rank (4) and unique $SU(5)$ GUT model can not explain proton decay nor yield a single intersection point for the "running" equations for the three coupling constants involved in the standard model; hence, if we do not wish to surrender the grand unification approach, we are compelled to consider a new "candidate" GUT group of rank 5 or higher. [An alternative approach is to consider a supersymmetric version of $SU(5)$ GUT [8.10]; this obviously permits more flexibility in phenomenological fitting than non-supersymmetric $SU(5)$ GUT but we prefer to explore the range of unification possibilities without introducing the new and as-yet-unverified supersymmetry principle (see §2.3a).] The straightforward approach then is to try the rank 5 candidate groups since — without supersymmetry — they involve the smallest number of intermediate mass scales (IMS's) between the grand unification and electroweak scales as well as the smallest number of "exotic" Weyl fermion states beyond the 15 Weyl states of the standard model; in a sense, ν_R is an "exotic" Weyl fermion but, to avoid confusion, we reserve the phrase "exotic" fermion for those fermions not included in the **16** just enumerated. There are only two "candidate" rank 5 groups: $SU(6)$ and $SO(10)$, and it is easy to show that $SU(6)$ GUT is not interesting since it does not break down to the standard group with the correct quark and lepton quantum numbers. Hence, we must turn to $SO(10)$ for a rank 5 "candidate" GUT group. Since the minimal (complex) spinorial representation of $SO(10)$ (the only type that accounts for chiral fermions) is **16**, it follows that the 15 Weyl fermion states of the standard model, augmented by a 16*th* Weyl fermion [which turns out to possess the quantum numbers $(1,1,0)$ of ν_R under G_{321}], must belong to the **16** representation of $SO(10)$. This, indeed, is the case and provides one of several good arguments for paying serious attention to $SO(10)$ GUT.

Let us first prove that the spinorial **16** representation of $SO(10)$ accommodates precisely the quarks and leptons of one generation and has room for ν_R; we then review the mathematical structure and physical implications of $SO(10)$ GUT. In order to establish the fermion content of the **16** representation of $SO(10)$, it is convenient to look at the maximal little groups of $SO(10)$ and, in particular, the maximal little groups that are composed only of unitary subgroups [compare our discussion of $SU(5)$ GUT in §8.3]. The two maximal little groups of $SO(10)$ that satisfy these two conditions are: $G_{51} = SU(5) \times U(1)$ and $G_{422} = SU(4) \times SU(2) \times SU(2)$ [we write $G_{n_1,n_2...} = SU(n_1) \times SU_{n_2} \times ...$]. It is easy to show that for both maximal little groups, the spinorial **16** representation of $SO(10)$ contains precisely the 15 Weyl quarks and leptons of one generation plus ν_R. Thus, the **16** representation of $SO(10)$ decomposes, under $SU(5) \times U(1)$, as follows [the numbers in parentheses are the $U(1)$ charges]:

$$\mathbf{16} = \mathbf{1}(-5) + \mathbf{\bar{5}}(3) + \mathbf{10}(-1) \qquad (8.20a)$$

It is seen that **16** contains precisely the $\mathbf{\bar{5}} + \mathbf{10}$ representations of $SU(5)$ and, in addition, a singlet $SU(5)$ representation, which is just ν_R. The decomposition of the **16** representation of $SO(10)$, shown in Eq. (8.20a), also makes it clear why the irreducible **16** representation of $SO(10)$ is triangular anomaly-free: the combination of the $\bar{\mathbf{5}}$ and **10** representations of $SU(5)$ is triangular anomaly-free (see §8.3) and a singlet representation is always triangular anomaly-free. While the SSB of $SO(10)$ GUT through the maximal little group $SU(5) \times U(1)$ yields the phenomenologically correct fermion content, it can be shown that this $SO(10)$ SSB path leads to a prediction for proton decay that can not be more favorable than the prediction for proton decay by $SU(5)$ GUT itself [8.11].

Since the descent of $SO(10)$ GUT through $SU(5) \times U(1)$ can not save proton decay, we must look at the descent of $SO(10)$ through the other maximal subgroup G_{422} $[\equiv SU(4)_C \times SU(2)_L \times SU(2)_R]$ to ascertain whether $SO(10)$ GUT is compatible with the lower limit on the proton decay lifetime. The decomposition of the **16** representation of $SO(10)$ under G_{422} is:

$$\mathbf{16} = (\mathbf{4, 2, 1}) + (\bar{\mathbf{4}}, \mathbf{1, 2}) \tag{8.20b}$$

where **4** and $\bar{\mathbf{4}}$ represent the 3 color states of each quark and antiquark plus the single "fourth color" states of each lepton and antilepton residing in one generation; with this understanding, it is easy to see that the first term on the R.H.S. of Eq. (8.20b) yields the lefthanded quark and lepton doublets whereas the second term on the R.H.S. of Eq. (8.20b) has merged the two righthanded quark singlets into a righthanded quark doublet and replaced the righthanded lepton singlet by a righthanded lepton doublet (thereby making provision for the 16*th* Weyl fermion in one generation, namely ν_R). The fact that the **16** representation of $SO(10)$ — with SSB through the G_{422} path — contains the additional ν_R is not surprising but the implicit replacement of the group $SU(2)_L \times U(1)_Y$ by $SU(2)_L \times SU(2)_R$ amounts to a significant change in the underlying physics of $SO(10)$ GUT compared to $SU(5)$ GUT and this must be explained in further detail. In view of the substantial amount of "new physics" that may enter $SO(10)$ GUT if its SSB proceeds through G_{422}, we digress to consider independent arguments that can be invoked to justify — through a series of steps — the enlargement of G_{321} to G_{422}.

Since the evidence is very strong that the confining gauge group for quarks is the unbroken rank 2 $SU(3)$ color group, the enlargement of the unified color-flavor (GUT) group from rank 4 to 5 must be sought through the enlargement of the $G_{21}[\equiv SU(2)_L \times U(1)_Y]$ flavor group from rank 2 to 3. Such an enlargement of G_{21} comes naturally if we allow for the existence of ν_R in each generation, as we demonstrate. We list in Table 8.2 the quark and lepton quantum numbers under G_{321}, together with an auxiliary chiral quantum number — weak flavor F — that will soon be described; we revert to the use of L and R representations since we are moving in the direction of the left-right-symmetric (LRS) enlargement of the standard group (when we include the ν_R). We see from Table 8.2 that while the electric charge Q is the same for each L and R chiral fermion (i.e. it is a non-chiral quantity), the weak hypercharge Y depends on whether we have an L or R chiral fermion, i.e. $Y_L \neq Y_R$ (reflecting the fact that the L fermions are in doublet representations of $SU(2)_L$ whereas the R fermions are in singlet representations). Nevertheless, as we have seen in §7.2c, the triangular anomaly due to ΣY_L^3 exactly cancels that due to ΣY_R^3, provided that the sums are taken over quarks and leptons of one complete generation.

Table 8.2: Electroweak qauntum numbers of quarks and leptons (including weak chiral flavor) for each generation.

	Q	$B-L$	$SU(2)_L$	I_{3L}	$Y_L = B - L + F_L$	F_L	$Y_R = B - L + F_R$	F_R
(q_L^+, q_R^+)	$\frac{2}{3}$	$\frac{1}{3}$	$(2,1)$	$(\frac{1}{2},0)$	$\frac{1}{3}$	0	$\frac{4}{3}$	1
(q_L^-, q_R^-)	$-\frac{1}{3}$	$\frac{1}{3}$	$(2,1)$	$(-\frac{1}{2},0)$	$\frac{1}{3}$	0	$-\frac{2}{3}$	-1
(ℓ_L^+, ℓ_R^+)	$(0,-)$	$(-1,-)$	$(2,1)$	$(+\frac{1}{2},-)$	-1	0	$-$	$-$
(ℓ_L^-, ℓ_R^-)	-1	-1	$(2,1)$	$(-\frac{1}{2},0)$	-1	0	-2	-1

We argue that this seemingly accidental triangular anomaly cancellation looks more natural if Y is expressed in terms of B, L and a global quantity F (really "weak chiral flavor") as follows [1.84]:

$$Y_{L,R} = B - L + F_{L,R} \tag{8.21}$$

so that:

$$Q = I_{3L} + \frac{Y_{L,R}}{2} = I_{3L} + \frac{B - L + F_{L,R}}{2} \tag{8.22}$$

Equation (8.21) tells us that for L fermions (quarks and leptons) $F_L = 0$ (so that $Y_L = B - L$) and for R fermions (quarks and leptons) $F_R = \pm 1$ (so that $Y_R = B - L \pm 1$) where $F_R = +1$ for the (weak isospin) "up" quark or "up" lepton and $F_R = -1$ for the (weak isospin) "down" quark or "down" lepton. The hidden "up-down" quark-lepton symmetry revealed by separating out $(B - L)$ in the hypercharge [Eq. (8.21)] confirms our suspicion that the triangular anomaly cancellation is not quite so accidental and basically reflects a "weak chiral flavor" quark-lepton symmetry. This statement only follows if the "up" righthanded chiral lepton (i.e. ν_R) is assumed to exist. To put it another way, "full" "weak chiral flavor" quark-lepton symmetry requires the existence of ν_R, and we can assign $F_R = +1$ to the hypothesized (weak isospin) "up" R lepton. [Within the framework of the standard electroweak group G_{21}, ν_R is "inert" with the quantum numbers $(1,0)$ and we see from Table 8.1 that these are precisely the quantum numbers listed under $SU(2)_L$ and Y_R.]

The last step — the insistence on "full' quark-lepton symmetry [when we speak of "full" quark-lepton symmetry, we mean "full" "weak chiral flavor" quark-lepton symmetry] and, *a fortiori*, on the existence of ν_R — constitutes the basic departure from the underlying assumption of the standard group (no ν_R) and leads to the enlargement of the rank 2 standard electroweak group to a rank 3 "electroweak" group. This follows if the global $(B - L)$ and F symmetries are gauged so that the $U(1)_Y$ of the standard group G_{321} becomes $U(1)_{B-L} \times U(1)_F$ and the rank 2 standard electroweak gauge group G_{21} has been enlarged to the rank 3 G_{211} [$\equiv SU(2)_L \times U(1)_{B-L} \times U(1)_R$] gauge group [8.12]. We have changed the notation from $U(1)_F$ to $U(1)_R$ because F_L is always 0 and only F_R takes on the values ± 1 (see Table 8.1). At the same time, the "weak" GNN relation (see §1.2d) has been changed from $Q = I_{3L} + \frac{Y}{2}$ to $Q = I_{3L} + \frac{B-L}{2} + \frac{R}{2}$ [i.e. Eq. (8.22)]. But now triangular anomaly cancellation for the enlarged "electroweak" group G_{211} follows directly from the vector-like character of $U(1)_{B-L}$ and the equality of the "up" values of R for quarks and leptons and, similarly, for the "down" values of R. There is a further advantage in highlighting the gauging of $B - L$ symmetry at the electroweak level [8.13] as a natural implementation of "full" quark-lepton symmetry: it sheds some light on the interesting result that $SU(2)_L$ flavor instantons separately break B and L global conservation [albeit in a miniscule way], while maintaining absolute global $B - L$ conservation at the electroweak level. [The maintenance of global $B - L$ conservation is an essential feature of the sphaleron-induced mechanism of baryon charge violation within the framework of QFD (see §10.3a)].

Equation (8.22) accepts parity violation in the weak interaction as given since the L fermions are still isodoublets of $SU(2)_L$ (in the enlarged "electroweak" group G_{211}) while the R fermions are still isosinglets of $SU(2)_L$. If nature makes provision for ν_R — consistent with "full" quark-lepton symmetry — one can argue that the status of electric charge quantization is less clearcut in G_{221} than it is in the enlarged LRS group $G_{422} = SU(2)_L \times SU(2)_R \times U(1)_{B-L}$ (see §7.7); when the R leptons and R quarks are placed in weak isodoublets to create G_{422}, more "new physics" is possible because parity violation can become an SSB phenomenon [8.14]. It is important that the LRS group G_{221} — although containing G_{211} as a subgroup — still possesses the rank 3. For G_{221}, the "weak" GNN relation becomes:

$$Q = I_{3L} + I_{3R} + \frac{B-L}{2} \qquad (8.23)$$

where the (weak) gauged hypercharge is now $(B - L)$, independent of the chiralities of the fermions, and is therefore a vector quantity. The SSB of G_{221} can induce parity violation — via the Higgs mechanism — to the standard electroweak group G_{21}, with or without the intermediate group G_{211}. In sum, G_{221} is initially non-chiral (LRS before SSB) as well as "full" quark-lepton symmetric, whereas G_{211} is only "full" quark-lepton symmetric. In either case, the rank of the standard electroweak gauge group has been increased from 2 to 3, and contains $U(1)_{B-L}$ as a gauge symmetry.

Both enlarged "electroweak" groups G_{211} and G_{221} have interesting physical consequences for the simple reason that G_{211} involves five gauge bosons and G_{221} seven gauge bosons, instead of the four gauge bosons associated with G_{21}. The additional gauge boson (or bosons) must acquire a mass (or masses) larger than the W and Z boson masses due to a larger SSB scale than the Λ_{QFD} scale responsible for the masses of Z and W, and, in principle, these new gauge bosons should be observable. Thus, if G_{211} exists, it follows that, above the M_W scale — since $\Delta Q = \Delta I_{3L} = 0$ — one obtains the selection rule [8.11]:

$$-\Delta R = \Delta(B - L) \qquad (8.24)$$

which relates the SSB scale of $U(1)_R$ breaking to the SSB scale of $U(1)_{B-L}$ breaking. For the larger group G_{221}, the selection rule becomes [from Eq. (8.23)]:

$$-2\Delta I_{3R} = \Delta(B - L) \qquad (8.25)$$

which now relates the SSB scale of $SU(2)_R$ breaking to the SSB scale of $U(1)_{B-L}$ breaking. The selection rule (8.25) is particularly interesting because it can be shown — in the "minimal" version of the LRS model [1.147] — that Eq. (8.25) implies (since $\Delta I_{3R} = 1$):

$$\Delta(B - L) = 2 \qquad (8.26)$$

Thus, Eq. (8.26) provides the rationale — within the framework of G_{221} or of a group of which G_{221} is a subgroup (see below) — for Majorana neutrinos $(\Delta L = 2)$ and neutron oscillations $(\Delta B = 2)$.

Since a grand unification group is a unification of the color and flavor groups as well as the quark and lepton representations, we must consider the consequence of adjoining $SU(3)_C$ to either the enlarged "electroweak" flavor group G_{211} or G_{221}. Since G_{211} is a subgroup of G_{221}, we content ourselves with adjoining $SU(3)_C$ to G_{221}, which we denote by G_{3221}. But, if we examine $G_{3221} [\equiv SU(3)_C \times SU(2)_L \times SU(2)_R \times U(1)_{B-L}]$, it is most natural to take the partial "unifying" step of considering $(B-L)$ as the "fourth" color, and of "unifying" the non-Abelian gauge group $SU(3)_C$ with the Abelian gauge group $U(1)_{B-L}$ into the rank 3 simple gauge group $SU(4)_C$ of four "colors" [three quark colors plus one "$B-L$ color" [8.15]]. $SU(4)_C$, together with the $SU(2)_L \times SU(2)_R$ part of the "electroweak" LRS group, then defines a rank 5 semi-simple gauge group $G_{422} [\equiv SU(4)_C \times SU(2)_L \times SU(2)_R]$ that happens to be the "Pati-Salam" group (i.e. G_{422} PUT [1.148]) and is one of the two maximal (unitary) subgroups [8.4] of the rank 5 simple gauge group $SO(10)$ [the other being $SU(5) \times U(1)$].

We have thus arrived at the interesting result that the "full" quark-lepton symmetry principle (with its ν_R) has led us to the $SO(10)$ GUT group with the maximal subgroup G_{422} [not $SU(5) \times U(1)$] as an intermediate stage. Another virtue of the $SO(10)$ group is that it is the only simple group that contains $SU(2)_L$ and $U(1)_{B-L}$ as subgroups, without requiring mirror fermions (i.e. fermions with opposite chirality) [8.16]. [Parenthetically, the only simple groups that contain $SU(2)_L$ and $U(1)_Y$ as gauge symmetries and do not contain "exotic" or "mirror" fermions are $SU(5)$ and $SO(10)$ [8.17]; however, $SO(10)$ GUT has the added property that it contains $U(1)_{B-L}$ as a gauge symmetry.] If we lift the condition that the unification group is simple and allow semi-simple groups, it can also be shown [8.16] that G_{422} is the only semi-simple group that contains $SU(2)_L$ and $U(1)_{B-L}$ as subgroups, without "mirror" fermions. We see now why the inclusion of $U(1)_{B-L}$ as a subgroup (of the electroweak group) raises the rank of the unification group by one unit but no more, and makes provision for just two viable $(B-L)$ "unification" groups $SO(10)$ GUT and G_{422} PUT.

§8.4a. SSB of $SO(10)$ GUT via the left-right-symmetric (G_{422}) path

We next spell out the "new physics" implied by the path of descent $SO(10) \rightarrow G_{422}$ and the subsequent stages as SSB proceeds from G_{422} downward. For this purpose, it is useful to write down a possible G_{422}-type pattern of intermediate stages in the symmetry-breaking chain between $SO(10)$ and

G_{31}:

$$SO(10) \xrightarrow[M_U]{54} G_{422} \xrightarrow[M_C]{45} G_{3221} \xrightarrow[M_{W_R}]{45} G_{3211} \xrightarrow[M_{Z_R}]{126} G_{321} \xrightarrow[M_W]{10} G_{31} \qquad (8.27)$$

where the reader is reminded that: $G_{422} = SU(4)_C \times SU(2)_L \times SU(2)_R$, $G_{3221} \equiv SU(3)_C \times SU(2)_L \times SU(2)_R \times U(1)_{B-L}$, $G_{3211} \equiv SU(3)_C \times SU(2)_L \times U(1)_R \times U(1)_{B-L}$, $G_{321} \equiv SU(3)_C \times SU(2)_L \times U(1)_Y$, and $G_{31} \equiv SU(3)_C \times U(1)_{EM}$. The number above each arrow in Eq. (8.27) denotes the lowest-dimension Higgs representation [of $SO(10)$] responsible for the SSB at that stage [other Higgs representations may be chosen to achieve certain well-defined objectives — see below] and satisfying the "Michel conjecture", and the symbols below the arrows represent the masses acquired by the gauge bosons at the indicated level of SSB. For example, **54** is the lowest-dimension Higgs representation that spontaneously breaks $SO(10)$ down to its maximal "little" group G_{422} and M_U is the "superheavy" mass acquired by **24** of the 45 gauge bosons of $SO(10)$ [the **45** representation of $SO(10)$ is the analog of the **24** representation of $SU(5)$ and the 24 "superheavy" gauge bosons in $SO(10)$ GUT are the analog of the 12 "superheavy" gauge bosons in $SU(5)$ GUT. The remaining 21 gauge bosons of G_{422} stay massless down to the next level of SSB, etc.] The situation is summarized by the chain of inequalities [see Eq. (8.27) for the definition of the M's]:

$$M_U \geq M_C \geq M_{W_R} \geq M_{Z_R} \geq M_W \qquad (8.28)$$

In contradistinction to $SU(5)$ GUT, we see [from Eq. (8.28)] that for $SO(10)$ GUT, there are, in general, three intermediate mass scales (IMS) between the unification mass scale M_U and the scale for the SSB of the electroweak group (M_W). In contrast to the "desert" of $SU(5)$ GUT (where there is no intermediate mass scale between M_U and M_W — see §8.3), the three possible IMS's of $SO(10)$ GUT open up a plethora of "new physics" beyond the standard model and before the GUT scale. In what follows, we try to identify the "new physics" associated with the three possible IMS's in $SO(10)$ GUT and to indicate the experimental feasibility of testing for the presence of each of the $SO(10)$ IMS's. To decide whether the intermediate mass scale M_C or M_{W_R} or M_{Z_R} (or more than one) is present, we must sharpen up our definition of each of them; in order of decreasing mass scale, we can say: (1) M_C characterizes the SSB of $SU(4)_C$ down to $SU(3)_C \times U(1)_{B-L}$, creating a distinction between the (color triplet) quarks and the (color singlet) leptons. It is the lack of distinction between quarks and leptons that makes possible proton decay in $SO(10)$ GUT although there is a "hidden" selection rule that prohibits proton decay within the framework of G_{422} PUT [1.147]; (2) M_{W_R} is a possible R charged weak boson mass — associated with the SSB of $SU(2)_R$ down to $U(1)_R$, which signals the spontaneous breaking of parity conservation. However, it will be seen later that the SSB of parity need not be associated with the SSB of $SU(2)_R$ down to $U(1)_R$ but can be induced (see §8.4c) by an "inner automorphism"-type of discrete parity (called the D parity) contained within the $SO(10)$ group but not G_{422}; and, finally, (3) M_{Z_R} is the R neutral weak boson mass — which is associated with the SSB of $U(1)_{B-L} \times U(1)_R$ down to $U(1)_Y$ — and which can have a very different value than M_{W_R} [e.g. M_{Z_R} can be related to a large Majorana mass of ν_R and — through the "see-saw" mechanism — to the Majorana mass of ν_L (see §6.4b)]. What is common to these three intermediate mass scales — M_C, M_{W_R} and M_{Z_R} — is that they are all associated with interme- diate groups between $SO(10)$ and G_{321} which contain $U(1)_{B-L}$ as a gauge subgroup [8.16]; this is in contrast to the more limited global $B - L$ symme- try of G_{321} and $SU(5)$ GUT. It should be understood that the quark-lepton dynamics may be such (we can only engage in pure speculation on this score) that some of the "elementary" Higgs representations are suppressed and one or more of the three intermediate stages is bypassed on the way from $SO(10)$ GUT to the standard gauge group G_{321}; for pedagogical reasons, we have allowed for the maximum number (3) of intermediate stages.

As we have pointed out, there is potential "new physics" associated with each IMS but before we comment on these possibilities, we must show that $SO(10)$ GUT is capable of giving larger values of the proton decay lifetime and of $\sin^2 \theta_W$ than $SU(5)$ GUT when the SSB of $SO(10)$ GUT proceeds through the maximal "little" group G_{422}. What we must prove is that the values of τ_p and $\sin^2 \theta_W$ — as determined by $SO(10)$ GUT through its de- pendence on the intermediate mass scales M_{Z_R}, M_{W_R} and M_C — does not rule out $SO(10)$ GUT. This turns out to be the case; indeed, it can be shown, on fairly general grounds [8.11], that:

$$M_{10} = M_5 \left(\frac{M_5}{M_{W_R}} \right)^{\frac{1}{2}} \tag{8.29}$$

where M_{10} and M_5 are the $SO(10)$ and $SU(5)$ grand unification masses and M_{W_R} is defined by Eq. (8.27). Since the ratio M_5/M_{W_R} can be substantially larger than unity, it is possible to have $M_{10} >> M_5$ and, *a fortiori*, $\tau_{10} >> \tau_5$ [the $SO(10)$ and $SU(5)$ predictions for τ_p respectively — see Eq. (8.19)]. Note that the unification mass, M_{10}, is independent of M_C and M_{Z_R} in Eq. (8.29), depending only on M_{W_R}. One can also deduce a simple relation between $\sin^2 \theta_W$ predicted by $SO(10)$ GUT in terms of the $SU(5)$ GUT prediction; one gets [8.11]:

$$\sin^2 \theta_{10} = \sin^2 \theta_5 + 11\alpha_e/(6\pi)[\ln(M_U/M_C) + \frac{1}{2}\ln(M_5/M_{W_R})] \qquad (8.30)$$

It is obvious from Eq. (8.30) that $\sin^2 \theta_{10} = \sin^2 \theta_5$ when $M_{W_R} = M_C = M_U$, i.e. there is no intermediate mass scale. More importantly, the expression for $\sin^2 \theta_{10}$ [Eq. (8.30)] is again independent of M_{Z_R} but is no longer independent of M_C nor M_{W_R}. [The fact that τ_{10} and $\sin^2 \theta_{10}$ are independent of M_{Z_R} is due to M_{Z_R} originating from the SSB of one non-semi-simple group (G_{3211}) to another such group (G_{321}) [8.11].]

Hence, the two key parameters of $SO(10)$ GUT — τ_p and $\sin^2 \theta_W$ — do not impose an undetectable (see below) lower limit on M_{Z_R}; that is, a relatively low mass of M_{Z_R} (but still higher than M_Z) is compatible with the observed value of $\sin^2 \theta_W$ and the measured lower limit on τ_p. Unfortunately, Eq. (8.30), while implying a larger $\sin^2 \theta_W$ for $SO(10)$ GUT than for $SU(5)$ GUT, requires too large a value of M_{W_R} (and, *a fortiori*, of M_C) to permit detectability of M_{W_R} and M_C by means of laboratory-based experiments. If we work with the "traditional" $SO(10)$ GUT model [see §8.4c for comments on "non-traditional" $SO(10)$ GUT], we must adjust the value of M_{W_R} in Eqs. (8.29) and (8.30) so that $\tau_p \geq 6 \times 10^{32} y$ and $\sin^2 \theta_W = 0.232$. A trivial calculation soon shows [even if we take $M_C = M_U$ so that the G_{422} group is eliminated from the "downward" SSB path shown in Eq. (8.27)] that M_{W_R} must be in the range of 10^{11} GeV, a value which guarantees undetectability of M_{W_R} and M_C within the framework of "traditional" $SO(10)$ GUT.

In sum, "traditional" $SO(10)$ GUT can — through the intervention of up to three intermediate mass scales that are absent in $SO(5)$ GUT — overcome the deficiencies of $SU(5)$ GUT, namely the prediction of values of τ_p and $\sin^2 \theta_W$ that are compatible with experiment. Moreover, the constraints on τ_p and $\sin^2 \theta_W$ of $SO(10)$ GUT are compatible with a relatively low mass for Z_R. This opens the door to the whole subfield of massive neutrino physics, which is excluded for rank 4 $SU(5)$ GUT (because of the presumed absence of ν_R). The phenomenological compatibility of "traditional" $SO(10)$ GUT with τ_p and $\sin^2 \theta_W$ exacts a price: two of the three IMS's, M_{W_R} and M_C, are much too large to be measured in laboratory experiments. Under these circumstances, it would seem pointless to discuss a sampling of phenomenological tests for the existence of M_{W_R} and M_C were it not that there is an intriguing modification of "traditional" $SO(10)$ GUT — through the utilization of an "inner automorphism" of $SO(10)$ — so-called

D parity (see §8.4c) — which is compatible with much lower values of M_{W_R} and M_C. We therefore review a sampling of phenomenological tests of all three intermediate mass scales — M_{W_R} and M_C, in addition to Z_R — in the next section (§8.4b), in preparation for §8.4c where we discuss $SO(10)$ GUT in combination with D parity.

§8.4b. Phenomenological tests of intermediate mass scales in $SO(10)$ GUT

There is general agreement that a fully satisfactory unified theory of all particle interactions would include gravity in addition to the other three basic interactions. But, short of a TOE that would descend from the Planck scale (see §1.4c) and a possible role for a new preon scale at the multi-TeV scale, it seems reasonable to search for significant departures from the standard model within the traditional conceptual framework of further enlargement of the standard group through partial (not necessarily grand) unification of the strong and electroweak interactions among the three generations of quarks and leptons (with the possible inclusion of ν_R). For reasons already stated, it seems to us that the SSB of $SO(10)$ GUT through G_{422} — giving rise, in general, to intermediate mass scales — serves as a good model to identify classes of significant departures from the standard model. Since rank 5 $SO(10)$ GUT is the first grand unification group with intermediate mass scales between the grand unification scale M_U and the electroweak scale M_W, our comments may be useful in considering the merits of E_6 GUT and various modifications of the smaller "partial unification" (PUT) groups.

With this in mind, we give a brief summary of the experimental limits on M_{Z_R}, M_{W_R} and M_C [it is more profitable to start the phenomenological discussion with the lowest anticipated detectable IMS M_{Z_R}] and comment on the significance of these numbers for the "traditional" $SO(10)$ GUT model; these numbers will also serve as the "testing ground" of non-traditional $SO(10)$ GUT [i.e. $SO(10)$ GUT with D parity violation (see §8.4c)]. For reasons alluded to in §8.4a, the lowest anticipated detectable IMS mass, of the three IMS's, is M_{Z_R}, which need not be much heavier than the well-known Z weak boson mass (~ 100 GeV) of the standard model. [Thus far, an analysis of the neutral current weak interaction data (e.g. the backward-forward asymmetry in $e^+e^- \to \mu^+\mu^-$ scattering [6.9]) yields a lower limit on M_{Z_R} of 350 GeV.] The existence of a relatively low mass weak boson Z_R (that interacts with the righthanded quark and lepton currents) has some interesting consequences. If we examine the path of descent from $SO(10)$

GUT to G_{31}, as given by Eq. (8.27), we note that the **126** Higgs representation of $SO(10)$ [with its decomposite Δ_R possessing the quantum numbers $(\overline{10}, 1, 3)$] under G_{422}] is the smallest $SO(10)$ representation that achieves the SSB of G_{3211} to G_{321} and, in the process, generates the possibly-not-so-heavy Z_R. The same Higgs Δ_R can be used to generate a Majorana mass for ν_R, which we denote by $M_{\nu_R} \simeq M_{\Delta_R} \simeq M_Z$; with the help of the "see-saw" mechanism (see §6.4b), one predicts $m_{\nu_L} \simeq m_\ell^2/M_{Z_R}$. Using the values of m_ℓ given in Table 1.6 and the lower limit of 350 GeV for M_{Z_R}, we obtain $m_{\nu_e} \lesssim 0.8$ eV, $m_{\nu_\mu} \lesssim 30$ KeV and $m_{\nu_\tau} \lesssim 8$ MeV. These neutrino masses are all consistent with the upper limits on the neutrino masses of the three generations given in Table 1.6 but, if Z_R is found in the multi-hundred GeV (or TeV) range, the resulting neutrino masses would be incompatible with the range of values: $\Delta m^2 \simeq 10^{-6} - 10^{-4} \text{eV}^2$ required to resolve the "solar neutrino problem" on the basis of matter-induced neutrino oscillations (see §6.4b). However, in view of the uncertainties connected with the "solar neutrino problem", it is highly desirable to search for a relatively light Z_R on the highest energy accelerators to decide whether there is any merit to the notion that the discovery of Z_R would herald the first-step enlargement (in the form of G_{3211}) of the standard group on the upward path to $SO(10)$ GUT or one of its PUT subgroups.

As expected, the story gets more complicated for the charged partner of Z_R, namely W_R. The righthanded charged gauge boson W_R plays a role in a variety of rare physical processes [8.18], e.g. the $K_L - K_S$ mass difference, neutrinoless double beta decay, $\mu \rightarrow e\gamma$ decay, muon conversion into electrons through interaction with nuclei, etc. A good example is the (righthanded) branching ratio B_R for $\mu \rightarrow e\gamma$ decay; a characteristic Feynman diagram is shown in Fig. 8.3 where N_1 and N_2 are the heavy righthanded neutrinos that are linear combinations of the first and second generation (N_e and N_μ) neutrinos coupled with the e and μ charged leptons to W_R. More precisely, the two mass eigenstates N_1 and N_2 can be written in terms of the mixing angle θ_R between N_e and N_μ (see §6.4b), thus:

$$N_1 = \cos\theta_R \, N_e + \sin\theta_R \, N_\nu; \quad N_2 = -\sin\theta_R \, N_e + \cos\theta_R \, N_\nu \quad (8.31)$$

Then the branching ratio B_R becomes:

$$B_R = \frac{3\alpha_e}{32\pi}\left(\frac{M_W^2}{M_{W_R}^2}\right)^4 \left[\sin\theta_R \cos\theta_R \frac{(m_{N_2}^2 - m_{N_1}^2)}{M_W^2}\right]^2 \quad (8.32)$$

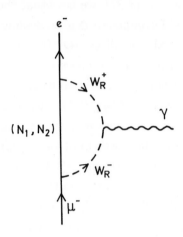

Fig. 8.3. Diagram for W_R-induced $\mu \to e\gamma$ decay.

The counterpart of Eqs. (8.31)–(8.32) for the light lefthanded neutrinos (denoting the lefthanded neutrinos of the first two generations by ν_e and ν_μ) is:

$$\nu_1 = \cos\theta_L\, \nu_e + \sin\theta_L\, \nu_\mu; \quad \nu_2 = -\sin\theta_L\, \nu_e + \cos\theta_L\, \nu_\mu \qquad (8.33a)$$

$$B_L = \frac{3\alpha}{32\pi}[\sin\theta_L \cos\theta_L \frac{(m_{\nu_2}^2 - m_{\nu_1}^2)}{M_W^2}]^2 \qquad (8.33b)$$

where ν_1 and ν_2 are the mass eigenstates of the lefthanded neutrinos ν_e and ν_μ. If we substitute the approximate value of $\Delta m^2 \simeq 10^{-6} \text{eV}^2$, derived from the presumed solution of the "solar neutrino problem" (see §6.4b) and take the maximum value of the mixing angle θ_L, Eq. (8.33b) yields a miniscule value (10^{-60}) of the branching ratio for $\mu \to e\gamma$ decay [to be compared with the observed upper limit of 4.9×10^{-11}]. However, if we look at B_R, given by Eq. (8.32), and take $M_{W_R} \simeq 400$ GeV and $m_{N_2}^2 - m_{N_1}^2 \simeq M_W^2$, the predicted branching ratio $B_R \simeq 4 \times 10^{-12} (\sin\theta_R \cos\theta_R)^2$ which comes within striking range (depending, of course, on θ_R) of the observed upper limit for $\mu \to e\gamma$ decay.

Unfortunately, Eq. (8.32) confirms the general expectation [since $M_{N_R} \simeq M_{W_R}$] that rare processes driven by W_R vary inversely as a high power of M_{W_R} and quickly become undetectable when M_{W_R} exceeds several tens of TeV. The example of $\mu \to e\gamma$ decay just discussed does not give us a true estimate of M_{W_R} because of our lack of knowledge of θ_R. Such an estimate can be obtained from the neutral kaon mass difference and one obtains $M_{W_R} \simeq 1.6$ TeV [8.19]; other phenomenological estimates of M_{W_R}

from laboratory-based experiments always lead to a value for the intermediate mass scale M_{W_R} many orders of magnitude lower than is compatible with the value of M_{W_R} ($\simeq 10^{11}$ GeV) derived on the basis of "traditional" $SO(10)$ GUT from a match with the observed value of $\sin^2 \theta_W$ (0.232 — [6.12]). Of course, we can "abort" the "upward" unification path [from G_{321} — see Eq. (8.27)] at, say, G_{3221} [with a number of elegant features remaining, including the possibility of giving a more natural explanation of parity and CP violation], allow M_{W_R} to be in a detectable range, and forego "traditional" $SO(10)$ GUT. However, "aborting" the "upward" unification path in "traditional" $SO(10)$ GUT amounts to surrendering the "grand unification" objective and settling for very partial predictive power *vis-à-vis* experiment. Before accepting this reduced objective, one might inquire whether there is a theoretically attractive "non-traditional" $SO(10)$ GUT that possesses the virtues of "traditional" $SO(10)$ GUT but none of its "phenomenological" defects. As we have already suggested, there is such a version of $SO(10)$ GUT that is based on exploiting the properties of an "inner automorphism" (called D parity) that is present in the $SO(10)$ group and can reconcile detectable ranges for M_{W_R} and M_C (see below) with the full $SO(10)$ group. Henceforth, we refer to $SO(10)$ GUT — with explicit use of D parity in the SSB chain — as "non-traditional" $SO(10)$ GUT, and we speak to this after some brief remarks about the M_C scale.

Physically, the third intermediate mass scale M_C arises from the SSB of the "four-color" gauge group down to the mixed "three-color" — $(B - L)$ gauge group $SU(3)_C \times U(1)_{B-L}$; in terms of the groups entering into the SSB pattern (8.27) for $SO(10)$ breaking down to G_{31}, M_C is associated with the SSB of G_{422} to G_{3221} [8.20] by means of the Higgs decomposite Σ [possessing the quantum numbers $(15, 1, 1)$ under G_{422}] of the Higgs 45 representation of $SO(10)$ [the Higgs **210** representation of $SO(10)$ also contains the decomposite $\Sigma(15, 1, 1)$ and it is the **210** representation of $SO(10)$ that is useful in "non-traditional" $SO(10)$ GUT — see §8.4c]; our discussion of the phenomenological role of M_C is, of course, affected by whether M_C is determined by the "running" equations of "traditional" $SO(10)$ [as contemplated in Eq. (8.27)] or by the "running" equations of "non-traditional" $SO(10)$ GUT. If we focus on Eq. (8.27), we expect $M_C \geq M_{W_R}$ and, hence, that the level of detectability of rare decay processes, involving the mediation of M_C, is comparable to that which might arise from the mediation of M_{W_R} (see Fig. 8.3). Probably, the two most interesting processes depending on the intermediate mass scale M_C are: $K_L \to \mu \bar{e}$ decay and $n - \bar{n}$ oscillations. The Feynman diagram for $K_L \to \mu\bar{e}$ decay is shown in Fig. 8.4, where it is seen to be driven by the M_C gauge boson. This decay process clearly involves the conversion of quarks (comprising K_L) into leptons as a consequence of the breaking of G_{422} [wherein quarks and leptons are placed in the same $(\mathbf{4}, \mathbf{2}, \mathbf{1})$ representation] to G_{3221} [wherein quarks and leptons are in separate irreducible representations of G_{3221}]. As we know, $SO(10)$ GUT permits proton decay [which is possible because quarks and leptons are in the same irreducible $\mathbf{16}$ representation of $SO(10)$]; however, there is a "hidden" selection rule that prohibits proton decay in G_{422} PUT [8.13] so that $K_L \to \mu\bar{e}$ decay is the most clearcut gauge-dependent process permitted by G_{422}. The branching ratio for $K_L \to \mu\bar{e}$ decay has been estimated to be [8.21] $\sim 10^{-8}(100 \text{ TeV}/M_C)^4$ and comparison with experiment — which gives an upper limit of 6×10^{-11} (see Table 6.8) — already places a lower limit on M_C of 3×10^2 TeV. Since the value of M_C increases with the inverse fourth power of the branching ratio, the branching ratio for $K_L \to \mu\bar{e}$ quickly becomes undetectable (in the foreseeable future) if M_C exceeds 10^3 TeV.

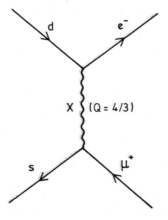

Fig. 8.4. Tree diagram that leads to $K_L \to \mu\bar{e}$ decay in models with $SU(4)_C$ unification (X is lepto-quark).

A second rare and unusual process that depends on the SSB of G_{422} down to G_{3221} (and thereby on M_C) is neutron oscillations (i.e. $n \to \bar{n}$) and its "bound nuclear" counterpart: $\Delta B = 2$ nuclear instability. We have already pointed out that the existence of the LRS group G_{221} above the M_W scale implies the relation $\Delta(B - L) = 2$ [see Eq. (8.26)] and consequently the probable appearance of $\Delta L = 2$ ($\Delta B = 0$) and $\Delta B = 2$ ($\Delta L = 0$) processes; as we have seen, the most interesting manifestation of the "new physics" associated with the $\Delta L = 2$ $(\Delta B = 0)$ "selection rule" is the possibility of a small finite mass for a Majorana-type ν_L and a much larger finite mass Majorana ν_R (with $m_{\nu_R} \simeq M_{W_R}$). One might think that the same Higgs $\Delta_R(\overline{\mathbf{10}}, \mathbf{1}, \mathbf{3})$ with a non-vanishing VEV, that gives rise to $m_{\nu_R} \simeq M_{W_R}$ would also give rise to baryon charge non-conservation [in accordance with the selection rule $\Delta B = 2$ $(\Delta L = 0)$]. However, the LRS group from which this selection rule is derived contains a "hidden" global symmetry $Q \to e^{i\alpha/3} Q$ (α is a constant), which is unbroken by the vacuum, and is most conveniently identified with the baryon charge. That is why a baryon charge-violating process like $n \to \bar{n}$ can only occur at the level of the larger gauge group G_{422} while the Majorana neutrino phenomenon can take place at the lower level of the LRS group G_{3221}. We can appreciate the difference in applying the same Higgs boson representation $\Delta_R(\overline{\mathbf{10}}, \mathbf{1}, \mathbf{3})$ [the quantum numbers of Δ_R refer to the group G_{422}] to the Majorana neutrino case and the neutron oscillation case, by examining Eq. (8.34) which exhibits the decomposition of $\Delta_R(\overline{\mathbf{10}}, \mathbf{1}, \mathbf{3})$, under G_{3221}, thus:

$$\Delta_R(\overline{\mathbf{10}}, \mathbf{1}, \mathbf{3}) = \Delta_{\ell\ell}(\mathbf{1}, \mathbf{1}, \mathbf{3}, -2) + \Delta_{\ell q}(\mathbf{3}, \mathbf{1}, \mathbf{3}, -\frac{2}{3}) + \Delta_{qq}(\mathbf{6}, \mathbf{1}, \mathbf{3}, \frac{2}{3}) \quad (8.34)$$

The point now is that for the Majorana neutrino case, the VEV of the dilepton decomposite $\Delta_{\ell\ell}(\mathbf{1}, \mathbf{1}, \mathbf{3}, -2)$ is given a non-vanishing VEV whereas, for the neutron oscillation case, it is the diquark decomposite $\Delta_{qq}(\mathbf{6}, \mathbf{1}, \mathbf{3}, 2/3)$ that receives the non-vanishing VEV; henceforth, we denote the G_{3221} decomposite relevant for the $n \to \bar{n}$ process by Δ'_R instead of Δ_R as we did for the Majorana neutrino case.

With the above understanding, we sketch the present situation with regard to neutron oscillations. While the M_C-dependent process $K_L \to \mu\bar{e}$ decay results primarily from an exchange of "lepto-quark" gauge bosons in the transition from G_{422} to G_{3221}, as shown in Fig. 8.3 [we use the same scale M_C for all the gauge bosons and Higgs bosons involved in the transition $G_{422} \to G_{3221}$], the dominant contribution to the $n \to \bar{n}$ process is the Higgs-driven six-quark diagram shown in Fig. 8.5. If the transition probability for the transition $n \to \bar{n}$ is denoted by $\delta m_{n\bar{n}}$ ($\delta m_{n\bar{n}}$ is the inverse of the mixing time $t_{n\bar{n}}$ for $n \to \bar{n}$) [8.22], one gets for $\delta m_{n\bar{n}}$:

$$\delta m_{n\bar{n}} \simeq \frac{h^3 \lambda \Delta'_R}{M_C^6} \quad (8.35)$$

where h is the Yukawa coupling constant (see Fig. 8.5), λ is the Higgs coupling constant, and Δ'_R has already been defined. Inserting reasonable values for h, λ and Δ'_R, Eq. (8.35) requires M_C to be of the order of 10^3 TeV in order to predict a detectable value of $t_{n\bar{n}}$, say, of the order of 10^8sec [1.147]. The above estimate of M_C for a detectable $t_{n\bar{n}}$ is rather crude and, in view of the sensitive dependence on M_C (which is basically a M_C^5 dependence), failure to detect neutron oscillations at the 10^8sec level would be, unfortunately, no more decisive than the failure to detect the branching ratio for $K_L \to \mu\bar{e}$ decay at the 10^{-12} level. It should be noted that experiments searching for $\Delta B = 2$ nuclear instability [i.e. the search for a "virtual" neutron oscillation effect inside a nucleus leading to the emission of energetic pions with a total energy of ~ 2 GeV] claim that "free" neutron oscillations at the $t_{n\bar{n}} \sim 10^8$sec level are just about ruled out [8.23]; this conclusion depends on plausible nuclear physics assumptions about the effective "optical potential" in which the "virtual" $\bar{n}$ moves inside a nucleus [8.24] but is not conclusive; an attempt to directly measure "free" neutron oscillations at the $t_{n\bar{n}} \sim 10^8$ sec (or better) is still worthwhile [8.25].

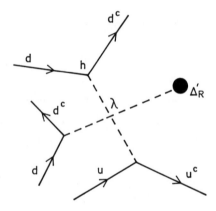

Fig. 8.5. The tree graph that induces the six-fermion $\Delta B = 2$ Higgs-type vertex that leads to $n \leftrightarrow \bar{n}$ oscillation.

Both experiments — the rare decay $K_L \to \mu\bar{e}$ and "bound" $\Delta B = 2$ nuclear instability — have led to a range of values of M_C that is consistent with the condition $M_C \geq M_{W_R}$ and the lower limits on M_{W_R} deduced from the rare weak processes — depending on the intermediation of the W_R boson — mentioned earlier. This is flimsy support for the notion that the two intermediate "$B-L$" gauge groups (G_{422} and G_{3221}) are embedded in $SO(10)$ GUT since the need to reconcile the predictions of "traditional" $SO(10)$ GUT for τ_p and $\sin^2\theta_W$ [see Eqs. (8.29) and (8.30)] with the observed lower limit on $\tau_p (6 \times 10^{32} \text{yr})$ and the measured value $\sin^2 \theta_W = 0.232$ require $M_{W_R} \simeq 10^{11}$ GeV and, *a fortiori*, $M_C \gtrsim 10^{11}$ GeV, thereby rendering hopeless the detectability of rare processes depending on M_{W_R} and M_C. [Parenthetically, similarly undetectable values of M_{W_R} and M_C result when the "$B - L$"-type unification stops at the Pati-Salam group G_{422}: the largest "$B - L$"-type group that can be reconciled with detectable values of M_{W_R} (M_C is no longer relevant) is the LRS group G_{3221}, as already noted.] We must conclude that, while the detectable range for M_{Z_R} is compatible with the "traditional" $SO(10)$ GUT model, the detectable ranges of the other two IMS's, M_{W_R} and M_C, are definitely incompatible. Consequently, if we wish to populate the $SU(5)$ GUT "desert" with more than one additional IMS, then we must scrutinize the rather large $SO(10)$ group to ascertain whether there is some "hidden" symmetry (discrete or continuous) that can open the door to laboratory-based experiments that could find direct or indirect evidence for at least one more IMS in addition to M_{Z_R}. It turns out that there is such a "hidden" discrete symmetry in $SO(10)$ GUT that can be exploited to convert traditional $SO(10)$ GUT into a non-traditional version that allows for detectable values of the intermediate mass scales M_{W_R} and M_C; this is D parity and its ramifications are summarized in the next section (§8.4c) on "non-traditional" $SO(10)$ GUT.

§8.4c. $SO(10)$ GUT with D parity

The key to "non-traditional" $SO(10)$ GUT [i.e. "traditional" $SO(10)$ GUT with the "inner automorphism" D parity taken into account] is the observation that the source of the undetectably large values of M_{W_R} and M_C in "traditional" $SO(10)$ GUT is the implicit assumption that the "running" gauge coupling constants g_L and g_R, associated respectively with the $SU(2)_L$ and $SU(2)_R$ subgroups of $SO(10)$ [i.e. G_{422} and G_{3221} — see Eq. (8.28)] are equal to each other at M_{W_R} and that the only way to secure a sufficient reduction (to 0.232) in the value of $\sin^2 \theta_W$ (from its grand unification value of 0.375) is to insert large values for M_{W_R} and M_C into Eq. (8.30). If a mechanism can be found whereby the "running" coupling constant equations — down from $SO(10)$ — produce the inequality $g_L(M_{W_R}) > g_R(M_{W_R})$, it is possible to secure the necessary reduction [from 0.27 to 0.231 — see Eq. (8.30)] with detectable values of M_{W_R} and M_C. It turns out that the inequality $g_L(M_{W_R} > g_R(M_{W_R})$ can be realized by exploiting the presence of the discrete symmetry — called D parity — that is embedded in the simple $SO(10)$ group [but not in the semi-simple Pati-Salam group G_{422}].

The presence of the D (parity) operation — to be defined below — creates the possibility of one more intermediate mass scale (M_D), larger than M_C, which can serve as a "surrogate" (so to speak) to reconcile the predictions of the new ("non-traditional") $SO(10)$ GUT for τ_p and $\sin^2\theta_W$ with detectable values of M_{W_R} and M_C. What is this "magical" property of D parity that achieves this reconciliation?

We have already noted that in "traditional" $SO(10)$ GUT, parity-breaking is coupled to $SU(2)_R$-breaking and this leads to the equality $g_L(M_{W_R}) = g_R(M_{W_R})$ with its negative consequences for discovering "new physics" in the multi-TeV region. The new approach succeeds in decoupling parity-breaking and $SU(2)_R$-breaking [8.26] by distinguishing the $SO(10)$ Higgs representations that are "odd" under the embedded discrete "D parity" operator and those that are "even". The formal definition of the "D parity" operator is: $D = \Sigma_{23}\Sigma_{67}$ [where $\Sigma_{\mu\nu}$ ($\mu = 1, 2, ..., 10$) are the 45 totally anti-symmetric generators of the $SO(10)$ group] so that D must be an "inner automorphism" of $SO(10)$ and play a role similar to charge conjugation on the fermions; for example, at the G_{422} level of $SO(10)$, the "D parity" operator transforms the fundamental fermion representation $(\mathbf{4}, \mathbf{2}, \mathbf{1})$ into $(\mathbf{\bar{4}}, \mathbf{1}, \mathbf{2})$, thus:

$$D(\mathbf{4}, \mathbf{2}, \mathbf{1}) = (\mathbf{\bar{4}}, \mathbf{1}, \mathbf{2}) \tag{8.36}$$

The subtlety in Eq. (8.36) is that not only does "D parity" complex-conjugate the $SU(4)_C$ fundamental representation but the $SU(2)_L$ and $SU(2)_R$ representations are interchanged.

The advantage of identifying "D parity" as a discrete operator embedded in the $SO(10)$ group is that one can distinguish the renormalization group effects of the singlet decomposites of $SO(10)$ Higgs representations that are "odd" under "D parity" from those that are "even" under "D parity". Thus, if one examines the decomposition of the $\mathbf{54}$ and $\mathbf{210}$ Higgs representations of $SO(10)$, one finds that both $\mathbf{54}$ and $\mathbf{210}$ lead to the maximal little group G_{422} because both possess a purely singlet G_{422} decomposite $\eta\,(1, 1, 1)$ (we denote the singlet decomposite by η); however, whereas $\eta\,(1, 1, 1)$ is "D-even" [i.e. $D\eta\,(1, 1, 1) = +\eta(1, 1, 1)$] for the Higgs $\mathbf{54}$ representation, it is "D-odd" [i.e. $D\eta\,(1, 1, 1) = -\eta\,(1, 1, 1)$] for the Higgs $\mathbf{210}$ representation. Since the $\mathbf{54}$ and $\mathbf{210}$ Higgs representations both break $SO(10)$ to G_{422}, we can first break $SO(10)$ with the "D-even" $\mathbf{54}$ Higgs representation, ending up with what we denote as G_{422D} [we add D to the subscript to differentiate G_{422D} from G_{422} where "D parity" is broken] and then break G_{422D} to G_{422} with the help of the "D-odd" Higgs $\mathbf{210}$ representation of $SO(10)$. The overall effect is to interpose the G_{422D} group into the $SO(10)$ chain of Eq. (8.27), thus [8.26]:

$$SO(10) \xrightarrow[M_U]{54} G_{422D} \xrightarrow[M_D]{210} G_{422} \xrightarrow[M_C]{45} G_{3211} \xrightarrow[M_{W_R}]{45} G_{3211} \xrightarrow[M_{Z_R}]{126} G_{321} \xrightarrow[M_W]{10} G_{31} \tag{8.37}$$

It is seen from Eq. (8.37) that there is a new intermediate mass scale M_D upon which to place the burden of parity-breaking [without breaking $SU(2)_R$, i.e. parity-breaking is decoupled from $SU(2)_R$-breaking] at a high mass scale; that is to say, while the equality $g_L(M_D) = g_R(M_D)$ must hold, it is now possible to find SSB chains in "non-traditional" $SO(10)$ GUT wherein the "running" gauge coupling constants g_L and g_R satisfy the inequality $g_L(M_{W_R}) > g_R(M_{W_R})$. Another advantage of exploiting D parity in the SSB chain down from $SO(10)$ is the removal of a troublesome obstacle in the realization of the "see-saw" mass matrix mechanism for neutrinos in the $SO(10)$ model. It was already pointed out in ref. [8.14] that in left-right — or $SO(10)$-type models — the "see-saw" mechanism can be compromised by the presence of an intrinsic $\nu_L \nu_L$ mass term of the order $\simeq \lambda M_{W_L}^2/V_{BL}$ [λ is the coupling parameter in the quartic part of the Higgs potential and V_{BL} is the VEV of the Higgs boson breaking $U(1)_{B-L}$], which converts the "see-saw" mass formula for light neutrinos into the form:

$$m_{\nu_\ell} \simeq \lambda \frac{M_W^2}{V_{BL}} - \frac{m_\ell^2}{V_{BL}} \tag{8.38}$$

Clearly, for values of $V_{BL} \leq 10^{12}$ GeV ($\lambda \simeq 1$), Eq. (8.38) contradicts the known upper limits on neutrino masses. However, it was shown that, in theories with broken D parity, Eq. (8.38) becomes:

$$m_{\nu_\ell} \simeq \lambda \frac{M_W^2}{M_D^2} V_{BL} - \frac{m_\ell^2}{V_{BL}} \tag{8.39}$$

Thus, for $M_D \gg V_{BL}$, the first term in Eq. (8.39) can be made negligible, thereby restoring the "see-saw" formula for neutrino masses [8.27]. It is therefore fair to say that in realistic "non-traditional" $SO(10)$ GUT models, D parity must be broken close to the GUT scale. Such solutions are also needed for compatibility with the inflation scenario and it is seen below that they do exist.

Once it can be shown that SSB chains exist satisfying the condition $g_L(M_{W_R}) > g_R(M_{W_R})$, it is not surprising that SSB chains in "non-traditional" GUT can be found predicting substantially smaller values of the intermediate mass scales M_C and M_{W_R}. However, it is gratifying that despite a very comprehensive search [8.28], few solutions with detectable IMS's exist if one insists on the value 0.23 for $\sin^2 \theta_W$ and the lower limit 6×10^{32}yr for τ_p. Interestingly enough, the imposition of a minimality condition (that the lowest dimensional Higgs is used at each symmetry-breaking stage) only allows a single chain (with detectable M_{W_R} and M_C) to survive, namely [8.27]:

$$ SO(10) \xrightarrow[54]{M_U} G_{422D} \xrightarrow[210]{M_D} G_{422} \xrightarrow[210]{M_C = M_{W_R}} G_{321} \xrightarrow[126]{M_{Z_R}} G_{321} \xrightarrow[10]{M_W} G_{31} \qquad (8.40) $$

One notes from Eq. (8.40) [in comparison to Eq. (8.37)] that the two-stage breaking: $G_{422} \to G_{3221} \to G_{3211}$, with two intermediate mass scales M_C and M_{W_R}, is replaced by the one-stage breaking: $G_{422} \to G_{3211}$, with one intermediate mass scale identified as: $M_C = M_{W_R}$; this should not be taken too seriously because the numbers given below are intended only to be illustrative of what is possible within the framework of "non-traditional" $SO(10)$ GUT when D parity is explicitly incorporated into the $SO(10)$ GUT model.

With this caveat, Eq. (8.40) tells us that there is a rather unique SSB chain in "non-traditional" $SO(10)$ GUT that has the following observable consequences: (i) detection of a "live" Z_R boson with a mass in the range 350 GeV to 1 TeV is possible; (ii) $K_L \to \mu\bar{e}$ decay to test the $SU(4)_C$ gauge sector of the theory — could be had with a branching ratio as high as 10^{-12}; (iii) $n - \bar{n}$ oscillations, with a mixing time of the order $10^8 - 10^{10}$sec — to test the Higgs sector of the model — could occur. The rather unique chain (8.40) also predicts that the grand unification mass M_U is in the narrow range $10^{16.0 \pm 0.2}$ GeV so that the predicted value of $\tau_p \simeq 1.5 \times 10^{36.0 \pm 1.0} (\frac{\Lambda_{\overline{MS}}}{150 \text{ MeV}})^4$ may be barely within reach of experiment. The predicted value of M_D falls in the range $10^{14.3 \pm 1.0}$ GeV, consistent with cosmological constraints [the "see-saw" mechanism for neutrino masses and the SSB of "D parity" generate unacceptable domain walls [8.29], which disappear — in the inflationary scenario [6.54] — for $M_D > 10^{12}$ GeV]. Finally, the chain of Eq. (8.40) requires $M_{W_R} \sim M_C$ so that if $n - \bar{n}$ oscilations and/or $K_L \to \mu\bar{e}$ decay are seen, it is supportive of the idea that the observed CP-violating effects in $K_L \to 2\pi$ decay are due to R (righthanded) currents in a left-right-symmetric model [8.30]. Incidentally, the same "LRS" model could provide an explanation for an observable electric dipole moment of the neutron — say, at the $10^{-27}e$ cm. level — in contrast to the unobservable value of $\sim 10^{-32}e$ cm predicted by the standard electroweak theory (see §6.3f).

We have thus seen that, while "traditional" $SO(10)$ GUT [wherein parity-breaking and $SU(2)_R$ breaking are coupled] does not lead to any

measurable consequences for weak processes associated with the intermediate mass scales M_{W_R} and M_C, "non-traditional" $SO(10)$ GUT [wherein
parity-breaking is achieved by D-parity-breaking so that parity-breaking and
$SU(2)_R$ breaking are decoupled] may lead to a completely different picture
for these two intermediate mass scales in the $SO(10)$ GUT model. The "non-
traditional" approach to the $SO(10)$ GUT model is, as far as we know, the
only way to establish whether the $SU(5)$ GUT-type "desert" — except for
the detectable consequences associated with a possible low mass Z_R gauge
boson — is "fructified" by enlarging rank 4 $SU(5)$ GUT to rank 5 $SO(10)$
GUT [with its built-in $(B - L)$ and left-right gauge symmetries] and taking into explicit account the D parity "inner automorphism" embedded in
$SO(10)$. While "non-traditional" $SO(10)$ GUT has its attractive features,
there is, of course, the possibility that the grand unification idea is only
implemented in a GUT group with rank greater than 5; in that case, the
most interesting GUT candidate, and the one most extensively studied, is
E_6 GUT and to this remaining "candidate" GUT model we now turn.

§8.5. E_6 grand unification

In §8.2 of this chapter, we pointed out that the only 1acceptable simple
compact Lie (GUT) groups were to be found within the $SU(N)$ $(N \geq 5)$,
$SO(4k + 2)$ $(k \geq 2)$ and E_6 groups. The lowest rank (4) candidate GUT
group, $SU(5)$, was examined first and its attractive features of simplicity
and uniqueness sadly brought it into conflict with the measured lower limit
on the proton lifetime and a precision measurement of $\sin^2 \theta_W$ [or, equivalently, with the recent determination that $SU(5)$ GUT — at least in its
minimal form — is incapable of furnishing a single intersection point for the
three "running" equations for the separate coupling constants [8.2]]. This
disagreement compelled us to move on to the next higher rank (5) candidate GUT group, of which the most promising is the $SO(10)$ group, breaking down spontaneously through its maximal G_{422} subgroup. $SO(10)$ GUT
(because of its increased rank) easily passes the proton decay and "single intersection point" tests and possesses several additional virtues: (1) its single
spinorial fermion representation **16** is triangular anomaly-free and accommodates all 15 (lefthanded) Weyl states of one generation (in the standard
model) plus one ν_R (a very desirable feature if the neutrino mass is finite);
(2) it places "full" quark-lepton symmetry on a natural footing through the
gauging of $B - L$ symmetry; and (3) it makes provision for an SSB mechanism for parity on a par with the other SSB-induced effects in the modern

gauge theory of particle interactions. Moreover, if $SO(10)$ GUT makes use of the embedded "D-parity" operator (that decouples parity-breaking from $SU(2)_R$-breaking), "non-traditional" $SO(10)$ GUT opens up a broad range of "new physics" associated with its three (or four if M_D is counted) intermediate mass scales, including the incorporation of the weak CP violation effects in the LRS subgroup of $SO(10)$.

However, it must be admitted that the price of exploiting D parity in $SO(10)$ GUT is very steep and "non-traditional" $SO(10)$ GUT may not work. In that case — and if one is still intrigued by the grand unification idea — it may be necessary to move on to the next higher rank (6) GUT groups to try to establish whether the hope of grand unification with observable consequences can still be maintained. Evidently, moving on to a rank 6 GUT model brings into play an even larger number of IMS's and of "exotic" fermions than $SO(10)$ GUT and, until recently, only a modest effort was devoted to a study of the most interesting candidate for a rank 6 GUT group, namely E_6 [1.153]. The problem with E_6 GUT is that not only is there greatly increased arbitrariness in the E_6 SSB chain [compared to $SO(10)$], but even the fundamental representation of E_6, **27**, already contains eleven additional "exotic" fermions [beyond the ν_R of $SO(10)$ GUT], whose absence thus far must somehow to be explained (see below).

The revival of E_6 GUT is primarily due to the surge of interest in superstring theories [1.185]. Phenomenologically, the most promising of these superstring theories in the recent past is the 10-dimensional (space-time) "heterotic string" $E_8 \times E_8'$ group with six of its nine space dimensions compactified in a so-called Calabi-Yau manifold (see §1.4c). It is claimed that the multiply-connected nature of a Calabi-Yau manifold provides an automatic mechanism for breaking the E_8 gauge symmetry (E_8' is relegated to a "shadow world"!) down to a E_6 subgroup whose structure, "in principle", can be predicted [1.185]. Since a graviton is contained in the "heterotic string" theory and the number of quark-lepton generations can be predicted from the difference between the **27** and $\overline{\textbf{27}}$ matter representations, the "heterotic string" theory generated a great deal of interest in "superstring-inspired" E_6 GUT models when it was first proposed. The mathematical discovery that there are about 10^4 Calabi-Yau possible manifolds in which the $D = 10$ space-time of the "heterotic string" theory can be compactified has dampened the enthusiasm for "superstring-inspired" E_6 but it still may be worthwhile to review briefly the phenomenological prospects of E_6 GUT.

The **27** representation of E_6 — to which reference has already been made — is the fundamental (complex) triangular anomaly-free representation of E_6 and, if it is used for Weyl fermions, it must, as a minimum, accommodate the 15 or 16 (if we include ν_R) (lefthanded) Weyl quark-lepton states of one generation. To exhibit the quantum numbers of the Weyl fermions of the **27** representation of E_6, we consider the three most promising maximal little groups (as usual, we satisfy the "Michel conjecture" — see §8.2) into which E_6 can break down spontaneously:

$$E_6 \xrightarrow{\textbf{78}} SO(10) \times U(1)$$
$$\xrightarrow{\textbf{650}} SU(3) \times SU(3) \times SU(3)$$
$$\xrightarrow{\textbf{650}} SU(6) \times SU(2) \tag{8.41}$$

where the number above the arrow indicates the lowest Higgs representation that breaks E_6 to the designated maximal little subgroup. We next write down the decomposition of **27** under the three subgroups of Eq. (8.41):

$$\textbf{27} \rightarrow \textbf{1}(4) + \textbf{10}(-2) + \textbf{16}(1) \quad \text{under } SO(10) \times U(1) \tag{8.42a}$$
$$\rightarrow (\bar{\textbf{3}}, \textbf{3}, \textbf{1}) + (\textbf{3}, \textbf{1}, \textbf{3}) + (\textbf{1}, \bar{\textbf{3}}, \bar{\textbf{3}}) \quad \text{under } [SU(3)]^3 \tag{8.42b}$$
$$\rightarrow (\bar{\textbf{6}}, \textbf{2}) + (\textbf{15}, \textbf{1}) \quad \text{under } SU(6) \times SU(2) \tag{8.42c}$$

Somewhere, among these **27** states, in each of the decompositions, we must find the **15** (or 16) quark-lepton states that are known to us and then figure out the quantum numbers of the remaining eleven "exotic" fermions.

This exercise is most easily carried out with the $SO(10) \times U(1)$ subgroup since we can immediately identify the **16** of $SO(10)$ as the same lowest-dimensional (complex) spinorial **16** representation that we have already discussed in connection with $SO(10)$ GUT. It can easily be seen that the singlet of $SO(10)$ represents another righthanded neutrino, and the **10** representation of $SO(10)$ [1.153] contains an "exotic" lepton doublet and its conjugate doublet — under $SU(2)_L$ (four states) — plus an "exotic" d-type quark and its conjugate (six states), totaling to the ten states of the (10) representation in Eq. (8.42a). The **1** and **10** decomposites of **27** under $SO(10) \times U(1)$ are real representations of $SO(10)$ and it has been argued (on the basis of the "survival hypothesis" [8.31]) that these ten fermions acquire large masses in E_6 GUT and hence need not have been observed. However, this is not a clearcut prediction and it can also be argued that there are many possible SSB patterns which can yield observable masses of these "exotic" fermions with the next round of accelerators.

What about the gauge bosons in E_6 GUT? The gauge bosons are in the adjoint **78** representation of E_6 but we only consider the decomposition of **78** under the two maximal subgroups $SO(10) \times U(1)$ and $[SU(3)^3]$ [since the maximal subgroup $SU(6) \times SU(2)$ is expected to break down to $SU(5) \times [U(1)]^2$ and the prediction for the proton decay lifetime for this E_6 SSB path can not exceed the $SU(5)$ GUT prediction, which is ruled out]; we have for the decomposition of the **78** representation under $SO(10) \times U(1)$ and $[SU(3)^3]$:

$$\mathbf{78} = (\mathbf{45}, 0) + (\mathbf{16}, -3) + (\overline{\mathbf{16}}, 3) + (\mathbf{1}, 0) \quad \text{under } SO(10) \times U(1) \quad (8.43\text{a})$$

$$\mathbf{78} = (\mathbf{8}, \mathbf{1}, \mathbf{1}) + (\mathbf{1}, \mathbf{8}, \mathbf{1}) + (\mathbf{1}, \mathbf{1}, \mathbf{8}) + (\mathbf{3}, \mathbf{3}, \overline{\mathbf{3}}) + (\overline{\mathbf{3}}, \overline{\mathbf{3}}, \overline{\mathbf{3}}) \text{under } (SU(3))^3$$
$$(8.43\text{b})$$

Equation (8.43a) tells us that the $(\mathbf{45},0)$ decomposite of **78** yields essentially the same IMS's and physical results as were obtained with the **45** Higgs of $SO(10)$ GUT. The novel maximal subgroup is $[SU(3)]^3$ and it is worth saying a few words about a typical SSB chain of E_6 GUT that proceeds through $[SU(3)]^3$; one such chain is:

$$E_6 \xrightarrow[\substack{\mathbf{650} \\ M_U}]{} SU(3)_C \times SU(3)_L \times SU(3)_R \xrightarrow[\mathbf{78}]{} SU(3)_C \times SU(2_L) \times SU(2)_R$$

$$\times U(1)_L \times U(1)_R \xrightarrow[\mathbf{27}]{} SU(3)_C \times SU(2)_L \times SU(2)_R \times U(1)_{B-L}$$

$$\xrightarrow[\mathbf{351'}]{} SU(3)_C \times SU(2)_L \times U(1)_Y \quad (8.44)$$

It is seen from Eq. (8.44) that, before reaching G_{321}, the SSB chain from E_6 passes through the LRS G_{3221} despite the fact that the LRS group does not originate with the SSB of $SO(10)$ or its maximal subgroup G_{422}. It is also evident from Eq. (8.44) that there is a (rank 6) intermediate group $G_{32211} \equiv SU(3)_C \times SU(2)_L \times SU(2)_R \times U(1)_L \times U(1)_R)$ whose SSB can generate another low-mass neutral gauge boson [in addition to the Z_R of $SO(10)$ GUT].

From the vantage point of the "heterotic superstring" group, it can be shown that discrete symmetries (achieved by means of "Wilson loops" [8.32]) can break the E_6 group down to the G_{32211} subgroup. Discrete groups are introduced into superstring theory in order to obtain a sufficiently low number of predicted quark-lepton families at low energies and it is considered a "bonus" that a second low-mass gauge boson (in addition to Z_R) is predicted. More precisely, the use of an Abelian discrete group in the compactification process yields the rank 6 group G_{32211} whereas the use of a non-Abelian discrete group in the compactification process breaks E_6 to a rank 5 subgroup G'_{3221} or G'_{3211} [which need not be the G_{3221} nor the G_{3211} of $SO(10)$ GUT]. Unfortunately, it is not possible to obtain the rank 4 standard group G_{321} directly through the use of a non-Abelian discrete group in the compactification process. Furthermore, since E_6 is spontaneously broken by the use of discrete groups in the compactification process, the problem remains of reaching the superstring-predicted subgroup in a way that is consistent with the observed lower limit on proton decay lifetime. This turns out to be difficult. In conclusion, it is fair to say that while the rank 6 E_6 GUT group is naturally much richer in SSB chains than the rank 5 $SO(10)$ GUT group and it does appear that promising SSB chains in E_6 GUT can proceed through the maximal little group $[SU(3)]^3$, the phenomenological implications of "superstring-inspired" E_6 GUT are still rather ill-defined.

Looking back, the history of the grand unification strategy in modern particle physics has fluctuated from tremendous hope and enthusiasm: when $SU(5)$ GUT was conceived with its unique predictions for proton decay and $\sin^2 \theta_W$ — to disappointment and deep skepticism about the GUT approach when more accurate experiments did not confirm the $SU(5)$ GUT predictions. The larger rank $SO(10)$ and E_6 GUT models were known in the mid-1970's when $SU(5)$ GUT was first formulated, but the disillusionment that set in after the failure of $SU(5)$ GUT undermined the theoretical interest in pursuing grand (or partial) unification. The more explicit awareness — in the early 1980's — of the capability of $SO(10)$-embedded local $B - L$ symmetry [and its (SSB)] to generate new physical effects "beyond the standard model" revived interest in $SO(10)$ GUT. More recently, the advent of 10-dimensional superstring theory and the probing into its phenomenological consequences have led to a renewal of interest in E_6 GUT. But it must be admitted that, as of now, not a single experiment has supported the grand unification approach and, what is particularly troubling, is that all three GUT models — $SU(5)$, $SO(10)$ and E_6 — share the common feature that they are "one-fermion-generation" models and shed very little light on why there are precisely three fermion generations [with quark and charged lepton masses varying over a range of 10^5], and why all the masses but one (the t quark mass) are extremely small compared to the electroweak breaking scale Λ_{QFD}. The "fermion generation problem" and the puzzle of the fermion mass spectrum are still two of the major unresolved problems of the standard model despite the increasing attention that has been accorded them. Of the two, a modicum of progress has been made on the "fermion generation problem" and a status report is given in Chapter 9. Unfortunately, very little progress can be reported in our understanding of the peculiar character of the quark-lepton mass spectrum of the three generations; the anomalously massive t-quark — whose existence is expected but still remains to be confirmed — only underlines our ignorance of the dynamics of mass generation of matter at the present time.

References

[8.1] H. Georgi, H.R. Quinn and S. Weinberg, *Phys. Rev. Lett.* **33** (1974) 451.

[8.2] U. Amaldi, preprint CERN-PPE 91-44, Geneva (1991); *ibid. Phys. Rev.* **D36** (1987) 1385.

[8.3] L. Michel and L. Radicati, *Ann. Phys. (N.Y.)* **66** (1971) 758; J. Buizlaff, T. Murphy and L. O'Raifeartaigh, *Phys. Lett.* **154B** (1985) 159.

[8.4] R. Slansky, *Phys. Reports* **79** (1981) 1.

[8.5] Cf. P. Langacker, *Phys. Reports* **72** (1981) 185.

[8.6] P. Langacker, *Ninth Workshop on Grand Unification*, AIX-LES-BAINS (Savoie), France (1988).

[8.7] S. Weinberg, *Phys. Rev. Lett.* **43** (1979) 1566; F. Wilczek and A. Zee, *Phys. Lett.* **88B** (1979) 311.

[8.8] C. Quigg, *Gauge Theories of Strong, Weak and Electromagnetic Interactions*, Benjamin-Cummings, California (1983).

[8.9] A. Zee, *Unity of Forces in the Universe*, Vol. I, World Scientific, Singapore (1982).

[8.10] S. Dimopoulos, S.A. Raby and F. Wilczek, *Physics Today*, October 1991.

[8.11] Y. Tosa, G.C. Branco and R.E. Marshak, *Phys. Rev.* **D28** (1983) 1731.

[8.12] A. Davidson, *Phys. Rev.* **D20** (1979) 776.

[8.13] R.E. Marshak and R.N. Mohapatra, *Phys. Lett.* **91B** (1980) 222.

[8.14] R.N. Mohapatra and G. Senjanović, *Phys. Rev. Lett.* **44** (1980) 912; *ibid. Phys. Rev.* **D23** (1981) 165.

[8.15] R.E. Marshak and R.N. Mohapatra, Festschrift in honor of Y. Ne'eman: *From SU(3) to Gravity*, ed. E. Gotsman and G. Tauber, Cambridge Univ. Press (1985).

[8.16] Y. Tosa, R.E. Marshak and S. Okubo, *Phys. Rev.* **D27** (1983) 444.

[8.17] Y. Tosa and S. Okubo, *Phys. Rev.* **D23** (1981) 2486; **D23** (1981) 3058.

[8.18] Riazuddin, R.E. Marshak and R.N. Mohapatra, *Phys. Rev.* **D24** (1981) 1310.

[8.19] G. Beall, M. Bander and A. Soni, *Phys. Rev. Lett.* **48** (1982) 848.

[8.20] J.M. Gipson and R.E. Marshak, *Phys. Rev.* **D31** (1985) 1705.

[8.21] R.N. Mohapatra, *Proc. of 8th Workshop on Grand Unification*, Syracuse (1987).

[8.22] R.N. Mohapatra and R.E. Marshak, *Phys. Lett.* **B94** (1980) 183.

[8.23] T.W. Jones *et al.*, *Phys. Rev. Lett.* **52** (1984) 720.

[8.24] C. Dover, A. Gal and J. Richard, *Phys. Rev.* **D27** (1983) 1090.

[8.25] M. Baldo-Ceolin and D. Dubbers, private communications (1988–89).

[8.26] D. Chang, R.N. Mohapatra and M.K. Parida, *Phys. Rev. Lett.* **52** (1984) 1072; D. Chang and R.N. Mohapatra, *Phys. Rev.* **D32** (1985) 1248.

[8.27] R.N. Mohapatra and M.K. Parida, U. of Maryland preprint (1992).

[8.28] D. Chang, R.N. Mohapatra, J.M. Gipson, R.E. Marshak, and M.K. Parida, *Phys. Rev.* **D31** (1985) 1718.

[8.29] T.W.B. Kibble, G. Lazarides, and Q. Shafi, *Phys. Rev.* **D26** (1982) 435.

[8.30] R.N. Mohapatra, *CP Violation*, ed. C. Jarlskog, World Scientific, Singapore (1989).

[8.31] H. Georgi, *Nucl. Phys.* **B156** (1979) 126.

[8.32] Y. Hosotani, *Phys. Lett.* **129B** (1983) 193.

Chapter 9

FERMION GENERATION PROBLEM
AND PREON MODELS

§9.1. Non-preon approaches to the fermion generation problem

We described in Chapter 8 the three major attempts to achieve a grand unification of the strong and electroweak groups using a single irreducible Weyl fermion representation (or at most two irreducible representations) of the quark and lepton fields of a single generation, namely the $SU(5)$, $SO(10)$ and E_6 GUT models. In the case of rank 5 $SO(10)$ GUT, we considered two versions — the "traditional" $SO(10)$ GUT and the "non-traditional" $SO(10)$ GUT with D parity — to illustrate the difference in the "new physics" that can result by taking cognizance of an "inner automorphism" in the $SO(10)$ group. Some consideration was also given to two "partial unification" (PUT) subgroups of $SO(10)$, namely the "four-color" G_{422} (Pati-Salam) PUT group and the LRS (left-right-symmetric) group. The $SU(5)$ GUT group is unable to reconcile its unique predictions of the proton decay lifetime and $\sin^2 \theta_W$ with experiment. The two other GUT groups [$SO(10)$ and E_6], as well as G_{422} PUT, have more flexibility than $SU(5)$ GUT and can be made consistent with the observed lower limit on τ_p and the measured value of $\sin^2 \theta_W$, but otherwise have received no definitive experimental support for the "new physics" that would result. Even the partial LRS (G_{3221}) unification group — which is a subgroup of G_{422} — is not supported by any experiment; even the latest Gallex experiment does not provide clear evidence for a massive neutrino nor for neutrino oscillations (see §6.4b). Apart from the lack of experimental support for a larger group than the standard group, it must be reiterated that the "superfluous family replication" problem and the fermion mass problem in general would still remain unsolved because none of the unification theories (partial or grand) pretends to "unify" more than one fermion generation nor to offer some credible rationale for the bewildering range and mass patterns of the quarks and leptons in the three generations.

The fermion mass problem that confronts the standard model can be considered under two headings: (1) the particularities of the quark and lepton masses of the three generations both as regards their intra-generational and inter-generational behavior; and (2) the physical significance of family replication, i.e. the three-fold generational replication of quantum numbers despite the vast range of masses within the three generations (see Tables 1.5 and 1.6). With regard to the "particularities" of the quark and lepton masses, their theoretical calculation is in a primitive state and basically resides within the standard model framework where the quark and charged lepton masses are arbitrary parameters, determined by the choice of the Yukawa coupling constants of the Higgs SSB doublet to the Weyl quarks and leptons — thrice repeated — of the electroweak group.

Empirically, one can make several observations: (1) all known quark and lepton masses are small compared to the electroweak symmetry-breaking scale $\Lambda_{QFD} \simeq 250$ GeV with the single exception of the thus-far-unobserved t quark (whose mass is > 92 GeV [6.9]). Calculations of radiative effects due to the t quark suggest that $m_t < \Lambda_{QFD}$; (2) the average quark-lepton mass in each generation increases very rapidly (by at least a factor of 10) from one generation to the next; furthermore, within each generation, there is a comparably rapid increase in going from the average lepton mass to the average quark mass. The surprisingly large t quark mass appearing on the horizon only accentuates the rapidity of both the inter-generational and intra-generational fermion mass increases. It also casts doubt on GUT predictions about quark-lepton mass ratios within each generation even when the "running" coupling constant effects are taken into account [8.5]; and, finally (3) there is the curious phenomenon of "mass inversion" between the first and the last two generations, i.e. the reversal of the relative sign of the mass difference between the (weak isospin) "up" and "down" quarks with respect to the mass difference between the (weak isospin) "up" and "down" leptons. The "mass inversion" seen in the first generation is probably not significant since it involves the smallest known quark and lepton masses; accepting the relative sign of the mass difference between the "up" and "down" quarks and leptons of the second and third generations, the explanation may be straightforward and due to the opposite sign of the baryon and lepton charges in the expression for the "weak" hypercharge of quarks and leptons (see §8.4a). However, no reliable calculations have been performed of the "mass inversion" effect. Overall, progress has been slow in explaining the "particularities" of the quark and lepton masses [9.1] and it is unlikely that the fermion mass spectrum can be fully understood until the t quark and Higgs masses are known and the question of a dynamical origin of either or both of these particles is settled (see §6.4c).

Under these circumstances, we focus on the more limited "fermion generation problem" and review in this chapter the attempts that have been made to explain the three-fold replication in quark numbers of the observed quarks and leptons, assuming only that no mass exceeds the electroweak scale Λ_{QFD}. Unfortunately, the numerous attempts — which we classify as the "non-preon" and "composite preon model" approaches — have not succeeded in truly solving the "fermion generation problem"; however, the attempts to do so raise issues and provide insights that should give us a deeper understanding of the conceptual foundations of the standard model. We first review several typical non-preon attempts to explain the observed fermion family replication and then treat the composite preon model approach. The non-preon attack on the "fermion generation problem" may be considered under three headings: (1) family group approach; (2) radial (or orbital) excitation approach; (3) "enlarged group plus representation" approach. The first two approaches are straightforward but quite arbitrary whereas the third approach is also fairly straightforward but its rationale has a somewhat deeper group-theoretic basis.

(1) *Family group*

We have seen in §7.5 that the 15 Weyl fermion representations of the standard group G_{321} and hypercharge quantization are direct consequences of the three chiral gauge anomaly-free constraints in four dimensions on the standard group; moreover, electric charge quantization then follows from the application of the Higgs SSB mechanism. However, while we can understand any number of copies of the generational quantum numbers on the basis of the three anomaly-free conditions, the precise number of observed generations is not explained. One of the obvious approaches to the "fermion generation problem" is to adjoin a global "family group" G_f to G_{321} — so that the operative group is $G_{321} \times G_f$, with a three-dimensional representation to explain the three-fold replication of fermion generations. One can suppose the SSB of G_f gives rise to different (average) masses of the three generations.

However, if G_f is a continuous global group, the SSB of G_f gives rise to N-G bosons, which have dubbed "familons" and these objects should be detected; but, if a "familon" field f exists, it should be coupled (like a N-G boson) to a current j_μ^f which generates the family symmetry, as follows:

$$\mathcal{L} = -\frac{1}{\Lambda_f} \partial^\mu f j_\mu^f \tag{9.1}$$

where Λ_f is the symmetry-breaking scale of the global family group. Indeed, if family symmetry exists, "familons" ought to appear in many processes, e.g. $K^+ \to \pi^+ + f$, with the branching ratio [9.2]:

$$\frac{\Gamma(K^+ \to \pi^+ f)}{\Gamma(K^+ \to \pi^+ \pi^0)} = \frac{1.3 \times 10^{14}}{\Lambda_f^2} \text{ GeV}^2 \tag{9.2}$$

If one uses $j_f^\mu = \bar{s}\gamma^\mu d$ for the current, one finds that the known upper limit on the branching ratio [9.3] corresponds to $\Lambda_f > 10^{10}$GeV. Other "intra-generational" processes give similarly large values for the family-breaking scale and, while not excluded, the "familon" solution to the "fermion generation problem" is, at this stage, based on completely arbitrary assumptions.

In general, a similar statement holds for a family gauge group which would predict extremely massive (and unobservable) family gauge bosons resulting from the SSB of a family gauge group. There exists a fairly imaginative proposal — already mentioned (see §7.7) — to combine an $SU(2)_f$ family gauge group with the standard group G_{321}, to impose the three chiral gauge anomaly-free constraints on the enlarged group $G_{321} \times SU(2)_f$, and to require asymptotic freedom to hold for the new group [7.18]. The trouble with this model is that the choice of family group is arbitrary, an additional eight "exotic" fermions have to be invented, and the condition of asymptotic freedom only leads to an upper bound on the number of generations. It is fair to say that while the family group approach to the fermion generation problem is not excluded, no formulation, until now, looks particularly promising.

(2) Radial or orbital excitation approach

In this approach, one follows the "non-relativistic" atomic analogy where one writes down a radial potential of the form $V(r) = ar^\nu$ and argues that the mass m of the radially excited states increases as n^α (n is the radial quantum number) where $\alpha = 2\nu/(\nu + 2) \leq 2$. For an orbital excitation, again using the (non-relativistic) atomic analogy, the dependence of m on the orbital angular momentum is $m \sim \ell^2$. In these crude "non-relativistic" approximations, the predicted mass increase with each succeeding generation — for both radial and orbital excitations — is much slower than the empirical variation of quark and lepton masses from generation to generation. And it is unlikely that "relativistic corrections" will alter this conclusion.

(3) *"Enlarged-group plus representation approach"*

With no supporting evidence for the family group approach nor the radial and/or orbital excitation approach, it is worthwhile to mention one more non-preon approach, which we call the "enlarged group plus representation" approach to explain the "superfluous replication" of quark-lepton generations. This third approach — which was articulated soon after the grand unification models were proposed [9.4] — tried to explain the observed repetitive family structure by means of a sufficiently large fermion representation R' of an enlarged GUT group G' such that when G' breaks down spontaneously to either the $SU(5)$, $SO(10)$ or E_6 GUT group G, with representation(s) R, the irreducible representation R' decomposes into exact replicas of the one-fermion-family representation R. It turns out at once that this aesthetically pleasing approach can not be applied to $SU(5)$ because the (anti-symmetric) tensorial properties of the $SU(N)$ groups are such that the triangular anomaly-free condition — obeyed by the $\mathbf{5} + \mathbf{10}$ representations of $SU(5)$ (see §8.3) — is not "replicated" in enlargements of the $SU(5)$ GUT group; for example, for three fermion generations, it would be natural to consider the $SU(8)$ group as a candidate for the "super-group", select a sufficiently large irreducible representation of $SU(8)$ and then postulate the SSB of $SU(8)$ to $SU(5) \times SU(3)$. To express the objection in another way, the cancellation of the triangular anomalies for the combination of $\bar{\mathbf{5}}$ and $\mathbf{10}$ irreducible representations of $SU(5)$ does not carry over to any suitable representation R' of the enlarged group $SU(8)$.

In proceeding from rank 4 $SU(5)$ GUT to rank 5 $SO(10)$ GUT, it seemed for a while that the "enlarged group plus representation" approach to the "fermion generation problem" could be solved in an elegant fashion because of a special property of the spinorial representations of the $SO(4k+2)$ ($k \geq 2$) class of orthogonal groups. It happens that the spinorial representations of $SO[2(n+m)]$ groups decompose into direct sums of 2^m spinorial representations of $SO(2n)$. This feature makes automatic provision for exact replicas of the smaller group spinorial representations in the decomposition of the spinorial representation of the larger orthogonal group. However — because of maximal parity violation in the weak interaction — we must use irreducible chiral fermion representations for the larger family group since each fermion family is a chiral multiplet. This means that if we wish to end up with exact replicas of the spinorial **16** representation of $SO(10)$, we must have recourse to the class of orthogonal groups $SO(10 + 4k)\,(k > 0)$, for which the fundamental spinorial representation is complex.

Unfortunately, the decomposition of the complex fundamental representation of $SO(10 + 4k)\,(k > 0)$ yields an equal number of righthanded (i.e. "mirror") $\overline{\mathbf{16}}$ representations and lefthanded **16** representations of $SO(10)$ and one ends up with real representations of $SO(10)$. More explicitly, the complex spinorial representation (R') of $SO(10 + 4k)\,(k > 0)$ decomposes into the sum of representations:

$$R' \to (2^{2k-1}, \mathbf{16}) + (2^{k-1}, \overline{\mathbf{16}}) \tag{9.3}$$

The $\overline{\mathbf{16}}$ representation in Eq. (9.3) represents **16** righthanded "mirror" fermions, in contrast to the **16** representation of the lefthanded "regular" fermions; Eq. (9.3) leads to trouble because there is an even number of **16** and $\overline{\mathbf{16}}$ representations so that the complex spinor representation R' of $SO(10+4k)$ essentially decomposes into 2^{2k-1} real representations of $SO(10)$. Thus, the $SO(14)\,(k = 1)$ group is ruled out because the predicted number of families is two and one must move on to the $SO(18)\,(k = 2)$ group where, in principle, the group G' and the spinorial representation R' are sufficiently large. For the $SO(18)$ group, G', and the spinorial representation $R' = \mathbf{256}$, the number of predicted families is, *ab initio*, much too large: eight lefthanded families [each in the **16** representation of $SO(10)$] and eight righthanded families [each in the $\overline{\mathbf{16}}$ representation under $SO(10)$]. It is possible to explain away half of the lefthanded and half of the righthanded families — by means of the "survival" hypothesis [8.31] — and even make plausible much heavier masses for the righthanded families compared to the lefthanded ones, but the predictions in this $SO(18)$ theory [9.5] — that eight light neutrinos should contribute to the Z width and that there should be four light lefthanded families — are now untenable (since there are just three neutrino species and, *a fortiori*, three generations — see §6.3b). It is evident that the "enlarged group plus representation" approach to family replication does not succeed within the framework of $SO(10 + 4k)\,(k > 0)$ super-GUT models.

The remaining GUT group, E_6, does not fare any better than $SO(10)$ GUT. The most plausible enlargement of the exceptional group E_6 is the exceptional group E_8, which can break spontaneously into $E_6 \times SU(3)$. However, E_8 only possesses real representations and, in particular, the fundamental (real) representation of E_8, **248**, decomposes under $E_6 \times SU(3)$, into:

$$\mathbf{248} = (\mathbf{1}, \mathbf{8}) + (\mathbf{78}, \mathbf{1}) + (\mathbf{27}, \mathbf{3}) + (\overline{\mathbf{27}}, \overline{\mathbf{3}}) \tag{9.4}$$

The last two terms in Eq. (9.4), at first sight, look interesting since it could be argued (in the absence of knowledge about the SSB mechanism from E_8 to E_6) that the **27** representation of E_6 can be assigned to the fermions of one generation and that the **3** representation can be thought of as the fundamental representation of an $SU(3)$ family group; however, one is again confronted with an equal number of "mirror" **27** representations. [Amusingly enough, the decomposition given in Eq. (9.4) has been exploited to derive an "index theorem" that relates the difference between the number of **27** fermion representations and the number of "mirror" **27** fermion representations to the Euler number of the 6-dimension Calabi-Yau space in superstring theory [1.185]; but it is clear that the superstring application of Eq. (9.4) is quite unrelated to the "enlarged group plus representation" approach to family replication.] Unfortunately, we must come to the conclusion that the "enlarged group plus representation" approach to the "fermion generation problem" does not work for any of the three GUT models: $SU(5)$, $SO(10)$ or E_6, when the fundamental representation is taken for R' of the enlarged group G'.

One might inquire whether one could make progress with the "fermion generation problem" by dealing with one of the candidate GUT groups (i.e. $SU(5)$, $SO(10)$ or E_6) but going to larger (rather than fundamental) representations R' of G'. For example, the **1728** representation of E_6 breaks down — under $SO(10) \times U(1)$ — as follows [the number in parenthesis is the U(1) charge]:

$$\mathbf{1728} = \mathbf{1}(4) + \mathbf{10}(-2) + \mathbf{16}(1) + \mathbf{16}(1) + \overline{\mathbf{16}}(7) + \mathbf{45}(4) + \mathbf{120}(-2)$$
$$+ \mathbf{126}(-2) + \mathbf{144}(1) + \overline{\mathbf{144}}(-5) + \mathbf{210}(4) + \mathbf{320}(-2) + \mathbf{560}(1)$$

$$(9.5)$$

It is seen from Eq. (9.5) that **1728** does contain two lefthanded **16** representations of $SO(10)$ but it also contains one $\overline{\mathbf{16}}$ "mirror" representation and an enormous number of "exotic" fermions. It is conceivable that an extremely large representation of E_6 would contain a "net" set of three **16** (lefthanded) Weyl fermion representations of $SO(10)$ but then one would have to concoct a myriad of dynamical arguments for discarding the huge number of remaining "exotic" fermions. Thus, the idea of explaining the three observed generations of quarks and leptons by enlarging a GUT group and its associated representation appears to be doomed to failure.

In view of the failure of any of the non-preon approaches to explain the three-fold replication of one-generation standard group quantum numbers, we scrutinize what appears to be a natural generalization of the composite quark flavor model of hadrons, namely the composite preon model of quarks and leptons. The hope is that just as the quark flavor model successfully predicts the multiple groupings of hadronic states, so might the preon model successfully predict the number (three) of quark-lepton generations. Details concerning the quark flavor model of hadrons are given in §3.2c and in §5.1, where it is shown that the quark model is grounded in the combined gauged quark color-global quark flavor group $\bar{G} = SU(3)_C \times SU(3)_F \times U(1)_B$, where $SU(3)_C$ is the confining color group for quarks, $SU(3)_F$ is the global quark flavor group that flows from the triplet of light quarks (u, d, s), and $U(1)_B$ is the global group for the baryon charge of the quarks. The lowest mass hadronic quark composites of Table 1.6 are all color singlets [under gauged $SU(3)_C$], flavor singlets, octets and/or decuplets [under global $SU(3)_F$], and possess 0 or 1 baryon charge [under global $U(1)_B$]. The important observation is that the quark group $\bar{G}$ is the product of a gauge group $[SU(3)_C]$ and a global group $[SU(3)_F \times U(1)_B]$ that is never gauged. It is this last statement which must be in sharp contrast to the preon group — whose composite representations are supposed to define the quarks and leptons of each generation and where the corresponding group is ultimately the totally gauged group $SU(3)_C \times SU(2)_L \times U(1)_Y$ of the standard model. We will see that most preon models of quarks and leptons, while bearing a strong resemblance to the quark flavor model of hadrons, differ not only in the "gauge-global" nature of the underlying group but in other fundamental ways.

In the next section (§9.2), we describe the generic similarities and differences between composite preon models of quark-lepton generations and the composite quark model of hadrons. We then review — in §9.3 — the phenomenological lower limits on the preon "metacolor" (MC) scale Λ_{QMCD} [Λ_{QMCD} is the analog — in the preon model — of Λ_{QCD} in QCD]. The following section (§9.4) is devoted to preon models with gauged "metacolor" and gauged "metaflavor" ["metaflavor" is the analog — in preon models — of flavor in the standard group]. This class of preon models differs significantly from the quark model of hadrons because the hypothesis of gauged "metaflavor" *ab initio* places constraints on the preon model and happens to predict precisely three fermion generations with the phenomenologically correct (quark-lepton) quantum numbers; unfortunately, this type of preon model suffers from an excess of "exotic" (multi-preon) fermions. The undesired "exotic" fermions can be eliminated if the initial preon group is taken to consist of a gauged "metacolor" group and a global "metaflavor" group [this is closer to the quark model of hadrons with its gauged color and global flavor groups — but, of course, the global "metaflavor" group must ultimately be gauged] which is subjected to the 't Hooft anomaly-matching condition [1.159]. Examples of models with gauged "metacolor" and global "metaflavor" are considered in §9.5, making extensive use of the 't Hooft anomaly-matching condition, first without (see §9.5a) and then with (see §9.5b) the use of the "tumbling complementarity" condition. Some candidate composite preon models are identified in this fashion but no claim can be made for the uniqueness of any model and we conclude §9.5 with some comments on the present status of the composite preon approach to the fermion generation problem.

§9.2. Similarities and differences between the composite preon and composite quark models

In order to compare in some depth the composite preon model (of quark-lepton generations) and the composite quark model (of hadrons), we must review the salient features of the quark model of hadrons, including the consequences of the 't Hooft triangular anomaly matching condition [1.159] between massless (Weyl) constituent quarks and massless three-quark composites (baryons). This exercise — which compares the compatibility of the 't Hooft anomaly matching condition between massless constituent quarks and massless (three-quark) composites for two and three quark flavors — is most instructive in clarifying the power and limitations of the 't Hooft anomaly matching condition in its application to preon models. The best starting point in dealing with the composite quark flavor model is to examine the quantum numbers corresponding to the QCD Lagrangian (for n flavors of massless quarks — see §5.1) for $q, \bar{q}$ and the quark condensates $<q\bar{q}>$ and $<qq>$, as in Table 9.1:

Table 9.1: Quark, anti-quark and quark condensate quantum numbers in massless QCD (with n arbitrary flavors).

	$SU(3)_C$	$SU(n)_L$	$SU(n)_R$	$U(1)_L$	$U(1)_R$
	3	n	1	1	0
	$\bar{3}$	1	$\bar{n}$	0	-1
$\langle q\bar{q}\rangle$	$(1+8)$	n	$\bar{n}$	1	-1
$\langle qq\rangle$	$(\bar{3}+6)$	$\frac{n(n+1)}{2} + \overline{\frac{n(n-1)}{2}}$	$\frac{n(n-1)}{2} + \overline{\frac{n(n+1)}{2}}$	2	0

With the two fundamental (Weyl) quark and antiquark representations given in Table 9.1, it is easily shown that, to lowest-order, the two-quark interaction (for both $q\bar{q}$ and qq) is given by:

$$V(r) \sim \frac{g^2}{2r}(C_R - C_i - C_j) \tag{9.6}$$

where g_s is the color gauge coupling constant and $C_R = 0, 3, \frac{4}{3}$ and $\frac{10}{3}$ are the second-order Casimirs for the corresponding representations $\mathbf{1}, \mathbf{8}, \bar{\mathbf{3}}$ and $\mathbf{6}$ of the two-quark $[SU(3)_C]$ $\mathbf{3} \times \bar{\mathbf{3}}$ and $\mathbf{3} \times \mathbf{3}$ systems. The resulting two-quark interaction V is respectively proportional to $-\frac{8}{3}g^2$, $\frac{1}{3}g^2$, $-\frac{4}{3}g^2$, and $-\frac{2}{3}g^2$ for the $SU(3)_C$ singlet, octet, antitriplet and sextet two-quark composites listed in Table 9.1.

Thus, the $q\bar{q}$ singlet and the qq anti-triplet are in the most attractive channels of the respective $q\bar{q}$ and qq systems (and hence capable of forming two-quark condensates) and, for obvious reasons, the $q\bar{q}$ singlet is called the maximally attractive channel (MAC) [9.6]. Indeed, the quantum numbers of the MAC $q\bar{q}$ condensate are precisely those required to explain the empirical observations. As discussed in §5.1, the confining color group $SU(3)_C$ and the global baryon charge group $U(1)_B[= U(1)_{L+R}]$ are unbroken by the MAC $q\bar{q}$ condensate (equivalent to a "quasi-Higgs boson") whereas the global chiral quark flavor group $SU(n)_L \times SU(n)_R$ is broken down to $SU(n)_F[= SU(n)_{L+R}]$ and the axial charge group $U(1)_{L-R}$ is broken down — with the help of the $SU(3)_C$ instantons — to the discrete group Z_{2n}. Putting aside the discrete symmetry group Z_{2n} until later, we can say that the resulting unbroken gauge-global group $SU(3)_C \times SU(n)_F \times U(1)_B$ is vector-like and allows for finite quark masses. While $n = 3$ flavors is the maximum for QCD, we continue our discussion in terms of an arbitrary number of n flavors in order to highlight the similarities and differences between the preon and quark models within the framework of the 't Hooft triangular anomaly-matching condition. For this purpose, we introduce a more compact notation and restrict ourselves to baryonic three-quark composites.

. We begin by invoking the Cartan notation for a two-component Weyl fermion since the constituent quarks in the quark model of hadrons are Weyl (massless) fermions and, later, the constituent preons in the preon model of quarks and leptons, are also Weyl (massless) fermions. We call a spin $\frac{1}{2}$ fermion a Weyl fermion that transforms as either a L (lefthanded) $(\frac{1}{2},0)$ or a R (righthanded) $(0,\frac{1}{2})$ representation under the Lorentz group. To find the spin of a composite state of three Weyl fermions, we use the decomposition rule for products of representations to find the Lorentz properties of composites. In particular, we can write:

$$LR = (\frac{1}{2}, 0) \times (0, \frac{1}{2}) = (\frac{1}{2}, \frac{1}{2}); \quad LL = (\frac{1}{2}, 0) \times (\frac{1}{2}, 0) = (1 + 0, 0) \quad (9.7a)$$

$$LLL = 2(\frac{1}{2}, 0) + (\frac{3}{2}, 0); \quad LRR = (\frac{1}{2}, 0) + (\frac{1}{2}, 1) \quad (9.7b)$$

The important points to note in Eqs. (9.7a) and (9.7b) are that LR yields a spin $J = 1$ non-chiral boson, LL a lefthanded spin $J = 1$ boson and a spin $J = 0$ boson, LLL two lefthanded $J = \frac{1}{2}$ and one lefthanded $J = \frac{3}{2}$ fermions and, finally, LRR one lefthanded $J = \frac{1}{2}$ fermion and a mixed state. Thus, insofar as three-Weyl fermion composites are concerned, both LLL and LRR contain lefthanded $J = \frac{1}{2}$ (Weyl) fermions [although the non-relativistic picture only allows spin $\frac{3}{2}$ for LLL]. We next denote by $T(\bar{3}, n, 1)_L$ and $V(3, 1, \bar{n})_L$ the "constituent ("constituent" in the sense of "preonic") quark" and the "constituent antiquark" Weyl representations under the QCD group for n flavors of quarks: $SU(3)_C \times SU(n)_L \times SU(n)_R$. In this T and V notation, the $U(1)_{L+R}$ group implicit in Table 9.1 becomes $U(1)_{T-V}$ and we will have occasion to speak about the $U(1)_{T-V}$ charge when we deal with the 't Hooft anomaly matching condition (see below). Thus, the fermionic baryons are the three-quark composites: TTT, T($\bar{V}\bar{V}$), V($\bar{T}\bar{T}$), VVV, which generate color-singlet lefthanded $J = \frac{1}{2}$ and $J = \frac{3}{2}$ three-quark composites — via TTT and T($\bar{V}\bar{V}$) — as well as the charge conjugate (righthanded) three-quark composites through VVV and $V(\bar{T}\bar{T})$.

Since the wavefunction associated with a color singlet three-quark (baryonic) composite is antisymmetrical, the rest of the wavefunction [for flavor, orbital angular momentum and chirality] must be symmetric; furthermore, in analogy with the strategy of the $SU(6)$ flavor-spin baryonic model (see §3.2d), flavor-chirality (assuming zero orbital angular momentum) must be symmetric under the larger simple group $SU_T(2n)$, rather than merely under the product flavor-chirality group $SU(n)_T \times SU(2)_{\text{chir}}$ [with a similar statement for $SU_V(2n)$]. When this generalized Pauli principle is applied to the color-singlet three-quark composites TTT, T($\bar{V}\bar{V}$), V($\bar{T}\bar{T}$, VVV, it turns out that the only surviving lefthanded color-singlet three-quark composites are those given by TTT, and, correspondingly, the surviving righthanded color-singlet three-quark composites are given by VVV.

The next step is to separate the $J = \frac{1}{2}$ and $J = \frac{3}{2}$ three-quark composites. Since TTT (and VVV) satisfies the generalized Pauli principle, we must decompose the totally symmetric part of TTT in terms of the symmetric (S), mixed (M), and antisymmetric (A) representations of $SU_T(n) \times SU(2)_{\text{chir}}$

(see Fig. 9.1):

$$(TTT)_S = S_3 \longrightarrow (M_3, M_3) + (S_3, S_3) \tag{9.8}$$

with the associated flavor dimensions:

$$2n(2n+1)(2n+2)/6 \rightarrow [n(n^2-1)/3, 2] + [\frac{n(n+1)(n+2)}{6}, 4] \tag{9.9}$$

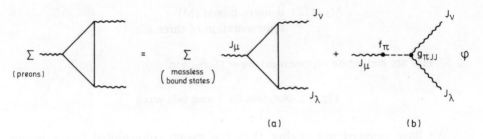

Fig. 9.1. 't Hooft triangular axial anomaly matching with (a): massless three-fermion composites and/or (b): N-G bosons.

The first term on the R.H.S. of Eq. (9.9) yields the number of $J = \frac{1}{2}$ three-quark composites, and the second term the number of $J = \frac{3}{2}$ three-quark composites. [For example, for $n = 3$, one obtains the $J = \frac{1}{2}$ flavor octet and $J = \frac{3}{2}$ flavor decuplet to explain the lowest-mass baryon $SU(3)_F$ multiplets.] The TTT composites are lefthanded Weyl fermions but, together with their righthanded partners from VVV, they combine to generate the Dirac fermions that must be used when the composites acquire mass. Parenthetically, if the color singlet part of the wavefunction were to give a symmetric contribution to the total wavefunction [we will find that the confining gauge group E_6 in the composite preon model leads to such a requirement], the flavor-spin contribution to the wavefunction would have to give an antisymmetric contribution. This would mean that the wavefunction would have to be antisymmetric under $SU_T(2n)$ and one would get for the decomposition of TTT under $SU_T(n) \times SU(2)_{\text{chir}}$ (see Fig. 9.2):

$$(TTT)_A = A_3 \longrightarrow (M_3, M_3) + (A_3, S_3) \tag{9.10}$$

where the dimensions of the irreducible representations are:

$$2n(2n-1)(2n-2)/6 \longrightarrow [n(n^2-1)/3, 2] + [\frac{n(n-1)(n-2)}{6}, 4] \tag{9.11}$$

Equations (9.10) and (9.11) will be useful for the composite preon model considerations.

If $F \equiv \square$ denotes fundamental representation of G, then

$S_2 \equiv \square\square$ denotes symmetric   ; $A_2 \equiv \begin{array}{c}\square\\\square\end{array}$ denotes antisymmetric
(S) representation of   (A) representation
two F's   of two F's

$S_3 \equiv \square\square\square$ denotes S representation ; $A_3 \equiv \begin{array}{c}\square\\\square\\\square\end{array}$ denotes A representation
of three F's   of three F's

$M_3 \equiv \begin{array}{c}\square\square\\\square\end{array}$ denotes mixed (M)
representation of three F's

$\bar{\square}$, $\bar{S}_2$, etc. are conjugate representations of $\square$, S_2, etc.

Fig. 9.2. Notation for Young tableaux.

We have pointed out earlier that the gauge color-global flavor group that governs n flavors in QCD is $SU(3)_C \times SU(n)_F$ [we omit $U(1)_B$ because the baryon charge of the quark is always $1/n$] and that the finite quark masses are consistent with the vector-like character of this group for the constituent quarks. Of course, vanishing mass for all n flavors of the quarks is permissible and the interesting question that then arises is whether it is possible — within the framework of the composite quark flavor model of hadrons — for the three-quark composite (baryonic) states to remain massless if one starts with masslesss constituent quarks, as we have done in writing down Eqs. (9.7). The answer to this question is crucial, as we will see, for the composite preon model of quarks and leptons (which must be a chiral gauge theory) but it is helpful to know the answer for the quark model because it highlights an important difference between the quark and preon models. For this purpose, we derive (and then apply) 't Hooft's triangular anomaly-matching condition which places a restriction on the number of chiral quark flavors that can yield massless three-quark composites (baryons) in QCD. We will see in §9.5 that satisfying 't Hooft's anomaly-matching condition establishes the precondition for the existence of realistic candidate preon models based on gauged "metacolor" and global "metaflavor". This is important because the existence of massless (to a first approximation) three-preon composites is the essential first step in preon model building, and it is interesting to learn what the 't Hooft anomaly matching condition teaches us in QCD.

But first, we must derive the 't Hooft anomaly matching condition; we start with the equation that exhibits the anomalous Ward-Takahashi identity for the Green's function of the product of three chiral gauge quark currents ($m = 0$), thus [see Eq. (7.39) for notation where we have replaced k by q and we have omitted the coefficient of D^{abc} (since it will be the same for constituent and composite fermions)]:

$$q^\lambda T^{abc}_{\mu\nu\lambda}(q_1, q_2, q_3) = D^{abc}_{\text{quark}}\epsilon_{\mu\nu\sigma\rho}q_1^\sigma q_2^\rho \qquad (9.12)$$

The anomaly coefficient D^{abc}, arising from the triangle diagrams in Fig. (7.1) [for the constituent (lefthanded) chiral quarks], is given by:

$$D^{abc}_{\text{quark}} \sim \sum \text{Tr}\, Q_a\{Q_b Q_c\} \qquad (9.13)$$

where Q_a is the chiral charge corresponding to the chiral quark current. We recall (from §7.2c) that Eq. (9.12) encodes the "short wavelength" character of the triangular chiral gauge anomaly but that the triangular anomaly also possesses a "long wavelength" limit which is best seen by exhibiting the explicit dependence of the Green's function $\Gamma^{abc}_{\mu\nu\lambda}$ on the "anomaly pole" singularity, thus [see Eq. (7.15)]:

$$T^{abc}_{\mu\nu\lambda}(q_1, q_2, q_3)\Big|_{q_i^2 = q^2} = \frac{D^{abc}_{\text{quark}}}{q^2}[\epsilon_{\mu\nu\sigma\rho}q_1^\sigma q_2^\rho q_{3\lambda} + \text{cyclic perm.}] \qquad (9.14)$$
$$+ \text{non-sing. terms}$$

With the standard definition of renormalized quantities [4.24], the "anomaly pole" form given in Eq. (9.14) requires that the L.H.S. [of Eq. (9.14)] is evaluated at the symmetry point $q_1^2 = q_2^2 = q_3^2 = q^2$.

Equation (9.14) enables us to invoke one of the basic principles of relativistic quantum field theory [9.7], namely that the $q^2 = 0$ singularity should be reproduced for $T^{abc}_{\alpha\beta\gamma}$ at the bound-state (composite) level. At the bound-state level, there are just two ways in which the $q^2 = 0$ pole of Eq. (9.14) can be duplicated [9.7]: (a) the $q^2 = 0$ pole is due to massless composite three-quark bound states, or (b) the $q^2 = 0$ pole arises because there are massless N-G bosons in the theory that couple to the chiral quark currents. In case (a), the triangular anomalies associated with the three-quark baryonic composites must exactly match the triangular anomalies associated with the constituent quarks, i.e. $D_{\text{quark}} = D_{\text{bound states}}$ (see Fig. 9.2a) so that we have chirality conservation in the binding. This is 't Hooft's necessary (but not sufficient) triangular anomaly matching condition for maintaining the masslessness of the fermionic composite bound states. On the other hand, in case (b), where massless N-G bosons reproduce the triangular axial "anomaly pole" [i.e. the pole singularity in $T^{abc}_{\alpha\beta\gamma}$] and signal the SSB of chiral symmetry, the Green's function $T^{abc}_{\alpha\beta\gamma}$ can be written in the form [9.7]:

$$\Gamma^{abc}_{\mu\nu\lambda}(q_1, q_2, q_3)|_{q_i^2 = q^2} = \frac{f_\pi g^{abc}_{\pi JJ}}{q^2}[\epsilon_{\mu\nu\sigma\rho}q_1^\sigma q_2^\rho q_{3\gamma} + \text{cyclic perm.}] \qquad (9.15)$$

where $g_{\pi JJ}$ is the coupling of the N-G "pion" to two chiral fermion currents and is given by (see Fig. 9.2b):

$$g^{abc}_{\pi JJ} = \frac{D^{abc}_{\text{quark}}}{f_\pi} \qquad (9.16)$$

with f_π the usual pion decay amplitude.

We will see that case (a) is of dominant interest for preon models because the masses of the three-preon composites (out of which quarks and leptons are supposed to be constructed) are small compared to the confining "meta-color scale", and it is hoped that the 't Hooft triangular anomaly matching conditions are satisfied so that three generations of light quarks and leptons [massless on the "metacolor" scale] are predicted by the preon model under consideration. This is in contrast to the quark flavor model of hadrons in QCD where the masses of the three-quark baryonic composites are comparable to the confining color scale, Λ_{QCD}, and it is preferable for the 't Hooft anomaly matching condition not to be satisfied. Indeed, for $n = 3$ (three quark flavors) in QCD, the 't Hooft anomaly matching condition is not satisfied so that baryonic three-quark composite states can not be massless and massless N-G bosons ("pions") are needed to cancel the non-vanishing triangular axial anomalies associated with the constituent quarks. These are the same N-G bosons (the pseudoscalar octet of "pions") that are associated with the SSB of the global chiral quark flavor group $SU(3)_L \times SU(3)_R$ to the vector diagonal sum group $SU(3)_F$ that can accommodate finite mass baryonic three-quark composite states. Curiously enough, the 't Hooft anomaly matching condition in QCD is satisfied for $n = 2$ flavors but, as we know from experiment, the N-G bosons play the same role in connection with the SSB of $SU(2)_L \times SU(2)_R$ down to $SU(2)_I$ ($n = 2$) as they do in the SSB of $SU(3)_L \times SU(3)_R$ down to $SU(3)_F$ ($n = 3$). The difference between $n = 2$ and $n = 3$ flavors in QCD does not create a problem for the composite quark model of hadrons because, as is implied by Eq. (9.15), the 't Hooft anomaly matching condition is necessary but not sufficient.

The striking difference in the behavior of the 't Hooft anomaly matching condition when the number of quark flavors is increased from 2 to 3 is not reflected in the experimental situation. The fact is that the three-quark composite nucleon, which basically involves the two lightest (u and d) "valence" quarks, is not substantially less massive than the two-quark composite ρ meson, say. This observation makes it most likely that the 't Hooft triangular anomaly generated by the constituent quarks in the two-flavor case is reproduced by the N-G "pions", just as it must be in the three-quark flavor case. This is a clearcut demonstration that 't Hooft anomaly matching between constituent fermions and fermion composites is a necessary but not a sufficient condition for the existence of massless composite fermions. We will see that the composite preon model of quarks and leptons strives mightily to justify the existence of massless quark and lepton states on the "metacolor" scale and the experience with the composite quark flavor model in QCD serves both as a model and as a cautionary warning.

One further lesson can be learned from the vector-like QCD confining theory of constituent quarks, and that is the proof [9.8] that the two-quark composites $(q\bar{q})$ [comprising the octet of pseudoscalar mesons ("pions") in QCD] must be lighter than the lightest three-quark (qqq) baryonic $(J = \frac{1}{2}^{+})$ composites. The basic assumptions that go into this useful theorem — which we call the WNW theorem after the three authors [9.8] — are that: 1) the constituent quark-confining group [i.e. $SU(3)_C$] is vector-like, so that the functional measure (in the path integral approach) is positive definite; and 2) the flavor part of the quark Lagrangian is invariant under $SU(n)_{L+R} \times U(1)_{L+R}$. If these two conditions are fulfilled, it can be shown rigorously — by examining the asymptotic behavior of the appropriate Green's functions describing the two-quark and three-quark bound states in n-flavor lattice QCD — that the following mass inequality between the two-quark and three-quark (composite) bound states must hold:

$$M(\bar{q}q) \lesssim kM(qqq) \qquad (9.17)$$

where $M(\bar{q}q)$ is the mass of the lightest two-quark "pion" composite, $M(qqq)$ is the mass of the lightest three-quark baryonic $(J = \frac{1}{2}^{+})$ composite, and k is a constant of the order of unity. The WNW theorem clearly requires composite $J = \frac{1}{2}^{+}$ fermions to be accompanied by less massive composite pseudoscalar mesons. Empirically, the mass inequality given by Eq. (9.17) holds in vector-like QCD for the entire pseudoscalar meson octet since the lowest baryonic mass is that of the nucleon although the theorem given by Eq. (9.16), in principle, only requires the charged pion to be less massive than the nucleon.

We will see that the WNW theorem is of fundamental importance for any composite preon theory of quarks and leptons which is modeled on the quark theory of hadrons with a confining "metacolor" interaction of the preon theory replacing the confining color interaction of quark theory and, similarly, the quark-lepton three-preon composites of the preon theory replacing the baryonic three-quark composites of quark theory. [A preon model of quarks and leptons based on three-preon composites is the most promising type of preon model because of the important role that chirality invariance and the Pauli exclusion principle can play in constraining the number of possible massless representations for a given "candidate" group; for example, a one fermion-one boson preon model becomes much too arbitrary [9.9]. Henceforth, we deal only with three-preon models of quarks and leptons.] If quarks and leptons are three-preon composites, the phenomenological absence of pseudoscalar bosons as light as quarks and leptons immediately eliminates vector-like three-preon models and strongly suggests that the confining "metacolor" group must be chiral. Since the difference between a vector-like confining quark group and a chiral confining preon group is so critical (e.g. in determining the type of gauged "metacolor" group that is possible — see §9.4), it seems reasonable to search for phenomenological arguments that favor the choice of a chiral gauged "metacolor" group in the preon model. The next section adduces phenomenological evidence for a preon "metacolor" scale that is large compared to the quark-lepton mass scale and thereby supports the choice of a chiral gauged "metacolor" preon group.

§9.3. Phenomenological lower limit on the preon metacolor scale

Not unrelated to the WNW theorem is the well-known fact that the three-quark baryonic composites possess masses comparable to the confining scale Λ_{QCD} in vector-like QCD. The situation is quite different in the quark-lepton world and we proceed to justify the choice of a confining metacolor scale Λ_{QMCD} (which we identify with the preon scale) that is larger than 1 TeV, and hence much larger than all the quark and lepton masses of the three generations [including the t quark mass]. The strong inequality $\Lambda_{QMCD} >> m_q, m_l$ places a major constraint on preon-model building and requires a successful preon theory — as a first step — to predict precisely three generations of massless quarks and leptons on the "metacolor scale". [Throughout this chapter, we consider each quark-lepton generation to consist of the 15 Weyl fermions in the standard model but allow for a possible "inert" ν_R (see §7.5) for a total of 16 Weyl fermions; three-preon composites with quantum numbers other than those of one quark-lepton generation (and ν_R) are called "exotic" (composite) fermions.]

We now justify the statement: $\Lambda_{QMCD} >> m_q, m_\ell$. For this purpose, we estimate lower limits on Λ_{QMCD} by examining several precision experiments which could give deviations from the standard model if quarks and leptons are composite structures; the upper limits on the deviations — based on the measured accuracies of the experiments — yield lower limits on Λ_{QMCD}. It is clear that the standard model works very well up to energies of at least $M_W \simeq$ 100 GeV (i.e. quarks and leptons are point-like down to at least 10^{-16} cm) so that the composite scale must be above 100 GeV and, as we will see, appreciably higher. One can next argue that deviations from the standard model in the energy range between M_W and Λ (we write Λ for Λ_{QMCD} in the remainder of this section) should not depend on the new interactions at the scale Λ and can be represented by additional terms of higher "canonical" dimension than 4 (the dimension of renormalizable gauge terms in the total Lagrangian denoted by $\mathcal{L}_0$) [9.10]; these higher-dimensional terms in the total Lagrangian are non-renormalizable "point-interaction" terms and can be written in the form:

$$\mathcal{L}_{\text{eff}} = \mathcal{L}_0 + \frac{1}{\Lambda}\mathcal{L}_1 + \frac{1}{\Lambda^2}\mathcal{L}_2 + ... \qquad (9.18)$$

where $\mathcal{L}_0$ is the standard model Lagrangian of dimension 4, $\mathcal{L}_1$ is a term of dimension 5 (with the factor $1/\Lambda$), $\mathcal{L}_2$ is of dimension 6 (with the factor $1/\Lambda^2$), and so on (see §5.4d). It is essential that the additional terms $(\mathcal{L}_1, \mathcal{L}_2, ...)$ are invariant under the standard group and obey all the conservation laws (baryon conservation, etc.) that are expected to hold in the energy region $M_W \to \Lambda$. A general analysis of these higher-dimensional terms soon indicates that the dimension 5 term can not be constructed out of the "standard model" fermions, gauge bosons and scalars without violating some of the conservation laws. The first serious higher-dimensional candidate, that can be made invariant under the standard group and obey the conservation laws, is the dimension 6 term in Eq. (9.18), with the factor $1/\Lambda^2$. All estimates of the lower limit on Λ given below are made in the context of a dimension 6 contribution to the total Lagrangian.

A large number of "dimension 6" processes are available to derive lower limits on Λ but we restrict ourselves to a sample of three flavor-conserving (also called "flavor-diagonal") and two flavor-changing (also known as "flavor-non-diagonal") processes. The flavor-diagonal processes that we consider are: (1) Bhabha scattering — a possible "composite" contribution to

Bhabha scattering would lead to an additional term to $\mathcal{L}_{\text{eff}}$ of the form:

$$\frac{g^2}{2\Lambda^2}(\bar{e}_L\gamma_\mu e_L)(\bar{e}_L\gamma_\mu e_L) \tag{9.19}$$

Assuming that $g^2/4\pi = 1$ (the usual assumption to make a conservative estimate of Λ) leads to $\Lambda \geq 750$ GeV [9.10]; (2) $g - 2$ for the muon: The "composite" contribution to this effect would add an additional term to $\mathcal{L}_{eff}$ of the form:

$$\frac{em_\mu}{\Lambda^2}(\bar{\mu}_L\sigma^{\mu\nu}\mu_R)F_{\mu\nu} \tag{9.20}$$

In Eq. (9.20), we note that μ_R must be used because the $\sigma^{\mu\nu}$ operator does not conserve chirality. From the error in the very precise measurement of $(g - 2)$ for the muon, one derives $\Lambda \geq 900$ GeV (see §4.5a); and (3) U_{CKM} quark mixing matrix: a lower limit on Λ can be derived from knowledge of the three elements of the U_{CKM} mixing matrix, namely $V_{ud} = 0.973 \pm 0.001$, $V_{us} = 0.231 \pm 0.003$, $V_{ub} = 0.01 - 0.007$, and the use of the unitary condition (see §6.3e),

$$V_{ud}^2 + V_{us}^2 + V_{ub}^2 - 1 = \pm 0.003 \tag{9.21}$$

Equation (9.21) leads to the lower limit $\Lambda \geq 4.9$ TeV.

The two "flavor-non-diagonal" processes that we consider for placing a lower limit on Λ are: (1) $K_L \rightarrow \mu + e^-$ decay: the "composite" contribution to this process would give an additional contribution to $\mathcal{L}_{\text{eff}}$ of the form:

$$\frac{g^2}{2\Lambda^2}(\bar{s}_L\gamma_\mu d_L)(\bar{e}_L\gamma_\mu \mu_L) \tag{9.22}$$

It is assumed in Eq. (9.28) that the "generation number" is conserved (see below for comment on this assumption) as are the regular conservation laws. From the measured branching ratio of this process, B.R.$\leq 6 \times 10^{-11}$ (see Table 6.8) one derives $\Lambda \geq 300$ TeV; and (2) $\mu \rightarrow e\gamma$ decay: The "composite" contribution to this process would give an additional term in $\mathcal{L}_{eff}$ of the form:

$$\frac{em_\mu}{\Lambda^2}(\bar{\mu}_L\sigma^{\mu\nu}e_R)F_{\mu\nu} \tag{9.23}$$

where the chirality non-invariance of $\sigma^{\mu\nu}$ again requires the combination of e_R with μ_L. As in $K_L \rightarrow \mu\bar{e}$, "generation number" conservation is assumed in Eq. (9.23). The upper limit on the measured branching ratio for this process is now $\leq 5 \times 10^{-11}$ [1.99], which leads to a lower limit on $\Lambda \geq 240$ TeV, essentially the same as the lower limit derived from the other "flavor-non-diagonal" process $K_L \rightarrow \mu\bar{e}$.

Evidently, the lower limits on Λ (i.e. Λ_{QMCD}) derived from "flavor-non-diagonal" processes are about 100 times greater than the lower limits derived from the "flavor-diagonal" processes but we must remember that the former values are more model-dependent (e.g. dependent on the assumption of "generation number" conservation) and, hence, it is safer to accept the lower limit $\Lambda_{QCMD} \gtrsim$ several TeV, which is still much larger than any of the quark and lepton masses. Thus, it is a good working hypothesis to take $\Lambda_{QMCD} >> m_q, m_\ell$, and we can now proceed with our discussion of, what we like to call, the "paradigmatic" preon model of quarks and leptons. We use the term "paradigmatic preon model" for the initial preon model which we discuss because it represents a class of preon models that are closely modeled on the composite quark model [which, as we know, is governed by the group $SU(3)_C \times SU(3)_F \times U(1)_B$] in at least three respects: (1) the constituent preons are $J = \frac{1}{2}$ fermions so that the quarks and leptons are three-preon composites; (2) the gauge bosons are "elementary", i.e. they are not preon composites; and (3) supersymmetry is not introduced into the theory. We can not say at the present time whether the Higgs boson is elementary or composite (quasi-) in the (standard) quark model; however, the discussion of preon models assumes — since it is so natural to do so — compositeness for the Higgs boson. The differences between the composite quark and preon models will become apparent as we take account of the absence of bosonic preon composites lighter than the fermionic ones (i.e. the WNW theorem) and make use of the fact that the quark and lepton masses are much smaller than the "metacolor scale".

§9.4. Paradigmatic preon model of quarks and leptons (based on gauged metacolor and gauged metaflavor)

A good starting point for the construction of the "paradigmatic preon model" is the observation, already made, that the failure to observe charged scalar particles as light as quarks or charged leptons eliminates vector-like three-fermion preon models of quarks and leptons. Since this observation is a consequence of the WNW theorem [see Eq. (9.17)] — which, in turn, depends on making use of Dirac fermions in a vector-like confining theory — it follows that it is possible to escape this constraint by constructing a three-fermion preon model based on the coupling of chiral preons to "metagluons" (the gauge fields of the preon theory). Indeed, there exist no two-preon condensates — necessary for the QCD result of Eq. (9.17) — in chiral three-fermion preon models. The chiral character of the composite preon model of

quarks and leptons — together with the requirement that the quark and lepton masses are small compared to the confining metacolor scale — drives us
to extensive use of the 't Hooft triangular anomaly matching condition. This
interconnectedness becomes apparent as we describe the essential ingredients
of the "paradigmatic preon model".

We make some simplifying assumptions — to be noted in due course —
to construct the "paradigmatic preon model" and, subsequently, we consider
several variations as we lift one or more of the simplifying assumptions. The
basic assumptions underlying the "paradigmatic preon model" are:

Assumption (1): the $m = 0$, $J = \frac{1}{2}$ preons transform under a single
irreducible representation R of a simple metacolor gauge group G_{MC}; Assumption (2): the metacolor sector is asymptotically free (ASF) as well as
being triangular chiral gauge anomaly-free (see §7.2c and §7.7); Assumption (3): since quarks and leptons are three-preon composites, $R \times R \times R$
must be a metacolor singlet. These three-preon composites are treated as
massless on the metacolor scale Λ_{MC}; and, finally Assumption (4): when all
gauge couplings — except metacolor coupling — are turned off, the metaflavor symmetry of the preons becomes a global metaflavor group denoted by
G_{MF}. In order to yield a "realistic" theory of quarks and leptons, the global
metaflavor group must contain a gauged metaflavor subgroup H_{MF} that
contains the standard gauge group as a subgroup. Thus, the initial mixed
gauged metacolor-global metaflavor group is $G_{MC} \times G_{MF}$, which becomes
$G_{MC} \times H_{MF}$ ($G_{MF} \supset H_{MF}$) upon the gauging of global metaflavor. We
next examine the extent to which Assumptions (1-4) above give a unique
determination of G_{MC}, G_{MF} and H_{MF}.

Let us start with the consequences of Assumption (1) by clarifying the
distinction between vector-like and chiral theories. As in the composite quark
model, we take all the massless preons to be lefthanded. Now suppose a preon
transforms under a representation (R, n) of $G_{MC} \times G_{MF}$. The theory is said
to be vector-like (with respect to G_{MC}) if, for every preon that transforms as
R, there exists another preon that transforms as $\bar{R}$ so that the "true" preon
Ψ is $(R, n) + (\bar{R}, n)$; the largest global chiral symmetry is then $U(n)_L \times U(n)_R$
— as in QCD (see §5.1). Since all such preon models are ruled out by the
constraint of Eq. (9.17), G_{MC} must be chiral and, hence R must be complex.
Consequently, G_{MC} is limited to those groups that have complex representations; in particular, if the metacolor group is simple — an assumption that
has been made in defining the "paradigmatic preon model" — then the simple
group G_{MC} must be $SU(N)(N \geq 3)$, $SO(4k + 2)(k \geq 2)$ or E_6 (see §8.2).

We next consider the triangular anomaly-free constraint on G_{MC} (Assumption 2), which must be taken into account if the chiral preon model is to be renormalizable. Since one of our assumptions is that there exists only one irreducible preon representation R under G_{MC}, the class of groups $SU(N)(N \geq 3)$ must be excluded [this assumption is lifted when we consider "non-paradigmatic preon models" in §9.5] because single complex irreducible representations with no triangular anomalies are very rare in $SU(n)$ and, indeed, their dimensions are so large as to violate the ASF condition [8.4] (Assumption 2). If we allow more than one irreducible representation of G_{MC}, the $SU(N)(N \geq 3)$ groups become "candidate" metacolor groups for preon models [e.g. $SU(5)$ with the 5 and $\overline{10}$ representations] and, indeed, the $SU(N)(N \geq 3)$ groups become interesting when we attempt to deal with some "unrealistic" consequences of the "paradigmatic preon model".

For $SO(4k+2)$ $(k \geq 2)$ groups, the spinorial representations are complex representations. The ASF condition on $SO(4k+2)$ $(k \geq 2)$ restricts the number of metaflavors n to be less than $11 \times k \times 2^{4-2k}$; thus, for $SO(10)$, $n < 22$, for $SO(14)$, $n < 9$, etc. For other representations, n is smaller, since the fermion contribution to the β function (see §5.2a) is proportional to the second-order index, which is proportional to the dimension of R. For three-fermion preon models of quarks and leptons, R is further restricted by the singlet condition, i.e. $R \times R \times R$ must contain a singlet. For those representations with congruence classes $(1,1),(1,3),(0,2)$ of $SO(4k+2)$ $(k \geq 2)$ [8.4], this is impossible. For the complex representations belonging to the $(0,0)$ congruence class, this is incompatible with the ASF condition. Thus, $SO(4k+2)$ $(k \geq 2)$ is also excluded for three-fermion preon models under Assumption 2.

Finally, the remaining G_{MC} candidate, E_6, is very promising: its fundamental $R = 27$ is a complex representation that satisfies the singlet condition (i.e. $27 \times 27 \times 27 \supset 1$) and the ASF condition if the number of metaflavors $n < 22$. In sum, in our discussion of the consequences of Assumption (2), it has been shown that, if G_{MC} is simple and R is a single complex irreducible representation, then the only simple group G_{MC} obeying the ASF and metacolor singlet conditions is:

$$G_{MC} = E_6 \text{ with } R = 27 \text{ and } n < 22 \qquad (9.24)$$

Thus, the metacolor group E_6 with its fundamental representation $R = 27$ is the unique "paradigmatic preon model" under Assumptions (1)–(3). It should be noted that, in contrast to the confining group $SU(3)_C$, which makes an antisymmetric contribution to the color singlet three-quark composite wavefunction, the E_6 metacolor group in "quantum metachromodynamics" (QMCD) makes a symmetric contribution to the metacolor singlet wavefunction of a three-preon composite.

We next examine how far we can go in defining the metaflavor sector (both global and local) of our "paradigmatic preon model" by adding Assumption (4). Evidently, the largest global metaflavor group G_{MF} for chiral theories is $U(n)$. The global $U(n)$ group may be spontaneously broken, as happens in vector-like theories, and the same criterion must be used in deciding whether G_{MF} is broken or not, namely the 't Hooft triangular anomaly-matching condition. If the triangular anomalies associated with the constituent preons and three-preon composites do not match, then G_{MF} must be spontaneously broken [but, as already pointed out, the reverse does not necessarily hold] into a global metaflavor subgroup, G'_{MF}. Furthermore, if one restricts oneself to a single irreducible representation, $\Psi = (R, n)$, and fermionic composites consist of three fermionic preons, then no triangular anomaly-free solution exists for $n \geq 3$, and $G_{MF} = U(n)$ must break into some subgroup of $U(n)$, G'_{MF}, where the triangular anomalies match.

A further step in the argument is to show that for chiral three-fermion preon models, the unbroken global metaflavor subgroup G'_{MF} must be gauged in order to dispose, in principle, of the N-G bosons that arise from the SSB of G_{MF} to G'_{MF}. In vector-like theories, such N-G bosons acquire mass through the explicit introduction of mass terms for the preons in the fundamental Lagrangian that break the G_{MF} symmetry, but not the G'_{MC} symmetry. Thus, in QCD, when $G_{MF} = SU(n)_L \times SU(n)_R$ is spontaneously broken (by the quark-antiquark condensates) into the vector subgroup $SU(n)_F$, the SSB of the chiral quark flavor symmetry leads to the three-quark composites acquiring masses of the order of $\Lambda_{\text{vector}} \simeq \Lambda_{QCD}$. In chiral three-fermion preon models of quarks and leptons, one cannot introduce explicit fermion mass terms to make the N-G bosons heavy, since such terms are forbidden by the chiral invariance of G_{MC}. Hence, one must use gauge boson radiative corrections to make N-G bosons heavy [9.11] and so there must be, at least, a gauge subgroup H_{MF} of the unbroken global metaflavor group G'_{MF} ($G'_{MF} \supset H_{MF}$) to generate these radiative corrections. Indeed, if there was no SSB of G'_{MF}, it would be a mystery why a subgroup of the global symmetry group had to be gauged. As we have already remarked, the gauging of global metaflavor also serves the necessary purpose of permitting the three-preon composites — that comprise the quarks and leptons of each generation

— to possess the necessary gauged color and flavor degrees of freedom under the standard group $SU(3)_C \times SU(2)_L \times U(1)_Y$.

Let us then assume that the gauged metaflavor subgroup H_{MF} exists; it is obvious that the preons have quantum numbers with respect to H_{MF} and thus the preon representation(s) under H_{MF} must be triangular anomaly-free for the renormalizability of the gauge theory. Moreover, the preon representation(s) under the gauge group H_{MF} must be complex; otherwise, composites are "real" and the "survival hypothesis" tells us that they become heavy [8.31]. Hence, the constraints on H_{MF} and its representation(s) are very strong: the group must have a complex representation (of dimension less than 22) and it must be triangular anomaly-free. If we assume that the preons transform under a single irreducible representation r, then H_{MF} must be $SO(10)$ and r must be spinor (or antispinor). This statement is obvious if H_{MF} is simple since, as we already know, $SU(n)(n \geq 3)$ does not satisfy the triangular anomaly-free condition for a single irreducible representation and E_6 is eliminated by the ASF condition (since $n < 22$) so that only $SO(10)$ remains. Consequently, as long as we limit ourselves to a single complex irreducible representation of H_{MF}, the choice is uniquely fixed — without any reference to the resulting composites; that is to say:

$$G_{MF} = U(16); \quad H_{MF} = SO(10); \quad r = \mathbf{16} \text{ or } \overline{\mathbf{16}} \qquad (9.25)$$

It should be emphasized that a unique "paradigmatic preon group" $E_6 \times SO(10)$, with the representation $\Psi = (\mathbf{27}, \overline{\mathbf{16}})$, has emerged when Assumption (4) is combined with Assumptions (1)–(3) because we have insisted on a single irreducible metacolor representation of G_{MC} and a single irreducible gauged metaflavor representation under H_{MF}. We will explore the consequences of this "paradigmatic preon model" in some detail because it is unique (under the specified conditions), and does give rise — on the metacolor scale — to three generations of massless three-preon composites with "standard model" quantum numbers, and because the identification of its limitations is suggestive of additional and/or modified guidelines for more "realistic" preon model building.

Before proceeding with a closer examination of the chiral preon group $E_6(MC) \times SO(10)(MF)$ with the single irreducible representation $\Psi = (\mathbf{27}, \overline{\mathbf{16}})$, it is worth noting how the relaxation of the conditions, that leads uniquely to this "paradigmatic preon model", opens up a range of new possibilities. Thus, if we loosen the condition that the preons transform under a single irreducible representation of H_{MF}, there are clearly more choices.

For example, any subgroup of $SO(10)$ is a solution: e.g. $H_{MF} = SU(5)$ with $r = 5 + \overline{10}$ (or its conjugate), or $H_{MF} = G_{422}$ with $r = [(\overline{4}, 2, 1) + (4, 1, 2)]$ are solutions. However, if we consider the conditions imposed on the three-preon composites by the phenomenologically correct three-family structure, the first possibility is ruled out while the second possibility survives [9.11]. Thus, the model $E_6 \times G_{422}$ can also serve as a chiral preon model of quarks and leptons if we are willing to accept two irreducible representations. Once two irreducible representations of G_{MF} are allowed, it is natural to inquire whether other solutions with two irreducible representations exist and, if so, whether they give the correct family structure for ordinary quarks and leptons. It turns out that while it is possible to identify several other chiral preon models with two complex irreducible representations of G_{MF}, none of them, except $E_6 \times G_{422}$, meets the family-structure test [9.11]. For this reason, after we have reviewed the predictions of the preon model based on $E_6(MC) \times SO(10)$ (MF) with one irreducible complex representation, we comment briefly on the more physically perspicious preon model based on $E_6(MC) \times G_{422}(MF)$ with two irreducible complex representations.

§9.4a. $E_6(MC) \times SO(10)(MF)$ preon model

The completely gauged $E_6 \times SO(10)$ "paradigmatic" chiral preon model has emerged as a unique candidate for a composite preon theory of quarks and leptons under the plausible assumptions spelled out in §9.4. It is of interest to examine some consequences of the "paradigmatic" $E_6 \times SO(10)$ preon model in order to clarify the strengths and weaknesses of this type of approach to the "fermion generation problem" and to the gauge hierarchy problem associated with the Higgs SSB mechanism. We first point out that the metaflavor $SO(10)$ gauge group possesses the global metaflavor symmetry:

$$SU(16) \times Z_{16}, \tag{9.26}$$

where the naive $U(1)$ symmetry [arising from $U(16) \to SU(16) \times U(1)$] is broken by the metacolor instantons (see §5.1) and reduces $U(1)$ to the discrete Z_{16} symmetry. We repeat that the global metaflavor symmetry $SU(16)$ is supposed to be spontaneously broken down to the gauged metaflavor subgroup $SO(10)$, at which stage the triangular anomaly matches trivially. Associated with the SSB of $SU(16)_{MF}$ to $SO(10)_{MF}$ are $(255 - 45) = 210$ N-G bosons that transform as an irreducible representation **210** under $SO(10)$; however, due to radiative corrections, these "would-be" N-G (scalar) bosons acquire masses on the $SO(10)$ scale and hence create no problem for the model.

It is also easy to check that the ASF of the $E_6 \times SO(10)$ preon model is robust; in the one-loop approximation, the β function of the E_6 metacolor coupling is (the subscript 6 refers to E_6):

$$\beta_6 = -(g_6)^3 1/(16\pi^2)\{44 - 2 \times 16\} = -(g_6)^3 \, 12/(16\pi^2) \qquad (9.27)$$

where, in the first equality on the R.H.S. of Eq. (9.27), the metagluon and fermion contributions are shown separately. For comparison, the β function for six-flavor QCD is (see §5.2a):

$$\beta_s = -(g_s)^3 1/(16\pi^2)\{11 - 6/3\} = -(g_s)^3 \, 9/(16\pi^2) \qquad (9.28)$$

It is evident from Eqs. (9.27) and (9.28) that the "robustness" of asymptotic freedom in the preon model compares favorably with that in QCD.

We next move on to a consideration of the multi-preon composites contained in the $E_6 \times SO(10)$ preon model, not only three-preon composites that might explain the three generations of fermionic quarks and leptons, but also the even numbers of composites that could give rise to Lorentz scalars that can serve the purpose of quasi-Higgs bosons in the theory. We begin with the scalar composites which, like the fermionic composites, must transform as metacolor singlets. The first remark is that there are no two-preon metacolor singlets that are Lorentz scalars, since the only candidate $\bar{P} \not{\!\!D} P$ (P denotes the preon) vanishes by the equation of motion. Metacolor singlets that are Lorentz scalars first arise at the four-preon level and, if we omit the Lorentz structure, these scalar bosons can be written in the form:

$$\bar{P}\bar{P}PP \sim 1 + 45 + 54 + 210 + \cdots \qquad (9.29)$$

where the decomposition into the first few lower-dimensional irreducible representations under $SO(10)$ is given (see §8.4c). It can be shown that all of these composites are invariant under the discrete symmetry Z_{16} so that, even if some of these scalars acquire VEV's, the discrete symmetry is unbroken. It can also be shown that, if any of the scalars acquires a VEV, the rank is preserved and cannot give masses to three-preon fermions since they belong to the congruence class $(0,0)$, but not $(0,2)$, of $SO(10)$ [9.11]. The situation is quite different for the six-preon composites (see §8.4c):

$$PPPPPP \sim 10 + 120 + 126 + 210' + \cdots \qquad (9.30)$$

where it can be shown that the six-preon composites are not invariant under Z_{16} and that, if a six-preon composite scalar acquires a VEV, it must reduce the rank of the group and can give mass to the three-preon composite fermions.

The strikingly different behavior of the four- and six-preon condensates in the "paradigmatic" $E_6 \times SO(10)$ preon model confirms qualitatively the gauge hierarchy result in $SO(10)$ GUT (see §8.4a) which is that, at the higher mass scales, the (four-preon composite) Higgs **45** and **210** representations of $SO(10)$ break down the $SO(10)$ group — along the G_{422} path — into equal rank subgroups (containing local $B - L$ symmetry). It is only when the SSB goes to the standard G_{321} group, and then to the color-electromagnetic G_{31} group, that there is a reduction of rank (and the breaking of local B-L symmetry) by means of the (six-preon composite) **126** and **10** irreducible representations of $SO(10)$ respectively. Amusingly enough, the **126** and **10** irreducible representations, because they are contained in the six-preon scalar composites [see Eq. (9.30)], are expected to bind at a lower scale than the four-preon composites [see Eq. (9.29)] since they can be thought of as "quark-quark" pairs. It does appear that the simplest "paradigmatic preon model" [based on gauged metacolor E_6 and gauged metaflavor $SO(10)$] can account, in gross qualitative terms, for the gauge hierarchy that is character-istic of GUT theories — in addition to providing a rationale for the composite structure of Higgs scalars.

Let us complete our consideration of the $E_6 \times SO(10)$ preon model by spelling out the extent to which it provides a mechanism for generating three generations of composite fermionic states that can be identified with ordinary quarks and leptons. In this preon model, the quarks and leptons are supposed to be metacolor singlet three-preon composites and the mass terms of these fermion composites should appear as:

$$(PPP)^T C (PPP) \tag{9.31}$$

where C is the charge conjugation operator. Equation (9.31) violates both $SO(10)$ gauge symmetry and the discrete symmetry Z_{16}. The breaking of $SO(10)$ follows from the fact that the multiplication of six **16**'s of $SO(10)$ belongs to the congruence class $(0, 2)$ but not $(0, 0)$, to which a singlet belongs [9.11]. The role of the metacolor-instanton-induced discrete symmetry Z_{16} in maintaining the masslessness of composite fermions — through the winding number of the instanton — is more complicated but it turns out that the combination of gauge $SO(10)$ symmetry and the discrete symmetry Z_{16} keep the fermionic composites massless on the metacolor scale. On the other hand, once the six-preon composite scalars get VEV's, then both the discrete symmetry and the $SO(10)$ symmetry are broken and the fermions acquire masses of the order of the VEV's of the six-preon composite scalars. Since the mass terms for chiral fermions must break both $SU(2)_L$ and $SU(2)_R$, the mass scale for the six-preon composite scalars must be of the order of $M(W_L)$ or $M(W_R)$. Hence, all the masses of fermionic composites will be of that order. This order-of-magnitude prediction for the t quark mass is quite satisfactory since the experimental lower limit on m_t already exceeds M_W but it clearly does not explain the small masses ($<< M_W$) of the other quarks and charged leptons of the three generations. Moreover, we will see that the composite three-preon representations of the $E_6 \times SO(10)$ model contain large numbers of "exotic" fermions — fermions whose quantum numbers differ from those of the observed quarks and leptons — for which the quasi-Higgs (six-preon composite scalar) — generated masses are predicted to be of the same order, M_{W_L} or M_{W_R}, as that of the quarks and leptons. This last result proves to be a major stumbling block to the "paradigmatic" type of preon model but, before we turn to some "non-paradigmatic" alternatives — we sketch the surprising prediction — and try to understand it — of the $E_6 \times SO(10)$ preon model that there should be precisely three generations of Weyl fermions with the quantum numbers of the observed quarks and leptons, in addition to the unwanted "exotic" fermions.

To show how the family structure and quantum numbers of the massless composite three-preon fermions are generated in the E_6 (MC)$\times SU(16)(MF)$ preon model, we follow the analogy with the $SU(6)$ three-quark flavor-spin model of hadrons (see §3.2d); we generalize the Pauli principle to the meta-color degree of freedom — in addition to the metaflavor and chiral degrees of freedom — and then unify the global metaflavor group $SU(16)$ with $SU(2)_{\text{chir}}$ [where the subscript on $SU(2)$ reminds us that our Weyl spinors possess two chiralities]. What we call the meta-Pauli principle then tells us that the total (composite) three-preon wavefunction must be totally antisymmetric under E_6 metacolor $\times SU(32)$ metaflavor-chirality. Since the three-preon metacolor singlet of E_6 contributes a symmetric part to the total three-preon wave-function, we need only consider the totally antisymmetric representation of $SU(32)$. The decomposition of the totally antisymmetric representation of $SU(32)$ into $SU(16) \times SU(2)_{\text{chir}}$ is (see Fig. 9.1):

$$(RRR)_A = A_3 = (M_3, M_3) + (A_3, S_3) \tag{9.32}$$

or:

$$(\mathbf{32} \times \mathbf{32} \times \mathbf{32})_A = \mathbf{4960} \rightarrow (\mathbf{1360}, \mathbf{2}) + (\mathbf{260}, \mathbf{4}) \tag{9.33}$$

under $SU(16)_{MF} \times SU(2)_{\text{chir}}$. The first term on the R.H.S. of Eq. (9.32) denotes the $J = \frac{1}{2}$ lefthanded fermions whereas the second term yields the

$J = \frac{3}{2}$ lefthanded fermions; this is to be compared with QCD where the symmetric **56** representation under $SU(6)$ decomposes into $\{(\mathbf{8,2}) + (\mathbf{10,4})\}$ representations of $SU(3)_F \times SU(2)_{\text{SPIN}}$ (cf. §3.2d); i.e. an octet of $J = \frac{1}{2}$ composite three-quark baryons and a decouplet of $J = \frac{3}{2}$ composite three-quark baryons.

Since the quarks and leptons of the three generations possess $J = \frac{1}{2}$, we focus on the irreducible representation 1360 of $SU(32)$ on the R.H.S. of Eq. (9.33). Further, since we have used the global $SU(16)$ metaflavor group as a tool for applying the meta-Pauli principle, and the actual contents are gauged $SO(10)$ particles, we must examine the decomposition of the **1360** representation of $SU(16)$ into irreducible representations of $SO(10)$; this decomposition gives:

$$\mathbf{1360} \rightarrow \mathbf{16} + \mathbf{144} + \mathbf{1200} \tag{9.34}$$

Equation (9.34) appears to tell us that there is one family of quarks and leptons (plus ν_R) given by the (fundamental) **16** representation of $SO(10)$. However, this is misleading because, as we have already pointed out, the SSB of $SO(10)$ to G_{422} [the only SSB path that is viable for $SO(10)$ GUT (see §8.5c) — accomplished through the quasi-Higgs four-preon scalar composites] does not reduce the rank of $SO(10)$ nor generate fermion masses. The fact that fermion masses are only generated beyond the Pati-Salam group G_{422} allows us to count up the number of quark-lepton generations on the basis of the G_{422} quantum numbers arrived at after the three irreducible representations of $SO(10)$ on the R.H.S. of Eq. (9.34) are decomposed under the G_{422} group. The decomposition of the three $SO(10)$ representations are shown in Eq. (9.35):

$$\mathbf{16} \rightarrow (\mathbf{4,2,1}) + (\bar{\mathbf{4}},\mathbf{1,2}) \tag{9.35}$$

$$\mathbf{144} \rightarrow (\mathbf{4,2,1}) + (\bar{\mathbf{4}},\mathbf{1,2}) + (\bar{\mathbf{4}},\mathbf{3,2}) + (\mathbf{4,2,3}) + (\overline{\mathbf{20}},\mathbf{2,1}) + (\mathbf{20,1,2})$$

$$\mathbf{1200} \rightarrow (\mathbf{4,2,1}) + (\bar{\mathbf{4}},\mathbf{1,2}) + (\bar{\mathbf{4}},\mathbf{3,2}) + (\mathbf{4,2,3}) + (\overline{\mathbf{20}},\mathbf{2,1}) + (\mathbf{20,1,2})$$

$$+ (\overline{\mathbf{20}'},\mathbf{2,1}) + (\mathbf{20',1,2}) + (\overline{\mathbf{20}},\mathbf{4,1}) + (\mathbf{20,1,4}) + (\overline{\mathbf{20}},\mathbf{2,3})$$

$$+ (\mathbf{20,3,2}) + (\overline{\mathbf{36}},\mathbf{1,2}) + (\mathbf{36,2,1}) + (\mathbf{36,2,3}) + (\overline{\mathbf{36}},\mathbf{3,2})$$

It is clear that each of the three $J = \frac{1}{2}$ representations of $SO(10)$ contains a single generation of ordinary quarks and leptons [if it is recalled that the pair of representations $(\mathbf{4,2,1}) + (\bar{\mathbf{4}},\mathbf{1,2})$ (under the G_{422} group) constitutes a single generation of quarks and leptons (plus ν_R)] so that the $E_6 \times SO(10)$ preon group predicts exactly three generations of ordinary quarks and leptons (plus ν_R) — at the G_{422} level — plus a large number of "exotic" fermions with definite quantum numbers under G_{422} but all belonging to the same congruence class as the quarks and leptons. While there are no "mirror" fermions — a very desirable result — we are confronted with a large number of "exotic" fermions with masses — in accordance with our earlier discussion of fermion mass generation by the quasi-Higgs preon composites — of the same order as ordinary quarks and leptons; there is no evidence for these "exotic" fermions.

Putting aside the "exotic" fermion problem — to which we return — one might inquire whether the $E_6 \times SO(10)$ preon model in its simplistic ("paradigmatic") form predicts any correlation between the empirically observed rapidly increasing (average) mass of each succeeding generation of quarks and leptons and the rapidly increasing dimensionality of each of the three irreducible representations of $SO(10)$ in Eq. (9.34). The tantalizing answer is that there does seem to be a correlation of this type when one examines the mass matrix of the $J = \frac{1}{2}$ fermionic states in the $E_6 \times SO(10)$ preon model. To see this, we note that the "flavor-diagonal" Yukawa couplings for each generation can occur for the 10 quasi-Higgs representation ϕ_{10} [a similar statement can be made about the 126 quasi-Higgs representation ϕ_{126} but with smaller contributions to the masses], namely:

$$\phi_{10} \times \mathbf{16} \times \mathbf{16}; \quad \phi_{10} \times \mathbf{144} \times \mathbf{144}; \quad \phi_{10} \times \mathbf{1200} \times \mathbf{1200} \qquad (9.36)$$

The "non-flavor-diagonal" couplings — responsible for the mass mixings — are:

$$\phi_{10} \times \mathbf{16} \times \mathbf{144}; \quad \phi_{126} \times \mathbf{144} \times \mathbf{1200}; \quad \phi_{120} \times \mathbf{16} \times \mathbf{144};$$

$$\phi_{120} \times \mathbf{16} \times \mathbf{1200}; \text{ and } \quad \phi_{120} \times \mathbf{144} \times \mathbf{1200} \qquad (9.37)$$

where ϕ_{126} is responsible for breaking G_{3221} or G_{3211} to G_{321} [see §8.4a], ϕ_{10} is responsible for the last stage of symmetry breaking in $SO(10)$ — from G_{321} to G_{31} — and ϕ_{120} is the (antisymmetric) quasi-Higgs representation that connects generations [see Eq. (9.30)].

Equations (9.36)–(9.37) can not be taken too seriously but it is worth mentioning some qualitative consequences of these equations in order to indicate the possible bearing of even such a simplistic preon model as $E_6 \times SO(10)$ on the U_{CKM} mixing matrix (see §6.3e). If we use the old quark model argument [9.12] [purporting to explain why the higher-dimensional representations of SU(3) color give larger masses than the low-dimensional representations] — which consisted of arguing that the lowest-order one-loop radiative corrections give the correct qualitative dependence on the dimensionality of the representation — then we might expect the Yukawa couplings in the $E_6 \times SO(10)$ preon model to be roughly proportional to the second-order indices of the fermion representations for the "flavor-diagonal" terms in Eq. (9.36). Since the second-order indices of the **16**, **144** and **1200** $SO(10)$ representations are **2**, **34**, **470** respectively, it follows that the mass should increase rapidly with the dimensionality of the fermion representation if we identify — as seems natural — the 16, 144 and 1200 representations as the first, second and third generations, respectively, of quarks and leptons. The correlation between the (average) mass increase and the increasing dimensionality of the irreducible $SO(10)$ representation, as one goes from one generation to the next, is surprisingly good but dynamical calculations must be performed to confirm the correct qualitative behavior exhibited by the "paradigmatic" $E_6 \times SO(10)$ preon model. Such dynamical calculations are notoriously difficult to carry out; after all, it is no easier to make dynamical calculations of masses with a non-Abelian confining E_6 metacolor group in the preon model than it is to do so with the non-Abelian $SU(3)$ color group in the quark model!

Further, the qualitative consequences of the $E_6 \times SO(10)$ preon model are not bad for the "flavor-non-diagonal" properties of the mass matrix; thus, if, the "flavor-non-diagonal" couplings are dominated by the Higgs **10** representation (as is usually assumed), then neither the **16** nor the **144** representation can couple to the **1200** representation. Hence, one would expect from the first interaction in Eq. (9.37) that the first two generations would mix with each other but not with the third generation; similarly suggestive connections between the $E_6 \times SO(10)$ preon model and the U_{CKM} mixing matrix can be made for the other interactions listed in Eq. (9.37). However, it must be reiterated that dynamical calculations must be performed to confirm these qualitative predictions and to ascertain whether the predicted masses and mixing angles in the U_{CKM} mixing matrix bear any resemblance to reality.

We now recapitulate the strengths and weaknesses of the "paradigmatic" composite preon model $E_6(MC) \times SO(10)(MF)$, that incorporates important ingredients of the composite quark theory of hadrons. The $E_6 \times SO(10)$ model is closely modeled on the quark theory in using fermionic constituent fields, elementary gauge fields and no supersymmetry. The confining metacolor group E_6 for the preons is the analog of the confining color group $SU(3)_C$ for the quarks and the initial global metaflavor symmetry group

$SU(16)$ for the preons [which must break down spontaneously to the gauged metaflavor subgroup $SO(10)$] is the analog of the global chiral quark flavor symmetry $SU(3)_L \times SU(3)_R$ [which breaks down to the global quark flavor vector subgroup $SU(3)_F$ — which is never gauged!] The three-preon composites are supposed to give rise to the quark and leptons of the three known generations — accepting the t quark as a third generation weak isopsin "up" quark — in analogy to the lowest mass $SU(3)_F$ baryonic multiplets. The four basic assumptions [Assumptions (1)–(4)] listed in §9.4 exploit the similarities between the composite preon and quark models but also incorporate their differences as well: (1) the metacolor group E_6 for the preons is chiral as contrasted with the vector-like character of $SU(3)_C$ for the quarks [thereby satisfying the constraint of Eq. (9.17)]; (2) there is only one irreducible representation (**27**) of the metacolor group E_6 instead of the two representations (**3** and $\bar{\mathbf{3}}$) of the color $SU(3)$ group; and (3) the global metaflavor group is gauged (or at least the largest subgroup thereof) in the preon case whereas the global flavor group in the quark case is never gauged. The greatest deficiency of the "paradigmatic" $E_6 \times SO(10)$ preon model [apart from its prediction of the thus-far-absent righthanded neutrino] lies in its prediction of a super-abundance of relatively low mass fermions — the "exotic" fermions — to be discussed more fully in §9.5. Before turning to the "non-paradigmatic" preon models considered in §9.5, we comment briefly on the slightly modified "paradigmatic" $E_6 \times G_{422}$ preon model.

If one insists on only one irreducible representation of the gauged metaflavor subgroup, then the "paradigmatic preon model" is uniquely determined to be $E_6(MC) \times SO(10)(MF)$ and this is the justification for the rather detailed consideration of this model that has been given in this section. However, the restriction to preon models with only one irreducible representation of the gauged metaflavor subgroup (of the initial global metaflavor group) may be too severe since the only certain feature of the gauged metaflavor subgroup is that it must contain the one-generation standard gauged color-flavor group $SU(3)_C \times SU(2)_L \times U(1)_Y$, and we know from §8.4a that a one-generation (partial) unification group like G_{422} can be just as viable as $SO(10)$ GUT. The interesting result — noted earlier — is that allowing the gauged metaflavor subgroup of the preon model to have two complex irreducible representations only adds one "candidate" preon model to the pool of composite preon models patterned after the quark model, namely the $E_6(MC) \times G_{422}(MF)$ preon model. Indeed, it has been found [9.13] that, as long as one limits oneself to semi-simple groups (non-semi-simple groups

introduce too much arbitrariness), the only semi-simple group with two complex irreducible representations that satisfies our initial Assumptions (1)–(3)
and predicts precisely three generations of massless quarks and leptons on
the metacolor scale, is the $E_6 \times G_{422}$ group with the two representations
$(\mathbf{27}; \bar{\mathbf{4}}, \mathbf{2}, \mathbf{1})_L$ and $(\mathbf{27}; \mathbf{4}, \mathbf{1}, \mathbf{2})_L$.

Actually, the last result is not surprising because G_{422} is a maximal
subgroup of $SO(10)$ and the fundamental $\overline{\mathbf{16}}$ representation of $SO(10)$ decomposes into the above two G_{422} representations. Indeed, except for the
possibility of a smaller (partial) unification mass than that for $SO(10)$ GUT
and the predicted absence of proton decay for G_{422} PUT [in contrast to
$SO(10)$ GUT — see §8.4a], the predictions of the $E_6 \times G_{422}$ preon model
parallel closely those of the "paradigmatic" $E_6 \times SO(10)$ preon model with all
of its advantages and limitations. We do not repeat the attractive features of
the "paradigmatic"-type of preon model [$E_6 \times SO(10)$ or $E_6 \times G_{422}$] but we
do emphasize that the large number of "exotic" fermions that characterizes
this type of preon model is a serious deficiency because the "exotic" fermions
belong to the same congruence class as the quarks and leptons and therefore
lead to the absence of asymptotic freedom in the composite quark-lepton
sector. The hurdle of a super-abundance of "exotic" fermions emerging from
a "paradigmatic"-type preon model [whether $E_6 \times SO(10)$ or $E_6 \times G_{422}$] is
a compelling reason to turn away from "paradigmatic" preon models and to
examine whether "realistic" models [in the minimal sense of three generations of massless three-preon composite fermions — on the metacolor scale
— with quark-lepton-like quantum numbers (under the standard group)] can
be constructed without a plethora of "exotic" fermions. The conditions under which such realistic preon models can be constructed are discussed in
the next section.

§9.5. "Realistic candidate" preon models of quarks and leptons

Let us remind ourselves that the "paradigmatic" preon model was constructed in close analogy with the QCD-grounded quark model of hadrons
— with a minimal number of modifications — in order to ascertain whether
a simple preon model of quarks and leptons could offer insight into the three-
fold replication of quark-lepton generations and the nature of the Higgs
boson. In many ways, the predictions of the "paradigmatic" $E_6(MC) \times
SO(10)(MF)$ preon model are in suprising qualitative agreement with the
known quark-lepton-Higgs boson phenomenology: three families of ordinary
massless quarks and leptons (on the metacolor scale), a quasi-Higgs boson

hierarchy that roughly relates to the observed gauge hierarchy, and the prediction of the gauged subgroup $SO(10)$ as the leading contender for the one-generation GUT group (see §8.4a). However, the large number of (relatively low-mass) "exotic" composite three-preon fermions predicted by the $E_6 \times SO(10)$ (or the $E_6 \times G_{422}$) preon model is a major obstacle to its acceptance as the starting point for a fully successful preon model of quarks, leptons and possibly Higgs bosons. We now deal with the question as to whether it is possible to improve upon the "paradigmatic"-type of preon model by maintaining its overall positive features and eliminating its unwanted "exotic" fermions in each of the three generations; we call such preon models — which are driven by the need to exploit the 't Hooft anomaly matching condition for global metaflavor — "realistic candidate" preon models (of which several are identified) because, in principle, they could serve as the starting point for a calculation of the observed quark-lepton mass spectrum. We review the advantages and limitations of these "realistic candidate" preon models in the next two sections: §9.5a–Preon models with E_6 metacolor and $SU(N)$ $(17 \leq N \leq 21)$ global metaflavor; and §9.5b–Preon models satisfying "tumbling complementarity".

§9.5a. Preon models with E_6 metacolor and $SU(N)$ $(17 \leq N \leq 21)$ global metaflavor

In constructing the "paradigmatic" preon model $E_6(MC) \times SU(16)(MF)$, the selection of the confining metacolor gauge group E_6 and the global metaflavor group $SU(16)$ is mandated by the hypotheses that the constituent preons are contained in single irreducible representations of each group; the gauging requirement on a subgroup of the global metaflavor $SU(16)$ group immediately leads to the maximal $SO(10)$ metaflavor gauge group and hence we proceeded to work out the consequences of the "paradigmatic" gauged $E_6(MC) \times SO(10)(MF)$ preon model. When we allow for two representations of the global metaflavor group, it is possible to choose the global metaflavor group to be $SU(8) \times SU(8)$ [instead of $SU(16)$] and the maximal gauged subgroup to be G_{422} [instead of $SO(10)$], yielding the $E_6(MC) \times G_{422}(MF)$ preon model with essentially the same predictions as the "paradigmatic" $E_6 \times SO(10)$ preon model. The limitation in our preon model building, until now, to one irreducible representation of the metacolor group — requiring the unique choice of the E_6 metacolor group — may also be too restrictive. [Later, in §9.5b, we allow for more than one representation of the confining metacolor group for the preons and this permits consideration of the $SU(N)(N \geq 3)$ class of chiral preon models that were previously rejected.] Nevertheless, if we continue to insist on only one irreducible representation of the metacolor group, a possible way to eliminate the unwelcome "exotic" fermions of the "paradigmatic" preon model is to relax the assumption of "minimality" with regard to the metaflavor degrees of freedom of the preons.

To show how a "realistic candidate" preon model can be constructed by allowing the number of metaflavor degrees of freedom to exceed 16 (while still satisfying the asymptotic (ASF) condition), we enlarge the global metaflavor group from $SU(16)$ to $SU(N)$ ($16 < N < 22$) [N must not exceed 21 in order to retain ASF for the E_6 metacolor group [9.14] (see §8.5)]. The basic difficulty with the "minimal" global $SU(16)$ metaflavor group is that the 't Hooft anomaly-matching condition is not satisfied so that there is no "pre-selection" of massless three-preon fermion composites for use with the gauged $SO(10)$ metaflavor subgroup. This lack of "pre-selection" is responsible for the large number of "exotic" fermionic composite states that are generated in addition to the quark-lepton states, at the gauged $SO(10)$ level. To achieve this "pre-selection" by means of the 't Hooft anomaly-matching condition, it is necessary to start with the larger preon group $E_6(MC) \times SU(N)(MF)$ ($17 \leq N \leq 21$) and to postulate that an SSB mechanism exists to break $G_{MF} \equiv SU(N)$ to unbroken $G'_{MF} \equiv SU(16) \times SU(N-16)$ [the $U(1)$ symmetry that appears in this breakdown of $G_{MF} \rightarrow G'_{MF}$ is assumed to be dynamically broken by the metacolor forces] that terminates with the same gauged metaflavor subgroup $SO(10)$ or G_{422}. The trick is to apply the 't Hooft anomaly matching condition to unbroken G'_{MF} and to look for solutions of the matching conditions that simply exclude all (or most) of the "exotic" fermions. It turns out that this strategy of "pre-selection" of massless composite three-preon fermions yields a unique solution for $N = 18$ [but no solution for $E_6 \times SU(N)$ ($N = 17, 19 - 21$) [9.15]]; furthermore, the maximal gauged subgroup of $SU(18)$ is $SO(10)$ so that the interesting result is obtained that starting with $E_6(MC) \times SU(18)(MF)$, the final gauge group $E_6(MC) \times SO(10)(MF)$ now predicts precisely three generations of ordinary quarks and leptons unaccompanied by "exotic" fermions.

The method for deriving the predictions of the modified "paradigmatic" preon model — which is based on $E_6 \times SU(N)$ ($17 \leq N \leq 21$) with $SU(N)$ breaking spontaneously down to $SU(16) \times SU(N-16)$ — is sufficiently instructive to give some details. Assuming that $SU(N) \rightarrow SU(16) \times SU(N-16)$, the preons then belong to the representation $(\mathbf{27}; N)$ of $[E_6; SU(N)]$ so

that the three-preon composites are in the representations S_3, M_3 or A_3
(see Table 9.1), where $\square = N$ is the fundamental representation of $G_{MF} =$
$SU(N)$. At this point, the important role of the meta-Pauli principle must
be noted; indeed, if the Pauli principle is not extended to the metacolor
degree of freedom, it is possible to satisfy the 't Hooft triangular anomaly
matching condition for the initial global metaflavor group $SU(N)$ for all
values of $N \geq 3$ (including $N = 17 - 21$) so that we can treat $G_{MF} =$
$SU(N)$ $(N = 17 - 21)$ as the unbroken metacolor group G'_{MF}. This can
be seen from Table 9.2 where the three representations S_3, M_3 or A_3 (see
Fig. 9.1) are all considered as valid composite three-preon composites [9.16].
Thus, with the 't Hooft indices ℓ_i defined in Table 9.2, the 't Hooft triangular
anomaly matching equation [$\mathcal{A}(R)$ is the triangular anomaly coefficient for
the specified representation] becomes [9.16]:

$$\ell_+ \mathcal{A}(S_3) + \ell_0 \, \mathcal{A}(M_3) + \ell_- \, \mathcal{A}(A_3) = 27 \qquad (9.38)$$

or:

$$\frac{1}{2}(N+3)(N+6)\ell_+ + (N^2 - 9)\ell_0 + \frac{1}{2}(N-3)(N-6)\ell_- = 27 \qquad (9.39)$$

Equation (9.43a) has an N-independent solution (for $N \geq 3$) without the
meta-Pauli principle — with the 't Hooft indices:

$$\ell_+ = \ell_- = 1; \quad \ell_0 = -1 \qquad (9.40)$$

Table 9.2: 't Hooft anomaly matching in the $(E_6)(MC) \times SU(N)(MF)$ $(N = 17, \ldots 21)$
preon model.

Preons	Representation of E_6	Multiplicity	Global chiral symmetry	
P	27	N	$SU(N)$	
Composites	Representation of E_6		Global chiral symmetry	't Hooft "index"
PPP	1		⊞⊞⊞	ℓ_+
PPP	1		(middle Young diagram)	ℓ_0
PPP	1		(vertical Young diagram)	ℓ_-

The situation is quite different if we take account of the meta-Pauli principle; in that case, only one of the three composite representations is permitted
— namely, M_3 — since the three-preon metacolor singlet under E_6 is totally

symmetric and, hence, the composite three-preon contribution to the total wavefunction of $SU(N)(MF) \times SU(2)$ (chirality) [besides the metacolor and metaflavor degrees of freedom in the preon model, one must also take into account the "chirality" of the Weyl fermions — see §9.4] must be antisymmetric. Since the "chirality" contribution to the total wavefunction comes from the mixed (M_3) representation of $SU(2)$, the $SU(N)$ metaflavor contribution must also come from the mixed representation M_3 of $SU(N)$. Using the meta-Pauli principle, the 't Hooft anomaly matching equation reduces to:

$$(N^2 - 9)l_0 = 27 \qquad (9.41)$$

and it is obvious that there is only the solution, $N = 6$, which is of no interest. Hence, if we accept the meta-Pauli principle — as we must — $G_{MF} = SU(N)$ must spontaneously break down to some unbroken subgroup G'_{MF}. Only detailed dynamical calculations can fix the precise SSB pattern of G_{MF} to G'_{MF}; in the absence of that knowledge, the consequences of the simple SSB pattern for $SU(N) \to SU(16) \times SU(N-16)$ have been explored [9.15] for the entire range $N = 17 - 21$, assuming that no 't Hooft anomaly matching "index" (see Table 9.1) exceeds unity (because, otherwise, there would be an additional global symmetry) [9.17]. It turns out that no solution of the 't Hooft's anomaly matching condition exists for $N = 17, 19 - 20, 21$ but only for $N = 18$. We sketch the steps in the proof for the $N = 18$ case because they are useful in the discussion of other "realistic candidate" preon models.

For the $N = 18$ case, $SU(18) \to SU(16) \times SU(2)$ and the fundamental 18 representation of $SU(18)$ decomposes into $(\mathbf{16}, \mathbf{1}) + (\mathbf{1}, \mathbf{2})$. The preon representations transforming under $E_6 \times SU(16) \times SU(2)$ are therefore: $P_1 = (\mathbf{27}; \mathbf{16}, \mathbf{1})$ and $P_2 = (\mathbf{27}; \mathbf{1}, \mathbf{2})$; the particle content is summarized in Table 9.3. The 't Hooft anomaly matching condition in this case comes from the three $SU(16)$ currents [the $SU(2)$ currents give no contribution] and, without the meta-Pauli principle, we have:

$$209\, l_1^+ + 65\, l_1^- + 247\, l_1^0 + 40\, l_2^+ + 24\, l_2^- + 3\, l_3^+ + l_3^- = 27 \qquad (9.42)$$

With the meta-Pauli principle, Eq. (9.42) reduces to:

$$247\, l_1^0 + 40\, l_2^+ + 24\, l_2^- + 3\, l_3^+ + l_3^- = 27 \qquad (9.43)$$

Moreover, without the restriction $\ell_i \leq 1$, Eq. (9.43) gives rise to a large number of solutions; however, using the argument that $\ell_i > 1$ implies the presence of a "hidden" global symmetry not contained in the initial global metaflavor $SU(N)$ group, Eq. (9.47) possesses a unique solution:

$$\ell_3^+ = \ell_2^- = 1 \quad \text{all other } \ell_i = 0 \tag{9.44}$$

Thus, accepting the plausible restriction $\ell_i = 0$ or 1 [further substantiated in §9.5b in connection with the "tumbling complementarity" approach to preon model building], the application of the 't Hooft anomaly matching condition to the intermediate unbroken global metacolor group $G'_{MF} = SU(16) \times SU(2)$ predicts $\ell_3^+ = \ell_2^- = 1$ (all other $\ell_i = 0$), i.e. two massless composite fermions $(\square, S_2)$ and $(S_2, \square)$ under this group.

Table 9.3: 't Hooft anomaly matching in the $E_6 \times SU(16) \times SU(2)$ preon model.

Preons	E_6	$SU(16)$	$SU(2)$	
P_1	27	$\square$	1	
P_2	27	1	$\square$	
Composites				Indices
$P_1 P_1 P_1$	1	$\square\square\square\,,\,\square\!\!\square\,,\,\boxplus$	1	$\ell_1^+, \ell_1^-, \ell_1^0$
$P_1 P_1 P_2$	1	$\square\square\,,\,\square\!\!\square$	$\square$	ℓ_2^+, ℓ_2^-
$P_1 P_2 P_2$	1	$\square$	$\square\square\,,\,\square\!\!\square$	ℓ_3^+, ℓ_3^-
$P_2 P_2 P_2$	1	1	$\square\square\,,\,\boxplus$	ℓ_4^+, ℓ_4^0

The next step in working out the consequences of this preon model is to gauge a maximal subgroup of the $SU(16)$ part of the unbroken non-simple global metaflavor group $SU(16) \times SU(2)$ [the physical significance of the $SU(2)$ part will be explained below] and, not surprisingly, this maximal gauge symmetry group is $SO(10)$. Consequently, the two massless fermion composites fixed by the 't Hooft anomaly matching condition, namely $(\square, S_2)$ and $(S_2, \square)$, correspond to three "copies" (families) of the **16** representation of $SO(10)$ and two "copies" (families) of the **120** representation of $SO(10)$. But the **120** representation is real, and according to the "survival hypothesis" [8.31], fermions corresponding to real representations acquire large masses at the $SO(10)$ GUT scale. Hence, we are left with exactly three generations of the spinorial representation **16** of $SO(10)$ that are massless on the metacolor scale and possess the quantum numbers of the ordinary quarks and leptons (plus ν_R). Since the $SO(10)$ gauge group is a subgroup of $SU(16)$, the $SU(2)$ part of G'_{MF} stays global and serves as the "family" group while its uniquely-derived symmetric **3** (S_2) representation (from the 't Hooft anomaly matching condition) determines the number of quark-lepton generations. The most important consequence of the $E_6(MC) \times SU(18)(MF)$ preon model [with the SSB of $SU(18)$ leading to $SU(16) \times SU(2)$] is that the "exotic" composite fermionic states have disappeared through the use of the 't Hooft anomaly-matching condition at the unbroken $SU(16) \times SU(2)$ level; the fully gauged (in metacolor and metaflavor) $E_6 \times SO(10)$ preon model has thereby been recaptured without any "exotic" fermions. Hence, asymptotic freedom in the composite "ordinary" color-flavor sector is restored.

The above result appears to be a stunning triumph for this particular brand of "realistic candidate" approach to preon model building — which is based on the intervention of the "intermediate-stage" unbroken global meta-color group G'_{MF} and the use of the 't Hooft anomaly matching condition to guarantee the masslessness of three quark-lepton-type generations. The fact that the unwanted "exotic" fermions disappear in the solution and that $SU(2)$ emerges as the "family" group seems to restore the "paradigmatic" preon group $E_6(MC) \times SO(10)$ (MF) to being more than a "toy" model and worthy of serious dynamical calculations of the quark-lepton masses of the three generations. Unfortunately, there is a hitch: there is no dynamical jus-tification for the assumption that the initial global metacolor $SU(18)$ group breaks spontaneously to the subgroup $SU(16) \times SU(2)$. Without some dy-namical argument in support of the SSB path $SO(18) \rightarrow SU(16) \times SU(2)$, the $E_6 \times SO(10)$ preon model is still left in limbo despite its many qualitatively correct predictions. A promising quasi-dynamical method, called "tumbling complementarity" [9.6], has been suggested in order to select the proper SSB path from the broken global metaflavor group G_{MF} to the unbroken group G'_{MF}. We discuss the method of "tumbling complementarity" in the next section (§9.5b) and subject the $E_6(MC) \times SU(N)(MF)$ preon model to its strictures. It turns out that with "tumbling complementarity", the entire E_6 metacolor class of preon models is eliminated and only the $SU(N)(N \geq 3)$ metacolor class of models — with at least two Weyl preon representations — is tenable. Indeed, we show in §9.5b that it is possible to construct "realistic candidate" preon models satisfying "tumbling complementarity" within the framework of the $SU(N)$ $(N \geq 3)$ metacolor group, using at least two Weyl preon representations.

§9.5b. Preon models satisfying "tumbling complementarity"

In the previous section (§9.5a), we considered "realistic candidate" preon models of quarks and leptons which retained the same E_6 metacolor group as

the "paradigmatic" preon models of §9.4 but started with a global metaflavor group (instead of a metaflavor gauge group), to a subgroup of which the
't Hooft anomaly matching condition is applied before gauging the unbroken
residual global metaflavor subgroup. The class of such "realistic candidate"
preon models considered in §9.5a can overcome the superabundance of "exotic" fermions' problem but there is no convincing rationale for choosing
the SSB path from the initial global metaflavor group to the final unbroken
global metaflavor group to which the 't Hooft anomaly matching condition
is applied. Moreover, in deriving the promising $E_6(MC) \times SU(18)(MF)$
preon model, it is necessary to impose the additional constraint that the
't Hooft "index" l_i does not exceed unity (see §9.5a). Thus, two questions
still remained unanswered in §9.5a: (1) is there some plausible justifcation
for choosing the SSB path $G_{MF} \to G'_{MF}$ [e.g. $SU(18) \to SU(16) \times SU(2)$]
among many possibilities? and (2) can one justify the constraint on the
't Hooft anomaly matching index of $l_i \leq 1$?.

The method of "tumbling complementarity" (invented by Dimopolous,
Raby and Susskind (DRS) [9.6]) attempts to answer these two questions by
establishing a connection between the SSB of the gauged metacolor group
G_{MC} in the "Higgs phase" to the SSB pattern of the global metaflavor group
(G_{MF}) in the "confining phase". DRS make the observation that confining
gauge theories can break down in sequential steps as a "natural response"
of the vacuum to the evolution of the "running" coupling constants and
that furthermore, scalar fermion condensates (quasi-Higgs) trigger this "tumbling" phenomenon in the absence of "elementary" Higgs fields. [This type
of argument has partly been taken over by Nambu in his recent attempt to
explain the electroweak Higgs boson as a $t\bar{t}$ condensate [9.18] — see §6.4c].
"Complementarity" implies a one-to-one correspondence between the surviving massless fermions in the "tumbling" "Higgs phase" and the massless
composite fermions remaining from 't Hooft anomaly matching in the "confining phase". When the metacolor gauge forces between fermions in a given
channel are sufficiently attractive [i.e. are in the "most attractive channel"
(MAC)] and the resulting two-fermion (scalar) condensate is in the fundamental representation of G_{MC}, so the DRS argument goes, the two-fermion
condensate acts as the quasi-Higgs boson in the SSB of $G_{MC} \to G'_{MC}$. [The
DRS argument is taken from lattice gauge theory [9.19] where it is shown
that, if the Higgs scalar is in the fundamental representation of a gauge
group, there is no "phase change" between the "Higgs phase" (wherein the
gauge coupling is small and the SSB scale is large) and the "confining phase"

(wherein the gauge coupling constant is large and the SSB scale is small).] In the process of "tumbling", the MAC condensate also breaks G_{MF} to G'_{MF}, and those fermions under the combined symmetry group $G'_{MC} \times G'_{MF}$, which participated in the MAC condensate, or which can form mass terms (because of the "survival hypothesis" [8.31]), become massive. "Tumbling" continues until a final group $G^F_{MC} \times G^F_{MF}$ is reached, in which the surviving massless fermions are all G^F_{MC} singlets or no massless fermion survives. The concept of "complementarity" then tells us that "anything which is exactly true in the Higgs region must extrapolate into the confinement region so that the fermion mass spectra map into each other and one can assume a one-to-one correspondence between the surviving massless fermions in the tumbling Higgs phase and the massless composite fermions remaining from 't Hooft anomaly matching in the confining phase [1.159].

"Tumbling complementarity" is an attractive approach to preon model building because it seems to give a quasi-dynamical foundation for the SSB from the initial global metaflavor group G_{MF} to the final stage of gauging the global metaflavor subgroup. We review briefly several preon models satisfying "tumbling complementarity" — from "toy" models to "realistic candidate" models — and indicate the potentialities in the DRS approach, and its basic limitations, in leading ultimately to a dynamically successful composite preon theory of quarks and leptons. In examining preon models of quarks and leptons satisfying "tumbling complementarity", we first test the method on our familiar E_6 metacolor preon models to ascertain whether the choice of the global metaflavor group $SU(18)$ with its SSB path $SU(18) \rightarrow SU(16) \times SU(2)$ — which is crucial for achieving three generations without "exotic" fermions — is confirmed. We show that the answer is negative and that, while a "toy" preon model based on $E_6(MC) \times SU(18)(MF)$, yielding three families and satisfying "tumbling complementarity" exists, no "realistic candidate" preon model emerges because the quantum numbers of the surviving massless fermions are not those of the observed quarks and leptons. We give a short description of the E_6 preon model with "tumbling complementarity" [9.20] because it clearly exhibits the formidable constraints imposed by the "tumbling complementarity" approach to preon model building.

We recall that the fundamental representation of the E_6 metacolor group is **27** and therefore the MAC channel of the two-fermion (scalar) condensate is $\mathbf{27} \times \mathbf{27} \rightarrow \overline{\mathbf{27}}$ which is, indeed, a fundamental representation of E_6 so that "tumbling" can proceed. If we start with the group $G_{MC} \times G_{MF} = E_6(MC) \times SU(N)(MF)$, the $\overline{\mathbf{27}}$ condensate spontaneously breaks down this gauged metacolor — global metaflavor group to a group of lower symmetry, which we denote by $G'_{MC} \times G'_{MF}$. Since the subgroup G'_{MF} must contain a singlet in the branching of the condensate $\overline{27}$ into G'_{MC}, we limit ourselves to the maximal little groups [8.4] of E_6 and identify the following five possibilities:

$$E_6 \to F_4 : \mathbf{27} \to \mathbf{1} + \mathbf{26}; \tag{9.45}$$

$$E_6 \to SO(10) \times U(1) : \mathbf{27} \to \mathbf{1}(4) + \mathbf{10}(-2) + \mathbf{16}(1); \tag{9.45a}$$

$$E_6 \to SU(3) \times SU(3) \times SU(3) : \ \mathbf{27} \to (\mathbf{1}, \bar{\mathbf{3}}, \bar{\mathbf{3}}) + (\mathbf{3}, \mathbf{1}, \mathbf{3}) + (\bar{\mathbf{3}}, \mathbf{3}, \mathbf{1}); \tag{9.46}$$

$$E_6 \to G_2 \times SU(3) : \ \mathbf{27} \to (\mathbf{1}, \bar{\mathbf{6}}) + (\mathbf{7}, \mathbf{3}); \tag{9.47}$$

$$E_6 \to SU(2) \times SU(6) : \ \mathbf{27} \to (\mathbf{1}, \mathbf{15}) + (\mathbf{2}, \bar{\mathbf{6}}) \tag{9.48}$$

where we have listed the branching rules for the $\mathbf{27}$ of E_6 for each maximal little group. Note that $U(1)$ of (9.45a), two of the three $SU(3)$'s of (9.46), $SU(3)$ of (9.47) and $SU(6)$ of (9.48) are broken, lacking singlets [zero charge for $U(1)$] in the branchings. Hence, we can take G'_{MC} to be F_4, $SO(10)$, $SU(3)$, G_2, or $SU(2)$, respectively. The metaflavor group G'_{MF} is determined by the same principle. The MAC condensate corresponds to the representation S_2 or A_2 (see Fig. 9.1) under $G_{MF} = SU(N)$. The meta-Pauli principle, however, requires the use of the symmetric representation S_2; this follows from the fact that the metacolor part is symmetric, while the chirality ("spin") part for the two-fermion condensate (scalar) is antisymmetric. Consequently, the possible breakings of G_{MF} into G'_{MF} are:

$$SU(N) \to SU(N-j) \times O(j) \quad (1 \le j \le N) \tag{9.49}$$

It can now be shown that pursuing the "tumbling complementarity" method for the first four SSB's of the E_6 metacolor group [see Eqs. (9.45)–(9.47)] excludes even "toy" preon models for E_6 metacolor whereas the fifth SSB of the E_6 metacolor group [i.e. $E_6 \to SU(2) \times SU(6)$] allows for a "toy" preon model [i.e. it predicts three massless fermion generations on the metacolor scale but not with the correct quark-lepton quantum numbers] but not for a "realistic candidate" preon model. We do not give a full accounting and only work out a few details for the fifth SSB path from E_6 to its maximal little group $SU(2) \times SU(6)$. The path $E_6(MC) \to SU(2) \times SU(6)$ gives rise to a "toy" preon model with three generations of massless fermions through the use of the $SU(2)$ part of $SU(2) \times SU(6)$. Let us see how this comes about by choosing $SU(N)_{MF} = SU(6m)_{MF}$ ($m < 4$ from metacolor ASF). Consider the Higgs phase: by the meta-Pauli principle, the metaflavor part of the MAC condensate is S_2 of $SU(6m)_{MF}$, which breaks $SU(6m)_{MF}$. By considering the branching of $SU(6m)_{MF}$ into $SU(6)_{MF} \times SU(m)_{MF}$, we can identify the diagonal subgroup of $SU(6)_{MC}$ in Eq. (9.48) and $SU(6)_{MF}$ as the new unbroken global symmetry $SU(6)'_{MF}$. This is permissible only when $m > 1$, because for $m = 1$, the S_2 of $SU(6)_{MF}$ and the $\overline{15}(= \bar{A}_2)$ from the metacolor part of the condensate cannot combine into a singlet. On the other hand, for $m > 1$, we have the following branching rules of $SU(6m)_{MF}$ into $SU(6)_{MF} \times SU(m)_{MF}$ (see Fig. 9.1):

$$\square \to (\square, \square); \quad S_2 \to (S_2, S_2) + (A_2, A_2) \tag{9.50}$$

The (A_2, A_2) term of Eq. (9.50) can form an $SU(6)'_{MF}$ singlet with the $\overline{15}$ in the $(1, \overline{15})$ part of metacolor decomposition. Up to this point, the MAC condensate can be written as: $(1; 1, A_2)$ under $SU(2)_{MC} \times SU'(6)_{MF} \times SU(m)_{MF}$. If $m = 2$, we can stop here because $S_2 = 1$ under $SU(2)_{MF}$. In sum, the fermions $(27, \square)$ under $E_6 \times SU(6m)_{MF}$ branch as follows for $m = 2$:

$$(\mathbf{2}; \mathbf{1}, \mathbf{2}); \quad (\mathbf{2}; \mathbf{35}, \mathbf{2}); \quad (\mathbf{1}; \mathbf{20}, \mathbf{2}); \quad (\mathbf{1}; \mathbf{70}, \mathbf{2}) \tag{9.51}$$

under $SU(2)_{MC} \times SU(6)'_{MF} \times SU(2)_{MF}$.

But now we are in trouble with $m = 2$ because $\mathbf{35}$ and $\mathbf{20}$ are real representations of $SU(6')_{MF}$ and the fermions corresponding to the first three terms of Eq. (9.51) acquire mass, leaving only $(\mathbf{1}; \mathbf{70}, \mathbf{2})$ as massless fermions. Thus, there are only two copies (i.e. two generations) of $(\mathbf{1}; \mathbf{70})$ under $SU(2)_{MC} \times SU(6')_{MF}$ where the non-trivial quantum number 2 under $SU(2)_{MF}$ lifts the degeneracy. Actually, we are close to a "toy" preon model based on the E_6 metacolor group and we achieve it if we move on from $m = 2$ to $m = 3$. Repeating the series of steps in the "tumbling complementarity" method that result in Eqs. (9.50)–(9.51), we find that, under $SU(2)_{MC} \times SU(6')_{MF} \times SU(2)_{MF}$, the massless fermion representations:

$$(\mathbf{1}; \mathbf{70}, \mathbf{2}); \quad (\mathbf{1}; \mathbf{70}, \mathbf{1}) \tag{9.52}$$

are the surviving massless fermions. Note that the threefold degeneracy is lifted by the $SU(2)_{MF}$ "family" quantum numbers 2 and 1. Thus far, we have proved that, in the Higgs phase, "tumbling" down the $E_6 \to SU(2)_{MC} \times SU(6)_{MC}$ path, when combined with $SU(18)_{MF}$, yields three generations of massless fermions (on the metacolor scale), distinguished by a quantum number of the "family" group $SU(2)_{MF}$. We must still prove that the solution (9.52) coincides with that in the "confining phase" in order to claim that a "toy" preon model has been achieved satisfying "tumbling complementarity".

Table 9.4: Confining phase of the $E_6 \times SU(6)'_{MF} \times SU(2)_f$ preon model derived from $G_{MF} = SU(18)_{MF}$.

contents		E_6	$SU(6)'_{MF}$	$SU(2)_f$	't Hooft indices
preons	P_1	27	☐	☐	
	P_2	27	☐	1	
composites	$P_1 P_1 P_1$	1	⊞ , ⊟ , 𝌆	⊞	$\ell_1^+, \ell_1^-, \ell_1^0$
				𝌆	$\ell_2^+, \ell_2^-, \ell_2^0$
	$P_1 P_1 P_2$	1	⊞ , ⊟ , 𝌆	☐	$\ell_2^+, \ell_2^-, \ell_3^0$
				⊟	$\ell_4^+, \ell_4^-, \ell_4^0$
	$P_1 P_2 P_2$	1	⊞ , ⊟ , 𝌆	☐	$\ell_5^+, \ell_5^{-\prime}, \ell_5^0$
	$P_2 P_2 P_2$	1	⊞ , ⊟ , 𝌆	1	$\ell_6^+, \ell_6^-, \ell_6^0$

To establish "tumbling complementarity" in the $E_6(MC) \times SU(18)(MF)$ "toy" model, we consider Table 9.4 to show what happens in the "confining phase". Using a similar notation to that in Table 9.3, but allowing for the unbroken $SU(6)'_{MF}$ to replace $SU(16)$ and the well-defined "family group" $SU(2)_f$ to replace $SU(2)_F$, we find that the 't Hooft anomaly matching condition — due exclusively to the three $SU(6)'_{MF}$ currents — leads to the following relations among the 't Hooft "indices" defined by Table 9.4:

$$3 \times 27 = 54(4\ell_1^+ + 3\ell_3^+ + 2\ell_{25}^+ + \ell_{46}^+) + 27(4\ell_1^0 + 3\ell_3^0 + 2\ell_{25}^0 + \ell_{46}^0) \quad (9.53)$$

where:

$$\ell_{ij}^{+,0} = \ell_i^{+,0} + \ell_j^{+,0} \quad (9.54)$$

It is seen from Eq. (9.54) that there are three solutions [if we insist on the meta-Pauli principle and the physical condition $\ell_i = 0, 1$], of which $\ell_{25}^0 = 1$, $\ell_{46}^0 = 1$ corresponds to the solution (9.53) for the "Higgs phase". It follows that there does exist a "toy" preon model based on the gauged metacolor-global metaflavor group $E_6(MC) \times SU(18)(MF)$ satisfying "tumbling complementarity". This is the same initial preon group that was discussed in §9.5a but unfortunately, insisting on "tumbling complementarity", leads to a different SSB pattern for the original global metaflavor $SU(18)$ group that results in three massless fermion generations (on the metacolor scale) with the wrong quantum numbers.

We have achieved a "toy" preon model based on E_6 metacolor and satisfying "tumbling complementarity" but a "realistic" candidate preon model still eludes us. This is a disappointment because Assumption (1) in §9.4 [i.e. that only one irreducible representation is required for the gauged metacolor group in a preon model] seems so attractive. However, the requirement of "tumbling complementarity" for selecting a "realistic candidate" preon model is equally attractive. Having gone this far, it behooves us to surrender Assumption (1) — while retaining Assumptions (2)–(4) (see §9.4) — and to search for a "realistic candidate" preon model satisfying "tumbling complementarity" by allowing for two irreducible representations of the gauged metacolor group defining the preon model. This opens the door to the class of $SU(N)$ ($N \geq 3$) metacolor preon models that obey the plausible Assumptions (2)–(4) put forward (in §9.4) to construct the "paradigmatic" preon model.

In actual fact the first "toy" chiral preon model satisfying "tumbling complementarity" — the Georgi model [8.31] — was based on the group $SU(5)_{MC} \times U(1)_{MF}$, with two lefthanded preon representations **5** and $\overline{\mathbf{10}}$ under $SU(5)_{MC}$ and respective $U(1)_{MF}$ global charges 3 and -1. It is easy to show that the **5** representation in the decomposition of $\overline{\mathbf{10}} \times \overline{\mathbf{10}}$ is the MAC channel and since this is the fundamental representation of $SU(5)_{MC}$, "tumbling" can proceed; the MAC condensate gives rise to the SSB of $SU(5)_{MC}$ to $SU(4)_{MC} \times U(1)_{MC}$. The $U(1)_{MC}$ can now be combined with $U(1)_{MF}$ to give a resultant $U(1)'$ with zero charge for the MAC condensate. It also follows from the "survival hypothesis" [8.31] that the $SU(4)_{MC}$ decomposite fermion representations arising from the original $\overline{\mathbf{5}}$ and **10** representations acquire mass from the SSB of $SU(5)_{MC}$ to $SU(4) \times U(1)_{MC}$. This can be seen by writing out the $SU(4)_{MC}$ content of the original $SU(5)_{MC}$ fermion representations:

$$\mathbf{5} \to \mathbf{1}(4) + \mathbf{4}(-1); \qquad \overline{\mathbf{10}} \to \overline{\mathbf{4}}(-3) + \mathbf{6}(2) \quad [\text{under } SU(4) \times U(1)] \quad (9.55)$$

and noting that — under $SU(4)_{MC} \times U(1)'$ — the two decomposite $SU(4)_{MC}$ fermion representations can combine to give a Dirac mass. In addition, since the **6** representation of $SU(4)$ is real, the fermions in that representation [see Eq. (9.55)] also acquire mass so that only the singlet fermion [under $SU(4)_{MC}$] stays massless. Furthermore, the end of "tumbling" is reached in the next stage [since the MAC channel for the surviving massless fermion is a singlet (i.e. $\mathbf{4} \times \overline{\mathbf{4}} \to \mathbf{1}$)] with the "Higgs phase" containing one singlet lefthanded fermion under $SU(4)_{MC}$ with $U(1)_{MC}$ charge = 5.

We must now determine whether the preon composites in the "confining phase" match the "Higgs phase" when the 't Hooft anomaly-matching condition is applied. The possible preon composites in the "confining phase" of the Georgi model are listed in Table 9.5; all preon composites are metacolor singlets — as they must be — and the $U(1)_{MF}$ charge is 5. The 't Hooft anomaly matching condition is easily seen to be fulfilled (see Table 9.5 for the notation on the 't Hooft indices):

$$[U(1)_{MF}]^3: \qquad \mathbf{125} = \mathbf{125}(\ell_1 + \ell_2) \qquad (9.56)$$

and hence $(\ell_1 + \ell_2) = 1$, which is acceptable. The conclusion is that the Georgi model satisfies "tumbling complementarity" and predicts only one family of composite fermions with a well-defined $U(1)_{MF}$ charge of -5. However, these composite fermions possess quantum numbers very different from those of ordinary quarks and leptons so that we are again dealing with a "toy" preon model satisfying "tumbling complementarity". Thus, while the Georgi "toy" preon model — like the E_6 "toy" preon model — is instructive in delineating the salient features of the "tumbling complementarity" method — which provides one of the few credible quasi-dynamical constraints on preon models of quarks and leptons — it is necessary to carry out a systematic search, within the framework of $SU(N)$ metacolor theories, to find "realistic candidate" composite preon models.

Table 9.5: Confining phase of the Georgi preon model.

Preons	$SU(5)_{MC} \times U(1)_{MF}$		
P_1	5	3	
P_2	$\overline{10}$	-1	
Composites			Indices
$P_1 P_1 P_2$	1	5	ℓ_1
$P_1 \bar{P}_2 \bar{P}_2$	1	5	ℓ_2

It is not a simple matter to construct "realistic candidate" preon models satisfying "tumbling complementarity". After much effort, one class of "realistic candidate" preon models based on $SU(N)$ (gauged) metacolor and a suitably chosen global metaflavor group, has been identified [9.21]. One starts with the group $SU(N)_{MC} \times SU(N+4)_{MF} \times U(1)_{MF}$ and uses two irreducible representations: the fundamental $F^i(\square)$ $(i = 1, \ldots, N)$ and the two-index symmetric-bar $\bar{S}_{ij}(i, j = 1, ...N)$ representations of $SU(N)_{MC}$ as shown in Table 9.6. In the "Higgs phase", we have, for MAC (see Fig. 9.1):

$$F^{ia}\bar{S}_{ij} = [\square;\,\square,\,(N+2)] \times [\bar{A}_2;\,1,\,-(N+4)] \rightarrow (\bar{\square};\,\square,\,-2) \qquad (9.57)$$

under $SU(N)_{MC} \times SU(N+4)_{MF} \times U(1)_{MF}$. Since $\bar{\square}$ is in the fundamental representation, we may proceed with "tumbling"; the choice for the two-preon (scalar) to secure SSB into the maximum little group is:

$$\Phi^a_j = < F^{ia}\bar{S}_{ij} > = \Lambda\,\delta^a_{j+4} \qquad (\Lambda = \text{const}) \qquad (9.58)$$

But:

$$SU(N+4)_{MF} \rightarrow SU(N)_{MF} \times SU(4)_{MF} \times U(1)'_{MF} \qquad (9.59)$$

with:

$$\square \rightarrow \begin{cases} \square & 1 & 4, \\ 1 & \square & -N \end{cases} \qquad (9.59a)$$

under $SU(N)_{MF} \times SU(4)_{MF} \times U(1)'_{MF}$. Hence we get:

$$SU(N)_{MC} \times SU(N+4)_{MF} \times U(1)_{MF} \rightarrow \widetilde{SU}(N)_{MF} \times SU(4)_{MF} \times \tilde{U}(1)_{MF} \qquad (9.60)$$

where $\widetilde{SU}(N)_{MF}$ is the diagonal subgroup of $SU(N)_{MC}$ and $SU(N)_{MF}$ [so that $SU(N)_{MC}$ is completely broken in the "Higgs phase"] and $\tilde{U}(1)_{MF}$ is a linear combination of $U(1)'_{MF}$ and $U(1)_{MF}$ [$\tilde{U}(1)_{MF} = U(1)'_{MF} + 2\,U(1)_{MF}$], that gives zero charge to MAC. Under $\widetilde{SU}(N)_{MF} \times SU(4)_{MF} \times \tilde{U}(1)_{MF}$, the preon representations decompose into:

$$F^{ia} \rightarrow \begin{cases} S_2 & 1 & 2(N+4) \\ A_2 & 1 & 2(N+4) \\ \square & \square & N+4 \end{cases} ; \qquad \bar{S}_{ij} \rightarrow \{\bar{S}_2 \quad 1 \quad -2(N+4)\} \qquad (9.61)$$

It follows that the massless fermions are:

$$\{A_2, 1, 2(N+4)\}; \quad \{\square, \square, N+4\} \qquad (9.62)$$

Table 9.6: GUT-like "realistic candidate" preon models satisfying "tumbling complementarity".

	$SU(N)_{MC}$	$SU(N+4)_{MF}$	$U(1)_{MF}$	
F^{ia}	$\square$	$\square$	$(N+2)$	$\begin{aligned} i &= 1,2,\ldots N \\ a &= 1,2,\ldots (N+4) \end{aligned}$
$\bar{S}_{ij}$	$\boxed{\ \ }$	1	$-(N+4)$	

It is seen from Eq. (9.62) that the family group $SU(4)$ has emerged in a natural way in the "Higgs phase"; but we must now show that the massless fermions in the "Higgs phase" [four $\square$'s and one A_2 of $SU(N)$] correspond to the massless (composite) fermions in the "confining phase". This is accomplished through the application of the 't Hooft anomaly matching condition to the preons in the gauged metacolor-global metaflavor group $SU(N)_{MC} \times SU(N)_{MF} \times SU(4)_{MF} \times \tilde{U}(1)_{MF}$. The representations for the preons and all three-preon composites in this model are listed in Table 9.7 [9.21]. Using Table 9.7, we find for the 't Hooft anomaly-matching equations (five in number);

$$[SU(N)_F]^3 : \quad N = 4\ell_1 + (N+4)\ell_2 + (N-4)\ell_2';$$

$$[SU(N)_F]^2\tilde{U}(1)_F : \; 2(N+4)N = 4(N+4)\ell_1 + 2(N+4)(N+2)\ell_2$$
$$+ 2(N+4)(N-2)\ell_2';$$

$$[SU(4)_F]^3 : \quad N = N\ell_1 + 8\ell_3;$$

$$[SU(4)_F]^2\tilde{U}(1)_F : \; (N+4)N = (N+4)N\ell_1$$

$$[\tilde{U}(1)_F]^3 : \; N^2[2(N+4)]^3 + 4N(N+4)^3 - \frac{N(N+1)}{2}[2(N+4)]^3$$

$$= 4N(N+4)^3 l_1 + \frac{N(N+1)}{2}[2(N+4)]^3\ell_2 + \frac{N(N-1)}{2}[2(N+4)]^3\ell_2'$$
$$\tag{9.63}$$

It is seen that the only solution to these five equations is $l_1 = l_2' = 1$ with all other $l_i = 0$, in precise agreement with the "Higgs phase". We therefore conclude that a solution of the $SU(N)_{MC} \times SU(N+4)_{MF} \times U(1)_{MF}$ model (satisfying "tumbling complementarity") exists and that the massless (composite) fermions are:

$$\{A_2, 1, 2(N+4)\}; \qquad \{\square, 4, (N+4)\} \tag{9.64}$$

The final step in the "tumbling complementarity" method — when there is promise in the global metaflavor quantum numbers of the matched "Higgs" and "confining" phases — consists of gauging the maximal subgroup of the global $SU(N+4)_{MF} \times U(1)_{MF}$ metacolor group and decomposing the surviving massless (composite) three-fermion representations of $SU(N+4)_{MF} \times U(1)_{MF}$ into irreducible representations of the gauged metaflavor subgroup H_{MF}. Obviously, the choice of N is decisive and we briefly summarize the results for $N = 15$. From Eq. (9.64), the massless (composite) fermions satisfying complementarity become:

$$(A_2, 1, 38); \qquad (\square, \square, 19) \tag{9.65}$$

under $SU(15)_{MF} \times SU(4)_{MF} \times \tilde{U}(1)_{MF}$. When we gauge the global chiral symmetry $G'_{MF} = SU(15) \times U(1)$ down to its subgroup $H_{MF} = SU(5)$, the preons transform according to the $\bar{5} + 10$ representation of $SU(5)$ and we get:

$$\Box \rightarrow \bar{5} + 10; \quad A_2 \rightarrow 5 + \overline{10} + \overline{45} + 45 \tag{9.66}$$

It would not appear at first sight that Eq. (9.66) implies three families; however, the "survival hypothesis" disposes of the $\overline{45} + 45$ composites in the decomposition of A_2 and the $5 + \overline{10}$ decomposite of A_2 constitutes a "mirror" representation that cancels one of the four $\bar{5} + 10$ decomposites of $\Box$. Consequently, we end up with three $(\bar{5} + 10)$ generations of $SU(5)$ without "exotic" fermions, and a family group, $SU(4)$, has emerged quite naturally (see Table 9.7).

Table 9.7: 't Hooft anomaly matching between the Higgs and confining phases for the $SU(N)_{MC} \times SU(N)_{MF} \times SU(4)_{MF} \times \tilde{U}(1)_{MF}$ preon model.

	Preons	$SU(N)_{MC}$	$SU(N)_{MF}$	$SU(4)_{MF}$	$\tilde{U}(1)_{MF}$	
$F^{ia} \rightarrow$	P_1'	$\Box$	$\Box$	1	$2(N+4)$	
	P_1''	$\Box$	1	$\Box$	$(N+4)$	
$\bar{S}_{ij} \rightarrow$	P_2	$\overline{\Box\Box}$	1	1	$-2(N+4)$	
Composites						**Indices**
$P_1' P_1'' P_2$		1	$\Box$	$\Box$	$(N+4)$	ℓ_1
$P_1' P_1' P_2$		1	$\Box\Box , \begin{array}{c}\Box\\\Box\end{array}$	1	$2(N+4)$	ℓ_2, ℓ_2'
$P_1'' P_1'' P_2$		1	1	$\Box\Box , \begin{array}{c}\Box\\\Box\end{array}$	0	ℓ_3, ℓ_3'

We can finally claim success in identifying one "realistic candidate" preon model — $SU(15)_{MC} \times SU(19)_{MF} \times U(1)_{MF}$ — that satisfies "tumbling complementarity" and thereby gives a quasi-dynamical basis to the choice of the final massless composite fermion representations. Furthermore, this composite preon model provides an SSB mechanism for the conversion of the global metaflavor degrees of freedom into a plausible family group; however, the gauged subgroup of the embedding global metacolor group corresponds to a three-generation $SU(5)$ GUT model. While it is gratifying to uncover at least one "realistic candidate" preon model, it must be admitted that a three-generation $SU(5)$ GUT-like preon model satisfying "tumbling complementarity" does not appear to have much phenomenological promise. Another choice of N in Eq. (9.64) can produce a second "realistic candidate" GUT-like preon model: e.g. $N = 20$ gives rise to four generations of $SO(10)$ 16 representations. Clearly, as long as we insist on placing all preons in a single irreducible representation of the metaflavor group, we end up with GUT-like preon models. What has not been ruled out in preon model building, is the possibility of a "realistic candidate" preon model (satisfying "tumbling complementarity") that predicts three "copies" of the one-generation standard group or a partially unified one-generation group [such as the left-right-symmetric group G_{3221} or the Pati-Salam group G_{422}], all of which would permit a smaller metacolor scale than that which is implicit in a three-generation GUT-like preon model. Some work looking into the possibility of a lower metacolor scale for the composite preon model satisfying "tumbling complementarity" has been carried out at the expense of allowing three or even four preon representations. With four irreducible preon representations, and an $SU(3)$ metacolor confining group, it is possible to show [9.22] that a low metacolor scale preon group can give rise to three generations of three-preon composites with the correct quantum numbers of the standard quarks and leptons.

While it is true that the three-generation $SU(3)_{MC}$ preon model requires one fewer representation than the five representations of the standard model [one L quark doublet, two R antiquark singlets, one L lepton doublet and one R anti-lepton singlet!], the initial gauged metacolor-global metaflavor preon group of the low composite scale three-generation preon model is larger (and more complicated) than the standard group, and it is difficult to believe that the solution to the "fermion generation problem" will come from such an inelegant composite preon model. The situation could change drastically if deviations from the point-like structure of quarks and leptons is uncovered in the TeV region. If this were to happen, a low metacolor scale preon model would become very attractive and it might be possible to solve the "fermion generation problem" without having recourse to large metacolor scale (approaching the Planck scale) preon models which would, *prima facie*, predict undetectable composite effects. Moreover, if compositeness of quarks and leptons is observed in the TeV or multi-TeV region, it would justify trying to cope not only with the "fermion generation problem" within the framework of a preon model but also with the full complexities of the mass spectrum of the quarks and leptons of the three generations.

References

[9.1] B. Stech, *E. Majorana Intern. Series*, Phys. Sci., Vol. 20, Plenum Press (1984).

[9.2]   F. Wilczek, *Phys. Rev. Lett.* **39** (1977) 1304.

[9.3]   R.D. Peccei, Lectures given at the 17 ème École d'Été de Physique des Particules, Clermont Ferrand, France (1985).

[9.4]   F. Wilczek and A. Zee, *Phys. Rev.* **D25** (1982) 553.

[9.5]   J. Bagger and S. Dimopoulos, *Nucl. Phys.* **B244** (1984) 247.

[9.6]   S. Raby, S. Dimopoulos and L. Susskind, *Nucl. Phys.* **B169** (1980) 373; S. Dimopoulos, S. Raby and L. Susskind, *Nucl. Phys.* **B173** (1980) 208.

[9.7]   Y. Frishman, A. Schwimmer, T. Banks, and S. Yankielowicz, *Nucl. Phys.* **B177** (1981) 157.

[9.8]   D. Weingarten, *Phys. Rev. Lett.* **51** (1983) 1830; S. Nussinov, *ibid.* **51** (1983) 2081; E. Witten, *ibid.* **51** (1983) 2351.

[9.9]   Y. Tosa and R.E. Marshak, *Phys. Rev.* **D26** (1982) 303.

[9.10] I. Bars, *Phys. Lett.* **106B** (1981) 105; *Nucl. Phys.* **B208** (1982) 77; W. Buchmüller and D. Wyler, *Nucl. Phys.* **B268** (1986) 621.

[9.11] J.M. Gipson, Y. Tosa and R.E. Marshak, *Phys. Rev.* **D32** (1985) 284.

[9.12] M.Y. Han and Y. Nambu, *Phys. Rev.* **139B** (1965) 1006.

[9.13] Y. Tosa and R.E. Marshak, *Phys. Rev.* **D27** (1983) 616.

[9.14] V. Silveira and A. Zee, *Phys. Lett.* **157B** (1985) 191.

[9.15] Y. Okamoto and R.E. Marshak, *Phys. Lett.* **162B** (1985) 333; *Nucl. Phys.* **B268** (1986) 397.

[9.16] X.Y. Li and R.E. Marshak, *Nucl. Phys.* **B268** (1986) 383.

[9.17] I. Bars, *Nucl. Phys.* **B207** (1982) 77.

[9.18] Y. Nambu, Univ. of Chicago preprint EFI 89-08 (1989).

[9.19] K. Osterwalder and E. Sciler, *Ann. Phys.* **110** (1978) 440; E. Fradkin and S.H. Shenker, *Phys. Rev.* **D19** (1979) 3682; T. Banks and E. Rabinovici, *Nucl. Phys.* **B160** (1979) 349.

[9.20] J.M. Gérard, Y. Okamoto and R.E. Marshak, *Phys. Lett.* **169B** (1986) 386.

[9.21] C.Q. Geng and R.E. Marshak, *Phys. Rev.* **D35** (1987) 2278.

[9.22] C.Q. Geng and R.E. Marshak, *Zeit. f. Phys.* **C33** (1987) 513.

Chapter 10

TOPOLOGICAL CONSERVATION LAWS

§10.1. "Noether" and topological conservation laws

The usual "Noether" conservation law in physics is a consequence of the invariance of the Lagrangian (or action) of a physical system under a continuous symmetry transformation, whether relating to space-time or to an internal symmetry. For most purposes in this book, the physical system is a relativistic quantum field but "Noether" conservation laws [10.1] can arise equally well for classical and non-relativistic fields. "Noether" conservation laws are expressed in terms of conserved currents (called "Noether" currents) — defined with the help of the Lagrangian equations of motion — and the resulting conserved "Noether" charges apply to field theories with a single physical vacuum (see §3.4).

The "Noether" conservation laws played a dominant role in the development of modern particle physics — with little attention given to the "topologically-oriented" proposals of the Dirac magnetic monopole and the Skyrme model — until the early 1970s. When it was realized that the $U(1)$ triangular axial anomaly $[U(1)_A]$ encodes non-perturbative information concerning gauge theories (whether Abelian or non-Abelian) and that the degenerate vacua resulting from spontaneous symmetry breaking (SSB) of gauge theories open up new possibilities for the topological classification of solutions of these theories, intense interest developed in the application of topological methods to modern particle physics. We propose to give a status report in this chapter on topological conservation laws, introducing physical examples from particle physics (for $D = 4$ space-time dimensions) and from condensed matter physics (for $D < 4$ space-time dimensions) where realistic particle physics examples are unavailable.

While it is possible to derive topological conservation laws governing field-theoretic solutions without explicit use of SSB [as in the cases of two-dimensional "kinks" and four-dimensional "instantons"], it is the degenerate vacua arising from SSB that create the most favorable conditions for identifying topological solitons (which give rise to topological conservation laws). Topological solitons — in contradistinction to non-topological solitons — are the non-dissipative finite-energy solutions of the Lagrangian equations of motion for which the boundary conditions at (spatial) infinity are topologically different from those valid for the single physical vacuum. [Non-topological solitons possess the same boundary conditions at infinity as do the perturbative, single-vacuum solutions of the Lagrangian equations and can arise when non-linear interactions among the fields are present (the non-linearity makes it possible to generate non-topological solitons for weak coupling) [2.18].] Topological soliton states are classified by means of the topologically inequivalent boundary conditions at spatial infinity and the selection rule that prohibits the transition from one class of soliton states to another (or to the vacuum) is called a "topological conservation law" [3.26]. Thus, topological conservation laws — unlike "Noether" conservation laws — are not a consequence of some symmetry of the Lagrangian but arise from the possibility of defining topologically distinct classes of non-singular, finite-energy solutions of the Lagrangian equations, i.e. classes of soliton solutions that can not be deformed into each other by topological means. The importance of topological solitons in modern particle physics stems from the fact that: (1) degenerate vacuum states play a prominent role in the standard gauge theory of the strong and electroweak interactions and its possible enlargements; (2) non-Abelian gauge theories — as in the standard model — necessarily involve topologically non-trivial gauge transformations, which, under suitable conditions, give rise to topological conservation laws; and (3) the structure of the chiral gauge anomalies in the standard model is strongly influenced by the topological soliton solutions (e.g. instantons) that arise in four-dimensional Yang-Mills theories.

The typical topological soliton solution is not only of finite energy (and, *a fortiori*, of finite size) and non-dissipative but may be time-independent; however, this is not always true and exceptions will be noted. The topological conservation laws associated with topological solitons (or similar "particle-like" objects) are usually of the "additive" type (like baryon or lepton charge conservation) although there are topological conservation laws of the "multiplicative" kind (like parity) which result from the use of the "large" gauge transformation in the presence of chiral fermions [such as the global $SU(2)$ anomaly — see §10.4a]. In either case — whether one is dealing with an "additive" or "multiplicative" topological conservation law — the conservation law can be expressed in homotopic group language, thus: $\Pi_k(G) = \mathcal{Z}$ for "additive" topological conservation whereas $\Pi_k(G) = \mathcal{Z}_2$ [$\mathcal{Z}_2$ is defined by integers mod 2] for "multiplicative" topological conservation, where $\Pi_k(G)$ denotes the homotopy group corresponding to the mapping of the S^k sphere at infinity (k is the number of dimensions) onto the G group manifold in parameter space. As we will see (in §10.4), the language of homotopic groups not only differentiates between the various types of topological conservation laws in four-dimensional particle physics but also helps to clarify the interrelationship between topological solitons that exist (or are conjectured to exist) in four-dimensional particle physics and those that have been identified in the lower-dimensional non-relativistic many-body systems of condensed matter physics.

As a historical note, it should be pointed out that the first hint of the possible relevance of an "additive" topological conservation law in particle physics arose in connection with the Abelian electromagnetic gauge group when Dirac derived the quantization condition for the hypothesized magnetic monopole (see §4.6), namely:

$$eg = \frac{n}{2} \quad (n \text{ an integer}) \tag{10.1}$$

where e is the electric charge and g is the "magnetic charge" of the magnetic monopole. While Dirac's magnetic monopole has never been found — and hence his quantization condition (10.1) has never been tested — we did consider the Wu-Yang approach to the Dirac magnetic monopole [4.13], with the gauge function ϕ serving as the transition transformation (see Fig. 4.12a) from the "north cap" magnetic field (α) to the "south cap" magnetic field (β) and where the requirement of single-valuedness for ϕ yields Eq. (10.1). It follows that the mapping (transition) gauge transformation is $e^{in\phi}$ [see Eq. (4.134) where $\alpha = 2g\phi$], which is an element of a $U(1)$ group manifold; hence, we may think of the mapping as going from the equatorial closed loop C at infinity (S^1) — non-contractible because of the "Dirac string" — onto the $U(1)$ manifold with "winding number" n; in homotopic language (see Table 10.3), we can write $\Pi_1[U(1)] = \mathcal{Z}$. Consequently, the quantization condition for a Dirac magnetic monopole is really the simplest example of an "additive" topological conservation law, and it will be interesting to compare the similarities and differences between the Dirac magnetic monopole and the 't Hooft-Polyakov magnetic monopole — which is also governed by the homotopic group $\Pi_1[U(1)] = \mathcal{Z}$ (see §10.2c) even though there is no manifest SSB in the case of the Dirac magnetic monopole.

A second hint of the importance of topological conservation laws emerged in §7.3, in connection with the global $SU(2)$ chiral gauge anomaly resulting from the presence of chiral fermions in a background of an $SU(2)$ gauge field. The global $SU(2)$ anomaly must be taken into account in the standard model because an $SU(2)$ chiral gauge theory is rendered mathematically inconsistent in the presence of an odd number of chiral (Weyl) fermion doublets and is only acceptable for an even number of chiral fermion doublets (see §7.3). In essence, a two-valued ("multiplicative") topological conservation law emerges from the evaluation of the "partition functional" of an odd number of chiral fermion doublets in the presence of an $SU(2)$ background gauge field to which the "large" gauge transformation has been applied (see §10.4a). Again, using homotopic language, it is shown (in §10.4a) that the pertinent "mapping" is from S^4 (the sphere in Euclidean four-space at infinity) onto the $SU(2)$ manifold so that the global $SU(2)$ anomaly is governed by the homotopy group $\Pi_4[SU(2)] = \mathcal{Z}_2$.

Insofar as applications of topological conservation laws to particle physics are concerned, we limit ourselves to four-dimensional space-time where, in principle, it should be possible to test the topological predictions. It is true that novel and unexpected features of the topological conservation laws found in four-dimensional particle physics have encouraged the application of topological methods to higher-dimensional particle physics (e.g. superstrings). However, at this stage, it is impossible to test the great variety of topological conservation laws uncovered in higher-dimensional particle physics and we forego reviewing the elegant mathematics involved in superstring theory and refer the reader to several published books [10.2] and to the brief historical introduction to superstring physics in §1.4c of this book. On the other hand, it is possible to manipulate physical conditions in other branches of physics (e.g. condensed matter physics) so as to simulate fewer than four space-time dimensions and thereby to test topological conservation laws in non-relativistic many-body systems for $D < 4$ space-time dimensions, and we do consider several such examples (in §10.2a and §10.2b).

With these remarks in mind, we proceed to discuss (in §10.2) the classical non-dissipative finite-energy soliton solutions of increasing dimensional complexity that exist in field theory when degenerate vacuum states — or their equivalent — exist. We start in §10.2a with the simplest case of $D = 2$ $[(1 + 1)-]$ space-time dimensions — where the degenerate vacua are built into the scalar field theory without gauging and without explicit SSB — and continue in §10.2b and §10.2c with $D = 3$ $[(2+1)-]$ and $D = 4$ $[(3+1)-]$ space-time dimensions respectively — where the degenerate vacua are generated by the SSB of gauge symmetry groups [Derrick's theorem comes into play for $D \geq 3$ [10.3] — see §10.2b]. Physical examples — whether observed or hypothetical — of the resulting topological conservation laws are given wherever possible. The topological conservation laws that emerge for $D = 2$ and $D = 3$ space-time dimensions are supported by well-studied phenomena in condensed matter physics and we discuss several examples. In going to $D = 4$ space-time dimensions, the topological conservation laws become especially relevant for particle physics and we consider the interesting (but speculative) physical application to 't Hooft-Polyakov magnetic monopoles.

If Minkowski (3+1)-space is converted into four-dimensional Euclidean space, the classical soliton solutions of the Yang-Mills equations, corresponding to minimum Euclidean action, become instantons and a special section (§10.3) is devoted to these "pseudo-particle" objects in four-dimensional Euclidean space. The role of QFD instantons in causing B and L violation [while maintaining $(B - L)$ conservation] and the capability of "sphaleron" solutions (to the Yang-Mills-Higgs equations) to overcome the instanton-induced suppression of B (and L) violation are covered in §10.3a. In §10.3b, we show that the instantons in unbroken non-Abelian QCD are responsible for the global QCD anomaly (see §7.2b) and for the θ vacuum, whose properties can be exploited to solve the "$U(1)$ problem" in QCD. In §10.3c, it is seen how the same θ vacuum that solves the $U(1)$ problem in QCD creates the "strong CP" problem in QCD and we consider several proposed remedies, including the Peccei-Quinn symmetry and the possibility that the θ parameter vanishes identically when the non-perturbative QCD vacuum is properly defined. Homotopic groups and the "index" theorem are treated in §10.4 with the object of providing greater mathematical precision (in terms of fermion "zero modes") to the origin of the "multiplicative" topological conservation law associated with the global $SU(2)$ anomaly, as well as to establish the relation between cancellation of the triangular chiral gauge anomaly and the absence of the global $SU(2)$ anomaly for certain classes of non-Abelian chiral gauge groups. Finally, the last section (§10.5) is devoted to the intriguing application of topological conservation laws to the Skyrme model for the light it might shed on the low energy (long-distance) behavior of the strong interaction and the usefulness of QCD and the "Wess-Zumino term" in fixing the properties of the baryonic topological soliton which is regarded as the Skyrmion.

§10.2. Topological solitons in field theory in $D = 2$ to $D = 4$ space-time dimensions

In the next several sections (§10.2–§10.2c), we focus on the topological conservation laws that are associated with topological solitons in field theories in $D = 2$ to $D = 4$ space-time dimensions. We will see that $D = 2$ scalar field theories are capable of giving rise to topological solitons ("kinks"), without gauging and without SSB, provided appropriate boundary conditions are chosen at spatial infinity. For $D > 2$, it is necessary to add vector fields (gauging) or higher spin fields (Derrick's theorem [10.3]), and for SSB to take place in order to obtain topological soliton solutions; the topological solitons are called "vortices" for the $D = 3$ case and "'t Hooft-Polyakov magnetic monopoles" for the $D = 4$ case. Before proceeding with our discussion of the $D = 2$ to $D = 4$ topological solitons in field theory, we should point out that "solitary waves" (i.e. solitons) were first observed in 1844 [10.4] as non-dissipative, time-independent solutions of finite energy in water. The "soliton" phenomenon was not immediately apparent in field theories since such well-known linear field theories as Maxwell's equations of electromagnetism, the Klein-Gordon equation, etc. only possess non-singular solutions of finite energy (i.e. wave packets) that are dissipative. Soliton solutions only exist in the presence of non-linear couplings and were first identified in two-dimensional (i.e. $D = 2$) non-linear scalar field theories such as the $\lambda\phi^4$ and Sine-Gordon theories. Since the non-linear "Higgs potential" is responsible for the SSB of modern gauge theories of particle interactions, it is not surprising that soliton solutions have been found in spontaneously broken field theories operating in $D \geq 2$ space-time dimensions. It should be emphasized that the soliton is a classical phenomenon and that the topological conservation laws that are associated with the "lumpy", "kinky", "packet-like" solutions of finite extension that are found in classical field theories must be re-interpreted in quantum mechanical terms to be relevant to the many-body physical systems of particle and condensed matter physics [2.18].

Strictly speaking, classical field theory can give rise to non-singular, non-dissipative solutions of finite total energy that are time-dependent, periodic in time, or with an even more complicated time dependence. Since we wish to consider classical field theories that are Lorentz-invariant (or Galilean-invariant), we need only examine static soliton solutions and then let these "lumps" or "kinks" move with uniform velocity. In the next three sections (§10.2a–§10.2c), we consider static soliton solutions in one, two and three spatial dimensions. In each case, we first discuss classical field theory and then the modifications introduced by quantization. As promised in the introduction, illustrations of the topological conservation laws associated with the soliton solutions — taken from particle physics or condensed matter physics — are given.

As an introduction to the subject, we begin in §10.2a with a set of classical field theories involving a single (real) scalar field in one-dimensional space (i.e. $D = 2$ space-time dimensions) — the $\lambda\phi^4$ theory — whose soliton solutions are due to two quasi-degenerate vacua and effectively give rise to a "multiplicative"' topological conservation law. We also treat the case of the $D = 2$ Sine-Gordon equation whose soliton solutions are due to an infinite number of quasi-degenerate vacua and give rise to an "additive" topological conservation law. We then move on to §10.2b where soliton solutions in two spatial dimensions (i.e. $D = 3$ space-time dimensions) are discussed, with an application to the quantized magnetic flux tubes of Type II low temperature superconductivity. Finally, in §10.2c, we discuss the soliton solutions in three spatial dimensions (i.e. $D = 4$ space-time dimensions), with particular emphasis on magnetic monopoles of the 't Hooft-Polyakov type.

§10.2a. Topological solitons in $D = 2$ space-time dimensions

In treating two examples of topological solitons in $D = 2$ space-time dimensions [the $\lambda\phi^4$ field theory and the Sine-Gordon field theory], we reiterate that both are purely scalar field theories [since Derrick's theorem allows topological soliton solutions for $D = 2$ (see §10.2b)]. We start with the $\lambda\phi^4$ theory [3.26] and show that the soliton solutions give rise to a topological conservation law that involves the integral over the one spatial dimension of the time component of an appropriately defined "kink" current (i.e. the "topological charge") .

The Lagrangian density for the real $\lambda\phi^4$ theory is:

$$\mathcal{L} = \int [\frac{1}{2}(\partial_0\phi)^2 - \frac{1}{2}(\partial_x\phi)^2 - V(\phi)]dx \qquad (10.2)$$

where the potential is independent of derivatives of ϕ and is written in the form:

$$V(\phi) = \frac{\lambda}{2}(\phi^2 - a^2)^2 \qquad (10.3)$$

with $a^2 = -\mu^2/\lambda$ (λ and $-\mu^2$ are positive parameters). The potential (10.3) for a single real scalar field is a "Higgs-type" potential and builds in two (quasi-) degenerate vacua; as a consequence, there are two classes of solutions: the topologically trivial non-soliton solutions corresponding to a "normal" vacuum and the topologically non-trivial soliton solutions (i.e. the "kink" type of solution and the "anti-kink" type of solution which is related to it by virtue of the built-in discrete symmetry [see Eq. (10.2)] and below). The energy is given by:

$$H = \int [\frac{1}{2}(\partial_0\phi)^2 + \frac{1}{2}(\partial_x\phi)^2 + V(\phi)]dx \qquad (10.4)$$

and the ground-state configuration is two-valued:

$$\phi = \pm a = \pm\sqrt{-\frac{\mu^2}{\lambda}} \qquad (10.5)$$

Equation (10.5) tells us that for a time-independent solution, the "normal" vacuum state (i.e. with energy $E = 0$) corresponds to the topologically trivial solution $\phi = +a$ or $\phi = -a$ for all x, i.e. ϕ =constant. To obtain the (static) topologically non-trivial soliton solution to the equation of motion resulting from (10.2) requires finding a solution $\phi(x)$ that satisfies the conditions $\phi(x) = -a$ as $x \to -\infty$ and $\phi(x) = +a$ as $x \to +\infty$, or conversely [i.e. $\phi(x) = +a$ as $x \to -\infty$ and $\phi(x) = -a$ as $x = +\infty$); the two finite-energy soliton solutions are then:

$$\phi^\pm(x) = \pm a \tanh(|\mu|x) \qquad (10.6)$$

In Eq. (10.6), ϕ^+ is the "kink" solution (see Fig. 10.1) [because ϕ^+ increases monotonically from the $\phi^-(-\infty) = -a$ to $\phi^-(+\infty) = +a$] whereas ϕ^- is the "anti-kink" solution [because ϕ^- decreases monotonically from $\phi^-(-\infty) = +a$ to $\phi^-(+\infty) = -a$. The "kink" and "antikink" solutions possess the same value of energy $E = 4|\mu|^3/3\lambda$ and both solutions are stable with respect to small perturbations even though they do not yield the absolute minimum of the potential energy $V(\phi)$ (which corresponds to the "normal" vacuum). Since the "normal" vacuum state corresponds to the choice $\phi = +a$ or $\phi = -a$ for all x, it is only the existence of more than two quasi-degenerate vacua — corresponding to $\phi = \pm a$ — that makes it possible to topologically distinguish the "kink" and "antikink" solutions from the "normal" vacuum solution. It is clear from Fig. 10.1 why the "kink" (and "anti-kink") solutions in the $\lambda\phi^4$ theory are sometimes said to describe "lumps" or "pseudoparticles": these finite-energy soliton solutions to the equation of motion resemble particle-like objects of finite extension with finite rest mass and well-defined location (defined by the center of mass). Moreover, the Lorentz covariance of the equation of motion guarantees that a Lorentz "boost" produces a moving "kink" that obeys the traditional energy-momentum relation: $E = \sqrt{P^2 + M^2}$, where P is the momentum of the center of mass and M is the rest mass of the "kink".

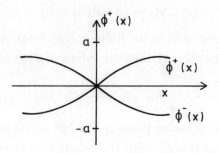

Fig. 10.1. "Kink" and "anti-kink" solutions in $\lambda\phi^4$ theory.

Beyond these interesting properties of the $\lambda\phi^4$ theory in $D = 2$ space-time dimensions, there is the important stabilizing property of these "kink" and "anti-kink" solutions that is due to an implicit topological conservation law. The topological conservation law follows from the finite-energy requirement at the spatial infinities $x = \pm\infty$, namely:

$$\phi^{\pm}(+\infty) - \phi^{\pm}(-\infty) = \pm(2a) \tag{10.7}$$

instead of the condition $\phi(+\infty) - \phi(-\infty) = 0$ for the "normal" vacuum. Evidently, Eq. (10.7) can be rewritten as:

$$\int_{-\infty}^{+\infty} (\partial_x \phi^{\pm}) dx = \pm(2a) \tag{10.8}$$

and, if we define the "topological" current $j_{\mu}^{\pm}(x) = \epsilon_{\mu\nu}\partial^{\nu}\phi^{\pm}$, this current is automatically conserved because of the antisymmetry of $\epsilon_{\mu\nu}$; the conserved "topological" charge Q_T follows from Eq. (10.8):

$$Q_T = \int_{-\infty}^{+\infty} j_0^{\pm}(x) dx = \int_{-\infty}^{+\infty} \partial_x \phi^{\pm} dx = \pm(2a) \tag{10.9}$$

Equation (10.9) implies, since the "kink" and "anti-kink" charges are conserved, that there can be no transition between a "kink" (or an "anti-kink") solution and the "normal" vacuum or between the "kink" and "anti-kink" solutions. This topological conservation law is not a consequence of the symmetry of the theory but rather results from the topological property of the space of non-singular finite-energy (soliton) solutions in the $\lambda\phi^4$ theory, namely that the space is not "connected". The mathematical language used to define this topological property is to denote the set of spatial infinities [which consists of the two discrete points $x = \pm\infty$] by S^0 and the set of minima of the potential $V(\phi)$ [which consists of the two discrete points: $\phi = \pm a$] by $\mathcal{M}_0$, thus:

$$\mathcal{M}_0 = \{\phi : V(\phi) = 0\} \qquad (10.10)$$

Since we are only interested in the finite-energy solutions to the equations of motion, it follows that the asymptotic values of $\phi(x)$ must be zeros of $V(\phi)$ and hence:

$$\lim_{x \to \pm\infty} \phi(x) = \pm a \in \mathcal{M}_0 \qquad (10.11)$$

The "mapping" that is taking place in the $\lambda\phi^4$ scalar theory is consequently the mapping of the points S^0 onto the points in $\mathcal{M}_0$ and encodes the topological result that the "kink" and "anti-kink" mappings are topologically distinct from each other (although one is the "reflection" of the other) and both mappings are distinct from the topologically trivial mapping for the "normal" vacuum state. The physical meaning of the last statement becomes clear if one tries, say, to continuously deform the ϕ^+ mapping [where $-\infty$ is mapped into $-a$ and $+\infty$ is mapped into $+a$] into the "normal" vacuum configuration [where $\pm\infty$ are both mapped into $+a$ or $-a$]; in trying to go from the first mapping to the second, it is necessary to penetrate the barrier around $\phi = 0$ over an infinite range of x and this would require an infinite amount of energy. Thus, the "kink" (and the "anti-kink") solution is the topologically non-trivial solution of the $\lambda\phi^4$ theory, in contradistinction to the topologically trivial "normal" vacuum solution. Since the "anti-kink" solution can be derived from the "kink" solution by parity inversion — although it is topologically distinct — we say in homotopic language (see §10.4) that the topological mapping for the $\lambda\phi^4$ theory is governed by $\Pi_0(\mathcal{M}_0) = \mathcal{Z}_2$ (see Table 10.3).

The overall conclusion is that non-dissipative finite-energy soliton solutions to the Lagrangian equations of motion in the real $\lambda\phi^4$ field theory exist in $D = 2$ space-time dimensions because there are two quasi-degenerate vacua produced by the non-linear ϕ^4 potential. The stability of these "particle-like" soliton states is ensured by the existence of a "multiplicative"-type topological conservation law, which forbids the soliton states from decaying into the vacuum (or into each other), i.e. altering their topological charges [see Eq. (10.9)]. These results for the $\lambda\phi^4$ field theory are easily generalized to other $D = 2$ field theories of which the real Sine-Gordon theory — which we discuss — is a particularly illuminating example that yields an "additive" topological conservation law.

The well-studied $D = 2$ Sine-Gordon model is based on a real scalar field with a sinusoidal "Higgs" potential which automatically gives rise to an infinite number of (quasi-) degenerate vacua possessing a discrete (not a continuous) symmetry. The Sine-Gordon Lagrangian is given by [10.5]:

$$\mathcal{L} = \frac{1}{2}\{(\frac{\partial\phi}{\partial t})^2 - (\frac{\partial\phi}{\partial x})^2\} - V(\phi); \quad V(\phi) = \frac{\alpha}{\beta}(1 - \cos\beta\phi) \qquad (10.12)$$

where α and β are positive constants. The equation of motion is:

$$\frac{\partial^2\phi}{\partial t^2} - \frac{\partial^2\phi}{\partial x^2} = -\alpha\sin\beta\phi \quad [\simeq -\alpha\beta\phi] \qquad (10.13)$$

where the mass $m = \sqrt{\alpha\beta}$ and the constant in $V(\phi)$ are chosen so that the ground state corresponds to zero energy. Plotting $V(\phi)$ in Fig. 10.2, it is obvious that the degenerate vacua belong to the set $\mathcal{M}_0$:

$$\mathcal{M}_0 = \{\phi: V(\phi) = 0\}; \quad \lim_{x\to\pm\infty}\phi(x) = \frac{2\pi n}{\beta}(n = 0, \pm 1, \pm 2, ...) \in \mathcal{M}_0 \qquad (10.14)$$

Equation (10.14) is consistent with the fact that the Lagrangian of Eq. (10.12) is invariant under the discrete translation:

$$\phi \to \phi + \frac{2\pi n}{\beta} \qquad (10.15)$$

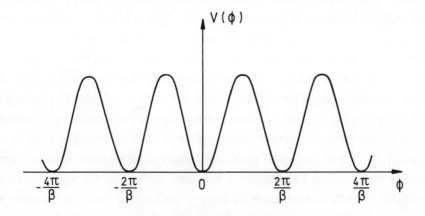

Fig. 10.2. Higgs potential in the Sine-Gordon model.

The finite-energy solutions must therefore satisfy the boundary condition at spatial infinity [see Eq. (10.7)]:

$$\phi(+\infty) - \phi(-\infty) = 2\pi n/\beta \quad \text{for } n \in \mathcal{Z} \tag{10.16}$$

Taking into account that, for a time-independent solution, the first integral of Eq. (10.12) is: $\frac{1}{2}(\partial\phi/\partial x)^2 - V(\phi) = 0$, we can derive the mass M of the $n = 1$ soliton by integrating over the first potential hill in Fig. 10.2, thus:

$$M = \int_0^{2\pi/\beta} [2V(\phi)]^{\frac{1}{2}} d\phi = 8m/\beta^2 \tag{10.17}$$

We anticipate our later notation by calling n the "winding number" of the soliton and hence Eq. (10.17) tells us that the mass of the soliton with "winding number" 1 is obtained by integrating over one potential hill in Fig. 10.2 and increases as the coupling constant decreases. To obtain the mass of the soliton with "winding number" n requires straightforward modification of Eqs. (10.16) and (10.17) (when $n > 1$).

Thus, the n in Eq. (10.16) serves as an "additive" topological quantum number and it follows that transitions between finite-energy solutions with different n are forbidden. Following the line of argument given for the $\lambda\phi^4$ theory, the corresponding conserved current can be written:

$$j^\mu = -i\frac{\beta}{2\pi}\epsilon^{\mu\nu}\partial_\nu\phi \tag{10.18}$$

with:

$$Q_T = \int_{-\infty}^{\infty} ij^0 dx = \frac{\beta}{2\pi}\int_{-\infty}^{\infty}\frac{\partial\phi}{\partial x}dx = \frac{\beta}{2\pi}[\phi(+\infty) - \phi(-\infty)] = n \tag{10.18a}$$

We note: (1) the topological nature of the conserved current in the Sine-Gordon theory follows from the fact that the conserved charge can be evaluated independently of the equation of motion; (2) j_0 has no canonically conjugate momentum and hence generates no continuous internal symmetry of the Lagrangian (10.13); and (3) j_0 is a spatial divergence and its spatial integral can only be non-vanishing if the scalar field satisfies the topologically non-trivial boundary condition (10.16). These remarks emphasize the basic point that a topological current — transporting a conserved quantity — requires a field theory with degenerate vacua and need not be connected with any internal symmetry of the field Lagrangian. Indeed, as we have already pointed out, the Sine-Gordon Lagrangian is invariant under the discrete transformation (10.15) (with n taking on all integral values) so that the "homotopic" statement for the Sine-Gordon topological solitons is: $\Pi_0(\mathcal{M}_0) = \mathcal{Z}$, to be contrasted with the "homotopic" statement $\Pi_0(\mathcal{M}_0) = \mathcal{Z}_2$ for the $\lambda\phi^4$ theory (see §10.2b). Thus, from the vantage point of homotopic group analysis, the topological solitons of both the $\lambda\phi^4$ theory and the Sine-Gordon theory are described by the zeroth homotopic group Π_0 but $\Pi_0(\mathcal{M}_0)$ is $\mathcal{Z}_2$ in the former case [because $\mathcal{M}_0$ represents two (quasi-) degenerate vacua] and $\Pi_0(\mathcal{M}_0) = \mathcal{Z}$ in the latter case [because $\mathcal{M}_0$ represents a discrete infinitude of (quasi-) degenerate vacua].

The two $D = 2$ scalar field theories that we have considered not only help to clarify the role of degenerate vacua in sustaining topological soliton solutions of both the "additive" and "multiplicative" types but are also useful for other purposes: (1) the Sine-Gordon theory, with its infinite number of soliton solutions, has served as a theoretical probe, particularly in connection with the Thirring model [10.6]. The equivalence of the bosonic Sine-Gordon theory with the fermionic Thirring model in certain essential respects not only sheds light on the relation between topological and "Noether" conservation laws but also provides the rationale for the fruitful application of topological concepts to combined bosonic soliton-fermion systems in condensed matter physics for $D = 2$ or $D = 3$ space-time dimensions; (2) building partly on the novel relationship between the bosonic soliton and the fermion field revealed by the comparison between the Sine-Gordon and Thirring models, the $\lambda\phi^4$ theory has been exploited to prove the existence of solitons with fractional fermion number which, in turn, has led the way to establishing the existence of stable fractionally charged excitations in a quasi-one-dimensional $(D = 2)$ conductor such as polyacetylene $(CH)_x$. We comment briefly on these interesting ramifications of the $D = 2$ Sine-Gordon and $\lambda\phi^4$ scalar theories before moving on to our discussion of the $D = 3$ and $D = 4$ topological solitons.

Consider then the Thirring model [10.6] which is a $D = 2$ fermion field theory described by the Lagrangian:

$$\mathcal{L} = \frac{1}{2}\bar{\psi}\gamma^\mu\partial_\mu\psi - m\bar{\psi}\psi - \frac{1}{2}g\bar{\psi}\gamma^\mu\psi\,\bar{\psi}\gamma_\mu\psi \tag{10.19}$$

where g is the coupling constant and the Lagrangian possesses the continuous internal symmetry:

$$\psi \to \psi' = \exp(i\alpha)\psi \tag{10.20}$$

As usual, the symmetry (10.20) gives rise to a conserved "Noether" current J^μ and associated fermion charge, Q_F, thus:

$$J^\mu = \bar{\psi}\gamma^\mu\psi \quad \text{with} \quad Q_F = \int_{-\infty}^{\infty} J^0 dx \tag{10.21}$$

where Q_F satisfies the commutation relation [see Eq. (2.13)]:

$$[Q_F, \psi(x)] = -\psi(x) \tag{10.21a}$$

The interesting observation now is that the respective currents of the Sine-Gordon and Thirring theories possess the same structure for the equal-time algebra; specifically, if we denote the common current by $\mathcal{J}^\mu$ [$\mathcal{J}^\mu$ is j^μ for Sine-Gordon and $\mathcal{J}^\mu = J^\mu$ for Thirring], one obtains for the equal-time C.R.:

$$[\mathcal{J}^0(\underline{x}, t), \mathcal{J}^0(\underline{y}, t)] = [\mathcal{J}^1(\underline{x}, t), \mathcal{J}^1(\underline{y}, t)] = 0;$$

$$[\mathcal{J}^0(\underline{x}, t), \mathcal{J}^1(\underline{y}, t)] = iC \frac{\partial}{\partial x} \delta(\underline{x} - y) \tag{10.22}$$

where $C = \beta^2/4\pi^2$ for Sine-Gordon and $C = \frac{\pi}{\pi+g}$ for the Thirring model. Furthermore, the energy-momentum tensors in the two theories have the common form:

$$T^{\mu\nu} = \frac{1}{C}(\mathcal{J}^\mu \mathcal{J}^\nu - \frac{1}{2}g^{\mu\nu}\mathcal{J}^2) + \sigma g^{\mu\nu} \tag{10.23}$$

where C is defined as before and $\sigma = (\alpha/\beta)(1 - \cos\beta\phi)$ and $m\bar{\psi}\psi$ for the Sine-Gordon and Thirring models respectively.

If the Sine-Gordon and Thirring parameters are properly adjusted, the equal-time algebra generated by J^μ and σ is the same in the two theories, guaranteeing the identity of the Green's functions for the currents of the two theories. Several aspects of this equivalence between the Sine-Gordon and Thirring models should be noted: (1) strong coupling (i.e. large g) in the fermionic Thirring model corresponds to weak coupling (i.e. small β^2) in the bosonic Sine-Gordon model; (2) conversely, strong coupling in Sine-Gordon corresponds to weak coupling in Thirring. Actually, $\beta^2/4\pi^2 = 1$ implies $g = 0$, corresponding to a "free" fermion theory!; and (3) since the "kinks" (solitons) in the boson theory correspond to weak coupling, they can be thought of as bound states of the fermions [i.e. the current is linear in the boson field in Sine-Gordon whereas it is quadratic in the fermion field in Thirring]. These novel connections between the $D = 2$ Sine-Gordon and Thirring models are intimations of the rich interplay between boson and fermion fields that has been found (and put to good use) in $D = 2$ and $D = 3$ dimension field theories. [There are subtle differences — within the framework of conformal field theory — between the Sine-Gordon and the massive Thirring model [10.7] and caution must be exercised in drawing conclusions from one model about the other.] The intriguing example of stable charged bosonic excitations carrying fractional quantum number (as applied to polyacetylene) still holds up and is now discussed.

The important step that made possible the polyacetylene application of the soliton in the $\lambda\phi^4$ scalar field theory was the paper by Jackiw and Rebbi [10.8]. In this paper, the authors add to the $\lambda\phi^4$ Lagrangian [Eqs. (10.2) and (10.3)] a $D = 2$ two-component massless, spinless Dirac term [which is possible since there is no spin-statistics theorem in $D = 2$ space-time dimensions] which interacts with the scalar field. The purely scalar part of the Jackiw-Rebbi (JR) Lagrangian gives rise to two degenerate vacua and associated soliton and anti-soliton solutions. The Dirac Hamiltonian is then (second-) quantized, with the scalar ϕ field taken as an external, c-number background field. With charge conjugation built into the theory, it is found that there is one "zero energy" fermion state ["zero mode" — see §10.4a] for each kink and that the change of fermion number in the vicinity of the kink is $\pm\frac{1}{2}$ when the "zero mode" state is full (and no change in fermion number when the "zero mode" state is empty). The fermion-generated soliton excitations with fractional fermion number is the key to an understanding of the experimental discovery of stable fractionally-charged excitations in quasi-one-dimensional polyacetylene [10.9].

It is interesting to sketch how nature simulates the conditions for fractionally-charged solitons in a $D = 2$ space-time régime. The substance in condensed matter physics that does the simulation is polyacetylene, a material that consists of parallel chains of CH_x groupings of atoms, in which electrons "hop" preferentially along each chain with very little crossing over to another chain. This strongly anisotropic behavior converts polyacetylene into a quasi one-dimensional conductor. In addition, there are two distinct patterns for bonding the electrons to the carbon atoms in the polyacetylene chain, which give rise to two degenerate vacua that possess a "reflection symmetry" in the coordinate displacement (u_n) along the symmetry axis of the nth CH group [this corresponds to the parity symmetry of $\phi(x)$ in the $\lambda\phi^4$ theory]. The two degenerate ground states (vacua) of polyacetylene are shown in Fig. 10.3; the configuration coordinate u_n for each CH group n describes a translation of the group along the symmetry axis (x) of the chain and is treated as a real, scalar boson field. There is a short ("double") bond between the $n-1$ and n groups, and a long ("single") bond between the n and $n+1$ groups in Fig. 10.3a whereas the situation is reversed in Fig. 10.3b; the ground state energies — by symmetry — are the same (i.e. we have degenerate vacua) and the electron "hopping" along the quasi-one-dimensional chain of CH groups is described by taking $\phi_n = (-1)^n u_n$. Finally, there is the electron-phonon (or electron-lattice displacement) interaction which is unstable with respect

to the SSB of the "reflection symmetry" $u_n \rightarrow -u_n$ (called the Peierls instability [10.10]) which gives rise to the "kink" ($D = 2$ soliton) and "anti-kink"
($D = 2$ anti-soliton) of the $\lambda\phi^4$ theory. [The "kink" in polyacetylene is also
called a "domain wall" since it is spatially one-dimensional: the "domain
wall' interpolates, so to speak, between two vacua.]

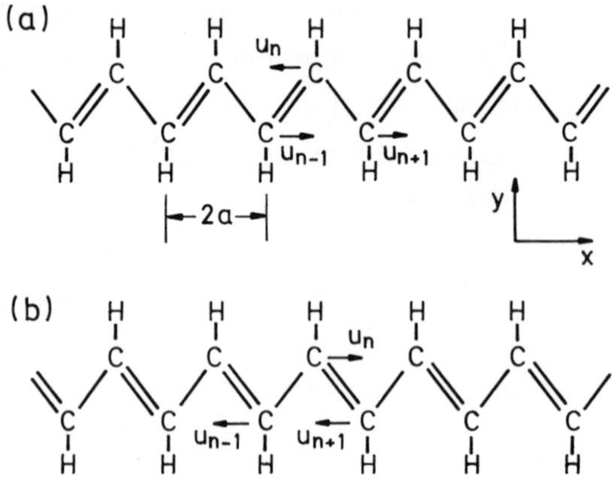

Fig. 10.3. Perfectly dimerized trans-polyacetylene showing the "dimerization coordinate"
u_n for the two degenerate ground states.

In first approximation, the two broken symmetry states $< \phi_n >$ are
equal to $\pm u_0$ (u_0 is a constant) and lead to an energy gap Δ in the electron
spectrum at the Fermi surface for the polyacetylene, as in a semi-conductor
(see Fig. 10.4). But now the JR result [10.8] informs us that the electrons
in the presence of the solitons (domain walls) acquire "zero mode" solutions
at the center of the energy gap Δ (as seen in Fig. 10.4). There is a complication in applying the JR result to the polyacetylene system because the
JR paper assumes zero spin for the electrons whereas the "real" electrons
and polyacetylene possess spin $\frac{1}{2}$; however, this complication can be taken
into account and the prediction is that there should exist stable fractionally-
charged excitations in polyacetylene in the form of spinless charged solitons
and spin $\frac{1}{2}$ neutral solitons. These anomalous properties of topological origin in $D = 2$ space-time dimensions have been confirmed experimentally in
polyacetylene through a great variety of experiments: nuclear and electron
spin resonance, infrared and optical absorption, transport experiments in
pure and chemically doped polyacetylene $(CH)_x$, and constitute an impressive triumph for the usefulness of topological concepts in condensed matter
physics [10.11].

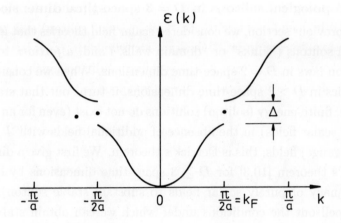

Fig. 10.4. Electron band structure in polyacetylene (see text).

Topological solitons and the related concepts of fractional charge and, more recently, fractional statistics (i.e. anyons [10.12]) also play a fruitful role in $D = 3$ space-time dimensional field theories; one need only mention the successful explanation of the fractional quantized Hall affect [10.13]. The success with the fractional quantized Hall effect [10.14] has encouraged further applications to condensed matter physics in $D = 3$ space-time dimensions and has even raised hopes that high temperature (cuprate) superconductivity can be explained in terms of stable so-called "anyonic" excitations of the "spin liquid" created by suitable doping of the anti-ferromagnetic parent compounds [10.15]. We do not enter into further discussion of fractional charge applications in condensed matter physics for $D = 3$ space-time dimensions, many of which are still speculative. Instead, we return to the more traditional case of vortex-type topological solitons in $D = 3$ space-time dimensions as they arise in low temperature Type II superconductivity. We find that in going from $D = 2$ to $D = 3$ (and higher) space-time dimensions, we are confronted by Derrick's theorem which states that spin 0 (scalar) fields must be augmented by higher spin fields in the Lagrangian in order to create the possibility of topological solitons. With Derrick's theorem derived below for an arbitrary number of D space-time dimensions, we first apply Derrick's result to the quantized magnetic flux realization of the $D = 3$ topological soliton due to the SSB of the $U(1)$ electromagnetic gauge group in a Type II superconductor [see §10.2b]. We then come back to particle physics where the relativistic quantum field theories live in $D = 4$ space-time dimensions and where the conventional "spin-statistics theorem" applies (see §2.2d).

§10.2b. Topological solitons in $D = 3$ space-time dimensions

In the previous section, we considered scalar field theories that give rise to topological solitons ("kinks" or "domain walls") and, *a fortiori*, topological conservation laws in $D = 2$ space-time dimensions. When we consider scalar field theories in $D > 2$ space-time dimensions, it turns out that stable, non-dissipative, finite-energy (soliton) solutions do not exist (even for an arbitrary number of scalar fields) in the absence of additional fields with $J > 0$, such as vector (gauge) fields; this is Derrick's theorem. We first give a direct proof of Derrick's theorem [10.3] for $D \geq 3$ space-time dimensions by exhibiting the inadequacy of purely scalar fields to substain stable soliton solutions; we then spell out the conditions under which we can obtain stable soliton solutions for a combination of scalar and vector (gauge) fields for arbitrary D space-time dimensions.

We start by writing down the time-independent Hamiltonian H for m scalar fields with symmetry group G for arbitrary D [note that there are $(D-1)$ space dimensions], namely:

$$H = \int d^{D-1}[\frac{1}{2}(\nabla\phi_i)^2 + V(\phi_i)] \tag{10.24}$$

where $V(\phi) = \frac{1}{4}\lambda(\phi_i^2 - a^2)^2$ $(i = 1, 2, \ldots m)$. The parameter space manifold $\mathcal{M}_0$ now denotes the set of constant values $\phi_i = \eta_i$ [the η_i's are assumed to be related by the same G symmetry group] that minimize $V(\phi_i)$, thus:

$$\mathcal{M}_0 = \{\phi_i = \eta_i : V(\eta_i) = 0\} \tag{10.25}$$

For the energy to be finite, the ϕ_i must tend to a point of $\mathcal{M}_0$ as we let $R\,\hat{r} \to \infty$ in any direction. The possible directions for going to ∞ are labelled by the unit vectors in the $(D-2)$-dimensional sphere:

$$S^{D-2}: \qquad \{\hat{r} : \hat{r}^2 = 1\} \tag{10.26}$$

It follows that the finite-energy solutions to the equation of motions have the property that:

$$\phi_i^\infty(\hat{r}) = \lim_{R\to\infty} \phi_i(R\,\hat{r}) \in \mathcal{M}_0; \quad \hat{r} \in S^{D-2} \tag{10.27}$$

In the notation of Eq. (10.27), we have for $D = 2$, $\hat{r} \in S^0$, i.e. a discrete set of points [in agreement with the $\lambda\phi^4$ and Sine-Gordon models]. On the other hand, for $D \geq 3$, S^{D-2} is a connected set and, assuming the ϕ_i^∞'s to be continuous, the requirement that the ϕ_i^∞'s attain the same values in all directions of S^{D-2} implies that the ϕ_i^∞'s must be constants on the connected set S^{D-2}. But this is the same trivial topology as that of the vacuum configuration and implies the absence of soliton solutions so that Derrick's theorem is proved.

In order to circumvent Derrick's theorem and create the possibility of topological solitons, it is necessary to introduce fields with $J > 0$ — in addition to scalar fields. Since $J = 1$ gauge fields play such a prominent role in modern particle physics (and in condensed matter physics where the electromagnetic gauge field is crucial), it is reasonable to inquire into the modifications of Derrick's theorem — for an arbitrary number D of space-time dimensions — when gauge fields are combined with scalar fields. It is possible to draw conclusions about the existence of topological solitons in combined scalar-gauge field theories, for arbitrary D, by the simple use of scale transformations on the scalar and gauge fields. It is instructive to give this derivation [10.5] before embarking on the more detailed discussion of the $D \geq 3$ cases in this and later sections. One starts by writing the total energy as the sum of three positive terms:

$$H = T_\phi + T_A + V \tag{10.28}$$

where:

$$T_\phi[\phi, A] = \int d^{D-1}x \; F(\phi) \, (\mathcal{D}^i \phi)^\dagger \mathcal{D}^i \phi; \quad T_A[A] = \frac{1}{4} \int d^{D-1}x \; F_{ij}^a F^{aij};$$

$$V[\phi] = \int d^{D-1}x \; U(\phi) \tag{10.28a}$$

The usual assumption is made that F and U are positive functions of ϕ involving no derivatives, $\mathcal{D}_i$ is the covariant derivative and F_{ij}^a is the non-Abelian anti-symmetric tensor field (a is the non-Abelian group index). If we apply the scale transformation $\underline{x} \to \lambda \underline{x}$, we get:

$$\phi(\underline{x}) \to \phi_\lambda(\underline{x}) = \phi(\lambda \underline{x}); \quad A(\underline{x}) \to A_\lambda(\underline{x}) = \lambda A(\lambda \underline{x}) \tag{10.29}$$

so that:

$$T_\phi[\phi_\lambda, A_\lambda] = \lambda^{3-D} \, T_\phi[\phi, A]; \quad T_A[A_\lambda] = \lambda^{5-D} \, T_A[A];$$
$$V[\phi_\lambda] = \lambda^{1-D} \, V[\phi] \tag{10.30}$$

In order to obtain a time-independent solution of the field equations, H of Eq. (10.28) must be stationary with respect to arbitrary field variations, including the scale transformations of Eq. (10.29). This is impossible if the separate terms T_ϕ, T_A and V in Eq. (10.28) all increase or all decrease when λ increases. Table 10.1 exhibits the behavior of the three terms in the total energy under a scale transformation as a function of D. The upward and downward arrows in Table 10.1 indicate "increase" and "decrease" (as λ increases) of the three terms in H. Table 10.1 is quite interesting as a function of D: (1) for $D = 2$ (the purely scalar case), it is seen that the T_ϕ and V contributions to H (there is no T_A contribution) can cancel and topological solitons are possible, as we found to be the case in §10.2a; (2) for $D = 3$, topological solitons are possible when only T_ϕ contributes to H or when both T_A and V contribute to H (and, of course, when all three terms contribute). When only T_ϕ contributes, one obtains a "Higgs model" leading to vortex lines [10.16] but otherwise one has available the types of models that have proved to be so useful in $D = 3$ condensed matter systems; (3) for $D = 4$, one requires — in contrast to the $D = 3$ case — that all three terms contribute to H; (5) for $D = 5$ space-time dimensions — which encompasses four-dimensional Euclidean space at constant time — the only possibility is a pure gauge theory leading to finite-energy instanton-type solutions; and, finally (6) for $D \geq 6$ space-time dimensions (i.e. $D - 1 \geq 5$ space dimensions), one can not obtain topological solitons within the class of scalar-gauge (vector) field theories under consideration.

Table 10.1: Behavior of terms in Hamiltonian under a scale transformation as a function of D space-time dimensions.

D	T_ϕ	T_A	V
2	↑	↑	↓
3	0	↑	↓
4	↓	↑	↓
5	↓	0	↓
$D \geq 6$	↓	↓	↓

Having established the general conditions — with the help of a scale transformation — under which topological solitons can be found in relativistic scalar-gauge quantum field theories for various values of D, we turn to a more detailed discussion of the $D > 2$ case and pinpoint how the presence of both scalar and vector fields enables one to find stationary finite-energy solutions of the field equations and to demonstrate that they possess nontrivial topologies. The basic point is that, without gauge fields, H in Eq. (10.28) becomes Eq. (10.24) and we get the inequality:

$$H \geq \int d^{D-1}x \left[\frac{1}{2}(\nabla\phi_i)^2\right] \tag{10.31}$$

Now, for $D > 2$, $(\nabla\phi_i)^2$ is the sum of radial and transverse terms and the transverse term must be of the order of r^{-2} (as $r \to \infty$) since ϕ_i^∞ can not be constant if topologically non-trivial solutions are to exist. But an r^{-2} behavior at ∞ makes the energy (H) diverge and destroys the possibility of a finite-energy solution unless the transverse contribution to H from $(\nabla \phi_i)^2$ is somehow attenuated. This can be accomplished by gauging the scalar field theory and replacing the gradient term, $\nabla_i \phi$, by the covariant derivative:

$$D_i \phi = \nabla_i \phi + ig(A_i^a \cdot T^a)\phi \qquad (10.32)$$

where A_i^a and T^a are respectively the gauge fields and transformation matrices belonging to the group G. The presence of the second term on the R.H.S. of Eq. (10.32) can then be exploited to arrange for $D_i \phi$ itself to decrease like r^{-2} (as $r \to \infty$) — even though A_i^a and $\nabla_i \phi$ both decrease like r^{-1} — so that $(D_i \phi)^2$ decreases as r^{-4} and topologically non-trivial finite-energy solutions become possible when the gauge group is spontaneously broken by the scalar (Higgs) fields.

We proceed to discuss the scalar-gauge field-theoretic model with SSB (and, *a fortiori*, degenerate vacua) in $D = 3$ space-time dimensions that has served as a paradigm for identifying topological solitons in $D > 3$ spontaneously broken gauge theories. The $D = 3$ model consists of gauging a complex scalar field theory that undergoes SSB. The well-known application of this $D = 3$ field theory to low temperature superconductivity makes use of the electromagnetic gauge field and associates the complex scalar field with the quasi-Higgs "Cooper pair" condensate which spontaneously breaks the $U(1)$ electromagnetic gauge group (see §4.4a); the topological solitons that emerge in the theory are translated into the quantized flux tubes in Type II superconductors. To see this, we write down the Lagrangian for the complex scalar theory with an Abelian $[U(1)]$ gauge symmetry, which is spontaneously broken; we have [see Eq. (4.27)]:

$$\mathcal{L} = -\frac{1}{4}F_{\mu\nu}F^{\mu\nu} + (D_\mu \phi^\dagger) \cdot (D^\mu \phi) - V(\phi) \qquad (10.33)$$

where $F_{\mu\nu}$ is the (Abelian) electromagnetic field, $D_\mu = (\partial_\mu + ieA_\mu)$ is the covariant derivative and the potential $V(\phi)$ becomes:

$$V = \frac{\lambda}{2}(\phi^\dagger \phi - a^2)^2 \qquad (10.34)$$

with λ and a positive numbers. Since the $U(1)$ gauge field is spontaneously broken, an infinite number of degenerate vacua is created — corresponding to the zeros of $V(\phi)$ which lie on a circle (see §3.4); we denote the zeros of

$V(\phi)$ by $\eta(\sigma)$ so that we can write:

$$\mathcal{M}_0 = \{\phi^\infty = \eta(\sigma) = ae^{i\sigma} : V[\eta(\sigma)] = 0\} \qquad (10.35)$$

where σ is a real number. Equation (10.35) imparts the information that the set of zeros of $V(\phi)$ belongs to the continuous $U(1)$ manifold [since the $U(1)$ gauge group has been completely broken], which is mapped onto the circle at ∞ in two-dimensional space, i.e. S^1; the statement that the mapping is $S^1 \to U(1)$ agrees with the general statement in Eq. (10.27) that, for arbitrary D, the mapping is $S^{D-2} \to \mathcal{M}_0$. From our discussion of the Dirac magnetic monopole in §10.1, we know that the mapping function for $S^1 \to U(1)$ is $e^{in\theta}$ (where n is the topological "winding number") and the homotopic group is $\Pi_1[U(1)] = \mathcal{Z}$, implying an "additive" topological conservation law.

We have invoked the homotopic group $\Pi_1[U(1)] = \mathcal{Z}$ on several occasions and since the mapping $S^1 \to U(1)$ is the simplest example of the mapping of a continuous manifold onto a (continuous) surface (at ∞), S^{D-2}, in D space-time dimensions, we pause for some mathematical definitions of homotopic classes and homotopic groups (to be supplemented by the discussion in §10.4) before continuing with our physical applications of topological conservation laws. We first define a homotopic class [3.26]: let X and Y be two topological spaces and let $f_0(x)$ and $f_1(x)$ be two continuous functions from X to Y. Let I denote the unit interval on the real line $0 \leq t \leq 1$; f_0 and f_1 are said to be homotopic if and only if there is a continuous function $F(x,t)$ which maps the direct product of X and I to Y such that $F(x,0) = f_0(x)$ and $F(x,1) = f_1(x)$. The continuous function $F(x,t)$ which deforms the function $f_0(x)$ continuously into $f_1(x)$ is called the homotopy and the ensemble of all such homotopic functions is called the homotopic class. From the definition of homotopic class, it is evident that the homotopic classes can be considered elements of a group, which is called the homotopic group.

From our brief encounter with homotopic groups, it is already clear that their classification is based on the dimension of the spatial surface which is mapped onto the group manifold consisting of the set of zeros of the "Higgs" potential [which is the coset space G/H, where G is the gauge group that is spontaneously broken to the subgroup H [2.25]]. In our notation — where D is the number of space-time dimensions — the dimension of the spatial surface is S^{D-2} so that the homotopic group corresponding to the coset space G/H is $\Pi_{D-2}(G/H)$. Thus, for $D = 2$ — which applies to the $\lambda\phi^4$ and Sine-Gordon theories — it is convenient to label the homotopic group by $\Pi_0(\mathcal{M}_0)$ where $\mathcal{M}_0$ is the "group manifold" consisting of the zeros of the equivalent "Higgs" potential in the scalar field Lagrangian; $\mathcal{M}_0$ takes the place of the coset space G/H in spontaneously broken gauge theories and, in the above notation, $\Pi_0(\mathcal{M}_0) = \mathcal{Z}_2$ (corresponding to two degenerate vacua) for the $\lambda\phi^4$ theory and $\Pi_0(\mathcal{M}_0) = \mathcal{Z}$ (corresponding to the discrete infinitude of degenerate vacua) for the Sine-Gordon theory (see §10.2a). It is only when we come to $D \geq 3$ space-time dimensions that we have the possibility of a continuous (internal) symmetry together with the SSB of a gauge group and the notation $\Pi_{D-2}(G/H)$ is, strictly speaking, appropriate. In the $D = 3$ field theory just considered — that of a complex scalar field with the SSB of a $U(1)$ gauge group — we can immediately identify the coset space $G/H = U(1)/I$ [since G is $U(1)$ and H is the completely broken $U(1)$ manifold denoted by the identity I] and hence the homotopic group corresponding to the mapping $S^1 \to U(1)$ is $\Pi_1[U(1)/I] = \Pi_1[U(1)]$. Once $\Pi_{D-2}(G/H)$ is determined for a specified scalar-gauge field theory, it is possible to look up the "Bott Table" (Table 10.3 in §10.4c) to decide whether the homotopic group is $\mathcal{Z}$, $\mathcal{Z}_2$ or 0 and to find out whether the topological soliton obeys an "additive" or "multiplicative" topological conservation law, or is absent. As we discuss the field-theoretic models with topological solitons in the remainder of this chapter, we identify the homotopic group involved in each example.

While homotopic group analysis is sufficient to specify the nature of the topological conservation law that obtains in a given field theory, in many cases we are also interested in knowing the explicit expression for the topological soliton solution and its consequences for the physical content of the theory. As a start, we return to the $D = 3$ gauge theory of a complex scalar field undergoing SSB, and prove that the homotopic classification of this $D = 3$ theory corresponds to the quantization of the magnetic flux tubes in Type II superconductivity. Since the mapping in this $D = 3$ theory is that of S^1 onto a $U(1)$ group manifold, the first topological space X consists of the points on a unit circle in ordinary space (S^1) labelled by the angle θ — with θ and $\theta + 2\pi$, etc. identified — and the second topological space Y consists of the set of unimodular complex numbers $u = [e^{i\sigma}]$ which represents the $U(1)$ group manifold. The mapping of $\{\theta\}$ onto $\{e^{i\sigma}\}$ is achieved by the continuous functions:

$$f_n(\theta) = \exp[i(n\theta + \alpha)] \tag{10.36}$$

which form a homotopic class for a fixed integer n and any real value of α.

This statement follows from the explicit construction of the homotopy:

$$F(\theta, t) = \exp\{i[n\theta + (1-t)\theta_0 + t\theta_1]\} \tag{10.37}$$

such that:

$$F(\theta, 0) = \exp(in\theta + \theta_0); \quad F(\theta, 1) = \exp[i(n\theta + \theta_1)] \tag{10.38}$$

Thus, each homotopic class is characterized by the "winding number" n (also called the Pontryagin index), which specifies the number of times points in $[\theta]$ are mapped into one point of $[e^{i\sigma}]$, i.e. the number of times points in S^1 "wind" around the point σ in the field parameter space $U(1)$. It follows that the "winding number" n can be expressed in terms of $f_n(\theta)$, thus:

$$n = \int_0^{2\pi} \frac{d\theta}{2\pi} [\frac{-i}{f_n(\theta)} \frac{df_n(\theta)}{d\theta}] \tag{10.39}$$

Consequently, the topological conservation law in the present instance consists of the statement that a soliton solution belonging to the nth homotopy class [as defined by Eq. (10.36)] is stable against transformation into a field configuration belonging to the mth homotopy class ($m \neq n$). It is evident that all of these homotopy classes with "winding number" n belong to the first homotopic group $\Pi_1[U(1)] = \mathcal{Z}$ (see Table 10.3), which confirms the "additive" character of the topological conservation law in question.

For later applications of homotopic groups to $D > 3$ space-time dimensional field theories, it is convenient to know the counterparts to the mapping function (10.36) in "Cartesian" coordinates and, for this purpose, we rewrite the mapping function for S^1 onto $U(1)$ in "Cartesian" coordinates, thus:

$$f(x, y) = x + iy \quad \text{with} \quad x^2 + y^2 = 1 \tag{10.40}$$

It is understood — in replacing Eq. (10.36) by Eq. (10.40) — that the mapping from the unit circle S^1 is replaced by the mapping from the whole real line $-\infty \leq x \leq \infty$, with the end-points $x = -\infty$ and $x = +\infty$ identified [this last condition is equivalent to the statement $f(x = -\infty) = f(x = +\infty)$]. Under these conditions, the topology is identical with that of the unit circle and the corresponding mappings belonging to the nth homotopic class can be represented by:

$$f_n(x) = \exp\{i\pi nx/(x^2 + \lambda^2)^{\frac{1}{2}}\} \tag{10.41}$$

where λ is an arbitrary parameter and the topological "winding number" n can be written as:

$$n = \frac{1}{2\pi} \int_{-\infty}^{+\infty} dx [\frac{-i}{f_n(x)} \frac{df_n(x)}{dx}] \tag{10.42}$$

The forms (10.41) and (10.42) are useful for mappings in $D > 3$ space-time dimensions, especially for instantons (see §10.3).

The power of homotopic group analysis is not needed to derive the quantization condition for magnetic flux tubes in Type II superconductivity for $D = 3$ space-time dimensions but utilization of it prepares us for the applications to topological solitons in particle physics. As a spontaneously broken field theory, the Ginzburg-Landau equation is a direct example of a gauged (complex) scalar field theory which undergoes SSB. It is well known that in Type II superconductors — to which the Ginzburg-Landau equation applies when the coherence length λ is less than $\sqrt{2}$ times the penetration depth d (see §4.4a) — the Meissner effect does not vanish suddenly at the first critical magnetic field (see §4.4a) but that, at first, quantized "magnetic flux tubes" [with their size determined by λ] appear in increasing number until the second critical magnetic field is reached when the Meissner effect disappears completely (see Fig. 4.2). To show that the magnetic flux tubes present in a Type II superconductor between H_{c_1} and H_{c_2} are quantized, we assume that the superconductor is sufficiently long so that the z dependence can be neglected; in that case, if we use polar coordinates in the expression for the Lagrangian given by Eq. (10.33), we find from Eq. (10.32) that the condition for a soliton solution is:

$$\frac{d\phi^\infty(\theta)}{d\theta} = ie \lim_{r \to \infty} r A_\theta \, \phi^\infty(\theta) \tag{10.43}$$

where $\phi^\infty(\theta) = e^{i\sigma(\theta)}$ [see Eq. (10.35)]. Taking account of the fact that the magnetic flux passing through the $x - y$ plane of the "magnetic flux tube" is given by:

$$\Phi = \lim_{r \to \infty} r \int_0^{2\pi} d\theta \, A_\theta(r, \theta) \tag{10.44}$$

we get:

$$\Phi = \int_0^{2\pi} d\theta e^{-1} \frac{d\sigma}{d\theta} = \frac{2\pi n}{e} \tag{10.45}$$

using Eq. (10.39). Thus, Eq. (10.45) demonstrates the quantization of the "magnetic flux tube" (vortex) in Type II superconductivity when the "Cooper pairs" are identified with the charged scalar field and the mapping is taken from the unit circle to the coset space $(U(1)/I)$ [so that $\Pi_1[U(1)/I] = \mathcal{Z}$]. It should be noted that, just as in the case of the Dirac magnetic monopole where the requirement of single-valuedness of the "transition" gauge transformation [between the "north-cap" and "south-cap" A_μ gauge fields —

see Fig. 4.12a] leads to an "additive" topological conservation law, so it is possible to derive the "vortex line" quantization condition (10.45) from the requirement of the single-valuedness of the "non-integrable phase" in the explicit solution of the vortex problem [10.16]. However, if one is only interested in establishing the existence of a topological conservation law (quantization), the homotopic group approach is more powerful and easily generalizable to field theories in $D > 3$ space-time dimensions. We will see in the next section how, under suitable conditions, the $D = 4$ topological soliton acquires ('t Hooft-Polyakov) magnetic monopole-type properties, of possible interest to particle physics.

§10.2c.   't Hooft-Polyakov magnetic monopoles in $D = 4$ particle physics

At the beginning (ca. 1975) of the Period of Consolidation and Speculation in modern particle physics — as we have termed it (see §1.4b) — there was a flurry of excitement in connection with the $SU(5)$ GUT model (see §8.3) and it was soon realized that the SSB of $SU(5)$ to the standard group $SU(3)_C \times SU(2)_L \times U(1)_Y$ generates topological solitons of the magnetic monopole type (since called 't Hooft-Polyakov monopoles [1.151]) that can have interesting physical and astrophysical consequences. Indeed, the need to suppress the over-abundant generation of extremely massive magnetic monopoles in the early universe by $SU(5)$ GUT was one of the factors that led to the inflationary universe scenario [10.17]. We do not dwell on the astrophysical status of the unseen GUT magnetic monopoles but the physics of the 't Hooft-Polyakov magnetic monopole, particularly vis a vis the Dirac magnetic monopole, is sufficiently interesting to justify a brief review of the subject. For this purpose, we first show how $D = 4$ topological soliton solutions acquire 't Hooft-Polyakov monopole properties in the simple "Georgi-Glashow" model [5.38], based on a triplet of scalar fields and the $SO(3)$ gauge group undergoing SSB. Fermion fields are not included since they are not needed to generate the topological solitons in this model.

The Lagrangian for the Georgi-Glashow model is:

$$\mathcal{L} = -\frac{1}{4}F^{a\mu\nu}F^a_{\mu\nu} + \frac{1}{2}(D^\mu \underline{\phi}) \cdot (D_\mu \underline{\phi}) - V(\phi) \qquad a = (1, 2, 3) \qquad (10.46)$$

where:

$$F^a_{\mu\nu} = \partial_\mu A^a_\nu - \partial_\nu A^a_\mu + g\epsilon^{abc} A^b_\mu A^c_\nu; \quad (D_\mu \phi)^a = \partial_\mu \phi^a - g\epsilon^{abc} A^b_\mu \phi^c;$$

$$V(\phi) = \frac{\lambda}{4}(\underline{\phi} \cdot \underline{\phi} - a^2)^2 \qquad (10.46a)$$

and $\underline{\phi}$ denotes the three real scalar fields ϕ_b ($b = 1, 2, 3$). It follows that the equations of motion are:

$$(D_\nu F^{\mu\nu})_a = g\epsilon_{abc}\phi_b(D^\mu\phi)_c; \quad (D^\mu D_\mu\phi)_a = -\lambda\phi_a(\underline{\phi}\cdot\underline{\phi} - a^2) \quad (10.47)$$

As usual, the values of $\underline{\phi}$ that minimize $V(\underline{\phi})$ in Eq. (10.46a) define $\mathcal{M}_0$, thus:

$$\mathcal{M}_0 = \{\underline{\phi} = \underline{\eta}; \ \underline{\eta}^2 = a^2 : \ V(\underline{\eta}) = 0\} \quad (10.48)$$

That is, to say, the zeros of $V(\phi)$ now lie on a two-sphere defined by $SO(3)$. Since all points in $\mathcal{M}_0$ are related to each other by $SO(3)$ transformations, it is possible to choose $\underline{\phi} = (0, 0, v)$ for the new ground-state configuration. With this choice, the new vacuum is invariant under the rotation about the 3-axis [i.e. under an $SO(2)$ transformation]; the SSB path becomes:

$$SO(3) \rightarrow SO(2) \sim U(1) \quad (10.49)$$

where the final unbroken $U(1)$ symmetry can be assigned to the electromagnetic interaction with the associated massless gauge field (photon).

If we wish to secure finite-energy (soliton) solutions in the Georgi-Glashow model, it is necessary that $\phi(\underline{r})$ approaches values in $\mathcal{M}_0$ [see Eq. (10.48)] as the spatial coordinate $\underline{r} \rightarrow \infty$. The possible directions in which $\underline{r} \rightarrow \infty$, for $D = 4$, are determined by the condition (10.26):

$$S^2 : \quad \{\hat{r} : \ \hat{r}^2 = 1\} \quad (10.50)$$

Since the spatial infinities also form a two-sphere S^2 with the same topology as the set $\mathcal{M}_0$, the mapping of S^2 to the $SO(3)$ group manifold $\mathcal{M}_0$ defined by Eq. (10.48) is expressed by writing:

$$\underline{\phi}^\infty = \underline{\eta} = a\hat{r} \quad (\underline{\eta}^2 = a^2) \quad (10.50a)$$

The choice of mapping given by (10.50a) — which is the "hedgehog" mapping (see Fig. 10.5b) and yields the magnetic monopole-type of topological

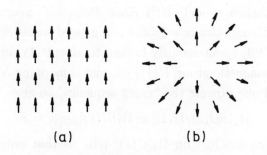

(a)   (b)

Fig. 10.5. Schematic ϕ configuration (a): for the vacuum; (b): for the soliton.

soliton — is not unique [10.18]; other choices are possible resulting in other kinds of topological solitons. What is true is that the "hedgehog" mapping (10.50a) can not be deformed continuously into the mapping for the vacuum configuration that requires S^2 to be mapped to a point. The striking difference in the ϕ configurations between the vacuum state and the soliton state corresponding to the "hedgehog" mapping is exhibited in Fig. 10.5 and it is evident that the "hedgehog" configuration (b) suggests some type of magnetic monopole.

We are now in a position to express our results in homotopic group language and then to demonstrate that the "hedgehog" solitons in the Georgi-Glashow model are "'t Hooft-Polyakov" magnetic monopoles [1.151]. For this purpose, we need several definitions: (1) a simply-connected group like $SU(2)$ satisfies the homotopy group condition $\Pi_1(SU(2)) = 0$ and, *a fortiori*, only contains "contractible" closed loops [in contrast to the Dirac magnetic monopole situation where the singularity requirement of the Dirac "string" (associated with a Dirac magnetic monopole) requires a "non-contractible" closed loop in $U(1)$ which yields the homotopic group condition $\Pi_1(U(1)) = \mathcal{Z}$]; (2) although the algebra of the $SO(3)$ group is identical with its $SU(2)$ "covering" group, its topological structure is different from that of $SU(2)$ because of the identification of the two "antipodal" points on the sphere defining $SO(3)$. The consequences are, that while $SO(3)$ is locally isomorphic to $SU(2)$, it is doubly-connected rather than singly-connected [like $SU(2)$], and its first homotopic group equation is: $\Pi_1(SO(3)) = \mathcal{Z}_2$ which contains both "non-contractible" as well as "contractible" closed loops in $U(1)$; (3) the "exact sequence" concept in homotopic group analysis [10.5], which relates the kth homotopic group of the coset space manifold G/H, i.e. $\Pi_k(G/H)$, to the $(k-1)^{th}$ homotopic group of H (the unbroken group manifold into which the G ground manifold breaks spontaneously) makes the subtle distinctions between isomorphic groups that help to gain a full understanding of the Georgi-Glashow model of "'t Hooft-Polyakov" magnetic monopoles. Thus, since the Georgi-Glashow model is a gauged scalar field theory starting with $G = SO(3)$ spontaneously broken to $H = U(1)$ [which is identified with the electromagnetic group $U(1)_{EM}$], one is mapping S^2 onto the coset $SO(3)/U(1)$ and one expects the "exact sequence" to give:

$$\Pi_2[SO(3)/U(1)] \simeq \Pi_1[U(1)]_{SO(3)} = \mathcal{Z} \qquad (10.51)$$

where the subscript $SO(3)$ on $\Pi_1[U(1)]$ tells us that only loops that are "non-contractible" in $U(1)$ but are contractible in part of $SO(3)$ contribute to the homotopic group equation $\Pi_1[U(1)] = \mathcal{Z}$. The physical significance of this statement is that the Georgi-Glashow model is capable of predicting a magnetic monopole of the non-singular "'t Hooft-Polyakov" type which acquires the property of a Dirac monopole as $r \to \infty$.

It is interesting to exhibit more explicitly the properties of a "'t Hooft-Polyakov" monopole. To show this, we revert to the usual notation and use A_μ^3 of the Georgi-Glashow model to denote the (photon) gauge field A_μ. If we are to obtain a non-trivial topology of the "hedgehog" form [i.e. $\phi_b^\infty = \eta_b = a\hat{r}_b$, $(b = 1, 2, 3)$ — see Eq. (10.51)], the existence of such finite energy solutions requires that $D_i\phi_b = 0$ is obeyed to order r^{-2}, which is possible since both $\nabla_i\phi_b$ and A_i^b decrease as r^{-1}. It is straightforward to show that the condition $D_i\phi_b = 0$ leads directly to the "hedgehog" form for the gauge field A_i^a; we can write, for $r \to \infty$:

$$D_i\phi_a = \partial_i\phi_a - g\epsilon^{abc}A_{bi}\phi_c = 0 \qquad (10.52)$$

Multiplying Eq. (10.52a) by ϵ^{ade} and evaluating the product of the ϵ^{abc}'s, we get:

$$\epsilon^{ade}\partial_i\phi_a - g(\delta_b^d\delta_c^e - \delta_c^d\delta_b^e)A_{bi}\phi_c = 0 \qquad (10.52a)$$

Taking account of Eq. (10.51), one obtains the desired result:

$$A_i^a = \frac{-\epsilon_{aij}}{e}\frac{\hat{r}_j}{r} \qquad (10.52b)$$

Equation (10.52c) gives us the asymptotic "hedgehog" behavior of the gauge field, and, with this knowledge, we can readily derive the lowest-energy soliton solution (corresponding to "winding number" $n = 1$) [3.26]:

$$\phi_b = \frac{r^b}{er^2}H(aer), \quad A_b^i = -\epsilon_{bij}\frac{r^j}{er^2}[1 - K(aer)], \quad A_b^0 = 0 \qquad (10.53)$$

where H and K are dimensionless functions with the asymptotic forms $H(\xi) \to \xi$ and $K(\xi) \to 0$ as $\xi(\equiv aer) \to \infty$. The total (time-independent) energy of the system is given — in terms of $H(\xi)$ and $K(\xi)$ — by:

$$E = \frac{4\pi a}{e}\int_0^\infty \frac{d\xi}{\xi^2}[\xi^2(\frac{dK}{d\xi})^2 + \frac{1}{2}(\xi\frac{dH}{d\xi} - H)^2 + \frac{1}{2}(K^2 - 1)^2$$
$$+ K^2H^2 + \frac{\lambda}{4e^2}(H^2 - \xi^2)^2] \qquad (10.54)$$

The $n = 1$ $H(\xi)$ and $K(\xi)$ functions have the shapes shown in Fig. 10.6 and, when substituted into the total energy Eq. (10.54), yield the following (classical) mass of the topological soliton [λ and a are the parameters of the

"Higgs" potential in Eq. (10.46a)]:

$$\text{Mass of "lightest" soliton (magnetic monopole)} = \frac{4\pi a}{e} f(\lambda/e^2) \quad (10.55)$$

where $f(\lambda/e^2)$ is of the order of unity for a wide range of values of λ/e^2. Thus, the "a" parameter, which is the VEV of the scalar field, sets the scale for the SSB of the $SO(3)$ group to the $SO(2)$ group and also sets the scale for the mass of the topological soliton. We expect the higher mass solitons to be associated with larger values of the "winding number" $(n > 1)$.

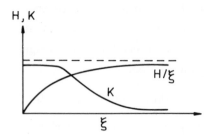

Fig. 10.6. Soliton functions in the Georgi-Glashow model; H is related to the scalar field, K to the electromagnetic field.

We have found a finite mass for the topological soliton in the Georgi-Glashow model and we must now prove that this topological soliton in $D = 4$ space-time dimensions is a magnetic monopole-type object, with an asymptotic magnetic field possessing the Dirac monopole form but differing in other essential respects. To see this, we insert the asymptotic values of H and K into Eq. (10.53); to fix the asymptotic behavior of the electromagnetic field, we find [see Eq. (10.52b)]:

$$F_b^{ij} \sim \frac{1}{er^4}\epsilon^{ijk}r_b r_k \sim \frac{\epsilon^{ijk}}{aer^3}r^k \phi_b \quad (10.56)$$

Equation (10.56) informs us that the asymptotic behavior of the magnetic field is:

$$\underline{B} \sim \frac{-1}{e}\frac{r}{r^3} \quad (10.57)$$

and confirms that the magnetic field of the topological soliton in the Georgi-Glashow model behaves asymptotically like the magnetic field due to a Dirac magnetic monopole but with "magnetic charge" $g = 1/e$ (in "Dirac" units); when this magnetic charge is inserted into the Dirac quantization relation (10.1), we note that the "'t Hooft-Polyakov winding number" $n = 2N$ (where we use N to distinguish it from the Dirac "winding number" n). The explanation of the "factor 2" difference in the Dirac quantization relation follows from the previous observation that $\Pi_1[SO(3)] = \mathcal{Z}_2$ so that in the "exact sequence" equation (10.51), only "half" of the closed loops — those that are contractible in $SO(3)$ — are available for $SO(3)$ in the generation of magnetic monopoles of the "'t Hooft-Polyakov" non-singular type. The Dirac quantization condition is not based on an "exact sequence" argument but follows directly from the homotopic equation $\Pi_1[U(1)] = \mathcal{Z}$.

There are other important differences between the "'t Hooft-Polyakov" and Dirac monopoles: in particular, the finite core spatial structure of the "'t Hooft-Polyakov" monopole is quite different from the "string" singularities associated with the Dirac monopole. The finite core of the 't Hooft-Polyakov monopole becomes apparent when one writes down the asymptotic behavior of the $H(\xi)$ and $K(\xi)$ functions, namely

$$H - \xi \sim e^{-|\mu|\xi/M} \simeq e^{-|\mu|r}; \quad K \sim e^{-\xi} \simeq e^{-Mr} \qquad (10.58)$$

where $|\mu| = (2\lambda)^{\frac{1}{2}}a$ and $M = ea$ are respectively the masses — after the SSB of the $SO(3)$ group — of the scalar and gauge bosons. If we compare Eqs. (10.55) and (10.58), we see that the mass of the 't Hooft-Polyakov monopole is of the order of a/α_e, i.e. $1/\alpha_e \simeq 10^2$ times the SSB scale for the breaking of $SO(3)$ to $U(1)$; this property is a general property of 't Hooft-Polyakov magnetic monopoles (see below). The finite core character of the 't Hooft-Polyakov monopole is highlighted further by pointing out that the "Dirac string" for the Dirac monopole case is replaced by the Higgs field for the 't Hooft-Polyakov monopole. Thus, combining the asymptotic expressions for A_a^i and ϕ_b derived from Eq. (10.53), we get:

$$A_a^i = \frac{1}{a^2 e} \epsilon^{abc} \phi^b \partial^i \phi^c \qquad (10.59)$$

so that the large-distance behavior of the magnetic field F_3^{ij} can be rewritten as:

$$F_3^{ij} = (\partial^i A_3^j - \partial^j A_3^i) + \frac{1}{ea^3} \underline{\phi} \cdot (\partial^i \underline{\phi} \times \partial^j \underline{\phi}) \qquad (10.60)$$

The second term on the R.H.S. of Eq. (10.60), involving the Higgs fields, replaces the additional singular term represented by the "Dirac string" in the case of the Dirac monopole [see Eq. (4.129)]; moreover, this second term explicitly exhibits its topological character because the integral over the surface at ∞ (the first term does not contribute) is nothing more than n, i.e.:

$$\frac{1}{ea^3} \int dS^i \frac{1}{2} \epsilon_{ijk} \underline{\phi} \cdot \partial^j \underline{\phi} \times \partial^k \underline{\phi} = \frac{1}{e} \qquad (10.60a)$$

which is precisely the equivalent magnetic charge of the "'t Hooft-Polyakov" monopole with the lowest "winding number", as it should be.

In many ways, the 't Hooft-Polyakov-type magnetic monopole is more attractive than the Dirac magnetic monopole since it arises quite naturally in a non-Abelian gauge theory with scalar and vector fields (Dirac fields can also be included) undergoing SSB, whereas the Dirac monopole is postulated for a special class of theories with "nodal lines of singularities" (see §4.6) and has never been detected. While the Georgi-Glashow model (with its special choice of scalar and gauge fields) gives rise to a 't Hooft-Polyakov-type monopole, it was soon realized that the essential conditions for the generation of the 't Hooft-Polyakov monopole can be duplicated with other choices of scalar and gauge (and fermion) fields that are closer to those in the standard group or its enlargements. Unfortunately — in terms of physical interest — the SSB of the standard electroweak group $G = SU(2)_L \times U(1)_Y$ to $H = U(1)_{EM}$ does not yield a 't Hooft-Polyakov monopole. With the knowledge that the electric charge Q is the sum of the I_{3L} generator in $SU(2)_L$ and the $U(1)_Y$ generator, this can be seen as follows: the $U(1)_{EM}$ group does not lie entirely in the $U(1)_Y$ group and hence any non-contractible closed loop in $H = U(1)_{EM}$ [remember that $G/H = [SU(2)_L \times U(1)_Y]/U(1)_{EM}$ is the coset space [2.25]] may be deformed in G to stay non-contractible in $U(1)_Y$. Thus, unless the deformation of the closed loop in $U(1)_Y$ is trivial, it cannot be deformed to a point in G and, consequently, the condition for a 't Hooft-Polyakov monopole [i.e. a contractible loop in G and non-contractible in H] is not satisfied. The "exact sequence" statement of this negative result is:

$$\Pi_2[SU(2)_L \times U(1)_Y/U(1)_{EM}] \simeq \Pi_1[U(1)]_{SU(2)_L \times U(1)_Y} = 0 \qquad (10.60b)$$

which tells us that no "'t Hooft-Polyakov" monopole can be generated from the SSB of the standard electroweak group.

Consequently, if one wishes to justify a search for 't Hooft-Polyakov monopoles in particle physics, it is necessary to consider simple (GUT) groups or semi-simple (PUT) groups (see §8.4a). In that case, it can be proved [2.25] that 't Hooft-Polyakov magnetic monopole solutions occur for any gauge theory in which a simple (or semi-simple) gauge group is broken down spontaneously to a factor group $H = H' \times U(1)$, where H' is a simple or semi-simple gauge group and $U(1)$ is identified with the electromagnetic gauge group. In general, the mass of the resulting monopole M_{mp} is fixed — in analogy to the Georgi-Glashow model — by the symmetry-breaking mass scale M_X for $G \rightarrow H' \times U(1)$ and is of order $M_{mp} \sim M_X/g^2$ [see

Eq. (10.58)]. Since the earliest candidate for grand unification, the $SU(5)$ group (see §8.3), must break down spontaneously to the standard group $SU(3)_C \times SU(2)_L \times U(1)_Y$, it satisfies the conditions for the creation of a magnetic monopole of *mass* $M_{mp} \sim M_X(\sim 10^{14}\ \text{GeV})/g^2 \sim 10^{16}\ \text{GeV} \sim 10^5$ grams. Unfortunately, the $SU(5)$ GUT-generated magnetic monopole has not been observed despite an extensive search [10.19]. This is not surprising since two other predictions of $SU(5)$ GUT — the proton decay lifetime and the value of $\sin^2 \theta_W$ (see §8.3) — have eliminated $SU(5)$ GUT as a viable theory and, as we pointed out in §8.3, one must move on to the next higher rank $SO(10)$ GUT group to keep open the option of grand unification "beyond the standard model". However, if the SSB path of $SO(10)$ GUT proceeds through $G_{422} = SU(4)_C \times SU(2)_L \times SU(2)_R$ — which it must do to satisfy the proton decay and $\sin^2 \theta_W$ conditions (see §8.4a) — no "super-massive" monopole is created [since G_{422} does not contain a $U(1)$ factor group]. Of course, the subsequent SSB of G_{422} to groups containing a $U(1)$ group which can be related to the $U(1)$ electromagnetic group, can, by the same token, produce 't Hooft-Polyakov monopoles, but presumably they would be less massive and more difficult to detect. This concludes our discussion of 't Hooft-Polyakov monopoles in $D = 4$ space-time dimensions and, while their "theoretical" existence has inspired substantial experimental searches and further theoretical work in particle physics, no evidence exists for the presence of such exciting objects in nature.

§10.3. Instantons in Euclidean four-space

This chapter has been devoted thus far to a discussion of the major types of finite-energy soliton solutions of the equations of motion in classical field theories (and their quantum counterparts) for $D = 2$ to $D = 4$ space-time dimensions, and to a partial delineation of their manifold topological properties through a suitable choice of physical examples. Starting with two $D = 2$ purely scalar field theories — the $\lambda\phi^4$ and Sine-Gordon theories — we identified the topological solitons in these theories ("kinks" associated with "built-in" degenerate vacua), exposed the character of the associated topological conservation laws, and commented on the highly successful applications to condensed matter physics problems (e.g. polyacetylene). As we moved on from $D = 2$ to $D = 3$ space-time dimensions, it was necessary to confront Derrick's theorem and to deal with gauged scalar field theories, their SSB and attendant degenerate vacua, in order to secure topological solitons. As in $D = 2$ space-time dimensions, the most fruitful use of topological solitons (which are vortex-type solutions) in $D = 3$ space-time dimensions turns out to be in condensed matter physics (e.g. the quantized magnetic flux tubes of Type II low temperature superconductivity). In homotopic group language (see Table 10.3), we moved from $\Pi_0(\mathcal{Z}_\ell) = \mathcal{Z}_\ell$ [where $\ell = 2$ for the $\lambda\phi^4$ theory to $\ell = \infty$ for the Sine-Gordon theory], to $\Pi_1[U(1)] = \mathcal{Z}$ [for $D = 3$ "vortices"]. It is clear that topological conservation laws have received impressive confirmation in non-relativistic many-body systems where physical conditions can be "engineered" to simulate $D = 2$ and $D = 3$ space-time dimensions.

One would have thought that when we come to four-dimensional spontaneously broken non-Abelian gauge theories (such as the standard model) — where topologically non-trivial gauge transformations are possible — the existence of topological solitons with interesting topological conservation laws would be a frequent occurrence. Unfortunately, as we have seen, the SSB of the standard gauge group does not give rise to 't Hooft-Polyakov magnetic monopoles and searches motivated by enlargements of the standard model have proved to be futile. This is where instantons enter the picture and raise the hope that the topologically non-trivial phenomena that they define in Yang-Mills theories are capable of giving rise to testable non-perturbative effects in the standard model. After all, instanton solutions are rigorous classical solutions of the Yang-Mills equations in Euclidean four-space and the translation from Euclidean to Minkowski space and quantization is presumed to be tractable!

The identification of classical soliton-like solutions of the Yang-Mills equations in Euclidean four-space was a significant discovery [1.136] and the interpretation of the topological solitons in Euclidean four-space — namely, the instantons — became a major undertaking from the very beginning. [Minkowski $D = 4$ [(3 + 1)-] space-time becomes Euclidean four-space when the time coordinate (t) in Minkowski space becomes the imaginary time coordinate (it) in Euclidean four-space and the term "instanton" is intended to signify that it possesses the finite spatial extension of a soliton but — unlike the solitons thus far considered — that it is also of finite temporal (albeit "imaginary") extension on account of rotational symmetry in Euclidean four-space.] The importance of instantons stems from the fact that the physical ("Minkowski") interpretation of the instanton solutions leads to some striking and unsettling consequences for relativistic quantum field theory and the standard model in our Minkowski world. The special character of instantons was already implied in our discussion of the dependence on D of the existence or non-existence of soliton solutions in scalar-gauge field theories (see Table 10.1). We found that only for $D = 5$ space-time dimensions (i.e. four "Euclidean" dimensions in the static limit of $D = 5$) could one obtain soliton solutions for pure Yang-Mills theories. Furthermore, since — as will be seen — instantons require the mapping of S^3 (three-space in Euclidean four-space) onto a group manifold with at least three generators, instanton solutions only arise in the presence of non-Abelian $SU(2)$ [or groups containing $SU(2)$ as a subgroup] gauge fields.

Instanton solutions are responsible for the non-uniqueness of the non-Abelian gauge theory vacuum — because they provide a mechanism for "vacuum tunneling" between what, at first sight, are topologically inequivalent vacua [the so-called n vacua where n is the "winding number" of the instanton]. This leads to the complicated "true" θ vacuum (see below) with its associated θ parameter and a number of unexpected physical consequences of the instanton solutions for particle physics. In many ways, the relevance of topological conservation laws to particle physics is being tested by our ability to make reliable calculations with instantons and to verify the phenomenological predictions when we believe these calculations to be correct. In reviewing the role of instantons in particle physics in this first section (§10.3), we delve into as much of the mathematical details as seem necessary to understand what is being said and to evaluate the rigor of the claims. We then comment, in the succeeding three sections (§10.3a–§10.3c), on the success — or lack thereof — in applying instanton concepts to QFD and QCD.

Let us start with the definition of "instantons" [or "pseudoparticle" solutions as they were originally called] by the authors in 1975 [1.136]: "by "pseudoparticle" solutions we mean the long range fields A_μ which minimize locally the Yang-Mills action S and for which $S(A) < \infty$" (S is taken over Euclidean four space). Thus we seek "minimal" finite Euclidean action solutions for the pure Yang-Mills fields which will then be reinterpreted in Minkowski space. We first treat the minimality condition on the Euclidean action and then take account of the finiteness requirement. Since time (t) in Minkowski space must be replaced by it in Euclidean space, we get for the Euclidean action S_E (in terms of the Minkowski action S):

$$S_E = \int d^4 x_E \mathcal{L}_E(x_E) = -iS; \quad \mathcal{L}_E = \frac{1}{2g^2} \text{Tr}(F_{\mu\nu} F_{\mu\nu}) \qquad (10.61)$$

where:

$$A_\mu^a \to \frac{i}{g} A_\mu^a \quad F_{\mu\nu}^a \to \frac{i}{g} F_{\mu\nu}^a; \quad F_{\mu\nu}^a = \partial_\mu A_\nu^a - \partial_\nu A_\mu^a + [A_\mu^a, A_\nu^a] \qquad (10.61a)$$

with:

$$A_\mu = \sum_a T^a A_\mu^a; \quad F_{\mu\nu} = \sum_a T^a F_{\mu\nu}^a \tag{10.61b}$$

The energy-momentum tensor is:

$$T_{\alpha\beta} = F_{\mu\alpha}^a F_{\mu\beta}^a - g_{\alpha\beta}\mathcal{L} = \frac{1}{4}(F_{\mu\alpha}^a + \tilde{F}_{\mu\alpha}^a)(F_{\mu\beta}^a - \tilde{F}_{\mu\beta}^a) \tag{10.62}$$

with:

$$\tilde{F}_{\mu\nu} = \frac{1}{2}\epsilon_{\mu\nu\alpha\beta}F_{\alpha\beta}$$

[In Euclidean space, the covariant notation can be dropped.] Evidently $T_{\mu\nu} = 0$ for a self-dual (or anti-self-dual) Yang-Mills field, i.e. when:

$$\tilde{F}_{\mu\nu}^a = \pm F_{\mu\nu}^a \tag{10.63}$$

It turns out that condition (10.63) guarantees the minimality of the Euclidean action as we show below; to prove this, we make use of the gauge and Lorentz-invariant quantity that represents the $U(1)$ triangular axial anomaly $\mathcal{A}$ (see §7.2a) which, in turn, is the divergence of the so-called Chern-Simons current K^μ [10.20], [we use the notation $A_\mu \to \frac{i}{g}A_\mu$, $F_{\mu\nu} \to \frac{i}{g}F_{\mu\nu}$], namely:

$$\mathcal{A} = 2 \, \text{Tr} \, (F_{\mu\nu}\tilde{F}_{\mu\nu}) = \partial_\mu K_\mu \tag{10.64}$$

where K_μ is:

$$K_\mu = 4\epsilon_{\mu\nu\alpha\beta} \, \text{Tr} \, (A^\nu \partial^\alpha A^\beta + \frac{2}{3}A^\nu A^\alpha A^\beta) \tag{10.64a}$$

It should be noted that K_μ in Eq. (10.64a), is not a gauge-invariant quantity in $D = 4$ space-time dimensions; however, the divergence of K^μ [see Eq. (10.64)] is a gauge-invariant quantity and, indeed, the integral of $\mathcal{A}$ over Euclidean four-space gives rise to the Pontryagin number [3.26], which is an integer and can be identified with the "winding number" of the instanton. [Parenthetically, the time-component of K_μ, i.e. K_0 (the true "Chern-Simons term" in the topological literature) is itself gauge-invariant under "small" (local) gauge transformations in $D = 3$ space-time dimensions where it can be used as a Lagrangian in $D = 3$ quantum gravity [10.21].

We write down the explicit expression for the Pontryagin number $\nu[A]$ for pure Yang-Mills fields which turns out to be precisely the "winding number" of the instanton; we have:

$$\nu[A] = \frac{1}{16\pi^2} \int_{S^4} d^4x \, \text{Tr}(F_{\mu\nu}\tilde{F}_{\mu\nu}) \tag{10.65}$$

where S^4 is the compactified four-space (since the points at infinity of $\mathcal{R}^4$ have been identified). To prove that the Pontryagin number is the "winding number" of an instanton, we must know the relation between $F_{\mu\nu}$ and

$\tilde{F}_{\mu\nu}$ when the (classical) action of a Yang-Mills field is minimized. The importance of the minimality condition is that the use of functional integral methods to quantize QCD requires the integration over all classical field configurations $A_\mu^a(x)$ in imaginary time (which interpolate between the vacua at $x_0 = \pm\infty$) and the dominant contribution to this integration results from those field configurations for which $S_E[A_\mu^a]$ is stationary [10.22]. To establish minimality, we consider the following expression in Euclidean space:

$$\text{Tr}[(F_{\mu\nu} \mp \tilde{F}_{\mu\nu})(F_{\mu\nu} \mp \tilde{F}_{\mu\nu}) \geq 0 \tag{10.66}$$

from which it follows that:

$$\text{Tr}[F_{\mu\nu}F_{\mu\nu} + \tilde{F}_{\mu\nu}\tilde{F}_{\mu\nu}] \geq \pm 2\,\text{Tr}[F_{\mu\nu}\tilde{F}_{\mu\nu}] \tag{10.67}$$

But:

$$\text{Tr}[\tilde{F}_{\mu\nu}\tilde{F}_{\mu\nu}] = \text{Tr}[F_{\mu\nu}F_{\mu\nu}] \tag{10.68}$$

and yields the inequality for the Euclidean action:

$$S_E = \frac{1}{2}\int d^4x\,\text{Tr}[F_{\mu\nu}F_{\mu\nu}] \geq \pm\frac{1}{2}\int d^4x\,\text{Tr}[F_{\mu\nu}\tilde{F}_{\mu\nu}] \tag{10.69}$$

Equation (10.69) can finally be expressed in terms of $\nu[A]$, thus:

$$S_E[A] \geq |\frac{1}{2}\int d^4x\,\text{Tr}[F_{\mu\nu}\tilde{F}_{\mu\nu}]| = \frac{8\pi^2}{g^2}|\nu[A]| \tag{10.70}$$

Clearly, the lower bound in Eq. (10.70) is attained when $\tilde{F}_{\mu\nu} = \pm F_{\mu\nu}$, corresponding to the self-dual and anti-self-dual solutions respectively.

With the condition of minimality established, we now prove that $\nu[A] \in \mathcal{Z}$ when the finiteness condition is imposed on the Euclidean action S_E. Finite S_E requires the Lagrangian density $\mathcal{L}_E = \frac{1}{2g^2}\text{Tr}\,(F_{\mu\nu}F_{\mu\nu})$ [see Eq. (10.61)] to vanish on $S^3 = S_\infty^3$, i.e.:

$$\lim_{|x|\to\infty} F_{\mu\nu}(x) \to 0 \tag{10.71}$$

Eq. (10.71) is certainly satisfied by letting the gauge field A_μ go to 0 as $|x| \to \infty$, but there is another possibility; we recall that under a non-Abelian gauge transformation U (U is unitary), the gauge field (if we take $A_\mu \to i/gA_\mu$) transforms as [see Eq. (4.19)]:

$$A_\mu' \to U^\dagger A_\mu U + U^\dagger \partial_\mu U \tag{10.72}$$

Hence, condition (10.71) holds if:

$$\lim_{|x|\to\infty} A_\mu(x) \to U^\dagger(x)\partial_\mu U(x) \tag{10.73}$$

The more restrictive condition: $\lim\limits_{|x|\to\infty} A_\mu(x) \to 0$ corresponds to the vacuum configuration, i.e. requiring the "winding number" n to vanish. When the less restrictive limiting condition (10.73) is used — which defines A_μ to be a so-called "pure gauge" field — topologically non-trivial solutions for the instantons can be identified. Thus, for self-dual (or anti-self-dual) solutions, the "winding number" n is directly related to the Euclidean action. To evaluate n, we take into account Eq. (10.73) and use the $A_0 = 0$ gauge; we get:

$$n = \frac{1}{6\pi^2} \int_\Sigma \epsilon_{ijk} \ \text{Tr} \ [(U^\dagger \partial_i U)(U^\dagger \partial_j U)(U^\dagger \partial_k U)] \ d^3x \qquad (10.74)$$

(where Σ is the 3-surface S^3_∞ at infinity). Equation (10.74) enables us to see clearly that the mapping function $U(x)$, which is an $SU(2)$ group element, takes us from the group manifold $SU(2)$ to Σ ; since the $SU(2)$ manifold is topologically the same as S^3, the mapping defines a set of homotopic classes that belong to the third homotopic group, thus (see Table 10.3):

$$\Pi_3[SU(2)] = \mathcal{Z} \qquad (10.75)$$

where, as usual, the integer $n \in \mathcal{Z}$ of Eq. (10.75) is the number of times the $SU(2)$ manifold is covered by the mapping of Σ. Moreover, the trivial gauge field $A_\mu = 0$ maps the Euclidean sphere Σ into a single point of the internal group space $SU(2)$ and is therefore defined by the identity with $n = 0$. There is another ("analogical") way to grasp the applicability of the homotopic group equation (10.75) to instantons. One can summon up the Dirac magnetic monopole case where the "non-contractible" closed loop C (see Fig. 4.11a) separates the two overlapping two-dimensional hemispheres Ω_α and Ω_β and the mapping of S^1 onto the Abelian gauge group $U(1)_{EM}$ gives rise to the homotopic group condition $\Pi_1[U(1)] = \mathcal{Z}$; in the instanton case, we can think of $\Sigma = S^3_\infty$ separating two overlapping four-dimensional "hemispheres" (in Euclidean space) and serving as the "non-contractible loop" which is mapped onto the non-Abelian $SU(2)$ gauge manifold and gives rise to the homotopic group equation (10.75).

We next evaluate $\nu[A]$ and then use the result to obtain an explicit expression for the instanton solution. For this purpose, we pursue yet another topological analogy — with the "quantized magnetic flux" topological conservation law (see §10.2b) [which is governed by the homotopic group $\Pi_1[U(1)] = \mathcal{Z}$] — because we feel that it is more instructive than the "brute force" method of calculating $\nu[A]$ with the help of the three-angle-parameterization of Σ and $SU(2)$ [of course, the result is the same]. Recall that the mapping function with "winding number" n for the "quantized magnetic flux" case in "Cartesian" coordinates could be written as [see Eq. (10.40)]:

$$f_n(x,y) = (x+iy)^n; \qquad x^2+y^2 = 1 \qquad (10.76)$$

Since the $SU(2)$ group element is $U = e^{i\underline{\epsilon}\cdot\underline{\tau}/2}$, we can write (by virtue of the properties of the Pauli matrices):

$$U = u_0 + i\underline{u}\cdot\underline{\tau} \; ; \qquad u_0, \underline{u} \text{ real with} \qquad u_0^2 + \underline{u}^2 = 1 \qquad (10.76a)$$

which defines the equation of a sphere in Euclidean four-space. Consequently, in mapping S_∞^3 to $SU(2)$, the generalization of Eq. (10.40) is:

$$f_n(x_0,\underline{x}) = x_0 + in\underline{x}\cdot\underline{\tau} \quad \text{with } x^2 = x_0^2 + \underline{x}^2 = 1 \qquad (10.77)$$

The nth mapping function given by Eq. (10.41) is replaced by:

$$f_n(x) = \exp\{i\pi n\underline{x}\cdot\underline{\tau}/(x^2+\lambda^2)^{\frac{1}{2}}\} \qquad (10.78)$$

and the expression (10.42) for the "winding number" n in the "quantized magnetic flux" case becomes for the instanton:

$$n = \frac{-1}{24\pi^2}\int_\Sigma d^3x \; \text{Tr}(\epsilon_{ijk}A_iA_jA_k) \qquad (10.79)$$

with:

$$A_i = f_n^{-1}(x_0,\underline{x})\partial_i f_n(x_0,\underline{x}) \qquad (10.79a)$$

where $f_n(x)$ has the form (10.78). It should be reiterated that the integration is carried out over the whole three-dimensional space in Eq. (10.79). That is, the instanton "winding number" n, given by Eq. (10.79), is identical with the Pontryagin number ν given by Eq. (10.65).

In sum, instantons are the localized, non-singular solutions of the classical Yang-Mills equations in Euclidean four-space, i.e. they are the topological solitons in Euclidean four-space. It is worth noting: (1) instantons are self-dual (or anti-self-dual) fields guaranteeing the minimality of the Euclidean action (and the vanishing of the energy-momentum tensor). The Euclidean action S_E for an instanton with "winding number" n is $S_E = 8\pi^2 n/g^2$; (2) instantons are due to the topologically non-trivial gauge transformations in Yang-Mills theories and the smallest gauge group that generates instanton solutions is the $SU(2)$ group. The Abelian gauge group does not give rise to instantons in $D=4$ Euclidean space although the flavor singlet $U(1)_A$ triangular axial anomaly is affected by instantons associated with the Yang-Mills groups contributing to the $U(1)_A$ anomaly; (3) the topological conservation law for instantons is "additive" since the homotopic group governing the topological behavior of instantons is $\Pi_3[SU(2)] = \mathcal{Z}$. It can be shown that any G containing $SU(2)$ as a subgroup [i.e. $G \supset SU(2)$] generates instanton solutions, which is consistent with the fact that $\Pi_3[SU(N)]$ $(N \geq 2) = \mathcal{Z}$; and, finally (4) since instantons are topological solitons in Euclidean four-space, they are of special relevance to the non-Abelian gauge theories of the strong and electroweak interactions (standard model and enlargements). The reinterpretation of instantons in Minkowski space has altered the conceptual structure of modern particle physics in ways that are not yet fully understood, as will be seen in §10.3a–§10.3c.

With these remarks in mind, we give a qualitative derivation of the $n = 1$ $SU(2)$ "instanton" expressions for the mapping function U and the resulting non-Abelian gauge field A_μ [see Eq. (10.73)]. We will then be in a position to draw out some of the physical consequences in Minkowski space due to the presence of instantons in both spontaneously broken non-Abelian QFD and unbroken non-Abelian QCD. Insistence on the self-duality (or anti-self-duality) of the non-Abelian gauge field gives the clue to the structure of instanton solutions when it is recognized that the pertinent algebra in Euclidean four-space is the $O(4)$ gauge algebra and that $O(4)$ is locally isomorphic to $SU(2) \times SU(2)$ so that one $SU(2)$ algebra can represent the $SO(3)$ symmetry of the spatial three-sphere at ∞, $\Sigma = S^3_\infty$, and the other $SU(2)$ algebra can represent the internal $SU(2)$ symmetry of the group manifold. The self-dual condition of (10.63) then imposes certain symmetry conditions on the charge and space-time indices (e.g. invariance under the "combined rotation" of space and isospin), leading to terms of the "hedgehog" type like $\underline{x} \cdot \underline{\tau}$.

The "hedgehog" solution for the 't Hooft-Polyakov magnetic monopole arose from the asymptotic condition on the Higgs field, i.e. $\underline{\phi} \to a\hat{\underline{r}}$ as $R\,\hat{r} \to \infty$; there is no SSB (and, *a fortiori*, no Higgs field) in the instanton situation but the topological conditions are similar. The simplest way to achieve the "hedgehog" form for the $n = 1$ mapping function $U_1(x)$ is to utilize Eq. (10.77) and write $U_1(x)$ in a manifestly unitary form, namely:

$$U_1(x) = \frac{x_0 + i\underline{x} \cdot \underline{\tau}}{\rho} \tag{10.80}$$

where $\rho^2 = x_0^2 + \underline{x}^2$. The corresponding $n = 1$ "instanton" solution for the gauge field itself is:

$$A_\mu(x) = \left(\frac{\rho^2}{\rho^2 + \lambda^2}\right) U_1^\dagger \partial_\mu U_1 \tag{10.81}$$

which translates into:

$$A_0(x) = \frac{-i\underline{\tau} \cdot \underline{x}}{\rho^2 + \lambda^2}; \qquad \underline{A}(x) = \frac{-i(\underline{\tau}\, x_0 + \underline{\tau} \times \underline{x})}{\rho^2 + \lambda^2} \tag{10.82}$$

with λ, an arbitrary scale parameter, called the "instanton size". It can easily be checked that the insertion of the expressions (10.82) for A_μ into Eq. (10.65) yields unit Pontryagin number so that the Euclidean action takes on its minimal value $8\pi^2/g^2$. It should be noted that implicit in Eq. (10.82) for the gauge field is the assumption that the topological charge density is concentrated around $x_\mu = 0$; but that is no problem because of the invariance of the theory under "four-translations". The "instanton size" is arbitrary because the pure Yang-Mills gauge field is scale invariant and the question of renormalization does not arise in connection with a global topological property of the Yang-Mills field. While it has been possible to treat instantons of "small" size in the so-called "dilute gas approximation" in various types of instanton-related calculations in QFD and QCD, the "large" instantons must be treated by non-perturbative methods and hence much less is known about their effects. Most of our knowledge about instantons is based on the "dilute instanton gas approximation" [10.23] where the mapping function U_n becomes $(U_1)^n$; it is then obvious that the Pontryagin number and Euclidean action acquire their predictable values for instantons with "winding number" n, namely $\nu[A] = n$ and $S_E = 8\pi^2 n/g^2$.

As we have already indicated, the determination of the instanton mapping function and associated gauge field in Euclidean four-space is only part of the task: it is necessary to find an interpretation for instanton solutions in Euclidean four-space (imaginary time) within the context of Minkowski $D = 4[(3+1)-]$ space-time (i.e. real time) in order to understand their physical significance. To achieve this understanding, one must define the vacuum states in Minkowski space and then establish how instantons (in Euclidean space) become "tunneling events" between two different Minkowski vacua. To find the structure of the "Minkowski vacuum" [which is defined by setting $F_{\mu\nu}^{\text{vac}} = 0$], we chose the temporal gauge $A_0^a = 0$ so that the gauge transformations are time-independent. It follows that $A_i(\underline{x})$ (we drop the superscript a on A_i and assume that it continues to contain the factor $\frac{i}{g}$) has the pure gauge form:

$$A_i(\underline{x}) = V^\dagger(\underline{x})\partial_i V(\underline{x}) \tag{10.83}$$

where we have written $V(\underline{x})$ to distinguish it from $U(x)$ [see Eq. (10.73)]. Furthermore, we have the gauge freedom to choose for the static "Minkowski" vacua the asymptotic behavior for $V(\underline{x})$:

$$V(\underline{x}) \to 1 \text{ as } |\underline{x}| \to \infty \tag{10.84}$$

But (10.84) is a familiar condition and implies that $V(x)$ defines a mapping of S^3 onto the $SU(2)$ group. The applicable homotopic group is the same as for the mapping function $U(x)$ for the instanton solution, namely $\Pi_3[SU(2)] = \mathcal{Z}$; however, in contrast to the V mapping, the U mapping is from the Euclidean three-dimensional surface S^3_∞ in the compactified Euclidean four-space S^4. Thus, the "Minkowski" vacuum is decomposed into an infinite number of topologically distinct vacua which are classified by an index $m \in \mathcal{Z}$ (again, to avoid confusion, we write m for the "winding number" of the "Minkowski" vacuum instead of the "winding number" n of the instanton). The perturbative "Minkowski" vacuum has $m = 0$ and all of the gauge transformations of the $m = 0$ vacuum are continuously deformable to the identity $V_0(\underline{x}) = 1$.

The novel feature of the non-perturbative "Minkowski" vacuum for Yang-Mills fields is that it is described by gauge fields A_i that are associated with gauge transformations $V_m(\underline{x})$ $(m > 0)$ that are not continuously deformable to the identity $V_0(\underline{x}) = 1$. Because of the asymptotic condition (10.84), the V_m's can immediately be written down, starting with $V_1(\underline{x})$ [10.24]:

$$V_1(\underline{x}) = \frac{\underline{x}^2 - \lambda^2}{\underline{x}^2 + \lambda^2} - i \int \lambda \frac{\underline{x} \cdot \underline{\sigma}}{\underline{x}^2 + \lambda^2} \tag{10.85}$$

Equation (10.85) gives rise to the following expression for $A_i(\underline{x})$:

$$\underline{A}_1(\underline{x}) = \frac{2i\lambda}{(\underline{x}^2 + \lambda^2)^2} [\underline{\sigma}(\lambda^2 - \underline{x}^2) + 2\underline{x}(\underline{\sigma} \cdot \underline{x}) + 2\lambda \, \underline{x} \times \underline{\sigma}] \tag{10.86}$$

The gauge (mapping) transformation for the mth homotopy class is given by:

$$V_m(\underline{x}) = [V_1(\underline{x})]^m \tag{10.87}$$

with expressions corresponding to (10.86) for $\underline{A}_m(\underline{x})$ (derived from Eq. (10.83)).

Equation (10.86) and its counterparts for $\underline{A}_m(\underline{x})$ $(m > 0)$ define the "Minkowski" m vacua and we must now show how an instanton with "winding number" n behaves as a "tunneling event" between two "Minkowski" vacua with "winding numbers" m and m', satisfying the condition $n = m' - m$. For this purpose, the "Chern-Simons charge" in Minkowski space Q_{CS} [see Eq. (10.64a)] plays the crucial role; we have:

$$Q_{CS}(t) = \frac{1}{16\pi^2} \int d^3x K_0(\vec{x}, t) \tag{10.88}$$

where $Q(t)$ is now a function of time. It should be emphasized that, in general, K_0 [which depends on the $A_\mu(\underline{x}, t)$'s in Minkowski 3-space at a fixed time] is not necessarily an integer; however, when $A_\mu(x)$ is a pure gauge field, $Q_{CS}(t)$ becomes an integer, i.e. $Q_{CS}(t) = n \in \mathcal{Z}$. This implies that the initial "Minkowski vacuum" [defined by $Q_{CS}[A]$ $(t = -\infty)$ with "winding number" m] and the final "Minkowski" vacuum [defined by Q_{CS} $[A]$ $(t = +\infty)$ with "winding number" m'] can be related to the Pontryagin number $\nu[A] \in \mathcal{Z}$ by using the equality [see Eq. (10.65)] [10.25]:

$$\nu[A] = Q_{CS}[A] \ (t = +\infty) - Q_{CS}[A] \ (t = -\infty); \quad n = m' - m \quad (10.89)$$

The significance of an earlier statement is now clear: the non-perturbative Euclidean instanton solution defined by $A_\mu(x_0, \underline{x})$ with "winding number" n interpolates between the real-time distant past and distant future "Minkowski" vacua — defined by $A_\mu(t = \pm\infty, \underline{x})$ — with respective "winding numbers" m and m'. The imaginary time $x_0 = it$ is the interpolating parameter and the "tunneling event" — accomplished by the instanton — does not take place if Q_{CS} $(t = -\infty)$ and $Q_{CS}(+\infty)$ are equal to each other, which is evidently the case when one reverts to the perturbative vacuum.

From Eq. (10.89), it follows that the path-integral representation of the ("Minkowski") vacuum-to-vacuum transition amplitude can be written in the form:

$$< n|e^{-iHt}|m >_J = \int [dA]_{n=m'-m} \exp\{-i \int d^4x(\mathcal{L} + JA)\} \quad (10.90)$$

where the path integral is taken over all gauge fields belonging to the same homotopic class with "winding number" $n = m' - m$, $\mathcal{L}$ is the QCD Lagrangian and J is the external source. The form (10.90) for the path integral suggests — when it is translated back to Euclidean time (so that the integrand of the path integral takes on the form e^{-S_E}) — that the instanton configurations with minimal finite action dominate the transition probability for "penetrating" the energy barrier between two m vacua. Thus, for an $n = 1$ instanton "tunneling" from the m vacuum to the $(m+1)$-vacuum, the vacuum-to-vacuum transition amplitude of Eq. (10.90) acquires the approximate form (in the semi-classical WKB approximation):

$$\lim_{x_0 \to \infty} <(m+1)| \exp(-Hx_0)|m> \to \int [dA_\mu] e^{-S_E} \sim e^{-\frac{8\pi^2}{g^2}} \quad (10.91)$$

where $S_E \sim 8\pi^2/g^2$ is the Euclidean action for an $n = 1$ instanton. Equation (10.91) yields an extremely small value for the transition probability for weak coupling [see §10.3a for the QFD example] although the situation can change drastically in the presence of strong coupling (e.g. QCD) or if the weak coupling contribution is balanced off by some other type of non-perturbative solution of a non-Abelian gauge theory [see §10.3a for the effect of "sphaleron" solutions on QFD predictions]. The moral is clear: perturbation-theoretic treatments of non-Abelian gauge theories — which assume the true vacuum to be $|m\rangle = 0$ — ignore the non-perturbative effects of quantum "tunneling" and may omit qualitatively significant effects, as will be seen below.

From what has been said, it is unwise to calculate non-perturbative effects in a Yang-Mills theory starting with the perturbative vacuum. What then is the non-perturbative vacuum (i.e. the "true" vacuum) in a Yang-Mills theory? To identify the "true" vacuum in a Yang-Mills theory, one must recognize that all the m vacuum states ($m \neq 0$) are degenerate in energy with the $m = 0$ vacuum state (i.e. that we have a discrete infinititude of "degenerate vacua"). Moreover, under a "large" gauge (instanton) transformation with $n = 1$ — call it $\bar{U}$ — we have the relation:

$$\bar{U}|m\rangle = |m+1\rangle \tag{10.92}$$

where, by gauge invariance, $\bar{U}$ commutes with the total Hamiltonian, i.e.:

$$[\bar{U}, H] = 0 \tag{10.93}$$

Equation (10.93) requires the "true" vacuum to be an eigenstate of all gauge transformations — both "large" and "small" — in contradiction with (10.92). It appears possible to construct the "true" vacuum — the so-called θ vacuum — by taking a superposition of the m vacua, as follows:

$$|\theta\rangle = \sum_m e^{-im\theta}|m\rangle \tag{10.94}$$

which (naively) possesses the property: $\langle \theta'|\theta \rangle = 2\pi\delta(\theta - \theta')$. The appearance of the δ function in the transition amplitude between the θ and θ' vacua is troublesome because it implies that the norm of the θ vacuum, i.e. $\langle \theta|\theta \rangle$ is infinite, and hence not normalizable. The usual interpretation of lack of normalizability [10.26] might lead one to believe that, in the presence of instantons, there are an infinite number of mathematically possible Hilbert spaces (corresponding to different θ vacua) and that the physically realizable Hilbert space is only one of them with a particular value of θ. In view of some undesirable consequences of a non-vanishing value of the θ parameter (see §10.3c), considerable effort has gone into attempting to prove that θ must vanish [10.27]. [It must be emphasized that the existence of the θ vacuum and the value of the θ parameter are distinct questions.]

Whatever the value of θ, the existence of the θ vacuum in Yang-Mills theories must be taken seriously and it is necessary to explore further the mathematical and physical basis of the non-perturbative θ vacuum and then to draw out some of the striking consequences, especially for QCD. Allowing the θ parameter to take on any value (including $\theta = 0$), we can say that an instanton with "winding number" n satisfies the eigenvalue equation:

$$\bar{U}_n |\theta> = \bar{U}_1^n |\theta> = e^{in\theta} |\theta> \qquad (10.95)$$

Since $\bar{U}_n$ commutes with the total Hamiltonian, the θ eigenvalue does not change under time evolution and hence, the Hilbert space of the theory is divided into sectors labelled by the continuous parameter θ and each $|\theta>$ is the ground state of an independent and physically inequivalent sector: different θ worlds do not "communicate" with each other since gauge-invariant disturbances are confined to a single sector [this is a more precise mathematical statement of our earlier remark about physically realizable Hilbert spaces]. The θ vacuum must now replace the n vacua and we must write for the path-integral representation of the vacuum-to-vacuum transition amplitude — in Minkowski space for QCD [see Eq. (10.90)] the following:

$$\lim_{x_0 \to \infty} <\theta' |e^{-Hx_0}|\theta> \sim \delta(\theta - \theta') \sum_n \int [dA]_n e^{-i \int d^4 x (\mathcal{L}_{\text{eff}} + JA)} \qquad (10.96)$$

where:

$$\mathcal{L}_{\text{eff}} = \mathcal{L}_{QCD} + \mathcal{L}_\theta; \quad \mathcal{L}_\theta = \frac{\theta}{16\pi^2} \text{Tr} \, G_{\mu\nu} \tilde{G}^{\mu\nu} \qquad (10.96a)$$

with θ the as-yet-undetermined parameter and $G_{a\mu\nu}$ the eight gluon fields of QCD. It is the parity-violating and CP-violating term $\mathcal{L}_\theta$ in Eq. (10.96a) — due to the presence of the topologically non-trivial color instantons — that is quite unexpected and has both reassuring and disturbing consequences in QCD.

Before considering the novel consequences resulting from the presence of instanton solutions in Yang-Mills theories [and, in particular, the implications of Eqs. (10.96) and (10.96a) for QCD], we list three essential differences between the θ vacuum-to-θ' vacuum transition amplitude [given by Eq. (10.96)] and the m vacuum-to-m' vacuum transition amplitude [given by Eq. (10.90)]: (1) for the m to m' vacuum transition amplitude, the path integral is taken over all gauge fields belonging to the homotopic class $n = m' - m$, whereas for the θ to θ' vacuum transition amplitude, the path-integral is taken over the discrete infinitude of homotopic classes n; (2) the Lagrangian for the m vacuum formulation, $\mathcal{L}_{QCD}$, is replaced by the effective Lagrangian density $\mathcal{L}_{\text{eff}}$ [Eq. (10.96a)] for the θ vacuum formulation; $\mathcal{L}_{eff}$ contains the additional term $\mathcal{L}_\theta$ that is both P- and CP-violating. It is true that the extra term possesses the familiar triangular axial anomaly structure (see §7.2a) but, in contrast to that situation — where the term $\text{Tr}\, G_{\mu\nu}\tilde{G}^{\mu\nu}$ is a (pseudoscalar) correction to the divergence of an axial vector current, the P- and CP-conserving term in Eq. (10.96a) represents a breakdown (if $\theta \neq 0$) of P and CP (or T) conservation (see §2.2d) that holds for $\mathcal{L}_{QCD}$ and is confirmed phenomenologically to high accuracy; and (3) the "cluster decomposition theorem" does not hold for the m vacua but does hold for the θ vacuum. The "cluster decomposition theorem" is a technical, albeit important, property of acceptable quantum field theories [3.26] and states that the expectation value for widely separated operators (in space-time) should vanish; however, it can be shown that the presence of "intermediate state" m vacua ($m \neq 0$) causes a violation of the theorem for widely-separated operators [with non-vanishing chirality — see §10.3c] when the expectation value is taken over the $m = 0$ (normal) vacuum state. This problem does not arise for the θ vacuum.

Thus, if we allow $\theta \neq 0$, the existence of instanton solutions for Yang-Mills theories in Euclidean four-space drastically alter the Yang-Mills vacua in Minkowski space. It is therefore not surprising that a number of puzzling predictions have ensued for QCD and QFD, bearing testimony to the importance of understanding the topological aspects of the standard model. In the following sections (§10.3a–§10.3c), we consider some of the surprising consequences that have emerged in QFD and QCD when account is taken of the instantons associated with these Yang-Mills theories. The spontaneous breaking of B and L conservation due to the $SU(2)_L$ "flavor" instantons in QFD is considered in §10.3a but the effects are minuscule due to the weak coupling; the possibility of enhancing B and L non-conservation effects due to non-perturbative "sphaleron" solutions — and the experimental consequences thereof — is also considered in this section. In §10.3b, we treat the "$U(1)$ problem" in QCD; one can say that the resolution of the "$U(1)$ problem" in QCD [consisting of the unexpectedly large branching ratio for $\eta \to 3\pi$ decay and the absence of a ninth N-G boson resulting from the SSB of the global chiral quark flavor symmetry $U(3)_L \times U(3)_R$ (see §5.1)] does not depend on the value of the θ parameter and is definitely facilitated by the color instantons of QCD. However, the resolution of the "strong CP" problem in QCD is closely tied to the value of the θ parameter and we must be satisfied with a status report on the salient features of the problem and the thus far unsuccessful axion approach — as well as other attempts — to deal with it.

§10.3a. Spontaneous breaking of B and L charge conservation by instantons and the sphaleron hypothesis

A good example of the power of instantons to break global conservation laws in quantum field theory in unexpected ways is the breaking of baryon and lepton charge conservation — while maintaining $B - L$ conservation — within the framework of QFD. We describe the nature of "flavor" instanton-induced B and L violation and then consider how the persistence of $B - L$ conservation, despite separate B and L violation, opens the door to a possible role for the topologically-related unstable "sphalerons". We start by recalling the quantum numbers of the Weyl fermion representations (for quarks and leptons) under the standard group (see Table 1.5); we remind the reader that all fermions in this table are given in lefthanded representations. The electroweak part of the standard group, $SU(2)_L \times U(1)_Y$, contains the non-Abelian $SU(2)_L$ factor subgroup, with which are associated the QFD "flavor" instantons. The "flavor" instantons are responsible for the topologically non-trivial $[SU(2)]^2 U(1)$ global triangular axial anomalies that give rise to the non-conservation of baryon and lepton charge in QFD. While these non-conservation effects are miniscule in QFD — because of the smallness of the weak coupling constant — it is illuminating to prove this result in the one-instanton approximation and, at the same time, to show that the separate breakdowns of global B and L cancel each other [i.e. $B - L$ is conserved despite the anomaly]. Comments about "sphalerons" follow.

In QFD, the anomalous four-divergence of the baryon (quark) current j_B^μ can be written in the form (the transcription to the lepton current is obvious):

$$\partial_\mu j_B^\mu = -\frac{N_B}{3} \frac{g^2}{16\pi^2} F_{\mu\nu}^a \, \text{Tr} \, F_{\mu\nu} \tilde{F}^{\mu\nu} \tag{10.97}$$

where N_B is the number of doublets of lefthanded quark fields, the baryon charge of each quark is $\frac{1}{3}$ and g is the weak coupling constant; the baryon current, j_B^μ, is given by:

$$j_B^\mu = \frac{1}{3} \sum_i \bar{\psi}_i \gamma^\mu \psi_i \tag{10.98}$$

where the sum extends over all right- and lefthanded quark fields. For the leptons, the lepton current, j_L^μ, replaces j_B^μ in Eq. (10.97) and N_L (the number of doublets of lefthanded lepton fields) replaces $N_B/3$ in the same equation (since the lepton charge of each lepton is 1). The analog of Eq.

(10.98) for leptons is:

$$j_L^\mu = \sum_j \bar{\psi}_j \gamma^\mu \psi_j \qquad (10.98a)$$

where the sum now extends over all right- and lefthanded lepton fields. To compute the changes in the baryon and lepton charges resulting from the instanton-induced anomalies, we calculate the changes in total baryon (lepton) charge $B(L)$ between the times $-\infty$ and $+\infty$, following the procedure that led to Eq. (10.89); we get [10.28]:

$$\Delta B = B(+\infty) - B(-\infty) = \int_{-\infty}^{\infty} dt\, \partial_0 \int d^3x\, j_B^0(\underline{x}, t)$$

$$= \int_{-\infty}^{\infty} dt \int d^3x\, \partial_\mu j_B^\mu \qquad (10.99)$$

For ΔB, Eq. (10.99) yields $-\frac{N_B}{3}\nu$ and for ΔL, $-N_L\nu$, where ν is the Pontryagin number of the $SU(2)_L$ Yang-Mills field, thus:

$$\nu = \int d^4x \frac{g^2}{16\pi^2} F_{\mu\nu}^a \tilde{F}^{a\mu\nu} \quad (a = 1, 2, 3) \qquad (10.100)$$

where $F_{\mu\nu}^a$ $(a = 1, 2, 3)$ is the weak gauge field tensor. For one family of quarks and leptons in the standard model, $N_B = 3$ (the number of colors) and $N_L = 1$, so that:

$$\Delta B = \Delta L = -\nu; \qquad \Delta(B - L) = 0 \qquad (10.101)$$

Several comments should be made about Eq. (10.101): (1) the Pontryagin number ν, as given by Eq. (10.100), is nothing more than the "winding number" of the $SU(2)_L$ "flavor" instanton which interpolates between two "Minkowski" vacua whose "winding numbers" differ by ν [see Eq. (10.89)]. Equation (10.101) tells us that the amount of baryon charge violation ΔB (and of lepton charge violation ΔL) is precisely equal to the instanton "winding number"; (2) the fermion charge violation (whether B or L), as given by Eq. (10.101), results from the fact that the instanton "winding number" is also — as will be proved in §10.4a — equal to the difference between the number of positive and negative chirality "zero modes" of the Dirac operator in the $SU(2)_L$ background gauge field; to put it another way, as the $SU(2)_L$ gauge field changes from one "non-perturbative" vacuum to another, the fermion (quark and lepton) energy levels in this background gauge field "cross over" (i.e. acquire zero energy eigenvalues) in an unsymmetrical way, thereby "transporting" net fermion charge (see §10.4a); (3) while the baryon decay predicted by Eq. (10.101) [not to be confused with the proton decay associated with the GUT groups discussed in §8.3] and the lepton decay predicted by Eq. (10.101) are qualitatively interesting examples of instanton-induced effects in QFD, the amounts of B and L decay are minuscule by virtue of the smallness of the weak coupling constant. An estimate of the amount of B violation (the amount of L violation is the same) is given by the transition probability for the "tunneling" of instantons with $\nu = 1$ and the result is $\exp(-16\pi^2/g_W^2) = \exp(-2\pi \sin^2 \theta_W/\alpha_e)$ [see Eq. (10.91)], where $\sin^2 \theta_W = 0.23$ (θ_W is the Weinberg angle). With these numbers, we get the tiny transition probability $\sim e^{-400}$, hardly a demonstration of the importance of topologically non-trivial gauge transformations in QFD. However, it should be pointed out that the above estimate is based on the assumption of an unbroken standard electroweak gauge group and that the energy barrier between the topologically distinct vacua through which the instanton "tunnels" may be substantially altered by the Higgs breaking of the electroweak group. Indeed, it turns out that under suitably exotic (but not impossible) conditions, the instanton prediction can be substantially modified by paying attention to the topologically non-trivial "sphaleron" solutions that are possible in Yang-Mills theories (such as QFD) with Higgs fields; and, finally (4) the result $\Delta(B-L) = 0$ contained in Eq. (10.101) — that the separate B and L violation induced by the $SU(2)_L$ flavor instantons is subject to the constraint of global $B - L$ conservation — informs us that, like, electric charge quantization, global $B - L$ conservation is reflective of the non-perturbative aspects of the standard electroweak group and does not require enlargement to a GUT group (see §8.3). As we will see, global $B - L$ charge conservation plays a key role in enabling sphaleron solutions (see below) to overcome the strongly suppressive effects of instanton solutions on baryon (and lepton) charge violation in QFD.

The first step in dealing with the modifications introduced into the topological aspects (instanton effects) of the Yang-Mills gauge theory is to take account of the SSB which introduces a mass scale otherwise absent in the Yang-Mills theory. In the case of the electroweak group, the SSB scale is M_W (strictly speaking, we should use Λ_{QFD} but we are only interested in rough numbers) and the height of the energy barrier between the topologically distinct vacua becomes [10.29]:

$$E_{sp} = C(\frac{M_H}{M_W})\frac{\pi M_W \sin^2 \theta_W}{\alpha_e} \sim 7 - 14 \text{ TeV} \qquad (10.102)$$

where we have written E_{sb} for the height of the energy barrier (since it is also the "sphaleron" energy — see below), C is a function of λ (in the Higgs potential for the electroweak group) and M_H is the Higgs mass; the value of E_{sp} changes from 7 to 14 TeV as λ increases from 0 to ∞. Whereas the instanton (with arbitrary "size" since there is no scale) is the static, classical, particle-like solution of the Yang-Mills equations in Euclidean four-space (without SSB) that minimizes the action, the "sphaleron" solution is the static, saddle point configuration of gauge and Higgs fields that sits on top of the energy barrier, given by Eq. (10.102) [10.30]; thus, the "sphaleron" solution establishes an energy scale in the neighborhood of which interesting physical phenomena might occur. From a realistic point of view, there are two possible laboratories for the study of "sphaleron"-induced baryon and lepton charge violation (although we focus on baryon charge violation) at energies of tens of TeV: the cosmic laboratory of the "big bang" universe where the temperatures — albeit fleetingly — were sufficiently high, and the earthly laboratory of sufficiently high accelerator energies. According to theory, the cosmological laboratory makes direct use of the "sphaleron" solution (because of the thermal equilibrium associated with the high temperatures) while the possible enhancement of baryon charge violation at extremely high accelerator energies depends more on cumulative contributions to the baryon charge-violating cross section due to multiple particle production (involving chiefly the W's, Z's and Higgs), which suppress the "low energy" instanton effect.

We comment first on electroweak baryon charge violation in the early universe and then consider the possibility of enhancing the minuscule value for baryon charge violation in QFD at high accelerator energies. At the extremely high temperatures of the early universe, the transitions between the topologically distinct vacua in non-Abelian QFD is no longer dominated by instantons ("tunneling") but, instead, by the thermal fluctuations in the neighborhood of the "sphaleron" energy, i.e. E_{sp} of Eq. (10.102); roughly speaking, the system can "roll over" the energy barrier instead of penetrating it [10.31]. At temperatures above E_{sp}, the system acquires a Boltzmann-type transition probability $\Gamma_{thermal}$ [which depends on the ratio $E_{sp}(T)/T$ — the dependence of $E_{sp}(T)$ on T is due to the dependence of M_W (or Λ_{QFD}) on T] instead of the instanton-type transition probability. Actually, as long as $T \gtrsim E_{sp}(T)$, the baryon charge-violating transitions are essentially unsuppressed and the estimate for the transition probability (per unit time per unit volume) is:

$$\Gamma_{\text{thermal}} = \kappa(\alpha_W T)^4 \tag{10.103}$$

where κ is of order 1 $(0.1 \sim 1)$ [10.31]. The rate (10.103) is to be compared with the expansion rate for the universe, which is of the order of T^2/M_{Planck}, and it can be shown that the electroweak mechanism for baryon charge violation is still dominant for $T \simeq \Lambda_{QCD}$. At first sight, this result implies that whatever baryon asymmetry of the universe (BAU) was generated during the GUT epoch $(T \simeq 10^{15}$ GeV) is washed out by the electroweak baryon charge-violating processes at later times (down to Λ_{QFD}). This is where $SO(10)$ GUT — which permits the SSB of gauged $(B - L)$ symmetry [in contrast to $SU(5)$ GUT] can come to the rescue by generating a sufficient amount of $(B - L)$ asymmetry at GUT temperatures which is maintained by the global $(B - L)$ conservation law. Needless to say, this new electroweak source of baryon charge violation — subject to the condition of global $(B - L)$ conservation — opens up new avenues for dealing with the BAU problem and even the possibility that the electroweak mechanism is the chief ingredient of its resolution [10.30].

Since much of the cosmological "sphaleron" work is still in the speculative stage, we pass on to the possibility of finding electroweak-induced baryon charge violation in the earthly laboratory of high energy collisions. The fairly confident prediction of appreciable baryon charge violation at sufficiently high temperatures (of the order of 10 TeV) raised the hope that a similar situation obtains in high energy collisions where the thermally-excited "sphaleron" transitions in the early universe are translated into their topological counterparts in the multi-particle scattering processes that take place at high energies. The difference — an important one — is that there is no thermal equilibrium in a high energy collision which is initiated, for example, by two particles (e.g. two quarks). This experimental limitation — at least in the foreseeable future — makes it extremely difficult to match the cosmic laboratory in its efficacy for generating appreciable baryon charge violation via the electroweak mechanism. The point is that the calculation of the cross section for $2 \to n$ collisions [where n denotes the final number of q's, ℓ's (satisfying the condition of $B - L$ conservation), W's, Z's and H's] uses the saddle-point approximation for the instanton contribution to the n-point Green's function (instead of calculating the effect of thermal fluctuations in the neighborhood of the "sphaleron" energy). This leads to a power correction to the usual instanton "penetrability" factor — instead of a Maxwell distribution in temperature; one finds [10.32]:

$$\sigma_{2 \to n} \simeq \frac{F(n, \alpha_W)}{V^2} e^{-\frac{4\pi}{\alpha_W}} \left(\frac{E}{V^2}\right)^{2n} \qquad (10.104)$$

where $F(n, \alpha_W)$ is a calculated function of n and α_W and V is the Higgs VEV.

We see from Eq. (10.104) that the huge instanton suppression factor is present and would eliminate the electroweak mechanism as a source of baryon charge violation in high energy collisions were it not for the discovery that, in the saddle-point approximation, the total cross section $\sigma_{\text{total}} = \sum_n \sigma_{2 \to n}$ increases exponentially with energy; one finds for the total cross section:

$$\sigma_{\text{total}} = \sum_n \sigma_{2 \to n} \simeq \exp\{\frac{4\pi}{\alpha_W}[-1 + \frac{9}{8}(\frac{E}{E_0})^{\frac{4}{3}} + ...]\} \qquad (10.105)$$

where $E_0 = \sqrt{6}\pi M_W / \alpha_W \simeq 17$ TeV. The particle multiplicity N becomes:

$$N \sim \frac{4\pi}{\alpha_W}(\frac{E}{E_0})^{\frac{4}{3}} \qquad (10.106)$$

If Eq. (10.105) was rigorous, the consequences would be very exciting: the inclusive cross section for electroweak baryon charge violation could become measurable at center-of-mass energies of tens of TeV and the events could be spectacular with $N \sim 4\pi/\alpha_W$ (mostly W's, Z's and H's) and the $(B-L)$ conservation law holding for the produced quarks and leptons. Furthermore, the instanton origin of these multiple events would mandate a spherically symmetric angular distribution. Unfortunately, there may be a flaw in Eq. (10.105), namely the violation of s wave unitarity for $E \simeq E_0$. The question must therefore remain open until the corrections to Eq. (10.105) are properly computed and determined to possess a power-law or exponential character [10.30].

§10.3b. $U(1)$ problem in QCD and its resolution with instantons

The discussion of instanton effects in QFD has been inconclusive: while theory predicts some qualitatively interesting instanton-like effects in spontaneously broken non-Abelian "weakly coupled" QFD — at the very high temperatures of the early universe and at very high accelerator energies — the predicted effects of significant baryon charge (and lepton charge) violation must still be tested. When we come to unbroken non-Abelian "strongly coupled" QCD, however, there are two striking instanton-induced effects that, in principle, should be observable at relatively low energies and could test our basic understanding of the non-perturbative QCD vacuum; the first instanton-related effect in QCD was given a great deal of attention soon after the discovery of the instanton [1.136] and was finally solved when it was recognized that the color instantons are responsible for the "hard" character of the $U(1)_A$ [flavor-singlet triangular axial] QCD anomaly and that this solution does not depend on the value of the θ parameter associated with the non-perturbative QCD vacuum [10.33]. The "$U(1)$ problem in QCD" is discussed in this section.

The second instanton-related effect in QCD — which depends more explicitly on our understanding of the θ vacuum and our knowledge of the θ parameter — received even more vigorous study, both theoretical and experimental, during the past fifteen years than the "$U(1)$ problem", and is still unsolved. The Peccei-Quinn (PQ) mechanism [1.138] is an ingenious attempt to explain the phenomenologically small value of the θ parameter [derived from measurements on the upper limit of the electric dipole moment of the neutron (see §6.3f)] by introducing a new chiral $U(1)$ symmetry — which can produce a vanishing value for θ. Unfortunately, a by-product of the SSB of the PQ $U(1)_A$ symmetry — which must occur — is the generation of a quasi-N-G boson, the axion, which has eluded detection. Our brief report on the present status of the instanton-created "strong CP problem in QCD" is given in §10.3c.

We define the precise nature of the "$U(1)$ problem" in QCD by recalling some key results from §5.1. It was noted there that a QCD Lagrangian with n_f flavors of massless quarks possesses the additional global chiral quark flavor symmetry:

$$SU(n_f)_L \times SU(n_f)_R \times U(1)_L \times U(1)_R \qquad (10.107)$$

It was also pointed out that the two-quark condensates — produced by the confining SU(3) color group (see Table 5.1) — spontaneously break down the global chiral quark flavor group $SU(3)_L \times SU(3)_R$ to the (global) vector "flavor" group $SU(3)_F$; at the same time, the condensates break down the global chiral quark group $U(1)_L \times U(1)_R$ to the global vector "baryon charge" group $U(1)_B$. In the process, the axial vector $SU(3)_A$ symmetry is completely broken and gives rise to an octet of pseudoscalar (massless) N-G bosons; a similar argument — naively applied — would lead to the conclusion that the quark condensates responsible for the SSB of the $U(1)_A$ symmetry (see Table 5.1) should give rise to a singlet N-G boson which — because of the finite quark masses — should become a singlet "quasi-N-G" boson with the quantum numbers of the η' meson. The scenario just described is followed with only one exception: the observed mass of the η' meson is 958 MeV, much larger than the pionic-type mass that is expected; while it is true that an early current algebra calculation [5.2], carried out before the discovery of instantons, accentuated the severity of the problem by placing an upper limit of $\sqrt{3}m_\pi$ on the mass of the $U(1)_A$-related N-G boson, the question of whether such a N-G boson should exist at all within the instanton framework, became the basic $U(1)$ problem.

We proceed to review the "instanton" solution to the "$U(1)$ problem" in QCD and then show that the "hard" character of the $U(1)_A$ anomaly in QCD ensures successful quantitative calculations of both parts of the "$U(1)$ problem" (i.e. the large η' meson mass and the large branching ratio for $\eta \to 3\pi$ decay). In treating the "instanton" solution to the "$U(1)$ problem" in QCD, we first prove that the color instantons spontaneously break $U(1)_A$ to Z_{2n_f} and then show that this breaking does not give rise to a N-G boson. We have the anomaly equation for arbitrary n_f [see Eq. (7.19)]:

$$\partial^\mu J_\mu^5 = 2n_f \frac{g^2}{16\pi^2} \text{Tr}(G_{\mu\nu}\tilde{G}^{\mu\nu}) \tag{10.108}$$

where $G_{\mu\nu}$ is the gluon tensor and $\tilde{G}$ its dual. Because of the anomalous term, the axial charge $Q_5 = \int d^3x J_5^0$ is not conserved and therefore can not serve as the generator of the chiral transformation $q_i \to e^{i\alpha\gamma_5}q_i$. However, by making use of the Chern-Simons current K^μ [see Eq. (10.64a)], we can extract a conserved (albeit gauge-variant) axial charge $\tilde{Q}_5 = \int d^3x \tilde{J}_5^0$ that can serve as the generator of the chiral transformation on the quarks; we get:

$$\partial_\mu \tilde{J}_5^\mu = 0; \qquad \tilde{J}_5^\mu = J_5^\mu - 2n_f K^\mu \tag{10.109}$$

so that $\tilde{Q}_5 = Q_5 - 2n_f Q_{CS}$ where the Chern-Simons charge Q_{CS}, defined by Eq. (10.88). The charge Q_{CS}, as given by Eq. (10.88), was shown in §10.3 to describe a "Minkowski" vacuum with "winding number" m when the static gauge field A_i [in Eq. (10.88)] acquires the pure gauge form: $A_i = U_m^\dagger \, \partial_i \, U_m$ [where U_m is the "mapping" function with "winding number" m].

Since we associate with Q_{CS} the "winding number" of the m vacuum, we have $Q_{CS}|m >= m|m >$ and we can write:

$$U_n^{-1}\tilde{Q}_5 U_n = Q_5 + 2m \, n_f \tag{10.110}$$

or:

$$[\tilde{Q}_5, U_m] = 2mn_f \tag{10.111}$$

The consequences of Eqs. (10.110) and (10.111) are rather interesting: first, since $U_m|0 >= |m >$ and $\tilde{Q}_5|0 >$, we have:

$$\tilde{Q}_5|m >= 2mn_f|m > \tag{10.112}$$

Equation (10.112) tells us that a vacuum state with definite "winding number" m also possesses a definite chirality (i.e. it is an eigenstate of $\tilde{Q}_5$).

Furthermore, since $\tilde{Q}_5$ commutes with the total Hamiltonian (i.e. $\tilde{Q}_5$ is conserved), the vacuum-to-vacuum transition amplitude vanishes unless the "winding number" is the same for both vacua. But this implies that, in the presence of massless fermions, "tunneling" between vacua with different "winding numbers" is suppressed. Finally, by virtue of Eq. (10.110), we get:

$$e^{-i\alpha\tilde{Q}_5}|\theta> = \sum_n e^{-i(\theta+2\alpha n_f)}|n> \qquad (10.113)$$

Hence, a chiral rotation in the amount α changes the θ vacuum state into the $\theta' = (\theta + 2n_f\alpha)$ vacuum state, thus:

$$e^{-i\alpha\tilde{Q}_5}|\theta> = |\theta'> \qquad (10.114)$$

From Eqs. (10.113) and (10.114), two important conclusions follow: (1) the θ' vacuum becomes identical with the θ vacuum if α is an integral multiple of $2\pi/2n_f$, i.e. the $U(1)_A$ symmetry is spontaneously broken to the discrete symmetry Z_{2n_f}. Thus, there is an incomplete breaking of the $U(1)_A$ symmetry [from $U(1)_A$ to Z_{2n_f}], which opens up the possibility that the SSB of $U(1)_A$ is not accompanied by a N-G boson, a possibility that is realized within the context of the θ vacuum, as we will see; (2) Eq. (10.114) tells us that, in the presence of at least one massless fermion, the theory is invariant under a chiral rotation that changes the vacuum state from $|\theta>$ to $|\theta'>$. Hence, since α is arbitrary, the parameter θ possesses no physical meaning and can be "rotated" away. Effectively, chiral (massless) fermions suppress instanton tunneling and eliminate the need for a θ vacuum (and, *a fortiori*, the parity- and CP-violating term in the Lagrangian) so that there is no "strong CP" problem in QCD (see §10.3c).

Following up on Eq. (10.113) and conclusion (1) above, we summarize the most rigorous available θ vacuum-type argument for the absence of the $U(1)$ problem in QCD [1.137]. Thus, consider an operator X with non-vanishing chirality χ and with non-vanishing VEV, to wit:

$$[\tilde{Q}_5, X] = \chi X; \qquad <\theta|X|\theta> \neq 0 \qquad (10.115)$$

[For example, if X is the "mapping" function U_m with "winding number" m $(m \neq 0)$, then $\chi = m \neq 0$ and $<\theta|U_m|\theta> \neq 0$.] It can then be shown — since $\tilde{J}_5^\mu$ is divergenceless — that the Green's function:

$$\tilde{G}^\mu(q) = \int d^4x e^{-iqx} <\theta|T(\tilde{J}_5^\mu(x)X)|\theta> \qquad (10.116)$$

possesses a $1/q^2$ singularity with a residue given by χ, i.e. :

$$\int d^4x \partial_\mu < \theta|T(\tilde{J}_5^\mu(x), X)|\theta >= \chi < \theta|X|\theta > \qquad (10.117)$$

However, since $\tilde{J}_5^\mu$ is not gauge invariant, it does not follow from Eq. (10.117) that a physical N-G boson is associated with the $1/q^2$ singularity contained in Eq. (10.116). A N-G boson is only present if the Green's function $G^\mu(q)$ [derived from Eq. (10.116) by replacing the gauge-variant $\tilde{J}_5^\mu$ by the gauge-invariant J_5^μ] possesses a non-vanishing residue; using Eq. (10.109), the residue of the q^2 pole in $G^\mu(q)$ becomes:

$$\int d^4x \partial_\mu < \theta|T(J_5^\mu(x), X)|\theta >= \chi < \theta|X|\theta >$$
$$+ 2n_f \frac{g^2}{16\pi^2} \int d^4x < \theta|T(\text{Tr } F_{\mu\nu}\tilde{F}^{\mu\nu}, X)|\theta >$$
$$(10.118)$$

It can now be shown that the R.H.S. of Eq. (10.118) does vanish [10.33] in the presence of an instanton, say, with "winding number" ν, because χ in the first term on the R.H.S. of Eq. (10.118) takes on the value $2n_f\nu$ [see Eq. (10.111)] and the second term [on the R.H.S. of Eq. (10.118)] acquires the same value with opposite sign [see Eq. (10.100)]. This result is perhaps not surprising since the same instanton field with "winding number" ν is used to evaluate both terms on the R.H.S. of Eq. (10.118). In any case, except for some caveats that have been raised [10.34], this is the basic argument for the generally accepted belief that the $U(1)_A$ group — in its instanton-induced breaking to Z_{2n_f} — is not accompanied by an N-G boson.

Implicit in the "instanton" solution to the "$U(1)$ problem" in QCD described above, is the use of the "instanton-driven" $U(1)_A$ QCD anomaly. As Witten [7.4] puts it, "the [QCD] anomaly term has non-zero matrix elements at zero momentum". This implies that the $U(1)_A$ symmetry is broken but is not accompanyied by an N-G boson. The point is that the presence of color instantons in QCD is responsible for the finite Chern-Simons charge, thereby breaking the $U(1)_A$ symmetry, and the non-perturbative QCD θ vacuum, originating with the instantons present in the theory, eliminates the unwanted N-G boson. (We note that the above proof nowhere depends on the value of the θ parameter.) The overall conclusion is that it should be possible to obtain good quantitative results for the two parts of the "$U(1)$ problem" — the large mass for the η' meson and the large branching ratio for $\eta \to 3\pi$ decay — by making explicit use of the "harmless" global QCD anomaly (see §7.2b) and without referring to the θ vacuum and the unknown value of the θ parameter. Such calculations have been performed, using current algebra and taking cognizance of the flavor-singlet triangular axial QCD anomaly at every stage of the calculation [7.7]; the result of the $m_{\eta'}$ calculation is summarized in §7.2b and gives good results. The calculation of the transition amplitude for $\eta \rightarrow 3\pi$ decay [7.7] also gives good agreement with experiment and the overall conclusion is that the two parts of the "$U(1)$ problem" in QCD can be understood in quantitative terms if the instanton-induced global QCD anomaly is fully applied.

§10.3c. Strong CP problem in QCD and the axion

The "strong CP" problem in QCD is not as tractable as the "$U(1)$ problem" precisely because the "strong CP problem" is defined by the P-violating and CP-violating term in the QCD Lagrangian [see Eq. (10.96a)] — required by the instanton-modified θ vacuum — and must cope with the phenomenological consequences of the θ parameter [1.137]. The "strong CP" problem arises in QCD because the upper phenomenological limit on the θ parameter [given in Eq. (10.96a)] is so much smaller than would be expected in a theory of the strong interaction (QCD). This extremely low value of θ_{QCD}, [we replace the θ of Eq. (10.96a) by θ_{QCD}] must somehow be explained and this section is primarily devoted to an abbreviated discussion of some of the attempts made to resolve the "strong CP" problem in QCD (including the Peccei-Quinn proposal and its corollary, the axion).

Before discussing the attempts to solve the "strong CP" problem, we must point out that there is an additional complication in connection with the "strong CP" problem arising from the higher order effects of the weak CP violation implicit in the U_{CKM} quark mixing matrix (see §6.3f). If we denote by θ_{QFD} the additional contribution to the QCD vacuum angle θ_{QCD}, the "strong CP" problem then consists of explaining why $\theta = \theta_{QCD} + \theta_{QFD} < 10^{-9}$. It is simple to show that θ_{QFD} is Arg det M (see §6.3e for notation); for this purpose, we write down the additional contribution to the QCD Lagrangian from finite quark masses:

$$\mathcal{L}_{\text{mass}} = -\bar{q}_{R_i} M_{ij} q_{L_j} - \bar{q}_{L_i} (M^\dagger)_{ij} q_{R_j} \qquad (i = 1, 2, 3) \qquad (10.119)$$

where q_L and q_R are the left- and righthanded quark states respectively, i denotes the ith quark generation and M_{ij} is the quark mass matrix (which, in general, is neither symmetric nor Hermitian). As we know (see §6.3e), the diagonalization of M_{ij} is carried out by means of separate unitary transformations on the left- and righthanded quark fields, which yields a real diagonal quark mass matrix with a common phase, say δ. To eliminate the phase δ

requires an axial rotation $q_i \rightarrow e^{i\frac{\delta}{2}\gamma_5}q_i$, i.e. a chiral $U(1)_A$ rotation, which gives rise to a triangular axial anomaly in QCD, namely $\frac{n_f\delta g^2}{16\pi^2}$ Tr $F_{\mu\nu}\tilde{F}^{\mu\nu}$; hence, the amount of chiral $U(1)_A$ rotation required to secure a γ_5-free quark mass matrix [see Eq. (10.119)] is given by:

$$\theta_{QFD} = \text{Arg det } M = n_f\delta \qquad (10.120)$$

One can now use Eq. (10.120) together [with θ_{QCD} (i.e. $\theta = \theta_{QCD} + \theta_{QFD}$)] in order to calculate the predicted value of the electric dipole moment of the neutron, d_n, in terms of the θ parameter. A good approximation is to consider only two quark flavors — the u and d quarks — and one finds for $\mathcal{L}_\theta$ the approximate form [1.137]:

$$\mathcal{L}_\theta \simeq -i\frac{m_u m_d}{(m_u + m_d)}\theta[\bar{u}\gamma_5 u + \bar{d}\gamma_5 d] \qquad (10.121)$$

With Eq. (10.121), one gets:

$$d_n \sim 10^{-16}\,\theta\,e\,\text{cm} \qquad (10.122)$$

which, when compared to the measured upper limit $d_n \lesssim 10^{-25}e$ cm [6.50] yields $\theta < 10^{-9}$. This tiny value of θ requires an almost exact cancellation of the "strong" QCD phase angle contribution, θ_{QCD}, by the "electroweak" vacuum angle contribution θ_{QFD} due to the quark mixing matrix. Such an "unnatural" requirement deepens the mystery of the "strong CP" problem and appears to imply a subtle linkage between QCD and QFD (see §8.1) that has not yet been understood.

We now undertake to comment on two "brute force" suggestions to solve the "strong CP" problem in QCD that are not ruled out as of this writing, before discussing at greater length the solution that has gained the widest currency, i.e. Peccei-Quinn symmetry. The first proposed "brute force" solution to the "strong CP problem" questions the accuracy of the current algebra calculations of the u, d and s quark masses (see §3.5c) and postulates the existence of at least one massless quark, e.g. the u quark. This proposal does not question the possibility of $\theta_{QCD} \neq 0$ and takes advantage of the fact that a chiral transformation — in the presence of just one massless quark — can rotate away both θ_{QCD} and θ_{QFD}. The basis for this statement with regard to θ_{QCD} is already implicit in Eq. (10.114) where the arbitrariness of α in the chiral rotation from the θ vacuum to the θ' vacuum implies that the theory is independent of the vacuum angle θ_{QCD} and hence one can set $\theta_{QCD} = 0$. It is also obvious that the existence of one massless quark in the quark mass matrix M [see Eq. (10.121)] implies the vanishing of the

θ_{QFD} = Arg det M contribution to the θ parameter. This would dispose of the "strong CP plus electroweak CP" problem in one fell swoop. However, it is difficult to believe that the current algebra calculations of the quark masses (see §3.5c) in terms of the observed meson masses are sufficiently inaccurate to permit a zero mass for the (lightest) u quark [1.137].

The second proposed "brute force" solution to the "strong CP problem" in QCD is of mathematical origin; the argument is that the topologically non-trivial structure of the QCD vacuum (i.e. the θ vacuum) is mathematically deficient and that a unique physical QCD vacuum is only possible when the θ parameter takes on the value 0 (or π) [10.26]. Other proposals for the choice $\theta_{QCD} = 0$ have recently been made justifying the choice $\theta_{QCD} = 0$ [10.27] but they are more complicated and less "brute force" and so we merely sketch the argument in ref. [10.26]. The argument in [10.26] stems from our earlier remark that the θ vacuum is not normalizable and, consequently, that there must be an infinite number of mathematically possible Hilbert spaces corresponding to different values of the θ_{QCD} parameter; however, there can be only one physically realizable "true" QCD vacuum. The real problem, so this argument goes, is to establish the value of θ_{QCD} that corresponds to the unique physical QCD vacuum. Ref. (10.26) suggests that the answer can be found in the local covariant and canonical operator formulation of non-Abelian gauge theory [10.35], which is modeled on the similar formulation of Abelian QED. In the old Bleuler-Gupta theory [2.4], it was shown that the use of the canonical formalism in covariant gauge in Abelian QED gives rise to unphysical "negative norm" states (the indefinite metric) which are eliminated from the physical Hilbert space by the subsidiary condition:

$$(\partial^\mu A_\mu)^\dagger(x)|\Psi_{\text{phys}}> = 0 \qquad (10.123)$$

where Ψ_{phys} is a state in the physical Hilbert space and A_μ is the electro-magnetic gauge field. As expected, the application of the covariant gauge and canonical formalism method to non-Abelian QCD gives rise to "negative norm" states that must be eliminated from the physical Hilbert space by a "non-Abelian" subsidiary condition. What the Kugo-Ojima paper [10.35] demonstrates is that a covariant formulation of the Yang-Mills theory is possible if one applies the global BRST transformation [10.36] to the total Yang-Mills Lagrangian (including the "gauge-fixing" and Faddeev-Popov terms — see §4.3), which yields all the Ward-Takahashi identities and effectively serves as the local gauge invariance for the Yang-Mills theory. Furthermore, it is proved that the "non-Abelian" counterpart of the subsidiary condition

(10.123) (for non-Abelian gauge theory) is the subsidiary condition:

$$Q_{BRST}|\Psi_{\text{phys}}>= 0 \qquad (10.124)$$

where Q_{BRST} is the so-called BRST charge [10.36]. The subsidiary condition $Q_{BRST}|\Psi_{\text{phys}}>= 0$ decouples all "negative norm" (non-physical) states from the physical Hilbert space and, when one takes account of the fact that Q_{BRST} commutes with the anti-unitary U_{CPT} operator (see §2.2d), one finds that $U_{CPT}|\Psi_{\text{phys}}>$ must also be a physical state; the uniqueness of the physical vacuum — proved in [10.35] — then requires:

$$U_{PCT}|\Psi_{\text{vac}}>= |\Psi_{\text{vac}}> \qquad (10.125)$$

apart from a trivial phase factor. But Eq. (10.125) can only hold when $\theta_{QCD} = 0$ if $|\Psi_{\text{vac}}>= |\theta>$. With the θ_{QCD} parameter vanishing, there would no longer be a "strong CP problem" in QCD.

The argument for $\theta_{QCD} = 0$ just given depends upon the rigor of the Kugo-Ojima formulation of QCD which appears to be the only consistent canonical formulation of QCD that is compatible with Lorentz invariance and the positivity of the physical Hilbert space. There are some "axiomatic" questions that still have to be settled in the Kugo-Ojima formulation (e.g. the correctness of the "asymptotic completeness" ansatz, the "cyclicity" assumption, etc.). But, even assuming that the choice of $\theta_{QCD} = 0$ is justified by a mathematically rigorous treatment of the non-perturbative QCD vacuum, one would still have to explain why $\theta_{QFD} = \text{Arg det } M \lesssim 10^{-9}$. The situation here looks promising: a three-loop calculation of the weak CP violation radiative effects on the basis of the standard model yields a value of θ_{QFD} many orders of magnitude less than 10^{-9} [10.37], in agreement with an earlier calculation that went as far as the seventh order in α_W where renormalization is required but the finite residual effect is small [10.38]. It seems that if the mathematical "brute force" approach can establish that $\theta_{QCD} = 0$, there will be no problem with θ_{QFD}.

With major uncertainties surrounding both "brute force" solutions to the "strong CP problem" in QCD, it is proper to search for other solutions that are more amenable to phenomenological tests. If one rejects the possibility of a massless quark and accepts the θ vacuum (with $\theta_{QCD} \neq 0$) as a legitimate signature of the topologically non-trivial property of Yang-Mills theories, the proposed Peccei-Quinn (PQ) symmetry $U(1)_{PQ}$ [1.138] — and its concomitant (pseudo-N-G) axion particle resulting from the SSB of chiral $U(1)_{PQ}$ [10.39] — is the solution of the "strong CP problem" most preferred at the present time. The basic PQ strategy is to seek a dynamical adjust-ment of $\theta = \theta_{QCD} + \theta_{QFD}$ — with its disparate contributions from QCD and QFD — to zero by imposing a new global chiral symmetry [the $U(1)_{PQ}$ sym-metry] on all Yukawa couplings and the Higgs potential and then to utilize the arbitrary phase associated with the $U(1)_{PQ}$ chiral anomaly to determine θ_{QFD} and then "rotate away" θ_{QCD} so that $\theta = 0$.

But there is a price to be paid for the PQ mechanism: the global chiral $U(1)_{PQ}$ symmetry is spontaneously broken by the same Higgs mechanism that generates the quark masses, thereby giving rise to an N-G boson that acquires a "quasi-N-G" mass by virtue of the $U(1)_A$-type triangular axial anomaly associated with the divergence of the $U(1)_{PQ}$ current. This quasi-N-G boson is the axion, which particle must be observed to validate the PQ solution to the "strong CP" problem. Unfortunately, the existence of a "visible" axion — the type of axion that must exist if the $U(1)_{PQ}$ symmetry-breaking scale Λ_{PQ} is of the order of Λ_{QFD} — as seems most natural — has been disallowed by numerous experiments carried out during the past decade [1.137]. However, the appeal of the PQ approach to the "strong CP" problem is so great that modified versions of the original PQ model — leading to an "invisible" axion (see below) — have been investigated, both to solve the "strong CP" problem and to shed light on other unsolved astrophysical and cosmological problems in which invisible axions are postulated to play a role. Because of the active interest in "axionology", it is instructive to present the salient features of the original visible axion model as a concrete example of the general PQ strategy and as a reference point for the non-standard "invisible" axion models still under consideration.

The original version of the Peccei-Quinn theory [1.138] introduced the $U(1)_{PQ}$ symmetry by postulating a second Higgs scalar doublet — in ad-dition to the usual Higgs doublet that breaks the electroweak symmetry and generates the quark masses. It is then natural to associate the scale of $U(1)_{PQ}$-breaking with Λ_{QFD} (the SSB scale of the standard electroweak group) and, in the process, to create a "visible" axion [10.39] to compensate for the elimination of the θ parameter. In the original PQ model, the Yukawa couplings are written in terms of two complex Higgs doublets, ϕ_1 and ϕ_2, as follows [see Eq. (6.65)]:

$$\mathcal{L}_Y = \sum_{i,j} h_{ij}^U (\bar{U}_i \bar{D}_i)_L \begin{pmatrix} \phi_2^0 \\ \phi_2^- \end{pmatrix} U_{jR} + \sum_{i,j} h_j^D (\bar{U}_i \bar{D}_i)_L \begin{pmatrix} \phi_1^+ \\ \phi_1^0 \end{pmatrix} D_{jR} \qquad (10.126)$$

In Eq. (10.127), U_i and D_i represent the up and down quarks respectively of the ith generation and the rest of the notation is standard [see Eq. (6.65)]. In the standard model, $\phi_2 = i\tau_2\phi_1^*$, but, in the PQ model, ϕ_2 and $i\tau_2\phi_1^*$ are allowed to differ. The next step is to consider chiral rotations on the quark fields and examine the conditions for the invariance of $\mathcal{L}_Y$ under such rotations; let the chiral rotations be:

$$\begin{bmatrix} U_i \\ D_i \end{bmatrix}_L \rightarrow e^{i\zeta_L} \begin{bmatrix} U_i \\ D_i \end{bmatrix}_L; \qquad U_{iR} \rightarrow e^{i\zeta_R^U} U_{iR}; \qquad D_{iR} \rightarrow e^{i\zeta_R^D} D_{iR} \quad (10.127)$$

and write the corresponding rotations of the ϕ_1 and ϕ_2 scalar fields as:

$$\phi_1 \rightarrow e^{i\zeta_1}\phi_1; \qquad \phi_2 \rightarrow e^{i\zeta_2}\phi_2 \qquad (10.128)$$

Then, assuming that the Higgs potential $V(\phi_1, \phi_2)$ is invariant under (10.127) and (10.128), the invariance of $\mathcal{L}_Y$ requires:

$$-\zeta_L + \zeta_1 + \zeta_R^D = 0; \qquad -\zeta_L + \zeta_2 + \zeta_R^U = 0 \qquad (10.129)$$

In the standard model, the condition $\phi_2 = i\sigma_2\phi_1^*$ implies $\zeta_2 = -\zeta_1$ and hence the relation between the phases:

$$\zeta_R^U + \zeta_R^D - 2\zeta_L = 0 \qquad (10.130)$$

Equation (10.130) brings us to the crux of the PQ argument: Eq. (10.130) is precisely the condition that yields a vanishing common phase δ [since $\delta = \zeta_R^U + \zeta_R^D - 2\zeta_L = 0$ for the common phase of the real diagonal quark matrix]; consequently, $\theta_{QFD} = 0$ [from Eq. (10.120)] and it is not possible to "rotate away" θ_{QCD} to eliminate the "strong CP problem". On the other hand, if there are two Higgs doublets satisfying the $U(1)_{PQ}$ symmetry, there is freedom to choose the arbitrary phase δ in the chiral rotation on the quark fields and to "rotate away" θ_{QCD}, thereby eliminating "strong CP" violation. Thus, the Peccei-Quinn theory, with two Higgs doublets and invariance of the standard model Lagrangian under the global chiral $U(1)_{PQ}$ symmetry, can, in principle, solve the "strong CP problem" in QCD.

But now one must deal with the consequences of the Peccei-Quinn procedure. Since the VEV's of ϕ_1 and ϕ_2 must be non-zero to generate the quark masses, the $U(1)_{PQ}$ symmetry is spontaneously broken and gives rise to a N-G boson (the axion); the axion, in turn, acquires mass due to the triangular axial anomaly associated with the global chiral $U(1)_{PQ}$ current. It is straightforward to derive an approximate expression for the axion mass within the framework of the two-Higgs-doublet PQ model. To find the explicit form of the axion field, one uses the usual reparametrizations of the Higgs scalar fields (see §4.4):

$$\phi_i(x) = (v_i + \rho_i)e^{i\theta_i(x)/v_i} \qquad (i = 1, 2) \qquad (10.131)$$

with $v = \sqrt{v_1^2 + v_2^2}$. One combination of the fields $\phi_1(x)$ and $\phi_2(x)$, namely the linear combination $\chi = \dfrac{1}{v}(v_1\phi_2 - v_2\phi_2)$, is "eaten" to give mass to the neutral weak boson Z [through the coupling term $Z^\mu \partial_\mu \chi$], while the orthogonal combination constitutes the axion field, i.e.:

$$a = \frac{1}{v}(v_1\phi_2 + v_2\phi_1) \tag{10.132}$$

Next, one writes down the $U(1)_{PQ}$ current associated with the axion field, the u and d quark fields (the more massive quark fields are neglected) and the charged lepton fields, to wit [1.137]:

$$J_{PQ}^\mu = -v\partial^\mu a + (v_2/v_1)\sum_i \bar{u}_{R_i}\gamma^\mu u_{R_i} + (v_1/v_2)\sum_i \bar{d}_{R_i}\gamma^\mu d_{R_i}$$
$$+ \left(\frac{v_1}{v_2}\right)\sum_i \bar{\ell}_{R_i}\gamma^\mu \ell_{R_i} \tag{10.133}$$

where $\sum\limits_i$ is the summation over the number N_g of fermion generations. The divergence of the chiral PQ current possesses a triangular axial anomaly, namely:

$$\partial_\mu J_{PQ}^\mu = N_g(v^2/v_1 v_2)\,\frac{g^2}{16\pi^2}\,\mathrm{Tr}\, G_{\mu\nu}\tilde{G}^{\mu\nu} \tag{10.134}$$

to which the instantons contribute. One then derives an approximate value of the axion mass by neglecting the charged lepton contribution and using standard current algebra techniques for the u and d quarks; one obtains [1.137]:

$$m_a = 2N_g m_\pi f_\pi \left(\frac{\sqrt{m_u m_d}}{(m_u + m_d)}\right) \cdot \left(\frac{1}{v\sin 2\alpha}\right) \tag{10.135}$$

where $\sin\alpha = v_1/v$ and $v = \Lambda_{PQ}$. It should be noted from Eq. (10.135) that $m_a \to 0$ as m_u and/or $m_d \to 0$ — a reassuring result. Using the mass values m_u and m_d, one gets:

$$m_a \simeq \frac{50\, N_g}{\sin 2\alpha}\,\text{KeV} \tag{10.136}$$

when $\Lambda_{PQ} \sim \Lambda_{QFD} \simeq 250$ GeV.

Equation (10.136) tells us that unless α is very small, the axion mass must lie below $2m_e$ and the dominant decay mode is: $a \to 2\gamma$; one can write $\tau(a \to 2\gamma)$ in terms of $\tau(\pi^0 \to 2\gamma)$ [1.137]:

$$\tau(a \to 2\gamma) = K_{a\gamma\gamma}^{-2}\left(\frac{m_\pi}{m_a}\right)^5 \tau(\pi^0 \to 2\gamma) \simeq 0.38 K_{a\gamma\gamma}^{-2}\left(\frac{100\,\text{KeV}}{m_a}\right)^5 \text{sec} \tag{10.137}$$

where $K_{a\gamma\gamma} = \sqrt{\frac{m_u}{m_d}} \simeq 0.75$. As an example, if $m_a = 100$ KeV, the two-photon decay lifetime is of the order of seconds. On the other hand, if $\sin 2\alpha$ is very small, it is possible that $m_a > 2m_e$ and then there is rapid decay of the axion into an electron pair (because there is direct axion coupling to lepton pairs) with the lifetime [1.137]:

$$\tau(a \to e^+ e^-) \simeq 4 \times 10^{-9} (\frac{1 \text{ MeV}}{m_a})[1 - \frac{4m_e^2}{m_a^2}]^{-\frac{1}{2}} \frac{v_2^2}{v_1^2} \text{ sec} \qquad (10.138)$$

Hence, the electron pair decay lifetime of the axion can be much shorter than the two-photon lifetime. Unfortunately, whether the "original" PQ axion is made visible via two photons or an electron pair, the "visible" axion [whether associated with the original PQ model or some "variant" model which retains the SSB of the $U(1)_{PQ}$ symmetry at the Λ_{QFD} scale] has been ruled out conclusively by many experiments [1.137] — such as searches for axions in $K^+ \to \pi^+ + a$ decay, heavy quarkonium decay (e.g. $\Upsilon \to a\gamma$ and $J/\psi \to a\gamma$), charged pion decay (e.g. $\pi^+ \to ae^+\nu_e$), and bremsstrahlung of axions in the process: $e^- + A \to e^- + a + A$. For example, from the last experiment, the mass region for the axion, $m_a > 10^4$ eV, is excluded.

Because of the attractiveness of the PQ approach to the solution of the "strong CP" problem in QCD, the failure to find "visible" axions [with easily measurable masses and reasonable coupling strengths (i.e. corresponding to $\Lambda_{PQ} \sim \Lambda_{QFD}$) has stimulated a succession of so-called "invisible" axion models (wherein $\Lambda_{PQ} >> \Lambda_{QFD}$) that might not only solve the "strong CP" problem but also shed light on a variety of cosmological and astrophysical problems. Since the axion mass and coupling go inversely as Λ_{PQ}, models with $\Lambda_{PQ} >> \Lambda_{QFD}$ are necessarily of the "invisible" axion type. Moreover, if $\Lambda_{PQ} >> \Lambda_{QFD}$, there is no need to express the $U(1)_{PQ}$ symmetry by means of an additional Higgs doublet and, indeed, most "invisible" axion models make use of a complex scalar field Φ which is an $SU(2)_L \times U(1)_Y$ singlet (rather than a doublet), carries PQ charge, and possesses a large value of $< \Phi >$:

$$< \Phi >= \frac{\Lambda_{PQ}}{\sqrt{2}} >> \Lambda_{QFD} \qquad (10.139)$$

We do not write out the detailed predictions for the masses and decay widths of the "invisible" axion models, except to note that since the axion mass varies inversely as Λ_{PQ} [and hence the mass of the axion is very light, possibly a fraction of eV] axions could play a significant astrophysical and/or cosmological role; thus, if $\Lambda_{PQ} \geq 10^8$ GeV, axions contribute importantly to the cooling loss of stars and distort normal stellar evolution. This leads to upper limits on the axion mass; for example, from the evolution of the sun, one obtains the limit $m_a < 1$ eV whereas the evolution of red giants limits the mass of the axion even further [1.137]. The strictest astrophysical upper limit obtained recently on the axion mass comes from the data on SN (supernova) 1987a which imposes the upper limit $m_a < 10^{-3}$ eV [10.40]. Such a "super-light", and therefore practically "invisible" axion, is an excellent candidate for the dark matter of the universe. If light axions are indeed the primary component of dark matter, they should have condensed into galaxies and, hence, it has been proposed that galactic axions might be detected through their conversion into microwave photons in a magnetic field [10.41]; unfortunately, available magnetic fields and other experimental limitations make it difficult to reach the level of the theoretically predicted coupling.

In sum, it appears that there are two remaining "windows" for the cosmic ("invisible") axions, namely: $m_a = 10^{-6}$ eV to 10^{-3} eV [$m_a = 10^{-6}$ eV is the lowest bound on the axion mass because otherwise the stored energy of the axions in the universe would exceed its critical value [10.36]], and $m_a \sim 1$ eV. It does seem that the odds of finding the "invisible" axion are rapidly diminishing and that the incentive to carry on the ingenious searches for "invisible" axions is fueled more by astrophysical and cosmological interest than by any hope to salvage the Peccei-Quinn-type solution of the "strong CP" problem in QCD. Consequently, despite the attractiveness of the PQ mechanism, the resolution of the "strong CP" problem may require more than the introduction of a new global symmetry, perhaps a deeper mathematical understanding of the topologically non-trivial QCD θ vacuum itself.

§10.4. Homotopy groups and the index theorem

Thus far, we can only claim modest phenomenological support for the usefulness of topological conservation laws in particle physics: while the topologically non-trivial instanton solutions of QCD have led to a solution of the "$U(1)$ problem", the same color instantons and their associated θ vacuum have created the "strong CP" problem in QCD. Moreover, neither the Dirac nor the 't Hooft-Polyakov magnetic monopole has been found to confirm the existence of topologically non-trivial structures in $D = 4$ spacetime dimensional particle physics; nor has any evidence been forthcoming as yet for the spectacular sphaleon-induced effects in QFD. Despite these disappointments, the study of the topological aspects of field theories — both classical and quantum — has given us greater insight into the conceptual foundations of the standard model and its likely future merging with a theory of quantum gravity. As a minimum, the impressive phenomenological successes of topological conservation laws registered in condensed matter physics in $D = 2$ and $D = 3$ space-time dimensions (where "topological engineering" is possible) has been most reassuring. In any case, throughout the first three sections of this chapter, we focused on the phenomenological status of various types of topological conservation laws and only introduced mathematical formalism as deemed necessary to define the arguments and describe the results. Along the way, we made reference to homotopic groups, "zero modes" of the Dirac operator in a background gauge field [10.42], the Pontryagin number, and other topological results.

In the next few sections, we provide more mathematical details concerning the basic topological concepts and reexpress several of the earlier topologically-related results in terms of the more precise mathematical language. As we proceeded earlier in this chapter, we identified the homotopic groups responsible for the topological conservation laws governing the physical examples that we considered; however, we try to give a more systematic account of the mathematics in §10.4a–10.4c, culminating in the "Bott Table" (Table 10.3). Specifically, in §10.4a, we use the global $SU(2)$ chiral gauge anomaly (see §7.3) as the springboard for introducing the concepts of spectral flow of eigenvalues and "zero modes" of the Dirac operator in a background gauge field and, in the process, derive the Atiyah-Singer "index theorem". This enables us, in §10.4b, to state more precisely the important relation between the perturbative triangular and non-perturbative global $SU(2)$ anomalies for chiral gauge groups in four space-time dimensions. Finally, in §10.4c, we list in what we call the "Bott Table" (Table 10.3) the homotopy groups of the classical Lie groups, and then indicate (in Table 10.4) the types of physical applications subject to these homotopy groups.

§10.4a. Global $SU(2)$ chiral gauge anomaly and the index theorem

In Chapter 7, we described the origin of the three chiral gauge anomalies in four dimensions — triangular, global $SU(2)$ and mixed gauge-gravitational anomalies — and examined the constraints placed on a chiral gauge theory by the conditions of anomaly cancellation in order to maintain the renormalizability and mathematical self-consistency of the theory. Two of the anomalies — the perturbative triangular and perturbative mixed anomalies — involve local (infinitesimal) gauge transformations and could be treated in a fairly straightforward fashion (in terms of Feynman diagrams); however, the non-perturbative global $SU(2)$ anomaly involves a "large" gauge transformation and its treatment was too qualitative and can be placed on a firmer mathematical footing with the help of the Atiyah-Singer "index theorem" [7.9]. It will be recalled that the global $SU(2)$ anomaly had its origin in the sign change of the fermion determinant for a single Weyl doublet, caused by a topologically non-trivial "large", global gauge transformation. We now justify the use of Eq. (7.66) by considering the "zero modes" of the Dirac operator in a background (instanton) gauge field and invoking the "index theorem". It will be seen that a single Weyl doublet in the background $SU(2)$ gauge field generates a single "zero mode" and that this odd number of "zero modes" is responsible — within the context of the "index theorem" — for the global $SU(2)$ anomaly. The connection of this result with the earlier conclusion that the homotopic group $\Pi_4[SU(2)] = \mathcal{Z}_2$ governs the global $SU(2)$ anomaly should also receive further clarification.

To define the problem [7.4], we recall that Eq. (7.66) expresses the fact the path integral is taken over over a single doublet of Weyl fermions and changes sign under the "large" gauge transformation U, i.e. $A_\mu^U = U^{-1}A_\mu U + U^{-1}\partial_\mu U$. This has the consequence that the partition function Z [in the presence of the $SU(2)$ gauge field] can be written as:

$$\mathcal{Z} = \int \mathcal{D}A_\mu \, (\det i\slashed{D})^{\frac{1}{2}} \, \exp(-\frac{1}{2g^2} \int_{\text{(Eucl.)}} d^4x \, \text{Tr} \, F_{\mu\nu}^2) \qquad (10.140)$$

From Eq. (10.140), it follows that Z vanishes identically because of the equal and opposite contributions of A_μ^U and A_μ when the integral is taken over gauge field space. For the same reason, the path integral vanishes identically when $\mathcal{O}$ is a gauge-invariant operator; hence the expectation value of $\mathcal{O}$ is indeterminate. Thus, the theory is ill-defined and the global $SU(2)$ chiral gauge anomaly must be eliminated. To justify this conclusion, it is necessary to establish the validity of Eq. (7.47), either by proving that the spectral flow of eigenvalues of the Dirac operator in the background $SU(2)$ gauge field involves an odd number of Dirac eigenvalue "crossovers" [10.42] or, returning to Eq. (7.44), by proving that the non-gauge invariance of the measure of the fermion path integral can be expressed by means of the "index theorem" [10.43]. The two methods, of course, are equivalent but, because each method illuminates a somewhat different topological aspect of the problem, we comment on both. Let us first exhibit the connection between the single Dirac eigenvalue "crossover" and the change of sign of $[\det i\slashed{D}(A_\mu)]^{-\frac{1}{2}}$ and, then, work out the connection between the "index theorem" and the number of "zero modes" associated with a general simple group $\supset SU(2)$. These results provide the rigorous underpinning for the existence of the global $SU(2)$ anomaly.

To show the relevance of spectral flow to the global $SU(2)$ anomaly, consider the expression $\det(i\rlap{/}{D})$, which is the value of the path integral for the Dirac fermion doublet [see Eq. (7.63)]; since $i\rlap{/}{D}(A_\mu)$ is Hermitian, it can be written as the product $\prod_n \lambda_n$ of all eigenvalues λ_n (the λ_n's are all real) of the Dirac operator in the background $SU(2)$ gauge field A_μ. Furthermore, for every eigenvalue λ_n, satisfying $i\rlap{/}{D}(A_\mu)\psi = \lambda_n\psi_n$, there is an eigenvalue $-\lambda_n$ satisfying $i\rlap{/}{D}(A_\mu)(\gamma_5\psi_n) = -\lambda_n(\gamma_5\psi_n)$. However, for a Weyl fermion doublet, we are interested in the square root of $i\rlap{/}{D}(A_\mu)$ and hence we must consider only half of the eigenvalues, which, for a given gauge field A_μ, can be chosen to be the product of only the positive eigenvalues. It is now possible to predict the relation of $[\det i\rlap{/}{D}(A_\mu^U)]^{\frac{1}{2}}$ to $[\det (i\rlap{/}{D}(A_\mu)]^{\frac{1}{2}}$ without knowing the λ_n's explicitly; for this purpose, one writes $A_\mu^t = (1-t)A_\mu+tA_\mu^U$, with t a real number, varying continuously from 0 to 1 as A_μ^t moves along a continuous path in field space from $A_\mu \rightarrow A_\mu^U$. Although the eigenvalue spectrum of $i\rlap{/}{D}$ must be exactly the same — because of gauge invariance — at $t = 1$ (corresponding to A_μ^U) as it is at $t = 0$ (corresponding to A_μ), the individual λ_n's may rearrange themselves as t goes from 0 to 1 (see Fig. 10.7) and the sign may change, reflecting the ambiguity in sign of $(\det i\rlap{/}{D})^{\frac{1}{2}}$. Figure 10.7 exhibits one change of sign because there is one crossover (at $t = t_0$) in the spectral flow of the eigenvalues λ_n from $t = 0$ to $t = 1$. It is possible to have more than one crossover but, for the simple case of the $SU(2)$ gauge group and a single Weyl fermion doublet (fundamental) representation, it can be shown [10.42] that there is just one crossover so that Eq. (7.47) follows and there is a global $SU(2)$ anomaly.

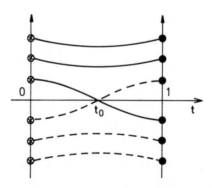

Fig. 10.7. The flow of eigenvalues as the $SU(2)$ gauge field is varied from A_μ (left of drawing) to A_μ^U (right of drawing) in the presence of a single Weyl doublet.

We now turn to the "index theorem" method of establishing the global $SU(2)$ anomaly and its analog in larger simple groups. The method is best described by starting with the explicit expression for the Weyl fermion path integral, that holds for an arbitrary group and an arbitrary Weyl fermion representation; we write for the generating functional:

$$\exp[-Z(A)] = \int \mathcal{D}\psi \mathcal{D}\bar{\psi} \, \exp-\{\int_{(\text{Eucl.})} d^4x \, \bar{\psi} \, i\!\!\not{D}\psi\} \tag{10.141}$$

In Eq. (10.141), $\not{D} = \gamma^\mu D_\mu = \gamma^\mu(\partial_\mu + igA_\mu)$, $A_\mu = A_\mu^a T^a$, $\mathcal{D}\psi\mathcal{D}\bar{\psi}$ is the measure of the Weyl fermion path integral and the integral in the action term is carried out over Euclidean four-space. The procedure is to inquire into the behavior of the Weyl fermion measure under (infinitesimal) global chiral transformations:

$$\psi \to (1 + i\alpha\gamma_5)\psi; \quad \bar{\psi} = \bar{\psi}(1 + i\alpha\gamma_5) \tag{10.142}$$

It is easy to show that the action integral transforms into:

$$\int_{(\text{Eucl.})} d^4x \, \bar{\psi}i\!\!\not{D}\psi \to \int d^4x \, \bar{\psi}i\!\!\not{D}\psi - \alpha \int d^4x \, \partial_\mu J_5^\mu(x) \tag{10.143}$$

where:

$$J_5^\mu = \bar{\psi}\gamma^\mu\gamma_5\psi \tag{10.143a}$$

It would appear from Eq. (10.143) that the invariance of $Z[A]$ under the chiral transformation gives naive conservation of the axial vector fermion current but, in fact, one must also take into account the change in the measure $\mathcal{D}\psi\mathcal{D}\bar{\psi}$ [10.43], which requires the addition of precisely the triangular chiral gauge anomaly term to the divergence of the axial vector fermion current. The anomaly term leads immediately to the Atiyah-Singer "index theorem" which, for the special case of the $SU(2)$ group with a single doublet of Weyl fermions, implies Eq. (7.47).

To calculate the change in measure, one defines the integration measure $\mathcal{D}\psi\mathcal{D}\bar{\psi}$ in terms of a complete set of eigenfunctions of the Hermitian operator $i\!\!\not{D}$:

$$i\!\!\not{D}\phi_n = \lambda_n\phi_n \tag{10.144}$$

with:

$$\int_{(\text{Eucl.})} d^4x\phi_n^\dagger(x)\phi_m(x) = \delta_{nm} \tag{10.144a}$$

One then expands ψ and $\bar{\psi}$ in terms of the eigenfunctions ϕ_n and $\phi_n^\dagger$:

$$\psi(x) = \sum_n a_n\phi_n(x); \quad \bar{\psi}(x) = \sum_n \phi_n^\dagger(x)\bar{b}_n \tag{10.145}$$

where a_n, $\bar{b}_n$ are independent elements of the "Grassmann (anti-commuting) algebra" [10.24]. With the help of Eq. (10.145), the integration measure becomes:

$$\mathcal{D}\psi\mathcal{D}\bar{\psi} = \prod_n da_n d\bar{b}_n \qquad (10.146)$$

The next crucial step is to prove that the integration measure is not invariant under the chiral transformations and that, indeed, the discrepancy is precisely the triangular chiral gauge anomaly; to see this, apply a global chiral transformation to ψ:

$$\psi' = \sum_n a'_n \phi_n = \sum_n a_n e^{i\alpha\gamma_5} \phi_n \qquad (10.147)$$

where:

$$a'_m = \sum_n \int_{(Eucl.)} d^4x \phi_m^\dagger e^{i\alpha\gamma_5} \phi_n a_n = \sum_n C_{mn} a_n \qquad (10.147a)$$

It follows that:

$$\Pi_m da'_m = (\det\ C)^{-1}\ \Pi_n a_n; \qquad C = e^{i\alpha\gamma_5} \qquad (10.148)$$

where:

$$(\det\ C)^{-1} = \exp\{-i\alpha \int d^4x \sum_n \phi_n^\dagger(x)\gamma_5\phi_n(x)\} \qquad (10.148a)$$

The Jacobian for $\mathcal{D}\bar{\psi}$ is the same as for $\mathcal{D}\psi$ and so we get:

$$\mathcal{D}\psi\mathcal{D}\bar{\psi} \rightarrow (\det\ C)^{-2}\mathcal{D}\psi\mathcal{D}\bar{\psi} \qquad (10.149)$$

Hence, the change in the Weyl fermion measure is given by:

$$\exp[-2i\alpha\tilde{Q}_5]; \qquad \tilde{Q}_5 = \int_{(Eucl.)} d^4x \sum_n \phi_n^\dagger \gamma_5 \phi_n \qquad (10.150)$$

where $\tilde{Q}_5$ turns out to be the familiar "Chern-Simons" charge when it is evaluated with care (see below).

Since the integrand in Eq. (10.150) is ill-defined, it must be regularized by means of a cutoff procedure involving a scale that is allowed to go to infinity after the integration; one gets [the Trace $\tilde{\mathrm{Tr}}$ is over group and Dirac indices — see [10.24]]:

$$\tilde{Q}_5 = \int d^4x \sum_n \phi_n^\dagger \gamma_5 \phi_n = \lim_{\Lambda\to\infty} \int d^4x \sum_n \phi_n^\dagger \gamma_5 \phi_n e^{-\lambda_n^2/\Lambda^2}$$

$$= \lim_{\Lambda\to\infty} \int d^4x \sum_n \tilde{\mathrm{Tr}}[\gamma_5 e^{-(i\not{D})^2/\Lambda^2}]\phi_n^\dagger \phi_n \qquad (10.151)$$

$$= \lim_{\Lambda\to\infty} \int d^4x \lim_{y\to x} \{\gamma_5 e^{-(i\not{D}_x)^2/\Lambda^2}\}_{kl}^{ab} \sum_n [\phi_n(x)]_\ell^b \ [\phi_n^*(y)]_k^a$$

Evaluating the last term in brackets on the R.H.S. of Eq. (10.151) in a plane wave basis yields:

$$\sum_n [\phi_n(x)]^b_\ell\, [\phi^*_n(y)]^a_k = \delta^{ab}\delta^{kl} \sum_n <x|n><n|y> = \delta^{ab}\delta^{kl} <x|y>$$

$$= \delta^{ab}\delta^{kl} \int \frac{d^4k}{(2\pi)^4} <x|k><k|y> = \delta^{ab}\delta^{kl} \int \frac{d^4k}{(2\pi)^4} e^{-ikx}e^{iky} \tag{10.152}$$

Inserting Eq. (10.152) back into Eq. (10.151) gives:

$$\tilde{Q}_5 = \lim_{\Lambda\to\infty} \int d^4x \int \frac{d^4k}{(2\pi)^4} \tilde{\mathrm{Tr}}\, [\gamma_5 e^{ikx}e^{-(i\!\not{D})^2/\Lambda^2}e^{-ikx}]$$

$$= \lim_{\Lambda\to\infty} \int d^4x\, \tilde{\mathrm{Tr}}\, \{\gamma_5([\gamma^\mu,\gamma^\nu]F_{\mu\nu})^2\}(\frac{1}{4\Lambda^2})^2\frac{1}{2!} \int \frac{d^4k}{(2\pi)^4} e^{k^\mu k_\mu/\Lambda^2}$$

$$= \frac{1}{16\pi^2} \int d^4x\, \mathrm{Tr}\,(F_{\mu\nu}\tilde{F}^{\mu\nu}) \tag{10.153}$$

The R.H.S. of Eq. (10.153) is recognized as the integral over the (triangular) "anomaly density" which becomes the "Chern-Simons" charge (or Pontryagin number) and is denoted by ν. If one now compares Eqs. (10.143) and (10.153), one finds that the Weyl fermion measure is multiplied by $e^{-2i\alpha\tilde{Q}_5}$ under the chiral transformations on ψ and $\bar{\psi}$ of Eq. (10.142). Hence, the invariance of the Weyl fermion path integral under the chiral transformations on ψ and $\bar{\psi}$ is guaranteed if the following equation holds:

$$\partial_\mu J^{\mu 5} = \frac{1}{16\pi^2} \mathrm{Tr}\,(F_{\mu\nu}\tilde{F}^{\mu\nu}) \tag{10.154}$$

Equation (10.154) constitutes the proof that the anomalous term is needed to preserve the gauge invariance of the Weyl path integral under chiral transformations. Equation (10.154) also leads directly to the "index theorem" — which relates ν to the difference between the numbers of positive and negative chirality "zero modes" of the Euclidean Dirac operator $i\!\not{D}(A_\mu)$.

To arrive at the "index theorem", we return to the quantity $\tilde{Q}_5$ of Eq. (10.150) and separate the sum (over n) on the R.H.S. into the terms corresponding to $\lambda_n \neq 0$ and those corresponding to $\lambda_n = 0$; the $\lambda_n = 0$ eigenfunctions of $i\!\not{D}$ are denoted by ϕ_n^0. It is easy to see that, in the case of $\lambda_n \neq 0$, the eigenfunction ϕ_n is orthogonal to the eigenfunction $\gamma_5\phi_n$ since $\gamma_5\phi_n$ is an eigenfunction of $i\!\not{D}$ with eigenvalue $-\lambda_n$. Consequently, terms with $\lambda_n \neq 0$ in $\tilde{Q}_5$ give vanishing contributions and we need only consider the $\lambda_n = 0$ ("zero mode") contributions; the "zero mode" eigenfunctions satisfy the equation:

$$i\not{D}\phi^0_{n_\pm} = 0; \qquad \phi^0_{n_\pm} = \frac{(1\pm\gamma_5)}{2}\phi^0_n \tag{10.155}$$

and hence are also eigenfunctions of the chirality operator γ_5, namely:

$$\gamma_5\phi^0_{n_\pm} = \pm\phi^0_{n_\pm} \tag{10.156}$$

We can now express the R.H.S. of Eq. (10.150) in terms of the numbers of positive and negative chirality "zero modes"; we get:

$$\nu(A_\mu) = \sum_{n(\lambda_n=0)} \int_{(\text{Eucl.})} d^4x\; \phi^\dagger_n(x)\gamma_5\phi_n(x)$$

$$= \sum_n \int_{(\text{Eucl.})} d^4x\; \{|\phi^0_{n_+}|^2 - |\phi^0_{n_-}|^2\} \text{ or } \nu(A_\mu) = n_+ - n_- \tag{10.157}$$

where n_+ and n_- are the numbers of positive and negative chirality "zero modes" respectively. Equation (10.157) is the Atiyah-Singer "index theorem" [7.11] and contains the interesting information that, for chiral fermions in the presence of Yang-Mills fields, the Pontryagin number ν is equal to the difference between the number of positive and negative chirality "zero modes" of the Dirac operator in the background of these Yang-Mills fields. We also learn from the "index theorem" that, when the background gauge field consists of instantons the "winding number" of the instanton [which is given by $\tilde{Q}_5$] can be identified with the net number of fermion "zero modes".

With the "index theorem" in hand, we are in a position to calculate the net number of "zero modes" (ν) for a given gauge group and Weyl fermion representation and to decide whether or not a global chiral gauge $SU(2)$ anomaly is present. If we return to the example with which we started — the $SU(2)$ gauge group with a single Weyl fermion doublet — we find $\nu = 1$, so that the net number of fermion "zero modes" in the space of $SU(2)$ gauge potentials — as determined by the Pontryagin number associated with the fundamental **2** representation — is 1, an odd number; it follows that a global $SU(2)$ anomaly is present [this is why it is sometimes said that the instanton with "winding number" 1 is responsible for the global $SU(2)$ anomaly]. This result can easily be generalized to an arbitrary Weyl fermion representation (of dimension k) of the $SU(2)$ background gauge field; the net number of "zero modes" ν, corresponding to the kth Weyl fermion representation of the $SU(2)$ gauge field, is given by:

$$\nu = 1/6k(k^2 - 1) \qquad (k = 1, 2, 3, ...) \tag{10.158}$$

Clearly, the (fundamental) chiral doublet representation ($k = 2$) yields $\nu = 1$, the triplet (adjoint) real representation ($k = 3$) gives the even number $\nu = 4$ and so on. Using "zero mode" language, Eq. (7.47) can be generalized to any $SU(2)$ representation, as follows:

$$\det{}^{\nu/2} [i\!\!\not{D}(A_\mu)] = (-)^\nu \det{}^{\nu/2} [i\!\!\not{D}(A_\mu^U)] \tag{10.159}$$

Thus, any representation of $SU(2)$ that yields $\nu = odd$ [i.e. the representations with dimension $m = 2(1 + 2\ell)\,(\ell = 0, 1, 2, ...)$] gives rise to a global $SU(2)$ anomaly and, *a fortiori*, to a mathematically inconsistent chiral gauge theory; all other representations of $SU(2)$ possess even ν and hence are global $SU(2)$ anomaly-free. Equation (10.159) is very useful — once the (net) number of "zero modes" is known — in determining whether a gauge group [of which $SU(2)$ is a subgroup] with a specified Weyl fermion representation, gives rise to a global $SU(2)$ anomaly. It is also useful to prove the "even-odd rule" that relates the triangular and global $SU(2)$ chiral gauge anomalies, to be discussed in the next section.

§10.4b. Relation between triangular and global $SU(2)$ chiral gauge anomalies: even-odd rule

As mentioned in the previous section, the chiral fermion "zero mode" language can be conveniently used to state an important theorem (called the "even-odd rule") — which relates the (perturbative) triangular and (non-perturbative) global $SU(2)$ chiral gauge anomalies [10.44] — that was exploited (see §7.7) in dealing with the uniqueness of the Weyl fermion representations for GUT groups. The "even-odd rule" states that, if the $SU(2)$ gauge group is embedded in a simple gauge group G, i.e. $G \supset SU(2)$, for which $\Pi_4(G) = 0$ [Π_4 is the fourth homotopy group — see §10.4c], freedom from the triangular chiral gauge anomalies associated with the Weyl fermion representations of G is a sufficient condition to ensure the absence of the global $SU(2)$ chiral gauge anomaly. It is seen from the "Bott Table" (Table 10.3) in §10.4c that $\Pi_4(G) = 0$ for all simple gauge groups except the symplectic group $Sp(N)$, for which $\Pi_4[Sp(N)](N \geq 1) \neq 0$. The absence of $Sp(N)(N \geq 1)$ from the condition $\Pi_4(G) = 0$ should not be surprising since $Sp(1)$ is isomorphic to $SU(2)$ and we have already noted (in §10.4a) that $\Pi_4[SU(2)] = \mathcal{Z}_2$; however, the exclusion holds for all $Sp(N)$ $(N > 1)$ groups. On the other hand, Table 10.3 informs us that $\Pi_4[U(N)] = 0$ for $N \geq 3$ and $\Pi_4[O(N)] = 0$ for $N \geq 6$ and that all five exceptional compact Lie groups: G_2, F_4, E_6, E_7 and E_8, satisfy the condition $\Pi_4(G) = 0$. Hence, the two conditions: $\Pi_4(G) = 0$ and $G \supset SU(2)$ are satisfied for all compact

Lie groups — except $Sp(N)$ — subject to the lower limits on the dimensions of the $U(N)$ and $O(N)$ groups.

If the two conditions: $\Pi_4(G) = 0$ and $G \supset SU(2)$ are satisfied, it is sufficient to demonstrate that if the net number of fermion "zero modes" in the background (instanton) gauge field is even, or equivalently, the Weyl fermion representation (or combination of representations) of G is triangular anomaly-free, the global $SU(2)$ anomaly is absent [10.44]. On the other hand, if the net number of "zero modes" is odd, the global $SU(2)$ anomaly is present; for this reason, the theorem is called the "even-odd rule". The proofs of the "even-odd rule" — which can be carried through either by means of group-theoretic [10.45] or topological methods [10.46] — are rather complicated and, therefore, we merely sketch the key points of the "group-theoretic" proof.

Let R be a Weyl fermion representation of a simple compact group G [strictly speaking, R is a representation of the Lie algebra of G] — satisfying the condition $\Pi_4(G) = 0$ — and let $N(R)$ be the number of "zero modes" associated with R, then it can be shown that [10.46]:

$$N(R) = kQ_2(R)N_{\min} \tag{10.160}$$

where $Q_2(R)$ is the second-order Dynkin index [10.47] of R, normalized to $Q_2(\square) = 1$ ($\square$ is the fundamental representation of G), k is a constant independent of R [but dependent on how $SU(2)$ is embedded in G], and $N_{\min}$ is the number of "zero modes" for the fundamental (Weyl fermion) representation of G; $N_{\min}$ is listed in Table 10.2 for the classical and exceptional Lie groups. It can then be proved by rigorous group-theoretic methods [10.46] that both k and $Q_2(R)$ are integers and, consequently, $N(R)$ is even for all simple Lie groups except $SU(N)$ and $Sp(N)$ (see Table 10.2). Putting aside $Sp(N)$ — which is the only G for which $\Pi_4(G) \neq 0$ — it follows from Table 10.2 and the "even-odd rule" that the global $SU(2)$ anomaly must be absent for all groups except $SU(N)$; indeed, all of these groups [e.g. E_6] — for which $N(R)$ is even — are known to be triangular anomaly-free for all representations R and the "even-odd rule" clearly holds. The remaining group in Table 10.2 that satisfies $\Pi_4(G) = 0$ is $SU(N)(N \geq 3)$, which is the only simple group with possible triangular chiral gauge anomalies. It must now be shown that any choice of representations of $SU(N)$ that together satisfy the triangular anomaly-free condition, is also free of the global $SU(2)$ anomaly.

Table 10.2: Numbers of "zero modes" for fundamental Weyl fermion representations of simple compact Lie groups.

$$N_{\min} = \begin{cases} 1 \text{ for } SU(N) \text{ and } Sp(N) \\ 2 \text{ for } G_2 \text{ and } SO(N) \ (N \geq 7) \\ 6 \text{ for } F_4 \text{ and } E_6 \\ 12 \text{ for } E_7 \\ 60 \text{ for } E_8 \end{cases}$$

The group-theoretic proof that triangular anomaly-free representations of $SU(N)$ $(N \geq 3)$ imply an even number of "zero modes" [and, *a fortiori*, the absence of the global $SU(2)$ anomaly] can be carried out by mathematical induction and leads to an explicit form of the "even-odd rule", namely [10.46]:

$$Q_3(R) = Q_2(R) \,(\mathrm{mod}\, 2) \qquad (10.161)$$

where $Q_3(R)$ is the third-order Dynkin index [10.47] (normalized to $Q(\Box) = 1$) and is, except for normalization, the usual triangular axial anomaly coefficient (see §7.2b). The "even-odd rule" of Eq. (10.161) then tells us that the cancellation of the triangular axial anomaly, i.e. $Q_3(R) = 0$, implies the absence of the global $SU(2)$ anomaly [since the R.H.S. of Eq. (10.161) must be even]. It is also evident from Eq. (10.161) that the inverse of this statement is not true; the global $SU(2)$ anomaly can vanish in the presence of a triangular axial anomaly as long as the number of "zero modes" corresponding to R is even but not necessarily zero (we have a sufficient but not a necessary condition). The "even-odd rule" is the basis for the statement in §7.7 that the global $SU(2)$ anomaly is redundant for all candidate GUT groups as long as the Weyl fermion representation (or combination of representations) is triangular anomaly-free.

§10.4c. Homotopy groups of simple compact Lie groups

In previous sections of this chapter on topological conservation laws, we made use of the homotopy groups of various types of gauge groups in order to characterize the topological properties of the mappings from suitably chosen k-dimensional Euclidean spheres at $|\underline{x}| \to \infty : S^k$ $(k = 0, 1, ...)$ to the group manifold G in question. Because we chose to focus our discussion of topological conservation laws on plausible physical examples taken from condensed matter physics and the standard model (or some of its "enlargements") in particle physics, we paid greatest attention to the homotopy groups of the classical (Lie) gauge groups; however, interest in the homotopy groups of exceptional Lie groups has grown with the construction of new "unification" and "super-unification" models in particle physics and it is useful to have readily available what we call the "Bott Table" of homotopic groups. Thanks to Bott [10.48], the homotopy groups of all classical and exceptional compact Lie groups have been identified for arbitrary k-dimensional Euclidean spheres (and, *a fortiori*, for arbitrary D space-time dimensions); this is not such an arduous task because "Bott periodicity" was uncovered. Table 10.3 contains (in abbreviated form) all this information about the classical and exceptional Lie groups in order of increasing "Euclidean" dimension; the comments that immediately follow Table 10.3 are intended to cover many of the special situations that are not included among the entries in Table 10.3. It should be noted that $\Pi_0(G)$ really speaks to the connectedness of the group.

Table 10.3: "Bott Table": homotopy groups of the compact Lie groups.

$G \rightarrow$ $k \downarrow \quad \Pi_k(G) \searrow$	$U(N)$ $(N > k/2)$	$SU(N)$	$0(N)$ $(N > k+1)$	$SO(N)$	$Sp(N)$ $(N > (k-2)/4)$	Exceptional groups $(G_2, F_4, E_6, E_7, E_8)$
0	0	0	$\mathcal{Z}_2$	–	0	–
1	$\mathcal{Z}$	$\mathcal{Z}$	$\mathcal{Z}_2$	$\mathcal{Z}_2$	0	0
2	0	0	0	0	0	0
3	$\mathcal{Z}$	$\mathcal{Z}$	$\mathcal{Z}$	$\mathcal{Z}$	$\mathcal{Z}$	$\mathcal{Z}$
4	0	0	0	0	$\mathcal{Z}_2$	0
5	$\mathcal{Z}$	$\mathcal{Z}$	0	0	$\mathcal{Z}_2$	0
6	0	0	0	0	0	0
7	$\mathcal{Z}$	$\mathcal{Z}$	$\mathcal{Z}$	$\mathcal{Z}$	$\mathcal{Z}$	–
8	0	0	$\mathcal{Z}_2$	$\mathcal{Z}_2$	–	–
"Bott" Periodicity	2	2	8	8	8	8

Comments on Table 10.3

1. $\Pi_0(S^k) = \mathcal{Z}$; $\Pi_k(S^k) = \mathcal{Z}$; $\Pi_k(S^n) = 0$ except for $k = 0, n$
2. $\Pi_1[SO(2)] = \mathcal{Z}$; $\Pi_k[U(1)] \simeq \Pi_k[SO(2)] = 0$ $(k \geq 2)$
3. $\Pi_k[O(3)] \simeq \Pi_k[Sp(1)] \simeq \Pi_k[SU(2)]$ $(k \geq 2)$
4. $\Pi_k[O(5)] \simeq \Pi_k[Sp(2)]$; $\Pi_k[O(6)] \simeq \Pi_k[SU(4)]$ $(k \geq 2)$
5. $\Pi_{4,5}[SO(3)] \simeq \Pi_{4,5}[SO(5)] = \mathcal{Z}_2$
6. $\Pi_{4,5}[SO(4)] \simeq \Pi_{4,5}[O(4)] = \mathcal{Z}_2 + \mathcal{Z}_2$
7. $\Pi_5[SO(6)] \simeq \Pi_5[O(6)] = \mathcal{Z}$

As we have already pointed out, many of the homotopy groups of the classical Lie groups listed in Table 10.3 — for low values of k — have found application in condensed matter physics and the standard parts of particle physics. It seems worthwhile to provide in Table 10.4 — again, in order of increasing "Euclidean" dimension k — a partial list of the relevant physical applications of homotopy groups to particle and condensed matter physics; there are, of course, useful physical applications of homotopy groups to other fields but those are outside the purview of this book. We do not list in Table 10.4 physical applications of homotopy groups of the exceptional Lie groups because of the speculative nature of these applications. Insofar as higher-dimensional superstring theory is concerned, this deficiency can easily be remedied with the help of the "Bott periodicity" rules.

Table 10.4: Some physical applications of homotopy groups to particle and condensed matter physics.

Homotopy group	Physical application
$\Pi_0[0(N)] = \mathcal{Z}_2$	parity
$\Pi_1[0(N)(N > 3)] = \mathcal{Z}_2$	spin
$\Pi_1[U(1)] = \mathcal{Z}$	Dirac magnetic monopole, polyacetylene, quantized magnetic flux in Type II superconductivity
$\Pi_2[SU(2)/U(1)] = \Pi_1[U(1)]$	't Hooft-Polyakov magnetic monopole
$\Pi_3[SU(N)](N \geq 2) = \mathcal{Z}$	instantons, standard Skyrme model
$\Pi_4[SU(2)] = \Pi_4[Sp(1)] = \mathcal{Z}_2$	global $SU(2)$ anomaly
$\Pi_5[SU(N)(N \geq 3)] = \mathcal{Z}$	Wess-Zumino term in extended Skyrme model

§10.5. Topological (baryonic) solitons in the Skyrme model

We introduced the subject of topological conservation laws in §10.1 of this chapter and then went on (in §10.2–§10.2c) to discuss topological solitons in $D = 2$ to $D = 4$ space-time dimensions, reviewing the necessary formalism as well as offering some physical examples from condensed matter and particle physics. We noted that, while purely scalar topological solitons can be defined for $D = 2$, it is necessary to introduce higher spin fields to generate topological solitons for $D \geq 3$. While quantum field theories of elementary particles must ultimately be formulated in $D = 4$ Minkowski space to acquire phenomenological interest, there are meaningful lower-dimensional ($D = 2$ and $D = 3$) physical examples of topological conservation laws in many-body non-relativistic systems. The verified topological conservation laws in condensed matter physics for $D < 4$ space-time dimensions (reviewed in §10.2a–§10.2b) are impressive and add a new dimension to our understanding of the non-perturbative aspects of classical and quantum field theories. Unfortunately, these successes are not duplicated in $D = 4$ particle physics; for example, the 't Hooft-Polyakov magnetic monopole is an application of topological soliton theory to $D = 4$ quantum field theory but, thus far, no one has found magnetic monopoles of any type in the universe. We did discuss — in §10.3 — the most striking application of topological soliton theory to particle physics, namely the identification of instantons in Euclidean four-space and their translation into "tunneling" events in the θ vacuum in Minkowski space, but our examination of the physical consequences of instantons, left the unsolved "strong CP" problem in its wake.

The review (in §10.4) of homotopy groups and the "index theorem" was more favorable to particle physics by providing a rigorous foundation for the non-perturbative global $SU(2)$ anomaly and the "even-odd" rule relating the absence of this anomaly to the cancellation of the triangular chiral gauge anomaly. We do think that the full potential of topological methods has not as yet been realized in particle physics and, in support of this view, we conclude this last chapter on the conceptual foundations of modern particle physics, with a status report on the phenomenologically-inspired use of topological solitons in low-energy particle physics in the form of the Skyrme model. The Skyrme model, in its modern "incarnation", augments the classical Skyrme Lagrangian [1.139] with a QCD-inspired Wess-Zumino term [10.49] and purports to show that the "winding number" of the topological soliton contained in its early version can be interpreted as the baryonic charge of fermionic hadrons (i.e. nucleons). The topological approach is deepened in the most recent version – the "extended Skyrme model" (ESM) as we call it — and we close our discussion of the Skyrme model with a review of the key topological arguments of ESM and of the extent to which ESM reflects the properties of QCD.

The Skyrme model has had a long history, having first been proposed in 1960 [1.139] and then revived during the past decade in the modern trappings of QCD and the "Wess-Zumino term" [10.49]. The early version of the Skyrme model — which we call the standard Skyrme model (SSM) — attempted to develop a theory of the low-energy strong pion-nucleon interaction modeled on Pauli's even earlier semi-classical strong coupling theory of pions interacting with nucleons [10.50]. Pauli's "symmetrical" (in spin and isospin) pseudoscalar strong coupling theory correctly predicted, among other things, the existence of the Δ resonance [e.g. its mass, its decay width, etc.]. Indeed, some of Skyrme's predictions bore a close resemblance to their counterparts in strong coupling meson theory [1.50]. Since SSM aimed to understand the low-energy behavior of the strong interaction processes involving pions and nucleons, it is not surprising that SSM was based on two flavors — the "proton flavor" ("up" isospin) and the "neutron flavor" ("down" isospin); the global two-flavor chiral symmetry group $SU(2)_L \times SU(2)_R$ of SSM is still very useful (because of the much smaller masses of the u and d quarks compared to the s quark) and is still favored for applications in nuclear physics. However, as we will see, the three-flavor ESM is essential for the inclusion of the "Wess-Zumino term" — thereby ensuring a better approximation to QCD and mandating the half-integral spin and Fermi statistics of the (baryonic) topological soliton already contained in SSM. A closer linkage to QCD is important because ESM may provide a useful approach to the long-wavelength-limit of QCD. It is small wonder that nuclear physicists have welcomed ESM as an alternative method for calculating some of the phenomenological consequences of QCD at low energies without invoking cumbersome lattice-gauge calculations. The revived interest in the Skyrme model is also due to an increasing awareness of the usefulness of the non-linear σ model — which is basic to the Skyrme model — as a theoretical probe of QCD properties at low energies.

We review the present status of the Skyrme model — as a phenomenologically-oriented exercise in the application of topological methods to particle physics — under three headings: §10.5a. Non-linear σ model; §10.5b. Topological solitons in the standard Skyrme model (SSM); and §10.5c. QCD and the Wess-Zumino term in the extended Skyrme model (ESM).

§10.5a. Non-linear σ model

In this section, we spell out the properties of the non-linear σ model — which is considered the first term in the Skyrme action — with particular emphasis on its replication of the global chiral quark flavor symmetry of the light quarks in QCD. In the succeeding sections, we first add (in §10.5b) a second ("quartic") term to the non-linear σ model action (to obtain SSM), that ensures a stable topological soliton solution (Skyrmion) and helps to define the baryonic current. In §10.5b, we also comment on some key experimental predictions of SSM and the status of these predictions. The final section (§10.5c) explains the need for the "third" Skyrme term (the "Wess-Zumino term") to enlarge SSM to ESM; ESM replicates the chiral gauge anomalies of QCD and succeeds in fixing the fermionic character of the Skyrmion.

In motivating the non-linear σ model, we recall that the linear σ model (with nucleon) (see §3.5b) served a useful pedagogical purpose, by exhibiting how the SSB of a global chiral-invariant $SU(2)_L \times SU(2)_R$ theory [with a massless quartet (isospin $I = 0$ and $I = 1$) of $J = 0$ bosons, and a massless doublet $(I = \frac{1}{2})$ of $J = \frac{1}{2}$ fermions] down to an $SU(2)_{L+R}$ vector (isospin) group takes place and generates masses for the $I = 0\,(\sigma)$ boson and the $I = \frac{1}{2}\,(N)$ fermion, while maintaining the masslessness of the $I = 1\,(\pi)$ boson. However, if one seeks a topological origin for the nucleon, one can forget about the nucleon field and write down the action for the linear σ model in the absence of nucleons as follows (see §3.5b):

$$S = \int d^4x \{ \frac{1}{2}[(\partial_\mu \sigma)^2 + (\partial_\mu \underline{\pi})^2] - V(\sigma^2 + \underline{\pi}^2)\}; \quad \underline{\pi} = (\pi_1, \pi_2, \pi_3) \quad (10.162)$$

where:

$$V(\sigma^2 + \underline{\pi}^2) = \frac{\mu^2}{2}(\sigma^2 + \underline{\pi}^2) + \frac{\lambda}{4}(\sigma^2 + \underline{\pi}^2)^2; \quad \mu^2 < 0 \quad (10.163)$$

The fermion-free Lagrangian of Eq. (10.162) is invariant under the "vector" (V) and "axial vector" (A) transformations ($\underline{\alpha}$ and $\underline{\beta}$ are constant isovectors):

$$V: \sigma \rightarrow \sigma; \quad \underline{\pi} \rightarrow \underline{\pi} + \underline{\alpha} \times \underline{\pi}; \quad A: \sigma \rightarrow \sigma + \underline{\beta} \cdot \underline{\pi}; \quad \underline{\pi} \rightarrow \underline{\pi} - \underline{\beta}\sigma \quad (10.164)$$

which give rise to "vector" and "axial vector" currents respectively:

$$J_\mu^a = \epsilon^{abc}\pi^b \partial_\mu \pi^c; \quad J_{5\mu}^a = (\partial_\mu \sigma)\pi^a - (\partial_\mu \pi^a)\sigma \quad (a = 1, 2, 3) \quad (10.165)$$

The associated charges $Q^a = \int i d^3x \ J_0^a(x)$ and $Q_5^a = \int i d^3x \ J_{50}^a(x)$ still satisfy the $SU(2)_L \times SU(2)_R$ algebra [see §3.3]. SSB occurs because $\mu^2 < 0$ in Eq. (10.163) so that the minimum of the potential is at (f replaces the usual a in the Higgs potential):

$$\sigma^2 + \underline{\pi}^2 = f^2; \quad f = (-\frac{\mu^2}{\lambda})^{\frac{1}{2}} \quad (10.166)$$

If we make the choice $< 0|\sigma|0 > = f$ [and $< 0|\pi|0 > = 0$], and substitute the shifted field $\sigma' = \sigma - f$ into $V(\sigma^2 + \underline{\pi}^2)$, we find that the isosinglet meson mass is $m_\sigma = \sqrt{2}|\mu|$ and the isotriplet meson mass m_π vanishes, i.e. the three "pions" are N-G bosons. These results follow from the discussion in §3.3 since the axial charges Q_5^a ($a = 1, 2, 3$) break the vacuum so that the surviving group is the isospin group $SU(2)_I$ with the three "vector" charges Q^a ($a = 1, 2, 3$); the resulting particle spectrum consists of the isosinglet massive σ meson and the isotriplet massless π meson.

Before deriving the action for the non-linear σ model — which is reached by taking the limit $m_\sigma \to \infty$ and placing $\underline{\pi}$ in a non-linear representation of the $SU(2)_I$ group (see below) — it is convenient to rewrite the action for the linear σ model in the form:

$$S = \int d^4x [\frac{1}{2}\text{Tr}\,(\partial_\mu\phi^\dagger\partial^\mu\phi) - \frac{\lambda}{n^2}(\text{Tr}\,\phi^\dagger\phi - nf^2)^2] \qquad (10.167)$$

where n is the number of flavors and the N-G boson field, $\phi(x)$, is now a complex $n \times n$ matrix. For $n = 2$, we may write $\phi = \sigma - i\underline{\tau} \cdot \underline{\pi}$ ($\underline{\tau}$ is the Pauli matrix) — to recapture the spontaneously broken action (10.162) (when $\mu^2 < 0$). We now examine Eq. (10.167) in the limit $m_\sigma \to \infty$ [this is reasonable since we are interested in a model for low-energy hadron phenomena and m_σ may be arbitrarily adjusted by the proper choice of μ]; since $m_\sigma \simeq \sqrt{\lambda}$, we can study this limit by considering what happens to Eq. (10.167) as $\lambda \to \infty$. To do this, we impose the constraint $\phi^\dagger\phi = f^2$ and define the unitary matrix:

$$U(x) = \phi(x)/f; \qquad U^\dagger U = \mathbf{1} \qquad (10.168)$$

By this simple device, Eq. (10.167) is transformed into the non-linear σ model action, thus:

$$S = \int d^4x \frac{f^2}{2} \text{Tr}[\partial_\mu U^\dagger \partial_\mu U]; \qquad U^\dagger U = \mathbf{1} \qquad (10.169)$$

We call Eq. (10.167) the linear σ model because the matrix ϕ sits in the general linear group of Hermitian matrices. On the other hand, the field theory of unitary matrices defined by Eq. (10.169) is called the non-linear σ model because $U(x)$ is an element of $SU(n)_{L+R}$, i.e. $U(x) = \exp(2iT^a\pi^a/F_\pi)$, with T^a and π^a ($a = 1, 2, 3...n^2 - 1$) respectively the generators and N-G bosons of $SU(n)_{L+R}$. Furthermore, $U(x)$ possesses a non-linear realization under $SU(n)_L \times SU(n)_R$, i.e. $U(x)$ transforms as $U \to V_L^\dagger U V_R$, where $V_L^\dagger$ and V_R are $n \times n$ unitary matrices under the global chiral group $SU(n)_L \times SU(n)_R$, operating respectively on the left and right of U. Actually, Eq. (10.169) possesses the global chiral symmetry:

$$U(n)_L \times U(n)_R / U(1) \qquad (10.170)$$

where $U(1)$ takes account of the fact that each $U(1)$ subgroup of $U(n)_{L,R}$ changes U by an overall phase.

The non-linear σ action, given by Eq. (10.169), is, by itself, not very interesting. The real question is what kinds of higher-order (interaction) terms should be added to Eq. (10.169) — as implied by the unitary constraint — in order to simulate the real world of hadronic interactions at low energies. A clue to the enlargement of the non-linear σ model action — which becomes SSM — comes from rewriting Eq. (10.169) in terms of currents associated with the symmetry of Eq. (10.170) and then comparing the resulting action with the appropriate QCD-inspired action in the low energy limit. Thus, we write for the "left" and "right" currents:

$$L_\mu = -iU^\dagger \partial_\mu U, \qquad R_\mu = -iU \partial_\mu U^\dagger \tag{10.171}$$

where L_μ and R_μ are both Hermitian and resemble Yang-Mills gauge fields with vanishing field strength. [Equation (10.171) suggests the potential usefulness of $U(x)$ as a "mapping" function because of the "gauge-like" structure of L_μ and R_μ [see Eq. (10.171)]]. The action can now be written in terms of L_μ:

$$S = \frac{f^2}{2} \int d^4x \ \mathrm{Tr} L_\mu L^\mu \tag{10.172}$$

One next tries to guess a suitable enlargement of Eq. (10.172) — by relating the action of this equation to the QCD action for massless quarks (see §5.1). One realizes, of course, that the full hadron spectrum and its effective couplings are very complicated, but one may hope to describe low-energy strong interaction phenomena by an effective chiral Lagrangian in terms of the $[U(n)_L \times U(n)_R]/U(1)$ chiral currents, $J_\mu^{(+)}$ and $J_\mu^{(-)}$ of QCD [see Eq. (5.1)], which one identifies with L_μ and R_μ respectively. However, one must understand the limitations of any non-linear σ model above some cutoff because it is intrinsically non-renormalizable — as compared to QCD. Nevertheless, the intractability of doing QCD calculations at low energy encourages one to go on and to imagine that one can split up the QCD degrees of freedom into (10.172) and "everything else" (which includes the ρ meson, Δ baryon, etc.). In principle, after integrating out "everything else" from the path integral, one should get an effective action in terms of L_μ, which can be expanded as a power series (a and b are constants):

$$S = \int d^4x \left[\frac{f^2}{2} \mathrm{Tr}\ (L_\mu L^\mu) - \frac{1}{32a^2} \mathrm{Tr}\ [L_\mu, L_\nu]^2 + \frac{1}{b^2} \mathrm{Tr}\ (L_\mu L^\mu)^2 \cdots \right] \tag{10.173}$$

Equation (10.173) can serve as a model for adding additional terms to the Skyrme action in terms of the currents defined by Eq. (10.171), e.g. the QCD theory of strong interactions suggests adding an even higher-order term (than the quartic) to the action of Eq. (10.169) (see §10.5c). Insofar as the quartic terms in Eq. (10.173) are concerned, the second (quartic) term predicts, together with the first (quadratic) term [which, in combination, constitute SSM], a topological soliton in three spatial dimensions that can be interpreted as a baryon; the third term in Eq. (10.173) does not lead to a stable soliton and so we discuss it no further. The standard Skyrme model (SSM) is treated in the next section (§10.5b).

§10.5b. Topological solitons in the standard Skyrme model (SSM)

We begin by rewriting the first two terms of Eq. (10.173) in terms of the current given by Eq. (10.171); the action for SSM becomes:

$$S = \int d^4x \{ \frac{f^2}{2} \text{Tr} \, (U^+ \partial_\mu U)^2 - \frac{1}{32a^2} \text{Tr}[U^\dagger \partial_\mu U, U^\dagger \partial_\nu U]^2 \} \qquad (10.174)$$

Before deriving the soliton solution contained in Eq. (10.174), we reiterate that the quadratic term is incapable of giving stable classical soliton solutions of the equations of motion. When the quartic term (in U) is included, the potential energy of a classical solution can describe a stable soliton if $U(x) = e^{\frac{2i}{F_\pi} \pi^a(x) T^a}$ ($F_\pi \sim 190$ MeV) satisfies the following necessary condition:

$$U(\underline{x}, t) = \bar{U}(\underline{x}) \to \bar{U}_0 \quad \text{as } r = \sqrt{\underline{x} \cdot \underline{x}} \to \infty \quad \text{(for all } t) \qquad (10.175)$$

where $\bar{U}_0$ is a constant matrix. To make sure that the topological soliton that emerges can be identified with the baryon, one must establish that the topological conservation law is "additive" and that this property does not depend on the number n of quark flavors as long as $n \geq 2$. It is not difficult to convince oneself that the mapping is given by

$$\bar{U}: \quad S^3_\infty \to \frac{SU(n)_L \times SU(n)_R}{SU(n)_{L+R}} \sim SU(n)_A \quad (n \geq 2) \qquad (10.175a)$$

and that the governing homotopy group is:

$$\Pi_3[SU(n)] = \mathcal{Z} \quad (n \geq 2) \qquad (10.176)$$

The arguments for Eqs. (10.175)–(10.176) are: (1) since a global chiral transformation can bring $\bar{U}_0$ to unity, we can identify all points at ∞ and so compactify the space $\mathcal{R}^3$ to the three-sphere S^3_∞ which is mapped onto the coset space; (2) the coset space is G/H where G corresponds to the original global chiral symmetry group $SU(n)_L \times SU(n)_R$ ($n \geq 2$ flavors) which breaks down spontaneously to $SU(n)_{L+R}$, generating $(n^2 - 1)$ N-G bosons ("pions") in the process. The coset space G/H is isomorphic to $SU(n)_{L-R} = SU(n)_A$; and (3) hence the "mapping" is from S^3 to $SU(n)(n \geq 2)$ and the associated homotopy group is precisely (10.176) and leads to an "additive" conservation law for the topological charge. It should be noted that Eq. (10.176) is the same homotopy group condition that obtains for instantons although the physical meaning of S^3 and $SU(n)$ is quite different [see Eq. (10.75)].

With the basic topological underpinning in place, we can comment on the explicit expressions for the "mapping" function $U(\underline{x})$, the Pontryagin number and mass of the topological soliton, and some of the key phenomenological predictions that follow from the standard Skyrme model (SSM). To start with, we write down the "mapping" function $\bar{U}(\underline{x})$ for $n = 2$ flavors and "winding number" m of the soliton [m is used instead of n to avoid confusion with the number of flavors]; it takes on the familiar "hedgehog" form:

$$\bar{U}_m(\underline{x}) = e^{im\hat{\underline{x}} \cdot \underline{\tau} \, \theta(r)} = \cos m\theta(r) + i\hat{\underline{x}} \cdot \underline{\tau} \sin m\theta(r); \quad \bar{U}_m(\underline{x}) \in SU(2)_f \tag{10.177}$$

with $\theta(0) = \pi$, $\theta(\infty) = 0$, and $\frac{d\theta}{dr} \equiv \theta'(r) < 0$. Because $\theta' < 0$, θ is monotonically decreasing as a function of r and is single-valued. For $m = 1$, we have, furthermore, that $\bar{U}(\underline{x})$ is single-valued and $\bar{U}_0(\underline{x})$ takes on every possible value in S^3; it is not hard to see that for arbitrary "winding number" m, $SU(2)$ is covered m times and, as expected from Eq. (10.176), m is conserved. However, before we can identify m with baryon charge, we must look more carefully at the properties of the soliton itself.

A key property of baryon charge — like any quantum field-theoretical charge — is its change of sign under charge conjugation; it is simple to see that m changes sign under charge conjugation, which is defined by $\bar{U} \to \bar{U}^\dagger$. Next, we must show that the "canonical" topological current associated with the "mapping" $S^3 \to SU(2)$ actually leads to the conserved charge m [10.51]; for this purpose, we use Eq. (10.171) to write down an expression for the topological current which is modeled on the "gauge-variant" (Chern-Simons) current K^μ introduced into instanton theory. This is reasonable because the applicable homotopic group for instantons is also $\Pi_3[SU(n)] = \mathcal{Z}$ [see Eq. (10.75)], namely:

$$b^\mu = \frac{i}{24\pi^2}\epsilon^{\mu\nu\alpha\beta} \, \mathrm{Tr} L_\nu L_\alpha L_\beta; \quad L_\mu = -iU^\dagger \partial_\mu U \tag{10.178}$$

It is easy to show that $\partial_\mu b^\mu = 0$ and, since $U(\underline{x})$ is independent of the time, that the topological charge $B = \int d^3x \, b^0$ becomes:

$$B = \int d^3x \, b^0 = \frac{i}{24\pi^2}\epsilon^{ijk} \int d^3x \, \mathrm{Tr} \, L_i L_j L_k \tag{10.179}$$

If we now use Eqs. (10.171) and (10.177) for L_k, Eq. (10.179) yields:

$$B = m \tag{10.180}$$

so that the topological charge B is the "winding number" m. We note that b_μ is not a total divergence and therefore can be localized arbitrarily in space; b_μ is a number current which implies that the topological (baryonic) charge is not coupled to any long-range field and its measurement is performed through simple counting. Hence, B is an "additive" global conserved quantum number and is independent of n for $n \geq 2$.

With the Pontryagin number of the topological soliton ("Skyrmion") identified as baryon charge, it is possible to show that the mass of the Skyrmion is bounded from below by a constant times the baryon charge B. This bound is similar to the bound that exists for 't Hooft-Polyakov magnetic monopole solutions of Yang-Mills fields spontaneously broken to $U(1)$ by scalars [cf. §10.2c]. We can write down the expression for the mass of the soliton by using Eq. (10.173) (without the third term on the R.H.S.) and taking account of the fact that $\partial_0 U = 0$; we get:

$$M_{\text{sol}} = \int d^3x [\frac{F_\pi^2}{8} L_i^a L_i^a + \frac{1}{8a^2} \tilde{L}_i^a \tilde{L}_i^a] \qquad (10.181)$$

where $\tilde{L}_i^a = \epsilon_{ijk} \epsilon^{abc} L_j^b L_k^c$. But:

$$\int d^3x [\frac{F_\pi^2}{8} L_i^a L_i^a + \frac{1}{8a^2} \tilde{L}_i^a \tilde{L}_i^a] \geq \pm \{\frac{F_\pi}{4a} \int d^3x \; L_i^a \tilde{L}_i^a\}^2 \qquad (10.182)$$

so that:

$$M_{\text{sol}} \geq \frac{F_\pi}{4a} | \int d^3x L_i^a \tilde{L}_i^a | \qquad (10.183)$$

It is now easy to show that the R.H.S. of Eq. (10.183) is related to the baryon charge by inserting the transformation matrices T_a's for the L_i^a's in Eq. (10.179); we get:

$$B = \frac{i}{24\pi^2} \int d^3x \epsilon^{ijk} \; \text{Tr} \; (L_i^a T^a L_j^b T^b L_k^c T^c) = \frac{1}{12\pi^2} \int d^3 L_i^a \tilde{L}_i^a \qquad (10.184)$$

Using:

$$\text{Tr} \; (T^a T^b T^c) = i\epsilon^{abc} \qquad (10.185)$$

the final result is:

$$M_{\text{sol}} \geq \frac{3\pi^2 F_\pi}{a} |B| \qquad (10.186)$$

Equation (10.186) indicates that the soliton mass depends linearly on the baryon charge, a reasonable prediction from the phenomenological point of view. Since Eq. (10.190) only yields a lower bound, it is worth mentioning that the variational estimate of the "baryonic" soliton mass (for $m = 1$) — using a spherically symmetric form for $\bar{U}(\underline{x})$ [see Eq. (10.177)] — yields a value of M_{sol} larger (by a factor of about 3) than the R.H.S. of Eq. (10.186).

Thus far, we have demonstrated that the enlargement of the non-linear σ model [given by Eq. (10.169)] into SSM [given by Eq. (10.174)] gives rise to topological solitons that can be assigned baryon charge. We have not proved that these topological solitons possess half-integral spin and obey Fermi statistics (a task that is carried out in §10.5c), nor have we shown that these topological solitons interact with the N-G "pions" in a physically plausible way. Before turning to ESM (§10.5c) to establish the fermionic character of the "Skyrmion", we pursue the physics implicit in SSM, at least to the extent of spelling out the spins, isospins and several other properties of the two lowest mass "Skyrmions" (N and Δ baryons). For this purpose, "collective coordinates" are introduced and it is their quantization that helps to define the physical content of SSM. The "collective coordinate" approach seems reasonable if we start with the premise that the baryon wave function is peaked about a one-soliton configuration. The classical energy of this configuration is then highly degenerate with respect to flavor rotations and we therefore expect these rotations to be the highest frequency modes in the quantum theory and to approximate them by treating the remaining modes as static. The simplest way to implement the "collective coordinate" strategy [10.52] — in the two-flavor case — is to perform a global chiral transformation on the soliton solution $\bar{U}(\underline{x})$ by means of an arbitrary time-dependent $SU(2)$ matrix $V(t)$ (the "collective coordinates") — which can be written as $V(t) = v_0(t) + i\underline{v}(t) \cdot \underline{\tau}$ (with $v_0^2 + \underline{v}^2 = 1$) — and then to treat v_0 and the v_i's ($i = 1, 2, 3$) as quantum mechanical operators obeying the canonical commutation relations. This procedure enables one to write down the eigenstates (of spin and isospin) of the two lowest mass baryons (N and Δ); if we write:

$$U(\underline{x}, t) = V^\dagger(t)\bar{U}(\underline{x})V(t); \qquad V(t) \in SU(2) \tag{10.187}$$

where $\bar{U}$ is given by (10.177) (with $m = 1$) and substitute Eq. (10.187) into Eq. (10.178) — assuming spherical symmetry — we find that the Lagrangian reduces to:

$$\mathcal{L} = -M_{\text{sol}} + \frac{\Omega}{2} \text{Tr}(\dot{V}^\dagger \dot{V})^2 \tag{10.188}$$

where Ω is the "moment of inertia" (with contributions from both spin and isospin — see below) given by:

$$\Omega = \frac{2}{3}\pi F_\pi^2 \int_0^\infty dr \, \sin^2\theta(r)[1 + 4(\theta'^2 + \frac{\sin^2\theta}{a^2 F_\pi^2 r^2})] \tag{10.189}$$

It is easy to convince oneself that Eq. (10.188) is the Lagrangian of a four-dimensional rigid rotator; to see this explicitly, we rewrite Eq. (10.188) in terms of v_0 and $\underline{v}$:

$$\mathcal{L} = -M_{\text{sol}} + \frac{\Omega}{2}(\dot{v}_\mu \dot{v}^\mu)^2 \tag{10.190}$$

Equation (10.190) is clearly a free particle Lagrangian, except for the constraint: $v_\mu v^\mu = 1$ (that the particle lies on the unit 3-sphere). The Hamiltonian is therefore proportional to the angular part of the Laplacian in four dimensions; thus:

$$H = M_{\text{sol}} + \frac{1}{2\Omega} K_{\mu\nu} K^{\mu\nu} \tag{10.191}$$

where the differential operators $K_{\mu\nu}$ are the orbital angular momentum generators in $\mathcal{R}^4$:

$$K_{\mu\nu} = v_\mu \frac{\partial}{\partial v_\nu} - v_\nu \frac{\partial}{\partial v_\mu} \tag{10.192}$$

The eigenstates and eigenvalues are classified by noting that the $K_{\mu\nu}$'s form the algebra $SU(2) \times SU(2)$ (see §2.1c), where one $SU(2)$ refers to spin, the other $SU(2)$ to isospin; we have:

$$K_{oi} = I_i - J_i, \quad K_{ij} = -I_i - J_i \tag{10.193}$$

where the I's and J's satisfy the well-known commutation relations (see §2.1c):

$$[I_i, I_j] = i\epsilon_{ijk} I_k; \qquad [J_i, J_j] = i\epsilon_{ijk} J_j; \qquad [I_i, J_j] = 0 \tag{10.194}$$

The meaning of the operators I_j, J_j in the Skyrme model is clarified by their commutation relations with $U(\underline{x})$; we find:

$$[J_i, U(\underline{x})] = -(\underline{x} \times \underline{\nabla})_i \, U(\underline{x}); \qquad [I_i, U(\underline{x})] = \frac{i}{2} \, [\tau_i, U(\underline{x})] \tag{10.195}$$

which confirm that $\underline{J}$ is the "Skyrmion" spin and $\underline{I}$ is "Skyrmion" isospin.

In principle, J and I can be chosen to be any pair of integers or half-integers. However, a valuable feature of the quantum mechanics of a rigid rotator system is the requirement that $J = I$; this still leaves open the question of whether $J = I$ is an integer or a half-integer, and nothing in the above analysis requires either possibility. In sum, SSM does not dictate the allowed values of spin and isospin for the topological soliton (only that $J = I$) and, within the framework of the two-flavor Skyrme model, we can only deduce that $J = I$ is a half-integer from experiment. If we do take half-integral values for the spin and isospin of the $B = 1$ "Skyrmion", SSM predicts that the lowest mass state (nucleon) possesses $J = I = \frac{1}{2}$ and that the first excited state (Δ resonance) corresponds to the $I = J = \frac{3}{2}$ topological baryon. Moreover, the theory predicts that the J (and, *a fortiori*, I) dependence of the mass $M_{I,J}$ of the $B = 1$ "Skyrmion" is given by:

$$M_{I,J} = M + \frac{1}{2\Omega}[J(J+1) - \frac{3}{4}] \qquad (I = J) \qquad (10.196)$$

Equation (10.196) tells us, not surprisingly, that $M_{I,J}$ is independent of the spin direction and isospin "direction" (i.e. the electric charge) but that the mass degeneracy between the $J = I = \frac{1}{2}(N)$ and $J = I = \frac{3}{2}$ (Δ) states is removed. It is instructive to compare Eq. (10.196) with the prediction of the old strong coupling theory of the pion-nucleon interaction [1.50] to see that the form (10.196) is precisely the same and that the pion-nucleon coupling constant f and "nucleon size" a [1.50] are replaced by the two parameters of SSM: F_π and a, in terms of which M and Ω of Eq. (10.196) are determined. In SSM, knowledge of M_N and M_Δ predicts $F_\pi = 129$ MeV (30% less than the experimental value — see Table 10.5) and $a = 5.45$ (no direct phenomenological comparison for a is possible). One can test the self-consistency of SSM by computing other low energy properties of the proton and neutron [10.53] such as charge distributions and magnetic moments; the comparison is shown in Table 10.5, where $< r^2 >_q$ and $< r^2 >_m$ refer to the electric charge and magnetic moment distributions, and μ_p and μ_n are the proton and neutron magnetic moments respectively. The overall agreement is about 30%, which is quite remarkable, considering that the SSM action (10.174) only refers to "pions".

Table 10.5: Comparison of standard Skyrme model with low energy hadron experiments.

Quantity	SSM prediction	Experiment
m_N	*input*	939 MeV
m_Δ	*input*	1232 MeV
F_π	129 MeV	186 MeV
$\langle r^2 \rangle^{\frac{1}{2}}_{q(I=0)}$	0.59 fm	0.72 fm
$\langle r^2 \rangle^{\frac{1}{2}}_{m(I=0)}$	0.92 fm	0.81 fm
μ_p	1.87	2.79
μ_n	-1.31	-1.91
$\left| \frac{\mu_p}{\mu_n} \right|$	1.43	1.46

We have seen how far SSM brings us with its ingenious interpretation of
the topological soliton solutions as finite-extension baryons ("Skyrmions").
Agreement with experiment is not bad and one is encouraged to try to improve SSM. Several "theoretical" limitations of SSM can be listed: (1) the
fermionic character of the baryon is not explained; (2) there are hidden discrete symmetries in SSM that are not present in the QCD action (see §10.5c)
so that certain observed processes in nature are forbidden; (3) no cognizance
is taken of the triangular axial anomalies inherent in QCD so that SSM does
not reflect the subtle differences between two-flavor and three-flavor global
chiral symmetries; and, finally (4) SSM is not linked to the special status
of the three colors in QCD (see §9.2). These limitations of SSM are overcome in an interesting way by the addition of the "Wess-Zumino term" to
three-flavor SSM, as discussed in the next section.

§10.5c. Extended Skyrme model (ESM) with the Wess-Zumino term

We have seen that in two-flavor SSM there is no guiding principle to
specify whether the "Skyrmion" should be quantized as a boson or a fermion.
In principle, the SSM action (10.174) should hold for an arbitrary number of
flavors — in particular, for three flavors (corresponding to the three known
light quarks), which leads to the successful $SU(3)_L \times SU(3)_R$ current algebra.
Indeed, the $SU(2)$ solution (10.177), corresponding to "winding number" m,
is easily generalized to $SU(3)$:

$$\bar{U}_m(\underline{x}) = \begin{bmatrix} \cos m\theta(r) + i\underline{x} \cdot \underline{\tau} \sin m\theta(r) & 0 \\ 0 & 1 \end{bmatrix}; \quad \bar{U}_m(x) \in SU(3)_f \quad (10.197)$$

Defining L_μ once again by (10.171), the baryon charge is again given by Eq.
(10.184) and is equal to m when the "mapping" function in (10.197) is taken
with "winding number" m. Topological solitons exist and the lower bound
on their masses is given by Eq. (10.186) as a function of the baryon charge.
This extension of SSM is fine but if nothing more could be done, the Skyrme
model would suffer from the four limitations enumerated above. Fortunately,
a closer study of QCD — which the Skyrme model is supposed to simulate as
an effective low energy theory — identifies two essential ingredients of QCD
that may be relevant for the Skyrme model: the odd number of colors and
the triangular axial anomalies, that can be incorporated into an additional
"Wess-Zumino term" [1.141] to overcome the four limitations of SSM. It
turns out that the "Wess-Zumino term" is proportional to the number of
colors, N_C, and only exists if the number of flavors, n_f, is at least three.

We now sketch a crude derivation of the "Wess-Zumino" addition to SSM — to convert it into ESM — and explain why this term requires "Skyrmions" to be fermions when the number of QCD colors is odd. While the "Wess-Zumino" action was originally invented to describe, in closed form, all axial contributions to the anomalous Ward-Takahashi identities (at low energies) in the presence of external Yang-Mills fields, the addition of the "Wess-Zumino" term to the ESM action is intended not only to deal with the anomalous Ward-Takahashi identities but also to cope with another deficiency of SSM, namely the presence of certain unwanted discrete symmetries in SSM that are absent in QCD (and the real world). This discrete symmetry problem is solved first because it helps to justify the structure of the "Wess-Zumino term" in ESM. Thus, consider that, for three flavors ($n_f = 3$), $U(\underline{x})$ can be written in the non-linear form:

$$U(x) = \exp[(2i/F_\pi)\lambda^a \pi^a(x)] \qquad (a = 1, 2, ..., 8) \qquad (10.198)$$

where the λ^a's are the eight $SU(3)$ generators (see §3.2c) and the π^a's are the eight N-G boson fields. It is easy to check that the Lagrangian of Eq. (10.174) possesses the two symmetries [10.49]:

$$P' : \quad U(\underline{x}, t) \to U(-\underline{x}, t); \qquad P'' : \quad U(\underline{x}, t) \to U^\dagger(\underline{x}, t) \qquad (10.199)$$

where the product $P'P''$ represents the usual parity operation P, i.e. $L_\mu(\underline{x}) \to R_\mu(\underline{x})$ under $P = P'P''$. The second discrete symmetry of Eq. (10.199) — by itself — yields a change in sign of the N-G boson field and corresponds to G conjugation for the $n_f = 2$ flavor case [see §2.2b]; thus, a state with ℓ "pseudoscalar mesons" changes phase by $(-1)^\ell$ under P''. If P' and P'' were separately good symmetries, the phase $(-1)^\ell$ would always be conserved. The fact that the vector ϕ meson can decay into either two or three pseudoscalar mesons, i.e. $K^+ K^-$ or $\pi^+ \pi^0 \pi^-$ in QCD [10.49] argues that P'' is an untenable symmetry of SSM [it must be noted that the process $K^+ + K^- \to 3\pi$ would also be in violation of the OZI rule (see §5.4b)].

It follows that another term must be added to the SSM action (10.174) in order to eliminate the two separate discrete symmetries P' and P'' while maintaining the usual parity conservation $P = P'P''$. It can be shown [10.49] that a Lorentz-invariant term that destroys P'' symmetry, must involve the Levi-Civita tensor, $\epsilon^{\alpha\beta\mu\nu}$, if the action contains no more than four L_μ's, i.e. is of the form $\int d^4x \, \epsilon^{\alpha\beta\mu\nu} \, \mathrm{Tr} \, L_\alpha L_\beta L_\mu L_\nu$. However, an action of this form must vanish because of the antisymmetry of $\epsilon^{\alpha\beta\mu\nu}$ and the cyclic symmetry of the trace. The addition to the SSM action which deals with the discrete symmetry problem and is only slightly more complicated, can be derived by expanding L_μ in powers of the "pseudoscalar meson" field, $\pi = \sum_a \pi^a \lambda^a$, to wit:

$$L_\mu = -iU^\dagger \partial_\mu U \simeq \frac{2i}{F_\pi} \partial_\mu \pi + O(\pi^2) \tag{10.200}$$

With the simplified form (10.200) for L_μ, we can write down a term, denoted by S_{WZ}, with the smallest number of π fields that manifestly allows processes like $K^+ K^- \to 3\pi$; we have:

$$S_{WZ} = \omega \int d^4x \, \epsilon^{\mu\nu\sigma\tau} \text{Tr}\{\pi \partial_\mu \pi \partial_\nu \pi \partial_\sigma \pi \partial_\tau \pi\} + 0(\pi^6) \tag{10.201}$$

where ω is a constant still to be determined. The action term (10.201) turns out to be the part of the "Wess-Zumino" action that simulates the triangular axial anomaly in the strong interactions (see below).

To show that Eq. (10.201) possesses the property claimed for it, we sketch the derivation of the "Wess-Zumino" action, which makes provision for all anomalous Ward-Takahashi identities for hadronic systems in the presence of external vector and axial vector fields subject to $SU(3)_L \times SU(3)_R$ chiral flavor symmetry (i.e. $n_f = 3$). We find that the "Wess-Zumino" expression for the total triangular axial anomaly not only contains a term that can be identified with (10.201) (and lead to a determination of ω in terms of a Pontryagin number ν') but that it also contains a term which accounts for the $\pi^0 \to 2\gamma$ decay process (and thereby fixes $\nu' = N_C = 3$). To obtain the full expression for the "Wess-Zumino" anomaly, one modifies the QCD action (for three massless quarks) by adding the interaction of the quarks with "background" vector and axial vector fields, as follows:

$$S_{QCD} \to S_{QCD} + \int d^4x [\bar{\psi}\gamma^\mu (V_\mu + \gamma_5 A_\mu)\psi]. \tag{10.202}$$

where:

$$V_\mu = \frac{\lambda_a}{2} V_\mu^a \quad \text{and} \quad A_\mu = \frac{\lambda^a}{2} A_\mu^a \tag{10.202a}$$

Equations (10.202) and (10.202a) lead to an anomalous term in the divergence of the axial vector current J_μ^5 that is evaluated with the help of a gauge-invariant regularization scheme invented long ago by Schwinger [10.54]. In carrying through the procedure for $n_f = 3$ flavors, one must take into account the triangular and box contributions to the axial anomaly; imposing

CVC on all the vector channels, one gets [7.2]:

$$\partial^\mu J_{5\mu}^a = \mathcal{A}_{WZ}^a = \frac{1}{4\pi^2}\epsilon_{\mu\nu\sigma\tau}\,\mathrm{Tr}[\frac{\lambda^a}{2}\{\frac{1}{4}V_{\mu\nu}V_{\sigma\tau} + \frac{1}{12}A_{\mu\nu}A_{\sigma\tau}$$
$$+ \frac{2}{3}i(A_\mu A_\nu V_{\sigma\tau} + A_\mu V_{\nu\sigma}A_\tau + V_{\mu\nu}A_\sigma A_\tau) - \frac{8}{3}A_\mu A_\nu A_\sigma A_\tau\}] \quad (10.203)$$

where:

$$V_{\mu\nu} = \partial_\mu V_\nu - \partial_\nu V_\mu - i[V_\mu, V_\nu] - i[A_\mu, A_\nu];$$
$$A_{\mu\nu} = \partial_\mu A_\nu - \partial_\nu A_\mu - i[V_\mu, A_\nu] - i[A_\mu, V_\nu] \quad (10.203a)$$

The total "Wess-Zumino" action W_{WZ} in terms of the total "Wess-Zumino" anomaly $\mathcal{A}_{WZ}$ of Eq. (10.203) can be written [10.24]:

$$W_{WZ} = \frac{1 - \exp(-\frac{\pi^a X^a}{F_\pi})}{\frac{\pi^a X^a}{F_\pi}} \cdot \frac{(\pi^a \mathcal{A}_{WZ}^a)}{F_\pi} \quad (10.204)$$

where:

$$X^a = -\partial_\mu \frac{\delta}{\delta A_\mu^a(x)} \quad (10.204a)$$

To obtain the part of W_{WZ} that contributes to the Skyrme action (so that we have ESM), one must expand the exponential in Eq. (10.204) for W_{WZ} to the fifth power of $(-\pi^a X^a/F_\pi)$ to match the last term in Eq. (10.203) for the "Wess-Zumino" anomaly; one gets:

$$W_{WZ}^{\mathrm{Skryme}} (=) \frac{1}{5!}\frac{1}{4\pi^2}\frac{8}{3}\frac{1}{F_\pi}(\frac{\pi^a X^a}{F_\pi})^4 \epsilon_{\mu\nu\sigma\tau}\,\mathrm{Tr}\,[\pi^a \frac{\lambda^a}{2}A_\mu A_\nu A_\sigma A_\tau] \quad (10.205)$$

When one then expresses the axial vector currents A_μ in terms of the $\bar{U}(x)$ of the Skyrme model, one finally obtains an expression with the same structure as Eq. (10.201) but with ω still to be determined.

The determination of ω in Eq. (10.201) turns out to require a topological discussion that leads to the determination of the Pontryagin number ν' (not to be confused with the topological charge of the soliton in SSM) implicit in the structure of the "Wess-Zumino" contribution to ESM. A good way to motivate the derivation of ω is to restate the derivation of Dirac's quantization condition (for the magnetic monopole) in sharply-defined homotopy language. We make use of the non-contractible "equatorial" closed curve (circle) C in Fig. 4.11a, the sphere S^2, the north-cap α (with surface area Ω_α) and the south-cap β (with surface area Ω_β). One first considers Dirac's non-integrable phase factor:

$$\exp i\Gamma = \exp\{ie\oint_C \underline{A}\cdot d\underline{x}\} = \exp\{ie\int_{\Omega_\alpha} F_{ij}d\sigma^{ij}\} = \exp\{-ie\int_{\Omega_\beta} F_{ij}d\sigma^{ij}\}$$
$$(10.206)$$

where F_{ij} is the electromagnetic field tensor and the normal to the surface is always taken outward. The consequence of the non-uniqueness of the surface is that we must have:

$$\exp\{ie \int_{\Omega_\alpha + \Omega_\beta} F_{ij} d\sigma^{ij}\} = 1 \qquad (10.207)$$

but $\int_{\Omega_\alpha + \Omega_\beta} F_{ij} d\sigma^{ij}$ in Eq. (10.207) is simply $4\pi g$ by Gauss' law (since F_{ij} is not singular over $\Omega_\alpha + \Omega_\beta$). That is to say, we get:

$$e^{4\pi i e g} = 1; \qquad eg = \frac{n}{2} \qquad (10.208)$$

which is Dirac's quantization condition.

This line of reasoning for the Dirac magnetic monopole is translated by Witten [10.49] into the "quantization" condition for the "Wess-Zumino" term in the Skyrme model. Taking cognizance of the fact that the "Wess-Zumino" action should contain at least the fifth power of the "pion" fields [see Eq. (10.205)] and recognizing that on the $SU(3)$ manifold — the minimum dimension corresponding to three quark flavors — there is a unique fifth-rank antisymmetric tensor s_{ijklm}^{WZ} that is invariant under the chiral $SU(3)_L \times SU(3)_R$ group, Witten chose for the analog of Γ in Eq. (10.206) the following (see Fig. 10.8):

$$\exp \ i\Gamma = \exp\{i\omega' \int_{S_\alpha^5} s_{ijklm}^{WZ} d\sigma^{ijklm}\} = \exp\{-i\omega' \int_{S_\beta^5} ds_{ijklm}^{WZ} d\sigma^{ijklm}\} \qquad (10.209)$$

where:

$$s_{ijklm}^{WZ} = \text{Tr} \ \{U^\dagger \partial_i U U^\dagger \partial_j U U^\dagger \partial_k U U^\dagger \partial_l U U^\dagger \partial_m U\} \qquad (10.210)$$

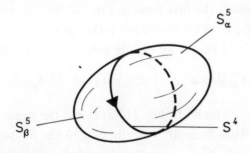

Fig. 10.8. Witten's quantization condition for the Wess-Zumino term in the Skyrme model.

Since the analog of Eq. (10.206) is Eq. (10.209), we have:

$$\exp\{i\omega' \int_{S^5} s_{ijklm}^{WZ} d\sigma^{ijklm}\} = 1 \qquad (10.211)$$

so that:

$$\omega' \int_{S^5} s^{WZ}_{ijklm} d\sigma^{ijklm} = 2\pi\nu' \qquad (10.212)$$

where ν' is a Pontryagin number so that Eq. (10.212) is the counterpart of Dirac's quantization condition (10.208). Equation (10.212) reflects the fact that the mapping is governed by the homotopy group: $\Pi_5[SU(3)] = \mathcal{Z}$ [instead of $\Pi_1[U(1)] = \mathcal{Z}$ as in the Dirac case] and the 2π on the R.H.S. of Eq. (10.212) is the normalization of the "one-soliton sector". Since the integral on the L.H.S. of Eq. (10.212) is the integral of a total divergence, it can be expressed — by "Stokes' theorem" — as an integral over the "boundary" S^4 (see Fig. 10.8); we get:

$$\Gamma_{WZ} = \nu' \frac{64}{15\pi^2 F_\pi^5} \int_{S^4} d^4 x \epsilon^{\mu\nu\sigma\tau} (\text{Tr } \pi \partial_\mu \pi \partial_\nu \pi \partial_\sigma \pi \partial_\tau \pi) \qquad (10.213)$$

where ν' is the Pontryagin number, $\pi = \Sigma_a \frac{\lambda^a}{2} \pi^a$ and the approximation of Eq. (10.200) has been used for $U^\dagger \partial_\mu U$. Equation (10.218) thus represents the "Wess-Zumino" contribution to the Skyrme action.

Witten's derivation of the "Wess-Zumino" contribution to the Skyrme action given by Eq. (10.213) has succeeded in incorporating the triangular axial anomalies of QCD and has eliminated the unwanted discrete symmetries in SSM. However, the most striking consequence of this topological derivation is that a Pontryagin "winding number" ν' enters the expression (10.213) for the "Wess-Zumino" action and a physical interpretation must now be found for ν'. It turns out that $\nu' = N_C$ (the number of colors in QCD) and to establish this, one returns to the total "Wess-Zumino" action W_{WZ}, given by Eq. (10.204), expands the exponent to the first power in $(-\pi^a X^a / F_\pi)$ to match the first term in Eq. (10.203) for the "Wess-Zumino" anomaly A^a_{WZ} [10.24]. This operation picks up the term in W_{WZ} that describes the $\pi^0 \to 2\gamma$ decay; if we denote this term by W^{EM}_{WZ}, we get:

$$W^{EM}_{WZ} = \nu'' \omega'' \int d^4 x \ \pi^0(x) \ \epsilon^{\mu\nu\alpha\beta} \ F_{\mu\nu} F_{\alpha\beta} \qquad (10.214)$$

where $\omega'' = \frac{e^2}{48\pi^2 F_\pi}$. The ν'' in Eq. (10.214) has the same topological origin as the ν' in the "Wess-Zumino" contribution to the Skyrme action (10.213) but we can now invoke the experimental result for the $\pi^0 \to 2\gamma$ decay width to establish $\nu' = \nu'' = N_C = 3$.

Accepting this last important result, it is still necessary to demonstrate that a "Skyrmion" with topological charge $B = 1$ is a fermion and to show that the oddness of N_C implies the fermionic character of the $B = 1$

"Skyrmion". This final step can be achieved by applying the "collective co-ordinate" approach to the total Skyrme action (ESM) when there are three flavors so that the "Wess-Zumino" action is included. The three-flavor "mapping" function $\bar{U}(\underline{x})$, given by Eq. (10.197), serves as the "classical background" field that is subject to the "collective coordinate" transformation and can be written:

$$\bar{U}(\underline{x}, t) = V(t)^{\dagger} \bar{U}(\underline{x}) V; \qquad V \in SU(3) \tag{10.215}$$

It is evident that $\bar{U}(\underline{x}, t)$ is invariant under:

$$V(t) = v(t) e^{iY\alpha(t)}; \qquad Y = \frac{1}{3} \begin{pmatrix} 1 & 0 & 0 \\ 0 & 1 & 0 \\ 0 & 0 & -2 \end{pmatrix} \tag{10.216}$$

where $\alpha(t)$ is a scalar function of time. It is then possible to show that the "collective coordinate" contribution to the time-dependent "Wess-Zumino" Lagrangian is given by:

$$\mathcal{L}_{WZ} = \frac{-iN_C}{2} B(\bar{U}) \operatorname{Tr} (Y V^{\dagger} \dot{V}) \tag{10.217}$$

where B is the baryon charge and ν'' has been replaced by N_C in accordance with Eq. (10.214). We note that $\alpha(t)$ is an arbitrary function of time, which means that it is equivalent to a $U(1)_Y$ gauge symmetry. Pursuing the "collective coordinate" method [10.55], one now quantizes an infinitesimal version of Eq. (10.215) with the v, α operators:

$$v \rightarrow v + iv Y \alpha(t) \tag{10.218}$$

The infinitesimal (quantum) generator of this transformation is $\hat{Y}$, satisfying the condition:

$$[\hat{Y}, v] = vY \tag{10.219}$$

Under such a variation, the Lagrangian (10.217) changes by;

$$\mathcal{L} \rightarrow \mathcal{L} + \frac{1}{3} N_C B \dot{\alpha} \tag{10.220}$$

and the condition on the physical states Ψ becomes:

$$\hat{Y} \Psi = \frac{1}{3} N_C B \Psi \tag{10.221}$$

This sort of condition always arises when there is a local gauge invariance.

We are now in a position to see explicitly why the $B = 1$ "Skyrmion" is a fermion; consider a 2π rotation of $\bar{U}(\underline{x})$ about the third axis in space:

$$\bar{U}(\underline{x}) \rightarrow \begin{pmatrix} -1 & 0 & 0 \\ 0 & -1 & 0 \\ 0 & 0 & 1 \end{pmatrix}^{\dagger} \bar{U}(\underline{x}) \begin{pmatrix} -1 & 0 & 0 \\ 0 & -1 & 0 \\ 0 & 0 & 1 \end{pmatrix} \tag{10.222}$$

This is equivalent to rotating v by:

$$v \to v \begin{pmatrix} -1 & 0 & 0 \\ 0 & -1 & 0 \\ 0 & 0 & 1 \end{pmatrix} = v e^{3\pi i Y} \qquad (10.223)$$

But, according to Eq. (10.221), this means that the wavefunction picks up a phase factor:

$$\Psi \to e^{i\pi N_C B} \Psi = (-1)^{N_C B} \Psi \qquad (10.224)$$

Thus, the $B = 1$ "Skyrmion" of ESM is a fermion if N_C is odd and a boson if N_C is even; moreover, since $N_C = 3$ on phenomenological grounds, the $B = 1$ "Skyrmion" must be a fermion and the goal of bringing the Skyrme model closer to the correct QCD theory has been achieved.

The triumph of ESM in bringing the Skyrme model so close to QCD naturally calls for further phenomenological scrutiny of the three-flavor Skyrme model. The qualitative consequences are favorable: for $n_f = 3$, the "Skyrmion" wavefunctions are $SU(3)$ generalizations of the $n_f = 2$ case, with an additional quantum number p labelling the $SU(3)$ representation. Moreover, for $N_C = 3$, we have from Eq. (10.221) that $Y = 1$. The two lowest $SU(3)$ representations with $Y = 1$ are $p = 8$ and $p = 10$. The 8 representation has states with $I = \frac{1}{2}$ and hence $J = \frac{1}{2}$, whereas the 10 representation has states with $I = \frac{3}{2}$ and hence $J = \frac{3}{2}$. Thus, ESM — the three-flavor Skyrme model with the "Wess-Zumino term"' — not only accounts for the fermionic character of the "Skyrmion" but recaptures for us the completely symmetric $\mathbf{56} = (\mathbf{8,2}) + (\mathbf{10,4})$ representation of the $SU(6) \supset SU(3)_{\text{flavor}} \times SU(2)_{\text{spin}}$ quark model of QCD (see §3.2d). In conclusion, it must be admitted that despite the major effort that has been expended to integrate the Skyrme model more fully into QCD with the help of the "Wess-Zumino term", the quantitative predictions of ESM *vis-à-vis* SSM are not much improved; there is still a 30% discrepancy between theory and experiment. The lack of experimental improvement can be excused by the fact that $n_f = 3$ is a substantially greater approximation to the real world than $n_f = 2$, and that approximations have been made [e.g. only pseudoscalar "pions" have been considered and one would think that improvements are possible with the help of "vector mesons"]. Nevertheless, it should be said that while one must still await the verdict in the phenomenological domain, one can only marvel at the topological successes of the Skyrme model and its potentially constructive role in the intractable low-energy domain of particle physics. It will be a great irony, indeed, if one gains a deeper insight into the long-distance properties of QCD physics from an application of higher-dimensional topological concepts to a model that was born three decades ago!

References

[10.1] E. Noether, *Nachr. Kgl. Geo. Wiss. Gottinger* **235** (1918).

[10.2] M.H. Green, J. Schwarz and E. Witten, *Superstring Theory*, Cambridge Univ. Press (1987); *String Theory in Four Dimensions*, North-Holland (1988).

[10.3] G.H. Derrick, *J. Mod. Phys.* **5** (1964) 1252.

[10.4] S. Russell, British Association Report (1844).

[10.5] P. Goddard and D. I. Olive, *Rep. Prog. Phys.* **41** (1978) 1357.

[10.6] W. Thirring, *Annals of Phys.* **3** (1958) 91.

[10.7] T.R. Klassen and E. Melzer, Cornell Univ. preprint CLNS-92/1149, ITP-SB-92-36 (1992).

[10.8] R. Jackiw and C. Rebbi, *Phys. Rev.* **D13** (1976) 3398.

[10.9] W.P. Su, J.R. Schrieffer and A.J. Heeger, *Phys. Rev. Lett.* **42** (1979) 1692; *Phys. Rev.* **B22** (1980) 2099.

[10.10] R.E. Peierls, *Quantum Theory of Solids*, Clarendon Press, Oxford (1955).

[10.11] J.R. Schrieffer, *Symposium on Anomalies, Geometry and Topology*, ed. W.A. Bardeen and A.R. White, World Scientific, Singapore (1985).

[10.12] D. Arovas, J.R. Schrieffer, F. Wilczek, and A. Zee, *Nucl. Phys.* **250** (1985) 117.

[10.13] D.D. Tsui, H.L. Störmer and A.C. Gossard, *Phys. Rev. Lett.* **48** (1982) 1559; R.B. Laughlin, *Phys. Rev. Lett.* **50** (1983) 1395.

[10.14] *The Quantum Hall Effect*, ed. R.E. Prange and S.M. Girvin, Springer-Verlag, New York (1987).

[10.15] Y.H. Chen, F. Wilczek, E. Witten, and B.I. Halperin, *J. Mod. Phys.* **B3** (1989) 1001.

[10.16] H.B. Nielsen and P. Olesen, *Nucl. Phys.* **B61** (1973) 45.

[10.17] A.H. Guth, *Phys. Rev.* **23** (1981) 347.

[10.18] J. Arafune, P.G.O. Freund and C.J. Goebel, *J. Mod. Phys.* **16** (1975) 433.

[10.19] B. Cabrera, *Phys. Rev. Lett.* **48** (1982) 1378.

[10.20] R. Jackiw, *Rev. Mod. Phys.* **52** (1980) 661.

[10.21] J.M.F. Labastida, M. Pernici and E. Witten, *Nucl. Phys.* **B310** (1988) 611.

[10.22] H.J. deVega, J.L. Gervais and B. Sakita, *Nucl. Phys.* **B139** (1978) 20.

[10.23] R. Jackiw and C. Rebbi, *Phys. Rev. Lett.* **37** (1976) 172.

[10.24] W. Dittrich, *Selected Topics in Gauge Theories*, Springer-Verlag (1986).

[10.25] R. Haag, *Kgl. Danske Videns. Selb. Mat. Fys. Medd.* **29** (1955) 12.

[10.26] R.E. Marshak and S. Okubo, *Prog. Theor. Phys.* **87** (1992) 1059.

[10.27] S. Samuel, *Mod. Phys. Lett.* **A7** (1992) 2007.

[10.28] N.H. Christ, *Phys. Rev.* **D21** (1980) 1591.

[10.29] R.F. Dashen, B. Hasslacher and A. Neveu, *Phys. Rev.* **D10** (1974) 4138; F.R. Klinghamer and N.S. Manton, *Phys. Rev.* **D30** (1984) 2212.

[10.30] V.A. Rubakov, *Proc. of 25th Intern. Conf. on High Energy Physics*, ed. K.K. Phua and Y. Yamaguchi, World Scientific, Singapore (1991).

[10.31] V. Kuzmin, V. Rubakov and M Shaposhnikov, *Phys. Lett.* **155B** (1985) 36.

[10.32] A. Ringwald, *Nucl. Phys.* **B330** (1990) 1.

[10.33] G. 't Hooft, *Phys. Rev. Lett.* **37** (1976) 5; *Phys. Rev.* **D14** (1976) 3432; ref. [10.36].

[10.34] R. Crewther, *Nuovo Cim.* **2** (1979) 63; G. 't Hooft, *Phys. Reports* **142** (1986) 357.

[10.35] T. Kugo and I. Ojima, *Prog. Theor. Phys. Suppl.* **66** (1979).

[10.36] C. Becchi, A. Rouet and R. Stora, *Annals of Phys.* **98** (1976) 287; Y. Tyupkin, Lebedev preprint FIAN n. 39 (1975) (unpublished).

[10.37] I.B. Khriplovich, *Phys. Lett.* **173** (1986) 193.

[10.38] J. Ellis and M.K. Gaillard, *Nucl. Phys.* **B150** (1979) 141.

[10.39] S. Weinberg, *Phys. Rev. Lett.* **40** (1978) 223; F. Wilczek, *Phys. Rev. Lett.* **40** (1978) 279.

[10.40] G. Raffelt and D. Serkel, *Phys. Rev. Lett.* **60** (1988) 1793; M. Turner, *Phys. Rev. Lett.* **60** (1988) 1797.

[10.41] P. Sikivie, *Phys. Rev. Lett.* **51** (1983) 1415; *Phys. Rev.* **D32** (1985) 1988.

[10.42] M. Atiyah, *Symposium on Anomalies, Geometry and Topology*, ed. W.A. Bardeen and A.R. White, World Scientific, Singapore (1985).

[10.43] K. Fujikawa, *Phys. Rev. Lett.* **42** (1979) 1195; *Phys. Rev.* **D21** (1980) 2848.

[10.44] C.Q. Geng, Z.Y. Zhao, R.E. Marshak, and S. Okubo, *Phys. Rev.* **D36** (1987) 1953.

[10.45] S. Okubo, C.Q. Geng, Z.Y. Zhao, and R.E. Marshak, *Phys. Rev.* **D36** (1987) 3268.

[10.46] S. Elizur and W.P. Nair, *Nucl. Phys.* **B243** (1984) 205.

[10.47] S. Okubo, Lecture notes on "Group Theory", ed. Y. Tosa, Univ. of Rochester (1980).

[10.48] R. Bott, *Ann. of Math.* **70** (1959) 313.

[10.49] E. Witten, *Nucl. Phys.* **B223** (1983) 422; *ibid.* **B223** (1983) 433.

[10.50] W. Pauli and S. M. Dancoff, *Phys. Rev.* **62** (1942) 85; W. Pauli, *Meson Theory of Nuclear Forces*, Interscience, New York (1946).

[10.51] A.P. Balachandran, V.P. Nair, S.G. Rajeev, and A. Stern, *Phys. Rev. Lett.* **49** (1982) 1124.

[10.52] J.L. Gervais and B. Sakita, *Phys. Rev.* **D30** (1984) 1795.

[10.53] G.S. Adkins, C.R. Nappi and E. Witten, *Nucl. Phys.* **B228** (1983) 552.

[10.54] J. Schwinger, *Phys. Rev.* **82** (1951) 664.

[10.55] I.J.R. Aitcheson, CERN-TH 4458/86 (1986).

INDEX

Robert E. Marshak began his lifelong devotion to theoretical physics with degrees from Columbia and Cornell. After participating in the Manhattan Project, he returned to the University of Rochester Physics Department, where he served as Chairman and Distinguished University Professor. In 1970 he became President of the City College of New York, where he initiated many programs to aid the economically disadvantaged. His career concluded as University Distinguished Professor at Virginia Polytechnic Institute. Prof. Marshak's research interests ranged from the nature of stellar interiors and high-temperature shock hydrodynamics to the character of strong and weak interactions in the sub-nuclear regime. Among his incisive contributions are the two-meson theory and the universal V-A theory (with G. Sudarshan). Prof. Marshak was dedicated to fostering international scientific ties in the interest of world peace, and he led the way in organizing major international conferences, notably the landmark "Rochester Conference" in high-energy physics. He served the physics community as president of the American Physical Society and chairman of the Federation of Atomic Scientists, and his achievements have been recognized by membership in the National Academy of Sciences and by numerous awards. He died on December 23, 1992, just after sending this book to press.